Elements

of

Ecology

Of Related Interest

GENERAL BIOLOGY

N. A. Campbell
Biology, Fourth Edition (1996)

N. A. Campbell, L. G. Mitchell, and J. B. Reece
Biology: Concepts and Connections, Second Edition (1997)

J. Dickey
Laboratory Investigations for Biology (1995)

R. J. Ferl and R. A. Wallace
Biology: The Realm of Life, Third Edition (1996)

J. Hagen, D. Allchin, and F. Singer
Doing Biology (1996)

A. Jones, R. Reed, and J. Weyers
Practical Skills in Biology (1994)

R. J. Kosinski
Fish Farm: A Simulation of Commercial Aquaculture (1993)

A. Lawson and B. D. Smith
Studying for Biology (1995)

M. C. Mix, P. Farber, and K. I. King
Biology: The Network of Life, Second Edition (1996)

J. G. Morgan and M. E. B. Carter
Investigating Biology: A Laboratory Manual for Biology, Second Edition (1996)

J. Pechenik
A Short Guide to Writing Biology, Third Edition (1997)

G. Sackheim
Introduction to Chemistry for Biology Students (1996)

R. M. Thornton
The Chemistry of Life (1998)

R. A. Wallace
Biology: The World of Life, Seventh Edition (1997)

R. A. Wallace, G. P. Sanders, and R. J. Ferl
Biology: The Science of Life, Fourth Edition (1996)

ECOLOGY AND EVOLUTION

J. Bowers
Sustainability and Environmental Economics (1998)

C. J. Krebs
Ecological Methodology, Fourth Edition (1989)

C. J. Krebs
Ecology, Fourth Edition (1994)

C. J. Krebs
The Message of Ecology (1987)

E. R. Pianka
Evolutionary Ecology, Fifth Edition (1994)

P. Skelton
Evolution: A Biological and Paleontological Approach (1998)

R. L. Smith
Ecology and Field Biology, Fifth Edition (1996)

MARINE BIOLOGY AND OCEANOGRAPHY

M. Lerman
Marine Biology: Environment, Diversity, and Ecology (1986)

J. W. Nybakken
Marine Biology: An Ecological Approach, Fourth Edition (1997)

D. A. Ross
Introduction to Oceanography (1995)

H. V. Thurman and H. H. Webber
Marine Biology, Second Edition (1991)

ENVIRONMENTAL SCIENCE (SUPPLEMENTS)

J. Graves and D. Reavey
Global Environmental Change: Plants, Animals and Communities (1995)

A. R. W. Jackson and J. M. Jackson
Environmental Science: The Natural Environment and Human Impact (1996)

W. Levy and C. Hallowell
Green Perspectives: Thinking and Writing About Nature and the Environment (1994)

W. J. Makofske and E. F. Karlin
Technology and the Global Environment Issues (1995)

I. F. Spellerberg
Conservation Biology (1996)

PLANT PHYSIOLOGY

M. G. Barbour, J. H. Burke, and W. D. Pitts
Terrestrial Plant Ecology, Second Edition (1987)

D. Dennis, D. H. Turpin, D. D. Lefebure, and D. B. Layzell
Plant Metabolism, Second Edition (1997)

ZOOLOGY

C. L. Harris
Concepts in Zoology, Second Edition (1996)

Elements

of

Ecology

fourth edition

Robert Leo Smith
West Virginia University, Emeritus

Thomas M. Smith
University of Virginia

An imprint of Addison Wesley Longman, Inc.

Menlo Park, California • Reading, Massachusetts • New York • Harlow, England
Don Mills, Ontario • Sydney • Mexico City • Madrid • Amsterdam

Publisher: Jim Green
Sponsoring Editor: Elizabeth Fogarty
Project/Developmental Editor: Evelyn Dahlgren
Publishing Assistant: Amy Dhillon
Production Editor: Lisa Weber
Art Supervisor: Carol Ann Smallwood
Artists: Robert L. Smith, Jr., Karl Miyajima
Photo Editor: Kathleen Cameron
Composition and Film Buyer: Vivian McDougal
Senior Permissions Editor: Ariane de Pree-Kajfez
Copy Editor: Kristin Zimet
Indexer: Nancy Kopper
Cover Designer: Yvo Riezebos
Cover Photo: © George Grall/National Geographic Society.

Library of Congress Cataloging-in-Publication Data
Smith, Robert Leo.
      Elements of ecology / Robert Leo Smith, Thomas M. Smith. — 4th ed.
            p.   cm.
      Includes bibliographical references and index.
      ISBN 0-321-01518-5
      1. Ecology.   I. Smith, T. M. (Thomas Michael), 1955-   .
II. Title.
QH541.S624   1998
577—dc21   97-43763
                                                              CIP

ISBN 0-321-01518-5
1 2 3 4 5 6 7 8 9 10—RNT—02 01 00 99 98

The Benjamin/Cummings Publishing Company, Inc.
2725 Sand Hill Road
Menlo Park, CA 94025

For Ben and Nate

BRIEF CONTENTS

CONTENTS

PREFACE

The first edition of *Elements of Ecology* appeared in 1976 as a short version of *Ecology and Field Biology*. It had evolved by the third edition into a text aimed at nonmajors. That evolution continues with the fourth edition, which has been rewritten and reorganized with the nonmajor completely in mind.

We believe that ecology should be part of a liberal education. It is essential that students who major in such diverse fields as economics, sociology, engineering, political science, law, history, English, languages, and the like have some basic understanding of ecology, for the simple reason that it impinges on their lives. They cannot appreciate or arrive at an informed opinion on such highly politicized environmental issues as clear air, clean water, wetland preservation, endangered species, biological diversity, logging, ozone depletion, global warming, old-growth forest, dams, flood control, and myriad other issues without a grounding in ecological concepts.

The changes in this edition mark a major departure from the organization and style of previous editions and, for that matter, from other ecology textbooks. Our goal in making this dramatic shift is to make ecology more accessible to nonmajors. For this reason we have employed a different pedagogical approach. We were impressed with the modular approach to biology in *Biology: Concepts and Connections*, by Campbell, Mitchell, and Reece. We have modified their approach in this book by discussing a key concept in each section and introducing it with a conceptual statement. These sentence headings become the focal point of the section. The organization of the chapter summaries is now a hierarchy of related topics identified by brief headings. Such groupings enable students to see how the concepts in the chapter fit together.

REVISIONS IN TEXT

We have tightened the text and shifted its emphasis to reflect the major objective of this book: to provide an accessible introduction to ecology for nonmajors. The chapters on natural selection and speciation have been revised, shortened, and moved to the end of Part III, because nonmajors need some background on population forces in order to grasp the natural selection process. In their place three new opening chapters ground the reader in essentials: the scientific method in ecology; homeostasis and adaptation, which keep the organism in tune with its environment; and the key processes of photosynthesis, assimilation, and decomposition, which are the basis of all ecological processes, from organisms to ecosystems. We have incorporated human disturbances, environmental problems such as pollution, and other topics of applied ecology into their appropriate conceptual chapters, rather than treating the material in "afterthought" chapters. We have added a new chapter on a topic of particular concern, global climate change. Part VI, "A Diversity of Ecosystems" in the third edition, we renamed "A Guide to Ecosystems." We retained the organization of this Part as it was in the third edition specifically to emphasize in a descriptive way the structure and functions of the various ecosystems and the human impacts on them.

Because it is a nonmajors text, we have deliberately held to a minimum the amount of mathematics, chemistry, and physics. Of course, in ecology you cannot completely escape math and chemistry. Where they are essential to the topic, we have tried to explain them clearly. In a number of places, especially in chapters on populations, we have placed this material in special boxes entitled Quantifying Ecology.

PEDAGOGY

Special features of this edition make learning easy:

- A list of objectives alerts students to what they should gain from each chapter.
- The modular approach flags major concepts as they arise.
- Topical summaries provide a hierarchical overview of the concepts presented in the chapter.
- Study Questions reinforce the objectives of the chapter. The questions are of two types. Review questions provide a guide to the study of the material. Special study questions marked with an asterisk (*) stimulate students to think about issues, investigate problems, and apply the concepts they have just learned.
- References for each chapter at the end of the text encourage further exploration. The references include the source material for the chapter as well as selected books and journal articles. Annotation helps the student choose among them.
- Key terms appear in bold face where they are first defined. Terminology is necessary in any

science, but we keep it secondary to the key concepts.

- A glossary of over 400 key terms used in the text provides an abridged dictionary of ecology.
- Cross-references throughout the text tie both concepts and chapters together, emphasizing the interrelated ecological principles.
- Boxed material enhances Chapters 1 through 26. It comes in two types. Focus on Ecology boxes showcase special topics. Quantifying Ecology boxes clarify mathematical or quantitative aspects of ecology.

ILLUSTRATION PROGRAM

This fourth edition introduces full-color illustrations. Retained in black and white are the outstanding original pen-and-ink drawings by the late Ned Smith that date back to the first edition as well as a number of pen-and-ink illustrations by Robert Leo Smith, Jr. that are most effective in their original format. All the remaining art has been redesigned, redrawn, and generated on the computer by Robert Leo Smith, Jr. All new color photographs have been carefully selected to supplement the text.

SUPPLEMENTAL MATERIAL

A set of supplementary materials supports the instructor who uses the fourth edition of *Elements of Ecology*.

- A combined Instructor's Manual by Robert Leo Smith and Test Bank by Edward Bedecarrax and Eugene Fenster.
- Computerized Test Banks for Macintosh and Windows for ease in developing tests.
- A set of 72 full-color overhead transparencies.

ORGANIZATION

We have divided this text into six related parts with numerous cross-references and an ever-broadening focus. Part I sets the stage. Chapter 1 explains what ecology is, how it relates to other sciences, and how ecologists use scientific methods. Chapter 2 introduces the concept of homeostasis and the intimate connection between organism and environment. Chapter 3 introduces the three processes basic to life: the fixation of carbon-based energy in photosynthesis and the use of that energy in assimilation and respiration. These chapters provide a conceptual foundation for the chapters to come.

Part II begins with individual organisms. It explores how organisms interact with their physical environment. Chapters 4 through 10 each focus on a significant condition for life.

Part III turns to the biological environment—the other organisms with which an individual shares the environment and interacts. Chapter 11 introduces the population and its major properties—density, distribution, and age structure. Chapter 12 looks at the relationships among individuals reflected in various life history patterns, including mating and reproductive strategies. Much of this chapter falls into the category of behavioral ecology. Chapter 13 explores population growth, tying it to mortality, natality, and survivorship. Regulation of population growth involves intraspecific competition, covered in Chapter 14. Introduced in this chapter are the concepts of density and growth in plants, dispersal, and social behavior. Organisms have to deal not only with individuals of their own species, but with other species as well. These relationships include interspecific competition, predation, parasitism, and mutualism, the topics of Chapters 15 through 17. The impact of humans on natural populations has a chapter of its own. Chapter 18 looks at exploitation, restoration, conservation, and pest control. It introduces the concepts of sustained yield and integrated pest management. Part II culminates in Chapter 19. It introduces natural selection, population genetics, and speciation, concepts crucial to the management of endangered species.

Part IV broadens the focus from the population to the community. A thin line separates population ecology from community ecology. Chapter 20 introduces the concept of the community, its vertical structure and horizontal patterns, and the concepts of dominance and diversity. Chapter 21 explores the spatial and temporal dynamics of the community, with an emphasis on the concepts of edge, succession, and island biogeography. Basic processes that affect community structure and that drive community change are explored in Chapter 22.

Part V, broader still, explores ecosystem dynamics. Chapter 23 presents the concept of the ecosystem and primary and secondary production. How energy flows through the ecosystem is the topic of Chapter 24. It discusses trophic levels, food chains, and food webs. Chapter 25 explores major biogeochemical cycles and examines how humans have intruded upon them. A more detailed look at the carbon cycle and human intrusions upon it forms the basis for Chapter 26, which deals with the looming problem of global climate change and associated environmental change.

Part VI covers the whole range of Earth's ecosystems. These chapters fall into related groups. Terrestrial ecosystems cover Chapters 27–31. The Human Impact sections in these chapters discuss the effects of overgrazing (Chapter 27), urbanization and increasing desertification (Chapter 28), mining, oil drilling, and timber exploitation (Chapter 29), timber harvest and fragmentation (Chapter 30), and deforestation (Chapter 31). Freshwater ecosystems occupy three chapters, Chapters 32–34. Special topics of interest are the effect of pollution on lakes and ponds (Chapter 32), the value and demise of wetlands (Chapter 33), and the impact of dams and channelization on flowing water ecosystems (Chapter 34). The last three chapters, Chapters 35–37, explore the marine environment and the effects of development, habitat destruction, and oil and toxic pollution on marine ecosystems.

The chapters in Part VI should in one way or another form an integral part of a nonmajors course. The student can bring everything in the book to bear upon them. They provide examples of the physical structure of communities, nutrient cycling, energy flow, and human intrusions. They equip the student to enter many of the ecopolitical debates today that will affect the ecosystems of tommorow.

ACKNOWLEDGMENTS

As you will note, a coauthor, my son Dr. Thomas M. Smith, Associate Professor, Environmental Science Department, University of Virginia, has joined this edition. He brings to this text global experience, fresh ideas, and a familiarity with the needs and problems of nonmajor students in an ecology course.

No textbook is a product of the authors alone. The material it covers represents the work of hundreds of ecological researchers who have spent lifetimes in the field and the laboratory. Their published experimental results, observations, and conceptual thinking provide the raw material out of which a textbook is fashioned.

Revision of a textbook depends heavily on the input of users who point out mistakes and opportunities. We took these suggestions seriously and incorporated many of them. We are deeply grateful to the following reviewers for their helpful comments and suggestions on how to improve this edition for a nonmajors audience: Edmund E. Bedecarrax, San Francisco City College; Mike Bell, Richland College; Leslie S. Bowker, California Polytechnic State University; Renee Brooks, University of South Florida; Donald Dahlsten, University of California-Berkeley; Gerald R. Dotson, Front Range Community College; Courtney Hackney, University of North Carolina-Wilmington; Ron Hofstetter, University of Miami; Norman Jensen, Milliken University; Michael Kutilek, San Jose State University; David V. McCalley, University of Northern Iowa; Larry Meisner, Concordia University; Bette H. Nybakken, Hartnell College; David Pimentel, Cornell University; Fred Smeins, Texas A & M University; Jack Stout, University of Central Florida; Robert A. Wright, West Texas A & M University; Richard Wunderlind, University of South Florida.

My first son, Robert Leo, Jr. rendered all the color graphics under considerable time pressure. His familiarity with the text, artistic ability, and skill at computer graphics allowed close collaboration and success.

After working 30 years with one publisher, Harper-Collins, we have had to adjust to a new publisher. Helping us with the new editorial policies and production techniques was a friendly and patient staff. Cathy Pusateri introduced us to our new publisher, its organization, and its staff and helped develop the approach the text would take. Lisa Weber, production editor, coordinated the myriad aspects of production and did a great job of keeping authors, editors, art, and proofs on track. Evelyn Dahlgren, project editor, steered the manuscript through its final stages, making sure captions, figures, tables, cross-references, and other parts of the manuscript fell into place, and rechecking copy-edited text and page proofs for inconsistencies. Kathleen Cameron, photo researcher, did an excellent job of selecting the many new photos. It was a real chore choosing photos to match my specific requirements for each subject, and she did so with much patience and care. Ecologically knowledgeable Kristin Zimet, copy editor for the fourth and fifth editions of *Ecology and Field Biology* and for the third editon of *Elements of Ecology*, did a good job of fine-tuning this fourth edition. Carol Ann Smallwood, art coordinator, took great care in keeping the art program organized and ensuring that all needed changes and corrections made it into the final art.

Through it all our wives, Alice and Nancy, had to endure the throes of book production. My wife Alice took care (and still does) of all the problems of living, while I devoted myself full time, including evenings and weekends, to working on this book. She has patiently endured book widowhood for years. In the midst of this revision Tom's responsibilities increased when he became the new father of twins, Ben and Nate, to whom this edition is dedicated.

Robert Leo Smith
Thomas M. Smith

WHAT IS ECOLOGY?

PART I

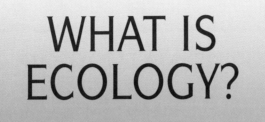

THE NATURE OF ECOLOGY

OBJECTIVES

On completion of this chapter, you should be able to:

- Define ecology and ecosystem.
- Tell why an ecosystem is a system.
- Define population and community.
- Relate ecology to other sciences.
- Describe how an ecologist does research.
- Define a hypothesis and discuss its importance in scientific studies.
- Explain the importance of models in ecology.

Zoologists examining the biotic community in a freshly cut tree in the Costa Rican rain forest.

Ecology—for years the term was familiar only to specialists in an obscure field. Overshadowed by molecular biology, the subject was scarcely recognized by the academic world. Then came the environmental movement of the late 1960s and early 1970s. Suddenly ecology became popular. The term appeared everywhere in newspapers, magazines, and books. Ecology became a household word, often misused. Even now people confuse it with the environment and with environmentalism. Ecology is neither one; yet sound environmental choices depend on understanding ecological issues and concepts.

1.1 Ecology is a science

So what is ecology? **Ecology,** according to the usual definition, is the scientific study of the relationship between organisms and their environment. That definition is good so long as you consider relationship and environment in their fullest meaning. *Environment* includes not only the physical but also the biological conditions under which an organism lives. *Relationship* includes interactions with the physical world and with members of other species and the same species.

The term *ecology* comes from the Greek words *oikos*, meaning "the family household," and *logy*, meaning "the study of." Literally, ecology is the study of the household. It has the same root word as economics, or "management of the household." We could consider ecology to be the study of the economics of nature. In fact, some economic concepts, such as resource allocation, and cost-benefit ratios, have crept into ecology. The term originally was coined by the German zoologist Ernst Haeckel in 1866. He called it *Oecologie* and defined its scope as the study of the relationship of animals to their environment.

1.2 The major unit of ecology is the ecosystem

Organisms interact with their environment within the context of the ecosystem. The *eco* part of the word relates to the environment. The *system* part implies that the ecosystem is a system. A system is a collection of related parts that function as a unit. The automobile engine is a system; subparts of the engine, such as the ignition, are also systems. Thus the ecosystem has interacting parts that support a whole. Broadly, the ecosystem consists of two basic interacting components, the living or biotic, and the physical or abiotic.

Figure 1.1 A forest ecosystem. Note the layers of vegetation and the diversity of plants.

Consider a natural ecosystem, a forest (Figure 1.1). The physical component of the forest consists of the atmosphere, climate, soil, and water. The many different plants and animals that inhabit the forest make up the biotic component. Each organism not only responds to the physical environment but also modifies it, becoming part of the environment itself. The trees in the canopy of a forest intercept the sunlight and use this energy to fuel the process of photosynthesis. In doing so, they modify the environment of plants below them, reducing the sunlight and lowering air temperature. Birds foraging on insects in the litter layer of the forest reduce insect numbers, modifying the environment for other organisms that depend on this shared food resource. Thus in ecosystems the living and the physical environment interact in ways that become complex, as succeeding chapters will show.

1.3 Ecosystem components form a hierarchy

The various kinds of organisms that inhabit our forest make up populations. The term *population* has many uses and meanings in other fields of study. In ecology, a **population** is a group of (potentially) interbreeding individuals occurring together in space and time. This definition implies that the individuals comprising the population are of the same species.

Populations of plants and animals in the ecosystem do not function independently of each other. Some populations compete with other populations for limited resources, such as food, water, or space. In other cases, one population is the food resource for another. Two populations may mutually benefit each other, each doing better in the presence of the other. All populations within an ecosystem are referred to as a **community** and have some connection to one another.

The community and the physical environment make up the **ecosystem.** We can now see the ecosystem has many levels. On one level, individual organisms, including humans, both respond to and influence the physical environment. At the next level, individuals of the same species form populations that we can describe in terms of number, growth rate, and age distribution. Further, individuals of these populations interact among themselves and with individuals of other species to form a community. Herbivores consume plants, predators eat prey, and individuals compete for limiting resources. When individuals die, their remains decompose, releasing nutrients that had been consumed and incorporated into their tissues back into the soil to be recycled. All this is the study of ecology—the interaction of organisms with their environment.

Combined, the ecosystems of Earth form the planetary ecosystem or **biosphere.** Organisms within the biosphere not only adapt to the environment but interact to modify and control chemical and physical conditions of the biosphere. This view of a self-sustaining biosphere, in which every organism is linked to the other, is known as the Gaia hypothesis. Although not all ecologists agree with this hypothesis, it does serve as a warning that to greatly disturb the biosphere can endanger our own survival.

1.4 Ecology has strong ties to other sciences

The complex interactions taking place within the ecosystem involve all sorts of physical and biological processes. To study these interactions, ecologists have to draw on other sciences. This dependency makes ecology an interdisciplinary science (Figure 1.2). (See Focus on Ecology 1.1: Ecology Has Complex Roots.)

In the following chapters we will discuss aspects of biochemistry, physiology, and genetics. We do so only in the context of understanding the interplay of organisms with their environment. The study of how plants take up carbon dioxide and lose water (Chapter 3), for example, belongs to plant physiology. Ecology looks at how these processes respond to variations in rainfall and temperature. This information is crucial to understanding the distribution and abundance

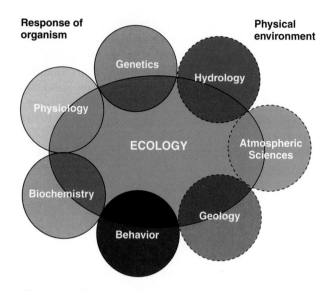

Figure 1.2 Ecology is an interdisciplinary science. It overlaps with many elements of physical and biological sciences.

of plant populations, and the structure and function of ecosystems on land. Likewise, we must dip into many of the physical sciences, such as geology, hydrology, and meteorology. They will help us chart other ways organism and environment interact. For instance, as plants take up water, they influence soil moisture and the patterns of surface water flow. As they lose water to the atmosphere, they increase atmospheric water content and influence regional patterns of precipitation. The geology of an area influences the availability of nutrients and water for plant growth.

In each of these examples, other scientific disciplines are critical to understanding how individuals respond to their environment and shape it. It is this connection with so many other fields that makes ecology such a rich area of study.

1.5 Ecologists use scientific methods

To investigate the relation of organisms to their environment, ecologists have to undertake experimental studies in the laboratory and in the field. All of these studies have one thing in common. They involve the collection of data to test hypotheses. A **hypothesis** is a statement of causation that can be tested, then either accepted or rejected. Usually a hypothesis is based on some observation in the field or on previous investigations.

For example, an ecologist might hypothesize that the availability of the nutrient nitrogen is the major

ECOLOGY HAS COMPLEX ROOTS

The genealogy of most sciences is direct. Tracing the roots of mathematics, chemistry, and physics is easy. The science of ecology is different. Its roots are complex.

You can argue that ecology goes back to the ancient Greek scholar Theophrastus, a friend of Aristotle, who wrote about the relations between organisms and the environment. On the other hand, ecology as we know it today has vital roots in plant geography and natural history.

In the 1800s, botanists began exploring and mapping the world's vegetation. One of the early plant geographers was Carl Ludwig Willdenow (1765–1812). He pointed out that similar climates supported vegetation similar in form, even though the species were different. Another was Friedrich Heinrich Alexander von Humboldt (1769–1859), for whom the Humboldt Current is named. He spent five years exploring Latin America, including the Orinoco and Amazon rivers. Humboldt correlated vegetation with environmental characteristics and coined the term *plant association.*

Among a second generation of plant geographers was Johannes Warming (1841–1924) at the University of Copenhagen, who studied the tropical vegetation of Brazil. He wrote the first text on plant ecology, *Plantesamfund.* In this book Warming integrated plant morphology, physiology, taxonomy, and biogeography into a coherent whole. This book had a tremendous influence on the development of ecology.

Early plant ecologists were concerned mostly with terrestrial vegetation. Another group of European biologists was interested in the relationship between aquatic plants and animals and their environment. They advanced the ideas of organic nutrient cycling and feeding levels, using the terms *producers* and *consumers.* Their work influenced a young limnologist at the University of Minnesota, R. A. Lindeman. He traced "energy-available" relationships within a lake community. His 1942 paper, "The Trophic-Dynamic Aspects of Ecology," marked the beginning of ecosystem ecology, the study of whole living systems.

Lindeman's theory stimulated considerable research on energy flow and nutrient budgets both in North America and Europe. The use of radioactive tracers to follow the fate of energy and nutrients enabled ecologists to develop systems ecology. **Systems ecology** applies general system theory and methods to ecology.

Meanwhile, activities in other areas of natural history were assuming an important role. One was the voyage of Charles Darwin (1809–1882) on the *Beagle.* Working for years on notes and collections from this trip, Darwin compared similarities and dissimilarities among organisms within and among continents. He attributed differences to geological barriers. He noted how successive groups of plants and animals, distinct yet obviously related, replaced one another.

Developing his theory of evolution and the origin of species, Darwin came across the writings of Thomas Malthus (1766–1834). An economist, Malthus advanced the principle that populations grow in a geometric fashion, doubling at regular intervals until they outstrip the food supply. Ultimately, the population would be restrained by a "strong, constantly operating force such as sickness and premature death." From this concept Darwin developed the idea of "the survival of the fittest" as a mechanism of natural selection and evolution.

Meanwhile, unknown to Darwin, an Austrian monk, Gregor Mendel (1822–1884), was studying the transmission of characteristics from one generation of pea plants to another in his garden. Mendel's work on inheritance and Darwin's work on natural selection provided the foundation for the study of evolution and adaptation, the field of **population genetics.**

Interest in Malthusian theory stimulated the study of population in two directions. One, **population ecology,** is concerned with population growth (including birthrates and death rates), fluctuation, spread, and interactions. The other, **evolutionary ecology,** is concerned with the natural selection and evolution of populations.

At the same time **physiological ecology** arose. It is concerned with the responses of individual organisms to temperature, moisture, light, and other environmental conditions.

Natural history observations also spawned **behavioral ecology.** Nineteenth-century behavioral studies included those on ants by William Wheeler and on South American monkeys by Charles Carpenter. Later Konrad Lorenz and Niko Tinbergen gave a strong impetus to the field with their pioneering studies on the role of imprinting and instinct in the social life of animals, particularly birds and fish.

Other observations led to investigations of chemical substances in the natural world. Scientists began to explore the use and nature of chemicals in animal recognition, trail-making, and courtship, and in plant and animal defense. Such studies make up the specialized field of **chemical ecology.**

Ecology has so many roots that it probably will always remain many-sided—as the ecological historian Robert McIntosh calls it, "a polymorphic discipline." Insights from these many specialized areas of ecology will keep enriching it.

factor limiting growth and productivity of plants in the prairie grasslands of North America. To test this hypothesis, the ecologist can gather data in a number of ways. The first approach might be a **field study.** The ecologist would examine the correlation between available nitrogen and grassland productivity across a number of locations or sites. Both factors vary across the landscape. If nitrogen is controlling productivity, grassland productivity should increase with nitrogen. The scientist would measure grassland productivity and nitrogen availability at a number of sites in the region. Then the relationship between these two variables, nitrogen and productivity, could be expressed on a graph.

The graph in Figure 1.3 shows nitrogen availability on the horizontal or x axis and plant productivity on the vertical or y axis. The reason for this arrangement is important. The scientist is assuming that nitrogen is the cause and that plant productivity is the effect. We call x the independent variable and y the dependent variable.

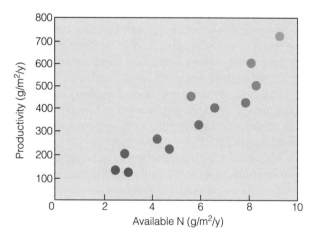

Figure 1.3 The response of grassland production to nitrogen availability. Nitrogen, the independent variable, goes on the x axis; grassland productivity, the dependent variable, goes on the y axis.

Although the data suggest that nitrogen does control grassland production, they do not prove it. It might well be that some other factor that varies with nitrogen availability, such as moisture or acidity, is actually responsible for the observed relationship. To test the hypothesis a second way, the scientist may choose to undertake an experiment. In designing the experiment, the scientist will try to isolate the presumed causal agent—in this case, nitrogen availability.

The scientist may choose to do a **field experiment,** adding nitrogen to some natural sites and not to others. The investigator controls the independent variable, levels of nitrogen, in a predetermined way and monitors the response of the dependent variable, plant growth. By observing the differences in productivity between grasslands that were fertilized with nitrogen and those that were not, the scientist tries to test whether nitrogen is the causal agent. However, in choosing the experimental sites, the scientist must try to locate areas where other factors that may influence productivity, such as moisture and acidity, are similar. Otherwise the scientist cannot be sure which factor is responsible for the observed differences in productivity among the sites.

Finally, the scientist might try a third approach, a series of **laboratory experiments.** The advantage of laboratory experiments is that the scientist has much more control over the environmental conditions. For example, the scientist can grow grass in the greenhouse under conditions of controlled temperature, soil acidity, and water availability. If the plants exhibit increased growth under higher nitrogen fertilization, then the scientist has evidence in support of the hypothesis. Nevertheless, the scientist faces a limitation common to all laboratory experiments. The results are not directly applicable in the field. The response of grass plants under controlled laboratory conditions may not be the same as their response under the natural conditions in the field. In the field, the plants are part of the ecosystem, interacting with other plants, animals, and the physical environment. Despite this limitation, the scientist now knows the basic growth response of the plants to nitrogen availability and goes on to design both laboratory and field experiments to explore new questions about it.

1.6   Experiments can lead to predictions

Scientists use the understanding derived from observation and experiments to make models. Data are limited to the special case of what happened when the measurements were made. Like photographs, data represent a given place and time. Models use the understanding gained from the data to predict what will happen in some other place and time.

Models are abstract, simplified representations of real systems. They allow us to predict some behavior or response using a set of explicit assumptions. Models can be mathematical, like computer simulations; or they can be verbally descriptive, like Darwin's theory of evolution by natural selection. Hypotheses are such word-based models. Our hypothesis about nitrogen availability is a model. It predicts that plant productivity will increase with increasing nitrogen availability. However, this prediction is qualitative—it does not predict how much. In contrast, mathematical models offer quantitative predictions. For example, from the data in Figure 1.3, we can develop an equation that predicts the amount of plant productivity per unit of nitrogen in the soil (Figure 1.4).

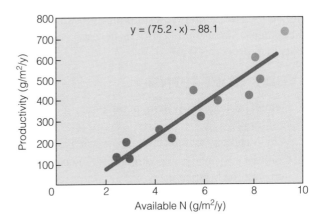

Figure 1.4 A mathematical model that predicts plant productivity (*y* axis) from nitrogen availability (*x* axis).

All of the approaches discussed above—observation, experimentation, hypothesis testing, and models—appear in the following chapters to back up concepts and relationships. They are the tools of science.

Ecology encompasses a broad area of science, ranging from individuals to ecosystems. There are many points from which we can depart to begin our study. We have chosen to begin with the individual organism, to examine the processes it uses and constraints it meets in maintaining life under varying environmental conditions. It is here that we can begin to understand the mechanisms that give rise to the diversity of life and of ecosystems on Earth.

CHAPTER REVIEW

SUMMARY

An Overview of Ecology (1.1–1.3)   Ecology is the study of the relationship between organisms and their environment **(1.1)**. This interaction takes place within the context of the ecosystem. A system is a collection of related parts that function as a unit **(1.2)**. The components of an ecosystem form a hierarchy. Organisms of the same kind that inhabit a given physical environment make up a population. Populations of different kinds of organisms interact with members of their own species as well as with individuals of other species. These interactions range from competition for shared resources to predation to mutual benefit. Interacting populations make up a biotic community.

Community plus the physical environment make up an ecosystem **(1.3)**.

An Interdisciplinary Science (1.4–1.6)   Ecology is an interdisciplinary science because the interactions of organisms with their environment and with each other involve physiological, behavioral, and physical responses. The study of these responses draws upon such fields as physiology, biochemistry, genetics, geology, hydrology, and meteorology **(1.4)**. The study of patterns and processes within ecosystems involves field study or experiments. Experimentation begins with formulating a hypothesis. A hypothesis is a statement about a causative agent that we can test experimentally **(1.5)**. From research data, ecologists develop models.

Models are abstractions and simplifications of natural phenomena. Such simplification is necessary to grasp ecosystem processes **(1.6).**

STUDY QUESTIONS

1. Define: ecology, ecosystem, population, community.
2. Why is ecology an interdisciplinary science?
3. Contrast a population, a community, an ecosystem, and the biosphere.
4. What is a system?
5. Why must ecosystems be considered systems?
6. What are the differences between a field study and a field experiment? What are the differences between a field experiment and a laboratory experiment? How are they related?
7. What is a hypothesis? How does it relate to a scientific study?
8. What is a model?
9. Of what value are models in ecology?

THE ORGANISM AND ITS ENVIRONMENT

OBJECTIVES

On completion of this chapter, you should be able to:

- Discuss the concept of adaptation.
- Explain why an organism's environment is so variable.
- Discuss the concept of homeostasis.
- Explain the meaning of environmental tolerance.
- Discuss the relationship between environment and the distribution and abundance of organisms.

A red fox *(Vulpes vulpes)* in its open field environment.

All living organisms are constantly interacting with their environment. Plants absorb carbon dioxide (CO_2) from the air through their leaves and water and mineral nutrients from the soil through their root system; they return water and oxygen to the atmosphere. Animals consume plants and other animals. They digest food, absorb nutrients, and discharge waste products.

For an organism to succeed, it needs to find essential resources and supporting conditions. If the organism can survive, grow, and reproduce under a given set of environmental conditions, we say it is **adapted** to that environment. If the environment does not offer the resources and conditions essential for its survival, the organism dies.

If the environment of the earth were homogeneous—constant in space and time—the adaptation of organisms to the environment would present only a single problem. An organism or group of organisms adapted to the constant conditions could inhabit the entire planet. But this is not the case. The environment varies, and with it, the set of characteristics an organism needs to survive, grow, and reproduce. In this chapter we will introduce some fundamental ways the organism relates to its environment.

2.1 Environmental conditions vary both in time and space

All organisms live in a varying physical environment of temperature, moisture, light, and nutrients. These factors differ from location to location—in latitude, region, and locality. In addition, at any location, the physical environment varies with time—yearly, seasonally, and daily. One important example of environmental variation is the flux of solar radiation reaching Earth's surface (Chapter 4). Solar radiation directly influences air temperature, atmospheric moisture, and light. To a large extent, it defines the general physical environment in which organisms live.

The amount of solar radiation reaching any point on Earth's surface and the resulting patterns of surface air temperature vary both spatially and temporally (see Figures 4.4 and 4.5). Organisms at any location on Earth's surface face both seasonal and daily variations in temperature. The variations are greatest in the temperate regions, where differences between average daily temperatures in the winter and summer can be extreme (see Chapter 6).

Even these variations do not define the conditions under which an organism lives (see Section 4.8). A bat in the canopy of a forest inhabits an environment quite different from that of a shrew on the floor of the same forest. A fox in a burrow in the desert experiences a different environment than a lizard on the desert surface. A daisy exposed to the open sunlight exists in a different environment than a violet growing nearby in the shade. Within the broad structure imposed by solar radiation is a wide range of microclimatic differences to which the organism must adapt. Although we humans are aware of variations in the physical environment, we shelter ourselves from the great variations that most plants and animals experience.

2.2 Organisms need a fairly constant internal environment

In an ever-changing physical environment, organisms must maintain a fairly constant internal environment, within narrow limits required by their cells, organs, and enzyme systems. For example, the human body must maintain internal temperatures within a narrow range around 37° C. An increase or decrease of only a few degrees from this value could prove fatal. Likewise, organisms must maintain certain levels of water, acidity, and salinity, to mention a few factors.

Maintaining these constant conditions requires a continuous exchange of energy and materials between the organism and the external physical environment. The organism must consume and digest food to ad-

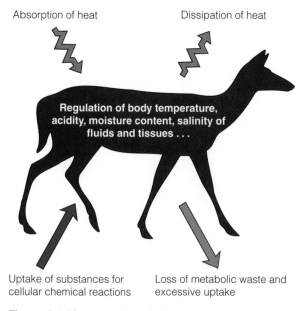

Absorption of heat

Dissipation of heat

Regulation of body temperature, acidity, moisture content, salinity of fluids and tissues . . .

Uptake of substances for cellular chemical reactions

Loss of metabolic waste and excessive uptake

Figure 2.1 Homeostasis calls for two-way exchange between the internal and external environments.

just its metabolism. Then it must excrete by-products and wastes from these chemical processes. The maintenance of conditions within the range that the organism can tolerate is called **homeostasis** (Figure 2.1).

What keeps this internal environment fairly constant is a feedback mechanism. It provides environmental information to which a system responds. An example is temperature regulation in humans (Figure 2.2). The normal temperature for humans is 37° C. We call such a norm a set point. When the temperature of the environment rises, sensory mechanisms in the skin detect the change. They send a message to the brain, which automatically relays the message to receptors that increase blood flow to the skin, induce sweating, and stimulate behavioral responses. Water excreted

through the skin evaporates, cooling the body. When the environmental temperature falls below a certain point, another reaction takes place. This time it reduces blood flow and causes shivering, an involuntary muscular exercise producing more heat. This type of reaction, which halts or reverses movement away from a set point, is called **negative feedback** (Figure 2.3).

If the environmental temperature becomes extreme, the homeostatic system breaks down. When it gets too warm, the body cannot lose heat fast enough to maintain normal temperature. Metabolism speeds up, further raising body temperature, until we die of heat stroke. If the environmental temperature drops too low, metabolic processes slow down, further decreasing body temperature, until death by freezing.

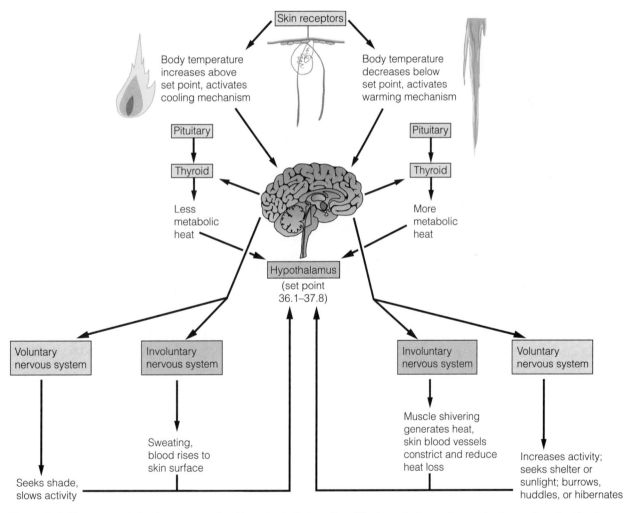

Figure 2.2 Thermoregulation is an example of homeostasis in action. The hypothalamus in your brain receives feedback on environmental temperature from the skin. It also senses the temperature of blood arriving from the body core. It responds to temperature change in two ways, activating the autonomic (or involuntary) and voluntary nervous systems and the endocrine system.

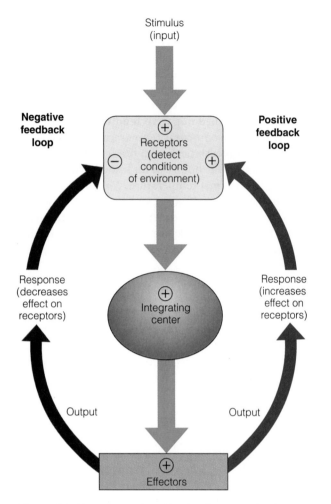

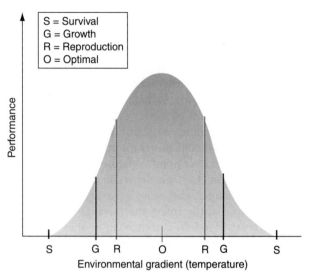

Figure 2.4 The response of an organism to an environmental gradient such as temperature. The end points of the curve represent the upper and lower limits for survival. Within this gradient are more restricted ranges within which the organism can grow and reproduce. This curve represents the law of tolerance.

Figure 2.3 Negative and positive feedback mechanisms. In negative feedback the response inhibits the mechanisms that brought it about. Positive feedback stimulates them. Negative feedback brings the mechanism back to the set point. Positive feedback leads away from the set point and can damage the system.

Such a situation, in which feedback reinforces change, driving the system to higher and higher or lower and lower values, is called **positive feedback.**

2.3 Homeostasis is possible only within a limited range of conditions

As the example of body temperature shows, there are limits to the range of environmental conditions over which homeostasis works. A graph illustrates this range (Figure 2.4). Let the x axis represent some feature of the physical environment. Because it is related to so many important aspects of homeostasis, as well as develop-

ment and survival, we will use temperature. Axis y will represent the response of the organism. There are a variety of responses we could use. Because survival is the most basic response, we will focus on it.

The response of an organism to the physical environment falls along a bell-shaped curve describing performance (in this case, the probability of survival). The point along the x axis where the response of the organism is the highest is called the optimum. As environmental conditions vary from this optimum, the probability of survival decreases. The two points (minimum and maximum) at which the curve intercepts the x axis represent the environmental conditions, in this case temperatures, beyond which the organism cannot survive. Within these two points is the range of environmental conditions under which an organism can survive, but not necessarily grow or reproduce. The minimum and maximum values of the environment are referred to as the **environmental tolerance** of the organism (see Focus on Ecology 2.1: Conditions for Reproduction).

The same axes can define other responses of the organism, particularly growth and reproduction. Note that the bell-shaped curves describing the responses of growth and survival lie within the curve describing survival. Although in this example the optimal environmental conditions are the same for these three processes, the range of conditions over which growth can

CONDITIONS FOR REPRODUCTION

The environmental conditions under which an organism can survive may be much broader than those required for reproduction. The brook trout *(Salvelinus fontinalis)* inhabits clear, cool, well-oxygenated streams of eastern North America. As an adult it can live within a temperature range from 0.5° to 25° C. The temperature range for egg development, however, is much lower, from 0° to 12° C.

The balsam fir *(Abies balsamifera)* is a conifer that grows in the cool moist climate of northern Canada and the northern United States. Popular as a Christmas tree

and ornamental, it is artificially grown and propagated outside its natural range. Although balsam fir will grow and survive where the mean annual temperature ranges between −4° and 7° C, it achieves optimum growth in that part of its range where the mean annual temperature is 2° to 4° C and the mean July temperature is 16° to 18° C. Balsam fir seeds require moisture and exposure to a low temperature of 5° C for a minimum of 30 days to germinate. This requirement precludes natural regeneration outside its natural range.

be maintained is smaller than that for survival. The range of conditions for reproduction is smaller yet, because reproduction depends on processes involved in growth.

Figure 2.4 represents the response of an organism to a range of values for a single environmental factor, temperature. However, organisms depend upon a wide range of environmental factors, each having an optimum and tolerances. To complicate things further, the factors interact. In the example of body temperature in humans, an important homeostatic response to rising body temperature is evaporative cooling, or sweating. This response requires water. Therefore, the water you need to survive is related to the temperature. When conditions are hot, our demand for water intake increases.

Because organisms respond to a variety of environmental factors, any one factor has the potential to limit survival, growth, and reproduction. In 1840, a German organic chemist named Justus von Leibig put forth a concept that has become known as **Leibig's law of the minimum.** In his study of the relation between soil and plants, Leibig noted that plants require certain kinds and quantities of nutrients. If one of those nutrients is absent, the plant dies. If it is present in minimal quantities only, the growth of the plant will be minimal. Stated in general terms, the law of the minimum says that the performance (survival, growth, and reproduction) of an organism will be a function of the most limiting environmental factor.

Organisms, then, are limited by a number of conditions and often by an interaction among them. Organisms live within ranges from too much to too little,

the limits of tolerance. This concept, that maximum and minimum conditions limit the presence and success of an organism, is called the **law of tolerance.**

2.4   An organism cannot do equally well in differing environments

The characteristics that enable an organism to do well under one set of conditions limit its performance under a different set of conditions. This important concept is plain to sports fans. Figure 2.5 is a photograph of two

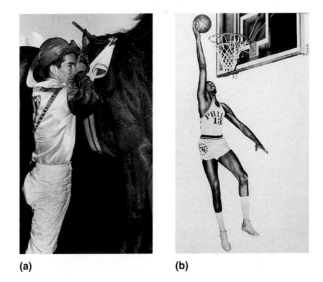

(a)   (b)

Figure 2.5 (a) Willie Shoemaker and (b) Wilt Chamberlain each excelled in a sport for which the other was physically unsuited.

great sports figures. Wilt Chamberlain was probably the greatest center in basketball. Willie Shoemaker was probably the best jockey to ride a racehorse. Now at the height of 1.49 m, Willie Shoemaker could never have played center for the Los Angeles Lakers, and at 2.15 m, Wilt Chamberlain could never have ridden to victory at the Kentucky Derby. The set of physical characteristics that enables you to excel at one of these sports precludes your ability to do well at the other. So, too, the characteristics of organisms constrain them.

For example, organisms specializing in different foods have developed distinctive mouthparts (Figure 2.6). The characteristics baleen whales require to filter small organisms from the water are quite different from those killer whales require to feed on larger prey. Hummingbirds, which feed on nectar and tiny insects, require a different type of bill than seed-eating cardinals. The grasshopper, which feeds on vegetation, possesses a set of chewing mouthparts, whereas the mosquito has piercing mouthparts with which to pen-

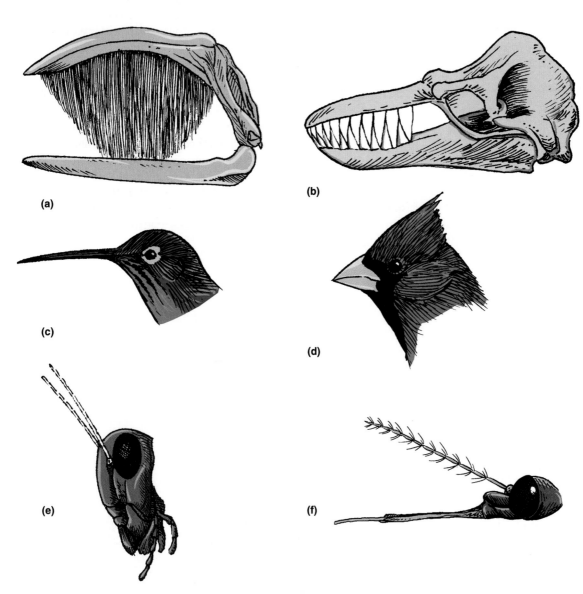

Figure 2.6 Mouthparts reflect how organisms gain their food. (a) Baleen, with which a baleen whale strains out krill and small fish. (b) Conical teeth of a predaceous toothed whale. (c) The long narrow bill of a hummingbird, a nectar feeder. (d) The strong, pointed conical bill of a cardinal, a seed eater. (e) The chewing mouthparts of a grasshopper. (f) The piercing mouthparts of a mosquito.

etrate skin and withdraw blood. The physical characteristics that permit each organism to efficiently use one of these feeding methods also prevent it from feeding in another manner.

2.5   The distribution of organisms reflects environmental variation

Figure 2.4 relates the performance of the organism to a range of values describing its physical environment. However, organisms in the real world are not distributed along an *x* axis. How does the distribution and abundance of organisms relate to variations over the landscape? By **distribution** we mean presence or absence. **Abundance** refers to numbers or population size.

To answer this question we must first see how features of the physical environment vary over the landscape. Suppose the feature of the environment described along the *x* axis is temperature. Then we might expect the geographic distribution of the organism to be limited to the region where temperatures fall within the range of tolerance of the organism. The red maple *(Acer rubrum)* is the most widespread of all

deciduous trees of eastern North America (Figure 2.7). Its northern limit coincides with the area in southeastern Canada where minimum winter temperatures drop to −40° C. Its southern limit is the Gulf Coast and southern Florida. Dry conditions halt its westward range. Within this geographic range it grows under a wide variation of soil types, soil moisture, acidity, and elevations, from wooded swamps to dry ridges. Thus red maple exhibits a high degree of tolerance to temperature and other environmental conditions. In turn, this high degree of tolerance allows a widespread geographic distribution.

Minimum and maximum temperature tolerances define the limits of species distribution. Although conditions close to the tolerances may be sufficient to maintain survival, growth, and reproduction, their values will be much below those that occur closer to the optimum. The nearer conditions approach the minimum and maximum tolerances of the organism, the fewer the individuals. We would expect the abundance of a species to increase as we move toward optimal environmental conditions.

Consider another example, the Carolina wren *(Thryothorus ludovicianus)*, a nonmigratory bird with

Figure 2.7 Red maple, one of the most abundant and widespread trees in eastern North America, thrives on a wider range of soil types, texture, moisture, acidity, and elevation than any other forest species in North America. The northern extent of its range coincides with the −40° C minimum winter temperature in southeastern Canada.

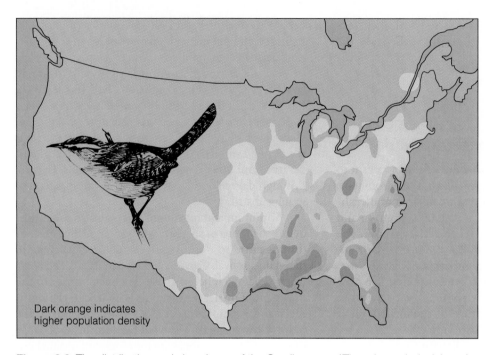

Dark orange indicates
higher population density

Figure 2.8 The distribution and abundance of the Carolina wren *(Thryothorus ludovicianus)* are strongly associated with temperature.

southern affinities, sensitive to cold winter temperatures (Figure 2.8). It is absent where the average minimum January temperature drops below −12° C. It occurs regularly only where the average minimum winter temperature is −7° C. It is most abundant in its optimal range, where the average minimum January temperature is over −4° C. Note how the abundance of the species decreases as it extends from the optimum to the edges of its range. The western edge of the wren's range appears to be limited by another aspect of its physical environment, moisture. The wren's western distribution ends where annual rainfall is less than 52 cm.

2.6   An organism lives in a habitat

An organism responds to a variety of environmental factors, and only when all of them are within the range of tolerance can it inhabit a location. The actual location or place where an organism lives is called its **habitat.** Because habitat describes a location, we can define it at many levels. Your habitat could be the country you live in, your state and city of residence, or the location of your home. Depending on your activity, such as eating, it could be your kitchen. The same is true for other organisms.

Consider the horned lark *(Eremophila alpestris)*, a bird of open grassland (Figure 2.9). Although it has a continent-wide distribution, its optimal environment is the midwestern prairie. Even within this optimal region, its distribution is influenced by tolerances. Because it requires a landscape of open, short grasses, it is much less abundant in taller, lightly grazed grassland. Within the short grassland, territory size limits colonization. Within each bird's territory, not all parts are usable for occupancy and specific activities. This hierarchy of suitable physical environments narrows the horned lark's habitat.

2.7   Constraints and trade-offs in habitat use define an organism's niche

The ideas of habitat and niche are closely related. The word *niche* in everyday terms means a recess in a wall where you place some object, or a place or activity for which a person or thing is best fitted. Joseph Grinnell,

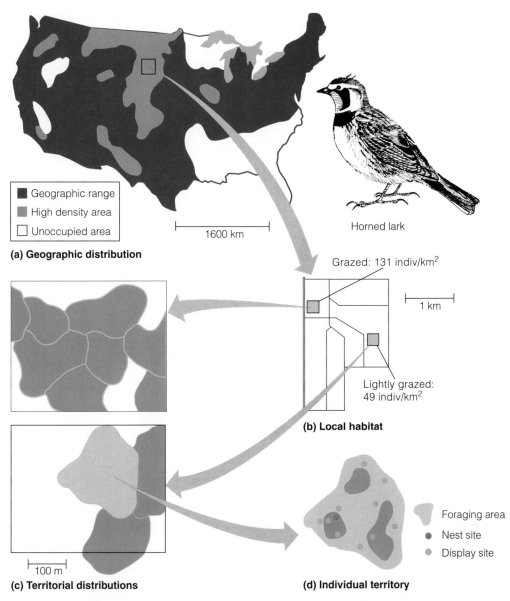

Figure 2.9 Population distribution on different scales for the horned lark *(Eremophila alpestris)*. (a) Although the horned lark is distributed over a wide breeding range, it inhabits areas ranging from high to low densities. (b) On a local scale the distribution of the bird is influenced by available grassland. (c) Within a plot of given habitat, the bird's distribution is influenced by territorial behavior. (d) Within each territory, the bird allocates space to different activities.

a California ornithologist, was the first to propose its use in ecology in 1917. He defined the niche as the ultimate distributional unit within which a species is restrained by the limitations of its physical structure and its physiology. What Grinnell had defined was a species's habitat. In 1927 an English animal ecologist, Charles Elton, considered the niche the basic role of an organism in the community—what it does, its relationship to its food and enemies. In other words, he defined the niche as the species's occupation.

In 1958 G. E. Hutchinson, a limnologist, expanded the idea for the niche to its current form. Now the **niche** includes all the physical and biological variables that affect an organism's well-being. Hutchinson's approach to the niche is similar to the graph in Figure 2.4. However, the niche does not fall along a single axis or environmental factor. All environmental factors to which an organism responds are part of it. Rather than a two-dimensional graph, a surface, Hutchinson's niche is a multidimensional response called a **hypervolume.**

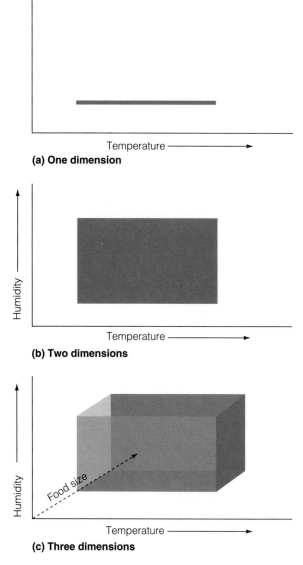

(a) One dimension

(b) Two dimensions

(c) Three dimensions

Figure 2.10 An illustration of niche dimension. Assume three elements comprising the hypothetical organism's niche: temperature, humidity, and food size. (a) A one-dimensional niche involving only temperature. (b) A second dimension, humidity, has been added. Enclosing that space, we have a two-dimensional niche. (c) Adding a third axis, food size, and enclosing all those points gives a three-dimensional niche space, or volume, for the organism. A fourth element would create a hypervolume.

We can begin to visualize a multidimensional niche by creating a three-dimensional one. Consider three niche-related variables for a hypothetical organism: temperature, humidity, and food size (Figure 2.10). If the organism can live only within a certain range of temperature, we plot that range on one axis. Temperature is one dimension of its niche. Next suppose that our organism can survive and reproduce only within a certain range of humidity. We plot humidity on a second axis. Enclosing that space, we have defined a two-dimensional niche. Now suppose our organism can eat only a certain range of food size. Food size is plotted on a third axis. Enclosing the new space, we come up with a volume, a three-dimensional niche. An example of a two-dimensional niche is the feeding niche of the blue-gray gnatcatcher *(Polioptila caerulea)*, a bird of open woods and brushy edges in the eastern United States (Figure 2.11). Two variables, foraging height and size of insect prey, define it.

Organisms with a wide range of tolerances occupy a large niche. Such organisms, like the red maple, we call generalists. Organisms with a narrow range of tolerances occupy a smaller niche. They are specialists.

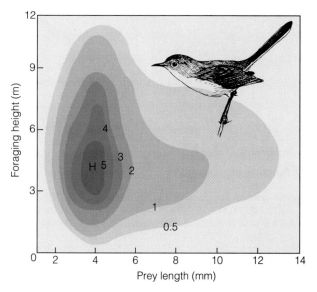

Figure 2.11 The feeding niche of the blue-gray gnatcatcher *(Polioptila caerulea)* is based on two variables, size of prey and foraging height. The contour lines map the feeding frequencies for adult gnatcatchers during the incubation period from July to August in California oak woodlands. The maximum response level is at H. Contour lines spreading out from this optimum represent decreasing response levels. The outermost line represents the boundary of the niche for these two variables.

CHAPTER REVIEW

SUMMARY

Adapting to a Variable Environment (2.1–2.4)   The physical environment, especially its temperature, moisture, and light, differs with latitude, region, and locality. It varies daily and seasonally, due to the variability of global solar radiation **(2.1)**. Despite this variable environment, organisms have to maintain a fairly constant internal environment. Doing so requires an exchange between the internal and external environments. This exchange involves negative and positive feedback **(2.2)**.

There are limits to the range of conditions under which an organism can live, described by a bell-shaped curve of tolerance. Minimum and maximum points on the curve define the limits beyond which organisms cannot survive. Within this range of survival are narrower ranges of conditions under which the organism can grow and reproduce **(2.3)**. Characteristics that enable an organism to survive, grow, and reproduce under one set of conditions are drawbacks under another **(2.4)**.

Environmental Patterns and Distribution of Organisms (2.5–2.7)   Distribution and abundance of organisms depend on both environmental variation and tolerances. Presence and absence define distribution, and the number of organisms defines abundance **(2.5)**. The place where an organism lives is its habitat **(2.6)**.

How it uses that habitat defines its niche, which includes the environmental and biological variables to which it responds. Organisms occupying wide niches are generalists; those occupying narrow niches are specialists **(2.7)**.

STUDY QUESTIONS

1. How does solar radiation influence the variability of an organism's physical environment?
2. Why must organisms maintain a fairly constant internal environment?
3. What is homeostasis, and how does it relate to survival in a variable environment?
4. Contrast negative feedback and positive feedback.
5. Describe the tolerance curve.
*6. An organism may have a wide range of tolerances for survival, yet be narrowly restricted in its distribution. Why? Provide some examples.
*7. How do the distribution and abundance of organisms relate to variability of the physical environment and the tolerances of those organisms?
8. Distinguish between an organism's habitat and its niche.

CHAPTER 3

KEY PROCESSES OF EXCHANGE

OBJECTIVES

On completion of this chapter, you should be able to:

- Contrast photosynthesis with respiration.
- Describe the process of C_3 photosynthesis.
- Discuss how C_4 and CAM photosyntheses are adaptations to warm and arid environments.
- Show the relationship between water loss and CO_2 uptake in photosynthesis.
- Discuss the process of decomposition.

Hygophorous sp. mushrooms and maple seedlings.

We built a framework for exploring the relationship between organisms and their environment in Chapter 2. Before we use this framework to examine how organisms respond to specific features of the environment, we must introduce three concepts—assimilation, respiration, and decomposition.

All living organisms require certain essential nutrients for synthesis, growth, and maintenance (Chapter 9). These substances are what we generally call food. Specific food requirements vary, but certain nutrients are essential for maintaining life processes in all organisms. One such nutrient is carbon.

3.1 Life on Earth is based on carbon

All life on Earth is carbon-based. What this means is that all living creatures are made up of complex molecules built on a framework of carbon atoms. The carbon atom is able to bond readily with other carbon atoms, forming long, complex carbon-containing molecules. The carbon needed to construct these molecules—the building blocks of life—derives from various sources. Humans, like all other animals, gain their carbon by consuming plant and animal materials. However, the ultimate source of carbon is carbon dioxide (CO_2) in the atmosphere.

Not all living organisms can use this abundant form of carbon directly. Only one process is capable of transforming carbon in the form of CO_2 into organic molecules and living tissue. That process, carried out by green plants, is **photosynthesis.** It is essential for the maintenance of life on Earth. All other organisms derive their carbon (and most other essential nutrients) from plant and animal tissue. In general, the process by which carbon and other essential nutrients transform into part of the organism is called **assimilation.**

The acquisition and assimilation of essential nutrients and the processes associated with life—synthesis, growth, reproduction, and maintenance—require energy. Chemical energy is generated in the breakdown of carbon compounds in living cells, a process called **respiration.**

The energy that fuels photosynthesis, the process of assimilation in green plants, comes from the sun. All other organisms in terrestrial and shallow water systems use energy that comes directly or indirectly from photosynthesis. The source from which an organism derives its energy is one of the most basic distinctions in ecology (Chapter 23). **Primary producers** derive energy from sunlight. **Secondary producers** derive energy from consuming plant and animal tissue and breaking down assimilated carbon compounds.

Let us take a closer look at the process so essential to life on Earth. Photosynthesis is "the lone reaction that counterbalances the vast expenditures of respiration, that reverses decomposition and death" (John Updike, 1968).

3.2 Photosynthesis converts carbon dioxide and hydrogen into carbohydrates

Photosynthesis is the process by which plants harness light energy from the sun to drive a series of chemical reactions. These reactions transform CO_2 into simple sugars and release oxygen (O_2). Photosynthesis is responsible not only for the carbon-containing molecules we animals consume, but also for the oxygen we breathe. We can express photosynthesis in simplified form:

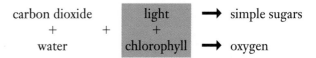

Plants use one unit of water (H_2O) and produce one unit of oxygen (O_2) for every unit of CO_2 they transform into simple sugars such as glucose $6(CH_2O)$. The initial products in the transformation are molecules that contain three carbon atoms; hence the process has been called **C_3** photosynthesis.

Notice the roles of light and chlorophyll. Chlorophyll is a pigment that acts as an antenna, absorbing energy from light. This energy splits the molecule of water. The oxygen is released. The plant uses the hydrogen in the transformation of CO_2 into simple sugars. Then it uses these sugars to make more complex carbon compounds, such as carbohydrates and proteins. These compounds build leaves, stems, roots, flowers, and seeds.

Like many chemical reactions in nature, the process of transforming CO_2 into sugars would occur at an extremely slow rate if it were not accelerated by the action of enzymes. Enzymes are chemical compounds that speed up or catalyze chemical reactions. The enzyme that catalyzes a crucial reaction in the transformation of CO_2 into sugars in photosynthesis is **rubisco** (ribulose biphosphate carboxylase-oxygenase). Rubisco is the most abundant enzyme on Earth.

Both the pigment chlorophyll and the enzyme rubisco are molecules that are constructed of carbon and nitrogen. In fact, over 50 percent of the nitrogen in a plant leaf is contained in these two compounds. Farmers use nitrogen-rich fertilizers to encourage production of these two compounds essential to photosynthesis.

3.3 Photosynthesis involves key exchanges

The uptake of CO_2 for photosynthesis is a crucial interaction between plants and the environment. In terrestrial (land) plants, CO_2 enters the plant through openings on the surface of the leaf called **stomata** (Figure 3.1). CO_2 moves through the stomata by diffusion. **Diffusion** is driven by a concentration gradient. Concentrations of CO_2 are often described in units of parts per million (ppm) of air. A CO_2 concentration of 355 ppm would be 355 units of CO_2 for every one million units of air. Substances flow from areas of high concentration to areas of low concentration until equilibrium is achieved—that is, until the concentrations in the two areas are equal. As long as the concentration of CO_2 in the air outside the leaf is greater than inside the leaf, CO_2 will continue to diffuse through the stomata.

As CO_2 is transformed into simple sugars, the concentration inside the leaf declines, pulling in more CO_2 from the atmosphere. As long as photosynthesis occurs, the gradient remains. If photosynthesis were to stop and the stomata were to remain open, then CO_2 would diffuse into the leaf until the internal CO_2 equalled the outside concentration. But this is not how plants work.

When photosynthesis and the demand for CO_2 are reduced for any reason, the stomata tend to close, reducing flow into the leaf. The reason for this closure is that stomata play a double role. They not only take up CO_2 but also lose water to the surrounding atmosphere. Water loss through the stomata is called **transpiration.** It represents a major constraint on the uptake of CO_2 by terrestrial plants.

The rate at which the leaf loses water to the surrounding air depends on the diffusion gradient of water vapor from inside to outside the leaf (see Section 7.5). Like CO_2, water vapor moves from areas of high concentration to areas of low concentration, from wet to dry. For all practical purposes, the air inside the leaf is saturated with water, so the outflow of water is a function of the amount of water vapor in the air—the humidity. The drier the air (lower the humidity), the more the water inside the leaf will diffuse through the stomata into the surrounding air. The leaf must replace this lost water with water that is taken up through the root system (see Section 7.5).

We can now see the tradeoff terrestrial plants face. To carry out photosynthesis, the plant must open its stomata; but when it does, it will lose water, which it must replace. If water is scarce, the plant must balance the opening and closing of the stomata, taking up enough CO_2 while minimizing the loss of water.

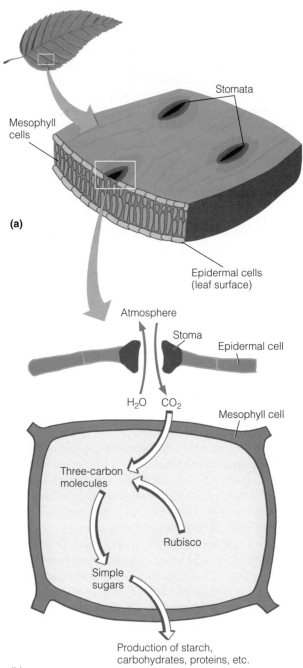

(a)

(b)

Figure 3.1 (a) Cross section of a leaf, showing stomata, mesophyll cells, and epidermal cells. (b) The C_3 pathway of photosynthesis. Carbon dioxide from the atmosphere diffuses into the leaf through the stoma to the mesophyll cells, where it is transformed into three-carbon molecules.

This balancing act affects the characteristics of plants and of ecosystems.

Photosynthesis in submerged aquatic plants is similar to that in terrestrial plants with one major exception. Submerged plants lack stomata. Dissolved CO_2

and bicarbonates (HCO_3^-) diffuse directly across the outer cell walls of the plant. Under water, the tradeoff between carbon uptake and water loss does not exist. However, other constraints related to oxygen and salinity come into play (Chapter 7).

3.4 Plants of warm, dry environments use a different photosynthetic path

Most plants carry on C_3 photosynthesis. However, many plants in warm, dry environments use a different photosynthetic pathway, called C_4. **C_4** plants possess a leaf anatomy different from C_3 plants (Figure 3.2).

In C_3 plants, the capture of light energy and the transformation of CO_2 into simple sugars occur in **mesophyll** cells. Then the products of photosynthesis move into the vascular bundles, part of the plant's transport system.

In contrast, plants possessing the C_4 pathway have vascular bundles surrounded by distinctive bundle sheath cells. A layer of mesophyll cells concentrates around each of the bundle sheaths. C_4 plants divide photosynthesis between the two types of cells, the mesophyll and the bundle sheath cells.

The first step takes place in the mesophyll cells. They have no enzyme rubisco, catalyzing the transformation of CO_2 into three-carbon molecules. Instead another enzyme, called PEP (phosphoenolpyruvate carboxylase), fixes CO_2 into four-carbon acids, malate and aspartate. Then malate and asparate are transported to the bundle sheaths (Figure 3.3). There enzymes break down the acids to form CO_2, reversing the process that is carried out in the mesophyll cells. In the bundle sheath cells the CO_2 is then transformed into sugars using the same enzyme that is used by C_3 plants—rubisco.

One of the advantages of C_4 photosynthesis is its effective use of CO_2. In the mesophyll, the CO_2 concentration will not go above the concentration of the outside air. This limits the CO_2 available to drive chemical reactions. However, carbon dioxide in the bundle

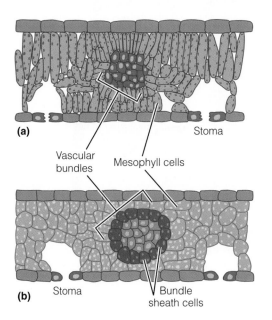

Figure 3.2 Comparison of leaf anatomy: (a) a C_3 plant leaf; (b) a C_4 plant leaf. In the C_4 plant, bundle sheath cells form a ring around the vascular bundle.

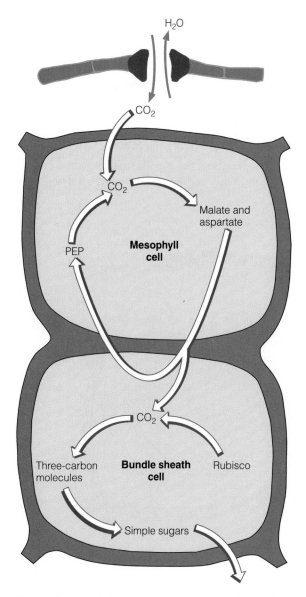

Figure 3.3 The C_4 pathway of photosynthesis. Different reactions take place in the mesophyll and bundle sheath cells. Compare to the C_3 pathway (Figure 3.1).

sheath cells can reach concentrations six times greater than that of the outside air. Under most conditions the rate of photosynthesis is limited by the concentration of CO_2 inside the leaf. Therefore, the maximum rate of photosynthesis is generally greater in C_4 plants.

To understand the second advantage of the C_4 pathway, we must go back to the tradeoff in terrestrial plants between the uptake of CO_2 and the loss of water through the stomata. The ratio of carbon fixed (photosynthesis) per unit of water lost (transpiration) is called the **water-use efficiency** (see Section 7.6). Due to the higher photosynthetic rate, C_4 plants exhibit greater water-use efficiency. That is, for a given degree of opening and water loss, C_4 plants typically fix more carbon. This increased water-use efficiency can be a great advantage in hot, dry climates where water is a major factor limiting plant production. However, it comes at a price. The C_4 pathway has a higher energy expenditure because of the need to produce the extra enzyme, PEP.

3.5   Desert plants modify photosynthesis to save water

In the hot deserts of the world, environmental conditions are even more severe. Solar radiation is high and water is scarce. To counteract these conditions, a small group of desert plants, mostly succulents in the families Cactaceae (cacti), Euphorbiaceae, and Crassulaceae, use a third type of photosynthetic pathway. It is the Crassulacean acid metabolism, CAM for short. The **CAM** pathway is similar to the C_4 pathway in that CO_2 is first transformed into the four-carbon acid malate using the enzyme PEP. The four-carbon product, malate, is later turned back into CO_2 and then transformed into sugars using the enzyme rubisco. Unlike C_4 plants, however, in which these two steps are physically separate (in mesophyll and bundle sheath cells), both steps occur in the mesophyll cells, at separate times (Figure 3.4).

CAM plants open their stomata at night, taking up CO_2 and converting it to malate using PEP. During the day the plant closes its stomata and reconverts the malate into CO_2, which it fixes into sugars. By opening their stomata at night, when temperatures are lowest and humidity is high, CAM plants conserve water. Although CAM plants have high water-use efficiency, the pathway is an inefficient use of energy. Under the extremely hot and dry conditions of the desert, however, this strategy allows the continuation of photosynthesis.

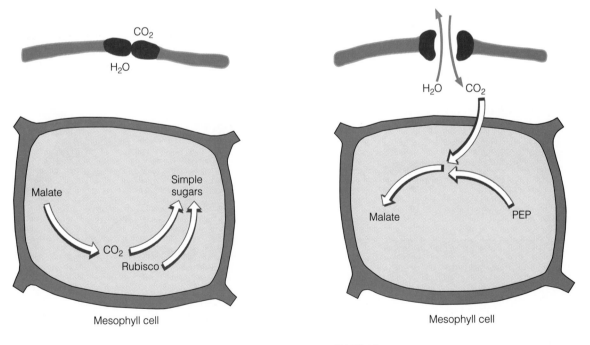

(a) Day   **(b) Night**

Figure 3.4 Photosynthesis in CAM plants. (a) By day the plant loses little water and takes in little CO_2. Photosynthesis takes place in the mesophyll cells. (b) At night the stomata open, the plant loses water through transpiration, and CO_2 diffuses into the leaf. CO_2 is stored as malate in the mesophyll, to be used in photosynthesis by day.

FOCUS ON ECOLOGY 3.1

STUDYING DECOMPOSITION

Ecologists study the process of decomposition by designing experiments that follow the decay of dead plant and animal tissues through time. Ecologists use litterbags to examine the decomposition of dead plant tissues (called plant litter). Litterbags are mesh bags constructed of synthetic material that does not readily decompose. The holes in the bag must be large enough to allow decomposer organisms to enter and feed on the litter, but small enough not to allow decomposing plant material to fall out of the bag. Ecologists most often use mesh bags with openings of 1–2 mm for plant leaf litter.

The graph shows the result of a litterbag experiment designed to examine the rate of decomposition of leaves from white oak. Ten grams of dry white oak leaves were placed in each of thirty litterbags. These bags were buried in the litter layer of the forest. At six intervals over the course of a year, five bags were collected and their contents dried and weighed in the laboratory.

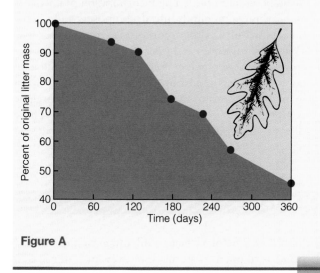

Figure A

Note the decreasing mass of litter remaining in the bags as time progresses. It shows that decomposer organisms were consuming the litter. The carbon was lost to the atmosphere as CO_2 as a result of respiration by decomposer organisms.

3.6 Respiration generates energy from carbon compounds

At the beginning of this chapter we introduced respiration. Respiration, recall, is the chemical breakdown of carbon-containing compounds to form energy. All processes associated with life require chemical energy in the form of **ATP** (adenosine triphosphate) generated by cellular respiration. Although the process of respiration uses other compounds, most often it is the breakdown of simple sugars and carbohydrates to yield carbon dioxide, water, and ATP. Breaking down these carbon compounds by aerobic respiration requires oxygen, so the chemical reaction is referred to as oxygenation. The same reaction occurs when you burn wood or fossil fuels such as gas and oil. Oxygen is used up and carbon dioxide is released. In your furnace, the energy takes the form of heat. In the cell, it is stored in chemical form as ATP or released as heat.

This reaction appears to be a reversal of photosynthesis, but it is not. There is an important difference. The energy that begins photosynthesis comes from sunlight, not from ATP. First energy from the sun is harnessed as chemical energy (ATP) and used to create organic compounds. Then respiration releases energy. The cell uses this energy to make new ATP to drive all metabolic processes in plants and animals.

3.7 Decomposition returns CO_2 to the atmosphere

Eventually all organisms die. Organic carbon compounds contained in once-living biomass break down, and the carbon returns to the atmosphere as carbon dioxide through decomposition.

Decomposition is the breakdown of energy-rich organic matter by consumers (secondary producers), mostly bacteria, fungi, and **detritivores**—organisms that feed on larger dead organic matter. Whereas photosynthesis incorporates solar energy, CO_2 and other nutrients into living tissue, decomposition releases heat and converts organic compounds into CO_2 and other inorganic nutrients (see Focus on Ecology 3.1: Studying Decomposition).

The greater part of the decomposition process begins with eating. Consumers assimilate the organic matter they eat, and their tissue becomes a source of food for still other consumers. Over varying periods of time organic matter ends up as CO_2 and other inorganic molecules, but by indirect routes.

Decomposition is a complex of many processes: (1) rainwater leaches out soluble compounds from dead organic matter; (2) detritivores fragment it; (3) bacteria and fungi break it down; (4) animals eat bacteria and fungi; and (5) organisms excrete organic and inorganic compounds. These processes link a diversity of organisms in highly tangled food webs (see Chapter 24).

All consumers function in some way as decomposers. As they digest food, they break down organic matter, change it, or release it as partially decomposed material, including feces, to other consumers. However, the true decomposers—bacteria, fungi, and detritivores—accomplish most of the decomposition. Detritivores open holes and break dead leaves and animal remains into smaller parts. They also oxidize organic

compounds, releasing energy and CO_2, and degrade them into smaller, simpler products. Bacteria and fungi convert this organic material into inorganic forms. This action gradually disintegrates dead organic matter into nutrients available to primary producers and microbes.

Some of these nutrients are incorporated into decomposer tissue. Feeding on bacteria and fungi are another group of organisms, known as microbivores (feeders on microbes). The incorporation of nutrients and carbon compounds into decomposers and microbivores withdraws them for a time from further use. When the microbes and microbivores themselves decompose, these nutrients return to the soil. The final outcome of decomposition is the return of CO_2 fixed by photosynthesis to the atmosphere and of inorganic compounds and elements to the soil and water. There they are used again, in never-ending cycles that you will meet in Chapter 25.

CHAPTER REVIEW

SUMMARY

Importance of Carbon (3.1) All life on Earth is carbon-based. The ultimate source of carbon is CO_2 in the atmosphere. Fixed by plants in photosynthesis, it is transferred directly or indirectly to all consumers. Carbon becomes part of organisms through assimilation. Assimilation requires energy derived from breakdown of carbon compounds in respiration. Because plants derive their energy from the sun, they are called primary producers. Organisms that derive their energy directly or indirectly from plants are called secondary producers.

An Overview of Photosynthesis (3.2–3.3) Photosynthesis harnesses light energy from the sun to convert CO_2 and H_2O into energy-rich carbohydrates. A nitrogen-based enzyme, rubisco, catalyzes the transformation of CO_2 into sugars. The first product of the reaction is two three-carbon compounds. For this reason, this photosynthetic pathway is called C_3 photosynthesis **(3.2)**.

Photosynthesis involves two key physical processes: diffusion and transpiration. Diffusion is the movement of molecules from more dense to less dense concentration. CO_2 diffuses from the atmosphere to the leaf through leaf pores or stomata. As photosynthesis slows down during the day and demand for CO_2 lessens, stomata close to reduce loss of water to the atmosphere. Water loss through the leaf is called transpiration. The amount of water lost depends upon the

humidity. The plant faces a dilemma: to pull in CO_2, it needs to open its stomata; to prevent water loss, it needs to keep its stomata closed **(3.3)**.

Photosynthesis in Warm and Arid Environments (3.4–3.5) Plants of warm environments use a C_4 pathway of photosynthesis. It involves two steps, made possible by a leaf anatomy that differs from C_3 plants. C_4 plants have vascular bundles surrounded by chlorophyll-rich bundle sheath cells. C_4 plants store carbon in the form of malate and aspartate in the mesophyll cells. They transfer these acids to the bundle sheath cells, where they release the CO_2 they contain. Photosynthesis now follows the C_3 pathway. Thus C_4 plants are able to store CO_2 as malate and carry on photosynthesis even if stomata close. C_4 plants have a high water-use efficiency **(3.4)**.

Succulent desert plants, such as cacti, have a third type of photosynthetic pathway, called CAM. CAM plants open their stomata to take in CO_2 at night, when the humidity is high. They convert CO_2 to a four-carbon compound, malate. During the day CAM plants close their stomata, convert malate back to CO_2, and follow the C_3 photosynthetic pathway **(3.5)**.

Releasing Carbon-Based Energy (3.6–3.7) Cellular respiration releases energy from carbohydrates to yield energy, H_2O, and CO_2 **(3.6)**.

Death and decomposition free carbon-based molecules and other nutrients sequestered in living biomass. Decomposition breaks down energy-rich molecules into simpler substances. All consumers become involved in decomposition to some degree, but bacteria, fungi, and detritivores are the major decomposers. Much of the carbon becomes incorporated into the living biomass of decomposer organisms. However, ultimately CO_2 is released by respiration back to the atmosphere to begin the cycle again **(3.7).**

STUDY QUESTIONS

1. What does it mean to say that life on Earth is carbon-based?
2. Distinguish between photosynthesis and assimilation. How are they related?
3. What is respiration? Is it the reverse of photosynthesis?
4. Distinguish between primary producers and secondary producers.
5. Describe the essential features of photosynthesis.
6. What are ATP and rubisco? What is the relationship between them?
7. How do diffusion and transpiration enter into photosynthesis?
8. How do C_3, C_4 and CAM plants differ in their photosynthetic processes?
9. What major obstacle do plants experience in the photosynthetic process? How do some plants get around it?
10. What is decomposition? How does it relate to respiration? What organisms are involved?

CONDITIONS
FOR LIFE

PART II

CLIMATE

OBJECTIVES

On completion of this chapter, you should be able to:

- Describe the fate of solar energy reaching Earth.
- Describe how the atmosphere heats and circulates.
- Tell how solar radiation influences seasonal temperatures.
- Explain the Coriolis effect on atmospheric circulation and ocean currents.
- Discuss the measures of atmospheric moisture.
- Describe microclimates and their ecological effects.

A fierce winter wind blows across the taiga
(Waterton Lakes National Park, Alberta, Canada).

The aspect of the physical environment that places the greatest constraint on organisms is climate. Climate is one of those terms we use loosely. In fact, people sometimes confuse climate with weather. **Weather** is the combination of temperature, humidity, precipitation, wind, cloudiness, and other atmospheric conditions at a specific place and time. **Climate** is the long-term average pattern of weather. We can describe the local, regional, or global climate.

Climate determines the availability of heat (Chapter 6) and water (Chapter 7). It influences the amount of solar energy that plants can capture (Chapter 3). Thus it controls the distribution and abundance of plants and animals (Chapters 2 and 22).

4.1   Earth intercepts solar radiation

Earth, immersed in sunlight, intercepts solar radiation on the outer edge of its atmosphere. The intercepted energy causes thermal patterns. Coupled with Earth's rotation and movement around the sun, it generates the prevailing winds and ocean currents. These movements of air and water in turn influence the distribution of rainfall.

Consider a simple energy "budget." Suppose Earth intercepts 100 units of solar energy (Figure 4.1). Of the 100 units, the atmosphere reflects 25 units directly into space and absorbs another 25 units. That leaves 50 units to reach Earth's surface. Of those 50 units, 5 units reflect into space, while the remaining 45 units are absorbed by the surface. The 45 units that are absorbed heat the land, oceans, and other bodies of water. Plants capture a small percent in photosynthesis (Chapter 3). Eventually these 45 units find their way back to the atmosphere, primarily as heat energy.

Solar radiation travels through space as waves. We describe them in terms of their wavelength. A very hot surface, such as that of the sun (6000° C), gives off

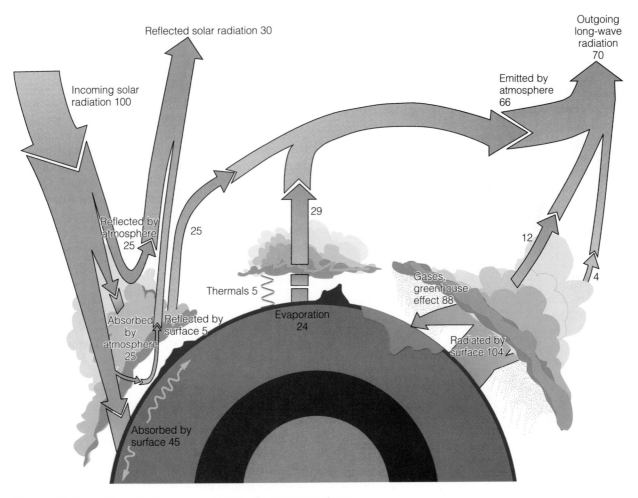

Figure 4.1 Disposition of solar energy reaching Earth's atmosphere.

primarily short-wave radiation. In contrast, cooler objects such as Earth's surface (average temperature of 15° C) emit radiation of longer wavelengths, or long-wave radiation.

Inbound short-wave radiation from the sun passes through the atmosphere with ease. Outbound long-wave radiation from Earth cannot readily escape. Instead certain gases in Earth's atmosphere, such as carbon dioxide and water vapor, absorb it and then send it back toward Earth. We call this process the **greenhouse effect** (see Chapter 26). The greenhouse effect is critical for maintaining the surface warmth of Earth. Without it, Earth would be a frozen planet.

4.2 Intercepted solar radiation varies over Earth's surface

The amount of solar energy intercepted by any point on Earth's surface varies markedly with latitude. Two factors influence this variation. First, at higher latitudes, radiation hits the surface at a greater angle, so it spreads over a larger area (Figure 4.2). Second, radiation that intercepts the atmosphere at an angle must travel through a deeper layer of air. It meets more particles in the atmosphere, so more of it bounces back to space. This difference explains why the temperature is higher in the tropics, near the equator, than at the poles.

Because the earth is tilted 23.5° from vertical on its axis with respect to the sun, solar radiation is perpendicular to different parts of the earth during the year (Figure 4.3). This situation causes seasonal varia-

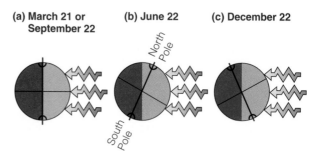

(a) March 21 or September 22 **(b) June 22** **(c) December 22**

North Pole

South Pole

Figure 4.3 Angle of the sun and circle of illumination at the equinoxes and winter solstices.

tion in temperature and daylength. Only at the equator are there exactly 12 hours of daylight and darkness every day of the year. At the spring (vernal) equinox and fall (autumnal) equinox (March 21 and September 22), the sun's rays fall directly on the equator (Figure 4.3a). At this time the equatorial region is heated most intensely, and every place on Earth receives the same daylength.

In the Northern Hemisphere at the summer solstice (June 22), the sun's rays fall directly on the Tropic of Cancer (23.5° north latitude). Then the Northern Hemisphere is heated most intensely, and daylength is at its longest (Figure 4.3b). In contrast, the Southern Hemisphere is experiencing winter at this time. The summer solstice in the Northern Hemisphere coincides with the winter solstice in the Southern Hemisphere. In the Northern Hemisphere at the winter solstice (December 22), the sun's rays fall directly on the Tropic of Capricorn (23.5° south latitude). During this period the Southern Hemisphere experiences summer, while the Northern Hemisphere has winter, with cold temperatures and short daylength (Figure 4.3c).

The seasonality of solar radiation, temperature, and daylength increases with latitude. At the Arctic and Antarctic Circles (66.5° north and south latitude), daylength varies from 0 to 24 hours over the course of the year. The days shorten until the winter solstice, a day of continuous darkness. The days lengthen with spring, and on the day of the summer solstice the sun never sets.

Figure 4.4 shows how annual, seasonal, and daily solar radiation vary over Earth. Although in theory every location on Earth receives the same amount of daylight over the course of a year, the tilt of the earth gives equatorial regions the most solar radiation. In high latitudes, where the sun is never directly overhead, the annual input of solar radiation is the lowest.

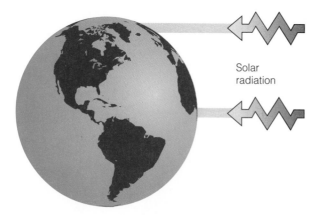

Solar radiation

Figure 4.2 Solar radiation striking Earth at high latitudes arrives at an oblique angle and spreads over a wide area. Therefore it is less intense than energy arriving vertically at the equator.

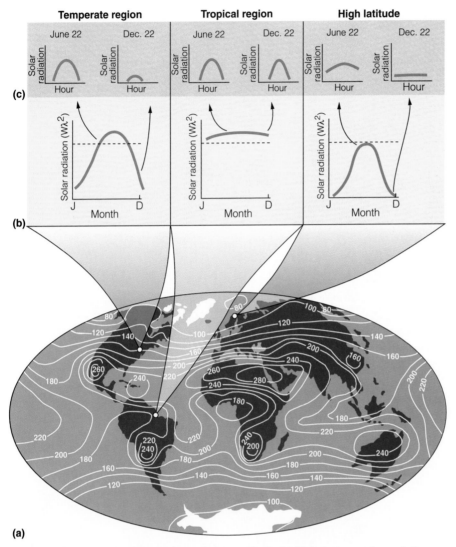

Figure 4.4 Annual variation in solar radiation on Earth. (a) Global mean solar radiation. (b) Variations in solar radiation from summer solstice to winter solstice at three locations: a temperate region, a tropical region, and a high latitude region. (c) Diurnal variations in solar radiation on two days in the year: the summer solstice and the winter solstice.

This pattern controls mean annual temperature around the globe (Figure 4.5). Like annual solar radiation, mean annual temperatures are highest in tropical regions and decline toward the poles.

4.3 Air temperature decreases with altitude

Variations in solar radiation explain latitudinal, seasonal, and daily changes in temperature, but they do not explain why air gets cooler with altitude. Mount Kilimanjaro is in tropical East Africa, but it is perma-

nently capped with ice and snow (Figure 4.6). The answer to this paradox lies in the physical properties of air.

Air molecules under pressure collide and heat up. As the warm air rises, the pressure on it decreases. The air expands. Now there are fewer collisions, so the air cools. This process is called **adiabatic cooling.** The same process works in an air conditioner, where a coolant is compressed. As the coolant moves from the compressor to the coils, the drop in pressure causes it to expand and cool.

The rate of adiabatic cooling depends on how much moisture is in the air. The adiabatic cooling of dry air

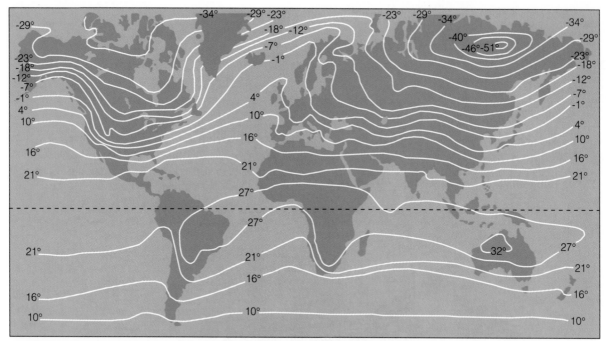

(a) January isotherms (lines of equal temperature) around the earth

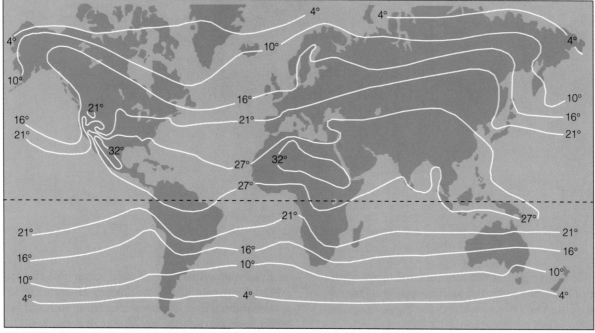

(b) July isotherms (lines of equal temperature) around the earth

Figure 4.5 Mean annual global temperatures change with latitude and season. (a) Mean sea-level temperatures (° C) in January. (b) Mean sea-level temperatures (° C) in July.

Figure 4.6 Although near the equator, Mount Kilimanjaro in Africa is snow-capped and supports tundralike vegetation near its summit. (See Figure 29.4)

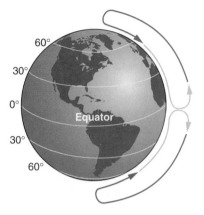

Figure 4.7 Circulation of air cells and prevailing winds on an imaginary, nonrotating Earth. Air heated at the equator rises and moves north and south. Cooling at the poles, it descends and moves back toward the equator.

is approximately 10° C/1000 m elevation. Moist air cools more slowly. The rate of temperature change with elevation is called the **adiabatic lapse rate.**

4.4   Air masses circulate globally

Now we can see how the global patterns of heating make the atmosphere circulate. As we saw in Figure 4.4, the equatorial region receives the largest annual input of solar radiation. Warm air rises because it is less dense than the cooler air above it. Air heated at the equatorial region rises to the top of the atmosphere, decreasing pressure down at the surface. More air rising beneath it forces the air mass to spread north and south toward the poles. As air masses approach the poles, they cool, become heavier, and sink over the Arctic and Antarctic regions. The sinking air raises surface air pressure. The cooled, heavier air then flows toward the equator, replacing the warm air rising over the tropics (Figure 4.7).

If Earth were stationary and without irregular land masses, the atmosphere would circulate as shown in Figure 4.7. Earth, however, spins on its axis from west to east, deflecting the pattern. We call this effect the **Coriolis force** after a nineteenth-century French mathematician, G. C. Coriolis, who first analyzed the phenomenon. The rotation of Earth causes all moving objects in the Northern Hemisphere, including air masses, to deflect to the right and those in the Southern Hemisphere to move to the left (Figure 4.8).

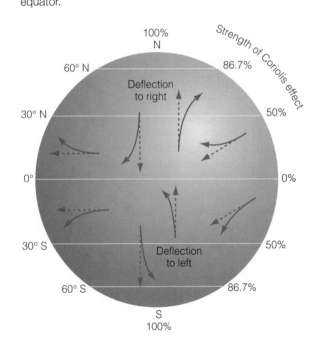

Figure 4.8 Effect of the Coriolis force on wind direction. The effect is absent at the equator, where the linear velocity is the greatest, 465 m/sec (1040 mi/hr). Any object on the equator is moving at the same rate. The Coriolis effect increases regularly toward the poles. If an object, including an air mass, moves northward from the equator at a constant speed, it speeds up because Earth moves more slowly (403 m/sec at 30° latitude, 233 m/sec at 60° latitude, and 0 m/sec at the poles). As a result the path of the object appears to deflect to the right or east in the Northern Hemisphere, and to the left or west in the Southern Hemisphere.

The Coriolis force, then, prevents a direct, simple flow from the equator to the poles. It creates a series of belts of prevailing winds, named for the direction from which they come. In the polar regions, they are the polar easterlies; and near the equator, they are the easterly trade winds. In the middle latitudes is a region of west winds known as the westerlies. These belts break the simple flow of air toward the equator and the flow aloft to the poles into a series of cells. They produce areas of high and low pressure (Figure 4.9).

The flow is divided into six cells, three in each hemisphere. The air that flows up from the equator forms an equatorial zone of low pressure, a region of calm called the **doldrums** by sailors. The equatorial air rises, cools, and spreads northward and southward away from the equatorial region. In the Northern Hemisphere the northward flow of air becomes nearly a westerly flow, and northward movement slows. The air cools, and at about 30° north latitude the air mass sinks, forming a semipermanent high pressure belt encircling Earth. This region of light winds is known as the subtropical high, or **horse latitudes.** The descended air warms and flows northward at the surface toward the pole or southward toward the equator. The northward-flowing air current turns right to become the prevailing **westerlies;** the southward-flowing air, also deflected to the right, becomes the **northeast trade winds** of the low latitudes. The air aloft gradually moves northward, continues to cool, and descends at the polar region. There it cools further at the surface, and flows southward. The flow of air deflected to the right becomes the polar **easterlies.** Southward-flowing air meets rising warm air moving toward the poles to produce a semipermanent low pressure area at about 60° north latitude. Similar flows take place in the Southern Hemisphere (see Figure 4.9).

4.5   Solar energy, wind, and Earth's rotation create ocean currents

This global pattern of winds initiates major patterns of surface flow in the oceans. These systematic patterns of water movement are called **currents.**

Each ocean is dominated by two great circular water motions, or **gyres.** Within each gyre the ocean current moves clockwise in the Northern Hemisphere and counterclockwise in the Southern Hemisphere (Figure 4.10). Trade winds push warm surface waters westward at the equator. As the waters encounter the continents, they split into north- and south-flowing

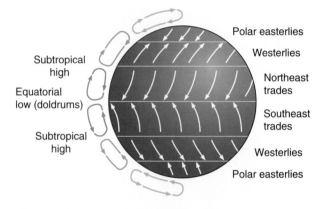

Figure 4.9 Belts and cells of air circulation about a rotating Earth. This circulation gives rise to the trade, westerly, and easterly winds.

currents along the eastern coasts, forming north and south gyres. As the waters move north and south, they cool and eventually encounter the westerly winds (30–60° N and S). The westerlies produce eastward-moving currents at the higher latitudes. These eastward-moving currents eventually encounter the continents, producing cool currents down the western margins. Ocean waters circulate unimpeded around the globe at the Antarctic continent.

4.6   Temperature influences the amount of moisture air can hold

Temperature plays yet another role in climate. The amount of water that can be held in a given volume of air is a function of its temperature. Warm air can hold more water than cold air. The amount of water vapor in the air is measured in units of pressure. The maximum amount of water vapor that can be held in a volume of air at a given temperature is called the **saturation vapor pressure.** This amount increases with increasing air temperature (Figure 4.11). **Relative humidity** is the amount of water in the air expressed as a percentage of the saturation vapor pressure. At saturation vapor pressure the relative humidity is 100 percent.

If the air cools while the amount of moisture it holds remains constant, the relative humidity increases, because cool air can hold less water than warm air. If the air cools beyond the saturation vapor pressure (100 percent relative humidity), the moisture condenses into clouds. When the particles of water or ice become too heavy to remain suspended in the air, rain

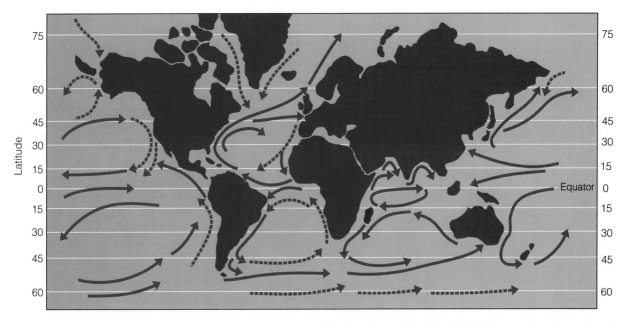

Figure 4.10 Ocean currents of the world. Notice how the circulation is influenced by the Coriolis force and continental land masses, and how oceans are connected by currents. Dashed arrows represent cool water and solid arrows warm water.

or snow falls. For a given water content of the air (vapor pressure), the temperature at which saturation vapor pressure is achieved is called the **dew point temperature.** Think of finding dew or frost on a cool morning. As nightfall approaches, temperatures drop and relative humidity rises. If cool night temperatures reach the dew point, water condenses and dew forms, lowering the amount of water in the air. As the sun rises, air temperature warms and the amount of moisture that the air can hold increases. The dew evaporates, increasing the vapor pressure of the air.

4.7   Precipitation has a global pattern

Bringing together the patterns of temperature, winds, and ocean currents, we are ready to understand the global pattern of precipitation (Figure 4.12). As the westerly winds move across the tropical oceans, they gather moisture. The warm air cools as it rises. When the dew point is reached, clouds form and precipitation falls as rain. This pattern accounts for high precipitation in the tropical regions of eastern Asia, South America, and Africa, as well as relatively high precipitation in southeastern North America.

As the winds move northward and southward, they cool. In the horse latitudes, where the cool air descends,

$$\text{Relative humidity} = \frac{\text{current VP}}{\text{saturation VP}} \times 100$$

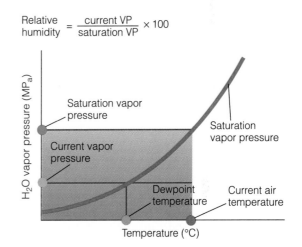

Figure 4.11 Saturation vapor pressure as a function of air temperature. For a given air temperature, the relative humdity is the ratio of actual vapor pressure to saturation vapor pressure. For a given vapor pressure, the temperature at which saturation vapor pressure occurs is called the dew point.

two belts of dry climate encircle the globe. The descending air warms and can therefore hold more moisture. The dry air draws water from the surface, causing arid conditions. In these belts the world's deserts have formed (Chapter 28).

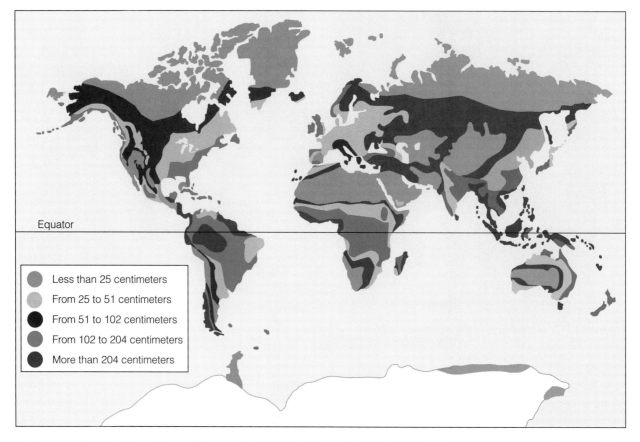

Figure 4.12 Annual world precipitation. Relate the wettest and driest areas to mountain ranges, ocean currents, and winds.

- Less than 25 centimeters
- From 25 to 51 centimeters
- From 51 to 102 centimeters
- From 102 to 204 centimeters
- More than 204 centimeters

Equator

4.8   Most organisms live in microclimates

The conditions in which most organisms live do not match the general climate. Their immediate surroundings modify the climate.

Today's weather report may state that the temperature is 28° C and the sky is clear. Nevertheless, environmental conditions will be quite different underground or on the surface, beneath vegetation or on exposed soil, on mountain slopes or on a ridgetop. Heat, moisture, air movement, and light all vary greatly from one part of the landscape to another to create a whole range of localized climates. These **microclimates** define the conditions under which organisms live. (See Focus on Ecology 4.1: Urban Microclimates.)

On a sunny but chilly day in early spring, flies may be attracted to sap oozing from the stump of a maple tree. The flies are active on the stump in spite of the near-freezing air temperature because during the day the surface of the stump absorbs solar radiation, heat-ing a thin layer of air above the stump. On a still day the air heated by the tree stump remains close to the surface. Temperatures decrease sharply above and below this layer. A similar phenomenon occurs when the frozen surface of the ground absorbs solar radiation and thaws. On a sunny late winter day, you walk on muddy ground, even though the air about you is cold.

By altering wind movement, evaporation, moisture, and soil temperatures, vegetation moderates the microclimate of an area, especially near the ground. Temperatures at ground level under shade are lower than in places exposed to the sun. On fair summer days a dense forest cover can reduce the daily range of temperatures at 25 mm (1 inch) above the ground by 7° to 12° C from the temperatures in the soils of bare fields. Within dense vegetation, such as heavy grass and low plant cover, the air is completely calm at ground level. This calm is an outstanding feature of the microclimate near the ground. It influences both temperature and humidity, creating a favorable environment for insects and other ground-dwelling animals.

URBAN MICROCLIMATES

Urban structures and the density and activity of their occupants create urban microclimates. In the urban complex, stone, asphalt and concrete pavement, and buildings share a high capacity for absorbing and re-radiating heat. They replace natural vegetation, with its low conductivity. Rainfall on impervious surfaces rapidly drains away, before evaporation can cool the air. Metabolic heat from masses of humans and waste heat from buildings, industrial combustion, and vehicles raise the temperature of the surrounding air. Industrial activities, power production, and vehicles pour water vapor, gases, and particulate matter into the atmosphere in great quantities.

These processes create a **heat dome** about cities large and small. In this dome the temperature may be 6° to 8° C higher than in the surrounding countryside. Heat domes have high temperature gradients. The highest temperatures are in areas of highest population density and activity, whereas temperatures decline markedly toward the periphery of the city. Although they are detectable throughout the year, heat domes are most pronounced during the summer and early winter. They are most noticeable at night, when heat stored by pavements and buildings reradiates into the air.

During summer, buildings and pavement of the inner city absorb and store considerably more heat than the vegetation of the countryside. In cities with narrow streets and tall buildings, the walls radiate heat toward each other instead of skyward. At night these structures slowly give off heat stored during the day.

In winter, solar radiation is considerably less because of the low angle of the sun, but heat accumulates from human and animal metabolism, industry, home heating, power generation, and transportation. In fact, heat from these sources is 2½ times that from solar radiation. Warming the atmosphere directly or indirectly, it makes winter milder in the city than in the countryside.

Throughout the year urban areas are blanketed with particulate matter, carbon dioxide, and water vapor. This haze reduces solar radiation reaching the city, which may receive 10 to 20 percent less than the surrounding countryside. At the same time, the blanket of haze absorbs part of the heat radiating upward and reflects it back; part of this heat warms the air, and part warms the ground. The higher the concentration of pollutants, the more intense is the heat dome.

Particulate matter has other microclimatic effects. Because of the city's low evaporation rate and the lack of vegetation, relative humidity is lower in the city than in surrounding rural areas. However, the particulate matter acts as condensation nuclei for water vapor in the air, producing fog and haze. Fog is much more frequent in urban areas than in the country, especially in winter.

Microclimates are a matter of scale. Consider two microclimatic extremes, north-facing and south-facing slopes in the Northern Hemisphere. South-facing slopes receive the most solar energy, whereas north-facing slopes receive the least energy. At other slope positions energy varies between these extremes, depending upon their compass direction.

The difference in solar radiation has a marked effect on moisture and heat on the two sites. High temperatures and associated high rates of evaporation draw out moisture from soil and plants. The evaporation rate is often 50 percent higher, the average temperature higher, the soil moisture lower, and the extremes more variable on south-facing slopes. Therefore the microclimate ranges from warm, dry, variable conditions on south-facing slopes to cool, moist, less variable conditions on north-facing slopes. Conditions are driest on the top of south-facing slopes, where air movement is the greatest, and dampest at the bottom of north-facing slopes.

The same microclimatic conditions occur on a much smaller scale on north-facing and south-facing slopes of large anthills, mounds of soil, dunes, and small ridges on the ground in otherwise flat terrain, and on the north-facing and south-facing sides of trees, logs, and buildings. The south-facing sides of buildings are always warmer and drier than the north-facing sides, a consideration for landscape planners, horticulturists, and gardeners. North sides of tree trunks are cooler and moister, a fact reflected in more vigorous growth of moss than on south sides. In winter the north-facing side of a tree may be below freezing while the south side, heated by the sun, is warm. This temperature difference may cause frost cracks in the bark as the sap, thawed by day, freezes at night. Bark beetles and other wood-dwelling insects seek cool, moist north-facing areas in which to lay their eggs. Flowers on the south side of tree crowns often bloom sooner than those on the north side.

Wide microclimatic extremes also occur in the concave surfaces of valleys and depressions in the ground. These places have lower temperatures at night, especially in winter, higher temperatures during the day, especially in summer, and higher relative humidity. Protected from the wind, the air becomes stagnant. It is heated by sunlight and cooled by terrestrial vegetation, in contrast to the wind-exposed, well-mixed layers of convex terrain. In the evening cool air from the higher slopes flows into the valley and on a smaller scale into surface depressions to form lakes of cool air (Figure 4.13). Moisture in this air often condenses to form valley fog. If the temperature drops low enough, frost pockets form in depressions. The microclimates of the frost pockets often support different kinds of plant life than the surrounding higher ground.

Mountainous topography influences local and regional microclimates by changing the pattern of precipitation. Mountains intercept air flow. As an air mass reaches the mountains, it ascends, cools, becomes saturated (because cold air holds much less moisture than warm air), and releases much of its moisture on the windward side (Figure 4.14). As the cool, dry air descends on the leeward side it warms and picks up moisture. As a result, the windward side of a mountain supports more vigorous, denser vegetation and different species of plants and associated animals than the leeward side, where in some areas dry, desertlike conditions exist. This phenomenon is called a **rain shadow.** Thus, in North America the westerly winds that blow over the Sierra Nevada and the Rocky Mountains drop their moisture on west-facing slopes to support vigorous forest growth, while on the eastern side desert or semidesert conditions exist.

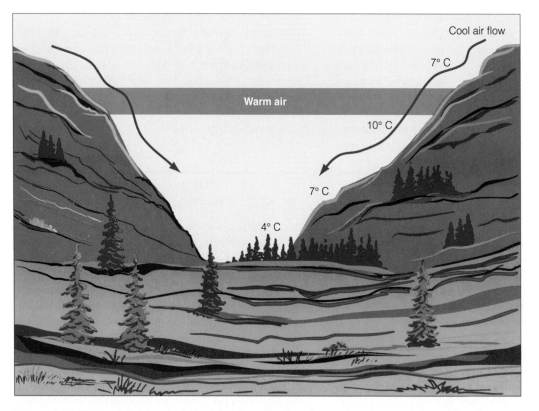

Figure 4.13 Topography can produce daily microclimatic extremes in valleys and depressions in the ground. At night air cools next to the ground, forming a weak surface inversion in which the temperature increases, rather than decreases, with height. At the same time, cool air moves downslope, deepening the inversion. When air is sufficiently cool and moist, fog forms in the valley. Smoke or air pollution released in such a situation will rise only until its temperature equals that of the surrounding air. Then it will flatten out just below the layer of warm air.

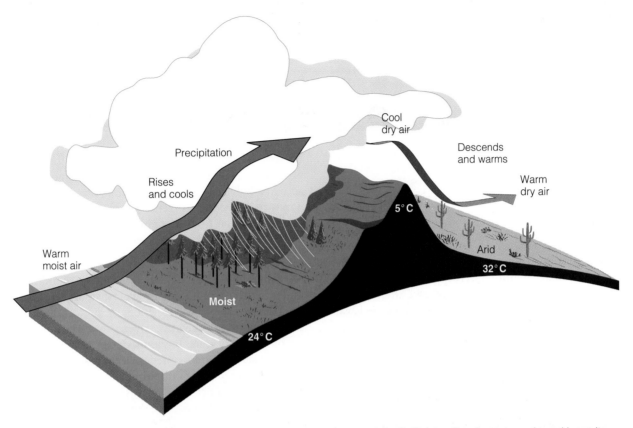

Figure 4.14 Formation of a rain shadow. Air is forced to go over a mountain. As it rises, the air mass cools and loses its moisture as precipitation on the windward side. The descending air, already dry, picks up moisture from the leeward side.

The same phenomenon occurs on a more local scale. Some of the most pronounced effects occur in the Hawaiian islands. Vegetation ranges from scrubby vegetation on the leeward side of an island to moist, forested slopes on the windward side (Figure 4.15). Within several square kilometers in the

(a)

(b)

Figure 4.15 Rain shadow on the mountains of Maui, Hawaiian islands. (a) The windward, east-facing slopes, intercepting the trade winds, are cloaked with wet forest. (b) The low-growing, scrubby vegetation of the dry western side.

central Appalachian Mountains, the western, wind-ward slopes support mesic forest vegetation dominated by yellow-poplar *(Liriodendron tulipifera)*, red oak *(Quercus rubra)*, white oak *(Q. alba)*, and black cherry *(Prunus serotina)*; whereas the dry, eastern, lee-ward slopes are dominated by scarlet oak *(Q. coccinea)*, black oak *(Q. velutina)*, and chestnut oak *(Q. prinus)*.

CHAPTER REVIEW

SUMMARY

Interception of Solar Radiation (4.1–4.2) Earth intercepts solar energy in the form of short-wave radiation that passes easily through the atmosphere. It emits much of it back as long-wave radiation that cannot escape readily outward. The atmosphere returns it to Earth, producing the greenhouse effect **(4.1)**. The amount of solar radiation intercepted varies markedly with latitude. Tropical regions receive the greatest amount of solar radiation, and high latitudes the least. Because Earth tilts on its axis, parts of Earth receive seasonal differences in solar radiation. These differences give rise to seasonal variations in temperature and rainfall. There is a global gradient in mean annual temperature. It is highest in the tropics and declines toward the poles **(4.2)**.

Dynamics of the Atmosphere (4.3–4.5) Heating and cooling cause air masses to rise and sink **(4.3)**. Vertical movements of air masses give rise to global patterns of atmospheric circulation. The spin of Earth on its axis deflects air and water to the right in the Northern Hemisphere and to the left in the Southern Hemisphere. This Coriolis effect produces three cells of global air flow in each hemisphere **(4.4)**. The global pattern of winds and the Coriolis effect cause major patterns of ocean currents. Each ocean is dominated by great circular water motions, or gyres. These gyres move clockwise in the Northern Hemisphere and counterclockwise in the Southern Hemisphere **(4.5)**.

Atmospheric Moisture and Precipitation (4.6–4.7) Atmospheric moisture is expressed in terms of relative humidity. The maximum amount of moisture the air can hold at any given temperature is called the saturation vapor pressure. It increases with temperature.

Relative humidity is the amount of water in the air expressed as a percentage of the maximum amount the air could hold at a given temperature **(4.6)**. Wind, temperature, and ocean currents produce global patterns of precipitation. They account for regions of high precipitation in the tropics and belts of dry climate in the horse latitudes **(4.7)**.

Microclimates (4.8) The actual climatic conditions under which organisms live vary considerably within one climate. These local variations or microclimates reflect topography, vegetative cover, exposure, and other factors on every scale. In mountainous areas the windward sides receive more precipitation than the leeward sides. Angles of solar radiation cause marked differences between north-facing and south-facing slopes, whether on mountains, sand dunes, or ant mounds.

STUDY QUESTIONS

1. What is the fate of the sun's energy when it reaches Earth's atmosphere?
2. Why does less solar energy reach the polar regions than the equator?
3. How does Earth's tilt influence the seasons?
4. How does Earth maintain a heat balance?
5. What is the Coriolis effect? How does it affect atmospheric circulation?
6. What causes ocean currents? What are gyres?
7. What are the relationships among saturation vapor pressure, relative humidity, and dew point?
8. What is a microclimate?
9. Why and how does a north-facing slope differ from a south-facing slope?

LIGHT

OBJECTIVES

On completion of this chapter, you should be able to:

- Describe the nature of light as it reaches Earth.
- Explain the fate of visible light in the plant canopy and in water.
- Describe how leaf area index influences the distribution of light.
- Examine how the process of photosynthesis responds to variations in light.
- Contrast shade-tolerant and shade-intolerant plants.
- Compare the light environment of aquatic plants to that of terrestrial plants.
- Discuss the significance of ultraviolet radiation for organisms.

Light streams through the canopy of a California coastal redwood *(Sequoia sempervirens)* grove.

Most of us associate natural light with visibility. Rarely do we think of light as a driving force of life. While we go about our daytime activities, plants are using visible light as an energy source to convert carbon dioxide and water to organic carbon compounds (Chapter 3). The hours of light and dark influence the daily and seasonal activities of terrestrial and shallow water organisms, a role we will explore in Chapter 8. Natural light is much more than light to see by.

5.1 Solar radiation includes visible light

Of the total range of solar radiation reaching Earth's atmosphere (Chapter 4), the wavelengths of approximately 400 to 700 nm (a nanometer is one-billionth of a meter) make up visible light (Figure 5.1). Collectively, these wavelengths are known as **photosynthetically active radiation (PAR)** because they include the wavelengths plants use in photosynthesis (Chapter 3). Wavelengths shorter than the visible range are ultraviolet or UV light. There are two types of ultraviolet light: UV-A with wavelengths from 315 nm to 380 nm, and UV-B with wavelengths from 280 to 315 nm. Radiation with wavelengths longer than the visible range is infrared. Near infrared has wavelengths of approximately 740 to 4000 nm, and far infrared or thermal radiation from 4000 to 100,000 nm.

Light that reaches the earth's surface is not quite the same light that arrives at the top of the earth's atmosphere (Figure 5.2). The ozone layer in the upper atmosphere (stratosphere) absorbs nearly all wavelengths, but especially the violets and blues of visible light. Molecules of atmospheric gases scatter these shorter wavelengths, giving a bluish color to the sky and causing Earth to shine in space. Water vapor scatters all wave-

lengths, giving clouds their white appearance. Dust scatters long wavelengths to produce reds and yellows in the sky. Because of the scattering, part of solar radiation reaches Earth as diffuse light from the sky, known as skylight. The relative contributions of direct light and skylight change not only over the course of a single day but globally as a function of latitude. Indirect light represents a larger proportion of total intercepted light at higher latitudes, as a result of low sun angles (see Section 4.2).

Light intercepted by Earth is reflected, absorbed, or transmitted through objects. Of greatest ecological interest is light reaching vegetation. Leaves reflect about 6 to 12 percent of photosynthetically active radiation, whereas they reflect about 70 percent of infrared light and only about 3 percent of ultraviolet light striking them directly. The degree of reflection varies with the nature of the leaf surface. Because leaves preferentially absorb violet, blue, and red light and strongly reflect green light, they appear green. The light not reflected or absorbed is transmitted through the leaf. How much light is transmitted depends upon the thickness and structure of the leaf. A leaf may transmit up to 40 percent of the light it receives, but 10 to 20 percent is more usual. Transmitted light is primarily green and far red. At ground level in a dense forest, even green light may be extinguished.

Light that enters water is absorbed rapidly (Figure 5.3). Only about 40 percent reaches 1 m into clear lake water. Moreover, water absorbs some wavelengths more than others. First to be absorbed are visible red light and infrared radiation in wavelengths greater than 750 nm. This absorption reduces solar energy by one-half. In clear water yellow disappears next, followed by green and violet, leaving only blue wavelengths to penetrate deeper water. A fraction of blue

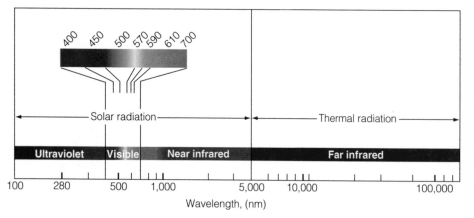

Figure 5.1 A portion of the electromagnetic spectrum, separated into solar and thermal radiation. Wavelengths on a spectrum are bunched irregularly. Ultraviolet, visible, and infrared light waves represent only a small portion of the spectrum. To the left of ultraviolet radiation are X rays and gamma rays (not shown).

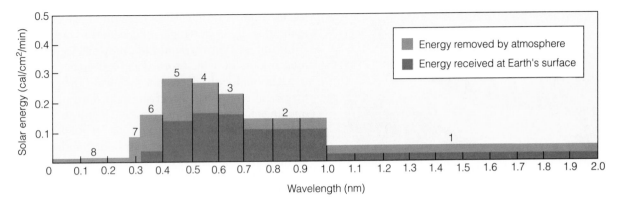

Figure 5.2 Energy in the solar spectrum before and after depletion by the atmosphere, given a solar altitude of 30°. Figures above the bars indicate (1) near infrared wavelengths over 1 micron; (2) near infrared, 0.7–1.0 μ; (3–5) visible light; (3) red; (4) green, yellow, and orange; (5) violet and blue; (6–8) ultraviolet. Note the strong reduction in ultraviolet. Nearly all the ultraviolet wavelengths are absorbed by the ozone. The region of peak energy shifts toward the red end of the spectrum. Visible light in the blue wavelength is scattered rather than absorbed, producing the blue light of the sky.

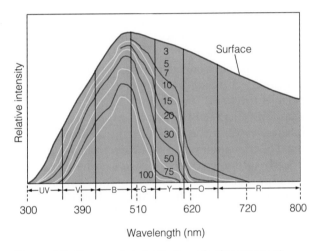

Figure 5.3 The spectral distribution of solar energy at Earth's surface and after it has been modified by passage through varying depths, measured in meters, of pure water. Note how rapidly red wavelengths are attenuated. At approximately 10 m, red light is depleted; but at 100 m, blue wavelengths still retain nearly one-half their intensity.

light is lost with increasing depth. In the clearest of seawater only about 10 percent of blue light reaches to more than 100 m in depth.

5.2 Plant cover intercepts considerable light

Walk beneath a stand of trees in summer and you notice a decrease in light (Figure 5.4a). If you could examine the lowest layer in a grassland or an old field,

you would observe much the same effect. The amount of light that does penetrate a stand of vegetation to reach the ground varies with both the quantity and position of the leaves.

The amount of light at any depth in the canopy is a function of the number of leaves above. As you move down through the canopy, the number of leaves above you increases, so the amount of light decreases. However, because leaves vary in size and shape, the number of leaves is not the best measure of quantity.

The quantity of leaves, or foliage density, is generally expressed in terms of leaf area. Because most leaves are flat, the leaf area is the surface area of one or both sides of the leaf. When the leaves are not flat, the entire surface area is sometimes measured. To quantify the changes in light environment with increasing area of leaves, we need to define the area of leaves per unit ground area (m^2 leaf area/m^2 ground area). This measure is the **leaf area index,** or LAI for short. A leaf area index of 3 (LAI = 3) would mean that there are three square meters of leaf area over each one square meter of ground area. The greater the leaf area index above any surface, the lower the quantity of light reaching that surface. This relationship can be seen in Figure 5.5. As you move from the top of the canopy to the ground in a forest stand, the cumulative leaf area and leaf area index increase. Correspondingly, there is a decrease in light. The general relationship between available light and leaf area index is described by Beer's law (see Quantifying Ecology 5.1: Beer's Law and the Attenuation of Light).

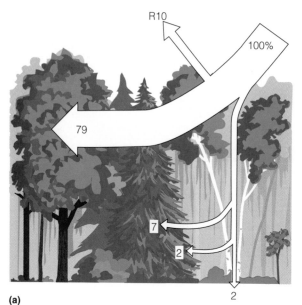

(a)

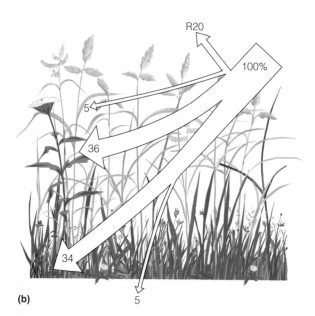

(b)

Figure 5.4 Thinning of light by the plant canopy. (a) A boreal mixed forest reflects (R) about 10 percent of the incident photosynthetically active radiation (PAR) from the upper crown, and it absorbs most of the remainder within the crown. (b) A meadow reflects 20 percent of the photosynthetically active radiation from the upper surface. The middle and lower regions, where the leaves are most dense, absorb most of the rest. Only 2 to 5 percent of PAR reaches the ground.

The arrangement of leaves on plants influences the attenuation of light with increasing leaf area. Densely grown plants with angled leaves and a high LAI will intercept more total light than those with horizontal leaves (Figure 5.6), because horizontal leaves shade each other. Thus leaf angle influences the vertical distribution of light through the canopy as well as the total amount of light absorbed and reflected.

Although light decreases downward through the forest canopy, some direct sunlight does penetrate openings in the crown and reaches the forest floor as sunflecks. Sunflecks can account for 70 to 80 percent of solar energy reaching the forest floor. Sunflecks and skylight enable many plants of the forest floor to endure the shaded conditions on the ground.

Only about 1 to 5 percent of the light that strikes the canopy of a typical temperate deciduous forest (LAI = 3–5) in the summer reaches the forest floor. More light travels through a stand of pine trees (LAI = 2–4)—about 10 to 15 percent. In a tropical rainforest (LAI = 6–10), only 0.25 to 2 percent gets through. Relatively open woodlands, with trees such as birch and oaks, allow light to filter through. There light attenuates gradually throughout the canopy. Likewise, in grassland, where the top of the canopy is relatively open, the middle the lower layers intercept the most light (see Figure 5.4b).

In many environments seasonal changes strongly influence leaf area. For example, in the temperate regions of the world many forest tree species are deciduous, shedding their leaves during the winter months. In these cases, the amount of light that penetrates a stand of vegetation varies with the season (Figure 5.7). In early spring in temperate regions, when leaves are just expanding, 20 to 50 percent of the incoming light may reach the forest floor. Spring-flowering plants take advantage of this flood of light, completing the reproductive phase of their life cycle before the canopy closes. When less than 10 percent of the light reaches the forest floor, flowering is over. In fall, when the leaves begin to drop, increased light again reaches the forest floor, and another surge of flowering, involving goldenrods and asters, takes place.

5.3 The light a plant receives affects its photosynthetic activity

Because plants live in a variable light environment, their photosynthetic activity changes too (Figure 5.8). The light level at which the rate of carbon dioxide uptake in photosynthesis is equal to the rate of carbon dioxide loss due to respiration is called the **light compensation point.** At that point photosynthesis proceeds so slowly that it produces only enough carbon compounds to meet the needs of respiration. As light levels exceed the light compensation point, photosynthetic

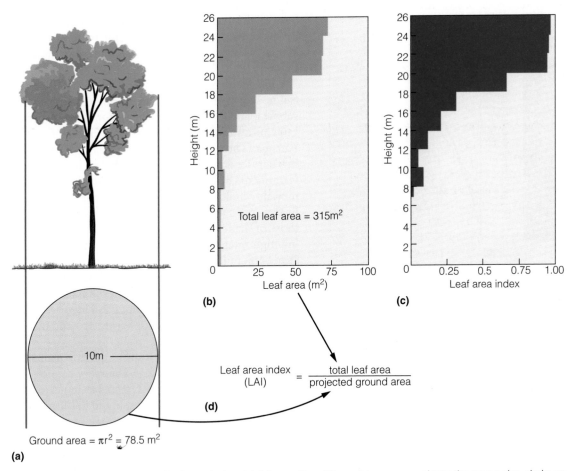

Figure 5.5 The concept of the leaf area index. (a) A tree with a 10-m wide crown projects the same size circle on the ground. (b) The foliage density of the crown at various heights above the ground. (c) The contributions of layers in the crown to the leaf area index. (d) Calculation of LAI. The total leaf area is 315 m². The projected ground area is 78.5 m². The LAI is 4.

rates increase. Eventually the light level no longer limits the rate of photosynthesis. Other factors, such as water (Chapter 7), temperature (Chapter 6), and nutrients (Chapter 9) become the limiting agents. At this point the curve levels off (see Figure 5.8).

The light level at which a further increase in light no longer results in an increase in the rate of photosynthesis is the **light saturation point.** In some plants adapted to extremely shaded environments, photosynthetic rates decline as light levels exceed saturation. This negative effect of high light levels is called **photoinhibition.**

5.4 Species of plants are adapted to either high or low light

The relationship between the availability of light and the rate of photosynthesis varies among plants (Fig-

ure 5.9). Plants adapted to shade tend to have a lower light compensation point, a lower light saturation point, and a lower maximum rate of photosynthesis than plants adapted to high light environments.

These differences, in part, relate to lower concentrations of the photosynthetic enzyme rubisco (Chapter 3) in shade-grown plants. Plants must expend a large amount of energy and nutrients to produce rubisco. Because low light, not the availability of rubisco to catalyze the fixation of CO_2, limits the rate at which photosynthesis can proceed, the plant produces less rubisco. Production of chlorophyll, the light-harvesting pigment in the leaves, increases. The reduced cost of producing rubisco and other compounds allows a lower rate of respiration. It is this lower rate of respiration that gives the lower light compensation point. However, this same reduction in enzyme concentrations limits the maximum rate at which photosynthesis can

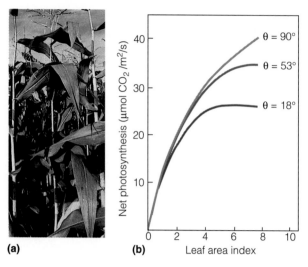

(a)   **(b)**

Figure 5.6 (a) The sharply angled leaves of corn plants with their high LAI allow dense planting. (b) The relationship of photosynthetic capacity of a barley plant to leaf angle ($\angle$) from the ground surface and to leaf area index.

occur when light is available. Thus it lowers the light saturation point and maximum rate of photosynthesis.

This shift in the characteristics of the light response curve is a tradeoff for the plant. The decreased enzyme concentrations and other reductions in metabolic costs enable the plant to decrease respiration rates and maintain photosynthesis under reduced light levels. This reduced metabolic activity, however, limits the rates of photosynthesis when light levels are high. Conversely, when a plant maintains a high level of enzyme concentrations, the cost of respiration is high. However, the plant can maintain high levels of photosynthesis when more light is available. The high costs of respiration do not allow the plant to continue to maintain a positive photosynthetic rate under low light conditions.

The patterns in photosynthetic response to light described above can occur among plants of the same species grown under different light conditions, and even among leaves on the same plant with different exposures to light. Such changes in the physiology or form of an organism in response to changes in environmental conditions is called **acclimatization.** These differences, however, are most pronounced between species of plants adapted to high and low light environments. These adaptations represent genetic differences in the potential response of the plant species to the light environment. Plant species adapted to high light environments are called **shade-intolerant species** or sun species. Plant species adapted to low

light environments are called **shade-tolerant species** or shade species. The names refer to their relative ability to continue photosynthesis under low light (see Figure 5.9).

Shade tolerance affects survival. These differences can be seen in an interesting experiment involving the survival, under high and low light conditions, of seedlings of a number of tropical rain forest species in Panama. The species fall into two major groups, those that survive equally well in low light and high light environments and those that experience high rates of mortality under shaded conditions (Figure 5.10). Seedlings whose survival rates are unaffected by shading are those of shade-tolerant species, whereas species showing high rates of mortality are shade-intolerant. The shade-tolerant species continued photosynthesis under reduced light. The shade-intolerant species could not meet the demands of respiration and therefore died. The reduced vigor of the seedlings also increased their susceptibility to predation (Chapter 16) and disease (Chapter 17).

The difference between sun and shade plants has practical importance. Look in a nursery catalog and you will find plants keyed by symbols to indicate their adaptation to full sun, partial shade, and full shade (Figure 5.11). A number of horticultural books are devoted to sun gardens and shade gardens. Foresters and landscape designers base a part of their management plans on the shade tolerance or intolerance of plants.

Suppose you do plant a sun plant in the shade. How will the plant respond? Adapted to high light and possessing a high rate of respiration, it tries to compensate for low light conditions. The decrease in photosynthesis per unit leaf area is countered by an increase in leaf area. It does not make more leaves, but thinner leaves. (See Focus on Ecology 5.1: Leaf Morphology and Light.) In addition, more of the carbon that is fixed is allocated to stem growth rather than to leaves and roots in an attempt to reach light. Because these plants allocate so much of their energy to stem and leaves, they rarely flower or fruit. Rapid stem growth and thin leaves increase vulnerability to drought (Chapter 7), and thin cell walls weaken the stems and make the plant susceptible to fungal infections (Chapter 17).

Sun or shade, each type of plant has unique characteristics (Figure 5.12). Sun plants grow best in open places receiving full sunlight. They establish themselves rapidly on disturbed sites such as abandoned fields and roadsides. They produce seeds at an early age, and their seeds tend to be small to aid in dispersal. They carry

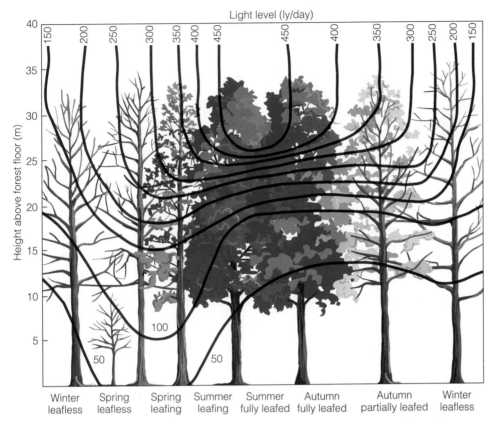

Figure 5.7 Light levels within and above a yellow-poplar *(Liriodendron tulipifera)* stand over a year. The greatest intensity of solar radiation occurs in summer, but the canopy attenuates most of the light, so little reaches the forest floor. Most illumination reaches the forest floor in spring, when trees are still leafless. The forest receives the least radiation in winter with its lower solar elevations and shorter daylengths. As a result, the amount of solar radiation reaching the forest floor is little more than that of midsummer.

on photosynthesis more efficiently in full sunlight and rapidly convert photosynthate to growth.

By contrast, shade plants have lower rates of photosynthesis and respiration. They grow more slowly in all environments. Many shade plants carry on photosynthesis at low light intensities and respond effectively to increased sunlight from sunflecks over a short period of time. Shade-tolerant tree species may live suppressed as seedlings or small trees for many years until the death of larger trees above provides sufficient light to grow to the canopy.

5.5 Aquatic plants live in a shaded environment

Because of the rapid attenuation of light with water depth (see Figure 5.3), most aquatic plants and algae live in the equivalent of a shaded environment. Aquatic plants exhibit the same light response shown in Figure 5.8. However, they more commonly experience inhibition of photosynthesis when exposed to higher light levels in the water column (Figure 5.13).

The depth at which most aquatic plants and algae grow corresponds to the light intensity that they find most favorable for photosynthesis. Aquatic organisms higher in the water column limit light for those below. Phytoplankton species can become photoinhibited at high light intensities at the surface, especially in sunny weather. They achieve highest photosynthetic rates at depths often well below the surface. Some species of phytoplankton are mobile and move up and down through the water column to escape inhibitory effects of high light and to reach the depth at which light intensity is most favorable. Photosynthetic

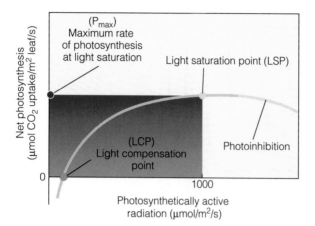

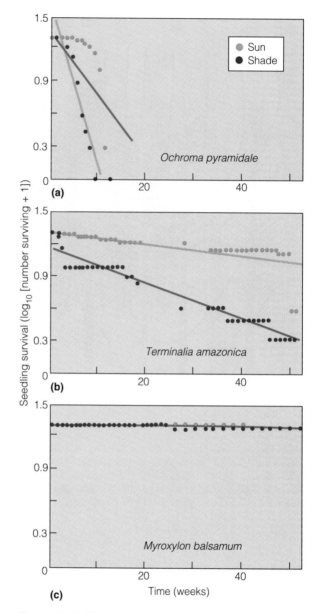

Figure 5.8 Response of photosynthetic activity to available light. The plant increases its rate of photosynthesis as the light level increases up to a maximum rate known as the light saturation point. After this point any increase in PAR results in a decline in photosynthesis or photoinhibition. The light compensation point is the light intensity at which the uptake of CO_2 for photosynthesis equals the loss of CO_2 in respiration.

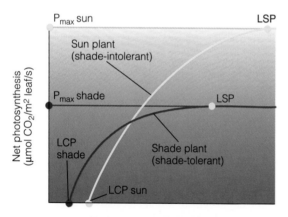

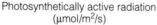

Figure 5.9 Patterns of photosynthetic response to light availability for shade-tolerant and shade-intolerant plants. Shade-tolerant plants have a lower light compensation point and a lower light saturation point than shade-intolerant plants.

Figure 5.10 Seedling survival during one year in sun and shade for three wind-dispersed tree species on Barro Colorado Island, Panama. (a) *Ochroma pyramidale,* a very shade-intolerant species. (b) *Terminalia amazonica,* which is shade-tolerant but survives better in sun than in shade. (c) *Myroxylon balsamum,* a shade-tolerant species that survives in both sun and shade.

ability is further influenced by seasonal variations in water temperature (Chapter 32) and light, and by adaptations for different wavelengths of visible light.

Red, green, and brown algae typically are found in different proportions with increasing depth (Figure

5.14). That pattern has been explained by the different absorptive spectra of their pigments. This theory holds that red algae are typical of the deepest water because they contain pigments that absorb green and blue-green wavelengths common there. Green algae,

QUANTIFYING ECOLOGY 5.1

BEER'S LAW AND THE ATTENUATION OF LIGHT

Equations can make a complicated picture easy to grasp. For instance, to describe the reduction, or attenuation, of light through a stand of plants we can use Beer's law:

$$AL_i = e^{-LAI_i * k}$$

The subscript i refers to the vertical height of the canopy. For example, a value of $i = 5$ refers to a height of 5 m above the ground. The value AL_i is the light reaching any vertical position i in the stand, expressed as a proportion of the light reaching the top of the plants (a value from 0 to 1.0); e is the natural logarithm; LAI_i is the leaf area index above height i; k is the light extinction coefficient. The light extinction coefficient is a measure of the degree to which leaves absorb and reflect light.

For the stand of yellow-poplar in Figure 5.7, we can construct a curve describing the available light at any height in the canopy. In Figure A, the light extinction coefficient has a value of $k = 0.6$. We label vertical positions from the top of the canopy to ground level on the curve. Knowing the amount of leaves (LAI) above any position in the canopy, we can use the equation to calculate the amount of light there.

The light levels and rates of light-limited photosynthesis for each of the vertical canopy positions are shown in the curve in Figure B. Light levels are expressed as a proportion of values for fully exposed leaves at the top of the canopy. As you move from

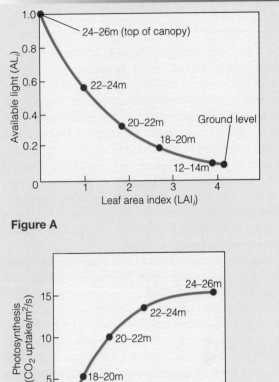

Figure A

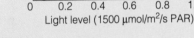

Figure B

the top of the canopy downward, the amount of light reaching the leaves and the corresponding rate of photosynthesis decline.

---

with chlorophyll *a* and *b* similar to that of terrestrial vegetation, mainly inhabit shallow water. Brown algae, with chlorophyll *a* and *c* and a special carotenoid pigment, fucoxanthin, have been considered the seaweeds of intermediate depths. Recent studies have shown that these seaweeds respond primarily to light intensity rather than spectral composition. Red algae do not have an exclusive claim to the deepest water and green algae may grow in greater abundance than red at depths of 90 to 100 m. The vertical distribution of these seaweeds in the water are more variable than first thought.

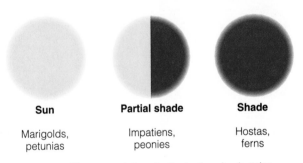

Figure 5.11 Nursery catalogs indicate the shade tolerance of annual, perennial, and woody plants they sell with these or similar symbols.

LEAF MORPHOLOGY AND LIGHT

The light under which a plant grows influences the morphology (size and shape) of its leaves. We find this effect both among individuals of the same species growing under different light conditions, and among leaves within the same plant. As the figure shows here, the leaves of a single red oak *(Quercus rubra)* tree vary in size and shape from the top to the bottom of the tree. Leaves at the top of the tree receive higher levels of solar radiation and experience higher temperatures than leaves at the bottom. Upper leaves are smaller and more lobed. The lobes increase the surface area of the leaf in contact with the air. This design lets more heat dissipate through convection (Chapter 6).

Leaves from the bottom of the canopy are larger, thinner, and much less lobed. Thinner leaves allow for a greater surface area of leaf per unit (weight) of carbon and other nutrients needed to construct the leaf. The amount of CO_2 that a plant can take up for photosynthesis and use for growth and maintenance depends upon both the rate of photosynthesis per unit of leaf area and the total leaf area over which photosynthesis

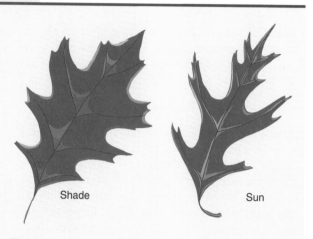

Shade   Sun

Figure A

occurs. The larger, thinner leaves partially compensate for the lower rates of photosynthesis per unit leaf area at lower levels of light. They increase the surface area for capturing light—the limiting resource.

(a)   (b)

Figure 5.12 (a) Wild-flowers, growing in open fields, are sun plants. (b) Woodland ferns are highly shade-tolerant. In midsummer ferns may be the only leafy ground plants visible in the forest.

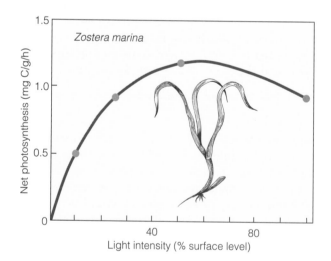

Figure 5.13 Net photosynthesis versus light intensity for the eelgrass *Zostera marina,* which grows in shallow estuarine and coastal waters. Note the inhibition of photosynthesis at higher light intensities.

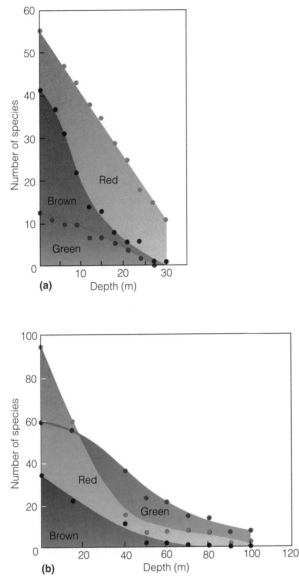

(a) Depth (m)

(b) Depth (m)

Figure 5.14 Variations in the distribution of red, green, and brown algae with depth. (a) In New England waters red algae show a gradual decline in the number of species with depth. Brown algae are more abundant at shallow depths than at greater depths. Green algae have more species adapted to light conditions in deeper waters. (b) In the clear tropical waters of the Caribbean, all three types of algae reach much greater depths. Below 50 m, however, green algae dominate.

5.6 Plants have evolved defenses against ultraviolet radiation

Of considerable ecological significance is solar ultraviolet radiation reaching Earth. Most of it is absorbed by ozone in the stratosphere. Solar UV-B radiation de-

creases with latitude from the tropics, where the ozone layer is the thinnest, to the poles, where the layer is the thickest. Solar UV-B radiation is greater at higher altitudes, increasing about 14 to 18 percent per 1000 m elevation. In recent years the stratospheric ozone layer has been diminishing as a result of increasing concentrations of human-made ozone-destroying chemicals such as chlorofluorocarbons. This depletion is most pronounced over the poles and the tropics. It allows an increase of ultraviolet radiation, especially UV-B radiation, to reach Earth's surface.

Increased UV-B radiation may have more impact on animals than on most plants. Especially vulnerable are light-pigmented organisms, especially humans, who are susceptible to ultraviolet light-induced skin cancer. Ultraviolet radiation accounts for 70 percent of the skin cancer in the United States, and the incidence of skin cancer has been increasing in most parts of the world. It is estimated that for every 1 percent decrease in the stratospheric ozone layer, there is a 1.4 percent increase in cancer-causing UV-B radiation.

Increases in ultraviolet radiation can also have an impact on plants. Laboratory and greenhouse experiments show that UV-B radiation can damage DNA, partially inhibit photosynthesis, alter the growth form, and reduce plant growth. These damaging effects, however, have not been clearly demonstrated for plants growing in the field.

Plants have evolved an array of defenses to inhibit UV-B radiation from reaching the interior of leaves. The major barriers to UV-B penetration are the epidermal (surface) cells of leaves. They contain certain substances that absorb UV-B radiation, yet transmit photosynthetically active radiation to the interior of the leaf.

Plants exhibit a wide range in their ability to screen ultraviolet radiation. Tropical and alpine plants, naturally exposed to high levels of ultraviolet radiation, more effectively block UV-B radiation than temperate species.

Faced with an increase of UV-B radiation, we know little about its effects. Research concentrated on crop plants indicates that about one-half of the three hundred species are sensitive to UV-B radiation. We know little about its effects on forest tree species and other native plants. We do not have a clear understanding of plant defenses against UV-B radiation. Can plants repair damage done to the interior of the leaf? Is resistance to UV-B radiation inherited? Will radiation cause inheritable genetic changes? Will UV-B radiation give a selective advantage (Chapter 19) to resistant species and thus alter plant diversity? These important questions need answers.

CHAPTER REVIEW

SUMMARY

Light: An Overview (5.1–5.3) Light has a major effect on almost all living organisms. Visible light, that part of the electromagnetic spectrum between the wavelengths of 400 to 700 nm, is known as photosynthetically active radiation (PAR). Short wavelengths between 280 and 380 nm are ultraviolet light; wavelengths longer than 740 nm are infrared. In addition to its spectral qualities, light also possesses intensity, duration, and directionality, all of which vary daily and seasonally. Light impinging on an object may be reflected, absorbed, or transmitted through it. Plants reflect green light most strongly and absorb violet, blue, and red wavelengths used in photosynthesis. Plants transmit the wavelengths they do not reflect or absorb **(5.1)**. Light passing through a canopy of vegetation or through water becomes attenuated. Certain wavelengths drop out before others. On a clear day in a forest, green and far red wavelengths pass through relatively unaltered. In pure water, red and infrared light are absorbed first, followed by yellow, green, and violet; blue penetrates the deepest **(5.2)**.

The density and orientation of leaves in a plant canopy influence the amount of light reaching the ground. Foliage density is expressed as leaf area index (LAI), the area of leaves per unit of ground area. The amount of light reaching the ground in terrestrial vegetation varies with the season. In forests only about 1 to 5 percent of light striking the canopy reaches the ground. Sunflecks on the forest floor enable plants to endure shaded conditions. The amount of light reaching a plant influences its photosynthetic rate. The light level at which the rate of carbon dioxide uptake in photosynthesis is equal to the rate of carbon dioxide loss due to respiration is called the light compensation point. The light level at which a further increase in light no longer produces an increase in the rate of photosynthesis is the light saturation point **(5.3)**.

Shade Tolerance and Intolerance (5.4–5.5) Plants are either sun plants (shade-intolerant) or shade plants (shade-tolerant). Each group is characterized by adaptations to a certain light regime. Shade-adapted plants have low photosynthetic, respiratory, metabolic, and growth rates. Sun plants generally have higher photosynthetic, respiratory, and growth rates, but have lower survival rates under shaded conditions. A plant's reaction to shade is reflected in its leaves. The leaves of some species of plants change structure and shape in response to light conditions. Leaves in sun tend to be smaller, lobed, and thick. Shade leaves tend to be larger and thinner **(5.4)**.

Aquatic plants live in a shaded environment because of the rapid depletion of red and far red light in the water. Shading by aquatic plants also reduces PAR. Phytoplankton can become photoinhibited by high light intensities at the surface. They achieve maximum photosynthesis at some depth below the surface. Large marine algae grow at depths where light is most favorable for photosynthesis **(5.5)**.

Impact of Ultraviolet Radiation (5.6) Depletion of the ozone layer in the stratosphere by human-introduced pollutants allows increased penetration of ultraviolet light in the wavelengths from 280 to 315 nm, known as UV-B. This narrow band of wavelengths can stunt the photosynthetic abilities and growth of plants as well as cause skin cancer in animals. Many plants, especially those of the tropics and alpine regions, possess defenses against ultraviolet light **(5.6)**.

STUDY QUESTIONS

1. What is photosynthetically active radiation?
2. What is leaf area index (LAI)? How does it relate to light penetration of the canopy?
3. How does the process of photosynthesis respond to varying levels of light?
4. Contrast light saturation point and light compensation point.
5. Contrast shade-tolerant and shade-intolerant plants.
6. How does light influence the size and shape of leaves?
7. Why is the aquatic environment shaded?
8. Why are levels of UV-B radiation reaching the earth's surface increasing?
9. What effect are increasing levels of UV-B radiation having on plant and animal species?

C H A P T E R   6

TEMPERATURE

OBJECTIVES

On completion of this chapter, you should be able to:

- Discuss the ways in which heat transfers between organisms and their environment.
- Describe the thermal balance of organisms.
- Discuss the metabolic and physiological means plants use to meet extremes of heat and cold.
- Distinguish among poikilothermy, homeothermy, and heterothermy.
- Explain how animal behavior, structure, and metabolism maintain body temperature.
- Distinguish among hibernation, aestivation, and torpor.
- Tell how hyperthermia can help some mammals tolerate heat.
- Describe the physiology of cold tolerance in some animals.
- Explain countercurrent circulation and its adaptive value.

A West African dwarf crocodile *(Osteolammus tetraspis)* basks on a log at the edge of a rain forest river.

Temperature has a pervasive influence on life. It affects rates of photosynthesis and energy storage in plants. It influences the need for moisture and the rates of chemical reactions in all living organisms. It is a key to climate, microclimate, and distribution of organisms. How do organisms adapt to their varied thermal environment? That question is explored in this chapter.

6.1 All organisms live in a thermal environment

All organisms live in and exchange energy with a thermodynamic environment—a world of heat and cold. They absorb solar radiation, which may be direct, diffused from the sky, or reflected from the ground, as well as thermal radiation from rocks, soil, vegetation, and the atmosphere (Figure 6.1). In addition, organisms produce heat during metabolic processes such as respiration and lose heat as infrared radiation. To maintain a constant body temperature, organisms must both lose heat to and gain heat from the environment.

Green plants convert a considerable amount of incoming solar radiation to chemical energy through photosynthesis. They store this energy and pass it on to animals when consumed as food. Other radiation is converted directly to heat. Whether produced by metabolism or absorbed, excess heat must be dissipated to the environment.

The transfer of heat between organisms and environment takes place in several ways. An important transfer is **evaporation,** the loss of moisture from a surface. Evaporation releases heat, the energy needed to convert water to vapor. Evaporation depends upon the difference in vapor pressure between the air and the object as well as the resistance of the surface to the loss of moisture (see Section 4.6). If the humidity of the air is high, little water evaporates and little heat is lost. As air becomes drier, the rate of evaporation and thus the rate of heat dissipation increases. You experience evaporative cooling when you step out of warm water into cooler air, or when you sweat. To dissipate heat by evaporation, an organism needs to take in water.

Heat also transfers between two solid objects, moving from the warmer object to the cooler one. This method of heat transfer is called **conduction.** How fast heat moves depends upon the degree of contact between the two objects, their temperature difference, and the resistance of the organism to heat transfer. The greater the temperature difference, the thinner the conducting layer, and the greater the conductivity (the ability to transfer heat), the faster heat moves. When you sit on a cold stone or walk across the hot sand on a beach, you experience heat loss or gain by conduction.

Convection takes place when fluids (air or water) move over an object. Heat transfers more rapidly by convection than by conduction for a given temperature difference. When you stand in front of a rotating fan, you cool yourself by convection.

A fourth important method of heat transfer is the emission of long-wave radiation, or **thermal radiation.** Radiant energy striking a surface is absorbed and then converted into heat. You experience that type of heat transfer when you stand in front of a fire.

When the surrounding or ambient temperature is lower than the temperature of the organism, the organism loses heat to the environment. When the surrounding temperature is higher, heat moves from the environment to the organism. The problem, then, is for the organism to balance heat gains with heat losses to maintain a constant internal temperature.

The heat balance of an organism may be summarized by the following expression:

Heat gain (solar radiation + thermal radiation + food energy storage + conduction + convection) = heat loss (thermal radiation + conduction + convection + evaporation)

6.2 Plant-temperature interactions can be complex

Most plants are fixed in place. Because they cannot move to a more favorable situation, they experience a wide range of temperatures (Figure 6.2). For example, in early spring in the northeastern United States, temperatures about the lower stems of clover plants 3 cm above the ground may be 21° C. Below the surface the temperature of the soil about the roots may be −1° C. Over a distance of 9 cm, the temperature range is 22° C.

To complicate the picture, an individual plant has many leaves, buds, and twigs. Some of these parts may be fully exposed to the sun, while others are shaded by twigs and leaves above. Because of these different exposures and different tolerances to heat and cold, some leaves, buds, and twigs may succumb to environmental extremes while others survive.

Plant metabolism contributes little to the internal temperature of the plant. Leaf temperature is influenced by the incident thermal radiation and, if the air temperature is warmer than the plant, by convection.

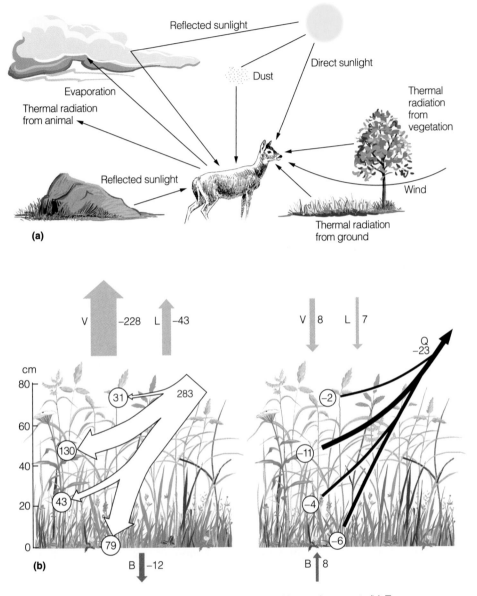

Figure 6.1 (a) Exchange of energy between a deer and its environment. (b) Energy exchange—absorption and emission—in a meadow on a sunny day (left) and at night (right). (Values are cal/cm².) The upper layer in the day absorbs 45 percent of net radiation. The lower layer absorbs 28 percent. By day 80 percent of the radiant energy goes to evaporation of water (V), 15 percent to heat convection (L), and 5 percent to raise soil temperature (B). At night net radiation (Q) as well as heat exchange is reversed. (Fig. 6.1b adapted from A. Cernusca, "Energy Exchange Within Individual Layers of a Meadow," *Oecologica* 23 (1976): 148, Fig. 1. Reprinted by permission of Springer-Verlag.)

Leaves absorb both short-wave (solar) and long-wave (thermal) radiation. Plants reflect some of this solar radiation and emit long-wave radiation back to the atmosphere. The difference between the radiation a plant receives and that which it radiates back is the net energy balance of the plant (R_n). Of the net radiation absorbed by the plant, some is used in photosynthesis and stored in plant biomass. This quantity is quite small, less than 5 percent of R_n. The remaining energy heats the leaves and the surrounding air. On a clear sunny day the amount of energy plants absorb can raise internal leaf temperatures 10° to 20° C above ambient. Internal

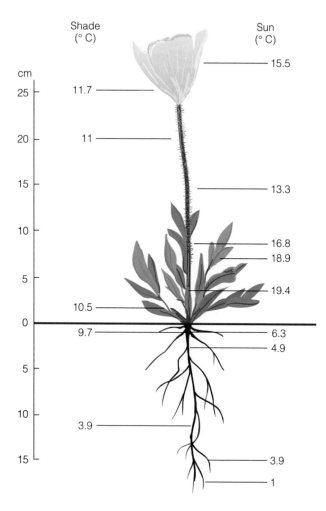

Figure 6.2 Temperature variation in a plant. Data are for the arctic plant *Novosieversia glacialis* on a sunny day with an air temperature of 11.7° C. Note the wide range in temperatures above and below ground.

leaf temperatures may go beyond the optimum for photosynthesis and possibly beyond critical levels.

How much heat a plant gains depends upon the reflectivity of leaves and bark, the orientation of the leaves to incoming radiation and wind, and the size and shape of its leaves. Surfaces perpendicular to the sun's rays absorb more heat than those lying at some other angle. Leaves sharply angled to 70° intercept little midday sun. Instead they intercept most of their solar radiation during the morning and evening hours. Some plants change the orientation of their leaves during the day relative to the sun's direction and intensity. Mature trees, big cacti, and other plants with large stem diameters experience higher temperatures on the sunny side than on the shaded side.

Terrestrial plants lose absorbed heat by convection and evaporation, and aquatic plants primarily by convection. Convective loss depends upon the difference between the temperature of the leaf and the sur-

rounding air. If the temperature of the leaf is higher than the surrounding air, the leaf loses heat to the air moving over it. Evaporation occurs during transpiration. As plants transpire water from their leaves to the surrounding atmosphere through the stomata, the leaves lose energy and their temperature declines.

Because plants lose heat by convection and evaporation, the size and shape of leaves are important. Deeply lobed leaves, like those of some oaks (see Focus on Ecology 5.1), and small, compound leaves, like those of black locust *(Robinia pseudoacacia)*, lose heat more effectively than broad, unlobed leaves. They expose more surface area per volume of leaf to the air for exchange of heat.

Temperatures within an individual leaf may vary because the edges can cool faster than the center. As a result, leaf margins may collect dew or experience frost damage while the midportion of the leaf is unaffected. Large trunks of trees have a higher temperature than ambient air because of heat storage and low conductance. This feature is exploited by birds and mammals that use cavities in trees for nesting and shelter.

6.3 Photosynthesis is temperature-sensitive

Leaf temperature relates to the ability of plants to carry on photosynthesis, which is temperature-sensitive. Both photosynthesis and respiration increase with temperature, and the net uptake of CO_2 by the leaf is a balance between these two processes (Sections 3.3–3.4). As Figure 6.3 illustrates, above some minimum temperature ($T_{minimum}$) photosynthetic rates increase with rising temperature. Eventually photosynthesis levels off at some optimum temperature ($T_{optimal}$) and then declines as temperatures continue to rise. At this point heat damages enzymes and proteins, breaking down the photosynthetic process. Eventually a temperature is reached at which the photosynthesis rate is zero ($T_{maximum}$).

The range of temperatures over which photosynthesis occurs ($T_{minimum}$ to $T_{maximum}$) and optimal temperature ($T_{optimal}$) vary both among species and among populations of a species growing in different temperature environments. In general, the temperature response of plants is closely related to the temperature of the environment in which they are found. This pattern suggests a high degree of acclimation. Nevertheless, there are inherent differences in the temperature responses of species characteristic of different environments. These differences are most pronounced between plants using the C_3 and C_4 photosynthetic pathways (Section 3.4). C_4 plants inhabit warmer, drier environments and exhibit higher optimal temperatures for photosynthesis (generally between 30° and 40° C) than do C_3 plants (Figure 6.4).

6.4 Plants have metabolic adaptations to heat and cold

Extremes of heat and cold kill plants by damaging tissues, inactivating enzymes, and denaturing proteins. Plants, however, have evolved ways to deal with both heat and cold.

Plants can tolerate subzero (−0° C) temperatures if the temperature decreases slowly, allowing ice to form in the cell wall. The effect is dehydration, which the plant can reverse when the temperature rises. If the temperature falls too rapidly for dehydration to take place, ice crystals form within the cell. The crystals may punc-

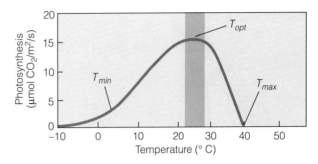

Figure 6.3 The relationship of photosynthetic rates to temperature. Here the optimal temperature is between 20° and 30° C.

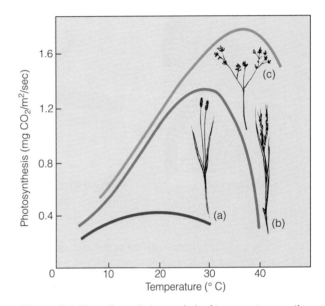

Figure 6.4 The effect of change in leaf temperature on the photosynthetic rates of C_3 and C_4 plants. (a) A C_3 plant, the north temperate grass *Sesleria caerulea,* exhibits a decline in the rate of photosynthesis as the temperature of the leaf increases. (b) A C_4 north temperate grass, *Spartina anglica.* (c) A C_4 shrub of the North American hot desert, Arizona honeysweet, *Tidestromia oblongifolia.* The C_4 species increase their rate of photosynthesis as the temperature of the leaf increases, up to a certain point.

ture cell membranes. When the plant thaws, the cell contents spill out, giving it a cooked, wilted appearance.

Cold tolerance is mostly genetic (Chapter 19), varying among species and among separated populations of the same species. In seasonally changing environments, plants develop frost hardening through the fall and achieve maximum hardening in winter (Figure 6.5). Plants acquire frost hardiness—the turning of cold-sensitive cells into hardy ones—through the formation or addition of protective compounds in the cell. They

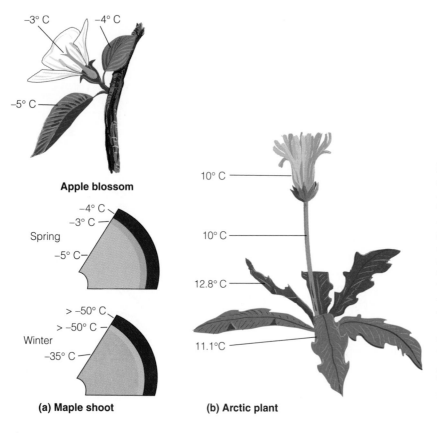

Apple blossom

Spring

Winter

(a) Maple shoot

(b) Arctic plant

Figure 6.5 (a) Frost resistance in an apple blossom in spring and in a maple shoot in spring and winter. The temperatures given are for 50 percent damage. The most sensitive tissue is shown in yellow. (b) Arctic plants live in a colder environment than temperate plants and their tolerance to heat is much lower. The temperatures given are those which will cause 50 percent injury to that plant structure. Note that the rosette leaves are more resistant than the shoot and the flower. (From W. Larcher, *Physiological Plant Ecology,* 2nd ed. (New York: Springer Verlag, 1980), pp. 46–49. Copyright © 1980 Springer-Verlag. Reprinted by permission.)

synthesize and distribute antifreeze substances, such as sugars, amino acids, and nontoxic protective compounds, that allow the **supercooling** of cell sap for short periods of time. Once growth starts in spring, plants lose this tolerance quickly, becoming susceptible to frost damage in late spring.

Insulation further increases resistance to freezing. Some species of arctic and alpine plants and very early-blooming species of temperate regions possess hairs that may act as heat traps and prevent cold injury. Cushion-type or ground-hugging leaves also trap heat. Their temperature may be 20° C higher than the surrounding air.

Most temperate plants succumb to heat damage at tissue temperatures of 44° to 50° C. Heat kills a plant by disturbing its protein metabolism and disrupting cell membrane functions. Heat resistance is lowest during main periods of growth. When exposed to heat, plants can initiate adaptive processes in cells, especially at midday in summer, when a short-term increase in heat resistance takes place. The plant loses this resistance by evening. During the growing season plants rely on transpiration to cool the leaves. Thus a plant's thermal tolerance relates to water availability.

6.5 Animals maintain temperature differently

Thermal relations for animals differ significantly from those for plants. The animal can produce heat by metabolism and can move. The structure of its body also makes a difference.

Consider a thermal model of an animal body (Figure 6.6). The deep interior or core of the body possesses a uniform temperature. The temperature of the environment surrounding the animal's body is variable. A thin layer of air called the **boundary layer** lies at the surface, just above and within hair, feathers, and scales. Therefore body surface temperature differs from both the air and core body temperatures. Separating the body core from the body surface are layers of muscle tissue and fat, across which the temperature gradually changes from the core temperature to the body surface temperature.

To maintain core body temperature the animal has to balance gains and losses of heat to the environment. It does so by changes in metabolic rate and by conduction. The core area exchanges heat, produced by metabolism and stored in the body, with the sur-

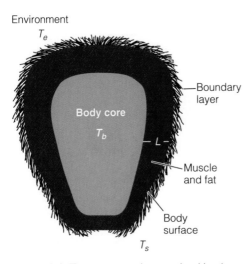

Figure 6.6 Temperatures in an animal body. Body core temperature is T_b, the environmental temperature is T_e, the surface temperature is T_s, and L is the thickness of the outer layer of the body.

face area by conduction. This exchange is influenced by the thickness and conductivity of fat and by the movement of blood near the surface. The surface layer exchanges heat with the environment by convection, conduction, radiation, and evaporation, all influenced by the characteristics of skin and body covering.

Physiology and environment heavily influence how animals confront thermal stress. Because air has a lower specific heat and absorbs less solar radiation than water, terrestrial animals face more radical and dangerous changes in their thermal environment than aquatic animals. Incoming solar radiation can produce lethal heat. Radiational loss of heat to the air, especially at night, can result in deadly cold. Aquatic animals live in a more stable energy environment, but they have a lower tolerance for temperature changes.

6.6   Animals fall into three physiological groups

Physiologically, animals can be divided into three groups, according to the way they maintain temperature.

One group, notably birds and mammals, relies primarily on stored energy to keep constant internal temperatures independent of external temperatures. This internal heat production is **endothermy,** meaning "heat from within." These animals are **homeotherms** (from *homeo,* "the same"). They are popularly called "warm-blooded."

A second group controls body temperature by external means. They gain heat through exposure to environmental sources and dissipate heat through conduction, convection, and evaporation. This means of maintaining body temperature is **ectothermy,** meaning "heat from without." The body temperature is variable. These animals are **poikilotherms** (from *poikilos,* "manifold" or "variegated"). They are often called "cold-blooded" because they are cool to the touch; actually, their body temperatures may get quite warm. These animals include invertebrates, amphibians, fish, and reptiles.

A third group regulates body temperature by endothermy at some times and by ectothermy at other times. These animals are **heterotherms** (from *hetero,* "different"). Heterotherms employ both endothermy and ectothermy, depending upon environmental situations and metabolic needs. Bats, bees, and hummingbirds belong to this group.

The terms *homeotherm* and *endotherm* are often used synonymously, as are *poikilotherm* and *ectotherm,* but there is a difference. *Ectotherm* and *endotherm* emphasize the mechanisms that determine body temperature. The other two terms, *homeotherm* and *poikilotherm,* represent the nature of body temperature.

6.7   Poikilotherms depend on environmental temperatures

Poikilotherms, such as amphibians, reptiles, and insects, gain heat easily from the environment and lose it just as fast. Environmental temperatures control the rates of metabolism and activity among most poikilotherms. Rising temperatures increase the rate of enzymatic activity, which controls metabolism and respiration. For every 10° C rise in temperature, the rate of metabolism in poikilotherms doubles (see Focus on Ecology 6.1: Body Size and Metabolism). They become active only when the temperature is sufficiently warm. Conversely, when ambient temperatures fall, metabolic activity declines, and poikilotherms become sluggish. Poikilotherms have an upper and lower thermal limit that they can tolerate. Most terrestrial poikilotherms can maintain a relatively constant daytime body temperature by behavioral means, such as seeking sunlight or shade. Lizards and snakes, for example, may vary their body temperature by no more than 4° to 5° C (Figure 6.7) and amphibians by 10° C when active. The range of body temperatures at which poikilotherms carry out their daily activities is the **operative temperature range.**

BODY SIZE AND METABOLISM

A close relationship exists between body size and metabolic rate. Basal metabolism (the rate when the body is resting) is proportional to body mass raised to the ¾ power. As body weight increases, the weight-specific metabolic rate decreases. Conversely, as body mass decreases, the weight-specific metabolic rate increases at the same rate. Within any given taxonomic group, small animals have a higher metabolism and require more food per unit of body weight than large ones (Figure A). Small shrews (*Sorex* spp.), for example, require daily an amount of food (wet weight) equivalent to their own body weight. It is as if a 150-pound human needed 150 pounds of food daily to stay alive. Therefore small animals are forced to spend most of their time seeking and eating food.

Part of the reason lies in the ratio of surface area to body mass (Figure B). A body loses heat to the environment in proportion to the surface area exposed. All other things being equal, large animals lose less heat to the environment than do small ones. To maintain homeostasis, small animals have to burn energy rapidly. The weight-specific rates of small endotherms rise so rapidly that below a certain size, they could not meet their energy demands. On the average, 5 g is about as small as an endotherm can be and still maintain a metabolic heat balance. A few are even smaller. The pigmy white-toothed

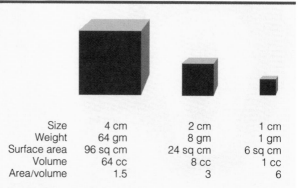

Size	4 cm	2 cm	1 cm
Weight	64 gm	8 gm	1 gm
Surface area	96 sq cm	24 sq cm	6 sq cm
Volume	64 cc	8 cc	1 cc
Area/volume	1.5	3	6

Figure B These three cubes—with sides of 4, 2, and 1 cm—point out the relationship between surface area and volume. A small object has more surface area in proportion to its volume than a large object of similar shape.

shrew *(Suncus etruscus)* of Africa, for example, weighs 2 g as an adult. Because of the conflicting metabolic demands of body temperature and growth, most young birds and mammals are born in an altricial state—blind, naked, and helpless—and begin life as ectotherms. They depend upon the body heat of their parents to maintain their body temperature. That allows young animals to allocate most of their energy to growth.

A similar relationship between size and metabolism exists among ectotherms. Among ectotherms, standard metabolic rates are a function of prevailing environmental temperatures and body size. Ectotherms smaller than 0.1 kg potentially experience a wide range of body temperatures that approach the daily range of temperature. They can heat up and cool down quickly. Small insects, for example, cannot elevate their body temperatures above air temperature because of convective heat loss, but can control their body temperature by movement to different microclimates. Ectotherms smaller than 1 kg can heat up quickly to temperatures at which they can be active. This ability is especially important for these ectotherms when the thermal environment is suitable for only a short period of time. When temperatures become extreme, they can retreat and cool down just as quickly.

Ectotherms larger than 10 kg are too large to lose heat or gain heat rapidly, because of a small body surface-to-volume ratio. Large ectotherms, especially large reptiles, retain some of their body heat. Losing

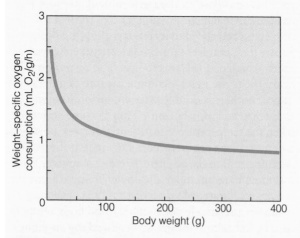

Figure A The weight-specific consumption of oxygen, a measure of metabolic rate, per gram of body weight. Small animals have a much higher metabolism than large animals.

heat slowly, large ectotherms are active over a longer period of the day and in more extreme environments.

Because they do not depend upon internally generated heat to maintain body temperature, larger ecto-therms, such as reptiles and amphibians, have a resting metabolic rate only 10 to 20 percent of birds. During periods of inactivity their temperatures drop to ambient, with a corresponding reduction in metabolic rate.

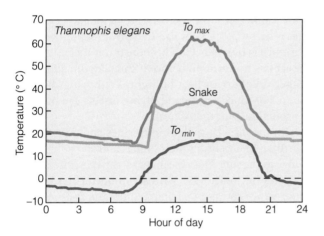

Figure 6.7 Daily temperature variation in the western terrestrial garter snake *Thamnophis elegans,* within its operative temperature range (T_o). Note that the snake maintains a fairly constant temperature during the day.

Biologists determine operative temperatures for an animal by measuring the temperature of an inanimate object, typically made of copper, of the same size, shape, and radiative properties as the animal and exposed to the same microclimates. Measurements on such a model represent the true environmental temperatures experienced by the animal and the maximum and minimum range of temperatures within which they can operate.

Poikilotherms have a low metabolic rate and a high thermal conductance between body and environment. Most of their energy production—50 to 98 percent—is from anaerobic respiration, in which oxygen is not used. This process depletes stored energy and accumulates lactic acid in the muscles. (Anaerobic respiration takes place in the muscles of marathon runners and other athletes, causing leg cramps.) Anaerobic respiration metabolism limits poikilotherms to short bursts of activity. It results in rapid physical exhaustion, often within 3 to 5 minutes.

Aquatic poikilotherms, completely immersed, do not maintain any appreciable difference between their body temperature and the surrounding water. Aquatic poikilotherms are poorly insulated. Any heat produced in the muscles moves to the blood and on to the gills and skin, where heat transfers to the surrounding water by convection. Because seasonal water temperatures are relatively stable, fish and aquatic invertebrates maintain a fairly constant seasonal temperature. They exhibit a low range of temperature variation in any given season.

Fish and aquatic invertebrates adjust seasonally to changing temperatures by acclimatization. They undergo physiological changes over a period of time. Poikilotherms have an upper and lower limit of tolerance to temperature that varies with the species. If they live at the upper end of their tolerable thermal range, poikilotherms will adjust their physiology, at the expense of being able to tolerate the lower range. Similarly, during periods of cold, the animals shift to a lower temperature range that would have been lethal before. Because water temperature changes slowly through the year, aquatic poikilotherms can make the adjustments slowly. Fish are highly sensitive to rapid change in environmental temperatures. If they are subjected to a sudden temperature change, they will die of thermal shock.

Nevertheless, being ectothermic has advantages. Poikilotherms can allocate more of their energy intake to biomass production than to metabolic needs. Because they do not depend upon internally generated body heat, ectotherms can curtail metabolic activity in times of food and water shortage and temperature extremes. Their low energy demands enable some terrestrial poikilotherms to colonize areas of limited food and water. Because they do not have the problem of metabolic heat loss, they are not restricted to any minimum body size or definite shape. Such characteristics enable poikilotherms to exploit resources and habitats unavailable to homeotherms. On the other hand, the same metabolic restrictions seem to impose an upper size limit. Ectotherms would not be able to absorb enough heat to warm a very large body. This fact has spurred debate over whether the large dinosaurs were ectothermic or endothermic.

6.8 Homeotherms escape the thermal constraints of the environment

Homeothermic birds and mammals meet the thermal constraints of the environment by being endothermic. They maintain body temperature by oxidizing glucose and other energy-rich molecules. They regulate the gradient between body and air or water temperatures by seasonal changes in insulation (the type and thickness of fur, structure of feathers, and layer of fat), which poikilotherms do not possess. They rely on evaporative cooling and on increasing or decreasing metabolic heat production. Homeothermy allows these animals to remain active regardless of environmental temperatures, although at high energy costs.

Maintenance of a high body temperature is associated with specific enzyme systems that operate within a high temperature range, with a set point about 40° C (Section 2.2). Homeotherms generally have a high metabolic rate and a low thermal conductance. Because efficient cardiovascular and respiratory systems bring oxygen to their tissues, homeotherms can maintain a high level of aerobic energy production. Thus, they sustain a high level of physical activity for long periods. What exhausts them is heat generated by the activity. Independent of external temperatures, homeotherms can exploit a wider range of thermal environments. They can generate energy rapidly when the situation demands, escaping from predators or pursuing prey.

The major disadvantage of homeothermy is the high cost of metabolism. It leaves a minimum of energy for biomass accumulation. A high resting metabolism is not compatible with a low-energy way of life. As a result, metabolic costs weigh heavily against smaller homeotherms, and place a lower limit on body size. (See Focus on Ecology 6.1.)

6.9 Heterotherms may or may not regulate body temperature

Species that sometimes regulate their body temperature and sometimes do not are called heterotherms. At different stages of their daily and seasonal cycle or in certain situations, these animals take on the characteristics of endotherms or ectotherms. They can undergo rapid, drastic, repeated changes in body temperature.

Insects are ectothermic and poikilothermic; yet in the adult stage most species of flying insects are heterothermic. When flying, they have high rates of metabolism, with heat production as great as or greater than homeotherms. They reach this high metabolic state in a simpler fashion than do homeotherms, because they are not constrained by cardiovascular and pulmonary pumping systems. Insects take in oxygen by demand through openings in the body wall and transport it throughout the body.

Temperature is critical to the flight of insects. Most cannot fly if the temperature of the thoracic muscles is below 30° C; nor can they fly if the muscle temperature is over 44° C. This constraint means that an insect has to warm up before it can take off, and it has to get rid of excess heat in flight. With wings beating up to 200 times per second, flying insects can produce a prodigious amount of heat.

Some insects, such as butterflies and dragonflies, warm up by orienting their bodies and spreading their wings to the sun. Most warm up by shivering their flight muscles in the thorax. Moths and butterflies vibrate their wings to raise thoracic temperatures above ambient. Bumblebees pump their abdomens without any external wing movements. They do not maintain any physiological set point, and they cool down to ambient temperatures when not in flight.

6.10 Torpor helps some animals conserve energy

To reduce metabolic costs during periods of inactivity, a number of small homeothermic animals become heterothermic and enter into daily torpor. **Daily torpor** is the dropping of body temperature to approximately ambient temperature for a part of each day, regardless of season.

A number of birds, such as hummingbirds (Trochilidae) and poorwills (*Phalaenoptilus nuttallii)*, and small mammals, such as bats, pocket mice, kangaroo mice, and even white-footed mice, experience daily torpor. Such daily torpor seems to have evolved as a means of reducing energy demands over that part of the day in which the animals are inactive. Nocturnal mammals, such as bats, go into torpor by day; and diurnal animals, such as hummingbirds, go into torpor by night. As the animal goes into torpor, its body temperature falls steeply. With the relaxation of homeothermic responses, the body temperature declines to within a few degrees of ambient. Arousal returns the body temperature rapidly to normal as the animal renews its metabolic heat production.

To escape the rigors of long cold winters, many terrestrial poikilotherms and a few heterothermic mammals go into a long seasonal torpor, called **hibernation.** Hibernation is characterized by the cessation of activity. Hibernating poikilotherms experience such physiological changes as decreased blood sugar, increased liver glycogen, altered concentration of blood hemoglobin, altered carbon dioxide and oxygen content in the blood, altered muscle tone, and darkened skin.

Hibernating homeotherms become heterotherms and invoke controlled hypothermia. They relax homeothermic regulation and allow the body temperature to approach ambient temperature. Heart rate, respiration, and total metabolism fall, and body temperature sinks below 10° C. Associated with hibernation are high levels of CO_2 in the body and acid in the blood. This state, called acidosis, affects cellular processes, lowers the threshold for shivering, and reduces the metabolic rate. Hibernating homeotherms, however, are able to rewarm spontaneously using only metabolically generated heat.

Entrance into hibernation is a controlled process difficult to generalize from one species of homeotherm to another. Some hibernators, such as the groundhog *(Marmota monax)*, feed heavily in late summer to lay on large fat reserves, from which they will draw energy during hibernation. Others, like the chipmunk *(Tamias striatus)*, lay up a store of food instead. All hibernators, however, acquire a metabolic regulatory mechanism different from that of the active state. Most hibernators rouse periodically and then drop back into torpor. The chipmunk, with its large store of seeds, spends much less time in torpor than species that store large amounts of fat.

Although popularly said to hibernate, black, grizzly, and female polar bears do not. Instead they enter a unique winter sleep, from which they easily rouse. They do not enter extreme hypothermia. The bears do not eat, drink, urinate, or defecate; yet they maintain a metabolism that is near normal. To do so, the bears recycle urea through the bloodstream. The urea is degraded into amino acids that are reincorporated in plasma proteins.

Hibernation provides selective advantages to small homeotherms. For them, maintaining a high body temperature during periods of cold is too costly. It is far less expensive to reduce metabolism and allow the body temperature to drop. To do so eliminates the need to keep warm and seek scarce food.

6.11 Animals exploit microclimates to regulate temperature

To maintain a tolerable and fairly constant body temperature during active periods, terrestrial and amphibious poikilotherms resort to behavioral means. They seek out appropriate microclimates (Figure 6.8). Insects such as butterflies, moths, bees, dragonflies, and damselflies bask in the sun to raise their body temperatures to the level necessary to become highly active.

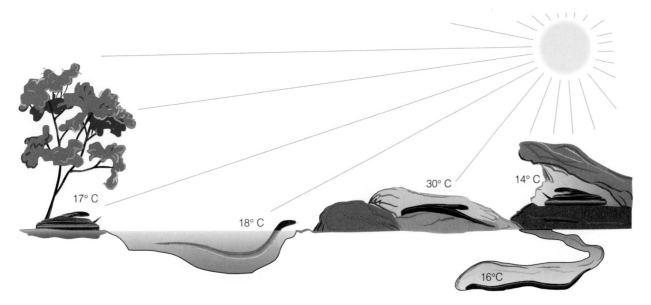

Figure 6.8 Microclimates a snake typically uses to regulate body temperature during the summer.

When they become too warm, these animals seek the shade. Salamanders, restricted to moist, shaded habitats, lack this option, but semiterrestrial frogs, such as bullfrogs and green frogs, exert considerable control over their body temperature. By basking in the sun, frogs can raise their body temperature as much as 10° C above ambient temperature. Because of associated evaporative water losses, such amphibians must either be near or partially submerged in water. By changing position or location or by seeking a warmer or cooler substrate, amphibians can maintain body temperatures within a narrow range.

Reptiles have their own response to temperature. Most are terrestrial and exposed to widely fluctuating temperatures. The simplest way for a reptile to raise body temperature is to bask in the sun. Snakes, for example, heat up rapidly in the morning sun (Figure 6.9). When they reach the preferred temperature, the animals move on to their daily activities, retreating when necessary to the shade to cool. In this manner they maintain a stable body temperature during the day. In the evening the reptile experiences a slow cooling. Its body temperature at night depends upon its location.

Lizards raise and lower their bodies and change body shape to increase or decrease the conduction of heat between themselves and the rocks or soil on which they rest. They also seek sunlight or shade or burrow into the soil to adjust their temperatures. Desert beetles, locusts, and scorpions exhibit similar behavior. They raise their legs to reduce contact between their

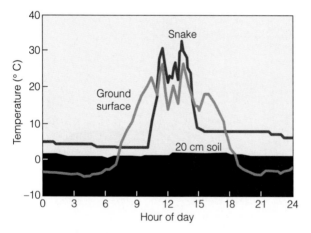

Figure 6.9 A snake warms up quickly in the morning, achieves a temperature plateau during the day, and cools off slowly in the evening to its nighttime temperature, which depends upon ambient temperature (T_0).

body and the ground, minimizing conduction and increasing convection by exposing body surfaces to the wind. Thus body temperatures of poikilotherms do not necessarily follow the general ambient temperature.

Endotherms also use microclimates to keep warm or cool. In the heat of a summer day, birds and mammals seek shady places. Desert mammals go underground by day and emerge at night. In winter some mammals, such as rabbits, go underground during periods of inactivity. Large mammals such as deer seek the thermal cover of conifers and rhododendron thickets. Mammals such as flying squirrels and birds such as penguins and quail huddle together during periods of cold, reducing individual surface area and conserving body heat.

6.12 Insulation reduces heat exchange

To regulate the exchange of heat between the body and the environment, homeotherms and certain poikilotherms use some form of insulation—a covering of fur, feathers, or body fat. For mammals, fur is a major barrier to heat flow, but its insulation value varies with thickness, which is greater on large mammals than on small ones. Small mammals are limited in the amount of fur they can carry, because a thick coat could reduce their ability to move. Mammals change the thickness of their fur with the season. Aquatic mammals, especially of arctic regions, and such arctic and antarctic birds as auklets and penguins have a heavy layer of fat beneath the skin. Birds reduce heat loss by fluffing the feathers and drawing the feet into them, making the body a round feathered ball. Some arctic birds, such as ptarmigan, have feathered feet; among most birds the feet are scaled and act as thermal windows, sites of heat loss.

Although the major function of insulation is to keep body heat in, it also keeps heat out. In a hot environment, an animal either has to rid itself of excess body heat or prevent heat from being absorbed in the first place. One means is to reflect solar radiation from light-colored fur or feathers. Another means is to grow a heavy coat of fur that heat does not penetrate. Large mammals of the desert, notably the camel, employ this method. The outer layers of hair absorb heat and return it to the environment.

A number of insects, notably moths, bees, and bumblebees, have a dense, furlike coat over the thoracic region, which serves to retain the high temperature of flight muscles during flight. The long, soft hairs of caterpillars, together with changes in body posture, act as insulation, reducing convective heat exchange.

When insulation fails, many animals resort to shivering, a form of involuntary muscular activity that increases heat production. Many species of small mammals increase heat production by **nonshivering thermogenesis**—burning highly vascular brown fatty tissue with a high rate of oxygen consumption. Brown fat, found about the head, neck, thorax, and major blood vessels, occurs in two groups of mammals: cold-acclimated adults, such as ground squirrels, and hibernators, such as bats and groundhogs.

6.13 Evaporative cooling in animals is important

Many birds and mammals, and even wasps and hornets, employ evaporative cooling to reduce the body heat load. Birds and mammals lose some heat by evaporation of moisture from the skin. They accelerate evaporative cooling by sweating and panting. Only certain mammals have sweat glands, particularly horses and humans. Panting in mammals and gular fluttering in birds increase the movement of air over moist surfaces in the mouth and pharynx. Many mammals, such as pigs, wallow in water and wet mud to cool down. Paper wasps maintain a rather constant temperature in their paper nest by fanning their wings. However, because of its gray color, the nest warms rapidly in the sunlight. If the nest temperature becomes excessive, over 35° C, wasps gather water from nearby sources, carry it to the nest, and fan their wings to speed evaporative cooling.

6.14 Some animals use unique physiological means for thermal balance

Storing body heat does not seem like a sound option to maintain thermal balance in the body, because of an animal's limited tolerance for heat. However, certain mammals, especially the camel, oryx, and some gazelles, do just that. The camel, for example, stores body heat by day and dissipates it by night, especially when water is limited. Its temperature can fluctuate from 34° C in the morning to 41° C by late afternoon. By storing body heat, these animals of dry habitats re-

duce the need for evaporative cooling and thus reduce water loss (see Section 7.11).

Many poikilothermic animals of temperate and arctic regions withstand long periods of below-freezing temperatures in winter through supercooling and developing a resistance to freezing. **Supercooling** of body fluids takes place when the body temperature falls below the freezing point without actual freezing. The presence of certain solutes in the body influences the amount of supercooling that can take place. Some arctic marine fish, certain insects of temperate and cold climates, and reptiles exposed to occasional cold nights employ supercooling.

Some intertidal invertebrates of high latitudes and certain aquatic insects survive the cold by freezing and then thawing out when the temperature moderates. In some species, more than 90 percent of the body fluids may freeze, and the remaining fluids contain highly concentrated solutes. Ice forms outside the shrunken cells, and muscles and organs are distorted. After thawing, they quickly regain normal shape.

Other animals, particularly arctic and antarctic fish, and many insects resist freezing by increasing the solutes, notably glycerol, in body fluids. Glycerol protects against freezing damage, increasing the degree of supercooling. Wood frogs (*Rana sylvatica*), spring peepers (*Hyla crucifer*), and gray tree frogs (*H. versicolor*) can successfully overwinter just beneath the leaf litter because they accumulate glycerol in their body fluids.

6.15 Countercurrent circulation conserves or reduces body heat

To conserve heat in a cold environment and to cool vital parts of the body under heat stress, a number of animals have evolved countercurrent heat exchangers (Figure 6.10). For example, the porpoise (*Phocaena* spp.), swimming in cold arctic waters, is well insulated with blubber. It could experience an excessive loss of body heat, however, through its uninsulated flukes and flippers. The porpoise maintains its body core temperature by exchanging heat between arterial and venous blood in these structures (Figure 6.11). Veins completely surround arteries carrying warm blood from the heart to the extremities. Warm arterial blood loses its heat to the cool venous blood returning to the body core. As a result, little body heat passes to the environment. Blood entering the flippers cools, while blood returning to the deep body warms. In warm waters, where the animals need to get rid of excessive body heat, blood bypasses the heat exchangers. Venous blood returns unwarmed through veins close to the skin's surface to cool the body core. Such vascular arrange-

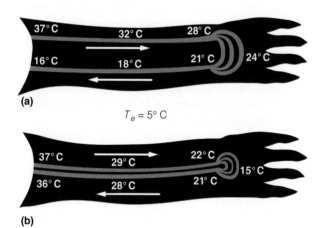

(a)

$T_e = 5° C$

(b)

Figure 6.10 A model of countercurrent flow in the limb of a mammal, showing hypothetical temperature changes in the blood (a) in the absence and (b) in the presence of countercurrent heat exchange.

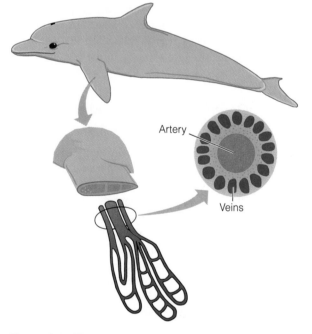

Figure 6.11 The porpoise and its relatives, the whales, use flippers and flukes as a temperature-regulating device. Arteries in the appendages are surrounded by several veins. Venous blood returning to the body core is warmed through heat transfer, retaining body heat.

ments are common in the legs of mammals and birds and the tails of rodents, especially the beaver.

Many animals have arteries and veins divided into small, parallel, intermingling vessels that form a discrete vascular bundle or net known as a **rete**. In a rete the principle is the same as in the blood vessels of the porpoise's flippers. Blood flows in opposite directions, and heat exchange takes place.

Countercurrent heat exchange can also dissipate heat. The oryx *(Oryx besia)*, an African desert antelope exposed to high daytime temperatures, and running African gazelles can experience elevated body temperatures yet keep the highly heat-sensitive brain cool by a rete in the head. The external carotid artery passes through a cavernous sinus filled with venous blood cooled by evaporation from the moist mucous membranes of the nasal passages (Figure 6.12). Arterial blood passing through the cavernous sinus cools on the way to the brain, reducing the temperature of the brain to 2° to 3° C lower than the body core.

Countercurrent heat exchangers are not restricted to homeotherms. Certain poikilotherms that assume some degree of endothermism employ the same mechanism. The swift, highly predaceous tuna *(Thunnus* spp.) and the mackerel shark *(Isurus tigris)* possess a rete in a band of dark muscle tissue used for sustained

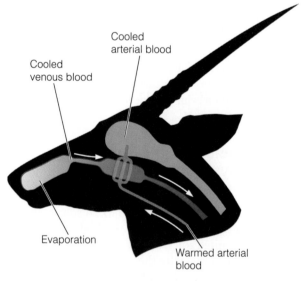

Figure 6.12 A desert gazelle can keep a cool head in spite of a high body core temperature by means of a rete. Arterial blood passes in small arteries through a pool of venous blood cooled by evaporation as it drains from the nasal region.

swimming effort. Metabolic heat produced in the muscle warms the venous blood, which gives up heat to the adjoining newly oxygenated blood returning from the gills. Such a countercurrent heat exchange increases the power of the muscles, because warm muscles contract and relax more rapidly. Sharks and tuna maintain fairly constant body temperatures, regardless of water temperatures.

CHAPTER REVIEW

SUMMARY

Heat Balance (6.1) All organisms live in a thermal environment. They must balance heat inputs and outputs between themselves and their environment. Heat gains come from direct and reflected sunlight, diffuse radiation, long-wave infrared radiation, convection, conduction, and metabolism. Organisms lose heat to the environment by infrared radiation, conduction, convection, and evaporation. Organisms can have a net loss of heat to the environment only when the current environmental temperature is less than body core temperature.

Plants and the Thermal Environment (6.2–6.4)
Plants not only experience a wide range of temperature between their roots and crown; their parts experience different temperatures during the day. Plant metabolism contributes little to internal plant temperature. The internal temperature of all plant parts is influenced by heat gained from and lost to the environment. Shoots absorb long-wave and short-wave radiation. They reflect some of it back to the environment. The difference between the amount reflected and the amount absorbed is the plant's net radiation balance. The plant uses some of the absorbed radiation in photosynthesis. Plants gain some control over heat absorbed by the reflectivity of leaves and bark, and leaf size, shape, and orientation toward the sun. Leaf temperatures affect photosynthesis (6.2). Plants have optimal temperatures for photosynthesis, beyond which photosynthesis declines (6.3).

Plants tolerate extremes of cold by frost hardening—the formation of or addition to protective antifreeze compounds in the cells of shoots, buds, roots, and seeds. Plants may also have insulation or lower the freezing point of their cells through supercooling. Plants are limited in their response to excessive heat, but cells can make short-term adaptive changes (6.4).

Poikilotherms, Homeotherms, and Heterotherms (6.5–6.9) Unlike plants, animals maintain a fairly constant internal body temperature, known as the body core temperature. They use behavioral and physiological means to maintain their heat balance in a variable environment. Layers of muscle fat and surface insulation of scales, feathers, and fur insulate the animal body core against environmental temperature changes. Terrestrial animals face a more changeable and often more threatening thermal environment than do aquatic animals (6.5).

Animals fall into three major groups: poikilotherms, homeotherms, and heterotherms. Poikilotherms, so named because they have variable body temperatures influenced by ambient temperatures, are ectothermic. They depend largely on the environment as a source of heat. Animals that depend upon internally produced heat to maintain body temperatures are endothermic. Because they maintain a rather constant body temperature independent of the environment, they are called homeotherms. Many animals are heterotherms. They function either as endotherms or ectotherms (6.6).

Poikilotherms gain heat from and lose heat to the environment. Poikilotherms have low metabolic rates and high thermal conductance. Environmental temperatures control their rates of metabolism. Poikilotherms are active only when environmental temperatures are warm and are sluggish when temperatures are cool. They have, however, upper and lower limits of tolerable temperatures. Most aquatic poikilotherms do not maintain any appreciable difference between body temperature and water temperature (6.7).

Homeotherms maintain high internal body temperature by oxidizing glucose and other energy-rich molecules. They have high metabolic rates and low thermal conductance. They are able to remain active regardless of environmental temperatures. For homeotherms, a high rate of aerobic metabolism comes at a high energy cost. These costs place a lower limit on body size (6.8).

Depending upon environmental and physiological conditions, heterotherms take on the characteristics of endotherms or ectotherms. Some normally homeothermic animals become ectothermic and drop their body temperature under certain environmental conditions. Many poikilotherms, notably the insects, need to increase their metabolic rate to generate heat before they can take to flight. Most accomplish this feat by vibrating their wings or wing muscles or by basking in the sun. After flight, their body temperatures drop to ambient temperatures **(6.9)**.

Special Animal Responses to the Thermal Environment (6.10–6.15) During environmental extremes some animals enter a state of torpor to reduce the high energy costs of staying warm or cool. They slow their metabolism, heartbeat, and respiration, and lower their body temperature. Birds such as hummingbirds and mammals such as bats undergo daily torpor, the equivalent of deep sleep, without the extensive metabolic changes of seasonal torpor. Seasonal torpor over winter is called hibernation. Hibernation involves a whole rearrangement of metabolic activity to run at a very low level. Heartbeat, breathing, and body temperature are all greatly reduced **(6.10)**.

Both poikilotherms and homeotherms resort to behavioral means to regulate body temperature. Mostly they exploit variable microclimates. Poikilotherms move into warm sunny places to warm and seek shaded places to cool. Many amphibians move in and out of water. Insects and desert reptiles raise and lower their bodies to reduce or increase conductance from the ground or convective cooling. Desert animals resort to shade or spend the heat of day in underground burrows **(6.11)**.

Body insulation of fat, fur, feathers, scales, and furlike covering on many insects reduces heat loss from the body. A few desert mammals employ heavy fur to keep out desert heat and cold. When insulation fails, many homeotherms resort to shivering and to nonshivering thermogenesis, the burning of brown fat **(6.12)**. For homeotherms evaporative cooling by sweating, panting, and wallowing in mud and water is an important way of dissipating body heat **(6.13)**.

Some desert mammals use hyperthermia to reduce the difference between body and environmental temperatures. They store up body heat by day, then by night unload it to the cool desert air. Hyperthermia reduces the need for evaporative cooling and thus conserves water. Some cold-tolerant poikilotherms use supercooling, the synthesis of glycerol in body fluids, to resist freezing in winter. Supercooling takes place when the body temperature falls below freezing without freezing body fluids. Some intertidal invertebrates survive the cold by freezing, then thawing with warmer temperatures **(6.14)**.

Many homeotherms and heterotherms employ countercurrent circulation, the exchange of body heat between arterial and venous blood reaching the extremities. This exchange reduces heat loss through body parts or cools blood flowing to such vital organs as the brain **(6.15)**.

STUDY QUESTIONS

1. How does heat move between organisms and their environment?
2. How do plants build up a tolerance to or adjust to cold? To heat?
3. Distinguish between poikilothermy and homeothermy. What are the advantages and disadvantages of each?
4. What are the relationships among body size, metabolic rate, and temperature regulation?
5. What is acclimatization, and how does it function among poikilotherms, particularly fish?
6. Speculate on how homeothermy might have evolved.
7. What behaviors help poikilotherms maintain a fairly constant body temperature during their season of activity?
8. How can insulation maintain an animal's thermal integrity?
9. How do homeotherms respond when ambient temperatures fall below body temperature? Poikilotherms?
10. How do homeotherms use evaporative cooling?
11. Explain the value of hyperthermia.
12. Compare supercooling in plants and animals.
13. How does countercurrent circulation work, and why is it important?
14. Distinguish between hibernation and torpor. Why is the black bear not a true hibernator?
*15. Consider a population of fish during winter living below a power plant discharging heated water. The plant shuts down for three days. What effect would that have on the fish?
*16. Bats hibernate in winter, becoming ectotherms for the season. But female little brown bats (*Myotis lucifugus*) shut off thermoregulation some 18 days before giving birth to their one young in late spring and during lactation. What is the advantage of such a physiological change? Consider the bat's summer habitat, energy demands, and daily activity patterns.

MOISTURE

OBJECTIVES

On completion of this chapter, you should be able to:

- Explain how the structure of water affects its properties.
- Discuss the ecologically important physical properties of water.
- Describe osmosis and osmotic potential.
- Trace the water cycle and the movement of water through soil.
- Explain how water moves from the soil through the plant to the atmosphere.
- Describe how plants meet the stresses of drought and flooding.
- Discuss how animals respond to environmentally induced moisture stress.
- Tell how plants and animals cope with a saline environment.

A rainstorm over the ocean—a part of the water cycle.

An organism's water balance is closely related to its thermal balance. It is difficult to discuss one without the other. Think of how marathon runners frequently drink water to replace that which they have lost through sweating. Sweating, which allows the evaporation of water, is one way animals reduce body heat generated metabolically during strong physical exertion. Likewise, evaporative cooling is a way in which plants maintain leaf temperatures within a range of tolerance. Terrestrial plants and animals could never maintain their thermal and moisture balance without the unique features of water that make life on Earth possible.

7.1 The structure of water is based on hydrogen bonds

Because of the physical arrangement of its hydrogen atoms and its hydrogen bonds, water is a unique substance. A molecule of water consists of one large atom of oxygen and two smaller atoms of hydrogen, bonded together covalently (Figure 7.1a). **Covalence** is the sharing of an electron. Each hydrogen proton shares its single electron with oxygen, leaving unbonded two pairs of oxygen electrons (Figure 7.1b). The repulsive force between these two pairs and between them and electrons in the O-H bonds pushes the bonds toward each other. Instead of a linear arrangement of the atoms of H-O-H, the arrangement is V-shaped (Figure 7.1a) with an angle of 105°. The shared hydrogen atoms are closer to the oxygen atom than they are to each other.

If the molecules were linear, then the two bonds' polarities—the pull they exert—would cancel each other out. Because of the V-shaped arrangement, the side of the molecule on which the H atoms are located has a positive charge and the opposite side has a negative charge, polarizing the water molecule (Figure 7.1c).

Because of its polarity, each water molecule becomes weakly bonded with its neighboring molecules (Figure 7.1d). The H or positive end of one molecule attracts the negative or opposite end of the other. The 105° angle of association between the hydrogen atoms encourages an open tetrahedral-like arrangement of water molecules. This situation, in which hydrogen atoms act as connecting links between water molecules, is **hydrogen bonding.** The simultaneous bonding of an H atom to two different water molecules accounts for the lattice arrangement of water.

In ice the lattice is complete. Each oxygen atom is hydrogen-bonded to two hydrogen atoms. Each hydrogen atom is covalently bonded to one oxygen atom and is hydrogen-bonded to another oxygen atom (Figure 7.1d). Ultimately, each oxygen atom connects to four other oxygen atoms by means of hydrogen atoms. One such unit built upon the other gives rise to a lattice with large open spaces (Figure 7.1e). Water molecules so structured occupy more space than they would in liquid form. As a result, water expands upon freezing and ice floats.

As ice melts, the hydrogen bonds break and the lattice-work partially collapses. The volume occupied

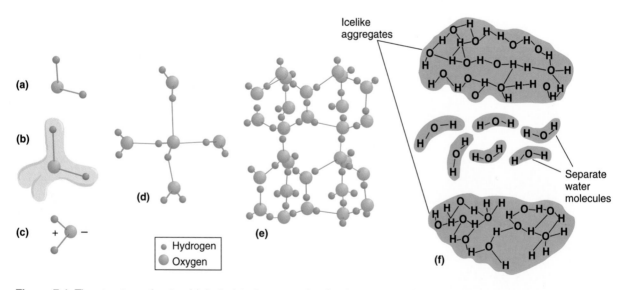

Figure 7.1 The structure of water. (a) An isolated water molecule, showing the angular arrangement of the hydrogen atoms. (b) Lone pairs of electrons in a water molecule. (c) Polarity of water. (d) Hydrogen bonds to one molecule of water in ice. (e) The open lattice structure of ice. (f) The structure of liquid water.

by the water molecules decreases and the density increases, until water achieves its greatest density at 3.98° C. At this temperature, contraction of the molecules, brought about by the partial collapse of the lattice structures, balances normal thermal expansion of warming molecules. As water heats, more hydrogen bonds break, converting the water to a liquid state, which is a mixture of individual and aggregated molecules (Figure 7.1f). Further heating results in a complete breakdown of the hydrogen bonds, separating aggregates of water molecules into individuals. At this point water enters the gaseous state, water vapor.

Seawater behaves somewhat differently. Seawater is a solution of salts dominated by sodium chloride. Although seawater is defined as containing 34.5 grams of salt per 1000 grams of water (34.5 ‰), it can vary in salinity. For this reason salt water has no definitive freezing point. Seawater begins to freeze at about −2° C. As pure water freezes out, the remaining unfrozen water becomes higher in salinity; the density of seawater increases, and its freezing point lowers. Finally a solid block of ice crystals and salt forms.

7.2 Water has important and unique properties

Water has a number of unique properties related to its hydrogen bonds. One property is high **specific heat**—that is, the number of calories necessary to raise one gram of water one degree Celsius. The specific heat of water is defined as a value of 1, and other substances are given a value relative to water. Water can store tremendous quantities of heat energy with a small rise in temperature. It is exceeded in this capacity only by ammonium, liquid hydrogen, and lithium.

Great quantities of heat must be absorbed before the temperature of natural waters, such as ponds, lakes, and seas, rises just 1° C. They warm up slowly in spring and cool off just as slowly in the fall. This behavior prevents wide seasonal fluctuations in the temperature of aquatic habitats, so characteristic of air temperatures, and moderates the temperatures of local and worldwide environments.

Because of the high specific heat of water, large quantities of heat energy must be removed before water can change from a liquid to a solid, and be absorbed before ice can convert to a liquid. Collectively, the energy released or absorbed in the transformation of water from one state to another is called **latent heat.** It takes approximately 80 calories of heat to convert 1 g of ice at 1° C to a liquid state. The same amount of heat would raise 80 g of water 1° Celsius.

Evaporation occurs at the interface between air and water at all ranges of temperature. Here, again, considerable amounts of heat are involved: it takes 536 calories to overcome the attraction between molecules and convert 1 g of water at 100° C into vapor. That is as much heat as is needed to raise 536 g of water 1° C. When evaporation occurs, the source of thermal energy may be the sun, the water itself, or objects in and around it.

Viscosity is the resistance of a liquid to flow. Because of the energy in hydrogen bonds, the viscosity of water is high. Imagine liquid flowing through a glass tube. The liquid behaves as if it consists of a series of parallel concentric layers, flowing over one another. The rate of flow is greatest at the center; because of internal friction between layers, the flow decreases toward the sides of the tube.

Viscosity is the source of frictional resistance to objects moving through water. This frictional resistance of water is 100 times greater than that of air. A mucous coating and streamlined body shape help fish reduce this frictional resistance. Replacement of water in the space left behind by the moving animal adds additional drag on the body. An animal streamlined in reverse, with a short, rounded front and a rapidly tapering body, meets the least resistance in the water. The acme of such streamlining is the sperm whale (*Physeter catodon*).

Within all substances similar molecules are attracted to one another. Water is no exception. Molecules of water below the surface are surrounded by other molecules. The forces of attraction are the same on all sides. At the water's surface there is a different set of conditions. Below the surface molecules of water are strongly attracted to one another. Above is the much weaker attraction between water molecules and air. Therefore, molecules on the surface are drawn downward, resulting in a surface that is taut like an inflated balloon. This condition, called **surface tension,** is important in the lives of aquatic organisms.

The surface of water is able to support small objects and animals, such as the water strider (Gerridae) and water spiders (*Dolomedes* spp.) that run across the pond's surface. To other small organisms, surface tension is a barrier, whether they wish to penetrate the water below or escape into the air above. For some the surface tension is too great to break; for others it is a trap to avoid while skimming the surface to feed or to lay eggs. If caught in the surface tension, a small insect may flounder on the surface. The nymphs of mayflies (*Ephemeroptera*) and caddisflies (*Trichoptera*) that live in the water and transform into winged adults find surface tension a handicap in their efforts to emerge from the water. Slowed down at the surface, these insects become easy prey for fish.

Surface tension is associated with capillary action, or capillarity—the rise and fall of liquids within narrow tubes. Capillarity affects the movement of water in soil and the transport of water to all parts of plants, as we shall see.

7.3   Osmosis and osmotic potential are essential to organisms

It seems paradoxical that in order to gain water from the soil, plants must lose water to the atmosphere. Plants take in water through the roots and lose it through the leaves and shoots in a process called transpiration (see Section 3.3). Transpiration is the evaporation of water from internal surfaces of leaves, stems, and other living parts and its diffusion from the plant into the atmosphere.

For water to move from soil solution into the roots it must pass through the cell membranes and cell walls of the root. Some membranes are permeable; they do not impede the movement of substances. Other membranes are selectively permeable; they allow some substances to pass through them, but not others. They are fairly permeable to water and are more or less permeable to various other substances.

The general tendency of molecules is to move from a region of high concentration to one of low concentration. This movement or diffusion accounts for the spread of a solute (dissolved substance) throughout a solvent (the medium that dissolves the solute). To understand the movement of water and solutes into the root, consider this example. Suppose we enclose a solute such as salt (sodium chloride) in high concentration (and water in low concentration) in a funnel sealed with a semipermeable membrane and lower it into a beaker of distilled water. The membrane is permeable to water but not to salt. The volume of the fluid within the funnel will increase and move up the tube as water moves across the membrane into the solution by a process called **osmosis** (Figure 7.2). Water continues to move across the membrane until the **osmotic pressure** of the solute, decreasing as the solute becomes more diluted by the pure water, is balanced by the physical pressure the fluid exerts on the tube.

Osmotic pressure accounts for the internal pressure (called turgor) plant cells achieve when the water supply is adequate. As plants lose water through transpiration, the concentration of water molecules in the cells decreases, so water, when available, moves from

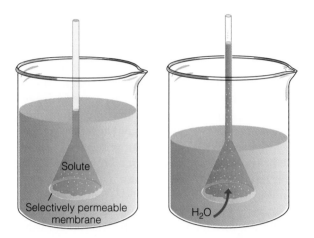

Figure 7.2 Demonstration of osmotic potential. Fluid within the funnel increases in volume and moves up the tube as water passes through the membrane into the solution by osmosis. Water continues to cross the membrane until the osmotic pressure of the solute, decreasing with dilution, is balanced by the gravitational pressure exerted by the fluid in the tube.

the soil solution into the plant (see Section 7.5). The tendency of solutes in a solution to cause water molecules to move from areas of high to low concentration is called **osmotic potential.** The osmotic potential of a solution depends upon its concentration. The higher the concentration of solutes, the lower is its osmotic potential and the greater is its tendency to gain water.

Osmosis and osmotic potential are involved in maintaining fluid balance not only in plants but in all living things from single-cell protozoans to vertebrates (see Section 7.10). They play a particularly important role in freshwater and marine life with different internal solute concentrations from the water around them.

7.4   Water cycles between Earth and the atmosphere

Water is the medium by which elements and other materials make their never-ending odyssey through the ecosystem. Without the cycling of water, decomposition and nutrient cycling could not proceed, ecosystems could not function, and life could not persist.

Solar energy, heating Earth's atmosphere (see Chapter 4) and evaporating water, is the driving force behind the water cycle (Figure 7.3). Precipitation sets the water cycle in motion. Water vapor, circulating in the atmosphere, eventually falls in some form. Some of the water falls directly on the soil and bodies of water. Some is intercepted by vegetation, plant litter on the ground, and urban structures and streets.

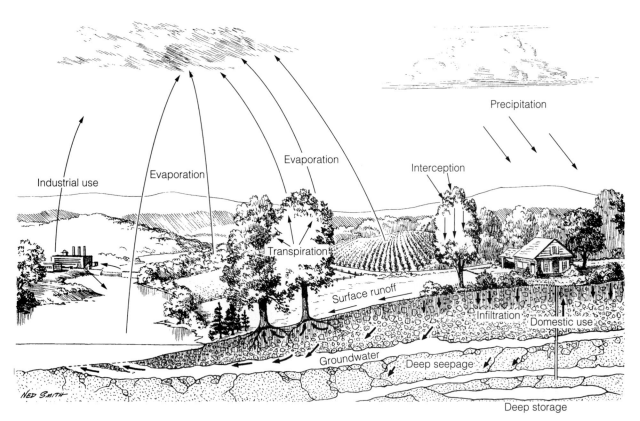

Figure 7.3 The water cycle on a local scale, showing the major pathways of water movement.

Because of interception, which can be considerable, various amounts of water never infiltrate the ground but evaporate directly back to the atmosphere. In urban areas a great portion of rain falls on roofs, sidewalks, roads, and other paved areas, which are impervious to water. Water runs down gutters and drainage ditches, to be directed off to rivers.

Precipitation that reaches the soil moves into the ground by infiltration. The rate of infiltration depends on the soil, slope, vegetation, and characteristics of the precipitation. In the soil, water seeps or percolates through the soil pores. After a heavy rain, the pore spaces fill with water. If the amount of water exceeds what the pore space can hold, we say the soil is saturated (Figure 7.4). Excess moisture drains freely from the soil. If water fills all the pore spaces and is held there by internal capillary forces, the soil is at **field capacity.** Field capacity is generally expressed as the percentage of the weight or volume of soil occupied by water compared to the oven-dried weight of the soil at a standard temperature. The amount of water a soil holds at field capacity varies with the soil's texture—the proportion of sand, silt, and clay (see Chapter 10). Coarse, sandy soil has large pores; water drains through it quickly. Clay soils have small pores and hold considerably more water. Water held between soil particles by capillary forces is **capillary water.**

As plants and evaporation from the soil surface extract capillary water, the amount of water in the soil declines. Eventually it reaches a level at which a plant cannot withdraw water faster than it loses water through its leaves. Soil water has now reached the **wilting point.** Although water still remains in the soil—filling up to 25 percent of the pore spaces—soil particles hold it tightly, making it difficult to extract. If the soil dries even further, the only moisture remaining adheres tightly to soil particles as a thin film. Known as **hygroscopic water,** this film of water is inaccessible to plants.

During heavy rains when the soil is saturated, excess water flows across the surface of the ground as **surface runoff.** At places it concentrates into depressions and gulleys, and the flow changes from sheet to channelized flow, a process you can observe on city streets as water moves across the pavement into gutters. Because of low infiltration, runoff from urban areas might be as much as 85 percent of the precipitation.

Some water entering the soil seeps down to an impervious layer of clay or rock to collect as **groundwater**

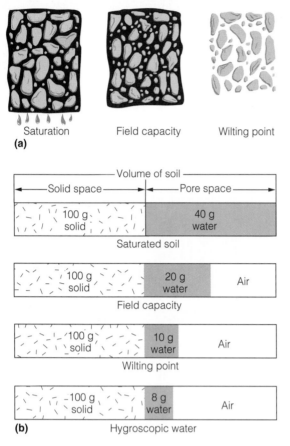

Figure 7.4 (a) Soil particles, water, and air at different moisture levels. (b) The proportions of solids, water, and air at the three levels and when only hygroscopic water remains. (From *Nature and Properties of Soil,* 8th edition, by Nyle C. Brady, © 1974. Reprinted by permission of Prentice-Hall, Inc., Upper Saddle River, NJ.)

(see Figure 7.3). From here water finds its way into springs, streams, and eventually rivers and seas. Humans draw a great portion of groundwater and surface water for domestic and industrial purposes, then discharge it into streams and rivers to reenter the water cycle.

Eventually all of this water returns to the atmosphere. Water evaporates from the surface of the ground and vegetation, as well as from streams, lakes, and oceans. Water in the soil evaporates directly or through plants by transpiration.

7.5 Water moves through plants from soil to the atmosphere

Terrestrial plants depend upon water in the soil. As a plant loses water by transpiration through the leaves, it

has to replace that water by uptake from the soil. To do so plants need a gradient along the pathway between soil and the atmosphere. From our discussion of osmosis, we might expect that the soil, with higher water and lower solutes, and the root, with lower water and higher solutes, would come to equilibrium. They never do. What prevents that equilibrium is a negative pressure gradient from the soil to the atmosphere (Figure 7.5), driven by transpiration. Plants pull water from the soil (where the water pressure is the highest) toward the atmosphere (where water pressure is the lowest).

When the relative humidity of the atmosphere drops below 100 percent, there is less water in the atmosphere than in the plant, where relative humidity is effectively 100 percent. As the atmosphere becomes drier, its capacity to accept water increases. Because of the difference in moisture outside and inside the plant, moisture escapes through the stomata and evaporates (Section 3.3). To replace the lost water, the plant has to transport water from roots to leaves through its conductive system (xylem). As the replacing water moves up, water pressure drops below it, all the way down to the fine rootlets in contact with soil particles and pores. This tension pulls more water from the soil to the root and up through the stem to the leaf. The movement is like sucking water up a straw.

Transpiration continues while the amount of energy striking the leaf is enough to evaporate water, the stomata are open, and moisture is available in the soil. As long as plants can maintain a gradient between roots and soil, they continuously remove water from the soil. Unless precipitation replenishes soil moisture, eventually only hygroscopic water remains.

The rate of water loss varies with daily environmental conditions such as humidity and temperature, and with the characteristics of plants. Opening and closing the stomata is probably the most important regulator of water loss through the plant. The stomatal openings are also entrances for CO_2. Whenever the stomata remain open to take in carbon dioxide, they also allow water to escape. For the plant this is no problem as long as water is available.

7.6 Plants reduce water loss in drought

When the atmosphere or soil is dry, plants respond by partially closing the stomata and opening them for shorter periods of time. In the early period of water stress, a plant closes its stomata during the hottest part of the day. It resumes normal activity in the afternoon. As its water balance deteriorates, the plant opens its

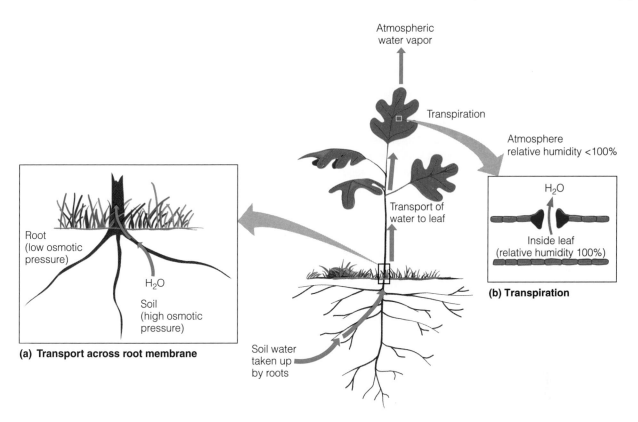

Figure 7.5 Transport of water along a pressure gradient from soil to leaves to air. (a) The solute concentration in the roots is higher than in the soil, so water moves into the roots. At the other end, moisture in the atmosphere is less than that in the leaves. (b) Water moves to the atmosphere through transpiration, decreasing the osmotic pressure within the plant, which now draws more water from the soil.

stomata only in the cooler, more humid conditions of morning. If drought conditions continue, the plant may keep its stomata closed. This closure reduces transpirational loss of water, but it may also raise the internal temperature of the leaf beyond the range of tolerance.

Stomatal closure also reduces CO_2 diffusion into the leaf. As a result the rate of photosynthesis declines. Reduced photosynthesis means less growth and smaller plants. Prolonged drought inhibits the production of chlorophyll, causing the leaves to turn yellow or, later in the summer, to show premature autumn coloration. As conditions worsen, deciduous trees may prematurely shed their leaves, the oldest ones dying first.

Some plants, such as rhododendrons, fold or roll their leaves to conserve water. Other species take on a wilted appearance as leaves lose turgor. Wilting reduces water loss by lessening the amount of solar radiation the leaf intercepts.

Not all water limitations occur in the heat of summer. Conifers and other evergreens can experience winter drought. When winter temperatures are warm enough to thaw the ice in their conductive tissues, the plants lose this water by transpiration. Because the soil is frozen, the plants cannot replace the water. Their needles turn brown and their twigs die back from dehydration.

7.7 Plants of arid regions possess other means of survival

As we move to regions of lower precipitation, we find plants possess adaptations to deal with the lower availability of water. Plants of arid and semiarid regions allocate more photosynthate to the production of roots and less to leaves. They need an extensive root system both to reach water and to gain maximum uptake when rains do fall. A smaller leaf area reduces water loss through transpiration. (See Focus on Ecology 7.1: Water, Roots, and Leaves.)

The angle of leaves to the sun affects a plant's heat load (see Section 6.2). When water is limited, many

FOCUS ON ECOLOGY 7.1:

WATER, ROOTS, AND LEAVES

Terrestrial environments vary widely in the availability of soil water, giving rise to ecosystems as different as deserts and tropical rainforests (Chapters 27–31). Plants may adapt to water availability by changing the relative production of roots and leaves. Plants with less water tend to decrease the production of leaves while increasing the production of roots. The reduction in leaf area decreases the amount of solar radiation the plant intercepts and the amount of water the plant loses by transpiration. In turn, the increased production of roots allows the plant to explore a larger volume of soil from which to extract water. Overall these changes increase the amount of root mass that provides water to the leaves.

Plant ecologists quantify these changes in morphology for plants growing under differing soil conditions. In general, the ratio of root mass to leaf area decreases with wetter conditions (Figure A).

Although an increase in root mass per leaf area improves water balance, it also decreases photosynthesis and growth. The surface area from which the plant loses water is smaller, but so is the surface area over which photosynthesis occurs. The overall uptake of CO_2 declines.

You can notice these changes in plant morphology if you grow plants of the same species under different soil-water conditions. They are most pronounced, however, among contrasting plant species adapted to wet and dry environments.

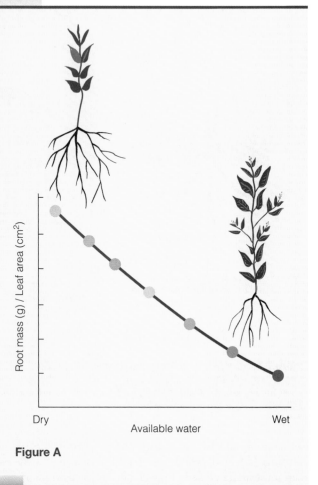

Figure A

species of plants turn leaf surfaces parallel to the sun's rays. This shift in position reduces leaf temperature and transpiration rates. When soil water is adequate, these plants orient their leaves perpendicular to the sun's rays.

In tropical regions with wet and dry seasons, deciduous trees lose their leaves at the onset of the dry season, a change similar to leaf fall in the cold season in temperate regions. Some desert shrubs, such as ocotillo *(Fouquieria splendens)*, shed their leaves four or five times a year and renew them with each rain. Other plants of dry areas possess extremely small leaves or lack leaves altogether. Those plants, such as palo-verde *(Cercidium floridum)*, carry on photosynthesis through green stem tissue rich in chlorophyll. Such

plants have the advantage of being able to start photosynthesis immediately, without a time lag for the formation of new photosynthetic tissue.

Some plants, notably cacti, store water in their tissues. Most of their photosynthate goes into the production of aboveground structures. Cacti have a minimal root system, but because their root systems are shallow, they can draw up considerable amounts of water during periods of rain. The plants swell rapidly as they store water in enlarged vacuoles (fluid-filled membrane-bounded structures) in the cells. During periods of drought, the cacti draw on this store of water.

Desert plants that employ the CAM type of photosynthesis close their stomata during the day. They open them at night to take in and fix CO_2 for use during the day (see Chapter 3). For this reason these plants experience minimal loss of water through transpiration.

Plants also endure aridity by means of leaf adaptations. In some plants the leaves are small, the cell walls are thickened, the stomata are tiny, and the vascular system for transporting water is dense. Many species have leaves covered with hairs that scatter incoming solar radiation and increase the depth of the boundary layer—a thin layer of still air close to the leaf surface. Others have leaves coated with waxes and resins that reflect light and reduce its absorption. All these structural features reduce the amount of heat energy striking the leaf and thus the loss of water through transpiration.

Root adaptations provide still other means of maintaining water balance. Plants of riverine and streamside habitats and deep depressions in arid and semiarid regions escape drought by tapping the deep groundwater supply. Known as **phreatophytes,** these plants have roots in constant contact with a fringe of capillary water just above the water table. They may send roots to a depth of 60 to 80 m to reach water, although 10 to 30 is more typical. Examples of such woody plants are cottonwood *(Populus deltoides)*, willows *(Salix* spp.), salt cedar *(Tamarix* spp.), and mesquite *(Prosopis* spp.). Such root systems enable them to secure the maximum amount of water when moisture is available.

Some plants adopt an ephemeral life cycle in which the population survives the dry period as dormant seeds ready to sprout quickly when the rains come. They complete the active stages of their life cycle—germination, growth, and production of seeds—within a short period before moisture is gone and then return to dormancy as seeds. The cycle may be based on annual wet and dry seasons; or the seeds may remain dormant for years, waiting for adequate moisture. Seeds that do not have a protective coat are subject to dehydration. To germinate successfully, they need to absorb large quantities of water. If they do germinate when water is inadequate, the seedlings may die early.

7.8 Plants respond to flooding in various ways

Too much water can place as much stress on plants as too little water. Plant species differ in their ability to deal with the stresses of flooding and waterlogged soils. Symptoms of flooding in plants intolerant of such conditions are similar to those of drought, including stomatal closure, yellowing and premature loss of leaves, wilting, and rapid reduction in photosynthesis. The causes, however, are different.

Growing plants need both sufficient water and a rapid gas exchange with their environment. Much of this exchange takes place between the roots and the soil atmosphere. When water fills soil pores, such gas exchange cannot take place. The plants, experiencing depressed O_2 levels, in effect asphyxiate; they drown. Roots depend upon gaseous O_2 to carry on aerobic respiration. Lacking oxygen, flooded roots are forced to shift to an alternative anaerobic metabolism. Further, anaerobic (oxygen-depleted) conditions in the soil allow chemical reactions that produce substances toxic to plants.

In response to these anaerobic conditions, some species of plants accumulate ethylene in their roots. Ethylene gas, a growth hormone, is highly insoluble in water. It is produced normally in small amounts in the roots. Under flooded conditions ethylene diffusion from the roots slows and oxygen diffusion into the roots virtually stops. Ethylene then accumulates to high levels in the root. Ethylene stimulates adjacent cells in the cortex of the root to self-destruct and separate to form interconnected gas-filled chambers called **aerenchyma** (Figure 7.6). These chambers, typical of aquatic plants, allow some exchange of gases between submerged and better aerated roots.

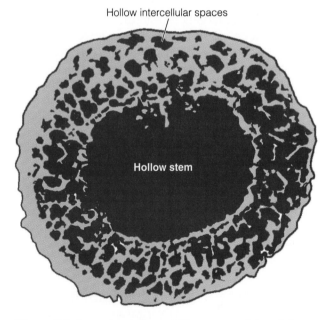

Hollow intercellular spaces

Hollow stem

Figure 7.6 Aerenchyma tissue with numerous intercellular spaces carries oxygen to the roots of aquatic plants.

In other plants, especially woody ones, the original roots, deprived of oxygen, die. In their place adventitious roots (roots that arise in positions where roots normally would not grow) emerge from the submerged part of the stem. Replacing the functions of the original roots, they spread horizontally along the oxygenated zone of the soil surface. Some plants found in poorly drained soils—for example, red maple (*Acer rubrum*) and white pine (*Pinus strobus*)—develop shallow, horizontal root systems to cope with flooding. These shallow root systems make them highly susceptible to drought and windthrow.

Prolonged flooding often causes dieback or death, particularly among woody species. The response of trees to flooding depends upon the season, duration, movement of water, and the species of tree. Floodplain trees and bottomland hardwoods are highly tolerant of seasonal, temporary inundation. Trees are injured more by standing in stagnant water than in oxygenated flowing water. Most trees die if their roots are flooded for more than half the growing season. For this reason you see standing dead trees above beaver dams and along highways where road construction has impeded soil drainage.

Plants associated with waterlogged and flooded soils have evolved certain adaptations. Most have gas-filled chambers called aerenchyma through which oxygen diffuses from the shoots to the roots (see Figure 7.6). In plants such as waterlilies, the aerenchyma ventilates the entire plant by taking air into the young leaves and discharging it from the old leaves. Interconnected gas spaces throughout the leaves and roots occupy nearly 50 percent of the plant tissue. Some plants of cold, wet, alpine tundra have similar spaces in leaf, stem, and root to carry oxygen to the roots.

Few woody species can grow on permanently flooded sites. Among them are baldcypress (*Taxodium distichum*), mangroves, willows, and water tupelo (*Nyssa aquatica*). Baldcypresses growing on sites with fluctuating water tables develop knees or **pneumatophores,** specialized growths of the root system (Figure 7.7). These growths may be beneficial, although they are not necessary for survival. Pneumatophores on mangrove trees help with gas exchange and provide oxygen to roots during tidal cycles (see Chapter 37).

Figure 7.7 Pneumatophores or "knees" are typical features of a baldcypress swamp.

7.9 Plants of saline habitats have physiologically dry environments

Plants of salt marshes and other saline habitats grow in a physiologically dry environment. Salinity limits the amount of water they can absorb (see Chapter 37). Known as **halophytes,** they take in water that contains high levels of solutes. Characteristically, they accumulate high levels of ions within their cells, especially in the leaves. That concentration, which may equal or exceed that of seawater, allows halophytes to maintain a high cell water content in the face of a low external osmotic potential. In addition, some plants have an internal osmotic pressure many times higher than freshwater and terrestrial plants. Other plants secrete a heavy waxy substance called cutin on their leaves to reduce water loss and are succulent. Succulent plants store water in their leaves.

Taking up water heavy in sodium and chloride, some halophytes dilute it with water they have stored in their tissues. Some plants have salt-secreting glands that deposit excess salt on the leaves to be washed away by rain. Others remove salts mechanically at the root membranes.

Plants of saline deserts encounter worse problems. Not only do they grow in salty soil; they must also endure dry conditions. A number of desert plants allow only certain ions to pass across root membranes and keep others out. That selectivity allows these plants to absorb essential nutrients from the soil and maintain osmotic pressure.

Halophytes, however, vary in their degree of tolerance to salt. Some plants, such as salt marsh hay grass (*Spartina patens*), grow best at low salinities. Others, such as salt marsh cordgrass (*S. alternifolia*), do best at moderate levels of salinity. A few, such as glasswort (*Salicornia* spp.), tolerate high salinities (see Chapter 37).

7.10 Animals have ways of maintaining water balance

Animals have a more complex and more energy-expensive problem than plants in maintaining water balance. All animals, however, possess a more or less universal mechanism, the excretory system. The system is simple in some animals and complex in others.

Remember that osmotic pressure moves water through cell membranes from the side of greater water concentration to the side of lesser water concentration. Aquatic organisms living in fresh water have a higher salt concentration in their bodies than in the surrounding water. Their problem is to prevent uptake or to rid themselves of excess water. Protozoans accomplish that task by means of contractile vacuoles, which collect and expel wastes. Freshwater fish maintain osmotic balance by absorbing and retaining salts in special cells in the body and by producing copious amounts of watery urine. Amphibians balance the loss of salts through the skin by absorbing ions directly from the water and transporting them across the skin and gill membranes. They store water from the kidneys in the bladder. If circumstances demand it, they can reabsorb the water through the bladder wall.

Terrestrial animals have three major means of gaining water and solutes: directly by drinking and eating, and indirectly by producing metabolic water in respiration (see Section 3.6). They lose water and solutes through urine, feces, evaporation from the skin, and respiration. Birds and reptiles have a salt gland and a cloaca, a common receptacle for the digestive, urinary, and reproductive tracts. They reabsorb water from the cloaca back into the body proper. Mammals possess kidneys capable of producing urine with high osmotic pressure and ion concentrations.

7.11 Animals of arid environments conserve water

In arid environments animals, like plants, face a severe problem of water balance. They can solve the problem in one of two ways: either by evading the drought or by avoiding its effects. Animals of semiarid and desert regions may evade drought by leaving the area during the dry season. That is the strategy employed by many of the large African ungulates. The spadefoot toad (*Scaphiopus couchi*) of the southern desert of the United States aestivates below ground and emerges when the rains return. Some invertebrates, such as the flatworm *Phagocytes vernalis*, which occupy ponds that dry up in summer, develop hardened cysts in which

they remain for the dry period. Other aquatic or semi-aquatic animals retreat deep into the soil until they reach the level of groundwater. Many insects undergo **diapause**, a stage of arrested development in their life cycle (see Section 8.6).

Other animals remain active during the dry season but reduce respiratory water loss. Some small desert rodents lower the temperature of the air they breathe out. Moist air from the lungs passes over cooled nasal membranes, leaving condensed water on the walls. As the rodent inhales, the warm, dry air is humidified and cooled by this water.

An African desert ungulate, the oryx (*Oryx besia*), reduces daytime losses of moisture by becoming hyperthermic (see Section 6.14). A substantial rise in daytime body temperature reduces the need for evaporation. The oryx also reduces water losses by suppressing sweating and by panting only at very high temperatures. Further, the oryx reduces its metabolic rate; by lowering the internal production of calories, the animal reduces the need for evaporative cooling. By night the oryx reduces evaporation across the skin by lowering its body temperature (Figure 7.8). With slower, more efficient respiration, the amount of water vapor it exhales is lower.

There are other approaches to the problem. Some small desert mammals reduce water loss by remaining in burrows by day and emerging by night. Many desert mammals, from the kangaroo to camels, produce highly concentrated urine and dry feces, and extract water metabolically from the food they eat. In addition, some desert mammals can tolerate a certain degree of dehydration. Desert rabbits may withstand water losses of up to 50 percent and camels up to 27 percent of their body weight.

7.12 Animals of saline environments have special problems

Animals in salty environments face problems opposite to those in fresh water. These organisms have to retain their body fluids. As discussed in section 7.9, when the concentration of salts is greater outside the body than within, organisms tend to dehydrate. Osmosis draws water out of the body into the surrounding environment. In marine and brackish environments, organisms have to inhibit the loss of water by osmosis through the body wall and prevent an accumulation of salts in the body.

There are many solutions to this problem. Invertebrates get around it by possessing body fluids that

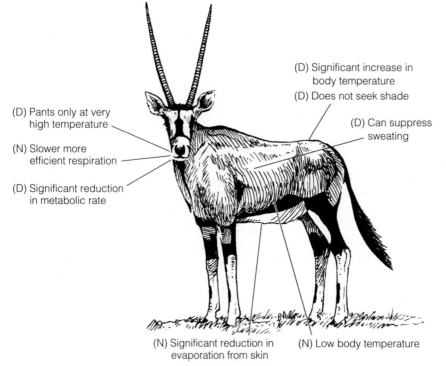

(D) Significant increase in body temperature

(D) Does not seek shade

(D) Pants only at very high temperature

(N) Slower more efficient respiration

(D) Can suppress sweating

(D) Significant reduction in metabolic rate

(N) Significant reduction in evaporation from skin

(N) Low body temperature

Figure 7.8 The physiological adaptations to aridity and heat of an African ungulate, the oryx. D = day; N = night.

have the same osmotic pressure as seawater. Marine bony (teleost) fish absorb salt water into the gut. They secrete magnesium and calcium through the kidneys and pass these ions off as a partially crystalline paste. Fish excrete sodium and chloride, major ions in seawater, by pumping the ions across special membranes in the gills. This pumping process is one type of active transport. Salts move across a concentration gradient at a cost of metabolic energy. Sharks and rays retain a sufficient amount of urea to maintain a slightly higher concentration of salt in the body than in surrounding seawater. Birds of the open sea can consume seawater because they possess special salt-secreting glands located on the surface of the cranium. Gulls, petrels, and other seabirds excrete from these glands fluids in excess of 5 percent salt. Petrels forcibly eject the fluids through the nostrils; other species drip the fluids out of the internal or external nares.

In marine mammals the kidney is the main route for the elimination of salt. Porpoises have highly developed kidneys to eliminate salt loads rapidly. In marine mammals the urine has a greater osmotic pressure (ion concentration) than blood and seawater; their physiology is poorly understood.

Vertebrates in the Arctic and Antarctic have special problems. As seawater freezes, it becomes colder and saltier. Increasing the solute concentrations in body fluids to lower body temperature is the only alternative for most organisms. Some species of fish in the Antarctic possess in their blood a glycoprotein antifreeze that enables the animals to survive in temperatures below the freezing point of their blood.

CHAPTER REVIEW

SUMMARY

Structure and Properties of Water (7.1–7.2)   Water has a unique molecular structure. One large oxygen atom and two smaller hydrogen atoms are bonded covalently. The two hydrogen atoms are separated by an 105° angle, which results in a V-shaped molecule. Because of this arrangement, the hydrogen atom side is electropositive and the oxygen side is electronegative, polarizing the water molecule. Because of their polarity, water molecules become coupled with neighboring water molecules to produce a lattice-like structure with unique properties **(7.1).**

Depending upon its temperature, water may be in the form of a liquid, solid, or gas. It absorbs or releases considerable quantities of heat with a small rise or fall in temperature. Water has a high viscosity that affects its flow. It exhibits high surface tension, caused by a stronger attraction of water molecules for each other than for the air above the surface. These properties are important ecologically and biologically **(7.2)**.

Osmosis and Osmotic Pressure (7.3) Water moves from the environment into organisms through permeable cell membranes or conductive tissues. Molecules move from areas of high concentration to low concentration until both sides are in equilibrium. This movement of molecules across membranes, called osmosis, generates pressure that slows the movement. The amount of pressure needed to counteract the movement is osmotic pressure. The tendency of a solution to attract water molecules from areas of high concentration to areas of low concentration is osmotic potential.

The Water Cycle (7.4) Water, on which life depends, follows a cycle from precipitation to interception, infiltration or surface flow, and evaporation. Water that infiltrates land moves into the soil, filling pore spaces. When water fills all pore spaces, the soil is saturated; it can hold no more water. When excess water drains away by gravity, the amount remaining is the soil's field capacity. Water held between soil particles by capillary forces is capillary water. Water held so tightly in a thin film on soil particles that it is unavailable to plants is hygroscopic water.

Water Movement from Soil to Plant to Atmosphere (7.5) Water moves from the soil into the roots, up through the stem and leaves, and out to the atmosphere. Pressure differences along a water gradient move water along this route. Plants draw water from the soil, where the water pressure is the highest, and release it to the atmosphere, where it is the lowest. Because of the lower amount of water in the atmosphere, water moves out of the leaves through the stomata in transpiration. Loss of water through the leaves reduces water pressure in the roots, so more water moves from the soil through the plant. This process continues as long as water is available in the soil. This loss of water by transpiration creates moisture conservation problems for plants. Plants need to open their stomata to take in CO_2, but they can conserve water only by closing the stomata.

Plant Responses to Water Problems (7.6–7.9) Plants respond to moisture deficits first by closing the stomata, reducing transpiration. As conditions worsen, leaves wilt. Severe drought decreases photosynthesis, causes leaves to turn yellow and shed, and may even kill the plant **(7.6)**.

Some plants of arid and semiarid regions avoid drought by being ephemeral, appearing only after rain. Other adaptations include increasing leaf thickness, possessing a waxy cuticle, storing water in cells (succulence), and shedding leaves during the dry season. Other plants extend roots widely. Phreatophytes have deep roots that reach a permanent water supply **(7.7)**.

Plants intolerant of flooding experience stress and show symptoms like those of drought. Flooded plants also undergo metabolic disturbances and changes in root growth. In extreme cases, prolonged flooding kills plants, especially in the growing season. In water-logged environments ethylene, a growth hormone in roots, may increase. It stimulates cells in the roots to form interconnected gas-filled chambers. These chambers permit the exchange of oxygen between submerged and aerated roots. Plants adapted to waterlogged environments have gas-filled chambers that carry oxygen from leaves to roots. Some woody plants have aerenchyma tissue that delivers oxygen to roots **(7.8)**.

Plants find saline environments physiologically arid. To maintain an osmotic balance with the environment, they must get rid of excess salt. Many species of plants accumulate and excrete salt through their leaves. Other plants, such as cacti, store water in their tissues **(7.9)**.

Animal Responses to Water Problems (7.10–7.12) Animals possess excretory systems from simple to complex to maintain an osmotic balance with the environment. Aquatic animals need to prevent the uptake of or rid themselves of excess water. Terrestrial animals gain water by drinking, eating, and producing metabolic water. They lose water through urine, feces, respiration, and evaporation **(7.10)**.

Animals of arid regions may reduce water loss by becoming nocturnal, producing highly concentrated urine and feces, becoming hyperthermic during the day, using only metabolic water, and tolerating dehydration **(7.11)**.

Animal responses to saline environments are complex. Most animals maintain a water balance by means of an excretory system. Many marine invertebrates maintain in their body cells the same osmotic pressure as seawater. Fish secrete excess salt and other ions through kidneys or across gill membranes. Birds and some reptiles have salt-excreting glands **(7.12)**.

STUDY QUESTIONS

1. How does the physical structure of water influence specific heat, latent heat, viscosity, and surface tension?

2. What are osmosis, osmotic pressure, osmotic potential, and transpiration?

3. What moves water through plants from soil to the atmosphere?

4. In what ways do plants cope with drought?

5. What are the advantages and disadvantages of closing stomata?

6. What adaptations enable land plants to live in an arid environment?

7. How do plants of saline habitats cope with salinity?

8. What are phreatophytes?

9. By what physiological means do animals maintain water balance with the environment?

10. How do animals of arid environments avoid dehydration?

11. What special problems must animals of salty environments overcome?

*12. Consider ways in which humans have interfered with the water cycle.

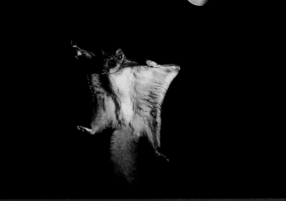

CHAPTER 8

PERIODICITY

OBJECTIVES

On completion of this chapter, you should be able to:

- Discuss the role of light in the daily and seasonal cycles of plants and animals.
- Describe circadian rhythms and their relation to the biological clock.
- Explain how the biological clock functions as a timekeeper.
- Discuss the relationship between biological clocks and critical daylengths.
- Describe the role of biological clocks in the rhythms of intertidal organisms.
- Show the relationship between biological clocks and seasonal periodicity.

The southern flying squirrel is a strongly nocturnal inhabitant of the northern hardwood and boreal forests. It does not emerge from its den until dark. Although abundant, it is rarely seen.

Daily and seasonal patterns govern life's activities. Bird song signals the arrival of dawn. Butterflies, dragonflies, and bees warm their wings, hawks begin to circle, and chipmunks and tree squirrels become active. Even the pattern of human activity quickens with daylight. At dusk, daytime animals retire, waterlilies fold, moonflowers open, and animals of the night appear. Foxes, raccoons, flying squirrels, owls, and luna moths take over niches others occupy during the day. Human activity also changes as evening begins.

As seasons progress, daylength changes, and activities shift. Spring brings migrant birds and initiates the reproductive cycles of many plants and animals. In fall the deciduous trees of temperate regions become dormant, insects and herbaceous plants disappear, summer birds return south, and winter birds arrive.

These rhythms are driven by the daily rotation of Earth on its axis and its 365-day revolution about the sun. Through time, life has become attuned to the daily and seasonal changes in the environment. At one time biologists thought that organisms were responding only to external stimuli such as light intensity, humidity, temperature, and tides. Laboratory investigations, however, have shown there is more.

8.1 Living organisms possess innate rhythms of activity

At dusk in the forests of North America, a small squirrel with silky fur and large black eyes emerges from a tree hole. With a leap the squirrel sails downward in a long, sloping glide, maintaining itself in flight with broad membranes stretched between its outspread legs. Using its tail as a rudder and brake, it makes a short, graceful upward swoop that lands it on the trunk of another tree. This is the flying squirrel *Glaucomys*

volans, perhaps the most common North American tree squirrel. (See the photograph that begins this chapter.) Because of its nocturnal habits, this mammal is seldom seen. Unless disturbed, it does not come out by day. It emerges into the forest world with the arrival of darkness; it retires to its nest before the first light of dawn.

The flying squirrel's day-to-day activities conform to a 24-hour cycle. The correlation of the onset of activity with the time of sunset suggests that light has a direct or indirect regulatory effect. If the flying squirrel is brought indoors and confined under artificial conditions of night and day, it will restrict its periods of activity to darkness and its periods of inactivity to light. Whether the conditions under which the squirrel lives are 12 hours of darkness and 12 hours of light or 8 hours of darkness and 16 hours of light, the onset of activity always begins shortly after dark.

Such behavior might not mean that the squirrel has any special timekeeping mechanism. The behavior could merely be a response to nightfall and daybreak. However, if we keep the same squirrel in constant darkness, it still maintains its pattern of activity and inactivity from day to day without any external cue. Under these conditions the squirrel's activity rhythm deviates from the 24-hour periodicity defined by the diurnal cycle (Figure 8.1). Its cycle of activity and inactivity in constant darkness varies from 22 hours 58 minutes to 24 hours 21 minutes; the average is less than 24 hours. Because the cycle length deviates from 24 hours, the squirrel gradually drifts out of phase with the external world. If we hold the squirrel under continuous light (a highly abnormal condition for a nocturnal animal), the activity cycle lengthens. Probably the animal, avoiding light, delays the beginning of activity as long as it can.

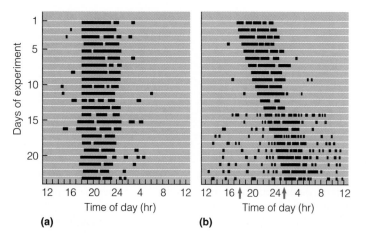

Figure 8.1 (a) Activity pattern of a flying squirrel under a cycle of 12 hours light and 12 hours dark. The solid black bars indicate extended activity. (b) Drift in phase of activity rhythm of a flying squirrel in continuous darkness at 20° C for 23 days. The red arrows indicate the beginning of greatest activity on the first and last days.

This innate rhythm of activity and inactivity covering approximately 24 hours is characteristic of all living organisms except bacteria. Because these rhythms approximate, but seldom match, the periods of Earth's rotation, they are called **circadian rhythms** (from the Latin *circa*, "about," and *dies*, "day"). The period of the circadian rhythm, the number of hours from the beginning of activity one day to the beginning of activity on the next, is its **free-running cycle.** In other words, the rhythm of activity exhibits a self-sustained oscillation under constant conditions of dark or light.

Circadian rhythms have a strong genetic component, transmitted from one generation to another. Temperature changes have little effect on them, and they are not learned from or imprinted upon the organism by the environment. They do not adapt to specific local or regional environmental conditions. Circadian rhythms influence not only the times of physical activity and inactivity, but also physiological processes and metabolic rates. They provide a mechanism by which organisms maintain synchrony with their environment.

Thus two daily periodicities—the external rhythm of 24 hours and the internal circadian rhythm of approximately 24 hours—influence the activities of plants and animals. If the two rhythms are to be in phase, some external cue or time-setter must adjust the internal rhythm to the environmental rhythm. The most obvious time-setters are temperature, light, and moisture. Of the three the master time-setter in the temperate zone is light. It brings the circadian rhythm of organisms into phase with the 24-hour photoperiod of their external environment. We call this effect **entrainment.**

8.2 Circadian rhythms operate the biological clock

The circadian rhythm and its sensitivity to light and dark are the major mechanisms that operate the biological clock—that timekeeper of physical and physiological activity in living things. Where is such a clock located in living things? Its position must expose it to its time-setter, light. In one-celled protists and plants, the clock appears to be located in individual cells. Light acts directly on photosensitive chemicals that activate cellular pathways. In multicellular animals the clock is within the brain.

Skillful surgical procedures have allowed circadian physiologists to discover the location of the physiological clock in some mammals, birds, and insects. In most insects studied, the photoreceptors—located

in cells at the base of the compound eyes—are connected by axons to the clock located in the optic lobe of the brain or in the tissues between the optic lobes. In birds and reptiles the clock is located in the pineal gland, functioning as a third eye, resting close to the surface of the brain. In mammals, including humans, the clock is in two clumps of neurons (suprachiasmatic nuclei) just above the optic chiasm. The optic chiasm is the place where the optic nerves from the eyes intersect. Operation of the clock involves a special hormone, **melatonin,** that serves to measure time. More melatonin is produced in the dark than in the light. The amount produced is a measure of changing daylength.

Whereas melatonin allows time measurement in animals, a colored protein, **phytochrome,** is the light detector in plants. Phytochrome comes in two forms, P_r that absorbs red light, and P_{fr} that absorbs far red light. When P_r absorbs red light, it converts to P_{fr}, and when P_{fr} absorbs far red light, it converts to P_r. The plant synthesizes phytochrome P_r. If the plant is kept in the dark, P_r remains in that form, and any P_{fr} remaining reverts to P_r in the dark. Thus after sunset each day any P_{fr} remaining converts to P_r. At sunrise P_r rapidly converts to P_{fr}. The conversion comes about because sunlight is richer in red than far red light. Because virtually all P_{fr} converts back to P_r within 3 to 4 hours, a time too short to act as a night-measuring clock, other hormones also seem to be an integral part of the clock.

To keep time, the clock has to have an internal mechanism with a natural rhythm of approximately 24 hours. Recurring environmental signals, such as changes in the time of dawn and dusk, should reset it. The clock has to be able to run continuously in the absence of any environmental time-setter and the same at all temperatures. Circadian rhythms fit all these criteria.

8.3 A model suggests how biological clocks measure time

Two basic models of biological clocks have been proposed. In 1960 E. Bünning proposed a single oscillation sensitive to light (Figure 8.2). Although dated, this model is helpful. The cycle or time-measuring process begins with the onset of light or dawn. The first half (12 hours) of the cycle requires light; the second half requires darkness. When light extends into the dark period, it triggers long-day or short-night effects, such as flowering or nonflowering in plants. The second model is a two-oscillator model. One oscillation is regulated by dawn and the other by dusk. There

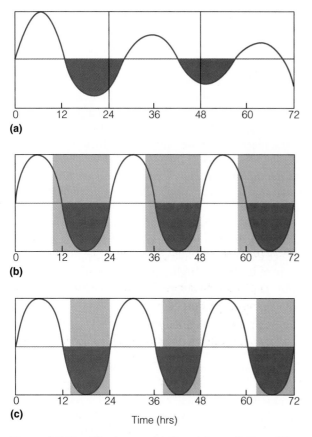

(a)

(b)

(c)

Time (hrs)

Figure 8.2 The Bünning model. One daily oscillation of the clock causes an alternation of half-cycles with different sensitivities to light. The dark blue portion represents the dark-sensitive part of the cycle, the white area the light-sensitive portion. The light blue areas represent periods of darkness or night. (a) The free-running clock in continuous light (or continuous dark) tends to drift out of phase with the 24-hour photoperiod. (b) Short-day conditions allow darkness to fall within the light-sensitive half-cycle. (c) In the long day, light falls during the dark-sensitive half-cycle.

are more complex models of the clock, but these basic models underlie most of them.

8.4 The biological clock is a hierarchy of clocks

Does one master clock drive the system, or are other clocks involved? It appears that the biological clock is a hierarchy of clocks (Figure 8.3). Subordinate clocks coupled to the master clock control rhythms of physiology and behavior. These clocks or pacemakers may be groups of cells within organs, where they have specific timekeeping functions.

When light enters the eye, the signal goes to the master clock. In turn, the master clock sends signals

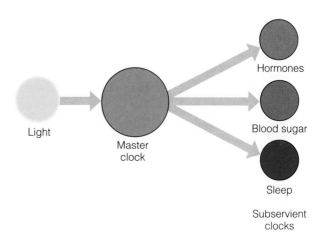

Figure 8.3 The hierarchy of clocks. The master clock, entrained to environmental changes in light, resets other clocks that control physiological rhythms.

to the other clocks. One clock may take longer to reset than another with a different period. Therefore, abrupt resetting of the master clock upsets its synchronization with the other clocks. These transient disturbances in the timing of various clocks put the physiological rhythms of the body out of phase. For this reason you feel jet lag when you fly across several time zones. Workers on night shift suffer effects similar to jet lag.

8.5 Circadian rhythms have adaptive value

How and why circadian rhythms and biological clocks function is the domain of the physiologist. Ecologists are more interested in the adaptive value of biological clocks. One adaptive value is that the biological clock provides the organism with a time-dependent mechanism. It enables the organism to prepare for periodic changes in the environment ahead of time. For example, trees of the African savanna begin leaf growth just prior to the onset of the rainy season.

Circadian rhythms help organisms with physical aspects of the environment other than light or dark. For example, the transition to night is accompanied by a rise in humidity and a drop in temperature. Wood lice, centipedes, and millipedes, which lose water rapidly in dry air, spend the day in the darkness and damp under stones, logs, and leaves. At dusk they emerge, when the humidity of the air is more favorable. These animals show an increased tendency to escape from light as the length of time they spend in darkness increases. On the other hand, their intensity of response to low humidity decreases with darkness. Thus, these invertebrates come out at night into places too dry for them during the day, and they quickly retreat to their dark hiding places as light comes.

The circadian rhythms of many organisms relate to biotic aspects of their environment. Predators such as insectivorous bats must match their feeding activity to the activity rhythm of their prey. Moths and bees must seek nectar when flowers are open. Flowers must open when insects that pollinate them are flying. The circadian clock lets insects, reptiles, and birds orient themselves by the position of the sun. Organisms make the most economical use of energy when they adapt to the periodicity of their environment.

8.6 Critical daylengths trigger seasonal responses

In the middle and upper latitudes of the Northern and Southern hemispheres, the daily periods of light and dark lengthen and shorten with the seasons (see Section 4.2). The activities of plants and animals are geared to the changing seasonal rhythms of night and day. The flying squirrel, for example, starts its daily activity with nightfall, regardless of the season. As the short days of winter turn to the longer days of spring, the squirrel begins its activity a little later each day (Figure 8.4).

Most animals and plants of temperate regions have reproductive periods that closely follow the changing daylengths of the seasons. For most birds, the height of the breeding season is the lengthening days of spring; for deer, the mating season is the shortening days of fall. Trilliums (*Trillium* spp.) and violets (*Viola* spp.) bloom in the lengthening days of spring while an abundance of sunlight reaches the forest floor. Asters (*Aster*

spp.) and goldenrods (*Solidago* spp.) flower in the shortening days of late summer.

The signal for these responses is **critical daylength.** When the duration of light (or dark) reaches a certain portion of the 24-hour day, it inhibits or promotes a photoperiodic response. Critical daylength varies among organisms, but it usually falls somewhere between 10 and 14 hours. Through the year plants and animals compare critical daylength with the actual length of day or night and respond appropriately. Some organisms can be classed as **day-neutral;** they are not controlled by daylength, but by some other influence such as rainfall or temperature. Others are short-day or long-day organisms. **Short-day** organisms are those whose reproductive or other seasonal activity is stimulated by daylengths shorter than their critical daylength. **Long-day** organisms are those whose seasonal responses, such as flowering and reproduction, are stimulated by daylengths longer than the critical daylength.

Horticulturists exploit short-day and long-day responses to force plants to bloom. When they hold plants under short-day and long-night conditions, short-day plants are stimulated to flower and long-day plants are inhibited from flowering (Figure 8.5a). When they increase daylength, short-day plants do not flower and long-day plants come into bloom (Figure 8.5b). If the dark period—the subjective night in the circadian rhythm—of a short-day and a long-day plant is interrupted, each plant responds as if it had been exposed to a long day. The long-day plant flowers and the short-day plant does not (Figure 8.5c). In reality, short-day and long-day plants respond not to the length of light, but to the length of darkness. The two might more accurately be called long-night and short-night plants.

Many organisms possess both long-day and short-day responses. Because the same duration of dark and light occurs two times a year in spring and fall, the organisms could get their signals mixed. For them the distinguishing cue is the direction from which the critical daylength comes. In one situation the critical daylength is reached as long days move into short, and at the other time as short days move into long.

Diapause, a stage of arrested growth over winter in insects of the temperate regions, is controlled by photoperiod. The time measurement in such insects is precise, usually between 12 and 13 hours of light. A quarter-hour difference in the light period can determine whether an insect goes into diapause or not. The shortening days of late summer and fall forecast the coming of winter and call for diapause. The

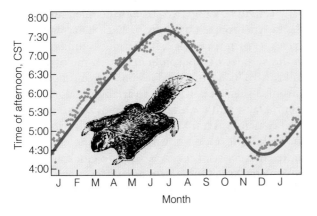

Figure 8.4 Seasonal variation in the time of day when flying squirrels become active.

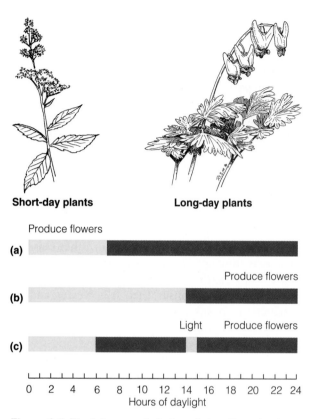

Short-day plants   **Long-day plants**

Produce flowers

(a)

Produce flowers

(b)

Light   Produce flowers

(c)

0 2 4 6 8 10 12 14 16 18 20 22 24
Hours of daylight

Figure 8.5 The influence of photoperiod on flowering in short-day and long-day plants.

lengthening days of late winter and early spring are the signals for the insect to resume development, pupate, emerge as an adult, and reproduce.

Increasing daylength induces spring migratory behavior, stimulates gonadal development, and brings on the reproductive cycle in birds. After the breeding season, the gonads of birds regress spontaneously. During this time light cannot induce gonadal activity. The short days of early fall hasten the termination of this period. The progressively shorter days of winter then become stimulatory. The lengthening days of early spring bring the birds back into the reproductive stage.

In mammals, photoperiod influences activity such as food storage and reproduction, too. (See Focus on Ecology 8.1: Deer Antlers.) Consider, for example, such seasonal breeders as sheep and deer. Melatonin initiates their reproductive cycle. More melatonin is produced when it is dark, so these animals receive a higher concentration of melatonin as the days become shorter in the fall. This increase in melatonin reduces the sensitivity of the pituitary gland to negative feed-

back effects of hormones from the ovaries and testes. Lacking this feedback, the anterior pituitary releases pulses of another hormone (called luteinizing hormone) that stimulates growth of ova in the ovaries and sperm production in the testes.

8.7 Activity rhythms of intertidal organisms follow tidal cycles

Along the intertidal marshes, fiddler crabs (*Uca* spp; the name refers to the enormously enlarged claw of the male, which he waves incessantly) swarm across the exposed mud of salt marshes and mangrove swamps at low tide. As high tide inundates the marsh, fiddler crabs retreat to their burrows, where they await the next low tide. Other intertidal organisms—from diatoms, green algae, sand beach crustaceans, and salt marsh periwinkles to intertidal fish, such as blennies and cottids—also obey both daily and tidal cycles.

Fiddler crabs brought into the laboratory and held under constant temperature and light, devoid of tidal cues, exhibit the same tidal rhythm in their activity as they show back in the marsh (Figure 8.6). This tidal rhythm mimics the ebb and flow of tides every 12.4 hours, one-half of the lunar day of 24.8 hours, the interval between successive moonrises. Under the same constant conditions, fiddler crabs exhibit a circadian rhythm of color changes, dark by day and light by night.

Is the clock in this case unimodal, with a 12.4-hour cycle; or is it bimodal, with a 24.8-hour cycle, close to the period of the circadian clock? Does one clock keep a solar-day rhythm of approximately 24 hours and another clock keep a lunar-day rhythm of 24.8 hours? J. D. Palmer and his associates at the University of Massachusetts did experiments to find out. The evidence suggests that one solar-day clock synchronizes daily activities, while two strongly coupled lunar-day clocks synchronize tidal activity. Each lunar-day clock drives its own tidal peak. If one clock quits running in the absence of environmental cues, the other one still runs. This feature enables tidal organisms to synchronize their activities in a variable tidal environment. Day-night cycles reset solar-day rhythms, and tidal changes reset tidal rhythms. Organisms, even at the cellular level, do not depend on one clock, any more than most of us keep a single clock at home. Organisms have built-in redundancies. Such redundancies enable the various clocks to run at different speeds, governing different processes with slightly differing periods.

cycle of 28 days or in some instances every semilunar cycle of 14 to 15 days. Among these species are the grunion *(Leurethes tenuis)*, a small California fish that swarms in from the sea to lay eggs on sandy beaches, and the intertidal midge *(Clunio marinus)*. These periodicities are so exact that these activities can be predicted well ahead of time. Laboratory studies confirm the entrainment of activity cycles to moonlight.

8.8 Phenology is the study of seasonal changes

Who in the temperate zones is not aware of the seasonal changes in plants and animals—the unfolding of leaves in spring and the dropping of leaves in fall, the blooming of flowers and the ripening of seeds, the migration of birds? In tropical regions, the dropping of leaves by some tree species and the fruiting by others mark the beginning of the dry season. Such collective biological events, recurring with the seasons, make up **seasonality.** The study of the causes of the timing of these events, of biotic and abiotic forces affecting them, and the relations among phases of the same or different species is called **phenology.**

Seasonality in temperate and arctic regions depends on changes in light and temperature. In a broad way, seasonal changes in temperature and light cause alternate warm and cold periods. The progression is gradual, however, and in temperate zones seasons can be identified as early or late spring, early or late fall, and so on. Seasonality in tropical regions is keyed to rainfall. The seasonal tropics have alternate wet and dry seasons, and their onset is abrupt. The beginning of the rainy season is a dependable environmental cue by which plants and animals become synchronized to seasonal changes. The onset of the rainy season, which may last up to six months, varies with the seasonal, latitudinal movements of tropical air masses. For this reason the alternating wet and dry seasons are predictable. In the wet tropics, the wet season is marked by monthly rainfall of 100 mm or more, and the dry season by several months of rainfall of 60 mm or less. The distinction between wet and dry seasons is less dramatic than in the seasonal tropical forests.

Seasonal responses reflect how light, temperature, and moisture change with altitude and latitude. The advance of spring in the temperate regions is marked by progressively later flowering of the same species of trees and herbs across a region, depending upon elevation. Although these progressive changes are most

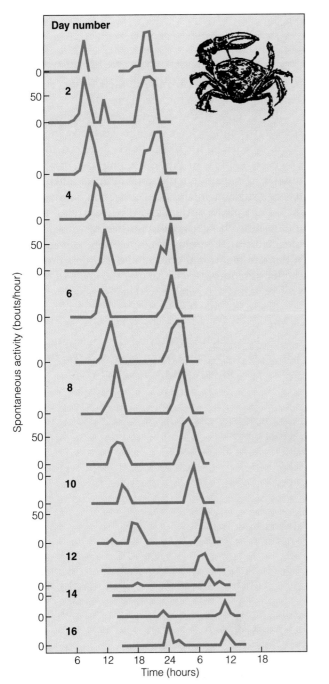

Figure 8.6 Tidal rhythm of a fiddler crab in the laboratory in constant light at a constant temperature of 22° C for 16 days. Because the lunar day is 51 minutes longer than the solar day, the tides occur 51 minutes later each solar day; thus peaks of activity appear to move to the right.

Reproduction in some marine organisms is restricted to a period that bears some relationship to tides. These rhythmic phenomena occur every lunar

FOCUS ON ECOLOGY 8.1

DEER ANTLERS

The familiar antlers you see on male deer—bucks and stags—in fall are not the same antlers they wore the fall before. They lose their antlers in early winter; they will produce new antlers the coming fall. The acquisition of new antlers and the associated reproductive cycle is controlled by the influence of daylength on the pituitary gland, located on the floor of the brain.

The lengthening days of spring stimulate the pituitary to increase secretion of growth hormones and prolactin, a hormone associated with lactation in females. In males these hormones stimulate the growth of antlers in the spring and early summer (Figure A). Fur-covered skin, called velvet, carries blood vessels and nerves to the growing antlers. During the shortening days of late summer, growth hormones and prolactin decrease. Under the influence of melatonin, testosterone secretion increases in the enlarging testes. The presence of testosterone inhibits the action of growth-stimulating hormones. Antler growth

ceases, and deer thrash and rub their antlers against vegetation to remove the shedding velvet.

By the onset of the season of sexual activity, called the rut, the antlers have become hardened and polished. The useful life of the newly acquired antlers, however, is short. In the shortening days of winter, pituitary stimulation of the testes declines, and testosterone, which maintains the connection between the dead bone of the antlers and the live frontal bone, diminishes. This decline of testosterone causes a loss of calcium at the point of connection between the antler and the frontal bone, and the antlers drop. In the lengthening days of spring, the cycle of antler growth and sexual resurgence begins anew.

Normally the deer is in velvet about one-third of the year. When the duration of that year is changed artificially by altering the daylength, deer may replace antlers as often as two, three, or four times a year, or only once every other year.

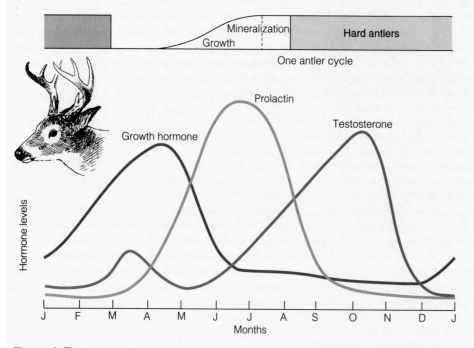

Figure A The seasonal course of hormonal levels in the white-tailed deer and its relationship to antler growth.

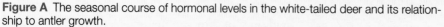

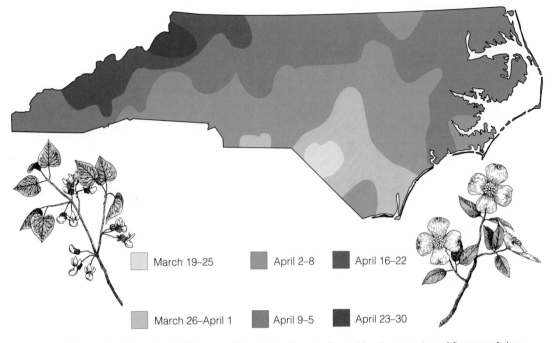

Figure 8.7 The arrival of spring 1970 across North Carolina, indicated by the opening of flowers of dogwood *(Cornus florida)* and redbud *(Cercis canadensis)*. (From H. Lieth, ed., *Phenology and Seasonality Modeling,* (New York: Springer-Verlag, 1974). © 1974 Springer-Verlag. Reprinted by permission.)

Legend:
- March 19–25
- March 26–April 1
- April 2–8
- April 9–5
- April 16–22
- April 23–30

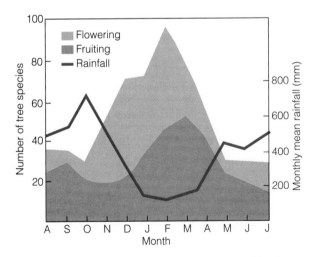

Figure 8.8 Synchronization of flowering and fruiting in rain forest trees with mean monthly rainfall in Golfito, Costa Rica. Flowering and fruiting are highest during the dry season, from January to March.

In tropical regions with a distinct wet-dry season, the sequence of flowering, fruiting, and leafy growth reflects the alternation of wet and dry seasons (Figure 8.8). The coming of the rainy season is marked by a flush of vegetative growth, just as warming spring temperatures trigger leafy growth in temperate regions. Over much of the seasonal tropics flowering and fruiting coincide with the dry season. Some species flower at the end of the rainy season when soil moisture is still high; other species flower at the end of the dry season.

Seasonal activities of animals center about reproduction and the availability of food. The reproductive cycle of the white-tailed deer (*Odocoileus virginianus;* Figure 8.9), for example, begins in fall, and the young are born in spring when the highest-quality food for lactating mother and young is available. In tropical Central America, home of numerous species of fruit-eating (frugivorous) bats, the reproductive periods track the seasonal production of food. The birth periods of frugivorous bats coincide with the peak period of fruiting. Young are born when both females and young will have adequate food. Insects and other arthropods reach their greatest biomass early in the rainy season in the Costa Rican forests. At this time the insectivorous bats give birth to their young.

pronounced across broad geographical areas (Figure 8.7), you can see distinct variations on a single mountain. Leaf emergence and flowering of trees advance upslope in spring, and fall coloration moves downslope in fall.

Peak of rutting

Necks of bucks swell

Velvet is rubbed off antlers

Fawns lose their spots

Fawns are raised

December

October

September

August–September

June–September

June

January

January–March

April–May

Antlers are shed

Food supply most important

New antlers appear

Peak of fawning

Figure 8.9 Seasonal reproductive cycle of the white-tailed deer. The cycle is attuned to the decreasing daylength of fall, when the breeding season begins, and to the lengthening days of spring, when antler growth begins.

CHAPTER REVIEW

SUMMARY

Circadian Rhythms (8.1–8.5) Living organisms, except bacteria, have an innate rhythm of activity and inactivity. This rhythm is free-running under constant conditions with an oscillation that deviates slightly from 24 hours. For that reason it is called a circadian rhythm. Under natural conditions the circadian rhythm is set or entrained to the 24-hour day by external time cues, notably light and dark (day and night). This setting synchronizes the activity of plants and animals with the environment. The onset and cessation of activity depend upon whether the organisms are light-active or dark-active **(8.1).**

Circadian rhythms operate the biological clocks of organisms. The biological clock is in the cells of plants and in the brain of multicelled animals. Animals produce more of a special hormone, melatonin, in the dark than in the light. Thus, melatonin becomes a device for measuring daylength **(8.2).** The basic clock has an oscillating circadian sensitivity to light and dark **(8.3).** The biological clock appears to be a hierarchy of clocks coupled to a master clock. Because of time lag, resetting the master clock can disturb the phase relationships among the clocks. An instance is jet lag (8.4). Circadian rhythms enable organisms to anticipate daily changes in the environment (8.5).

Seasonal Responses (8.6) Seasonal changes in activity are based on daylength. Lengthening days of spring and the shortening days of fall stimulate migration in animals, and reproduction and food storage in plants and animals alike. The signal for these responses is critical daylength, a specific duration of light or dark. Organisms that respond to daylengths shorter than critical daylength are short-day organisms. Those that respond to daylengths longer than critical daylengths are long-day organisms. Many organisms possess both long-day and short-day responses in their annual life cycle.

Tidal Cycles (8.7) Intertidal organisms are under the influence of two environmental rhythms, daylength and tidal cycles of 12.4 hours. Intertidal organisms appear to have two lunar-day clocks that set tidal rhythms, and one solar-day clock that sets circadian rhythms.

Seasonality (8.8) The biological clock also synchronizes seasonal activities among living organisms, reflected in the seasonal periodicities of flowering and fruiting of plants and in the migration and breeding cycles of animals. The study of the causes, timing, and relationship of these recurring seasonal biological events is phenology. In temperate regions, changes in light and temperature influence seasonality; in tropical regions, rainfall does the same. These seasonal rhythms bring living organisms into a reproductive state at the time of year when the probability of survival of offspring is the highest. It synchronizes within a population such activities as mating and migration, dormancy and flowering.

STUDY QUESTIONS

1. What is a circadian rhythm?
2. How does circadian rhythm relate to the 24-hour day?
3. What is the biological clock, and where is it located?
4. What conditions must a biological clock fulfill to function as a timekeeper?
5. Discuss four uses for a biological clock.
6. What is critical daylength? What are long-day and short-day organisms?
7. Explain diapause in insects.
8. Why do intertidal organisms need two kinds of clocks?
9. How does daylength influence the seasonal activity of plants and animals?
*10. What is the adaptive value of seasonal synchronization?

NUTRIENTS

OBJECTIVES

On completion of this chapter, you should be able to:

- Distinguish between macronutrients and micronutrients.
- Understand factors controlling the availability of nutrients to plants.
- Relate the characteristics of organic matter to rates of decomposition and nutrient release.
- Contrast plants adapted to low-nutrient and high-nutrient environments.
- Discuss the relationship between nutrients and the growth and reproduction of animals.

A black bear *(Ursus americanus)* feeding on the berries of mountain ash *(Sorbus americana)*.

Light, thermal energy, and moisture are not the only conditions of life. Nutrients are a fourth essential factor. Without them organisms could not survive, grow, and reproduce. What nutrients are essential? How do plants and animals acquire these nutrients? What happens if these nutrients are lacking in the diet? These questions and others are explored in this chapter.

9.1 Essential nutrients are either macronutrients or micronutrients

Living organisms require at least 30 to 40 chemical elements for growth, development, and metabolism (Table 9.1). Some of these elements are needed in large amounts. Known as **macronutrients,** they include among others carbon, oxygen, hydrogen, nitrogen, phosphorus, calcium, potassium, magnesium, sulfur, sodium, and chlorine. Other elements are needed in lesser, often minute quantities. These elements are called **micronutrients** or trace elements. They include iron, copper, zinc, iodine, boron, cobalt, molybdenum, manganese, and selenium.

The prefixes *micro-* and *macro-* refer only to the quantity in which the nutrients are needed, not their importance to the organism. If micronutrients are lacking, plants and animals fail as completely as if they lacked nitrogen, calcium, or any other macronutrient. Some micronutrients are essential to all organisms; others appear to be essential to only a few. All of the micronutrients, especially the heavy metals, can be toxic in quantities greater than needed.

9.2 Geology and climate affect availability of nutrients

The macro- and micronutrients plants and animals require come from either atmospheric or geological sources. The majority are released from rocks and minerals by weathering. **Weathering** is the process by which newly created or exposed geological substrates become soil (see Section 10.3). Two aspects of weathering are of importance in nutrient availability: the rate at which materials weather and the elements released.

The rate of weathering depends on the type of rock and on environmental conditions. Geologists define three major types of rock: igneous, sedimentary, and metamorphic. **Igneous rocks** are formed by the cooling of volcanic flows, surface or subterranean. The properties of these rocks depend on the rate and temperature at which they form. **Sedimentary rocks** are formed by the deposition of mineral particles (sediments). The properties of sedimentary rocks depend on the type of sediment from which they are formed. Some sediments are of biological origin; for example, shells of ocean invertebrates may fall to the sea floor. **Metamorphic rocks** are either igneous or sedimentary rocks that have been altered by heat and the pressure of overlying rock.

Rocks gradually break up and dissolve, releasing mineral nutrients into the soil solution (see Section 10.4). Igneous rocks first break into sediments and secondary minerals. Which type of sediments and secondary minerals form depends on environmental conditions and has an important effect on nutrients and water retention in the resulting soil (see Section 10.3).

Temperature, precipitation, and wind influence the rate at which rock weathers. In warmer, wetter climates, the depth to which weathering occurs is much greater than in cool, wet or in hot, dry regions (see Section 10.6). The depth of weathering influences the vertical distribution of nutrients in the soil. Extremely slow weathering will limit nutrient availability. High rainfall and temperatures can also be damaging. They can leach nutrients from the soil into groundwater, streams, and rivers.

9.3 Plant uptake influences local availability of nutrients

Plants acquire nutrients from the soil by the uptake of dissolved ions in the soil solution and through active transport. The uptake of dissolved ions occurs through diffusion from the soil to the roots (see Section 7.3). Where the concentration of nutrients in the roots is lower than in the soil solution, roots actively transport nutrients across their surfaces. This active transport expends a large amount of energy to move ions against the concentration (diffusion) gradient. As roots extract nutrients from the soil around them, they reduce the nutrients in the surrounding area. Water and nutrients diffuse through the soil from areas of higher concentration, eventually replacing them.

Different nutrients diffuse through the soil at different rates. Calcium and magnesium move readily through the soil in solution. Nitrogen and phosphorus diffuse slowly, and consequently take a much longer time to replenish in areas of the soil where roots have been. As a consequence, plants produce a large quantity of fine roots that readily absorb nutrients in a region of the soil and then die. The continuous production and dieback of fine roots provides a means of acquiring

TABLE 9.1

SOME ESSENTIAL ELEMENTS

Macronutrients		Micronutrients	
Element	**Role**	**Element**	**Role**
Carbon (C) Hydrogen (H) Oxygen (O)	Basic constituents of all organic matter.	Iron (Fe)	In plants, involved in the production of chlorophyll; is part of the complex protein compounds that activate and carry oxygen and transport electrons in mitochrondia and chloroplasts. In animals, iron-rich respiratory pigment hemoglobin in blood of vertebrates and hemolymph of insects transports oxygen to every organ and tissue. Synthesized into hemoglobin and hemolymph throughout life. Deficiency results in anemia.
Nitrogen (N)	Used only in a fixed form: nitrates, nitrites, ammonium. Component of chlorophyll and enzymes; building block of protein.		
Calcium (Ca)	In animals, needed for acid-base relation-ships, clotting of blood, contraction and relaxation of heart muscles. Controls movement of fluid through cells; gives rigidity to skeletons of vertebrates; forms shells of mollusks, arthropods, and one-celled Foraminifera. In plants, combines with pectin to give rigidity to cell walls; essential to root growth.	Manganese (Mn)	In plants, enhances electron transfer from water to chlorophyll and activates en-zymes in fatty acid synthesis. In animals, necessary for reproduction and growth.
		Boron (B)	Fifteen functions are ascribed to boron in plants, including cell division, pollen germination, carbohydrate metabolism, water metabolism, maintenance of conductive tissue, translocation of sugar. Deficiency causes stunted growth in leaves and roots and yellowing of leaves.
Phosphorus (P)	Necessary for energy transfer in living organisms; major component of nuclear material of cells. Animals require a proper ratio of Ca:P, usually 2:1 in the presence of vitamin D. Wrong ratio in vertebrates causes rickets. Deficiency in plants arrests growth, stunts roots, and delays maturity.	Cobalt (Co)	Required by ruminants for the synthesis of vitamin B_{12} by bacteria in the rumen.
Magnesium (Mg)	In all living organisms, essential for maximum rates of enzymatic reactions in cells. Integral part of chlorophyll; involved in protein synthesis in plants. In animals, activates more than 100 enzymes. Deficiency in ruminants causes a serious disease, grass tetany.	Copper (Cu)	In plants, concentrates in chloroplasts, influences photosynthetic rates, activates enzymes. Excess interferes with phos-phorus uptake, depresses iron concentra-tion in leaves, reduces growth. Deficiency in vertebrates causes poor utilization of iron, resulting in anemia and calcium loss in bones.
Sulfur (S)	Basic constituent of protein. Plants use as much sulfur as they do phosphorus. Excessive sulfur is toxic to plants.		
Sodium (Na)	Needed for maintenance of acid-base balance, osmotic homeostasis, formation and flow of gastric and intestinal secre-tions, nerve transmission, lactation, growth and maintenance of body weight. Toxic to plants along roadsides when used to treat icy highways.	Molybdenum (Mo)	In free-living nitrogen-fixing bacteria and cyanobacteria, a catalyst for the conversion of gaseous nitrogen to usable form. High concentration in ruminants causes teart, a disease characterized by diarrhea, debilita-tion, and permanent fading of hair color.
Potassium (K)	In plants, involved in osmosis and ionic balance; activates many enzymes. In ani-mals, involved in synthesis of protein, growth, and carbohydrate metabolism.	Zinc (Zn)	In plants, helps form growth substances (auxins); associated with water relation-ships; component of several enzyme systems. In animals, functions in several enzyme systems, especially the respiratory enzyme carbonic anhydrase in red blood cells. Deficiency in animals causes der-matitis, parakeratosis.
Chlorine (Cl)	Enhances electron transfer from water to chlorophyll in plants. Role in animals similar to that of sodium, with which it is associated in salt (NaCl).	Iodine (I)	Involved in thyroid metabolism. Defi-ciency results in goiter, hairlessness, and poor reproduction.
		Selenium (Se)	Closely related to vitamin E in function. Prevents white-muscle disease in rumi-nants. Borderline between requirement level and toxicity is narrow. Excess results in loss of hair, sloughing of hooves, liver injury, and death.

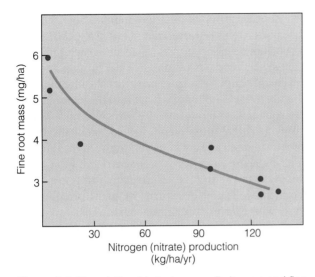

Figure 9.1 The relationship between soil nitrogen and fine root mass. When soil nitrogen is low, plants need a large mass of fine roots to scavenge nitrogen from the soil.

nutrients over a large volume of soil. The lower the availability of a less mobile nutrient such as nitrogen, the greater the production of fine roots to exploit a greater volume of soil (Figure 9.1).

9.4 Decomposers transform nutrients into a usable form

Decomposer organisms feed on dead organic material (see Section 3.7). This material contains energy in the form of carbon-containing compounds and minerals.

The energy value of the organic material is related to the types of carbon compounds present. Compounds such as simple fats, simpler carbohydrates, and proteins are easy to digest and of high quality. Compounds such as lignin and the complex carbohydrate cellulose are difficult to break down and low in energy. Cellulose is the source of fiber for paper and paper products, whereas lignin is the compound that makes wood "woody." Lignin is encrusted around cellulose in cell walls to provide rigidity and strength. In fact, lignin can be digested by certain fungi only, and it decomposes slowly. The amount of cellulose and lignin in dead plant materials determines the rate of decay.

The nutrient quality of dead organic material varies. We can use the macronutrient nitrogen as an example. Most dead leaf material, such as the leaves that fall from deciduous trees during the autumn in the temperate zone, has a nitrogen content in the range of 0.5 to 1.5 percent (of total weight). The higher the nitrogen content of the dead leaf, the higher the

nutrient value for the microbes and fungi that feed upon the leaf.

Plants cannot take up nutrients that are tied up in organic compounds such as proteins and enzymes. The organic compounds must first be broken down and transformed into inorganic compounds, or mineral nutrients. This process of transformation is carried out by microbial decomposers—bacteria and fungi. As they digest dead organic material, they release nutrients tied up in organic compounds into the soil in an inorganic form that plants can use. This process is called **mineralization.**

Like all living organisms, microbial decomposers have specific nutrient requirements. If these requirements cannot be met by the organic material on which they feed, they must take from the soil nutrients that have been mineralized already. We call the process by which microbial decomposers remove nutrients from the soil to meet their own nutritional needs and tie up these nutrients in their own biomass **immobilization.** This process decreases the availability of nutrients to plants. The net availability of nutrients is the difference between the rate at which nutrients are mineralized and the rate at which they are immobilized (Figure 9.2). This difference is called **net mineralization.** When the organic material is high in nutrients, decomposers meet their nutritional needs from their food. Immobilization is low, and net mineralization is high. If the nutrient content of the material is low, immobilization is high and net mineralization is low.

Nutrients move in a cycle. Plants take up nutrients, incorporate them into tissues, die, and decompose, releasing nutrients for uptake again. This process is **nutrient cycling** (Figure 9.3). We explore it in greater detail in Chapter 25.

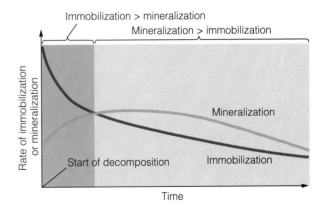

Figure 9.2 The relationship between immobilization and mineralization rates during the decomposition of plant material.

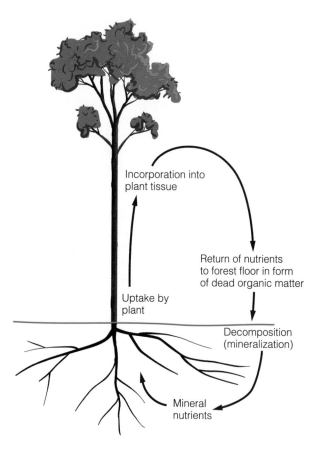

Figure 9.3 Nutrient cycling between plants and soil.

9.5 Microorganisms help plants take up some nutrients

Although plants take most nutrients directly from the soil or from the atmosphere, they need help with other nutrients, such as nitrogen. The source of nitrogen is nitrogen gas (N_2) and other nitrogen compounds in the atmosphere. Plants cannot use nitrogen gas directly from the atmosphere; it must first be fixed into a usable form. This task is accomplished by specific groups of bacteria in both terrestrial and aquatic environments.

In aquatic environments, free-living cyanobacteria carry out this transformation. In terrestrial environments, the free-living cyanobacteria *Rhizobium* and *Frankia* bacteria are the nitrogen-fixing bacteria. *Rhizobium* bacteria grow on the root systems of certain plant species, notably the legumes. *Frankia* grow on the roots of alder and other angiosperms. Both feed on carbon provided by the plant on which they live. In return, they give the plant nitrogen. This mutually beneficial relationship between organisms of different species is discussed in Chapter 17.

Another group of microorganisms associated with the root systems of terrestrial plants are special fungi. A root-fungus association is called a mycorrhiza. There are two major types of mycorrhizal fungi. **Ectomycorrhizal fungi** form large mats that sheathe the roots (Figure 9.4a). The fungi extend into the root between the root cells and out into the surrounding soil. **Endomycorrhizal fungi** do not form a mat or sheath but penetrate into the root cells and outward into the soil (Figure 9.4b). Both types of mycorrhizal fungi assist the plant in the uptake of nitrogen and phosphorus from the soil, and in return the plant provides the fungi with carbon-containing compounds, a source of food energy. As with nitrogen-fixing *Rhizobium* bacteria, the relationship between plant and fungi is mutually beneficial (Chapter 17).

9.6 Plants are adapted to low-nutrient or high-nutrient environments

The availability of nutrients has many direct effects on plant survival, growth, and reproduction. The best example of the direct link between nutrient availability and plant performance involves nitrogen.

Nitrogen plays a major role in photosynthesis. In Chapter 3, we examined two important compounds in photosynthesis—the enzyme rubisco and the pigment chlorophyll. Rubisco catalyzes the transformation of carbon dioxide into simple sugars, and chlorophyll absorbs light energy. Nitrogen is a large component of both of these compounds; the plant requires nitrogen to make them. In fact, over 50 percent of the nitrogen content of a leaf can be tied up in these two compounds. The more nitrogen in the leaf, the higher the observed rate of photosynthesis (Figure 9.5).

The uptake of a nutrient depends upon both supply and demand. Figure 9.6 illustrates the typical relationship between the uptake of a nutrient and its concentration in soil. Note that the uptake rate increases with the concentration until some maximum rate. Above this concentration no further increase occurs, because the plant meets its demands. In the case of nitrogen, low concentrations in the soil or water mean low uptake rates. A lower uptake rate decreases the concentrations of rubisco and chlorophyll in the leaf. Therefore lack of nitrogen will limit the growth of plants. A similar pattern holds for other essential nutrients.

We have seen that geology, climate, and biological activity alter the availability of nutrients in the soil.

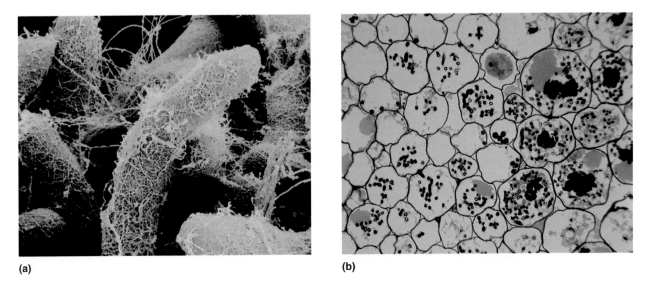

(a)   **(b)**

Figure 9.4 Mycorrhizae, which help plants take up nitrogen and phosphorus. (a) Coral-like growth of ectomycorrhizae. (b) Endomycorrhizae in the root cells of an orchid.

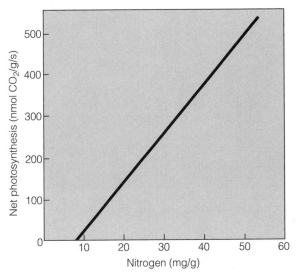

Figure 9.5 The relationship between light-saturated photosynthetic rate and the nitrogen content of leaves of plants from a variety of habitats.

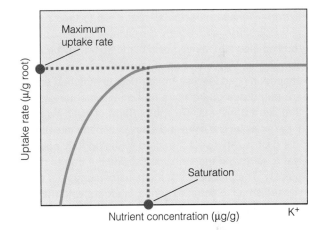

Figure 9.6 Uptake of nutrients increases with concentration in soil until the plant arrives at maximum uptake.

As a consequence, some environments are rich in nutrients, whereas others are poor. How do plants from low-nutrient environments succeed? (See Focus on Ecology 9.1: Carnivorous Plants.)

A plant's rate of growth influences its demand for a nutrient. In turn, the plant's uptake rate of the nutrient also influences growth. This relationship may seem circular, but the important point is that not all plants have the same rate of growth. In Chapter 5 we saw how shade plants have an inherently lower rate of photosynthesis and growth than sun plants, even under high light conditions. This lower rate of photosynthesis and growth means a lower demand for resources, including nutrients. The same pattern of reduced photosynthesis occurs among plants that are characteristic of low-nutrient environments. Figure 9.7 shows the growth responses of two grass species when soil is enriched with nitrogen. The species that naturally grows in a high-nitrogen environment keeps increasing its rate of growth with increasing nitrogen. The species native to a low-nitrogen environment reaches its maximum rate of growth at low to medium nitrogen availability. It does not respond to further additions of nitrogen.

Some plant ecologists suggest that a low natural growth rate is an adaptation to a low-nutrient environment. One advantage of slower growth is that

CARNIVOROUS PLANTS

Terrestrial and semiaquatic plants gain many of their nutrients from the soil. Some plants, however, that live in acidic, nutrient-poor communities—heaths, bogs, and swamps, freshwater marshes, and impoverished soil in forest openings—supplement their nutrition by feeding directly on animals. Out of one-quarter million or more species of plants, 400 resort to carnivory as a source of nitrogen and phosphorus.

Carnivorous plants trap their prey, which may vary from protozoans, small crustaceans, mosquito larvae, and minute water insects to small tadpoles, large insects, and small amphibians. Carnivorous plants fall into two groups. Active trappers employ rapid plant movements to open trap doors or to close traps. Passive trappers employ pitfalls or sticky adhesive traps.

Two examples of active trappers are the Venus flytrap *(Dionaea muscipula),* restricted to the coastal plain of North and South Carolina, and the ubiquitous aquatic and semiaquatic bladderworts (*Utricularia* spp.). The Venus flytrap (Figure A) employs clam-shaped, hinged leaves. Around their unattached edges are numerous guard hairs and minute nectar glands that attract insects. On the surface of each half are three small trigger hairs and a covering of minute digestive glands. An insect attracted to the brightly colored leaf touches the trigger hairs, causing the trap to close quickly. The bladderwort, as its name suggests, possesses small, elastic, flattened bladders with the entrance sealed by a flap of cells. When prey touch the

tactile cells on the flap, the trap door opens. The bladder walls spring apart, causing a sucking motion that sweeps a current of water into the bladder. Then the door closes, trapping the prey.

Pitcher plants (*Sarracenia* spp.) and the various sundews (*Drosera* spp.) are examples of passive trappers. Pitcher plants (Figure B) employ pitfalls. The leaves are modified into pitcherlike or funnel-like traps that arise from underground stems. Bright coloration and secretions of nectar along the hood and the rolled-up lip attract insects. When they alight and move down the

Figure B Pitcher plant

Figure A Venus flytrap

Figure C Sundew

leaf, the insects are unable to back up against the stiff, downward-directed hairs. They fall into watery fluid containing digestive enzymes secreted by the leaf. The sundews (Figure C) attract insects to sticky leaves by color, scent, and glistening droplets of adhesive. Sundews have two types of glands on the leaf surface that secrete adhesive droplets. Long-stalked glands on the edge of the leaf ensnare the insect. Shorter-stalked glands in the middle secrete digestive juices. The long-stalked glands slowly bend into the center of the leaf, securing the prey to the digestive area of the leaf.

The digestive glands of carnivorous plants vary considerably but share some features. The outer secretory cells of the leaf epidermis are specialized for enzyme synthesis. The capture of prey induces an osmotically driven outflow of fluid from the vascular system that flushes out digestive enzymes stored in the glands. After digestion is complete, the leaf reabsorbs the secretion pool, with its nitrogen-rich and phosphorus-rich products of digestion, through its cell walls. The plant moves the fluids through the vascular system to other parts of the plant.

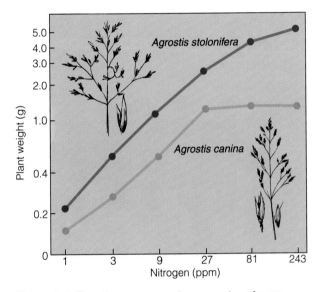

Figure 9.7 Growth responses of two species of grass—carpet bent grass *(Agrostis stolonifera),* found in high-nutrient environments, and velvet bent grass *(Agrostis canina),* found in low-nutrient environments—to the addition of different levels of nitrogen fertilizer. Note that *A. canina* responds to nitrogen fertilizer up to a certain level only.

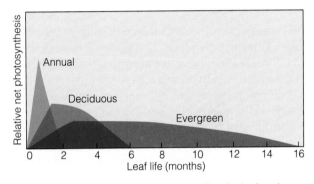

Figure 9.8 Relationship between the life of a leaf and photosynthetic capacity for annual plants, deciduous trees, and evergreen trees.

the plant can avoid stress under low-nutrient conditions. A slow-growing plant can still maintain optimal rates of photosynthesis and other processes critical for growth under low nutrient availability. In contrast, a plant with an inherently high rate of growth will show signs of stress.

A second adaptation to low-nutrient environments is leaf longevity (Figure 9.8). The production of a leaf has a cost to the plant. This cost can be defined in terms of the carbon and other nutrients required to make the leaf. At a low rate of photosynthesis, a leaf needs a longer time to "pay back" the cost of its production.

As a result, plants inhabiting low-nutrient environments tend to have longer-lived leaves. A good example is the dominance of pine species on poor, sandy soils in the coastal region of the southeastern United States. In contrast to deciduous tree species, which shed their leaves every year, pines have needles that live for up to three years.

Like water, nutrients are a belowground resource of terrestrial plants. Their ability to exploit this resource is related to the amount of root mass. One means by which plants growing in low-nutrient environments compensate is to increase the production of roots. This increase is one cause of their low growth rates. Just as was the case with water limitation (see Focus on Ecology 7.1), carbon is allocated to the production of roots at the cost of the production of leaves. The reduced leaf area reduces the total amount of carbon the plant fixes in photosynthesis.

Another distinctive feature of plants in low-nutrient environments is that their leaves, stems, and branches often contain lower concentrations of nutrients.

Therefore they have lower nutrient value to decomposer organisms (see Section 9.4). The decomposers must immobilize more nutrients in the soil, reducing net mineralization and the rate at which nutrients become available again to plants. Now the low availability of nutrients will reinforce low rates of uptake and low tissue concentrations in plants (see Figure 9.7). This positive feedback loop (see Section 2.2) is particularly pronounced for nitrogen, whose availability is largely dependent on rates of nutrient cycling (Chapter 25).

9.7 Soil acidity influences the availability and uptake of nutrients

Soil acidity can affect both the availability and the uptake of nutrients by plants. Soil acidity is measured by pH. By definition, pH is the logarithm (base 10) of the concentration of hydrogen ions in solution. An ion is a charged particle, and hydrogen carries a positive charge (H^+), making it a positive ion or **cation.** The logarithmic pH scale goes from 0 to 14, with a pH of 7 denoting a neutral solution. A pH greater than 7 denotes an alkaline (or basic) solution, and a pH of less than 7 an acidic solution. Because the scale is based on $\log_{10}$, a pH of 5 has ten times the hydrogen concentration of a solution with a pH of 6; and a solution of pH 4 has ten times more hydrogen ions than one of pH 5, and 100 times the hydrogen concentration of a solution of pH 6.

Hydrogen is not the only positive ion in the soil. Other macronutrients, such as calcium (Ca^{2+}), potassium (K^+), and magnesium (Mg^{2+}), also have a positive charge. These positively charged ions exist in the soil in two states: either in the soil water or attached to the negatively charged edges of clay particles and soil organic matter (see Chapter 10). When soils are acidic, the high concentration of hydrogen ions displaces other cations (such as calcium, magnesium, and potassium) from the soil surface, moving them into solution (Figure 9.9). If plants do not take up these nutrients over the short term, they can leach from the soil, lowering soil fertility. For this reason acidic soils (low pH) are often low in nutrient concentrations.

The reverse condition can also occur. When calcium is abundant due to the weathering of calcium-rich rocks (see Section 9.2), calcium can replace the hydrogen ions in soil solution and on the soil particles. These soils, high in calcium rather than hydro-

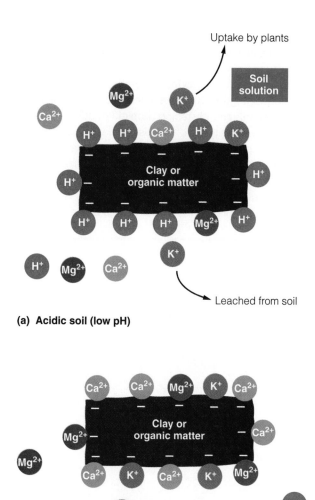

(a) Acidic soil (low pH)

(b) Neutral to basic soil (higher pH)

Figure 9.9 Cation exchange in soils. (a) In acidic conditions, hydrogen (H^+) ions replace calcium (Ca^{2+}), magnesium (Mg^{2+}), and potassium (K^+) ions that are subsequently leached from the soil, impoverishing it. (b) In neutral to basic soils, calcium, magnesium, and potassium ions replace the hydrogen ions.

gen, have a pH of 7, whereas acidic soils have a pH of 6.5 and less.

Many plants grow well in mildly acidic or acid soils, whereas others grow best in high calcium (often called high lime soils). Plants have been broadly classified as **calcicole** (lime-loving), **calcifuge,** (lime-intolerant), and **neutrophilus** (tolerant of either condition). This relationship between plants and calcium, however, is not a simple one. Calcium availability is only one of a number of pH-related soil conditions, including the toxicity of iron, aluminum, and other heavy metal ions, and deficiencies of nitrogen, phosphorus, and magnesium. The association between calcium deficiency and low pH gets the most attention only because calcium is the most abundant cation in the soil.

True calcifuges, such as rhododendrons and azaleas, are plants that have a low lime requirement and can live in soils with a pH of 4.0 and less. Some of their characteristics are the opposite of those used to describe a calcicole. Highly acidic soils associated with calcifuges invariably have a high concentration of aluminum and iron ions, toxic to most plants. Calcifuge plants are tolerant of aluminum. They are especially sensitive to calcium. If grown in calcareous soils, they suffer from lime chlorosis, a disease in which roots and leaves become stunted and the leaves turn yellow. It is the ability to grow in the presence of toxic ions, especially aluminum, that sets calcifuge plants apart from calcicoles.

These differences in the ability of plants to tolerate soil acidity can create a problem for gardeners and landscapers. Plantings of calcifuge azalea and rhododendron often grow adjacent to lawns of calcicole grasses such as fescue and bluegrass (*Poa pratensis*). Keeping lawn grasses healthy requires the application of lime. Maintaining the evergreen shrubs such as rhododendrons requires an acid fertilizer.

True calcicoles, such as alfalfa (*Medicago sativa*), blazing star (*Chamaelirium luteum*) and southern red cedar (*Juniperus silicicola*), are restricted to soils of high pH, not because they have any particular demand for calcium, but because they are susceptible to aluminum toxicity, acidity, and other factors influenced by calcium.

9.8 The quality of plants affects the nutrition of consumers

Consumers, notably vertebrates and arthropods, require mineral elements and some 20 amino acids, of which 14 are essential. Amino acids make up proteins. The need for these nutrients differs little among vertebrates and invertebrates. Insects, for example, have the same dietary requirements as vertebrates, although they need more potassium, phosphorus, and magnesium than vertebrates and less calcium, sodium, and chlorine. Directly or indirectly, the source of these essential nutrients is plants. For this reason, the quantity and quality of plants affect the nutrition of consumers. In the face of a limited quantity of food, consumers may suffer acute malnutrition, leave the area, or starve. In other situations, the quantity of food may be sufficient to allay hunger, but its low quality affects reproduction, health, and longevity.

The problem facing consumers is the conversion of plant tissue to consumer tissue. Plants and their consumers have different chemical compositions. Animals are high in fat and proteins, which they use as structural building blocks. Plants are low in proteins and high in carbohydrates. Much of their carbon is in the form of cellulose and lignins in cell walls (see Section 9.4). In plants the ratio of carbon to nitrogen (a major constituent of protein) is 40 to 1. In mammals the ratio is about 14 to 1. The plant-feeders or herbivores face the task of converting cellulose and a limited supply of plant protein into animal tissue. Few herbivores have the enzymes necessary to break down cellulose and lignin. Most of them require bacterial microbiota in the digestive system to break down these carbon compounds in plant tissue.

The highest quality plant food for herbivores, vertebrate and invertebrate, is high in nitrogen in the form of proteins. As the nitrogen content of food plants increases, assimilation of plant material improves, and herbivores show increased growth, reproductive success, and survival. Proteins concentrate in the growing tips, new leaves, and buds of plants. Its content declines as leaves and twigs mature and age. Herbivores have adapted to this period of new growth. Herbivorous insect larvae are most abundant early in the growing season and complete their growth before the leaves mature. Vertebrate herbivores give birth to their young at the start of the growing season, when the most protein-rich plant foods are available.

Although availability and season strongly influence food selection, herbivores exhibit a consistent preference for the most nitrogen-rich plants, which they probably detect by taste and color. Chemoreceptors in the nose and mouth of deer encourage or discourage consumption of certain foods. During drought, nitrogenous compounds concentrate in plants, making them more attractive and vulnerable to herbivorous insects. However, high preference for certain plants means little if they are unavailable. Food selection by herbivores is an interaction among quality, preference, and availability.

The need for quality foods differs among herbivores. Ruminant animals (see Chapter 24) such as sheep and cattle can subsist on rougher or lower-quality forage. Bacteria in their rumen, or first stomach, can synthesize vitamin B_1 and certain amino acids from simple nitrogenous compounds, allowing them to meet their nutritional requirements from plant materials that are too low in quality for nonruminants. Thus the caloric content and the nutrient status of a certain food item might not reflect its real nutritive value for a given herbivore. Nonruminant herbivores require forage containing more complex proteins, and they may carry bacterial microbiota in the caecum, a blind pouch arising off the intestine, that aid digestion. Other animals, such as seed-eating herbivores, specialize in plant tissues with an inherently high concentration of nutrients. Such animals are not likely to have dietary problems as long as the resource is available.

Among the carnivores, quantity is more important than quality. Carnivores rarely have a dietary problem because they consume animals that have resynthesized and stored proteins and other nutrients from plants in their tissues. Carnivores are simply converting animal tissue with the same chemical composition as their own into carnivore flesh.

9.9   Mineral availability affects animals' growth and reproduction

Mineral availability and deficiencies also appear to influence the abundance and fitness of some animals. One essential nutrient that has received attention is sodium, the most variable nutrient in forest and arctic ecosystems. In areas of sodium deficiency in the soil, herbivorous animals face an inadequate supply of sodium in their diets. The problem has been noted in Australian herbivores such as kangaroos, in African elephants *(Loxodonta africana)*, in rodents, in white-tailed deer, and in moose *(Alces alces)*.

Figure 9.10 A mineral lick used by white-tailed deer.

Sodium deficiency can influence the distribution, behavior, and physiology of mammals, especially the herbivores. The spatial distribution of elephants across the Wankie National Park in central Africa appears to be closely correlated with the sodium content of drinking water. The greatest number of elephants occurs at water holes with the highest sodium content. Three herbivorous mammals, the European rabbit *(Oryctolagus cuniculus)*, the moose, and the white-tailed deer, experience sodium deficiencies in parts of their range. In sodium-deficient areas in southwestern Australia, the European rabbit builds up reserves of sodium in its tissues during the nonbreeding season. These reserves become exhausted near the end of the breeding season, which ends abruptly. During the breeding sea-

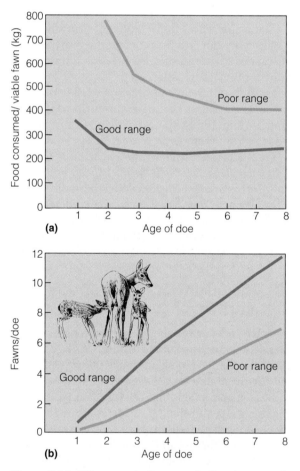

Figure 9.11 Differences in the reproductive success of female white-tailed deer on good and poor range in New York State. (a) Food consumed per viable fawn on poor range (Adirondack Mountains) was much greater than food consumed on good range (western New York). (b) Reproductive success was considerably greater on good range than on poor range.

son, the rabbits selectively graze on sodium-rich plants to the point of depletion.

Ruminants face severe mineral deficiencies in spring. Attracted by the flush of new growth, deer, bighorn sheep (*Ovis canadensis*), mountain goats (*Oreamnos americanus*), elk (*Cervus elaphus*), and domestic cattle and sheep feed on new succulent growth of grass, but with high physiological costs. Spring vegetation is much higher in potassium relative to calcium and magnesium than during the rest of the year. This high intake of potassium stimulates the adrenal gland to increase the secretion of aldosterone. Aldosterone is the principal mineral hormone that promotes retention of sodium by the kidney. Although aldosterone stimulates the retention of sodium, it also facilitates the excretion of potassium and magnesium. Because concentrations of magnesium in soft tissues and skeletal stores are low in herbivores, these animals experience magnesium deficiency. This deficiency results in a rapid onset of diarrhea and often a neuromuscular derangement or tetany. The deficiency comes late in gestation for females and at the beginning of antler growth for male deer and elk, a time when mineral demands are high.

To counteract this mineral imbalance in the spring, large herbivores seek mineral licks, places in the landscape where animals concentrate to satisfy their mineral needs by eating mineral-rich soil (Figure 9.10). Although sodium chloride is associated with mineral licks, animal physiologists hypothesize that it is not sodium the animals seek but magnesium, and in the case of bighorn sheep, mountain goats, and elk, calcium as well.

The size of deer, their antler development, and reproductive success all relate to nutrition. Other factors being equal, only deer obtaining high-quality foods grow large antlers. Deer on diets low in calcium, phosphorus, and protein show stunted growth, and bucks develop only thin spike antlers. Reproductive success of does is highest where food is abundant and nutritious (Figure 9.11).

CHAPTER REVIEW

SUMMARY

Macronutrients and Micronutrients (9.1–9.3) All living organisms require certain nutritive elements essential to survival, growth, and reproduction. Some they need in large quantities, including carbon, oxygen, nitrogen, calcium, phosphorus, potassium, and sodium. This group makes up the macronutrients. Organisms require other nutrients in lesser, often minute quantities. They are micronutrients or trace elements. They include copper, zinc, and iron. Without trace elements organisms will die or become impaired **(9.1)**. The essential nutrients are derived from atmospheric or geological sources. Most of them are released from rocks by weathering. Weathering is the gradual breakup and dissolution of rock materials, releasing mineral nutrients into the soil solution **(9.2)**. Terrestrial plants take up nutrients from soil through the roots. As roots deplete nearby nutrients, diffusion of water and nutrients through the soil replaces them. Plants continuously produce fine roots to reach an undepleted nutrient supply **(9.3)**.

Microorganisms and Nutrient Availability (9.4–9.5) Decomposer organisms derive their energy and nutrients from the dead organic matter on which they feed. Carbon compounds such as fats and simple carbohydrates are high sources of energy; cellulose and lignin are harder for them to digest. Decomposer organisms transform nutrients tied up in organic form into mineral nutrients available to plants. This process is called mineralization. If the nutrient quality of the dead organic matter is low, decomposers must take up nutrients from the soil, reducing the availability to plants. This process is called immobilization. The rate at which nutrients become available to plants during decomposition is the difference between the rates of these two processes **(9.4)**. Other microorganisms, notably the free-living cyanobacteria, *frankiae*, and *rhizobia*, fix atmospheric nitrogen, making it available to plants. Another group of organisms, the mycorrhizal fungi that live on and in the roots of plants, assists plants in the uptake of nitrogen and phosphorus from the soil **(9.5)**.

Plant Adaptations to Nutrient Availability (9.6–9.7) Availability of nutrients has a direct effect on a plant's survival, growth, and reproduction. Nitrogen is important because rubisco and chlorophyll are nitrogen-based compounds essential to photosynthesis. Uptake of nitrogen and other nutrients depends upon availability and demand. Plants with high nutrient demands grow poorly in low-nutrient environments. Plants with lower demands survive and grow, if slowly, in low-nutrient environments. Plants adapted to low-nutrient environments exhibit lower rates of growth and increased longevity of leaves. The lower nutrient concentration in their tissues means lower quality of food for decomposers **(9.6)**. The acidity of soil can influence the availability and uptake of nutrients. In acidic soils, hydrogen ions displace other nutrients, decreasing the nutrient content of the soil and increasing the loss of nutrients by leaching. Plants that grow well on acid soil, are tolerant of aluminum, and are sensitive to calcium are called calcifuges. Plants susceptible to aluminum toxicity grow best in soils high in calcium. They are called calcicoles **(9.7)**.

Animal Nutrition (9.8–9.9) Directly or indirectly, animals get their nutrients from plants. Low concentration of nutrients in plants can have adverse effects on the growth, development, and reproduction of plant-eating animals. Herbivores convert plant tissue to animal tissue. Among plant eaters, the quality of food, especially its protein content and digestibility, is critical. Carnivores consume nutrients already synthesized from plants and converted into animal flesh. Their major problem is securing a sufficient quantity of food **(9.8)**. Three essential nutrients that influence the distribution, behavior, growth, and reproduction of grazing animals are sodium, calcium, and magnesium. Grazers seek these nutrients from mineral licks and from the vegetation they eat **(9.9)**.

STUDY QUESTIONS

1. Distinguish between a macronutrient and a micronutrient.
2. What is weathering, and how does it influence nutrient availability?
3. What factors affect weathering?
4. What influences the rate at which nutrients are released during decomposition?
5. What parts do bacteria and fungi play in nutrition of plants and animals?
6. Contrast plants adapted to low-nutrient and high-nutrient environments.
*7. Discuss the differences between plants that grow best on high lime soils and on acid soils.
8. Why is the quality of food more important than the quantity of food for plant-eating animals?

CHAPTER 10

SOIL

OBJECTIVES

On completion of this chapter, you should be able to:

- Define soil and its general features.
- List the five major factors in soil development, and explain how soil develops.
- Discuss the role of living organisms in soil development.
- Describe the soil profile and its horizons.
- Describe humus.
- Explain the importance of cation exchange in the soil.
- List the distinguishing physical characteristics of soil.
- Relate climate and vegetation to the major soil orders.
- Describe the features of soil as an environment for life.
- State the causes and consequences of soil erosion.

Tree roots penetrate the rocky substrate, breaking it down into finer particles, which are then exposed to weathering.

Soil is the foundation upon which all terrestrial life and much freshwater aquatic life depend. It is the medium in which plants are rooted, a reservoir of minerals for plants upon which, in turn, animal life depends (see Chapter 9). It is the site of decomposition of organic matter and the staging area for the return of nutrients in the mineral cycle (see Chapter 25). Roots occupy a considerable portion of the soil, to which they tie the vegetation and from which they pump water and minerals in solution for photosynthesis (see Chapter 7) and other biogeochemical processes. Vegetation, in turn, influences the development of soil, its chemical and physical properties, and its organic matter content. Thus soil acts as a pathway between the organic and mineral worlds, a pathway easily altered or destroyed by human interference.

10.1   Soil is not easily defined

As familiar as it is, soil is difficult to define. One definition says that soil is a natural product formed and synthesized by the weathering of rocks and the action of living organisms. Another states that soil is a collection of natural bodies of earth, composed of mineral and organic matter and capable of supporting plant growth. Indeed, one eminent soil scientist, Hans Jenny, a pioneer of modern soil studies, will not give an exact definition of soil. In his book *The Soil Resource*, he writes:

> Popularly, soil is the stratum below the vegetation and above hard rock, but questions come quickly to mind. Many soils are bare of plants, temporarily or permanently, or they may be at the bottom of a pond growing cattails. Soil may be shallow or deep, but how deep? Soil may be stony, but surveyors (soil) exclude the larger stones. Most analyses pertain to fine earth only. Some pretend that soil in a flowerpot is not a soil, but soil material. It is embarrassing not to be able to agree on what soil is. In this pedologists are not alone. Biologists cannot agree on a definition of life and philosophers on philosophy.

Of one fact we are sure. Soil is not just an abiotic environment for plants. It is teeming with life—billions of minute and not so minute animals, bacteria, and fungi. The interaction between the biotic and the abiotic makes the soil a living system.

Soil scientists recognize soil as a three-dimensional unit or body, possessing length, width, and depth. A three-dimensional soil body large enough that we can study its physical and chemical properties is called a **pedon.** The pedon is the basic unit in the study of soils.

10.2   Soil formation involves five interdependent factors

Hans Jenny proposed that any individual soil or soil property results from the interaction of five interdependent soil-forming factors: parent material, climate, biotic factors, topography, and time.

Parent material is the unconsolidated mass from which soils form. It is derived from parent rock or from transported material. Rocks are residual or in-place parent material. They may be igneous, sedimentary, or metamorphic (see Section 9.2). The composition of all these rocks largely determines the chemical composition of the soil.

Other parent materials are transported by wind, water, glaciers, and gravity. Because of the diversity of materials, transported soils are commonly more fertile than soils derived from in-place parent materials.

Climate influences the development of soil. Temperature and rainfall, which vary with elevation and latitude, govern the rate of weathering of rocks, decomposition of minerals and organic matter, and leaching and movement of weathered material. Further, climate influences the plant and animal life in a region, both of which are important in soil development.

Biotic factors—plants, animals, bacteria, and fungi—all contribute to the formation of soil. Vegetation is largely responsible for the organic matter in the soil and the color of the surface layer. It reduces erosion, and it influences the nutrient content of the soil (see Section 9.3). Animals, bacteria, and fungi decompose organic matter, mix it with mineral matter, and aid in the aeration of soil material and the percolation of water through it.

Topography, the contour of the land, influences the amount of water that enters the soil. More water runs off and less enters the soil on steep slopes than on level land. Therefore soil is less developed on steep slopes, and layers are often indistinct and shallow. On low and flat land, extra water enters the soil, and the subsoil may be wet and grayish in color. Topography also influences the rate of erosion and downhill transport of soil material.

The weathering of rock material (see Section 9.2); the accumulation, decomposition, and mineralization of organic material; the loss of minerals from the upper surface; gains in minerals and clay in the lower layers; and layer differentiation—all require considerable *time*. The formation of well-developed soils may require 2000 to 20,000 years. Soil differentiation from parent material, however, may take place within 30 years. Certain acid soils in humid regions develop in

2000 years because the leaching process is speeded by acidic materials. Parent materials heavy in texture require a much longer time to develop into "climax" soils, because they impede downward percolation of water (see Section 7.4). Soils develop more slowly in dry regions than in humid ones. Soils on steep slopes often remain young regardless of geological age, because rapid erosion removes soil nearly as fast as it forms. Floodplain soils age little through time, because of the continuous accumulation of new materials. Young soils are not as deeply weathered as and are more fertile than old soils, because they have not been exposed to leaching as long. Old soils tend to be infertile because of long-time leaching of nutrients without replacement by fresh material.

10.3   Formation of soil begins with mechanical and chemical weathering

The formation of soil begins with the weathering of rocks and their minerals. Weathering includes both the mechanical destruction of rock materials into smaller particles and the chemical modification of primary minerals into new secondary minerals. These secondary minerals include a variety of clay minerals that may weather into still other kinds of clay minerals.

Mechanical weathering comes about through the interaction of several forces. Exposed to the combined action of water, wind, and temperature, rock surfaces flake and peel away. Water seeps into crevices, freezes, expands, and cracks the rock into smaller pieces. Growing roots of trees split rock apart.

Eventually, rocks break down into loose material. This material may remain in place, but more often than not, much of it is lifted, sorted, and carried away. Material transported by wind is known as *loess* and that transported by water as *alluvial, lacustine* (lake), or *marine* deposit. Material transported by glacial ice is *till*. In a few places soil materials come from accumulated organic matter, called *peat*. Materials remaining in place are called *residual*. This mantle of unconsolidated material is the **regolith**. It may consist of slightly weathered material with fresh primary minerals, or it may be intensely weathered and consist of highly resistant minerals such as quartz.

Accompanying this mechanical weathering and continuing long afterward is **chemical weathering,** brought about by the activities of soil organisms such as lichens and mosses, the acids they produce, and the continual addition of organic matter to mineral matter. Rain water falling upon and filtering through the accumulating organic matter picks up acids and minerals in solution and sets up a chain of complex chemical reactions in the regolith. This chemical activity decomposes primary minerals. Easily weathered, these minerals, particularly the aluminosilicates, convert to secondary minerals, particularly clays. Because iron is especially reactive with water and oxygen, iron-bearing minerals are prone to rapid decomposition. Iron may remain oxidized in the red ferric state, or it may be reduced to the gray ferrous state. The clay particles produced are shifted and rearranged within the mass by percolating water and on the surface by runoff, wind, or ice. Percolating water carries calcium, potassium, and other mineral elements, soluble salts, and carbonates deeper into the soil material or even carries them away into streams, rivers, and the sea. The greater the rainfall, the more water moves down through the soil material and the less moves upward. These localized chemical and physical processes in the parent material result in the development of layers or horizons in the soil, giving individual soils their distinctive profiles.

Because of variations in slope, climate, and native vegetation, many different soils can develop from the same parent materials. The thickness of parent material, the kind of rock from which it was formed, and the degree of weathering affect fertility and water relations of the soil.

10.4   Living organisms influence soil formation

Plants and animals have a pronounced influence on soil development. In time, plants colonize the weathered material. Plant roots penetrate and further break down the parent material. They pump nutrients up from its depths and add them to the surface. In doing so, plants recapture minerals carried deep into the soil by weathering processes. Through photosynthesis, plants capture the sun's energy and add a portion of it in the form of organic carbon to the soil. This energy source of plant debris enables bacteria, fungi, earthworms, and other soil organisms to colonize the area.

The decomposition of organic matter turns organic compounds into inorganic nutrients (see Section 9.4). Invertebrate animals in the soil—millipedes, centipedes, earthworms, mites, springtails, grasshoppers, and others—consume fresh material and leave partially decomposed products in their excreta. Microorganisms further reduce this material into water-soluble nitrogenous compounds and carbohydrates and an accumulation of insoluble waxes, resins, and lignins. Residual material makes up humus, which is eventually mineralized into inorganic compounds.

Humus is difficult to define and not fully understood. It is a dark, homogenous organic material made up of many complex compounds. It varies in nature, depending upon the vegetation from which it derives. Its decomposition proceeds slowly. The equilibrium between the formation of new humus and the decomposition of old determines the amount of humus in the soil.

Although there are a number of ways to describe humus, the most useful is based on physical features. The most familiar is raw or skeletal humus, whose appearance resembles the humus you purchase bagged at the garden center to add to flowerbeds or flowerpots.

Humus that develops under the cover of mixed and deciduous forests on fresh and moist soils is inseparable from the upper layer of mineral soil. A wide diversity of soil organisms fragment and decompose plant material into humic substances and mix it with the upper layer of mineral soil. This humic material gives forest topsoil its dark color.

In dry or moist acidic habitats, especially heathland and coniferous forests, the humus rests on top of the mineral soil as a matted or compacted deposit. An accumulation of litter slowly decomposes and remains unmixed with mineral soil. The reason is that soil organisms capable of living in the acidic conditions cannot mix humic materials with mineral soil. The main decomposer organisms there are acid-producing fungi.

The most prevalent type of humus is the one that is well mixed with mineral matter in the soil. Small arthropods, particularly springtails and mites, transform plant residues into feces and tiny fragments. The droppings, plant fragments, and mineral particles all form a loose, netlike structure held together by chains of small droppings that are easily separated.

10.5 The soil body has horizontal layers or horizons

A fresh cut along a roadbank or an excavation reveals bands and blotches of color. If you look closer, handling the material, you discover changes in texture and structure. Any vertical cut through a body of soil or a pedon is the **soil profile** (Figure 10.1). The apparent layers are called the **horizons.** Each horizon has a characteristic set of features, particularly color, that distinguishes it from other horizons. Each horizon has its own thickness, texture, structure, consistency, porosity, chemistry, and composition.

In general, soils have five major horizons: O, an organic layer, and A, E, B, and C, the mineral layers. Below the four may lie the R or nonsoil horizon. In some soils the horizons are quite distinct. In other soils the horizons form a continuum, with no clear-cut boundary between one horizon and another.

The O horizon is the surface layer, formed or forming above the mineral layer. It consists of fresh or partially decomposed organic material that has not been mixed into mineral soil. The O horizon is further subdivided into an O_i or litter layer and an O_a or humus layer (see Section 10.4). The O_i layer fluctuates seasonally. In temperate regions, it is thickest in the fall, when new leaf litter accumulates on the surface. It is thinnest in the summer after decomposition has taken place. It is usually absent in cultivated soils. This layer and the upper part of the next horizon, A, constitute the zone of maximum biological activity. They are subject to the greatest changes in temperature and moisture, contain the most organic carbon, and are the sites where most or all decomposition takes place.

The A horizon is the upper layer of mineral soil with a high content of organic matter. It is characterized by an accumulation of organic matter, and by the loss of some clay, inorganic minerals, and soluble matter. The E horizon (once labeled the A_2 horizon) is the zone of maximum leaching (eluviation, thus the label E). The downward movement of water and of

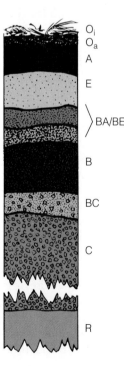

Figure 10.1 A generalized profile of the soil. Rarely does any one soil possess all of the horizons shown. Horizon designations are explained in the text.

suspended and dissolved materials alters its chemistry and structure. The E horizon has a granular, platy, or crumblike structure, often with a whitish color. Some soils possess both an A and an E horizon. Other soils lack one or the other, depending on weathering and the incorporation of organic matter with mineral soil.

The B horizon (the zone of illuviation) accumulates silicates, clay, iron, aluminum, and humus from the E horizon. There may be a transitional zone above it, BA or BE. The B horizon has blocky, columnar, or prismatic shapes (see Section 10.6). Below the B horizon in some soils there may be compact, slowly permeable layers of clay that become hard when dry and stiff when wet. Such layers of clay, called clay pans, interfere with root and water penetration in the soil.

The C horizon contains weathered material, either like or unlike the material from which the soil is presumed to have developed. Some active weathering takes place in this horizon, but it is little affected by soil formation. Below the C horizon is R, unweathered material or parent material.

10.6   Soils have certain distinguishing physical characteristics

Soils are distinguished by differences in their physical and chemical properties. Physical properties include color, texture, structure, moisture (see Chapter 7), and depth. All may be highly variable from one soil to another.

Color is one of the most useful characteristics for the identification of soil. It has little direct influence on the function of a soil, but when considered with other properties, it can tell a good deal about the soil. In temperate regions, brownish-black and dark brown colors, especially in the A horizon, generally indicate considerable organic matter. The B horizon of well-drained soils may range anywhere from pale brown to reddish and yellowish colors. Dark brown and blackish colors, especially in the lower horizons, indicate poor drainage. It does not always follow, however, that dark-colored soils are high in organic matter. Soils of volcanic origin, for example, are dark from their parent material of basaltic rocks. In warm temperate and tropical regions, dark clays may have less than 3 percent organic matter. Red and yellow soils derive their colors from the presence of iron oxides, the bright colors indicating good drainage and good aeration. The intensity of red and yellow colors increases from cool regions to the equator. Other red soils obtain their color from parent material such as red lava rock and not from soil-forming processes. Well-drained yellowish sands are white sands that contain a small amount of organic matter and such coloring material as iron oxide. Quartz, kaolin, carbonates of calcium and magnesium, gypsum, and various compounds of ferrous iron give whitish and grayish colors to the soil. Grayish colors indicate permanently saturated soils in which iron is in the ferrous form. Imperfectly and poorly drained soils are mottled with blotches of various shades of yellow-brown and gray. The colors of soils are determined by the use of standardized color charts, notably the Munsell color charts.

Soil texture is the proportion of different sized soil particles (Figure 10.2). Texture is partly inherited from parent material and partly a result of the soil-forming process.

Particles are classified on the basis of size into gravel, sand, silt, and clay. Gravel consists of particles larger than 2.0 mm. They are not part of the fine fraction of soil. Sand ranges from 0.05 to 2.0 mm, is easy to see, and feels gritty. Silt consists of particles from 0.002 to 0.05 mm in diameter, which can scarcely be seen by the naked eye and feel and look like flour. Clay particles are less than 0.002 mm, too small to be seen under an ordinary microscope. Clay controls the most important properties of soils. One is plasticity—the ability to change shape when pressure is applied and

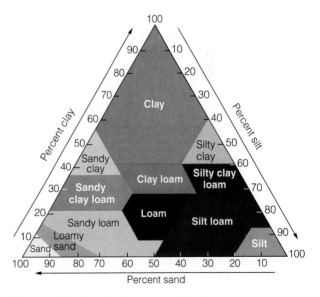

Figure 10.2 A soil texture chart, showing the percentages of clay (below 0.002 mm), silt (0.002 to 0.05 mm), and sand (0.05 to 2.0 mm) in the basic soil texture classes. For example, a soil with 60 percent sand, 30 percent silt, and 10 percent clay would be a sandy loam.

retain that shape when pressure is removed. The other is cation exchange—the exchange of ions between soil particles and soil solution. A soil's texture is the percentage (by weight) of sand, silt, and clay. Based on proportions of these components, soils are divided into texture classes.

Soil texture affects pore space in the soil, which plays a major role in the movement of air and water in the soil, penetration by roots, and field capacity. In an ideal soil, particles make up 50 percent of the total volume of the soil; the other 50 percent is pore space containing soil gases. Pore space includes spaces within and between soil particles, as well as old root channels and animal burrows. Coarse-textured soils possess large pore spaces that favor rapid water infiltration, percolation, and drainage. To a point, the finer the texture, the smaller the pores, and the greater the availability of active surface for water adhesion and chemical activity. Very fine-textured or heavy soils, such as clays, easily become compacted if plowed, stirred, or walked upon. They are poorly aerated and difficult for roots to penetrate. This undesirable condition can be alleviated by the addition of organic matter.

Soil particles are held together in clusters or shapes of various sizes, called **aggregates** or **peds.** The arrangement of these aggregates is called soil structure. There are many types of soil structure, including prismatic, columnar, blocky, platelike, and granular (Figure 10.3). Structure is influenced by texture, plants growing in the soil, other organisms, and the soil's chemical status.

Soil depth varies across the landscape, depending on slope, weathering, parent materials, and vegetation. Soils developed under native grassland tend to be several meters deep. Soils developed under forests are shallow, with an A horizon of about 15 cm and a B horizon of about 60 cm. On level ground at the bottom of slopes and on alluvial plains, soils tend to be deep. Soils on ridgetops and steep slopes tend to be shallow, with bedrock close to the surface. The natural fertility and water-holding capacity of such soils are low.

10.7 Cation exchange capacity is important to soil fertility

Chemical elements are dissolved in soil solution, bound up in organic matter, or adsorbed on soil particles. In soils ions are limited in their mobility, because they are closely held to particles of clay and humus. In aquatic systems the ions dissolve in water and obey the laws of diffusion and dilute solutions.

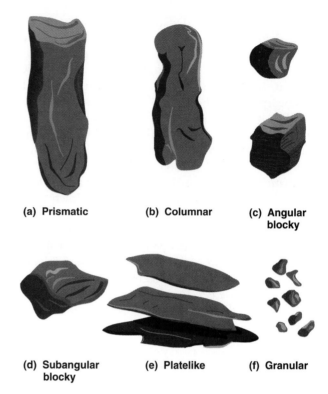

(a) Prismatic **(b) Columnar** **(c) Angular blocky** **(d) Subangular blocky** **(e) Platelike** **(f) Granular**

Figure 10.3 Some types of soil structure.

The key to the availability of nutrients in the soil is the clay-humus complex. Because of its unique structure and chemical properties, clay controls fertility. The negatively charged edges and sides of the platelike clay particles, termed **micelles,** are exchange sites. They attract positive ions known as cations, water molecules, and organic substances (see Section 9.7). The number of negatively charged exchange sites on clay and humus particles is the **cation exchange capacity.** The positively charged ions can be replaced by other ions in the soil solution (see Section 9.7). The negative charges enable a soil to prevent the leaching of its positively charged nutrient cations.

Not all positively charged ions are equally attracted to and held at exchange sites. Some bind more tightly than others. The decreasing order of binding is H^+, Al^{2+}, Ca^{2+}, Mg^{2+}, K^+, NH_4^+, and Na^+. Ions at the end of this list are more easily displaced, and aluminum and hydrogen ions may take their place. The percentage of sites occupied by ions other than hydrogen is the **percent base saturation.** Acidic soils may have a high cation exchange capacity but a low percent base saturation, because of a high number of exchangeable hydrogen ions. Soils with a high cation exchange capacity are potentially fertile. Soils with both a high cation exchange capacity and high base saturation are fertile, unless they are saline or contain toxic heavy metals.

Cations in soil solutions are continuously being replaced by or exchanged with cations on the clay and humus particles in response to changes in the concentration in the soil solution. For example, as roots remove certain cations, they reduce the concentration of those cations in soil solution, and pull more cations from clay particles. Hydrogen ions added by rain water, by acids from organic matter, and by metabolic acids from roots and microorganisms increase the concentration of hydrogen ions in the soil solution and displace other cations, such as Ca^{2+}, on the micelles. As more and more hydrogen ions replace other cations, the soil becomes increasingly acidic (see Section 9.7). In more acidic soils, aluminum and hydrogen ions become nonexchangeable, lowering the cation exchange capacity.

10.8 Climate and vegetation produce different soil orders

Broad regional differences in climate, vegetation, time, and degree of weathering produce major kinds of soils, called **soil orders.** Each order has distinctive features, summarized in Table 10.1 and its own distribution, mapped in Figure 10.4. Soil horizons vary tremendously among orders (Figure 10.5).

Two orders of soils lack B horizons—the **Vertisols** and **Entisols.** They have evolved under unusual conditions in many regions. Vertisols include soils high in clays. They swell in response to wetting and shrink while drying. Such wetting and drying causes frequent deep cracking. Entisols, found in a variety of climates, are typical of mountainous and sandy soil regions. They have a poorly developed E horizon.

In older mountain regions, such as the Appalachians, are the related **Inceptisols.** Inceptisols have a developing B horizon. Both Entisols and Inceptisols are considered young soils.

Soils of arid and semiarid regions are **Aridisols.** Because vegetation is sparse, little organic matter accumulates, and these soils are low in humus. They are, however, high in base content, and often have a gypsum, clay, or carbonate horizon. Scant precipitation results in slightly weathered and slightly leached soils high in plant nutrients. The horizons are faint and thin.

Within these regions are areas where soils contain excessive amounts of soluble salts, either from the parent material or from the evaporation of water draining in from adjoining land. Infrequent rains carry the salts to a certain depth. There water evaporates, leaving a more or less cemented horizon of sodium, calcium,

TABLE 10.1

THE ELEVEN MAJOR SOIL ORDERS (UNITED STATES SYSTEM)		
Order	**Derivation and Meaning**	**Description**
Entisol	Coined from "recent"	Dominance of soil materials; absence of distinct horizons; on floodplains and rocky soils.
Vertisol	Latin *verto*, "inverted"	Dark clay soils that exhibit wide, deep cracks when dry.
Inceptisol	Latin *inceptio*, "beginning"	Texture finer than loamy sand; little translocation of clay; often shallow; moderate development of horizons.
Aridisol	Latin *aridus*, "arid"	Dry for extended periods; low in humus; high in base content; may have carbonate, gypsum, clay horizons.
Mollisol	Latin *mollis*, "soft"	Surface horizons dark brown to black with soft consistency; rich in bases; in semihumid regions.
Spodosol	Greek *spodus*, "ashy"	Light gray, whitish E horizon on top of a black and reddish B horizon high in extractable iron and aluminum.
Alfisol	Coined from Al and Fe	Shallow penetration of humus; translocation of clay; well-developed horizons.
Ultisol	Latin *ultimus*, "last"	Intensely leached; strong clay translocation; low base content; in humid warm climate.
Oxisol	French *oxide*, "oxidized"	Highly weathered soils; red, yellow, or gray; rich in kaolinite, iron oxides, and often humus; in tropics and subtropics.
Histosol	Greek *histos*, "organic"	High content of organic matter; in bogs.
Andisol	Japanese *an*, "black," *do*, "soil"	Developed from volcanic ejecta; not highly weathered; upper layers dark; low density.

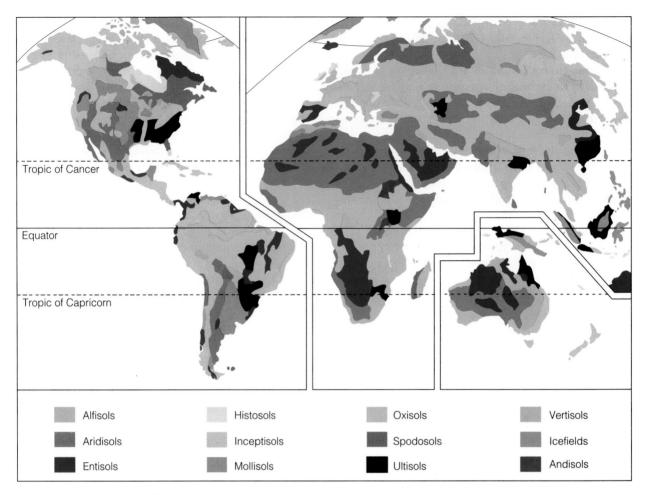

Figure 10.4 The world distribution of the 11 major soil orders and icefields.

Alfisols	Histosols	Oxisols	Vertisols
Aridisols	Inceptisols	Spodosols	Icefields
Entisols	Mollisols	Ultisols	Andisols

or magnesium salts below the surface. This cemented horizon is called **caliche.**

The subhumid-to-arid and temperate-to-tropical regions of the world—the plains and prairies of North America, the steppes of Russia, the veld and savannas of Africa, and the pampas of South America—support grassland vegetation. Dense root systems may extend many meters below the surface. Each year nearly all of the vegetative material above ground and a part of the root system go back to the soil as organic matter. Although the material decomposes rapidly, it never completely disappears before the next cycle of decay begins. Soil inhabitants mix the humus with mineral soil, creating a soil high in organic matter. There is a deep, dark, base-rich A horizon that is granular and soft and an indistinct B horizon full of calcium carbonate. These soils are **Mollisols.** Worldwide, much of their native grassland vegetation has been converted to grain.

Distinct A and B horizons characterize three other soils. All three occur in forested regions with ample rainfall but variations in temperature. As they form, bases are depleted, and insoluble iron and aluminum oxides combine in complex ways with organic acids and humus. Rain water carries these oxides and humic materials along with clays downward into the B horizon. This leaching creates an E horizon.

This soil-forming process, called **podzolization** (from *podzol,* meaning "ashy") is most intense in the highly acidic conditions of coniferous forests in cool humid climates. It leaves a light or ash-colored A horizon. The organic (O) horizon is a layer of fermented litter on top of a layer of humus unmixed with mineral soil. Such soils are called **Spodosols.**

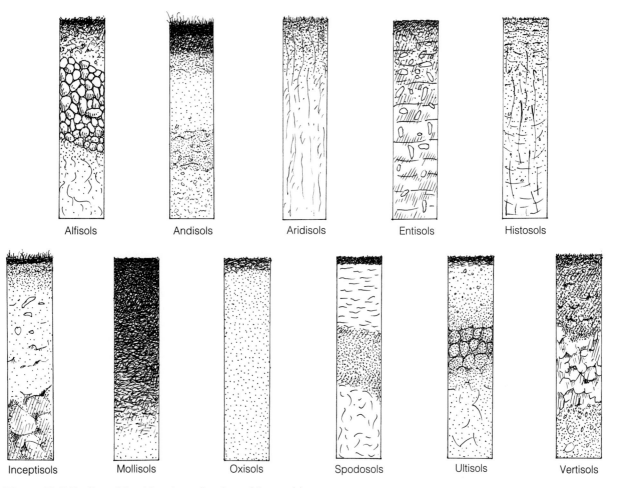

Figure 10.5 Profiles of the 11 major soil orders of the world.

Podzolization occurs in other soils, notably **Alfisols** and Ultisols. Alfisols are associated with humid, temperate forest regions. Like the Spodosols, these soils have accumulations of clay in the B horizon. Organic matter, however, tends to be well mixed by soil animals to form a darkened, organically enriched A horizon. The darkening is called melanization. In some soils the division between the A and B horizons is hard to distinguish and the E horizon is missing.

The related **Ultisols** are characteristic of warmer climates supporting coniferous and deciduous forest. Ultisols are more intensely weathered than Alfisols and less acidic than Spodosols. They have a noticeably reddish or strong yellow color because of the release of iron from the silicates. Such reddish and yellowish soils, characteristic of the southeastern United States and tropical regions, also are in part a product of an-

other soil-forming process, laterization, which produces true tropical soils, the highly weathered **Oxisols.**

In humid subtropical and tropical forested regions of the world, where rainfall is heavy and temperatures high, weathering and leaching are much more intense. They produce soils low in plant nutrients and intensely weathered to great depths. Iron and aluminum left after leaching of other minerals and bases become enriched as hydrous oxides, forming brilliant reddish colors in the E horizon. Some tropical soils, when deprived of vegetation and exposed to sun and air, tend to harden irreversibly. Although Oxisols can be farmed, they require careful management. Cutting tropical forests and mismanaging cleared land leaves vast areas eroded and unproductive.

The amount of water that passes through or remains in the soil determines the degree of oxidation and breakdown of soil minerals. Iron in soils where water stays near or at the surface most of the time is reduced to ferrous compounds. These compounds give a dull gray or bluish color to the horizons. The process, called **gleization,** may result in compact, structureless horizons. **Gley** soils are high in organic matter, because vegetation produces more organic matter than can be broken down. An absence of soil microorganisms and anaerobic conditions are responsible for a low rate of humification. Gleization is the characteristic soil-formation process in cold, wet situations, especially in the tundra; but it also is common in other regions where the water table remains above the B and C horizons. We say the water table is perched.

In places with a high water table and poor drainage and in areas of high precipitation and low evaporation, organic matter only partially decomposes. These dark soils of partially decomposed organic matter are **Histosols.** Histosols, typical of peatlands, occur in temperate regions, in the tropics, and in subarctic regions. When drained, some Histosols provide fertile soils for vegetables, but their properties change.

Regions subjected, past and present, to heavy volcanic activity have soils developed in volcanic ejecta. These soils, termed **Andisols,** are dark in color and dominated by silicate minerals. Highly fertile, they are important in Central America.

10.9   Soil supports diverse and abundant life

The soil is a radically different environment for life than the ones on and above the ground; yet the essential requirements do not differ. Like organisms that live outside the soil, life in the soil requires living space, oxygen, food, and water. Without the presence and intense activity of living organisms, soil development could not proceed. Soil inhabitants from bacteria and fungi to earthworms convert inert mineral matter into a living system.

Soil possesses several outstanding characteristics as a medium for life. It is stable structurally and chemically. The atmosphere remains saturated or nearly so, until soil moisture drops below a critical point. Soil affords a refuge from extremes in temperature, wind, light, and dryness. These conditions allow soil fauna to make easy adjustments to unfavorable conditions. On the other hand, soil hampers the movement of animals. Except to such channeling species as earth-worms, soil pore space is important. It determines the living space, humidity, and gaseous conditions of the soil environment.

Only a part of the upper soil layer is available to most soil animals as living space. Spaces within the surface litter, cavities walled off by soil aggregates, pore spaces between individual soil particles, root channels, and fissures—all are potential habitats. Most soil animals are limited to pore spaces and cavities larger than themselves.

Water in the pore spaces is essential. The majority of soil organisms are active only in water. Soil water is usually present as a thin film coating the surfaces of soil particles. This film contains, among other things, bacteria, unicellular algae, protozoa, rotifers, and nematodes. The thickness and shape of the water film restrict the movement of most of these soil organisms. Many small species and immature stages of larger centipedes and millipedes are immobilized by a film of water and are unable to overcome the surface tension imprisoning them. Some soil animals, such as millipedes and centipedes, are highly susceptible to desiccation and avoid it by burrowing deeper.

When water fills pore spaces after heavy rains, conditions are disastrous for some soil inhabitants. If earthworms cannot evade flooding by digging deeper, they come to the surface, where they often die from ultraviolet radiation and desiccation or are eaten.

A diversity of life occupies these habitats (Figure 10.6). The number of species of bacteria, fungi, protists, and representatives of nearly every invertebrate phylum found in the soil is enormous. A soil zoologist found 110 species of beetles, 229 species of mites, and 46 species of snails and slugs in the soil of an Austrian deciduous beech forest.

Dominant among the soil organisms are bacteria, fungi, protozoans, and nematodes. Flagellated protozoans range from 100,000 to 1,000,000, amoebas from 50,000 to 500,000, and ciliates up to 1000 per gram of soil. Nematodes occur in the millions per square meter of soil. These organisms obtain their nourishment from the roots of living plants and from dead organic matter. Some protozoans and free-living nematodes feed selectively on bacteria and fungi.

Living within the pore spaces of the soil are the most abundant and widely distributed of all forest soil animals, the mites (Acarina) and springtails (Collembola). Together they make up over 80 percent of the animals in the soil. Flattened, they wiggle, squeeze, and digest their way through tiny caverns in the soil.

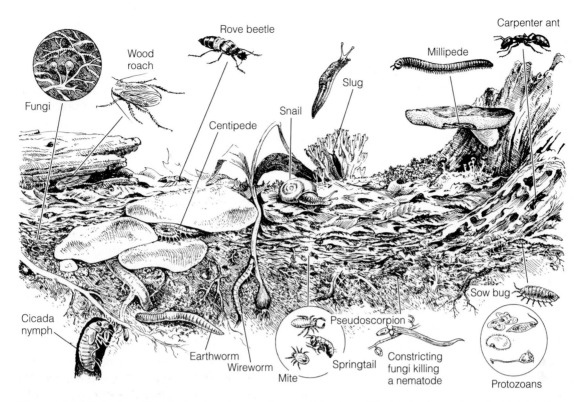

Figure 10.6 Life in the soil. This drawing shows only a small fraction of the kinds of organisms that inhabit the soil and litter. Note the fruiting bodies of fungi, which are consumed by animals, vertebrate and invertebrate.

They feed on fungi or search for prey in the dark interstices and pores of the organic mass.

The more numerous of the two, both in species and numbers, are the mites, tiny eight-legged arthropods from 0.1 to 2.0 mm in size. The most common mites in the soil and litter are the Orbatei. They live mostly on fungal hyphae that attack dead vegetation as well as on the sugars produced by the digestion of cellulose found in conifer needles.

Collembolae are the most widely distributed of all insects. Their common name, springtail, describes the remarkable springing organ at the posterior end, which enables them to leap great distances for their size. The springtails are small, from 0.3 to 1.0 mm in size. They consume decomposing plant material, largely for the fungal hyphae they contain.

Prominent among the larger soil fauna are the earthworms (Lumbricidae). Burrowing through the soil, they ingest soil and fresh litter and egest both mixed with intestinal secretions. Earthworms defecate aggregated castings on or near the surface of the soil or as a semiliquid in intersoil spaces along the burrow.

These aggregates produce a more open structure in heavy soil and bind light soil together. In this manner earthworms improve the soil environment for other organisms.

Feeding on the surface litter are millipedes. They eat leaves, particularly those somewhat decomposed by fungi. Lacking the enzymes necessary for the breakdown of cellulose, millipedes live on the fungi contained within the litter. The millipedes' chief contribution is the mechanical breakdown of litter, making it more vulnerable to microbial attack, especially by saprophytic fungi.

Accompanying the millipedes are snails and slugs. Among the soil invertebrates they possess the widest range of enzymes to hydrolyze cellulose and other plant polysaccharides, possibly even the highly indigestible lignins.

Not to be ignored are termites (Isoptera), white wingless, social insects. Except for some dipteran and beetle larvae, termites are the only larger soil inhabitants that can break down the cellulose of wood. They do so with the aid of symbiotic protozoans living in their gut. Termites dominate the tropical soil fauna. In the tropics termites are responsible for the rapid removal of wood, dry grass, and other materials from the soil surface. In constructing their huge and complex mounds, termites move considerable amounts of soil.

Detrital-feeding organisms support predators. Small arthropods are the principal prey of spiders, beetles, pseudoscorpions, predaceous mites, and centipedes. Protozoans, rotifers, myxobacteria, and nematodes feed on bacteria and algae. Various predaceous fungi live on bacteria-feeders and algae-feeders.

10.10 Soils have been severely disturbed

Over much of the landscape, soils have been highly disturbed. Soils have been buried under fill; overturned and moved about by excavations, surface mining, and road construction; and exposed to erosion by wind and water. Upper horizons have been mixed by agricultural plowing and tillage, and compacted by heavy machinery and trampling.

Soil protected by vegetation maintains its integrity. Vegetation breaks the force of the wind and disperses raindrops, breaking their force. Rain trickles slowly through the litter, infiltrating the soil. If rainfall exceeds the soil's capacity to absorb it, the excess runs across the surface, but vegetation slows its movement.

Stripped of its protective vegetation and litter by plowing, logging, grazing, road building, and urban and suburban construction, soil is highly vulnerable to soil **erosion**—the carrying away of particles by wind and water faster than new soil can form. (Soil forms at the rate of about 1 metric ton per hectare per year.) Loss of the upper layers of humus-charged, granular, absorptive topsoil exposes the humus-deficient, less stable, less absorptive, and erodible layers beneath. If the subsoil is clay, it absorbs water so slowly that heavy rains produce a highly erosive and rapid runoff.

Soil compaction intensifies the problem. Heavy machinery from large agricultural tractors to construction equipment compacts soil on large areas of ground. Trampling on lawns and playing fields, concentrated use of pathways and hiking and riding trails through woods and fields, and off-road use of all-terrain vehicles compact the soil on other sites. Soil compaction occurs when any weight pushes the soil particles together and reduces the size of the pores. Greatest compaction occurs under wet and moist conditions. Moist soil particles easily slide over one another. Compacted soil cannot absorb water, so the water flows across the surface.

Rain falling on bare ground hammers the soil's surface, removing lightweight organic matter, breaking down soil aggregates, and forming a seal on the surface. Unable to infiltrate the soil, the water moves across the surface as runoff, carrying soil particles with it. The least conspicuous type of soil erosion is **sheet erosion.** It is a more or less even removal of soil over a field. Soil compaction can increase sheet erosion. When runoff concentrates in small channels or rills instead of moving evenly over a sloping land, its cutting force increases. **Rill erosion** channels water rapidly downslope. On areas where concentrated water cuts the same rill long enough or where runoff concentrates in sufficient volume to cut deeply into the soil, highly destructive gullies result. **Gully erosion** often begins in wheel ruts made by off-road vehicles in fields and forests, on logging roads and skid trails, and on livestock and hiking trails.

Bare soil, finely divided, loose, and dry, as it often is after tillage, is ripe for wind erosion. Wind picks up fine particles of dust and carries them as dust clouds. Drastic erosion occurred in the Great Plains during the droughty Dust Bowl days of the 1930s. Wind erosion is increasing worldwide today, especially in arid and semiarid regions. Often dust particles are lifted high in the atmosphere and carried for hundreds and even thousands of kilometers. Wind erosion exposes roots of plants or covers vegetation with drifting debris. In the Great Plains wind erosion exceeds water erosion.

Erosion by wind and water ruins land. Worldwide, about 12 million hectares of arable land are destroyed and abandoned annually because of soil mismanagement. Such land is usually so degraded that natural vegetation has difficulty returning. Erosion becomes progressively worse unless extreme measures are taken to restore vegetation. (See Focus on Ecology 10.1: Fighting Soil Erosion.)

Effects of erosion are felt both on and off site. Erosion on agricultural and forest lands reduces organic matter and increases clay content. It reduces water-holding capacity of the soil, intensifying drought conditions in dry weather and flooding in wet weather. Erosion degrades soil structure and reduces plant nutrients and plant rooting depths, depressing crop yields. It also reduces the diversity and abundance of soil organisms, essential to soil productivity and water infiltration. The loss of 2.5 cm of topsoil reduces corn and

FOCUS ON ECOLOGY 10.1

FIGHTING SOIL EROSION

W. C. Lowdermilk, assistant chief of the newly created Soil Conservation Service (now called the National Resources Conservation Service) in the 1930s, wrote, "In a larger sense, a nation writes its record on the land . . . —a record that is easy to read by those who understand the simple language of the land." Past civilizations and nations have left a sorry record. The impoverished lands today—the near East, Egypt, Syria, Lebanon, North Africa, the Mediterranean—all once supported flourishing agriculture, extensive forests, and great empires. Because of overgrazing, deforestation, and poor land use, these ancient empires based on agriculture are gone, the remains buried beneath soil eroded from the hills.

In spite of past experience, soil erosion still is one of the world's major environmental problems. Worldwide loss of topsoil amounts to 23 billion tons a year, and is increasing as human population growth forces agriculture to move into marginal, erodible land. Each year the countries of the former Soviet Union abandon an estimated one-half million hectares of cropland because of erosion. Dust storms in North Africa carry up to 100 million tons of soil annually out to the Atlantic, speeding desertification. The Ganges in India and the Yellow River in China carry billions of tons of soil to the ocean, as does the Mississippi River.

Soil erosion was severe in the United States in the 1930s, when plowing of the grasslands of the Great Plains and drought resulted in the Dust Bowl, and continuous cropping of cotton and tobacco gullied most of the South. Alarmed, President Franklin Delano Roosevelt and Congress established the Soil Conservation Service, which took massive action. Solutions included planting windbreaks in the Midwest, grassland restoration, crop rotation instead of monoculture, terracing, contour farming, and gully control. Such approaches were working until after World War II, when family farms declined and corporations with large land holdings took over agricultural production. Corporate farmers tore out windbreaks, terraces, fencerows, and soil-holding features that interfered with the use of large-scale machinery. They abandoned crop rotation for the monocultural production of corn, soybeans, and cotton. Again soil erosion grew severe as corporate agriculture treated soil as an economic input, not as a renewable resource. Forty-four percent of United States cropland annually loses more than 11 metric tons per hectare through wind and water erosion. This amount is the maximum tolerable level if soil productivity is to be maintained. Twenty-three percent of cropland loses more.

Serious soil erosion is forcing agriculturists to come up with new techniques. Typically, crop production involves plowing land and cultivating crops in rows to control weeds. These expensive, time-consuming practices expose soil to severe erosion. New techniques, called conservation tillage systems, range from reduced tillage to no-tillage systems. No-tillage systems plant in the residue of the previous crop. An initial application of herbicides eliminates existing grass and weeds and the need for row cultivation. Conservation tillage significantly reduces soil erosion and the loss of soil nutrients. It increases the organic matter in soil and encourages higher populations of soil microbes and invertebrate fauna.

wheat yields by 6 percent. Costly, energy-demanding chemical fertilizers mask the ruinous effects of soil erosion on the inherent fertility of the soil. In the United States, the economic cost of soil erosion amounts to close to $27 billion annually, and the environmental costs are $17 billion.

The costs of off-site effects may be more than twice as high. Soil eroded by wind and water has to go somewhere. Erosion carries sediments into rivers, reducing light penetration and clogging navigation. Sediment fills reservoirs and hydroelectric dams, shortening their lives and polluting the water. Windborne soil contributes significantly to air pollution, illness, and damage to machinery.

All forms of soil erosion destroy the integrity of ecosystems and ecological cycles. They raise the cost of food, fostering hunger and famine. For humans, the ecological consequences could well be social disorder and degradation of life. There is a true saying, "Poor soils make poor people."

CHAPTER REVIEW

SUMMARY

Soil Defined (10.1)   Soil is a natural product of unconsolidated mineral and organic matter on Earth's surface. The medium in which plants grow, soil is the foundation of terrestrial ecosystems. It is the site of decomposition of organic matter and of the return of mineral elements to the nutrient cycle. It is the habitat of animal life, the anchoring medium for plants, and their source of water and nutrients.

Soil Formation (10.2–10.5)   Soil results from the interaction of five factors: parent material, climate, biotic factors, topography, and time. Parent material provides the substrate from which soil develops. Climate shapes the development of soil through temperature, precipitation, and its influence on vegetation and animal life. Biotic factors—vegetation, animals, bacteria, and fungi—add organic matter and mix it with mineral matter. Topography influences the amount of water that enters the soil and rates of erosion. Time is required for full development of distinctive soils (10.2).

Soil formation begins with the weathering of rock and minerals. In mechanical weathering, water, wind, temperature, and plants break down rock. In chemical weathering the activity of soil organisms, the acids they produce, and rain water break down primary minerals (10.3).

Living organisms influence soil development. Plants rooted in weathering material break it down, pump up nutrients from its depths, and add all-important organic material and nutrients to the surface. Soil organisms act on organic matter to produce humus. Humus is a dark-colored, noncellular, chemically complex organic material that affects soil structure and fertility (10.4).

Soils develop in layers called horizons. Five horizons are commonly recognized, although all are not necessarily present in any one soil: the O or organic layer; the A horizon, characterized by accumulation of organic matter; E, the zone of leaching of clay and mineral matter; the B horizon, in which mineral matter accumulates; and C, the underlying material, exclusive of bedrock. These horizons may divide into subhorizons (10.5).

Distinguishing Characteristics (10.6–10.7)   Soils and horizons within soils differ in texture, structure, and color. Texture is determined by the proportions of soil particles of different sizes—sand, silt, and clay. Texture is important in the movement and retention of water in the soil (10.6). Soil particles, particularly the clay-humus complex, are important to nutrient availability and the cation exchange capacity of the soil—the number of negatively charged sites on soil particles that can attract positively charged ions. Percent base saturation is the percentage of sites occupied by ions other than hydrogen. Soils with a high cation exchange capacity are potentially fertile (10.7).

Climate, Vegetation, and Major Soil Orders (10.8)   Vegetation and climate influence soil development over large areas. Similar vegetation and climate produce soils with similar characteristics. There are 11 major soil orders. In grassland regions, calcium accumulates in the lower horizons. In arid regions, salt accumulates close to the surface. In forest regions, podzolization takes place—the leaching of calcium, magnesium, iron, and aluminum from the upper horizons and the retention of silica. In tropical regions, silica is leached and iron and aluminum oxides are retained in the upper horizon. Gleization takes place in poorly drained soils. Organic matter decomposes slowly. Iron is reduced to the ferrous state, imparting a greyish or bluish color to the horizons.

Life in the Soil (10.9)   The soil provides a unique environment for diverse organisms that in turn influence soil development and structure. Bacteria, protozoans, and other microorganisms live in the thin film of water surrounding soil particles. Most microorganisms obtain their nourishment from organic matter. Larger soil invertebrates live in pore spaces and channels in the soil. They feed on fresh litter, other detrital material, bacteria, and fungi. These detritus feeders support an array of predators, from mites to spiders.

Soil Erosion (10.10)   Worldwide, soils have been and are being depleted by erosion in which topsoil loss exceeds formation. Deforestation, poor agricultural practices, urbanization, road building, and other disturbances expose the soil to the erosive forces of water and wind. Soil erosion destroys natural and agricultural ecosystems and fills rivers, lakes, reservoirs, and navigation channels with silt. Wind erosion carries soil far from its source in clouds of dust and increases particulate air pollution. Soil erosion in all its forms impoverishes regions and nations, reduces food production, and causes extensive economic losses.

STUDY QUESTIONS

1. Why is it difficult to define soil?
2. What five major factors affect soil formation?
3. What roles do mechanical and chemical weathering play in soil formation?
4. How do living organisms develop soil?
5. What is a soil profile? What are soil horizons?
6. Distinguish among the major soil horizons.
7. How do the major soil orders reflect climate and vegetation?
8. What gives soil its textures?
9. How does cation exchange capacity affect nutrient uptake?
10. What characteristics do soils possess that make them good habitats for living organisms?
*11. Discover soil texture. In a glass jar, mix several small samples of soil from your garden, a grass field, or forest with about twice as much water. Shake vigorously and allow the sediments to settle. After 24 hours measure the depth of sand (which settles out first), silt (which settles out second on top of the sand), and clay. Divide the depth of each layer by the total depth of all the settled soil and multiply by 100 to determine the proportions of sand, silt, and clay in your soil. Compare the results with the soil texture chart (Figure 10.2).
*12. From your soil conservation office or library, obtain a soil survey for your area. How do local soils relate to agricultural development, urban development, soil erosion problems, and forest distribution?
*13. What is the major cause of soil erosion in your local area? Note the degree of siltation in streams, rivers, dams, and lakes after heavy rains.
*14. What efforts are being made to reduce erosion in your area?

POPULATIONS

PART III

PROPERTIES OF POPULATIONS

OBJECTIVES

On completion of this chapter, you should be able to:

- Define population density, crude density, and ecological density.
- Summarize the types of population distribution.
- Discuss the problems and methods of determining the density of a population.
- Describe the age structure of a population.
- Explain the significance of different types of age pyramids.
- Summarize ways in which the ages of plants and animals can be determined.
- Compare the age structure of plants with that of animals.

Wildebeest *(Connochaetes taurinus)* graze on the savanna.

So far we have explored the relationship of individual organisms to their physical environment. Now we will turn to the biotic environment. How the individual interacts with others of its own species, and with competitors, predators, parasites, diseases, mutualists, and humans, are the subjects of the following chapters. First we will consider the group of individuals of which the organism is necessarily a part, the local population.

11.1 Populations possess unique features

As an individual, how do you perceive the world? Most of us regard a friend, a neighborhood maple tree, a daisy in a field, a squirrel in the park, or a bluebird nesting in the back yard as an individual. Rarely do we consider each as a part of a larger unit, a population. A **population** is a group of potentially interbreeding and interacting individuals of the same species living in the same place at the same time. It is reproductively isolated from other such groups.

Populations have unique features. They have an age structure, density, and distribution in time and space. They exhibit a birthrate, a death rate, and a growth rate. They respond in their own ways to competition, to predation, and to other pressures. Individuals that make up a population affect one another in various ways. The relations of one population with another influence the structure and function of whole ecosystems.

11.2 Plant populations are complex

The principles of population ecology developed around the study of animals. Only recently have ecologists begun to study plant populations. To do so, plant ecologists have had to modify the techniques animal ecologists use. The reasons lie in the nature of the plants themselves.

Consider a tree. The tree appears to be one unit; but if you look up into it, you realize that the tree is a collection of modules: buds, twigs, shoots, flowers, and leaves above ground, and roots and their extensions below ground (Figure 11.1). These modules have their own demography: birthrates, death rates, and growth rates. The births and deaths of these separate modules determine not only the rate of growth but also the form of the trees.

A tree or a shrub grown from a seed is an individual with its own genetic characteristics (see Chapter 19). It reproduces sexually, producing a new generation from seeds. If it is cut to the ground, new buds may develop along the root collar and send up sprouts or coppice growth. Although its original aboveground stem is dead, the individual lives on through new vegetative growth. Trees such as black locust (*Robinia pseudoacacia*) and aspen (*Populus tremuloides*), shrubs, and numerous perennial herbaceous plants grow root extensions that send up new shoots or suckers that may remain attached to root extensions or break off to live independently (Figure 11.2). These new "individuals" or clones may cover a considerable area and appear to be individuals of different sizes and ages. Because they are reiterations of the parent stem, the plant itself may be old although the cloned individuals are young. Some of the modules die while others live and new ones appear. Birth, senescence, and death occur at the level of the module as well as the whole organism. This growth behavior is typical of most plants, as well as protozoans and fungi. As a result, it is hard to quantify the life spans of plants, especially trees, shrubs, perennial herbs, and grasses. Clones, connected or living independently, complicate the definition of what an individual is and thus the studies of plant populations.

In contrast, most animals are unitary. Their form, development, growth, and longevity, barring an early death, are predictable. There is no question about what is an individual. Exceptions are groups of invertebrate modular animals, including the sponges, hydras, corals, bryozoans, and colonial ascidians. Experimental cloning of mammals is creating new exceptions. They are examples in the animal world in which the definition of an individual becomes blurred.

11.3 Density affects individuals in a population

Two outstanding attributes of a population are density and dispersion. Think of the change in human numbers from the concentrated masses of the city to the sparse populations of the countryside.

When you drive in city traffic or on lightly traveled country roads, you are aware of population density. You respond to that density in some way. You may seek out crowds, or you may escape to the quiet of home. High densities may frustrate or stimulate you.

Individuals in natural populations are also affected by density. Trees in crowded stands may grow more slowly, and some may succumb to a lack of water, nutrients, and light, unequally shared. Scarce food may

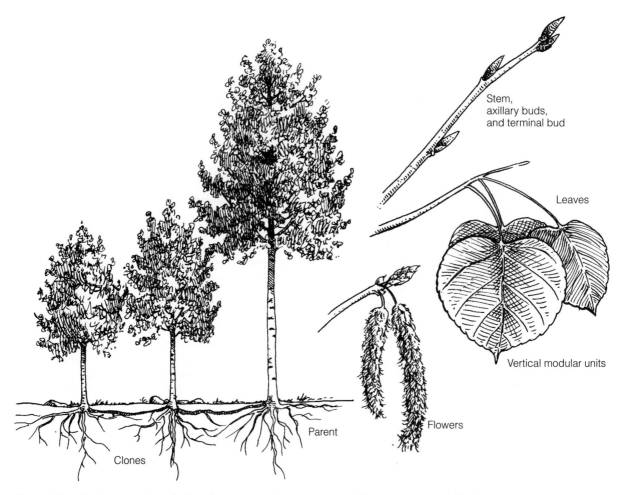

Figure 11.1 Horizontal and vertical modular growth in an aspen tree *(Populus tremuloides)*. Vertical growth involves modules of leaves, buds, and associated stems. Horizontal growth involves modules of roots and root buds, which give rise to clones. These clones are various ages, with the youngest individuals forming the leading edge of growth away from the parent.

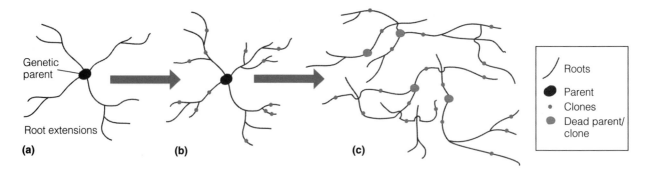

Figure 11.2 Connections between the original parent and its clones are lost in time, and some clones lead independent existences. (a) Genetic parent. (b) Clonal growth arising from root extensions. (c) Genetic parent is dead. Clonal growth arising from root extensions of the parent forms groups of new individual stems. Roots of some clones have separated from the parental and some clonal root extensions to establish new independent groups. However, all clones of the parent, independent or not, are identical genetically.

be denied to smaller or less aggressive mammals in a population. Some birds may deny others access to nest sites when not enough sites exist to meet the demand. Having too few individuals in a population may reduce the chances of finding a mate or inhibit behavior essential to the welfare of the population. Low population density may raise an individual's risk of succumbing to predation. Affecting the welfare of individuals in all these ways, density in part controls a population's birthrates, death rates, and growth.

11.4   Density is difficult to define

However important it may be, density is an elusive characteristic, difficult to define and to determine. **Density** can be characterized as the number of individuals per unit of space—as so many per square kilometer, per hectare, or per square meter. That measure is **crude density.** The trouble with this measure is that individuals do not occupy all the space within a unit, because not all of it is suitable habitat. A biologist might estimate the number of deer living in a square kilometer. The deer, however, might avoid half the area, because of human habitation, land use, and lack of cover and food. Goldenrods inhabiting old fields grow in scattered clumps because of soil conditions and competition from other old-field plants.

No matter how uniform a habitat may appear, it is usually patchy, because of microdifferences in light, moisture, temperature, exposure, or other physical conditions, or because of the lack of sites available for colonization. Each organism occupies only those areas that can meet its requirements.

To account for patchiness, density should refer to the number of individuals per unit of available living space. That measure would be **ecological density.** For example, in a study of bobwhite quail *(Colinus virginianus)* in Wisconsin, biologists expressed density in terms of the number of birds per mile of hedgerow rather than as birds per acre. Ecological densities are rarely estimated, because what portion of a habitat represents living space is difficult to determine.

11.5   Dispersion of individuals influences population density

How organisms are dispersed over space has an important bearing on density. Individuals of a population may be distributed randomly, uniformly, or in clumps (Figure 11.3). Individuals are distributed *randomly* if the position of each is independent of the others'. Forest trees of the same species in the canopy layer may be spaced at random. So are some invertebrates of the forest floor, particularly spiders.

By contrast, individuals distributed *uniformly* are more or less evenly spaced. In the animal world, uniform dispersion usually results from some form of competition, such as territoriality (see Section 14.9). Uniform distribution happens among plants when severe competition exists for crown or root space, as among some forest trees, or for moisture, as among desert plants.

The most common dispersion type is *clumped* dispersion, in scattered groups. Clumping results from responses to habitat differences, daily and seasonal weather changes, reproductive patterns, and social behavior. The distribution of human populations is clumped or aggregated because of social behavior, economics, and geography. There are various degrees and types of clumping. Groups may be randomly or nonrandomly distributed over an area. Aggregations may range from small groups to a single centralized group. If environmental conditions encourage it, populations may be concentrated in long bands or strips along some feature of the landscape such as a river, leaving the rest of the area unoccupied.

Because the environment is not homogeneous, individuals usually are not distributed as one large population across the landscape. Rather they exist as subpopulations linked together by movements of individuals into and out of them. These linked subpopulations are called **metapopulations.** Each subpopulation has its own birthrate, death rate, and probability of going extinct. Each habitat patch emptied by extinction has its own probability of being recolonized by individuals from other subpopulations. Such recolonization tends to stabilize populations across the landscape.

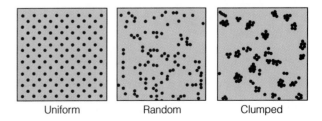

Figure 11.3 Patterns of dispersion.

11.6 Determining density and distribution requires sampling

To determine whether the distribution of individuals in a population is random, uniform, or clumped is difficult, requiring careful sampling techniques. Just as difficult is the determination of true population density. The accuracy of sampling can be influenced by the manner in which individuals are distributed over the landscape. The density figure can also be influenced by the choice of boundaries. If a population is concentrated in small areas and the population density is described in terms of individuals per square kilometer, you receive a false impression (Figure 11.4).

Fortunately, ecologists rarely need to know the exact abundance or density of a population. **Abundance** is the number of individuals in a given area, in contrast to density, which is the number of individuals expressed per unit area. Both can be determined only by a direct count. Except for unusual habitat situations, such as antelope living on an open plain or waterfowl concentrated in a marsh, direct counts are extremely difficult or impossible. Even if possible, direct counts may be too time-consuming or expensive.

For these reasons, ecologists use other means of gathering information on population density. One is sampling. Researchers divide the area of study into subunits, in which they count animals or plants of concern in a prescribed manner. From the data they determine the mean density of the unit sampled. They multiply the mean value by the total number of sampled and unsampled areas to arrive at the estimated population. They can calculate confidence limits on the estimate. Ecologists use this type of sampling most widely in the study of populations of plants and sessile (attached) animals, which cannot move in and out of an area.

For mobile populations, animal ecologists use other sampling methods. Sometimes they use mark and recapture techniques. (See Focus on Ecology 11.1: Capture-Recapture Sampling.) For most work with animals, ecologists find that relative abundance is sufficient. You might count the number of mourning doves seen along a prescribed length of road traveled year after year; you might record the number of drumming ruffed grouse heard along a stretch of trail; or you might tally the number of tracks of opossums crossing a certain dusty road. You can then convert the data to the number of individuals seen per kilometer or heard per hour. Such counts, called indexes, cannot stand alone. However, if you have a series of such index figures collected from the same area over a period of years, you can follow trends in density or abundance. You can also obtain counts from different areas during the same year to compare numbers from one habitat to another. Most population data on birds and mammals are based on indices of relative abundance rather than on direct counts.

11.7 Populations have an age structure

Unless each generation breeds and dies in a single season, not overlapping the next generation (like annual plants and grasshoppers), the population will have an age structure. Because reproduction is restricted to certain age classes and mortality is most prominent in others, the ratio of the age groups bears on how quickly or slowly populations grow.

Populations divide into three ecological periods: prereproductive, reproductive, and postreproductive. We might divide humans into young people, working adults, and senior citizens. In plants the prereproductive period is usually termed the juvenile period. The length of each period depends largely on the life history of the organism. Among annual species the length of the prereproductive period has little influence on the rate of population growth. In longer-lived plants and animals, the length of the prereproductive period has a pronounced effect on the population's rate of growth. Organisms with a short prereproductive period often increase rapidly, with a short span between generations. Organisms with a long prereproductive period, such as elephants and whales, increase slowly and have a long span between generations.

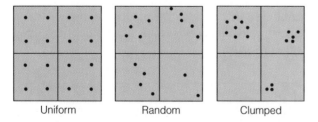

| Uniform | Random | Clumped |

Figure 11.4 The difficulty of sampling. Each area contains a population of 16 individuals. We divide each area into four sampling units and choose one at random. Our estimates will be quite different, depending upon which unit we select. For the uniform population, any sampling unit gives a correct estimate. For the random population, the estimates are 20, 20, 16, and 8. For the clumped population, the estimates are 32, 20, 0, and 12.

FOCUS ON ECOLOGY 11.1

CAPTURE-RECAPTURE SAMPLING

Capturing, marking, and recapturing animals from fish to mammals is the most widely used technique to estimate animal populations. There are many variations of this technique, ranging from single mark–single recapture to multiple capture–multiple recapture. Books are devoted to various methods of application and statistical analysis. Nevertheless, the basic concept is simple.

Capture-recapture or mark-recapture methods are based on trapping, marking, and releasing a known number of marked animals into the population. After an appropriate period of time (approximately one week for rabbits and mice), individuals are again captured from the population. Some of the individuals caught in this second period will be carrying marks, and others will not. An estimate of the population is computed from the ratio of marked to unmarked individuals in the sample. The ratio supposedly reflects the ratio of marked to unmarked animals in the population.

The simplest method, the single mark–single recapture, is known as the Lincoln index or Petersen index of relative population size. The basic model is:

$$N : M :: n : R$$

or

$$N = nM/R$$

where M is the number marked in the precensus period, R is the number of marked animals trapped in the census period, n is the total number of animals trapped in the census period, and N is the population estimate.

As an example, suppose that in a precensus period, biologists capture and tag 39 rabbits. During the census period they capture 15 tagged rabbits and 19 unmarked ones, a total of 34.

The following ratio results:

$$N : 39 :: 34 : 15$$

or

$$N = nM/R = (34)(39)/15 = 88$$

The estimate of 88 rabbits is exactly that, an estimate. To determine within what range the population lies, biologists estimate confidence limits. In this case, the confidence limits state that the chances are 95 out of 100 that the population of rabbits lies between 64 and 114 animals. Such results are typical in population studies of animals.

The capture-recapture method involves a number of assumptions:

- All individuals in a population have an equal chance of being captured. None are trap-shy and none are trap-happy.
- The ratio of marked to unmarked animals remains the same from the time of capture to the time of recapture.
- Marked individuals, once released, redistribute themselves throughout the population with respect to unmarked ones, as they were before capture.
- Marked animals do not lose their marks.
- The population is closed. No emigration or immigration takes place during the sampling period.

A number of conditions influence the probability of capture. They include time, behavioral responses to capture, mortality patterns, and the bias of sex and age on movement. Males and younger individuals are more susceptible to capture than others in the population.

11.8 Determining age structure requires finding individual ages

The **age structure** of a population is the ratio of the various age classes to each other at a given time. To determine age structure we need some means of obtaining the ages of members of a population. For humans, this task is not a problem, but it is with wild populations. Age data for wild animals can be obtained in a number of ways, the method varying with the species. The most accurate, but the most difficult method is to mark young individuals in a population and follow their survival through time. Such a method, however, requires both a large number of marked individuals and a lot of time. For this reason biologists may use other, less accurate methods. These methods include examining a representative sample of carcasses. A biologist might look for the wear and replacement of teeth in deer and other ungulates, growth rings in the cementum of the teeth of carnivores and ungulates, or annual growth rings in the horns of mountain sheep. Among birds, observations of plumage changes and wear can separate juveniles from subadults (in some species) and adults.

Studies of the age structure of plant populations are few. The major reason is the difficulty of determining the age of plants and whether the plants are genetic individuals or asexual modules or clones.

Foresters have tried to use age structure as a guide to timber management. They employ size (diameter of the trunk at breast height or dbh) as an indicator of age on the assumption that diameter increases with age. The greater the diameter, the older the tree. Such assumptions, foresters discovered, were valid for dominant canopy trees; but with their growth suppressed by lack of light, moisture, or nutrients, understory trees and even so-called seedlings and saplings add little to their diameter. Although their diameter suggests youth, small trees often are the same age as large individuals in the canopy.

You can determine the approximate age of trees in which growth is seasonal by counting annual growth rings. Attempts to age nonwoody plants, especially those with short, swollen underground stems called corms, which seem to exhibit rings of annual growth, have not been successful. The most accurate method of determining the age structure of short-lived herbaceous plants is to mark individual seedlings and follow them through their lifetimes.

11.9   Age pyramids portray age structure

Age pyramids (Figure 11.5) compare the sizes of age groups to help us visualize age structure. As the population changes with time, the number of individuals in each age class changes, and so do the ratios. A large number of young, which expands the base of the pyramid, characterizes a growing population. This large class of young eventually moves up into the reproductive age classes. If the young are as prolific as the parents, the young age class will expand even further. A higher proportion of individuals moving into the older age classes characterizes a declining population. With

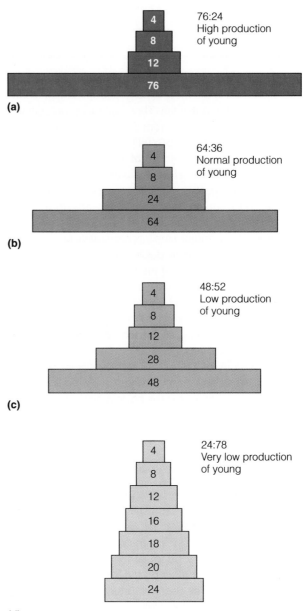

Figure 11.5 Theoretical age pyramids for mammals and birds. (a) Fast-growing populations have a large number of young and small older classes. The oldest potential age classes are missing. (b) A growing population has a broad base and a narrow, pinched top. Increasing numbers of young may enter the reproductive age and increase all age classes through time. (c) If a population is declining, a higher proportion may be in the older age classes (d) If the population is stationary, the number in each age class tends to remain the same. The pyramid is a narrow one with a narrow base of young.

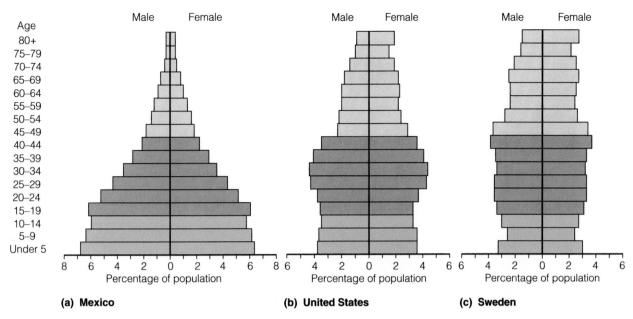

Figure 11.6 Age pyramids for three human populations. (a) Mexico shows an expanding population. A broad base of young will enter the reproductive age classes. (b) The age pyramid for the United States has a less tapered shape. The youngest age classes are no longer the largest. This pyramid reflects birth control, immigration, and declining fertility, which in time will further alter the age structure. (c) The age pyramid for Sweden is characteristic of a population that is approaching zero growth.

fewer young, fewer individuals will enter the reproductive age classes, further depressing the population. In this way, age structure changes over time.

These generalizations apply to human populations with long life span and low juvenile mortality (Figure 11.6). Age structure is much less predictive of population growth in wild animal populations. Population growth for wild animals is closely related to availability of resources, both habitat and food, and to competition for those resources. For example, a normally increasing population of a wild species should have an increasing number of young; a decreasing population, a decreasing number of young. Within this framework, however, there are a number of variations. One population may be decreasing, yet show an increasing percentage of young. Another population might remain static with an increasing percentage of young. Both situations result when too many adults are hunted. Only when we correlate such information as density, mortality, and reproduction with age structure can it provide insight into the dynamics of wild populations.

Age pyramids for plants take on a different configuration. Because the growth and survival of juve-

nile trees and the production of seedlings are often inhibited by dominant overstory trees, the distribution of age classes is often highly skewed (Figure 11.7). One or two age classes dominate the site until they die or are removed, allowing young age classes to develop and dominate in turn.

11.10 Sex ratios in populations may shift with age

Populations of sexually reproducing organisms tend toward a 1:1 sex ratio (the proportion of males to females). The primary sex ratio (the ratio at conception) also tends to be 1:1. This statement may not be universally true, and it is difficult to confirm. The secondary sex ratio (the ratio at birth) is often weighted toward males, but the population shifts toward females in the older age groups. This shift happens in most mammalian populations, including humans. Generally, males have a shorter life span than females. Among birds there is a tendency for males to outnumber females because of the loss of nesting females.

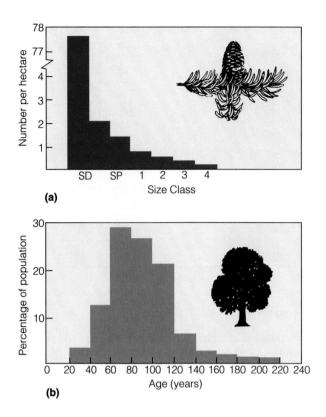

(a)

(b)

Figure 11.7 Age distributions for two plant populations. (a) A balsam fir *(Abies balsamea)* stand on Lake Superior in Ontario. Size classes are seedlings (SD), 1 to 3 cm diameter at breast height (dbh); saplings (SP) greater than 3 cm and less than or equal to 9 cm; and trees grouped in 7.62 cm diameter classes. This stand is young, with most of the population in younger age classes. (b) An oak *(Quercus)* forest in Sussex, England is dominated by trees in the 60- to 120-year age classes. There has been no recruitment for 20 years.

CHAPTER REVIEW

SUMMARY

Characteristics of Populations (11.1–11.2) A population is a group of individuals of the same species living in the same place at the same time, and reproductively isolated from other such groups. Populations are characterized by age structure, density, distribution in time and space, birthrate, death rate, and growth rate (11.1). Most animal populations are made up of unitary individuals with a definitive growth form and longevity. Plant populations are more complex. Individual plants consist of modules: buds, leaves, twigs, flowers, and roots. Plant populations may consist of sexually produced parent plants and modular subpopulations of asexually produced stems arising from roots (11.2).

Population Density and Dispersion (11.3–11.6) Density, the size of a population in relation to the area it occupies, affects individuals' access to living space, food, shelter, and mates (11.3). Because of a patchy environment, not all of the area is suitable habitat. The number of organisms in available living space is the true or ecological density (11.4).

Populations and individuals are dispersed in some pattern over the landscape. Some are randomly dispersed, with the spacing of each individual independent of the others; others are uniformly distributed; most are in aggregations, which in turn may be random, uniform, or clumped in distribution. Movements of individuals link subpopulations into metapopulations (11.5).

Determination of density and dispersion requires careful sampling and appropriate statistical analysis of the data. Plant researchers often use sample plots. Animal researchers use capture-recapture techniques or determine relative abundance by count surveys (11.6).

Age Structure (11.7–11.9) Individuals making up the population may be divided into three ecological periods: prereproductive, reproductive, and postreproductive. The distribution of individuals among age groups influences the birthrate, death rate, and growth rate. A large number of young about to enter the reproductive age suggests a potentially increasing population, whereas a high proportion of individuals in the postreproductive age classes suggest zero growth or a declining population **(11.7).**

To determine the age structure of a population, you must find out the ages of its members. Whereas obtaining data on age in human populations is fairly simple, determining the ages of animals and plants requires special techniques. It is difficult to determine age for plants. Although size is often used as a criterion, it can be misleading **(11.8).**

Age pyramids compare the percentages of the population in different age groups. Pyramids with a broad base of young suggest growing populations. Pyramids with a narrow base of young and even ratios suggest a declining or aging population. Between these two are many variations. Viewed over time, age pyramids depict changing dynamics of the population **(11.9).**

Sex Ratios (11.10)   Populations have a sex ratio that tends to be 1:1 at conception and birth but shifts toward females later.

STUDY QUESTIONS

1. Distinguish between ecological density and crude density.
2. Compare random, uniform, and clumped dispersion. What are some underlying reasons for each type of dispersion?
*3. Why is it hard to determine the density of a population? Do these same problems extend to a census of human populations?
4. Of what significance are the ratios of individuals in the three general age periods?
5. What is stable age distribution?
6. Why does age structure differ between animal and plant populations?
7. How does the modular nature of plant populations add complexity to the study of age structure in plants?
*8. Age pyramids, which represent the past of a population, are often used to predict its future. Think of drawbacks in this practice. What can age pyramids tell us?

LIFE HISTORY
PATTERNS

OBJECTIVES

On completion of this chapter, you should be able to:

- Discuss modes of sexual reproduction.
- Define a mating system and distinguish among monogamy, polygamy, polygyny, and polyandry.
- Explain the mechanisms in sexual selection.
- Discuss the importance of energy allocation to reproduction.
- Compare semelparity and iteroparity.
- Discuss the relationship between the number of young and parental investment.
- Explain the relationship among size, age, and fecundity.
- Contrast *r*-selection and *K*-selection.

Two bull elephant seals *(Mirounga angustirostris)* clash for possession of a section of beach and its attendant females.

Reproduction is the major drive of all living things. The role of the reproductive drive is to transmit genetic characteristics from one generation to another. The ability of an organism to accomplish that successfully is termed its **fitness.** Fitness is equated with the ability of an organism to leave behind reproducing offspring. Individuals that leave behind the most reproducing offspring are supposedly the fittest (see Chapter 19). Achieving fitness involves, among other things, fecundity and survivorship; physiological adaptations; modes of reproduction; age at reproduction; number of eggs, young, or seeds produced; parental care; size; and time to maturity. How organisms achieve fitness becomes the organism's life history pattern.

12.1 Reproduction may be sexual or asexual

How organisms form new individuals falls into two categories: asexual and sexual reproduction. Asexual reproduction, which takes many forms, creates new individuals genetically the same as the parent. The one-celled *Paramecium* reproduces by dividing in two; hydras, coelenterates that live in fresh water (Figure 12.1), reproduce by budding; strawberry plants spread by runners. Aphids produce eggs by normal cell division or mitosis that develop into female adults without fertilization, a process called **parthenogenesis.** However, organisms that rely heavily on asexual reproduction revert on occasion to sexual reproduc-

Figure 12.1 The freshwater hydra reproduces asexually by budding.

tion. Hydras at some time in their life cycle produce eggs and sperm. Strawberry plants bear flowers. At the end of the summer, aphids resort to sexual reproduction to make males.

Sexual reproduction is common in multicellular organisms. Two individuals produce haploid (one-half the normal number of chromosomes) gametes—egg and sperm—that combine to form a diploid cell or zygote. This halving and recombination of genes allow the gene pool to mix, producing genetic variability among offspring (see Chapter 19). For the individual, however, sexual reproduction is expensive. Each individual can contribute only one-half of its genes to the next generation. Because the success of that contribution depends upon the contribution from a member of the opposite sex, it follows that each individual should acquire the best possible mate.

12.2 Sexual reproduction takes several forms

Sexual reproduction takes a variety of forms. The most familiar involves separate male and female individuals. It is common to most animals. Plants with that characteristic are called **dioecious;** examples are holly (*Ilex* spp.) trees and stinging nettle (*Urtica* spp.). In such plants fertilization takes place with the fusion of a large nonmobile female egg and a small mobile male gamete in pollen.

Some individual organisms possess both male and female organs. They are **hermaphroditic.** In plants, individuals can be hermaphroditic either by having bisexual flowers with both male organs (stamens) and female organs (ovules), as lilies and buttercups do, or by being monoecious (Figure 12.2). A **monoecious** plant possesses separate male and female flowers on the same plant, as do birch (*Betula* spp.) and hemlock (*Tsuga* spp.) trees. Among animals, hermaphroditic individuals possess both testes and ovaries, a condition common in invertebrates such as earthworms (Figure 12.3) and some fish.

Some hermaphrodites are sequential. Sequential hermaphrodites are one sex when they are young or small and develop into the opposite sex when they mature or become larger. These individuals budget their reproductive costs by changing gender as they grow during their life cycle. Gender change, in which the individual switches between a costly female function—the production of eggs—to a less expensive male

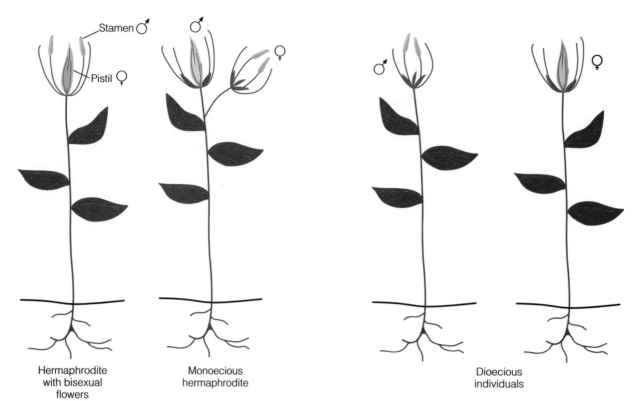

Figure 12.2 Floral structure in hermaphroditic and dioecious plants.

Figure 12.3 Hermaphroditic earthworms mating.

function—the production of sperm—maximizes their reproductive success.

Social change in the sex ratio of the population stimulates gender change among some animals, notably fish species. Removing individuals of the other sex initiates sex reversal among some species of marine fish. Among some coral fish, removal of females from a social group stimulates an equal number of males to change sex and become females. In other species, removal of males stimulates a one-to-one replacement of males by sex-reversing females. Among the mollusks, the Gastropoda (snails and slugs) and Bivalvia (clams and mussels) have sex-changing species. Almost all of these species change from male to female. Among these organisms male size has little effect on breeding success, but the most fertile females are large. In the presence of many males, a small individual could improve his fitness by switching and growing rapidly as a female.

Plants also can undergo gender change. On such plant is jack-in-the-pulpit *(Arisaema triphyllum)*, a clonal woodland herb (Figure 12.4). Jack-in-the-pulpit may produce staminate (male) flowers one year, an asexual vegetative shoot the next, and a carpellate (female) or monoecious shoot the next. Over its life span a jack-in-the-pulpit may produce both genders as well as an asexual vegetative shoot, but in no particular sequence. Usually an asexual stage follows a gender change. Gender change in jack-in-the-pulpit appears

Figure 12.4 The jack-in-the-pulpit becomes asexual, male, or female, depending on energy reserves. The plant gets its name from the flower stem or spadix enclosed in a hoodlike sheath. This fruiting plant is the female stage.

to be triggered by an excessive drain on photosynthate by female flowers. If the plant is to survive, one carpellate flowering cannot follow another.

12.3 Mating strategies take several forms

The behavioral mechanisms and the social organization involved in obtaining a mate make up a **mating system.** Systems differ in the number of mates males and females acquire and the manner in which they acquire them. After pair formation, mating systems vary in the nature of the pair bond and the pattern of parental care from each sex.

Monogamy is the formation of a pair bond between one male and one female. It is most prevalent among birds and rare among mammals, except several carnivores, such as foxes and mustelids, and a few herbivores, such as the beaver, muskrat, and the prairie vole.

Monogamy occurs mostly among species in which cooperation by both parents is needed to rear young successfully. Nearly 90 percent of all species of birds are seasonally monogamous, because in most birds the young are helpless at hatching and need food, warmth, and protection. The avian mother is no better suited to provide these needs than the father. Instead of seeking other mates, the male can increase his fitness more by continuing his investment in the young. Without

him, the young carrying his genes may not survive. Among mammals, the situation is different. The females lactate, providing food for the young. Males often can contribute little or nothing to the survival of the young, so it is to their advantage to mate with as many females as possible. Among the exceptions are the foxes, wolves, and other canids among which the male provides for the female and young and defends the territory. Both male and female regurgitate food for the weaning young.

Monogamy, however, has another side. Among some 100 species of monogamous birds, the female or male may "cheat" by engaging in extra-pair copulations while maintaining the reproductive relationship with the primary mate and caring for the young.

Polygamy is the acquisition by an individual of two or more mates, none of which is mated to other individuals. It can involve one male and several females or one female and several males. A pair bond exists between the individual and each mate. When one member of the pair is freed from parental duty, partly or wholly, the emancipated member of the pair can devote more time and energy to competition for more mates and resources. The more unevenly such critical resources as food or quality habitat are distributed, the greater is the opportunity for a successful individual to control the resource and several mates.

The number of the opposite sex an individual can monopolize depends upon the degree of synchrony in sexual receptivity. For example, if females in the population are sexually active for only a brief period, as with the white-tailed deer, the number a male can monopolize is limited. However, if females are receptive over a long period of time, as with elk, the size of a harem a male can control depends upon the availability of females and the number of mates the male has energy to defend.

Environmental and behavioral conditions result in various types of polygamy. In **polygyny,** an individual male gains two or more females. In **polyandry,** an individual female gains two or more males. A special form of polygamy is **promiscuity,** in which males and females copulate with one or many of the opposite sex and form no pair bonds.

Polyandry is interesting because it is the exception rather than the rule. The female rather than the male is the competitive individual. In resource defense polyandry, the female competes for and defends resources essential for the male. This system is best developed in two orders of shorebirds, Gruiformes (cranes and rails) and Charadriiformes (auks and gulls). As in polygyny, this mating system depends upon the distribution and defensibility of resources, especially quality habitat. The female produces multiple clutches of eggs, each of which requires the brooding services of a male. Females compete for available males. After the female lays a clutch, the male begins incubation. Then he becomes sexually inactive, resulting in a scarcity of available males.

12.4 Acquisition of a mate involves sexual selection

Regardless of the mating system, selection of a proper mate is essential if a plant or animal is to contribute genetically to the next generation. In any population, there are just so many males and females. Because males are not as selective with whom they mate, females usually have no trouble finding a partner. Females, however, are selective because of their high energy investment in production of eggs and young, and males must prove their fitness to females. There is intense rivalry among males for female attention. The males' success, however, depends not on the outcome of their conflicts, but on how the competition affects their acceptance by choosy females. In the end, the female determines which male will sire her offspring.

Supposedly, males, and in some situations females (see Section 12.3) that win this competition are the fittest. By selecting a mate from among competing males on the basis of some specific characteristic during courtship, females attempt to ensure their own fitness. Making such a choice is known as **sexual selection.**

How does a female select the mate with greatest fitness? That question still intrigues behavioral ecologists. Charles Darwin observed that the elaborate and often outlandish plumage of birds and the horns, antlers, and large sizes of polygamous males seemed incompatible with natural selection. To explain them, Darwin developed a theory of sexual selection. He hypothesized that competition among males selects for size of weapons and body and for striking plumage patterns. Females choose males from among the winners, based on their appearance and behavior. This hypothesis assumes that mate competition is a male trait and that mate choice is a female trait.

Another hypothesis focuses on the reproductive interests of the female. The basic strategy for both male and female is to ensure their own maximum fitness. What increases male fitness is not necessarily what improves female fitness. Sperm are cheap. With little investment, males should mate with as many females as possible to achieve maximum fitness. Eggs are costly, so it pays females to select the best mate, who will pass on the best genes to the next generation.

A third concept of sexual selection is the handicap hypothesis. It postulates the evolution of three characters: a male handicap, a female mating preference for the handicap, and a general viability trait. The handicap is any secondary male characteristic, such as bright plumage, elaborate tails as in peacocks and lyrebirds, or any trait that could make the male more vulnerable to predation or in any other way reduce the male's probability of survival, the viability trait. If a male can carry these handicaps and survive, it is an honest appraisal of the male's genetic superiority. Females showing preference for the handicapped male will produce offspring that will carry genes for high viability. Thus the driving force behind the evolution of exaggerated secondary sexual characteristics in males is selection by females.

12.5 Females attempt to acquire mates with the highest fitness

Selection is of two types: intrasexual and intersexual. **Intrasexual selection** (see Section 12.4) involves male-to-male (or female-to-female, as in some shorebirds) competition for the opportunity to mate. It leads to exaggerated secondary sexual characteristics, such as bright or elaborate plumage in male birds and antlers in deer. **Intersexual selection** is mostly female choice of a male.

There are two grounds for choosing a mate. In one the female bases her choice on resources such as habitat or food that will improve her fitness. In the other she selects characteristics not associated with resources.

For monogamous females the criterion for mate selection appears to be acquisition of a resource, usually a defended high-quality habitat or territory (see Section 14.9). Does the female select the male and accept the territory that goes with him, or does she select the territory and accept the male that goes with it? There is some evidence from the laboratory and the field that female songbirds base their choice, in part, on the variety of the male's song. In aviary studies, females respond with stronger solicitation displays to more varied songs (Figure 12.5). In the field, males with more complex song repertoires appear to hold the higher-quality territories, so the more complex song may convey that fact to the female. None of

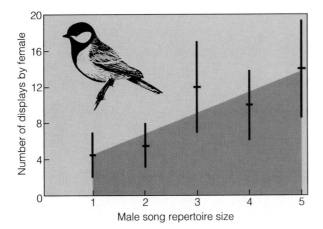

Figure 12.5 Mean number of copulation-solicitation displays given by 11 female great tits *(Parus major)* as a function of repertoire size of male songs.

these studies, however, determined the fitness of females attracted to these males.

Another criterion of fitness female birds may use is plumage. One hypothesis proposes that only healthy males can develop bright plumage. There is some evidence that the male with the lowest parasitic infection has the brightest plumage. Females pick these males and in doing so, select the fittest. This hypothesis is controversial, to say the least.

Among polygamous species, the question becomes more complex. In those cases in which females acquire a resource along with the male, the situation is similar to that of monogamous relationships. Among polygamous birds, for example, the females show strong preference for males with high-quality territories. On such territories, with superior nesting cover and an abundance of food, females can attain reproductive success even if they share the territory and the male with other females.

Among those fish species in which parental care of the eggs is an exclusively male function, females prefer to spawn with males already caring for another female's eggs, although the male gives less care to adopted eggs. Males of some fish, such as the fantail darter *(Etheostoma flabellare)*, capitalize on such female preferences by resorting to egg mimicry. Territorial males excavate nests beneath flat rocks and care for eggs attached by the female to the ceiling of the nest. Male darters have each of the seven to eight spines of the distal end of the first dorsal fin tipped with fleshy knobs that look like eggs when the males are resting beneath the rocks. Female darters prefer egg mimics over males without mimicry.

12.6 In polygamy females have limited information for mate choice

When polygamous males offer no resources, it seems that the female has limited information to act upon. She might select a winner from among males that best others in combat—as in bighorn sheep, elk, and seals. She might select mates by intensity of courtship display or some morphological feature that may reflect a male's genetic superiority and vitality.

Among some polygamous species, it is hard to see female choice at work. In elk and seals, a dominant male takes charge of a harem of females. Nevertheless, the females express some choice. Protestations by female elephant seals over the attention of a dominant male may attract other large males nearby, who attempt to replace the male. Such behavior ensures that females will mate only with the highest-ranking male.

In other situations females seem clearly in control. Males put on an intense display for them. Such advertising can be costly for courting polygamous males. In the breeding season some male three-spined sticklebacks *(Gasterostens aculeatus)*, a small fish of marine and freshwater habitats in North America, Europe, and Pacific Asia, have red throats and some do not. D. E. Semler gave females a choice between two genetically non-red males, one of which had been given an artificial red throat painted with nail polish. Females selected the artificially colored male. However, red-throated males are subject to intense predation by rainbow trout. The gene that attracts the female endangers the male. Thus a cost accrues to the red-throated males that acts against their reproductive advantage.

Extreme examples of genes-only female choice appear in lek species. These animals aggregate into groups on communal courtship grounds called **leks** or arenas. Males on the lek defend small territories that hold no resources and advertise their presence by colorful vocal and visual displays. Females visit the leks of displaying males, select a male, mate, and move on. Although few species engage in this type of mating system, it is widespread in the animal world, from insects to frogs to birds and mammals. Males defend small, clustered mating territories, whereas females have large overlapping ranges that the males cannot economically defend.

At least three hypotheses have been advanced to explain lek behavior. One hypothesis, female choice, is that females show preference for a particular mating territory because it is the safest place to mate or it forces males to cluster. Leks provide an unusual opportunity for females to choose a mate among the displaying males. This advantage is especially great among lekking insects, such as certain tropical *Drosophila*, in which females are widely dispersed.

A second, related hypothesis is the "hotspot" model. Males cluster in places where encounters with females are potentially high. If females have large overlapping ranges, males can ensure encounters with them by aggregating in overlap zones. For example, sage grouse *(Centrocercus urophasianus)* males display in leks near the females' wintering ranges (Figure 12.6). As females abandon the winter range for nesting areas, the males establish new lek areas nearby.

A third hypothesis, the "hotshot" model, emphasizes male-male interactions. This model envisions a strong hierarchy among the males, with the dominant male, in smaller leks at least, displacing all others and leaving no opportunity for female choice. In other words, the visiting females select an arena and mate with the dominant males. This model may be correct in the case of tropical manakins, prairie chickens, and birds of paradise. Among these birds, one dominant male on the lek may perform over 90 percent of successful copulations. In spite of the odds of not mating, males congregate on the lek because, by displaying together, they draw in females from a larger area. Congregating about "hotshot" males with the most effective displays, subdominant and satellite males may steal mating opportunities. A majority of matings on the lek, however, is done by a small percentage of the males in a dominance hierarchy formed in the absence of females.

12.7 Organisms budget time and energy to reproduction

Organisms spend their energy like income in a household or a business to meet many needs (see Chapter 23). Some energy must go to growth, to maintenance, to acquiring food, to defending territory, and to escaping predators. Some must go to reproduction. To achieve optimal fitness, an organism has to budget its energy and time in reproduction. Allocations of time and energy make up an organism's **reproductive effort.**

The more energy an organism spends on reproduction, the less it retains to spend for growth and maintenance. For example, reproducing females of the terrestrial isopod *Armadillidium vulgare* have a lower rate of growth than nonreproducing females. Nonre-

Figure 12.6 A male sage grouse on a strutting ground or lek. The males strut, inflate air sacs in their breast, and spread their pointed tails in courtship display. Here a dominant male has attracted a circle of hens.

producing females devote as much energy to growth as reproducing females devote to both growth and reproduction.

How much energy organisms invest in reproduction varies. The investment includes not only the weight of the progeny but also the energy expended to nourish it. Herbaceous perennials invest between 15 and 20 percent of annual net production in new plants, including vegetative propagation. Wild annuals that reproduce only once expend 15 to 30 percent; most grain crops 25 to 30 percent; and corn and barley, 35 to 40 percent. The lizard *Lacerta vivipara* invests 7 to 9 percent of its annual energy assimilation in reproduction. The female Allegheny Mountain salamander *Desmognathus ochophaeus* spends 48 percent of its annual energy flow on reproduction, including energy stored in eggs and energy costs of brooding.

How often can an organism afford to reproduce? One approach is to invest all energy into growth, development, and energy storage, make one massive reproductive effort, and die. An organism may sacrifice future prospects by expending all its energy in one suicidal act of reproduction. That mode of reproduction is **semelparity.**

Semelparity is employed by most insects and other invertebrates, by some species of fish, notably salmon, and by many plants. It is common among annuals, biennials, and some bamboos. Many semelparous plants, such as ragweed, are small, short-lived, and found in ephemeral or disturbed habitats. For them it would not pay to hold out for future reproduction, for the chances are slim. They gain their maximum fitness by expending all their energies in one bout of reproduction.

Other semelparous organisms, however, are long-lived and delay reproduction. Mayflies (Ephemeroptera) may spend several years as larvae before emerging from the surface of the water for an adult life of several days devoted to reproduction. Periodical cicadas spend 13 to 17 years below ground before they emerge as adults to stage an outstanding exhibition of single-term reproduction. Some species of bamboo delay flowering for 100 to 120 years, produce one massive crop of seeds, and die. The Hawaiian silverswords (*Argyroxiphium* spp.) live 7 to 30 years before flowering and dying. In general, species that evolve semelparity have to increase fitness enough to compensate for the loss of repeated reproduction.

Organisms that choose to produce fewer young at one time and repeat reproduction throughout their lifetime are **iteroparous.** Iteroparous organisms include most vertebrates, perennial herbaceous plants, shrubs, and trees. For an iteroparous organism the problem is timing reproduction—early in life or later. Whatever the choice, it involves trade-offs. Early reproduction means less growth, earlier maturity, reduced survivorship, and reduced potential for future reproduction. Later reproduction means increased growth, later maturity, and increased survivorship, but less time for further reproduction. In effect, if an organism is to make a maximum contribution to future generations, it has to balance the profits of immediate reproduction against future prospects, including the cost to fecundity (total offspring) and its own survival. Some timing is the best strategy for **optimal lifetime reproductive success,** the sum of present and future reproductive success.

12.8 Parental investment relates to the number of young

The same energy can produce many small young or one or two large ones. The number of offspring affects the investment each receives. If the parent produces a large number of young, it can afford only minimal investment in each one. In such cases, animals provide no parental care, and plants store little food in seeds. Such organisms usually inhabit disturbed sites, unpredictable environments, or places such as the open ocean, where opportunities for parental care are difficult at best. By

dividing energy for reproduction among as many young as possible, these parents increase the chances that some young will successfully settle somewhere. By such a strategy, parents increase their own fitness but decrease the fitness of the young. (See Focus on Ecology 12.1: Life History Strategies.)

Parents that produce few young are able to expend more energy on each individual. The amount of energy will vary with the number of young, their size, and their maturity. Some organisms expend less energy during incubation. The young are born or hatched in a helpless condition and require considerable parental care. These animals, such as young mice or nestling robins, are **altricial.** Other animals have longer incubation or gestation, so the young are born in an advanced stage of development. They are able to move about and forage for themselves shortly after birth. Such young are called **precocial.** Examples are gallinaceous (chickenlike) birds and ungulate mammals such as cows and deer.

The degree of parental care varies widely. Some species of fish, such as cod (*Gadus morhua*), lay millions of floating eggs that drift freely in the ocean with no parental care. Other species, such as bass, lay eggs in the hundreds and provide some degree of parental care. Among amphibians, parental care is most prevalent among tropical toads and frogs and poorly developed in salamanders. Among reptiles, crocodiles are an exception. They actively defend the nest and young for a considerable time. Parental care is not well developed among invertebrates, except for the social insects. Social insects perform all functions of parental care, including feeding, defense, heating, and sanitation.

12.9 Environmental conditions influence the number of young

The conditions under which organisms evolve and to which they adapt influence the number and the size of young. Among plants, short-lived annuals have numerous small seeds. High production of numerous offspring ensures that some will survive and germinate the next growing season. Optimum seed size and number for perennial plants relate to dispersal ability, colonizing ability, and the need to escape predation. Plants that colonize disturbed environments that are widely distributed or available for only a short period of time may have small wind-blown seeds that carry great distances. Because they invest minimal energy in each seed,

FOCUS ON ECOLOGY 12.1

LIFE HISTORY STRATEGIES

There are broad differences among reproductive patterns. Some species, such as weeds and insects, are small, have high reproductive rates, and lead short lives. Others, like trees and deer, are large, and have low reproductive rates and long lives. The population ecologists R. MacArthur, and E. O. Wilson, and later E. Pianka, called these types *r* species and *K* species.

The theory of *r*- and *K*-selection predicts that species in different environments will differ in life history traits such as size, fecundity, age at first reproduction, number of reproductive events during a lifetime, and total life span. Species popularly known as *r*-strategists are typically short-lived. They have high reproductive rates at low population densities, rapid development, small body size, large number of offspring (but with low survival), and minimal parental care. They make use of temporary habitats. Many inhabit unstable or unpredictable environments that can cause catastrophic mortality independent of population density. For these species environmental resources are rarely limiting. They exploit noncompetitive situations. Tough and adaptable, *r*-strategists, such as weedy species, have means of wide dispersal, are good colonizers, and respond rapidly to disturbance.

K-strategists are competitive species with stable populations of long-lived individuals. They have a slower growth rate at low populations, but they maintain that growth rate at high densities. *K*-strategists can cope with physical and biotic pressures. They possess both delayed and repeated reproduction, and have a larger body size and slower development. They produce few seeds, eggs, or young. Among animals, parents care for the young; among plants, seeds possess stored food that gives the seedlings a strong start. *K*-strategists' mortality relates more to density than to unpredictable environmental conditions. They are specialists, efficient users of a particular environment, but their populations are at or near carrying capacity and are resource-limited. These qualities, combined with their lack of means of wide dispersal, make *K*-strategists poor colonizers of new and empty habitats.

Although it is tempting to classify organisms as either *r*-strategists or *K*-strategists, it is difficult to force

species into such a classification. At some time in its life history, the same species may exhibit *r* traits or *K* traits. Meadow mice living in environments where dispersal can easily take place exhibit characteristics of an *r* species, whereas those with no place to disperse assume the characteristics of a *K* species. White-tailed deer exhibit *r* traits when a population spreads into new habitat or is greatly reduced by hunting. At high densities, however, the population exhibits *K* traits, with low overall recruitment of young. The concept of *r* species and *K* species is most useful to compare organisms of the same type, individuals within a population, or populations within a species.

The plant ecologist J. Grime has proposed a three-strategy system. In his system, there are ruderal or *R*-strategists, competitive or *C*-strategists, and stress-tolerant or *S*-strategists. *R*-strategists, typically weedy species, occupy uncertain or disturbed habitats, have a short growth form, reproduce early in life, possess high fecundity, experience one lethal reproduction (semelparity), and have well-dispersed seeds. *C*-strategists and *S*-strategists occupy more stable environments and live longer, which often drastically reduces the opportunity for seedlings to grow. Beyond these two characteristics, the two have evolved quite different life history strategies. *C*-strategists, such as grasses in an ungrazed grassland, live in competitive but productive and undisturbed environments, attain maximum vegetative growth, reproduce early, and repeatedly use an annual expenditure of energy stored prior to seed production. *S*-strategists live in stressed environments, such as highly disturbed sites or forest understory. They have delayed maturity, intermittent reproductive activity, and long-term energy storage.

Within these extremes are intermediate types. Competitive ruderals *(C-R)* live in environments of low stress and moderate disturbance, such as fertilized, moderately grazed grasslands; stress-tolerant ruderals *(S-R)* live in extreme environments such as rock crevices; and *C-S-R* plants occupy habitats where competition is reduced by the combined effects of stress and disturbance, such as old fields.

these plants can afford heavy losses of seeds. Plants subject to heavy predation may have small seeds that provide a less attractive source of food. Again an abundance of seeds is an insurance that some will escape predation (see Chapter 16).

Plants associated with more stable environments may produce fewer and larger seeds, with a large store of energy that the seedling can use to become established. Such plants may invest a considerable amount of energy in toxins or heavy seed coats to reduce predation. Some plants adjust their production of seeds to meet current environmental conditions. For example, in moist environments purslane speedwell, *Veronica peregrina*, produces few but heavy seeds. Where moisture is limited and competition from grass is keen, the same species grows taller and produces lighter seeds that disperse better. In other species the adjustment may involve a plastic response. On harsh, open environments plants allocate more energy to reproduction than in more moderate habitats.

Animals make similar adjustments. "Annual" species, such as insects, that overwinter in the egg stage produce enormous numbers of small eggs. Birds such as quail and mammals such as rabbits and mice, which are subject to heavy predation, produce large numbers of young.

12.10 Food supply affects the production of young

Within a given region, production of young may reflect the abundance of food. The normal clutch or litter size may not be optimal for that region. Parents cannot predict available food at the time of nesting, so they may have to reduce the number of young. The parent does not make a conscious decision to reduce a brood. Asynchronous hatching and **siblicide,** the killing of one sibling by another, take care of it.

In asynchronous hatching the young are of several ages. The older siblings beg more vigorously for food, forcing the harried parents to ignore the calls of the younger, smaller sib, who perishes. For example, the common grackle (*Quiscalus quiscula*) begins incubation before its clutch of five eggs is complete, a pattern of hatching ensuring survival of some young under adverse conditions. The eggs laid last are heavier, and the young from them grow fast. However, if food is scarce, the parents fail to feed these late offspring be-

cause of more vigorous begging by siblings. The last-hatched young then die of starvation. Thus asynchronous hatching favors the early-hatched young.

In other situations, the older or more vigorous young simply kill their weaker sib. A number of birds, including raptors, herons, egrets, gannets, boobies, and skuas, practice siblicide. The parents normally lay two eggs, possible to insure against infertility of a single egg. The larger of the two hatchlings kills the smaller sibling or runt, and the parents redirect all resources to the surviving chick. These birds are not alone in siblicidal tendencies. The females of some parasitic wasps lay two or more eggs in a host, and the larvae fight each other until only one survives.

12.11 The number of offspring may vary with latitude

David Lack, an English ornithologist, proposed that clutch size in birds has evolved to equal the largest number of young the parents can feed. Thus clutch size is an adaptation to food supply. Temperate species, he argued, have larger clutches, because increasing daylength provides a longer time to forage for food to support large broods. In the tropics, where daylength is roughly 12 hours, foraging time is limited.

An American ecologist, Martin Cody, modified Lack's ideas by proposing that clutch size results from different allocations of energy to egg production, avoidance of predators, and competition. In temperate regions, periodic local climatic catastrophes can hold a population below the level that the resources could support. Organisms respond with a higher rate of increase and larger clutches. In tropical regions, with predictable climate and increased probability of survival, there is no need for extra young.

A third hypothesis states that clutch size varies in direct proportion to the seasonal variation in resources, especially food. Population density is regulated primarily by mortality in winter, when resources are scarce. Greater winter mortality offers more food to the survivors during the breeding season. This abundance is reflected in larger clutches. Thus geographical variation in mean clutch size and the size of the breeding population inversely relates to winter food supply.

All three hypothesis predict that clutch size should be greater in high latitudes than low. Field examples support the prediction. Birds in temperate regions have larger clutch sizes than they have in the tropics (Figure 12.7), and mammals at higher latitudes have larger litters than they do at lower latitudes. Lizards exhibit a similar pattern. Those living at lower latitudes have smaller clutches, have higher reproductive success, reproduce at an earlier age, and experience higher adult mortality than those living at higher latitudes.

Insects, too, such as the milkweed beetle (*Oncopeltus* spp.), support the hypotheses. Temperate and tropical milkweed beetles have a similar duration of the egg stage, egg survivorship, developmental rate, and age at sexual maturity. Although the clutch sizes are the same, the temperate species lay a larger number of eggs because they lay more clutches. Total egg production of the tropical species is only 60 percent of that of the temperate species.

Plants also follow the general principle of allocation on a latitudinal basis. For example, three species of cattail grow in North and Central America. The common cattail (*Typha latifolia*) is a climatic generalist that grows from the Arctic Circle to the equator. The narrow-leaved cattail (*T. angustifolia*) is restricted to

the northern latitudes. The southern cattail (*T. domingensis*) grows only in the southern latitudes. These three cattails show a climatic gradient in the allocation of energy to reproduction. *T. angustifolia* and the northern populations of *T. latifolia* grow earlier and faster than *T. domingensis*, and produce a greater number of rhizomes. The southern populations produce larger rhizomes.

Although reproductive output among organisms does appear to be greater at higher latitudes, the number of comparable species for which there are data is too few to confirm the hypotheses. Many more studies along a latitudinal gradient are needed.

12.12 Fecundity relates to age and size

For many species clutch size and fecundity relate to the age and size of the parent. This relationship is especially strong in plants and ectothermic animals. Among plants, perennials delay flowering until they have attained a sufficiently large leaf area to support seed production. Many biennials in poor environments also delay flowering beyond the usual two-year life span, until environmental conditions become more favorable. Annuals show no such relationship between leaf area and the percentage of energy devoted to reproductive output, but size differences among annuals do result in differences in the number of seeds produced. Small plants produce fewer seeds, even though the

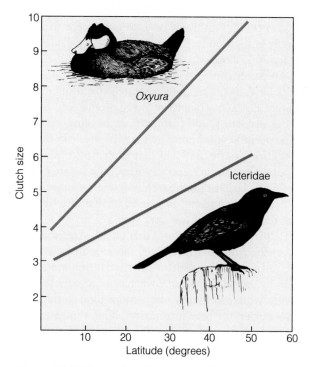

Figure 12.7 The relationship between clutch size and latitude in birds. Represented are the subfamily Icteridae (blackbirds, orioles, and meadowlarks) in North and South America and the worldwide genus *Oxyura* (ruddy and masked ducks) of the subfamily Anatinae.

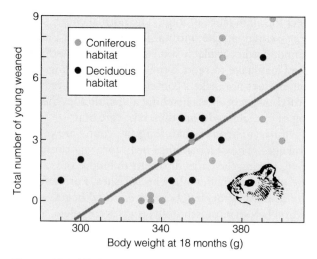

Figure 12.8 Lifetime reproductive success of the European red squirrel according to body weight in the first winter as an adult.

plants may be contributing the same share of energy to reproduction as larger plants.

Similar patterns exist among poikilothermic animals. Fecundity in fish increases with size, which increases with age. Because early fecundity reduces both growth and later reproductive success, fish obtain a selective advantage by delaying sexual maturation until they grow larger. Gizzard shad (*Dorosoma cepedianum*) reproducing at two years of age produce about 59,000 eggs. Those delaying reproduction until the third year produce about 379,000 eggs. Among the gizzard shad, only about 15 percent spawn at two years of age and about 80 percent at three years.

The clutch size of the loggerhead sea turtle (*Caretta caretta*) is constrained by the female's egg-carrying capacity, which is related to body size. Because of the high energy cost of laying eggs on land, sea turtles may be under pressure to maximize clutch size.

An apparent relationship also exists between body size and fecundity among endotherms. Heavier females are more successful in reproduction and more of their young survive. For example, the body weight of female European red squirrels (*Sciurus vulgaris*) in Belgian forests strongly correlated with lifetime reproductive success (Figure 12.8). Few squirrels weighing less than 300 g reproduced.

CHAPTER REVIEW

SUMMARY

Asexual and Sexual Reproduction (12.1–12.3) The ability of an organism to leave behind reproducing offspring is its fitness. Organisms that contribute the most offspring to the next generation are the fittest. Reproduction can be asexual or sexual. Asexual reproduction or cloning results in new individuals genetically the same as the parent. Sexual reproduction combines egg and sperm in a diploid cell or zygote. Sexual reproduction produces genetic variability among offspring **(12.1)**.

Several mating strategies are associated with sexual reproduction. Plants with separate males and females are called dioecious. An organism possessing both male and female sex organs is hermaphroditic. Plant hermaphrodites have a bisexual flower or, if they are monoecious, separate male and female flowers on the same individual. Some plants and animals change gender **(12.2)**. Mating systems include two basic types, monogamy and polygamy. In polygyny the male acquires more than one female; in polyandry the female acquires more than one male. The potential for competitive mating and sexual selection is higher in polygamy than in monogamy **(12.3)**.

Sexual Selection (12.4–12.6) Selection of a proper mate is essential if an organism is to contribute to the next generation. An important component of mating strategy is sexual selection. In general, males compete with males for the opportunity to mate with females, but females finally choose a mate. Sexual selection favors traits that enhance mating success, even if they handicap the male by making him more vulnerable to predation **(12.4)**. Male competition is intrasexual selection. In intersexual selection, females choose males offering defensible resources or supposedly superior genes. By choosing the best males, females ensure their own fitness **(12.5)**. Among polygamous species females have limited information on male fitness. Among lek species females may come to a safe communal breeding ground and choose among displaying males. Another hypothesis is that males collect in hotspots where female numbers are high. A third hypothesis allows females little choice: a dominant hotshot male accomplishes most of the matings **(12.6)**.

Energy Investment in Reproduction (12.7–12.8) To achieve optimal fitness, an organism has to balance immediate reproductive efforts against future prospects. One alternative, semelparity, is to invest maximum energy in a single reproductive effort. The other alternative, iteroparity, is to allocate less energy each time to repeated reproductive efforts **(12.7)**. The number of young produced relates to the amount of parental investment. The amount of time and energy parents allot to reproduction is reproductive effort. Organisms that produce a large number of offspring have a minimal investment in each offspring. They can afford to send a large number into the world with a chance that a few will survive. By so doing, they increase parental fitness but decrease the fitness of the young. Organisms that produce few young invest considerably more in each one. Such organisms increase the fitness of the young at the expense of the fitness of the parents **(12.8)**.

Environmental Influences (12.9–12.11) Organisms living in variable or ephemeral environments or facing heavy predation produce numerous offspring, ensuring that some will survive. A large number of young is characteristic of annual plants, short-lived mammals, insects, and semelparous species. Having few young is a characteristic of long-lived species. Iteroparous species may adjust the number of young in response to environmental conditions and the availability of resources (12.9). Production of young often reflects the availability of food. In times of food scarcity, parents may fail to feed some offspring. In other situations, vigorous young kill their weaker sibs (12.10).

In general, clutch and litter sizes increase from the tropics to the poles. This gradient may reflect length of daylight, which influences foraging time, or the more stable climate in the tropics (12.11).

Size and Fecundity (12.12) A direct relationship between size and fecundity exists among plants and ectotherms. The larger the size, the more young are produced. Among endotherms, too, heavier females are more successful in reproduction.

STUDY QUESTIONS

1. What is fitness?
2. Distinguish between sexual and asexual reproduction.
3. What are dioecy, monoecy, and hermaphroditism?
4. What are some advantages of hermaphroditism?
5. What is a mating system? Distinguish among monogamy, polygamy, polygyny, and polyandry.
6. What is sexual selection? In what ways does it differ for males and females?
7. Distinguish between intrasexual selection and intersexual selection.
8. What is the driving force in the evolution of secondary sexual characteristics in males?
9. What is involved in reproductive effort?
10. How does energy allocation affect reproduction and growth?
11. What are semelparity and iteroparity, and what conditions favor each one?
12. What are the advantages of early and late reproduction?
*13. Under what conditions should parents have many young or few young?
14. Compare investments in atricial and precocial young.
15. How might resource availability affect reproduction?
16. Explain the relationship between size and fecundity.
*17. What is the difference between r-selected and K-selected organisms?

CHAPTER 13

POPULATION GROWTH

OBJECTIVES

On completion of this chapter, you should be able to:

- Express mortality as the probability of dying or surviving.
- Construct a life table.
- Explain the derivation and meaning of life expectancy.
- Distinguish among several kinds of life tables.
- Plot mortality and survivorship curves.
- Distinguish among different types of survivorship curves.
- Explain how net reproductive rates are determined.
- Distinguish between exponential and logistic growth.
- Show the relationship between net reproductive rate and annual rate of increase.
- Explain the relationship of carrying capacity to the logistic growth curve.
- Discuss cycles and irregular fluctuations in populations.
- Explain why populations become extinct.

A cedar waxwing *(Bombycilla cedrorum)* at the nest with its hungry, begging young.

Demography—the study of populations—reviews the ways populations change in size. Births (natality) and deaths (mortality) account for most changes in a population. The difference between the two rates determines its growth or decline.

13.1 Mortality is the probability of surviving

In any population, what is an individual's chance of living or dying? The number of individuals dying in a given time period is called the **mortality rate** or probability of dying. To calculate mortality, q_x, we divide the number of individuals that died during an interval of time, d_t, by the number alive at the beginning of the period, N_t. The formula is $q_x = d_t/N_t$.

The complement of the probability of dying is the **probability of survival,** the number of survivors at the end of a time period divided by the number alive at the beginning. Because the number of survivors is more important to a population than the number dying, mortality is better expressed either as the probability of surviving or as **life expectancy,** e_x, the average number of years to be lived in the future by members of the population.

13.2 Life tables give a systematic picture of mortality and survival

To obtain a clear and systematic picture of mortality and survival, we can construct a **life table.** The life table is simply an account book of death. First developed by students of human populations, it is widely used by life insurance companies. It consists of a series of columns headed by standard notations, each of which describes certain age-specific mortality relationships within a population. It begins with a **cohort,**

a group of individuals of a certain size born in the same period of time. The cohort size at birth is usually expressed as 1000 or as a proportion of 1. A cohort of 1000 is obtained by converting data collected in the field to the equivalent numbers had the starting density of the cohort been 1000 (Table 13.1). (See Quantifying Ecology 13.1: Constructing a Life Table.) At one time data for life tables could be obtained only for laboratory animals and humans. As population sampling methods and age determination techniques became more refined, biologists began to construct life tables for wild animals and plants.

There are two basic kinds of life tables. One is the *cohort* or *dynamic life table.* It records the fate of a group of individuals, all born in a single short period of time, from birth to death. A modification of the dynamic life table is the *dynamic composite life table.* It takes as a cohort individuals born over several time periods instead of just one. For example, you might follow the fate of individuals born in 1955, or pool the data for all the individuals born in the fifties. The other type is the *time-specific life table.* It is constructed by sampling the population in some manner to obtain a distribution of age classes during a single time period. You assume that you sampled each age class in proportion to its numbers in the population and age at death, and that the birthrate and the death rate are constant. For example, you could age the carcasses of a particular species, using some age marker such as tooth wear in deer or growth rings in the horns of bighorn sheep.

Life tables for vertebrates deal with long-lived animals. Generations overlap, and different groups are contributing to reproduction at the same time. Many animal species, especially insects, live through only one breeding season. Because their generations do not overlap, all individuals belong to the same age class. We obtain the survivorship values, l_x, by observing a natural population several times over its annual sea-

TABLE 13.1

GRAY SQUIRREL LIFE TABLE

x	n_x	l_x	d_x	q_x	L_x	T_x	e_x
0–1	530	1.000	0.747	0.747	0.626	1.008	1.008
1–2	134	0.253	0.147	0.581	0.179	0.382	1.509
2–3	56	0.106	0.032	0.302	0.090	0.203	1.915
3–4	39	0.074	0.031	0.418	0.058	0.113	1.527
4–5	23	0.043	0.021	0.488	0.033	0.055	1.279
5–6	12	0.022	0.013	0.591	0.015	0.022	1.000
6–7	5	0.009	0.006	0.666	0.006	0.007	1.285
7–8	2	0.003	0.003	01.000	0.001	0.001	0.333

x = Age (years)
n_x = Raw data
l_x = Survivors at beginning of age class x
d_x = Deaths
q_x = Mortality rate
L_x = Average life of all individuals (years)
T_x = Time units left for all individuals to live from age x onward
e_x = Further expectation of life (years)

son, estimating the size of the population each time. For many insects, the l_x value can be obtained by estimating the number surviving from egg to adult. If records are kept of weather, abundance of predators and parasites, and the occurrence of disease, death from various causes can also be estimated.

Table 13.2 represents the fate of a cohort from a single egg mass of gypsy moth. The age interval or x column indicates the life history stages, which are of unequal duration. The l_x column indicates the number of survivors at each stage. The d_x column gives a breakdown of deaths by causes in each stage; d_{xf} is the

cause. In this population, dispersal and predation account for most of the losses. Note that life expectancy is not calculated because there is none. All adults in the population will die in late summer.

13.3 Plant life tables are more complex

Mortality and survivorship in plants are not easily condensed into life tables. To begin with, age is difficult to determine. Mortality of individuals usually stimulates the growth of survivors, increasing the living tissue

---

QUANTIFYING ECOLOGY 13.1

CONSTRUCTING A LIFE TABLE

There is no better way to understand a life table than to construct one. The life table consists of a series of columns related to one other. The columns include x, the units of age; l_x, the number of organisms in a cohort that survive to age x, $x + 1$, and so on; and d_x, the number or fraction of a cohort that dies during the age interval x, $x + 1$. The column d_x can be summed to give the number dying over a particular period of time. If l_x and d_x are converted into proportions—that is, if the number of organisms that died during the interval x, $x + 1$ is divided by the number of organisms alive at the beginning of age x, the result is q, the age-specific mortality rate. The q_x column, however, cannot be summed to give the overall mortality rate at any specific age.

The next two columns are L_x, the average years lived by all individuals in each age category, and T_x, the number of time units left for all individuals to live from age x onward. These two columns provide the data needed to calculate e_x, the life expectancy at the end of each age interval. The values for L_x, are obtained by summing the number alive at age intervals x and $x + 1$ and dividing the sum by 2. T_x is calculated by summing all values of L_x from the bottom of the table upward. Life expectancy is obtained by dividing T_x for a particular age class x by the l_x value for that age class.

Consider as an example Table 13.1, the life table for the gray squirrel. The table has another column, n_x, the actual data on which the life table is based. The data were obtained by following cohorts of squirrels marked as nestlings. The raw data are

converted to a cohort of 1000 by simple proportion: divide the number in each age category by 530, which is the equivalent of 1000: 530/530 = 1.000; 134/530 = 0.253; and so on. The proportionality can be multiplied by 1000 to obtain whole numbers for each age category. However, because the numbers will have to be converted back for other calcuations, it is more convenient to keep the data as proportionality.

Obtain the d_x column by subtracting the number of survivors at $x + 1$ from the survivors at previous age x: 1.000 − 0.253 = 0.747, the mortality at age 0–1; 0.253 − 0.106 = 0.147, mortality at age 1–2, and so on.

Divide the mortality (d_x) for age x by the value of l_x for the same age interval to obtain the value q_x for the same age class: 0.747/1.000 = 0.747 for age class 0–1; 0.147/.253 = 0.581 for age class 1–2; and so on.

Obtain the values for the L_x column by adding the survivorship of x and $x + 1$ and dividing by 2. For age class 0–1, (1.000 + 0.253)/2 = 0.626; for age class 2–3, (0.106 + 0.074)/2 = 0.090; and so on.

Calculate the T_x column by adding the values in the L_x column upward: for age class 6–7, 0.001 + 0.006 = 0.007; for age class 2–3, 0.001 + 0.006 + 0.015 + 0.033 + 0.058 + 0.090 = 0.203.

Obtain life expectancy, e_x, for each age class by dividing the T_x value by the l_x value. For age class 0–1, life expectancy is 1.008/1.000 = 1.008; for age class 4–5, life expectancy is 0.055/0.043 = 1.279.

In other words, an individual grey squirrel's life expectancy at birth is 1.008 years. If the individual lives to 4 years, on the average it can be expected to live another 1.279 years.

TABLE 13.2

LIFE TABLE OF A SPARSE GYPSY MOTH POPULATION IN SOUTHEASTERN NEW YORK				
x	l_x	d_{xf}	d_x	$1000q_x$
Eggs	450.0	Parasites	67.5	15
		Other	67.5	15
		Total	135.0	30
Instars I–III	315.0	Dispersal, etc.	157.5	50
Instars IV–VI	157.5	Parasites	7.9	5
		Disease	7.9	5
		Other	118.1	75
		Total	291.4	135
Prepupae	23.6	Dessication, etc.	0.7	3
Pupae	22.9	Vertebrate predators	4.6	20
		Others	2.3	10
		Total	7.6	33
Adults	16.0		5.6	35
Entire generation			439.6	233

TABLE 13.3

LIFE TABLE FOR A NATURAL POPULATION OF *SEDUM SMALLII*						
x	l_x	d_x	$1000q_x$	L_x	T_x	e_x
Seed produced	1000	160	160	920	4436	4.4
Available	840	630	750	525	756	0.9
Germinated	210	177	843	122	230	1.1
Established	33	9	273	28	109	3.3
Rosettes	24	10	417	19	52	2.2
Mature plants	14	14	1000	7	14	1.0

or biomass and the size of the modular populations of buds, leaves, and stems. Seedlings make up a large numerical proportion of individuals but an extremely small portion of the biomass. Further, it is difficult to separate and even identify individuals. The parent plant may die, yet live on in sprouts and suckers. The plant demographer has to deal with mortality (and natality) on two levels, the individual and the clones.

In plant demography the life table is most useful in studying three areas: (1) seedling mortality and survival; (2) population dynamics of perennial plants marked as seedlings; and (3) life cycles of annual plants. An example of the third type is Table 13.3. The time of seed formation is the initial point in the life cycle. The l_x column indicates the number of plants alive at the beginning of each stage and the d_x column the

number dying. The L_x column gives the mean number of plants alive during the life cycle. The T_x column gives the total number of plants remaining at the beginning of each life cycle stage. Life expectancy of these annual plants dropped rapidly in the seed stages and returned to a high level after seedlings were established. Although individuals that became established had a good chance of surviving, the high early mortality resulted in a low mean life expectancy.

Another approach to the life table of plants is the yield table developed by foresters (Table 13.4). The yield table considers age classes and the density of trees in each age class, with additional columns giving diameters and basal area (cross-sectional area). Yield tables chart the mortality of trees by the reduced number of individuals in each age class. However, as the

TABLE 13.4

YIELD TABLES FOR DOUGLAS-FIR ON FULLY STOCKED HECTARE (SITE INDEX 200)			
Age (years)	Trees per ha	Average dbh (cm)	Basal area (m²)
20	1427	14	23
30	875	23	35
40	600	30	46
50	440	38	53
60	345	46	58
70	282	53	62
80	242	59	66
90	210	65	69
100	187	70	72
110	172	75	75
120	157	80	77
130	147	83	79
140	137	87	81
150	127	91	83
160	120	94	85

numbers decline, basal area and biomass increase. Mortality may not indicate a declining population but a maturing one. Like life tables, yield tables are not constant for a species. We can construct them for different environmental conditions, such as soils and moisture availability. The conditions are called site classes.

13.4 Life tables provide data for mortality and survivorship curves

From the life table we can plot two kinds of curves: a *mortality curve* based on the q_x column, and a *survivorship curve* based on the l_x column. A mortality curve (Figure 13.1) plots mortality rates in terms of 1000 or 1.000 q_x against age. It consists of two parts: a juvenile phase, in which the rate of mortality is high; and a post-juvenile phase, in which the rate decreases with age until mortality reaches some low point, after which it increases again. For mammals a roughly J-shaped curve results. For plants the mortality curve may assume a number of patterns, depending on whether it is annual or perennial and how we plot the data.

Survivorship curves can be plotted in a number of ways. The usual method is to plot the logarithmic number of survivors, the l_x column, against time or age (*x*). The time interval is on the horizontal axis and survivorship is on the vertical axis (Figure 13.2).

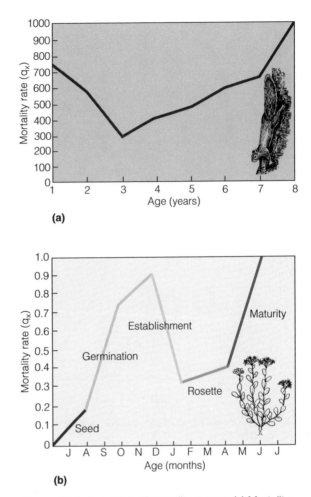

(a)

(b)

Figure 13.1 Examples of mortality curves. (a) Mortality curve for the gray squirrel population. (b) *Sedum* has two J-shaped curves, one for seeds through germination and another for plants through "youth," the rosette stage, and maturity.

The accuracy of survivorship curves depends upon the accuracy of the life table. Life tables, and so survivorship curves, are based on data obtained from one population of the species at a particular time and under certain environmental conditions. They are like snapshots. For this reason survivorship curves are useful for comparing one time, area, or sex with another. They show us, for example, differences in survival between sexes.

Survivorship curves fall into three general types (Figure 13.3). When individuals tend to live out their physiological life span and when there is a high degree of survival throughout the life span followed by heavy mortality at the end, the curve is strongly convex, or Type I. Such a curve is typical of humans and other

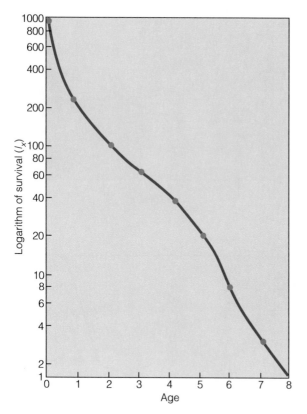

Figure 13.2 Survivorship curve for the gray squirrel, based on Table 13.1.

mammals and some plants. If mortality rates are constant at all ages, the survivorship curve will be straight, or Type II. Such a curve is characteristic of adult birds, rodents, and reptiles, as well as many perennial plants. If mortality rates are extremely high in early life—as in oysters, fish, many invertebrates, and some plants— the curve is concave, called Type III.

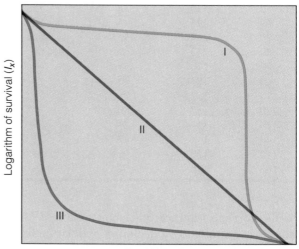

Figure 13.3 The three basic types of survivorship curves.

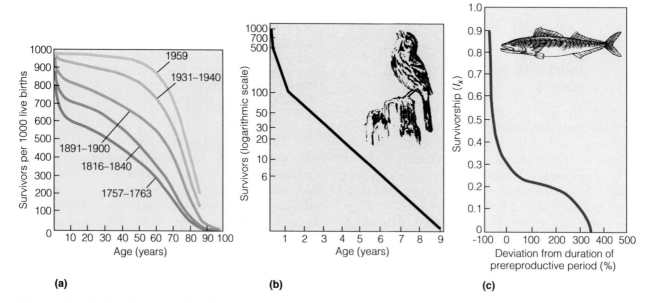

(a) **(b)** **(c)**

Figure 13.4 Survivorship curves for animals. (a) Type I curves for the human population of Sweden over several centuries. As health conditions improved, survivorship began to approach physiological longevity. (b) Survivorship curve for the song sparrow *(Melospiza melodia)*. The curve is typical of birds. After a period of high juvenile mortality (Type III), the curve becomes linear (Type II). (c) Atlantic mackerel *(Scomber scombrus)* has a Type III curve with high juvenile mortality.

These generalized survivorship curves are idealized models to which survivorship of a species can be compared (Figures 13.4 and 13.5). Most survivorship curves are intermediates between models.

13.5   Natality is age-specific

Birth or natality rates are usually expressed as births per 1000 population per unit time. This figure is obtained by dividing the number of births per unit time by the estimated population size at the beginning of the unit or period of time and multiplying it by 1000. This figure is the **crude birthrate.**

A better way of expressing birthrate is the number of births per female of age x per unit of time, because reproductive success varies with age. If we arbitrarily divide females of reproductive age into age classes and tabulate the number of births for each age class, we get an **age-specific schedule of births.** Because population increases are a function of the female, the age-specific birth schedule can be further modified by determining only the mean number of females born in each age group, m_x. This information is the **gross reproductive rate.** It contrasts with the **net reproductive rate**, R, the number of females left during a lifetime by a newborn female or the mean number of females born in each age group. Because it is calculated by multiplying the reproductive rate, m_x, by the survival of each age class, l_x, it includes adjustments for mortality of females in each age group.

13.6   Natality and survivorship determine reproductive rates

How are net reproductive rates determined? Let us make the gray squirrel population described in Table 13.1 the basis for the construction of a fecundity table (Table 13.5). The fecundity table uses the survivorship column, l_x from the life table and an m_x column, the mean number of females born to females in each age group. At age 0–1 females produce no young; therefore their m_x value is 0. The m_x value for females

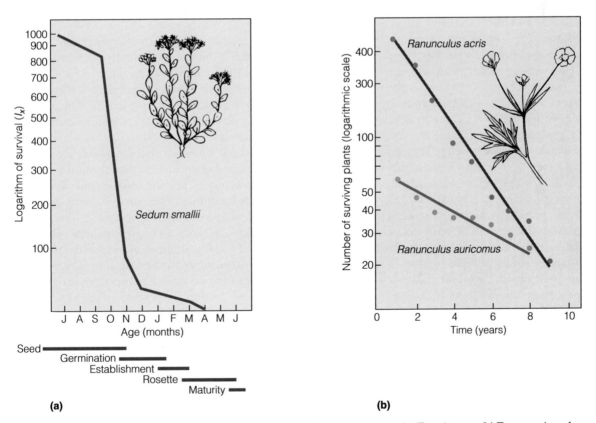

(a)   **(b)**

Figure 13.5 Survivorship curves for plants. (a) *Sedum smallii* has elements of a Type I curve. (b) Two species of buttercup *(Ranuculus)* exhibit Type II curves.

TABLE 13.5

GRAY SQUIRREL FECUNDITY TABLE

x	l_x	m_x	$l_x m_x$	$x l_x m_x$
0–1	1.000	0.0	0.0	0.0
1–2	0.253	1.28	0.324	0.324
2–3	0.106	2.28	0.242	0.484
3–4	0.074	3.24	0.240	0.720
4–5	0.043	3.24	0.139	0.556
5–6	0.022	2.48	0.055	0.275
6–7	0.009	2.28	0.021	0.126
7–8	0.003	2.28	0.007	0.049
			$R_0 = 1.027$	

Total weighted age = 2.60

at age 1–2 is 1.28. Before age 5–6, the m_x values increase; then they decline. Although m_x may increase with age, survivorship in each age class declines. To adjust for mortality, multiply the m_x values by the corresponding l_x, or survivorship, values. The resulting value, $l_x m_x$, gives the mean number of females born in each age group adjusted for survivorship.

Thus, for age class 1–2, the m_x value is 1.28; but when adjusted for survival l_x, the value drops to 0.324. For age 5–6, the m_x is 2.48 but $l_x m_x$ drops to 0.055, reflecting poor survival of adult females. The adjusted m_x values, $l_x m_x$, are summed over all ages at which reproduction occurs. The result represents the number of females that will be left during a lifetime by a newborn female, designated as R_0. If the R_0 value is 1, females have replaced themselves. If the value is less than 1, the females are not replacing themselves. If the value is much over 1, they are leaving additional offspring behind. Our gray squirrel, with an R_0 value of 1.027, is just maintaining its population.

Natality in plants, like mortality, is perplexing because plants reproduce both sexually and asexually. If you consider only the genetic individual, then natality is restricted to sexual reproduction. There are two separate populations, seeds and seedlings, and two separate processes, the production of seeds and the germination of seeds. Except for annuals and biennials, which have one reproductive effort resulting in the death of the parent plant, seed production by individual plants is hard to estimate. Woody plants and other perennials, even within a population, vary in longevity, in seed production over the years, and in the ability of seeds to germinate.

The formal equivalent of birth in plants is germination. Before "birth," seeds usually undergo a varying period of dormancy, often necessary before they can sprout. The seeds of some plants remain dormant for years, buried in soil or mud as a seed bank, until they are exposed to conditions conducive to germination. Once the seed has germinated, the seedling is subject to mortality. Thus, the plant population at all times consists of two parts, one growing and producing seeds, and the other stored as seeds in a dormant state.

13.7 Net reproductive rate is an estimator of population growth

Mortality and natality are the two major forces influencing population growth. Births minus deaths $(b - d)$ equals the rate of increase. When births exceed deaths, the population increases. When births equal deaths, the population remains the same. When deaths exceed births, the population declines.

Two additional influences on population growth are immigration *(i)*, an influx of new individuals into a population, and emigration *(e)*, the dispersal of individuals from a population. To account for those gains and losses, a general formula for the rate of increase (or decrease) is $(b + i) - (d + e)$.

The rate at which populations change can be estimated from R_0, the net reproductive rate, as determined from the fecundity table (Table 13.5). R_0 is the expected number of female offspring an average newborn female will produce during her lifetime, assuming discrete generations. It takes into account both births and mortality.

If generation times are discrete, as they are among annual plants and many insects, then the unit time and the generation time, *T*, are one and the same. But many populations, including invertebrates, plants, and all vertebrates, have overlapping generations. The parental generation continues to contribute new offspring to population growth, although at a reduced rate, while their older offspring reproduce. In this case, we have to convert generation time *T* to mean cohort generation time T_c by adding the product of age *(x)* and the number of expected offspring produced per age $(l_x m_x)$ to give a total weighted age $(x l_x m_x)$, as we did in Table 13.5. This total weighted age divided by R_0, the net reproductive rate, gives us the **mean cohort generation time,** T_c. For the gray squirrel population described in Table 13.1, the mean generation time is 2.60/1.027 = 2.53. Thus the gray squirrel produces an average of 1.027 offspring in an average of 2.5 years.

Because it is much more useful to compute population growth by year than by generation, we convert

R_0 to an annual finite rate of increase, designated by the Greek letter lambda (λ)

$$\lambda = R_0^{1/T_c}$$

which describes the geometric rate of increase by discrete time intervals. For the gray squirrel:

$$\lambda = 1.027^{1/2.5} = 1.027^{0.4} = 1.01$$

The annual rate of increase for the gray squirrel population is 1.01. It is lower than the generation rate of increase. When $R_0 = \lambda = 1$, females are replacing themselves, and the population remains the same. When λ is greater than 1, the population increases. How fast it increases is revealed by how much λ exceeds 1. When λ is less than 1, the population is declining. The gray squirrel population is growing very slowly.

13.8 Exponential growth is like compound interest

The general equation for population growth is $N_t = N_0\lambda$ in which N_t is the population size at some given time in the future, N_0 is the initial population, and λ is the annual rate of increase. To determine the population growth of the gray squirrel, you might start with an initial population of $N_0 = 20$ and $\lambda = 1.01$.

The equation can be stated as $N_{t+1} = N_t\lambda$, or $N_t = N_0\lambda^t$, in which λ is raised to the power of the appropriate time interval. For example $N_4 = N_0\lambda^4$. The population of the gray squirrel in four years would be $N_4 = 20 \times 1.01^4 = 21$.

If you lack sufficient life table data to construct a fecundity table, you can estimate annual rate of increase, λ, from the ratio of numbers at successive time intervals, provided you can obtain sufficient census data:

$$\lambda = N_{t+1}/N_t$$

For example, a mule deer herd in Colorado over three years had annual populations of 10,449, 10,702, and 11,153. Lambda in this case would be $N_2/N_1 = 10,702/10,449 = 1.02$; and $N_3/N_2 = 11,153/10,702 = 1.04$.

The equation $N_t = N_0\lambda^t$ describes a population that grows exponentially, like compound interest. (See Quantifying Ecology 13.2: Exponential Growth and Compound Interest.) Such growth can occur when λ is greater than 1, the environment remains constant, and resources are excessive. Some hypothetical examples of exponential growth are in Figure 13.6. Notice how the shapes of the curves vary with the value of λ.

QUANTIFYING ECOLOGY 13.2

EXPONENTIAL GROWTH AND COMPOUND INTEREST

If there were no movement into or out of a population and no mortality, then birthrate alone would account for population changes. Under this condition population growth would accrue like compound interest. If you refer to a mathematics handbook, you will learn that if interest is compounded annually,

$$A = P(1 + r)^n$$

where A is the new amount at some given time, P is the original amount or principal, r is the rate of interest expressed as a decimal, and n is the number of years.

If interest is compounded several times a year, then

$$A = P\left(1 + \frac{r}{q}\right)^{nq}$$

where q is the number of times interest is compounded during the year.

If interest is compounded continuously, then q approaches infinity. Letting r/q equal x, the expression can be written

$$A + P(1 + x)^{rn/x}$$

When q approaches infinity, x approaches 0. From calculus it can be shown that

$$\lim_{x\to 0}(1 + x)^{1/x} = e$$

where e is the base of natural logarithms, which is approximately 2.7183. Thus, if interest is compounded continuously, the expression reads:

$$A = Pe^{rn}$$

Using symbols common in population ecology, A, the amount at some time t, becomes N_t; the principal, P, becomes N_0, the initial population; r remains the growth rate or rate of increase; and t is the unit of time. Then our compound interest formula will read

$$N_t = N_0e^{rt}$$

This formula expresses logarithmic population growth, the accumulation of compound interest (births) in the population.

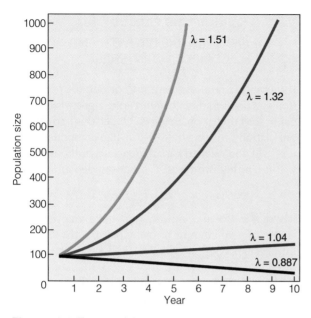

Figure 13.6 Exponential growth for four hypothetical populations with different values of λ.

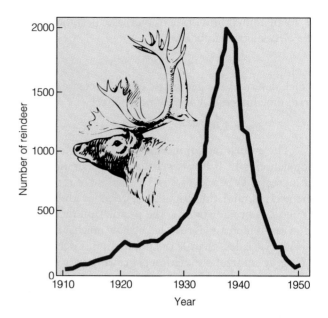

Figure 13.7 Exponential growth of the St. Paul reindeer (Rangifer tarandus) herd and its subsequent decline.

The closer λ comes to 1, the slower the growth. The population with value of λ = 1.04 is barely replacing itself, and the population with the value below 1 is declining. These curves suggest several features of population growth. It is influenced by life history features, such as age at the beginning of reproduction, the number of young produced, survival of young, and length of the reproductive period. A population may increase at an exponential rate until it overshoots the ability of the environment to support it. Then the population declines sharply from starvation, disease, or emigration.

The J-shaped or exponential curve is characteristic of some invertebrate and vertebrate populations introduced in a new or unfilled environment. An example of an exponential growth curve is the rise and decline of the reindeer herd introduced on St. Paul, one of the Pribilof Islands, Alaska (Figure 13.7). Introduced on St. Paul in 1910, reindeer expanded rapidly from 4 males and 22 females to a herd of 2000 in only 30 years. So severely did the reindeer overgraze their range that the herd plummeted to 8 animals in 1950. The decline produced a curve typical of a population that exceeds the resources of the environment. Growth stops abruptly and declines sharply in the face of environmental deterioration. From a low point the population may recover to undergo another phase of exponential growth. It may decline to extinction, or it may recover and fluctuate about some level far below the original high level.

13.9 Rate of increase is used in population studies

The finite rate of increase, lambda, can be expressed as the **rate of increase,** r, which describes instantaneous population growth. The population is considered to breed continuously, rather than possessing a discrete breeding season. Such an approach best describes population growth in organisms that reproduce throughout the year. In populations with overlapping generations, however, members of different generations may give birth or die during the same time interval. In such situations it is more usual to express population growth as instantaneous rate of change in births and deaths in the form of differential equations.

The rate of increase is obtained by taking the natural log of λ. Thus, for the gray squirrel $r = l_n\ 1.01 = 0.01$. Lambda, then, is often expressed as e^r where e is the base of the natural logarithm, 2.71828. For the gray squirrel $e^r = 2.71828^{0.01} = 1.01$.

The determination of r by this method only approximates the actual rate of increase. To determine the actual value of r_{max}, the intrinsic rate of increase, which is a continuous and not a discrete variable, requires a more complex calculation.

The rate of increase depends upon the exponential rate at which a population grows if it has a **stable age distribution** appropriate to the current life table and fecundity table. A stable age distribution is one in which the proportions of each age class remain the same, even though the population is growing. The rate of increase also depends upon mean fecundity and mean survival at each age in the population, which involves age at first reproduction, average litter size, and length of reproductive period. Because age structure is seldom stable and fecundity and survival vary over time, r is continually changing. Thus r, like λ, reflects the past and not the present.

Nevertheless, the use of r, the rate of increase, has certain advantages. It enables biologists to compare growth of populations living under different environmental conditions. Further, when population growth is measured as r, it has the same value as an equivalent rate of population decrease. Consider the declining population with $\lambda = 0.887$. If the population were increasing at the same rate, λ would be 1.127. It is hard to see that $\lambda = 0.887$ is the inverse of $\lambda = 1.127$. It is easy to see the connection between $r = +0.120$ and $r = -0.120$. Thus, r allows a direct comparison of rates and converts easily from one form to another. In addition, r allows us to compute the doubling time of a population.

Doubling time is the time required for a population to double its size. If $N_t/N_0 = 2$; then $e^{rt} = 2$; $rt = \log_n = 0.6931$. Therefore doubling time = $0.6931/r$. The doubling time for the gray squirrel population, $0.6931/0.01$, is 69 years. Demographers use the same method to determine the doubling time of human populations.

Now, with some of the mystery removed from r, we will use the term from this point on. To present exponential growth in terms of r, you would use the equation $N_t = N_0 e^{rt}$ to project 8 years into the future. For the size of a gray squirrel population 8 years hence, assuming an initial population at N_0 as 100, we calculate:

$$N_8 = 100\,(2.71828^{0.01 \times 8})$$
$$= 100\,(2.71828^{0.08}) = 108$$

13.10 Population growth is limited by the environment

In the real world, the environment is not constant and resources are limited. As the density of a population increases, competition among its members for available resources also increases. With shrinking resources and with an unequal distribution of those resources, mortality increases, fecundity decreases, or both. As a result, population growth declines with increasing

density, eventually reaching a level at which population growth ceases. That level is called **carrying capacity,** or K. Theoretically, at K the population is in equilibrium, neither increasing nor decreasing with respect to its resources or environment. In other words, population growth is density-dependent, in contrast to exponential growth, which is independent of population density.

Inhibitions on the growth of a population by competition among its members for available resources can be described mathematically. We can express the exponential growth model $N_t = N_0 e^{rt}$ as a differential equation:

$$\frac{dN}{dt} = rN$$

By adding the variable K to account for the effects of density that slows population growth, we can define the logistic model of population as:

$$\frac{dN}{dt} = rN\left(\frac{K-N}{K}\right)$$

in which dN/dt represents the instantaneous rate of change in population density N, K is carrying capacity, and $(K-N)/K$ is the further opportunity for population growth. As the population grows, this unutilized opportunity declines. This equation describes the logistic (sigmoidal or S-shaped) growth curve (Figure 13.8). The expression $(K-N)/K$ slows population growth. Note that in this formula, as N approaches K, the value of the expression decreases toward 0.

As an example of the logistic growth rate, consider a hypothetical population with an initial size of 100, a rate of increase r of 0.412 ($\lambda = 1.51$), and a carrying capacity of 400:

Year	Size
0	100
1	134
2	173
3	214
4	253
5	289
7	342
9	372
12	391
16	398
20	399.6
25	399.9
36	400

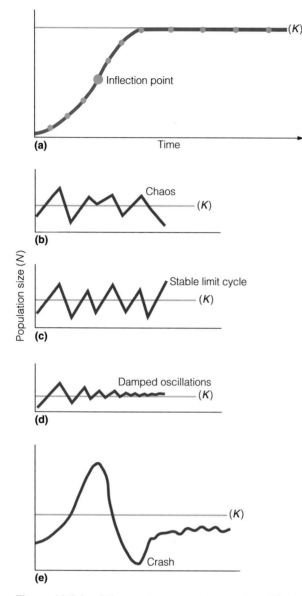

Figure 13.8 Logistic growth curve and examples of fluctuations around *K*.

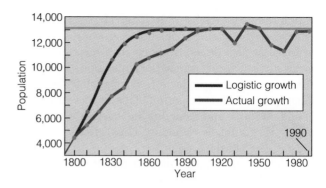

Figure 13.9 Actual and logistically predicted growth of population in Monroe County, West Virginia.

Note how the rate of increase is slow at first, accelerates, and then slows. The point in the logistic growth curve where growth is maximal is *K*/2, known as the inflection point. From this point on population growth slows. As population density approaches carrying capacity, *N* approaches *K* and the rate declines. Populations rarely remain at *K*; they may decline below *K*.

The logistic growth curve is theoretical, a mathematical model of how populations might grow under favorable conditions. Although natural populations might appear to grow logistically, they rarely do.

For example, Figure 13.9 considers the growth of a human population in Monroe County, West Virginia, which has always had a stable economic base of agriculture and small industry. It was settled in the early 1700s, and was well established in 1800, when the first United States census was taken. The population grew most rapidly between 1800 and 1850, so those years provided the data to estimate *r*. The population reached 13,200 in 1900 and fluctuated about that number since that year. The rate of increase *r* was calculated as 0.074 and *K* was set at 13,200. Although the actual growth curve of the population mimics the logistic, the calculated logistic growth curve rose much more steeply than the actual growth curve and predicted that the population would reach *K* around 1870, 30 years before it actually did.

The reasons for nonconformity are obvious. The age structure was not stable; birthrates and death rates varied from census period to census period; immigration and emigration were common to the population. The most surprising feature of the population is its stability after reaching *K*.

13.11 Populations fluctuate about some upper and lower limits

The logistic equation suggests that populations function as systems, regulated by positive and negative feedback. Positive feedback promotes growth (as illustrated by the exponential curve); negative feedback of competition and resource availability slows it. As the population approaches carrying capacity, it theoretically responds instantaneously as density-dependent reactions set in.

Rarely does such feedback work as smoothly in practice as the equation suggests. Often adjustments lag, and available resources may be sufficient to allow the population to overshoot equilibrium. Unable to sustain itself, the population then drops to some point below carrying capacity, but not before it has altered resource availability for future generations. The density of the previous generation and the depletion of resources, especially food, build a time lag into population recovery.

Time lags make a population fluctuate, sometimes widely. Such populations may be influenced by some powerful outside force such as weather or by some chaotic changes inherent in the population. A population may fluctuate about the equilibrium level, K, rising and falling between upper and lower limits (see Figure 13.8). Such fluctuations are **stable limit cycles.** Some populations oscillate between high and low points in a manner more regular than we would expect to occur by chance. Such fluctuations are **population cycles.**

The two most common oscillation intervals in animal populations are nine- to ten-year cycles, typical of the snowshoe hare (Figure 13.10) and three- to four-year cycles, typical of lemmings (Figure 13.11). These cyclic fluctuations are confined largely to simpler ecosystems, such as northern coniferous forest and tundra. Cycles in the snowshoe hare involve an interaction among the hare, its overwinter food supply (mostly small aspen twigs), and predators (see Chapter 16). A growing population of hares reduces the ability of plants to recover from pruning. Once decreased plant growth triggers an overwinter food shortage, a weakened hare population becomes highly vulnerable to predation. Now the hare population is low and the vegetation recovers, stimulating a resurgence of the hare and initiating another cycle.

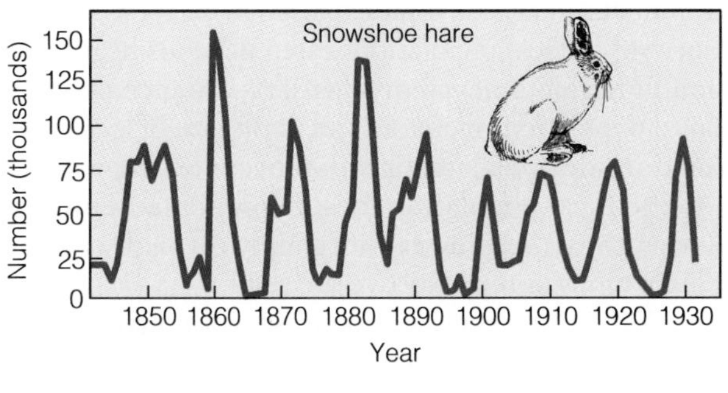

13.12 Low population can decline toward extinction

When R_0 is less than 1 or r is negative, the population declines. Increasing sparseness is associated with a reduction in the rate of increase. The population may become so low that it declines toward extinction (see Figure 13.6).

The extinct heath hen (*Tympanuchus cupido cupido*) is a classic example. Formerly abundant in New England, this eastern form of the prairie chicken was driven eastward by habitat destruction and excessive hunting to Martha's Vineyard, off the Massachusetts coast, and to the pine barrens of New Jersey. By 1880 it was restricted to Martha's Vineyard. Two hundred birds made up the total population in 1890. Conservation measures increased the population to 2000 by 1917; but then fire, gales, cold weather, and excessive predation by goshawks reduced the population to 50. The number of birds rose slightly by 1920, then declined to extinction in 1925. The last bird died in 1932.

The case of the heath hen points out the major causes of extinction, hastened by human interference. The overriding cause is habitat destruction. Loss of habitat forces remnant populations to exist in small, fragmented patches of habitat. These small populations are highly vulnerable to chance environmental

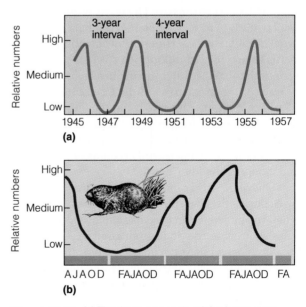

Figure 13.11 (a) The four-year cycle of the brown lemming *(Lemmus sibiricus)* near Barrow, Alaska. (b) A single four-year oscillation, showing subordinate fluctuations. The letters stand for April, July, August, October, December, and February.

Figure 13.10 The nine- to ten-year cycle of the showshoe hare *(Lepus americanus)* in Canada.

catastrophes and predation. The fewer the animals, the greater the individual's chances of succumbing to a predator. Loss of even a few individuals can severely impair the population's viability. Small populations may not be large enough to stimulate the social behavior necessary for successful reproductive activity. Heath hens are communal or lek breeders (see Section 12.6). The species requires a minimal number of displaying males to attract females and stimulate reproductive activity. The population was too small to carry off successful reproduction.

Extinction is a natural process, albeit a selective one. Species differ in their probability of extinction, depending on their characteristics as well as on random factors. Some of the qualities of a species that favor a high rate of extinction are large body size, small or restricted geographic range, habitat specialization, lack of genetic variability to cope with a changing environment, and inability to switch to alternate food. However, the recent extinction and rapid decline of populations are unnatural, the result of ever increasing pressures of human population.

13.13 Past extinctions have been clustered in time

Extinctions are not spread evenly across Earth's history. Most extinctions are clustered in geologically brief periods of time (less than several tens of millions of years). One mass extinction occurred in the late Permian period, 225 million years ago, when 90 percent of the shallow-water marine invertebrates disappeared. Another occurred in the Cretaceous period, 65 to 125 million years ago, when the dinosaurs vanished. That extinction perhaps was brought about by asteroids striking Earth, interrupting oceanic circulation, altering the climate, and accompanying volcanic and mountain-building activity.

One of the great extinctions of mammalian life took place during the Pleistocene, when such species as the woolly mammoth, giant deer, mastodon, and giant sloth vanished from Earth. Some students of the Quaternary (that geological period from the end of the Tertiary period to the present time) believe that climate changes as ice sheets advanced and retreated caused the extinctions. Others argue that Pleistocene hunters overkilled large mammals, especially in North America, as human populations swept through North and South America between 11,550 and 10,000 years ago. Per-

haps the large grazing herbivores could not withstand the combined predatory pressure of humans and other large carnivores. The greatest number of present-day extinctions have taken place since A.D. 1600. Humans have caused well over 75 percent of these extinctions through habitat destruction, introduced predators and parasites, and exploitative hunting and fishing.

13.14 Extinction of species begins locally

We usually think of extinction as taking place simultaneously over the full range of the species. Actually, it begins with isolated local extinctions when local conditions deteriorate or habitat disappears. Eventually, one local extinction after another adds up to total extinction.

The most important cause of extinction is habitat destruction or alteration, which is a local phenomenon. Cutting and clearing a forest, draining and filling a wetland, converting grassland to cropland, constructing highways and industrial complexes, and spreading cities and suburbs, with new housing and malls, greatly reduce available habitat for most species. When a habitat disappears, its unique plant and animal life also disappear (see Sections 18.9, 21.5).

Because of the rapidity of habitat destruction, no evolutionary time exists for a species to adapt to changed conditions. Forced to leave, dispossessed animals usually find the remaining habitats filled and face competition from others of their own kind or from different species. Restricted to marginal habitats, animals may persist for a while as nonreproducing members of a population or succumb to predation or starvation. As the habitat becomes more and more fragmented, the affected animal populations are fragmented into small isolated populations out of contact with others of the same species. As a result, isolated populations have less genetic variation, making them less adaptable to environmental change, a topic explored in Chapter 15. The survival of local populations often depends heavily on immigration of new individuals. As distance between local populations increases and as the size of local populations declines, immigration becomes impossible. As the local population falls below some minimum level, it may become extinct simply through random fluctuations in reproductive success.

Much the same situation exists with plants. They, too, face habitat destruction brought about by agriculture, mining, and urban and suburban developments.

Such activities result in the mass elimination of whole populations, many of them restricted to certain habitats. Unlike animals, most plants have poor dispersal abilities and cannot escape to favorable habitats, nor can they adapt quickly to changed environmental con-

ditions. Added to habitat destruction is the invasion of habitats by alien species humans introduce, which take over and crowd out native plant species (see Chapter 15).

CHAPTER REVIEW

SUMMARY

Mortality, Survivorship, and Life Tables (13.1–13.4) Mortality, concentrated in the young and the old, is the greatest reducer of populations. Mortality is measured by dividing the number dying in a given period by the number alive at the beginning of the period **(13.1)**. Mortality and its complement, survivorship, are best analyzed by means of a life table, an age-specific summary of mortality. Life tables for animals can be derived from data on mortality and survivorship **(13.2)**. Life tables for plants are more difficult to develop because plants have modular growth and two life cycle stages, germination and seedling development **(13.3)**. From the life table we derive both mortality curves and survivorship curves. They are useful for comparing demographic trends within a population and among populations under different environmental conditions, and for comparing survivorship among various species. In general, mortality curves assume a J shape. Survivorship curves fall into three major types: Type I, in which individuals tend to live out their physiological life span; Type II, in which mortality, and thus survivorship, is constant through all ages; and Type III, in which survival of young is low. Survivorship curves follow similar patterns in both plants and animals **(13.4)**.

Natality and Fecundity Tables (13.5–13.6) Birth is the greatest influence on population increase. Like death, births are age-specific. Certain age classes contribute more to the population than others **(13.5)**. Natality in plants is complicated because they have both vegetative and sexual reproduction. Seed production and germination are true natality, because they produce new genetic individuals. Modular populations of clones—leaves, buds, and twigs—exhibit their own "birth" rates. The contribution of various age classes can be determined by a fecundity table. It considers the gross reproduction of each age class, m_x, and survivorship, l_x, of each age, the sum of the

products of which gives the net reproductive rate, R_0 **(13.6)**.

Reproductive Rates and Population Growth (13.7–13.11) Influencing population size is the number of individuals added to the group by birth and immigration and the number leaving by death and emigration. When additions exceed removals, the population increases. The difference between birthrates and death rates $(b - d)$ when measured as an instantaneous rate is the rate of increase, r. It is derived either from R_0, determined from the fecundity table or from the annual rate of increase, λ, derived from R_0 **(13.7)**. In an unlimited environment, populations expand geometrically or exponentially. Exponential growth, described by a J-shaped curve, is like compound interest. Such growth may occur when a population is introduced into an unfilled habitat **(13.8)**. Finite rate of increase can be expressed as the rate of increase in which growth is considered continuous rather than discrete. This approach is most useful in studies of populations with overlapping generations or ones that reproduce throughout the year. The rate of increase is useful in comparing population growth under different environmental conditions **(13.9)**. Because resources are limited, geometric growth cannot be sustained indefinitely. Population growth eventually slows and arrives at some point of equilibrium with the carrying capacity (K) **(13.10)**.

Natural populations rarely achieve a stable level. They fluctuate within upper and lower limits about a mean. Some fluctuations have peaks and lows that occur more regularly than we would expect by chance. The two most common intervals are three to four years, as in lemmings, and nine to ten years, as in snowshoe hares **(13.11)**.

Extinction of Populations (13.12–13.14) A population may decline to extinction **(13.12)**. Past extinctions have been clustered in time. Extinction is a natural process. Over long geological periods of time, old species disappear and new ones evolve **(13.13)**.

Current extinctions, however, are not brought about by natural processes but by human activity. At the present time habitat fragmentation and destruction and overexploitation of populations are accelerating extinctions at an alarming rate (13.14).

STUDY QUESTIONS

1. Explain the meanings of mortality, natality, survivorship, and fecundity. How are they related?

2. What is a life table? What information do you need to construct one?

3. What are the advantages and weaknesses of a life table in a study of population dynamics?

4. What is the difference between a mortality curve and a survivorship curve? From what columns of the life table are they derived?

5. Why are natality and mortality more difficult to study in plants than in animals?

6. What are the differences between exponential and logistic population growth? How are they related?

7. How does R, net reproductive rate, relate to λ, the finite rate of increase, and r, the rate of increase? When do you use each one?

8. How do you determine the carrying capacity of a habitat? What causes fluctuations around carrying capacity?

*9. What are the causes of extinction? Relate them to some currently endangered species, such as the whooping crane, California condor, manatee, spotted owl, and desert tortoise.

*10. How do the concepts in this chapter apply to the worldwide problem of human population growth? Review the population growth rates of some less developed countries. For information, consult the *Population Bulletin* published by the Population Reference Bureau.

INTRASPECIFIC POPULATION REGULATION

OBJECTIVES

On completion of this chapter, you should be able to:

- Define competition.
- Distinguish between scramble and contest competition.
- Discuss the effects of intraspecific competition on growth, mortality, and natality.
- Explain the significance of dispersal.
- Describe a social hierarchy.
- Explain how social dominance may regulate populations.
- Define territoriality and explain its possible role in population regulation.
- Discuss the relationship between space capture and plant density.
- Explain how density-independent forces affect density-dependent responses.

A subdominant individual assumes a submissive pose when confronted by the dominant wolf in the pack.

No population continues to grow indefinitely. Even those with exponential growth confront the limits of the environment. Most populations, however, do not behave in an exponential fashion. As the density of a population changes, interactions set in among members of the population that tend to regulate its size.

14.1 Population regulation involves density dependence

Implicit in the concept of population regulation is **density dependence.** Density-dependent effects influence a population in proportion to its size. At low density there is no influence. Above that point, the larger the population becomes, the greater is the proportion of individuals affected. Density-dependent mechanisms act largely through competition for abundant or meager resources. If the effects of a particular influence do not change with population density, or if the proportion of individuals affected is the same at any density, then the influence is density-independent (Figure 14.1).

14.2 When resources are limited, competition results

One aspect of population regulation is competition among individuals of the same species for environmental resources. Individuals compete only when a resource is in short supply relative to the number seeking it. As long as resources enable each individual to survive and reproduce, no competition exists. When resources are insufficient to satisfy all individuals, the means by which they are allocated has a marked influence on the welfare of the population.

When resources are limited, a population may exhibit one of two responses, scramble competition and contest competition. **Scramble competition** occurs when no individual receives enough of the resource for growth and reproduction, as long as the population remains dense. **Contest competition** takes place when some individuals claim enough resources while denying others a share. Generally, under the stress of limited resources, a species will exhibit only one type of competition. Some are scramble species and others are contest species. One species may practice both types at different stages in the life cycle. For example, the larval stages of some insects endure scramble competition until the population declines, and the adult stages face contest competition.

The outcome of scramble and contest competition differs. Scramble competition can produce chaotic

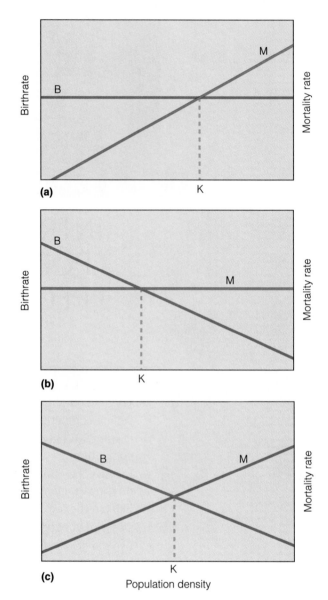

Figure 14.1 Regulation of density in three situations. (a) Birthrate (*B*) is independent of population density, as indicated by the horizontal line. Only the death rate (*M*) increases with density. At *K* equilibrium is maintained by increasing mortality. (b) The situation is reversed. Mortality is independent, but birth declines with density. At *K* a decreasing birthrate maintains equilibrium. (c) Full density-dependent regulation. Both birthrate and mortality are density dependent. Fluctuations in either one hold the population at or near *K*.

oscillations in the population over time. It limits the average density of the population to below that which the resources could support if they were available to only part of the population. For that reason, scramble competition can waste resources, from the point of view of population growth. In contest competition, a fraction of the population suffers, the unsuccessful individuals. That eliminates or greatly reduces the wastage of resources, permits the maintenance of high population density, and maintains some numerical constancy.

14.3 Intraspecific competition retards growth and reproduction

Because the intensity of intraspecific competition is density-dependent, it increases gradually, at first affecting just the quality of life. Later it affects individual survival and reproduction. (See Focus on Ecology 14.1: The African Buffalo.)

As population density increases toward a point at which resources are insufficient, individuals in scramble competition reduce their intake of food. That diet slows the rate of growth and inhibits reproduction. Examples of this inverse relationship between density and rate of body growth may be found among populations of ectothermic vertebrates (Figure 14.2). Tadpoles reared experimentally at high densities experienced slower growth, required a longer time to change from tadpoles to frogs, and had a lower probability of completing this transformation. Those that did reach threshold size were smaller than those living in less dense populations. Fish living in overstocked ponds exhibit a similar response to density. Bluegills *(Lepomis macrochirus)*, for example, normally grow to the size of a dessert plate, but in overstocked and underharvested farm ponds, many do not grow beyond the size of a silver dollar and never reproduce.

Other vertebrate groups, too, experience a density-dependent response, especially in fecundity, to increasing numbers. The timing of the response depends upon the nature of the population. Among large mammals with a long life span and low reproduction, regulating mechanisms do not function until the population approaches carrying capacity. At this point density mechanisms set in and tend to overcompensate. The birthrate of bison *(Bison bison)* shows such a response (Figure 14.3a), but the birthrates of other large mammals, such as the grizzly bear *(Ursus arctos)*, appear to be linear (Figure 14.3b). Among animals with high reproductive rates and short life spans, response to density may occur much earlier.

A close relationship exists between plant density and growth of individual plants as measured by accumulated living tissue or **biomass.** This biomass may be contained in many small individuals or a few large ones. For a time, all plant seedlings exhibit an increase in biomass. As their size increases, plants interfere with

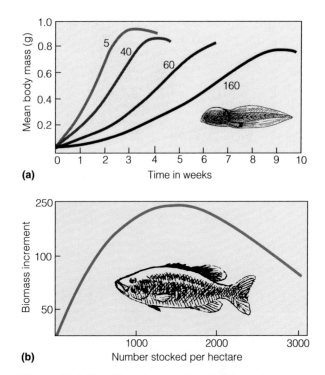

(a)

(b)

Figure 14.2 The effect of population density on the growth of individuals in scramble species. (a) The growth rate of the tadpole *Rana tigrina* declines swiftly as density increases from 5 to 160 individuals confined in the same space. (b) Growth of bass declines rapidly with density above 1500/ha.

each other, competing for the same resources—light, moisture, and space. Up to a point, plants are successful; then the population begins to experience density-dependent mortality.

Response takes place earliest in dense seedling populations, but in time it also occurs in less dense populations as plant size increases. As more biomass accumulates in fewer individuals, fewer plants per unit area are needed for competition to develop. As a result, plants starting at different densities converge toward some common value, decreasing through time. As mortality thins the population, more biomass accumulates in the remaining individuals. Therefore an old growth forest has a high biomass but fewer individuals than a young stand (see Table 13.4).

14.4 High density is stressful to populations

As a population reaches a high density, individual living space becomes restricted. Often aggressive contacts among individuals increase. One hypothesis of population regulation is that increased crowding and

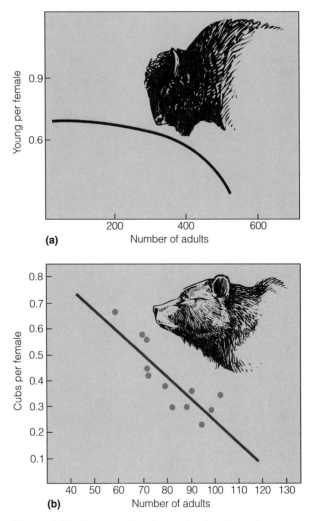

(a)

(b)

Figure 14.3 Linear and nonlinear density-dependent change in large mammal populations. (a) The birthrate of the American bison begins to decline sharply after the population reaches a certain density. (b) The grizzly bear shows a constant decrease in young as the population increases. (Figure 14.3b from D. R. McCullough, "Population Dynamics of the Yellowstone Grizzly," in *Dynamics of Large Mammal Populations,* C. W. Fowler and T. D. Smith, eds. (New York: John Wiley & Sons, 1981), p. 177. © 1981 John Wiley & Sons, Inc. Reprinted by permission.)

social contact cause stress. Such stress triggers hyperactivation of the system that controls the endocrine glands. Profound hormonal changes suppress growth, curtail reproductive functions, and delay sexual activity. They may also suppress the immune system and break down white blood cells, increasing vulnerability to disease. Social stress among pregnant females may increase intrauterine mortality and cause inadequate lactation, stunting nurslings. Thus stress results in decreased births and increased infant mortality. Such

population-regulating effects have been confirmed in confined laboratory populations of several species of mice and to a lesser degree in enclosed wild populations of woodchucks *(Marmota monax)* and old world rabbits *(Oryctolagus cuniculus).* Evidence of effects of stress in free-ranging wild animals, however, is difficult to obtain.

Pheromones, or chemical releasers, present in the urine of adult rodents can encourage or inhibit reproduction. One study involved wild female house mice *(Mus musculus)* confined to grassy areas within a highway cloverleaf. One group lived in a high-density population and the other in a low-density population. Urine from females of each population was absorbed onto filter paper and placed in laboratory cages with juvenile test females. Exposing juvenile females to urine from high-density populations delayed puberty, whereas exposing females to urine from low-density populations did not. The results suggest that pheromones in adult females in high-density populations delay puberty and help slow further population growth. Juvenile female mice exposed to urine of dominant adult males experience earlier onset of puberty.

14.5   Dispersal may or may not be density-dependent

Instead of coping with stress, some animals seek vacant habitats. Although dispersal is most apparent when population density is high, it goes on all the time. Some individuals leave the parent population whether it is crowded or not. However, there is no hard-and-fast rule about who disperses.

When a lack of resources forces some individuals to disperse, the ones to go are usually subadults driven out by adult aggression. The odds are that such individuals will perish, although a few may arrive at some suitable area and settle down. Because such dispersal follows overpopulation, it does not prevent it. More important to population regulation is dispersal when density is low or increasing, well before the population reaches a point at which it overexploits its resources. Individuals who participate are not a random selection of the population. Usually they are in good condition, are of either sex and any age, have a good chance of survival, and show a high probability of settling in a new area. Some evidence suggests that such individuals are genetically predisposed to disperse.

Such individuals can maximize their probability of survival and of having offspring only if they leave their birthplace. When intraspecific competition at home is intense, dispersers can relocate in habitats where resources are more accessible, breeding sites are more available, and competition is less. Further, the disperser reduces the risk of inbreeding (see Chapter 19). At the same time, dispersers incur risks that come with living in unfamiliar terrain.

Dispersers, according to recent studies, travel no farther than necessary. How far they go depends on the density of surrounding populations and the availability of suitable habitat. Individuals may travel in a straight line from their birthplace, or they may make a number of exploratory forays before leaving, and then occupy the first uncontested site they locate (Figure 14.4). How well they fare in their new location depends upon the quality of habitat.

Dispersal requires a source and a **sink,** an empty or unfilled habitat, or even marginal or unsuitable habitat where the animals can survive for a time. The dispersal sink must permanently remove animals from the resident or source population. Many dispersers die during their travels. Some dispersers discover and settle into patches of optimal habitat. Others move into areas where conditions such as predation, poor nesting sites, and lack of protective cover preclude successful reproduction. In such habitats reproduction

does not balance mortality. Even though some sink habitats support large populations, their populations would disappear without continual immigration. Because a species may actually be more abundant in a sink than in the source habitat, the sink appears to be optimal habitat. In reality, the habitat may be luring dispersers into danger or reproductive failure.

Such population sinks can be distinguished from optimal habitats only with some knowledge of the reproduction and survival of the species within it. Such knowledge is critical in the conservation of species. Because of population abundance, we may mistakenly select a sink habitat as a wildlife refuge. Such a decision could lead both sink and source populations to local extinction.

Does dispersal actually regulate a population? Although dispersal is positively correlated with population density, no relationship exists between the proportion of the population leaving and its increase or decrease. Dispersal may not function as a regulatory mechanism, but it contributes strongly to population expansion and aids in the persistence of local populations.

Successful dispersal of plants may depend upon the frequency of dispersal. The more often a plant has a large seed crop, the more often it will have new seeds available for colonization, even though seed losses will be enormous. Regardless of the means of dispersal, most seeds drop near the parent plant. Because they furnish a concentrated source of food, seeds near the parent plant undergo more intense predation by seed-eating animals than seeds scattered some distance away. Therefore the probability of seed survival increases with the distance from the parent plant.

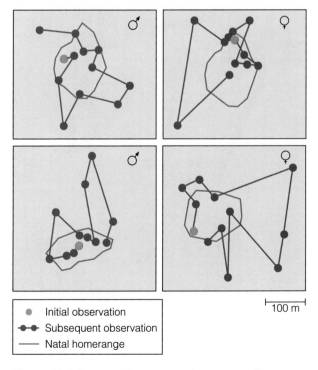

	Initial observation
●—●—●	Subsequent observation
——	Natal homerange

100 m

Figure 14.4 Dispersal forays made by young radio-collared red squirrels *(Tamiasciurus hudsonicus)* in Assiniboine, Alberta, Canada.

14.6 Social behavior may limit populations

Intraspecific competition expresses itself in social behavior, the degree to which individuals of the same species tolerate one another. Social behavior appears to be a mechanism that limits the number of animals living in a particular habitat, having access to a common food supply, and engaging in reproductive activities. It excludes the others.

To prove that social behavior limits populations in a density-dependent fashion, population ecologists have to show that (1) a substantial portion of the population consists of surplus animals that do not breed because they either die or attempt to breed and fail; (2) such individuals are prevented from breeding by dominant individuals; (3) nonbreeding individuals are

THE AFRICAN BUFFALO

The African buffalo *(Syncerus caffer),* studied in detail on the Serengeti in East Africa by A. R. C. Sinclair, is a large bovine ungulate of the African savanna. Its food is grass, preferably the protein-rich leaves. In 1894, rinderpest, a measleslike disease of cattle, spread from domestic cattle to the African buffalo and wildebeest *(Connochaetes taurinus).* The disease decimated the herds in the early 1900s. Veterinarians eliminated rinderpest in cattle, which reduced its incidence in wild bovines. The disease continued to cause high mortality among juvenile buffalo as late as 1964. Released from heavy juvenile mortality, the African buffalo expanded dramatically, reaching an apparent equilibrium density in the 1970s.

The critical time of year for the buffalo population is the dry season, when food may become scarce. Rainfall determines the productivity of grass. The greater the rainfall, the more vigorously the grass grows, increasing the amount of forage available in the dry season. Equilibrium density varies with the mean annual rainfall. The greater the rainfall, the more grass is available, and the greater is the density of buffalo (Figure A).

During the wet season on the African savanna, food is abundant, but during the dry season, the quality of food declines as the grasses dry. Buffalo become more selective, seeking green leaves and moving to the moist riverine habitat. They break into smaller units and use different areas. As the dry season progresses, buffalo become less selective, eating dry leaves and stems they otherwise would reject. As food quantity and quality decline, scramble competition becomes keener. Individuals in any one area reduce the food available to neighboring animals. The more buffalo present, the less food is available for each individual. Eventually the animals use up their fat reserves.

Undernourished and lacking the protein intake necessary to maintain their immunity to disease and the

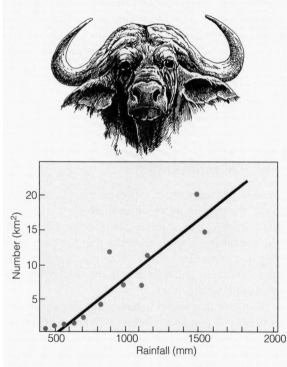

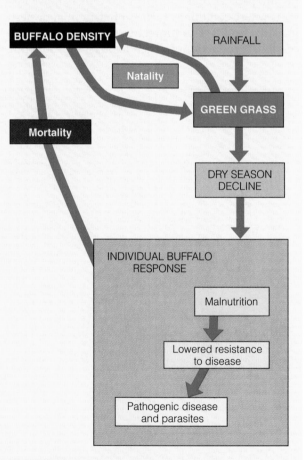

Figure A The relationship between rainfall and buffalo abundance.

Figure B Population regulation in the African buffalo.

parasites they normally harbor, adults become ill. Older animals are the most vulnerable. The number dying depends upon the rapidity with which adults use up their energy reserves before the coming of the rainy season and new growth. If the next season sees more rainfall and that rainfall extends sporadically into the dry season, the mortality of adults the following year declines (Figure B).

Thus mortality of adults, influenced by rainfall and poor resources, determines a variable equilibrium density and regulates the population. By contrast, juvenile mortality appears to be density independent. One or several environmental variables cause fluctuations in the population. Density-dependent adult mortality compensates for the disturbances and dampens fluctuations.

capable of breeding if dominant individuals are removed; and (4) breeding animals are not completely using food and space.

14.7 Social dominance can affect reproduction and survival

Many species of animals live in a group with some kind of social organization. This organization is based on aggressiveness, intolerance, and the dominance of one individual over another. Two opposing forces are at work. One is mutual attraction of individuals; the other is a negative reaction against crowding, the need for personal space.

Each individual occupies a position in the group based on dominance and submissiveness. In its simplest form, the group includes an alpha individual dominant over all others, a beta individual dominant over others except the alpha, and finally an omega individual subordinate to all others. Individuals settle social rank by fighting, bluffing, and threatening at initial encounters between any given pair of individuals or a series of such encounters. Once individuals establish social rank, they maintain it by habitual subordination of those in lower positions. They reinforce this relationship by threats and occasional punishment meted out by those of higher rank. Such organization stabilizes and formalizes intraspecific competitive relationships and resolves disputes with a minimum of fighting and wasted energy.

Social dominance plays a role in population regulation when it affects reproduction and survival in a density-dependent manner. An example is the wolf. Wolves live in small groups of 6 to 12 or more individuals, called packs. The pack is an extended kin group consisting of a mated pair, one or more juveniles from the previous year who do not become sexually mature until the second year, and several related, nonbreeding adults.

The pack has two social hierarchies, one headed by an alpha female and the other headed by an alpha male, the leader of the pack, to whom all other members defer. Below the alpha male is the beta male, closely related, often a full brother, who has to defend his position against pressure from males below.

Mating within the pack is rigidly controlled. The alpha male (occasionally the beta male) mates with the alpha female. She prevents lower-ranking females from mating with the alpha and other males, while the alpha male inhibits other males from mating with her. Therefore, each pack has one reproducing pair and one litter of pups each year. They are reared cooperatively by all members of the pack.

The size of the packs, which is heavily influenced by the availability of food, governs the level of the wolf population in a region. Priority for food goes to the producing pair. At high pack density, individuals may be expelled or leave the pack. Unless they have the opportunity to settle successfully in a new area and form a pack, they may not survive. Thus, at high wolf densities mortality increases and birthrates decline. When the population of wolves is low, sexually mature males and females leave the pack, settle in unoccupied habitat, and establish their own packs with one reproducing female in each. In this case, nearly every sexually mature female reproduces, and the wolf population increases. At very low densities, however, females may have difficulty locating males with whom to establish a pack and so fail to reproduce or even survive.

14.8 Social interactions influence activities and home ranges

Social interactions among individuals influence the movement, distribution, and reproductive activity of animals over an area. The space that an animal normally uses during a year is its **home range.** The two sexes may have the same or different home ranges (Figure 14.5), which may overlap. Often the home range of a male includes those of several females. Although the home range is not defended, aggressive interactions may influence the movements of individuals within another's home range. Some species, however, defend a core area of the home range against others. If the animal defends any part of its home range, then we define that part as a **territory**, a defended area. If the animal defends its entire home range, then its home range and territory are the same.

Home range is highly variable, even within a species. Seldom is a home area rigid in its use, its size, or its establishment. The home range may be compact and continuous, or it may be broken into discontinuous parts reached by trails. Irregularities in spatial and temporal distribution of food produce corresponding irregularities in home range and frequency of visitation.

Overall size of the home range varies with the available food resources, mode of food gathering, body size, and metabolic needs. In general, carnivorous animals require a larger home range than herbivorous and omnivorous animals of the same size. Males and adults

have larger home ranges than females and subadults. Weight alone is sufficient to account for differences within a species without invoking any competitive interactions. The home range of herbivores and omnivores increases at a nearly constant rate as body weight increases. Among carnivores, home range increases at a greater rate as body weight increases (Figure 14.6). The more concentrated the food supply, the smaller is the home range. Home ranges of some mammals and the center of activity within them appear to be in a constant state of flux, determined by such variables as the location of food and the reproductive condition of neighboring males and females.

The degree of aggressive behavior among individuals may limit the size of a home range. Dominant individuals hold the largest home ranges, with some overlap, while subdominant individuals occupy home ranges within those of dominant individuals. The dominant animal is able to range more freely over a larger area and have access to more resources than subdominant individuals. Usually a dominant male can control highly desirable locations, especially in relation to food and females, and force subdominant animals to move less freely in the area.

Restriction of activities to a home area confers certain advantages. Animals become familiar with the area

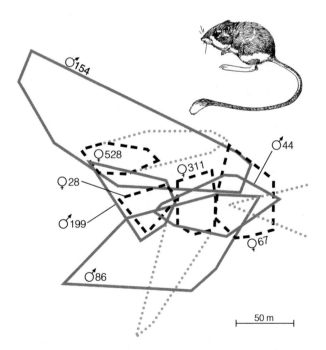

Figure 14.5 Home ranges of eight adult Merriam's kangaroo rats *(Dipodomys merriami)* in Arizona. Males are shown by solid lines, females by dashed lines. Dotted lines show excursions outside the usual home ranges. Note the overlap of all home ranges forming the greater home range. The males' ranges are larger; females' ranges lie within the home ranges of males.

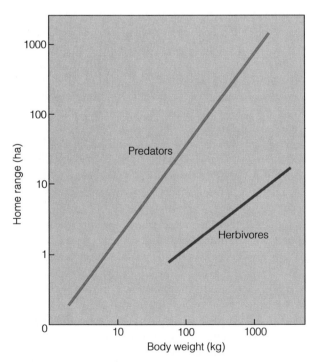

Figure 14.6 Relationship between the size of home range and body weight of North American mammals.

and the location of food, shelter, and escape cover. The animal is more secure than it would be if it were forced into unfamiliar areas.

14.9 Territoriality may regulate populations

Territoriality is a situation in which an animal defends an exclusive area not shared with rivals (Figure 14.7). Defense follows definitive behavioral patterns—song and call, intimidation displays such as spreading wings and tail in birds and baring fangs in mammals, attack and chase, and marking with scents that evoke escape and avoidance in rivals. As a result, territorial individuals tend to occur in more or less regular patterns of distribution (see Section 11.5).

Why should a cardinal or a wolf pack defend a territory? The proximate reasons vary. For some it is the acquisition and protection of food, a nesting site, or a mating area, or attraction of a mate. The ultimate reason is an increased probability of survival and reproductive success. By defending a territory, the individual forces others into suboptimal habitat, reducing their reproductive success. At the same time it increases the proportion of its own offspring in the population.

Defending a territory is costly in energy and time and can interfere with courtship, mating, feeding, and rearing young. Situations exist in which a territory is economically defensible and in which it is not. To be worthwhile, a territory must meet its owner's needs. If resources are unpredictable or patchy, it may be advantageous not to defend any space.

A feature of territorial ownership is its size. As the size of a territory increases, the cost of defense increases. In general, territory size tends to be no larger than required. That size may vary from year to year (see Figure 14.7) and from locality to locality, depending upon resources and number of animals seeking space.

For some animals, birds in particular, it is not the size of the territory that counts, but its quality. Some males, perhaps the most aggressive, claim the best territories, usually measured by features of vegetation that make them superior nesting sites. Less successful males occupy suboptimal territories, and some males secure no territories at all. The most successful males are assured of a mate, whereas male birds settled on a poor territory may be unable to attract one.

The total area available divided by the minimum size of the territory determines the number of territorial owners a habitat can support. When the available area is filled, owners evict excess animals or deny them

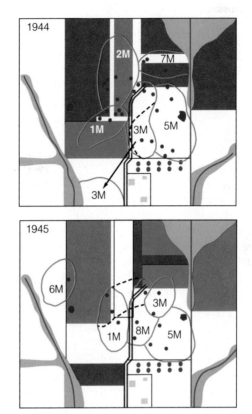

Figure 14.7 Territories of the grasshopper sparrow *(Ammodramus savannarum)*, determined by observations of male banded birds, indicated by 1M, 2M, etc. Dots indicate song perches. Note how they are distributed near the territorial boundaries. Shaded areas represent crop field. Dashed lines indicate territory shifts prior to second nesting. The return of the same males to nearly the same territory the second year is termed *philopatry*.

access. These individuals make up a floating reserve. The existence of such a reserve of potentially breeding adults has been described for a number of species, including the red grouse *(Lagopus lagopus)* in Scotland, the Australian magpie *(Gymnorhina tibicen)*, Cassin's auklet *(Ptychoramphus aleuticus)*, and the white-crowned sparrow *(Zonotrichia leucophrys)* of California. Studies of a banded population of white-crowned sparrows indicated a surplus of potentially breeding individuals.

In fact, 24 percent of the territory holders entered the population two to five years after banding. Twenty-five percent of the nestlings that acquired territories did so two to five years after hatching. Floaters quickly replaced territory holders that disappeared during the breeding season.

Few data exist on the social organization of the floaters. Floaters may form flocks with a dominance hierarchy on areas not occupied by territory holders, as do red grouse and Australian magpies. They may live singly outside occupied territories, as white-crowned sparrows do; or they may be accepted within the territory as nonrivals. They may spend much time on the breeding territories of others, as do the rufous-collared sparrows (*Zonotrichia capensis)* of Costa Rica.

Territoriality can function as a mechanism of population regulation. If all pairs that settle on an area get a territory, then territoriality only spaces a population and does not regulate it. By contrast, if territorial size has a lower limit, then the number of pairs that can settle on an area is limited, and individuals that fail to do so have to leave. In such a situation, territoriality might regulate the population, but only if an excess of nonterritorial males and females of reproductive age live in the area, as is the case for the red grouse and white-crowned and rufous-collared sparrows. Then reproduction is limited by territoriality, and we have density-dependent population regulation.

The size of home ranges and territories is of more than academic interest. It is an indicator of carrying capacity and of how much habitat is necessary to support viable populations. Animals with small home ranges or territories can be contained in smaller landscape patches than large mammals and predatory birds, the populations of which may require thousands of square kilometers. Management of such expansive habitats is a major ecological and economic problem.

14.10 Plants capture and "defend" space

Plants are not territorial in the same sense that animals are territorial; but plants can capture and hold onto space. This phenomenon in the plant world is analogous to territoriality in animals, especially if you accept the alternative definition of territoriality—individual organisms spaced out more than we would expect from a random occupancy of suitable habitat.

Plants from dandelions to trees do capture a certain amount of space and exclude individuals of their own species and other species from the same and smaller size classes. When a dandelion plant spreads its rosettes of leaves on the ground, it eliminates all other plants from the area it covers. The faster-growing trees in a forest achieve a height dominance, shading others. Plants with expansive crowns or rosettes of leaves intercept light, eliminating competition. They restrict occupancy of the ground beneath to shade-tolerant species. Because the root systems of trees more or less mirror in the ground the expanse of their crowns, dominant trees are also in a superior competitive position for nutrients and moisture. Because of their longevity, some plants, especially trees, occupy space for a long time, preventing invasion by individuals of the same or other species. Plants successful in capturing space increase their fitness at the expense of others. The survival rate of a few adults is high; others are eliminated.

Plants also capture space by the release of organic toxins that reduce competition for light, nutrients, and space. A variety of phenolic compounds released by roots and by leaves from litterfall and rain water throughfall accumulate in the soil. These compounds inhibit the germination of seeds and the growth of other plants, both herbaceous and woody (see Focus on Ecology 15.1).

14.11 Density-independent influences can be strong

We have seen that population growth and fecundity are heavily influenced by density-dependent responses. However, there are other, often overriding influences on population growth. These influences are density independent. By themselves, density-independent influences on the rate of increase do not regulate population growth. Regulation implies feedback. Density-independent influences, however, can have considerable impact on birthrates and death rates. They can mask or even eliminate density-dependent regulation.

A cold spring may kill the flowers of oaks, causing a failure of the acorn crop. Because of the failure, squirrels may suffer widespread starvation the following winter. Although the proximate cause of starvation is the density of squirrels and the low food supply, weather was the ultimate cause of the decline. In general, annual and seasonal changes in the environment tend to cause irregular population fluctuations. Conditions beyond an organism's limits of tolerance can have disastrous effects. They can affect growth, maturation, reproduction, survival, and movement. They can even eliminate local populations.

The influence of weather is irregular and unpredictable. It functions largely by influencing the availability of food. Pronounced changes in population growth often correlate directly with variations in moisture and temperature. For example, outbreaks of spruce budworm (*Choristoneura fumiferana*) are usually preceded by five or six years of drought, characterized by low rainfall and high evaporation. Outbreaks end when wet weather returns. Such density-independent effects can take place on a local scale where topography and microclimate influence the fortunes of local populations.

In desert regions a direct relationship exists between precipitation and rate of increase in certain rodents and birds. Merriam's kangaroo rat (*Dipodomys merriami*) occupies lower elevations in the Mojave Desert. The kangaroo rat has the physiological capacity to conserve water and survive long periods of aridity. However, it does require in its environment a level of moisture sufficient to stimulate the growth of herbaceous desert plants in fall and winter. The kangaroo rat becomes reproductively active in January and February when plant growth, stimulated by fall rains, is green and succulent. Herbaceous plants provide a source of water, vitamins, and food for pregnant and lactating females. If rainfall is scanty, annual plants fail to develop and the production of kangaroo rats is low. This close relationship to seasonal rainfall and success of winter annuals is also apparent in other rodents and birds occupying similar desert habitats.

CHAPTER REVIEW

SUMMARY

Density-Dependent Regulation (14.1) Populations do not increase indefinitely. As resources become less available to an increasing number of individuals, birthrates decrease, mortality increases, and population growth slows. If the population declines, mortality decreases, births increase, and population growth speeds up. Between positive and negative feedback, the population arrives at regulation.

Competition for Resources (14.2–14.3) Intraspecific competition occurs when resources are in short supply. Competition can take two forms: scramble and contest. In scramble competition, resources are shared equally. No individual receives a sufficient amount for growth and reproduction. In contest competition, dominant individuals claim sufficient resources for growth and reproduction. Others produce no offspring or perish. **(14.2)**. Competition for scarce resources can decrease or retard growth and delay reproduction in animals. Up to a point plants respond to competition by modifying form and size. After that they experience density-dependent mortality **(14.3)**.

Responses to Competition (14.4–14.5) The stress of crowding may cause delayed reproduction, abnormal behavior, increased adrenal activity, reduced ability to resist disease and parasitic infections in animals, and reduced growth and production of seeds in plants **(14.4)**. Stress might also lead animals to disperse. Dispersal is a constant phenomenon in populations at presaturation levels. Individuals seem to be genetically programmed to disperse. Many end up in submarginal habitats, but some succeed in new or unfilled habitat. A population sink may keep drawing immigrants but fail to sustain them. At saturation levels, dispersal is a response to overcrowding, and the dispersers are surplus to the population. There is no strong indication that dispersal regulates populations, but it does help them expand **(14.5)**.

Social Behavior (14.6–14.7) Intraspecific competition may be expressed through social behavior. The degree of tolerance can limit the number of animals in an area and some animals' access to essential resources **(14.6)**.

A social hierarchy is based on dominance. Dominant individuals secure most of the resources. Shortages are borne by subdominant individuals. Such social dominance may be a mechanism of population regulation **(14.7)**.

Home Range and Territoriality (14.8–14.10) Social interactions influence the distribution and movement of animals. The area an animal normally covers in its life cycle is its home range, which can overlap the home ranges of others. The size of a home range is influenced by body size, metabolic needs, food resources, and interactions within a species. Dominant individuals can strongly influence movements and home range size of subdominant individuals. If the animal or a group of animals defends a part or all of its home range as its exclusive area, it exhibits territoriality. The defended area is its territory **(14.8)**. Animals defend territories by songs, calls, displays, chemical scents, and fighting. Owners can afford a territory only if benefits exceed costs of defense. Territoriality is a form of contest competition in which a portion of the population is excluded from reproduction. These nonreproducing individuals act as a floating reserve of potential breeders, available to replace losses of territory holders. In such a manner, territoriality can act as a population-regulating mechanism (14.9).

Plants are not territorial in the same sense as animals, but they do hold on to space, excluding other individuals of the same or smaller size. Plants capture and hold space by intercepting light, moisture, and nutrients or by releasing organic toxins (14.10).

Density-Independent Effects (14.11) Density-independent influences affect but do not regulate populations. They can reduce populations to the point of local extinction. Their effects, however, do not vary with population density. Regulation implies feedback.

STUDY QUESTIONS

1. What is competition? How are scramble and contest competition important in population regulation?

2. How might density-dependent effects influence growth and size of individuals?

3. How might stress influence population growth in mammals? How do plants respond to stress?

4. How is dominance expressed in a social hierarchy? Does it regulate population size?

5. Distinguish between home range and territory.

6. How can territorial behavior affect reproductive success?

7. How can territoriality function in population regulation? What is the role of the floater?

*8. Comment on the remark, "If this woods is cut, the animals will just move elsewhere." What is the probable fate of the displaced individuals, and why?

*9. Look up the home ranges of such large mammals as the elephant, rhino, grizzly bear, and others. Discuss the problem of maintaining viable populations on such limited areas as national parks.

*10. What are some problems associated with the release of animals into unfilled or partially filled habitat?

INTERSPECIFIC COMPETITION

OBJECTIVES

On completion of this chapter, you should be able to:

- Describe the types of relationships between species.
- Contrast exploitation and interference competition.
- Describe four theoretical outcomes of inter-specific competition.
- Explain how potentially competing species may coexist.
- Explain resource partitioning.
- Relate the concept of the niche to interspecific competition.

Spotted hyaenas *(Crocuta crocuta)*, jackals *(Canus aureus)*, and vultures vie for the carcass of a wildebeest.

Although the most intense relationships exist between them, individuals of the same species do not live apart from individuals of other species. Living in close association, different species interact. They may compete for a shared resource, such as food, light, space, or moisture. One may depend upon the other as a source of food. They may provide mutual aid, or they may have no direct effect on each other at all.

15.1 Species have positive, adverse, or zero effects on each other

If we designate the positive effect of one species on another as +, a detrimental effect as −, and no effect as 0, we can express the different ways in which populations of two species interact (Table 15.1). When neither of the two populations affects the other, the relationship is (00), or neutral. If the two populations mutually benefit, the interaction is (++), or positive; the relationship is **mutualism.** If the relationship is not essential for the survival of either population, it is *nonobligatory mutualism.* If the relationship is essential for the survival of both populations, it is *obligatory mutualism* (Chapter 17). When one species maintains or provides a condition necessary for the welfare of another, but does not affect its own well-being, the relationship (+0) is **commensalism.** For example, the trunk or limb of a tree provides the substrate on which an epiphytic orchid grows. The arrangement benefits the orchid, which gets nutrients from the air and moisture from aerial roots, while the tree is unaffected.

When the relationship is detrimental to the populations of both species (−−), the interaction is **com-**

petition. In some situations, the interaction is (−0). One species reduces or adversely affects the population of another, but the affected species has no influence in return. This relationship is **amensalism.** Many ecologists consider it a form of competition.

Relationships in which one interactant benefits at the expense of the other (+−) are **predation, parasitism,** and **parasitoidism.** Predation is killing and eating prey (Chapter 16). Predation always has a negative effect on the individual. However, often at the population level predation has a mutualistic effect. Predators cull the diseased and less fit members of the prey population, thereby benefiting both predator and prey populations. In parasitism one organism feeds on the other, rarely killing it outright. The two, parasite and host, live together for some time. The host survives, although its fitness is reduced. Parasitoidism is like predation in that it kills the host eventually. Parasitoids, which include certain wasps and flies, lay eggs in or on the body of the host. When the eggs hatch, the larvae feed upon the host. By the time the larvae reach the pupal stage, the host has succumbed (Chapter 17).

15.2 Interspecific competition affects two or more species

The relationship in which the populations of both associated species are affected adversely (−−) is **interspecific competition.** In interspecific competition, as in intraspecific competition, individuals seek a resource in short supply, but they are of two or more species. Both kinds of competition may take place simultaneously. Gray squirrels, for example, compete among themselves for acorns during a poor crop year. At the same time, white-footed mice, white-tailed deer, wild turkey, and blue jays vie for the same crop. Because of competition, individuals within a species may be forced to broaden the base of their foraging efforts. Populations of various species may be forced to turn away from acorns to food less in demand. Thus intraspecific competition selects for a broadening of the use of the resource base, or generalization, whereas interspecific competition favors a reduction of the use of the resource base, or specialization.

Like intraspecific competition, interspecific competition takes two forms. **Interference competition,** like contest competition, is direct or aggressive (see Section 14.2). One competitor interferes with another's access to a resource (see Focus on Ecology 15.1: A Chemical Solution to Competition in Plants). **Ex-**

TABLE 15.1

POPULATION INTERACTIONS, TWO-SPECIES SYSTEMS		
	Response	
Type of Interaction	**A**	**B**
Neutral	0	0
Mutualism	+	+
Commensalism	+	0
Competition	−	−
Amensalism	−	0
Predation	+	−
Parasitism	+	−
Parasitoidism	+	−

A CHEMICAL SOLUTION TO COMPETITION IN PLANTS

A particular form of interference competition among plants is **allelopathy,** the production and release of chemical substances by one species that inhibit the growth of other species. These substances range from acids and bases to simple organic compounds that reduce competition for nutrients, light, and space. Produced in profusion in natural communities as secondary substances, most compounds remain innocuous. A few, however, influence community structure. For example, broom sedge *(Andropogon virginicus)* produces chemicals that inhibit the invasion of old fields by shrubs and thus maintains its dominance. Bracken fern *(Pteridium aquilinum),* the most widely distributed vascular plant in the world, produces plant poisons that accumulate in the upper surface of the soil. These phytotoxins kill the germinating seeds of many plants, especially conifers, and reduce growth of seedlings. These allelopathic effects, along with the heavy, smothering overwinter accumulation of dead fronds, allow bracken ferns to dominate large areas of ground. Likewise, the black walnut *(Juglans nigra)* of the eastern North American deciduous forest is antagonistic to many plants.

In desert shrub communities a number of shrubs *(Larrea, Franseria,* and others) release more or less toxic phenolic compounds to the soil through rain water. Under laboratory conditions, at least, these substances inhibit germination and growth of seedlings of annual herbs. In southern California two shrubs, sagebrush *(Artemisia)* and sage *(Salvia),* that commonly invade annual grasslands release aromatic terpenes such as camphor to the air. These terpenes are adsorbed from the atmosphere onto soil particles. In certain clay soils these terpenes accumulate during the dry season in quantities sufficient to inhibit the germination and growth of herb seedlings. As a result, invading patches of shrubs are surrounded by belts devoid of herbs and by wider belts in which the growth of grassland plants is reduced. Allelopathy may not be the only reason for belts devoid of vegetation. Studies of plant-animal interactions suggest that although plants do produce toxins, the absence of plants may result from grazing by hares and consumption of seeds by rodents, birds, and ants.

ploitative competition, similar to scramble competition, reduces the abundance of shared resources. Each species indirectly reduces the abundance of the other species. The outcome depends on how effectively each of the competitors uses the resource.

The concept of interspecific competition is one of the cornerstones of evolutionary ecology. Darwin based his idea of natural selection on competition, the struggle to survive. Because it is advantageous for a species to avoid it, competition has been regarded as the major force behind species divergence and specialization. Nevertheless, the role of interspecific competition in community structure is a controversial area of ecology.

15.3 There are four possible outcomes of interspecific competition

In the early part of the twentieth century two mathematicians, the American Alfred Lotka and the Italian Vittora Volterra, independently arrived at mathematical expressions to describe the relationship between two species using the same resource. Both began with the logistic equation for population growth that you may remember from Chapter 13:

$$\frac{dN}{dt} = rN\left(\frac{K-N}{K}\right)$$

Next each modified the logistic equation for each species by adding to it a coefficient to take into account the competitive effect of one species on the population growth of the other. For species 1 this coefficient is αN_2. N_2 is the number of individuals of species 2 and α is the competitive impact per individual of species 2 on species 1. This constant, in effect, converts the number of members of one species's population into an equivalent number of members of the other. Similarly, for species 2, the coefficient is βN_1. Now we have a pair of equations, which consider both intraspecific and interspecific competition:

$$\frac{dN_1}{dt} = r_1 N_1 \left(\frac{K_1 - N_1 - \alpha N_2}{K_1} \right) \qquad \text{Species 1}$$

$$\frac{dN_2}{dt} = r_2 N_2 \left(\frac{K_2 - N_2 - \beta N_1}{K_2} \right) \qquad \text{Species 2}$$

As you can see, the absence of interspecific competition—either α or $N_2 = 0$ in equation 1 and β or $N_1 = 0$ in the equation 2—the population of each species grows logistically to equilibrium at K or carrying capacity. In the presence of competition the picture changes.

For example, the carrying capacity for species 1 is K_1, and as N_1 approaches K_1 the population growth (dN_1/dt) approaches zero. However, species 2 is also vying for the limited resource that determines K_1, so we must consider the impact of species 2. Since α is the per capita effect of species 2 on species 1, the total effect of species 2 on species 1 is αN_2. The population density of species 2 that will exactly equal K_1 can be calculated as K_1/α. If the population of species 2 (N_2) is equal to K_1/α, the population of species 1 can never increase. Conversely, the population density of species 1 that exactly equals K_2 is K_2/β. If the population density of species 1 (N_1) is equal to K_2/β, the population of species 2 can never grow.

We consider both effects in calculating population growth. As the combined population effect ($N_1 + \alpha N_2$) approaches K_1, the growth rate of species 1 will approach zero. The greater the density of the competing species (N_2), the greater the influence on reducing the growth rate (dN_1/dt) of species 1. The lower the growth rate of species 1, the lower its density (N_1). The lower the density of species 1, the less its competitive effect on species 2 (βN_1). The less its competitive effect on species 2, the higher the growth rate (dN_2/dt). The higher the growth rate, the greater the negative influence of species 2 on species 1 (αN_2).

The outcome of competition, then, depends upon the relative values of K_1, K_2, α, and β. If $N_2 = K_1/\alpha$, N_1 can never increase; and if $N_1 = K_2/\beta$, N_2 can never increase. To state it differently, the presence of species 1 decreases the carrying capacity for species 2 at a certain rate. Similarly, species 2 decreases the carrying capacity for species 1 at a certain rate. The reason, of course, is that each species has to share limited resources with the other.

Depending upon the combination of values for the Ks and for α and β, the Lotka-Volterra equations predict four different outcomes. In two situations one species wins out over the other. In one case species 1 inhibits further increase in species 2 while continuing

to increase itself. In this case species 2 is driven to extinction. In the other case species 2 inhibits further increase in species 1 while continuing to increase itself, and species 1 disappears. In the third situation each species, when abundant, inhibits the growth of the other species more than it inhibits its own growth. Both species hang on in an unstable equilibrium. In the long run one wins. The outcome depends upon which species is the most abundant and upon which adapts better to environmental change. In the fourth situation neither population can achieve a density capable of eliminating the other. Each species inhibits its own population growth more than that of the other species. Figure 15.1 depicts all the possible outcomes with graphs.

In each graph the x axis represents the population size of species 1 and the y axis the population size of species 2. Two lines are plotted on each graph, one representing each of the two species. The diagonal line for species 1 represents the combined population densities of species 1 and 2 that equal K_1 and therefore $dN/dt = 0$. The line for species 2 is the combination of $N_1 + N_2$, equal to K_2. For any point on the species 1 line, $N_1 + N_2 = K_1$. When $N_1 = K_1$, then N_2 must be zero. When $N_2 = K_1/\alpha$, then N_1 must be equal to zero. For the line describing species 2, the values where the line crosses the axes will be K_2 and $N_1 = K_2/\beta$.

The diagonal lines are called zero growth isoclines. The zero growth isoclines for species 1 and 2 are shown in Figures 15.1a and 15.1b, respectively. In the space below the line, or isocline, for species 1 (Figure 15.1a), combinations of N_1 and N_2 are below carrying capacity; therefore, the population is increasing (dN_1/dt greater than zero). This increase is represented by the arrows that are parallel to the x-axis and point in the direction of increasing values of N_1. In the space above the isocline, combinations of N_1 and N_2 are greater than carrying capacity; therefore, the population is declining (dN_1/dt is negative). In this region the arrows point in the direction of decreasing population size. Figure 15.1b depicts the analagous situation for N_2.

Figures 15.1c–f represent the possible outcomes when the two isoclines are combined. In Figure 15.1c, the isocline of species 1 is parallel to, and lies outside, the isocline of species 2. In this case, even when the population of species 2 is at its carrying capacity (K_2), its density cannot stop the population of species 1 from increasing. As species 1 continues to increase, species 2 eventually becomes extinct. In 15.1d, the situation is reversed. Species 2 wins, leading to the exclusion of species 1.

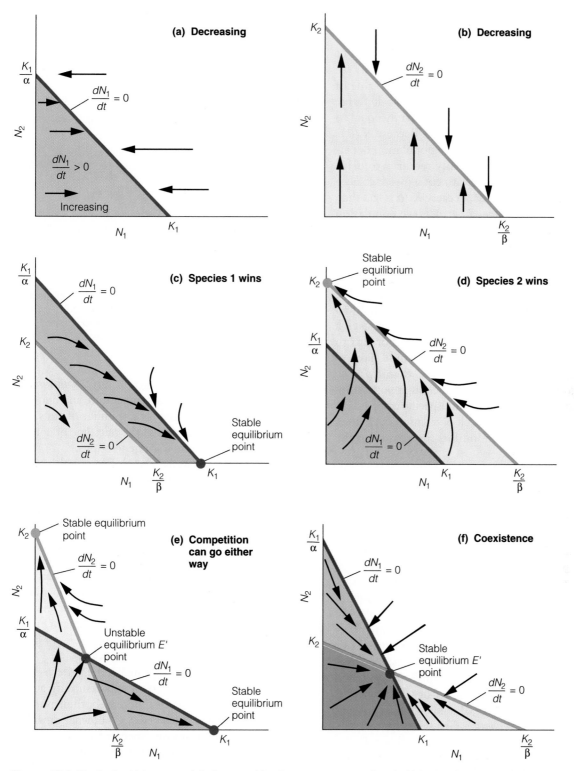

Figure 15.1 The Lotka-Volterra model of competition between two species. (a–b) In the absence of competition, populations of species 1 and 2 come to equilibrium at some point along the diagonal isocline of zero population growth ($dN/dt = 0$). In the shaded area below the line the population growth is positive and the population increases (as indicated by the arrows), whereas above the line population decreases. (c) The isocline of species 1 falls outside the isocline of species 2. Species 1 always wins, leading to the extinction of species 2. (d) The situation is the reverse of (c). (e) The isoclines cross. Each species inhibits the growth of the other more than its own growth. The more abundant species often wins. (f) Each species inhibits the growth of its own population more than that of the other by intraspecific competition. The species coexist.

In Figures 15.1e and 15.1f, the isoclines cross, but the outcomes of competition are quite different. In Figure 15.1e, note that along the x axis, the value of K_2/β is less than the value of K_1. Recall that the value $N_1 = K_2/\beta$ is the density of species 1 exactly equal to the carrying capacity of species 2 (K_2). Because K_2/β is less than K_1, species 1 can achieve population densities that would exceed the density required to drive the population of species 1 to extinction (produce negative growth rates). Likewise, because K_1/α is less than K_2 along the y axis, species 2 can achieve population densities high enough to drive species 1 to extinction. Which species "wins" the competition depends upon the initial populations of the two species.

In Figure 15.1f, the result of competition differs in that neither species can exclude the other, leading to coexistence. Note that in this case, K_1 is less than K_2/β. The population of species 1 can never reach a density sufficient to eliminate species 2. For this to happen, the population of species 1 would have to reach $N_1 = K_2/\beta$. Likewise, K_2 is less than K_1/α, so species 2 can never achieve a high enough population density to eliminate species 1. Intraspecific competition (K) inhibits the growth of each population more than interspecific competition inhibits the growth of the other species.

15.4   Laboratory experiments support the Lotka-Volterra equations

The theoretical Lotka-Volterra equations stimulated studies of competition in the laboratory, where under controlled conditions the outcome is more easily determined than in the field. One of the first to study the Lotka-Volterra competition model experimentally was the Russian biologist G. F. Gause. He used two species of *Paramecium*, *P. aurelia* and *P. caudatum*. *P. aurelia* has a higher rate of increase than *P. caudatum* and can tolerate higher population density. When he introduced both into one tube containing a fixed amount of bacterial food, *P. caudatum* died out (Figure 15.2). In another experiment, Gause reared the loser, *P. caudatum*, with another species, *P. bursaria*. These two species coexisted, because *P. caudatum* fed on bacteria suspended in solution, whereas *P. bursaria* confined its feeding to bacteria at the bottom of the tubes. Each used food unavailable to the other.

In the 1940s and 1950s Thomas Park at the University of Chicago conducted a number of classic competition experiments with laboratory populations of flour beetles. He found that the outcome of competi-

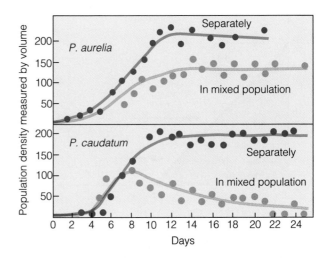

Figure 15.2 Competition experiments with two ciliated protozoans, *Paramecium aurelia* and *P. caudatum,* grown separately and in a mixed culture. In a mixed culture *P. aurelia* outcompetes *P. caudatum,* and the result is competitive exclusion.

tion between *Tribolium castaneum* and *T. confusum* depended upon environmental temperature and humidity and on fluctuations in the total number of eggs, larvae, pupae, and adults. Often the outcome of competition was not determined for generations.

In a much later study, David Tilman of the University of Michigan and his associates grew laboratory populations of two species of diatoms, *Asterionella formosa* and *Synedra ulna*. Both of these species require silica for the formation of cell walls. The researchers monitored not only population growth and decline but also the level of silica. When grown alone in a liquid medium to which the resource (silica) was continually added, both species kept silica at a low level. When grown together, *S. ulna* took silica to a level too low for *A. formosa* to survive and reproduce (Figure 15.3). By reducing resource availability, *Synedra* drove *Asterionella* to extinction.

In these laboratory experiments the patterns were similar. The two species grew exponentially at low densities. As their densities increased, their population growth declined. Each began to influence the other. Under these conditions one species had a greater fitness than the other, which declined toward extinction.

15.5   Studies suggest the competitive exclusion principle

In three of the four situations predicted by the Lotka-Volterra equations, one species drives the other to extinction. The results of laboratory studies tend to

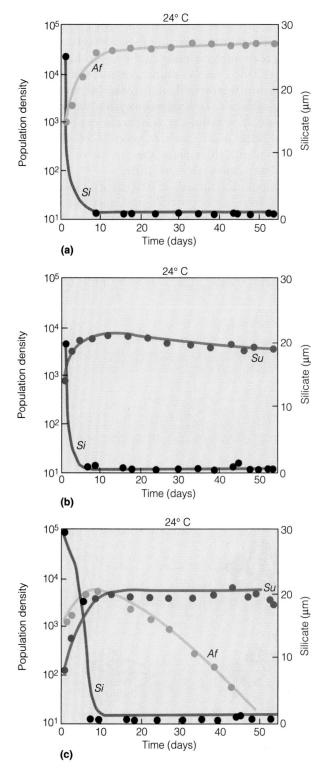

Figure 15.3 Competition between two species of diatom, *Asterionella formosa* (Af) and *Synedra ulna* (Su), for silica. (a–b) Grown alone in a culture flask, both species reach a stable population that keeps silica (Si) at a constant low level. *Synedra* draws silica lower. (c) When the two are grown together, *Synedra* reduces silica to a point at which the other species dies out.

support the mathematical models. These observations have led to the concept called Gause's principle, although the idea was far from original with him. More recently called the **competitive exclusion principle,** it states that complete competitors cannot coexist. Basically, if two noninterbreeding populations possess exactly the same ecological requirements and live in exactly the same place, and if population A increases the least bit faster than population B, then A eventually will occupy the area completely. B will become extinct.

Competitive exclusion, then, invokes more than competition for a limited resource. The competitive exclusion principle assumes that both competitors remain genetically unchanged, that immigrants from other areas with different environmental conditions do not move into the population of the losing species, and that environmental conditions remain constant. Such conditions rarely exist. Because they are not ecologically identical, two or more species can compete for an essential resource without being complete competitors (see Section 15.12). The idea of competitive exclusion, however, has stimulated a more critical look at competitive relationships in natural situations, including studies to determine what ecological conditions are necessary for coexistence of species.

15.6 Competition experiments in the field are difficult

Demonstrating interspecific competition in laboratory "bottles" is one thing; demonstrating competition under natural conditions is another. In the field, you (1) have no control over the environment; (2) have difficulty knowing whether the populations are at or below carrying capacity; (3) lack full knowledge of the life-history requirements or the subtle differences between the species. Further, the competitive outcome among associated species may have already been settled through evolution.

In reviews of more than 100 field experimental studies involving caged or enclosed populations of assumed competitors, the importance of competition in controlling populations was not universally demonstrated. However, results do suggest that interspecific competition has a large overall effect that varies widely among organisms. For example, the studies show strong competition among toads, frogs, and arthropods of flowing water. Among herbivores interspecific competition is less significant than intraspecific competition in controlling populations. Among most organisms the effects of intraspecific and interspecific competition are equally strong.

15.7 Exclusion may involve more than competition

The better field studies of exclusion are not experimental but observational. Many studies observe the impact of introduced species on native fauna and flora (see Focus on Ecology 15.2: Invasion of the Aliens). In the United States the aggressive common starling and the house sparrow compete strongly with the cavity-nesting bluebird and flicker for nesting sites and drive the native species away. (A sharp reduction in the populations of the two invaders has considerably lessened the competition.) An introduced, highly competitive fish, the carp *(Cyprinus carpo)*, present in virtually every type of freshwater habitat, easily outcompetes other fish, such as bass and trout, for both space and food. Uprooting and consuming aquatic plants as they feed, they eliminate spawning cover for other fish and cause nesting fish to desert the nest. They create such turbidity in the water that visually foraging fishes, such as sunfish, cannot see. The colorful, exotic weedy wetland perennial from Europe, purple loosestrife *(Lythrum salicaria)*, has invaded prime wetlands throughout the temperate regions of the United States and Canada. It has crowded out native wetland plants, threatening the survival of the plants and animals dependent upon them.

Chipmunks furnish a striking example of exclusion. In this case physiological tolerance, aggressive behavior, and restriction to habitats in which one organism has competitive advantage all play a part. On the eastern slope of the Sierra Nevada Mountains live four species of chipmunks: the alpine chipmunk *(Eutamias alpinus)*, the lodgepole pine chipmunk

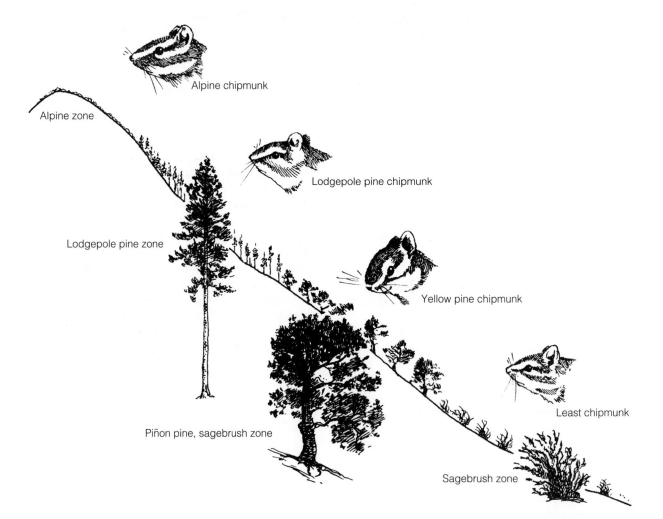

Figure 15.4 A transect of the Sierra Nevada Mountains in California, 38° north latitude, showing vegetation zonation and the altitudinal ranges of four species of chipmunks *(Eutamias)* on the east slope.

(E. speciosus), the yellow pine chipmunk *(E. amoenus)*, and the least chipmunk *(E. minimus)*, at least three of which have strongly overlapping food requirements. Each species occupies a different altitudinal zone (Figure 15.4).

The line of contact is determined partly by interspecific aggression. Aggressive behavior by the dominant yellow pine chipmunk determines the upper range of the least chipmunk. Although the least chipmunk can occupy a full range of habitats from sagebrush desert to alpine fellfields, it is restricted in the Sierras to sagebrush habitat. Physiologically, it is more capable of handling heat stress than the others, enabling it to inhabit extremely hot, dry sagebrush. If the yellow pine chipmunk is removed from its habitat, the least chipmunk moves into vacated open pine woods. However, if the least chipmunk is removed from the sagebrush habitat, the yellow pine chipmunk does not invade the habitat. The aggressive behavior of the lodgepole pine chipmunk in turn determines the upper limit of the yellow pine chipmunk. The lodgepole pine chipmunk is restricted to shaded forest habitat because it is vulnerable to heat stress. Most aggressive of the four, the lodgepole pine chipmunk also may limit the downslope range of the alpine chipmunk. Thus, the range of the least chipmunk is determined both by aggressive exclusion and by its ability to survive and reproduce in habitat hostile to the other species. The other species, too, are restricted to certain habitats by their own physiology and by aggressive exclusion.

Plant defenses may influence the competitive outcome among some herbivores. Lambs-quarter *(Chenopodium* spp.) is attacked by sap-feeding aphids that never actually come in contact with each other, although they share the same resource. One species, *Hayhurstia atriplicis*, feeds above ground, where it forms leaf galls. The other species, *Pemphigus betae*, feeds underground on the roots. *Pemphigus* has no apparent effect on its host. *Hayhurstia*, however, can reduce the biomass of the host by 54 percent and seed set by 60 percent. Some lambs-quarters are resistant to leaf galling by *Hayhurstia*; others are not. N. A. Moran and T. G. Whigham found that on susceptible plants, *Hayhurstia* reduced the number of *Pemphigus* by an average of 91 percent and often eliminated them entirely. Thus on susceptible plants, *Pemphigus* experiences competitive exclusion (Figure 15.5). On plants resistant to galling, colonies of *Hayhurstia* were smaller and did not affect *Pemphigus* infesting the roots. Although *Hayhurstia* had a negative effect on *Pemphigus*, the latter had no measurable impact on *Hayhurstia*, which suggests amensalism.

In these examples competitive exclusion does not fit the theoretical framework described by the Lotka-Volterra equations. Instead exclusion is determined by physiological tolerances and by aggressive behavior.

15.8 A variable environment upsets equilibrium

One species excludes another when it exploits a shared limiting resource more effectively, resulting in an increase in its population at the expense of its competitor. However, as we saw in Chapter 2, each organism does best under particular conditions. Low temperature, for example, may favor one competitor. If temperatures rise from time to time, the advantage may shift to another species. When environmental conditions vary, the competitive advantage may shift. As a result no one species will reach sufficient density to displace its competitors.

Variations in a shared limiting resource influence carrying capacities directly. Variations in other factors affecting growth, survival, or reproduction change the competition as well. In this manner, environmental variations allow the coexistence of competitors, where under constant conditions, one would exclude the other.

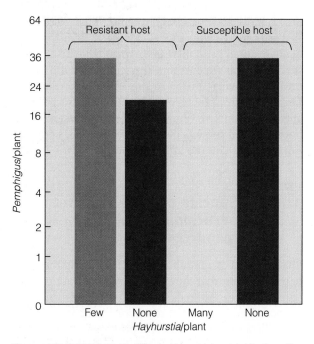

Figure 15.5 Exclusion of *Pemphigus betae* by *Hayhurstia atriplicis* in a screen cage experiment. The same competitive exclusion occurs in natural stands.

15.9   Coexistence limits the population of each species

Two or more competing species can coexist, although such competition reduces the fitness of all parties. Among some animals, notably birds, competing species may behave as if they were the same species. This interspecific territoriality reduces the number of breeding individuals of each species that occupy a given area of shared habitat.

Competing species belonging to different taxonomic groups also may coexist. Consider the competitive relationship between rodents and ants in the desert scrub of Arizona. Both groups feed on seeds of desert plants; both overlap considerably in the size of seeds they consume. Desert rainfall strongly influences the production of those seeds. A strong correlation between the number of species of rodents and ants and mean annual rainfall suggests that ants and rodents may engage in exploitative competition for the limited amount of seeds.

James Brown and Diane Davidson undertook field experiments to determine to what extent rodents and ants compete for a limited seed supply. They established six experimental plots. From two of the plots they removed rodents by trapping and excluded immigrants by fencing. From two plots they eliminated the ants by repeated applications of an insecticide. From the remaining two plots that were to serve as controls, they removed both ants and rodents. They discovered that ants and rodents indeed compete. Removing rodents resulted in an increase in ants, who ate as many seeds alone as rodents and ants combined did previously. Removing ants had a similar effect on rodents. On the control plots from which they removed both ants and rodents, the amount of seeds increased. Thus, both ants and rodents depressed each other's populations and perhaps the fitness of the seed plants as well.

15.10   Species coexist by partitioning available resources

Observations of a number of species sharing the same habitat suggest that they coexist by partitioning available resources. Animals use different kinds and sizes of food, feed at different times, or forage in different areas. Plants occupy a different position on a soil moisture gradient, require different proportions of nutrients, or have different tolerances for light and shade. Each species exploits a portion of the resources un-

available to others. We regard such partitioning as an outcome of interspecific competition.

Consider a species that feeds on seeds. The pattern of seed selection can be expressed as a bell-shaped curve on a graph with seed size as the x axis and proportion of total diet as the y axis (Figure 15.6a). In the absence of any competitor, species 1 uses a range of seeds of different sizes. Most of the diet is composed of intermediate-size seeds, with fewer large and small seeds selected. As population size increases, the range of seeds taken may expand, as intraspecific competition forces some individuals to seek food at the extremes. Such intraspecific competition fosters expanded use of the resource base.

When a second species, 2, enters the area, 1 and 2 show considerable overlap in the size of seeds they select. Interspecific competition forces both species 1 and species 2 to narrow their seed size range. Competition will favor the selection of seeds in the range of sizes not available to the competing species. Ultimately, the two species will narrow their ranges of resource use (Figure 15.6b). They will diverge, moving to the left and to the right on the graph. Direct interspecific competition will be reduced and the two species will coexist. Thus, whereas intraspecific competition favors expanded use of the resource base, interspecific competition narrows it.

Enter a third species, 3, which selects seeds in the narrow area of overlap between 1 and 2 (Figure 15.6c). By selecting the largest proportion of seeds in that part of the gradient, species 3 will force 1 and 2 to restrict further the range of resources they use (Figure 15.6d). The result is a partitioning of resources along the resource gradient. Such a group of functionally similar species whose members interact strongly with each other, exploiting the environment in a similar manner, is called a **guild.**

Field studies provide many examples of apparent resource partitioning. One example involves three species of annual plants growing together on prairie soil abandoned one year after plowing. Each plant exploits a different part of the soil resource (Figure 15.7). Bristly foxtail *(Setaria faberii)* has a fibrous, shallow root system that draws on a variable supply of moisture. It recovers rapidly from drought, takes up water rapidly after a rain, and carries on a high rate of photosynthesis even when partially wilted. Indian mallow *Abutilon theophrasti)* has a sparse branched taproot extending to intermediate depths, where moisture is adequate during the early part of the growing season but is less available later on. The plant is able to carry on photosynthesis at a low water potential (see Section 7.6). The

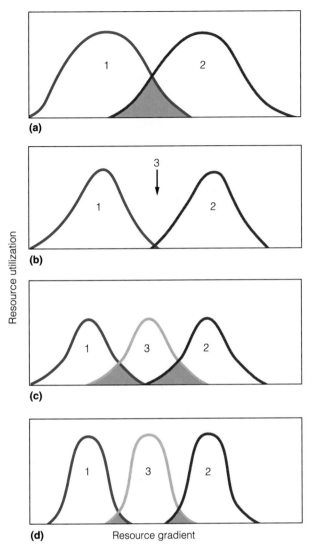

Figure 15.6 Theoretical resource gradient used by three competing species 1, 2, and 3. (a) 1 and 2 compete for the same resource—for example, food—along a segment of the resource gradient (such as food size). (b) As a result of competitive interaction, 1 and 2 narrow the range of the food resource they use. A third species (3) arrives, whose optimal use of the food resource lies in that area 1 and 2 use at less than optimal level. (c) 1 and 2 now compete with 3 along a certain segment of the food resource gradient. (d) Competition of 3 with 1 and 2 force all three species to narrow the range of their resource use to their optimal food size.

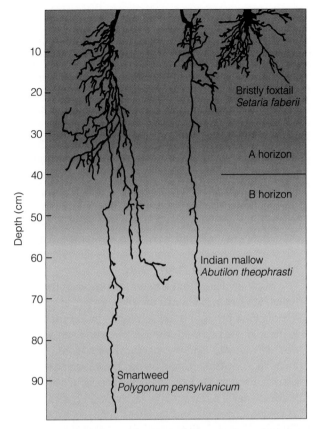

Figure 15.7 Partitioning of the prairie soil resource at different levels by three species of annual plants one year after disturbance.

avoids direct competition with the others. The different rooting patterns may have evolved in response to severe competition for resources in the past, although this hypothesis is difficult to test. Nevertheless, ecologists regard competition as instrumental in directing the evolutionary divergence of species (Chapter 19).

15.11 The niche relates to interspecific competition

Almost inseparable from the concept of interspecific competition is the niche. The niche, you will recall, was introduced in Chapter 2. There we defined it in terms of the adaptation of an organism to its total environment. We did not stress the relationship of one organism's niche to that of another. Because an organism's niche involves the full range of conditions and resources within which it can survive and reproduce, it will come into contact with or infringe on the niches

third species, smartweed *(Polygonum pensylvanicum)*, possesses a taproot that is moderately branched in the upper soil layer and develops mostly below the rooting zone of other species, where it has a continuous supply of moisture.

This pattern of soil use is an example of resource partitioning. By dividing the resource, each species

INVASION OF THE ALIENS

Aliens—nonnative plants and animals—have invaded every continent and nearly every island on Earth. They come in all forms: viruses, bacteria, fungi, seed plants, insects and other invertebrates, amphibians, fish, birds, and mammals. Many arrive insidiously, without notice. They come hidden as viruses and bacteria inhabiting living plants and animals, as stowaways on ships' ballast, in agricultural products and exotic timbers. Humans introduce many nonnative species, including exotic garden plants, aquarium fish and plants, fish for stocking in natural waters, and game animals. Escaping their confines, these nonnative species begin their often rapid invasion of the country.

Such invasions have been taking place for centuries. Stowaway black rats on ships coming in from the Middle East brought bubonic plague to Europe in the fourteenth century. European explorers and settlers carried diseases to Native Americans who were not immune to them. Sailing ships carried brown rats to remote islands. Escaping to land, they multiplied and destroyed predator-naive birds and mammals. Alien plants, such as purple loosestrife and Japanese honeysuckle (*Lonicera japonica*), are displacing native plants. In a world shrunken by rapid transportation and widespread movement of humans, invaders have been increasing at an unprecedented rate, destroying populations of native species and altering ecosystems to their own needs.

Not all would-be invaders are successful, but those that are gain an early foothold. Their local population remains low and undetected for a long period; then they suddenly expand and move out from their beachhead. The invaders are mostly *r*-strategists. They reproduce rapidly, producing a large number of spores, seeds, and young that are excellent dispersers. Scattered by the wind, carried by insects, birds, and mammals, and transported on automobiles, trucks, trains, and riverboats, the invaders take advantage of disturbed sites and empty niches. If the abiotic environment is appropriate for the invaders, they are likely to succeed, regardless of what biota is present.

Once established in an area, the invaders displace native species. The Great Lakes have been invaded by 139 nonnative aquatic species that are changing ecological relationships in them. The San Francisco Bay area is occupied by 96 nonnative invertebrates, from sponges to crustaceans. Exotic fish, introduced purposefully or accidentally, have been responsible for 68 percent of fish extinctions in North America during the past 100 years and for the decline of 70 percent of the fish species listed as endangered.

Other invaders have changed whole ecosystems. The chestnut blight, introduced through imports of the Chinese chestnut, has all but exterminated the chestnut (*Castanea mollissimi*), the dominant deciduous tree in the eastern hardwood forests of North America, and converted the oak-chestnut forest ecosystem to oak. The gypsy moth, accidentally escaped from a lot brought into Massachusetts for experimental silkworm production, has spread through northeastern oak forests, endangering oaks as commercial timber species and displacing many moths native to hardwood forests. Multiflora rose (*Rosa multiflora*), introduced and widely planted as a living fence in the 1950s, now covers pastures and forest edges in the eastern United States.

The most notorious recent invader is the zebra mussel (*Driessena polymorpha*). A European species that arrived in the ballast waters of ships, it was discovered in 1988 in Lake St. Claire, Ontario, north of Lake Erie. By 1997 the zebra mussel had spread to each of the Great Lakes and the major river systems of the United States. Its exploding populations are having a detrimental effect on aquatic ecosystems. The zebra mussel attaches itself to native mussels, covering them to the point of suffocation. Feeding on plankton, the zebra mussel, with its tremendous water-filtering capacity, reduces the population upon which larval fish such as alewife and smelt depend. Their filtering clarifies water so much that it increases the growth of aquatic plants.

The zebra mussel also is causing great economic damage. Attaching themselves to any hard surface, zebra mussels clog water intake pipes of power plants, water treatment plants, and industrial facilities, costing millions of dollars to replace and maintain. They foul boat hulls and engine cooling systems. Hitchhiking on the hulls of barges, recreational boats, and boat trailers, and drifting downstream with the current, the zebra mussel has the potential of affecting two-thirds of the nation's waterways ecologically and economically for years to come.

of others. This infringement might result in competitive interactions.

An organism free from interference from another species could use the full range of conditions and resources to which it is adapted. We call this range the species's **fundamental niche.** Competition from other species often restricts a species to a portion of its fundamental niche. The portion of the fundamental niche that a species actually exploits in the presence of competitors is its **realized niche** (Figure 15.8). Like the fundamental niche, the realized niche is an abstraction. In their studies, ecologists usually confine themselves to examining one or two niche dimensions, such as feeding behavior or tolerance of humidity (see Section 2.7).

The fundamental and realized niches of an organism can change with different stages of development. Insects with a complex life cycle occupy one niche as larvae and an entirely different niche as adults. Caterpillars, the larval form of butterflies, feed on the leaves of specific plants. The adult butterflies feed on the nectar of flowers.

Niche overlap occurs when two or more organisms use a portion of the same resource, such as food, simultaneously (Figure 15.9). The amount of niche

overlap is assumed to be proportional to the degree of competition for that resource. Niche overlap, however, does not always mean high competitive interaction. In fact, resources may not be in short supply. Extensive niche overlap may indicate that little competition exists and resources are abundant.

For simplicity, we think of a niche as one-dimensional. In reality, a niche includes many types of resources: food, a place to feed, cover, space, and so on. Rarely do two or more species possess exactly the same combination of requirements. Species may overlap on one gradient but not on another. The total competitive interaction may be less than that suggested by the niche overlap on one gradient alone (Figure 15.10).

Niche differentiation can also occur among individuals within a species. Division of feeding space, food size, and other niche components may exist between sexes of the same species. The male red-eyed vireo (*Vireo olivaceus*), for example, obtains its insect food in the upper canopy, where he sings, the female in the lower canopy and near the ground, where she nests (Figure 15.11).

When we plot the range of resources an organism uses, the observed range represents the **niche width** (also called niche breadth and niche size). Theoretically, niche width is the sum total of the different resources used by an organism. On the practical side, measurement of a niche usually involves the measure of a single ecological variable such as food size or habitat space (see Figure 2.11).

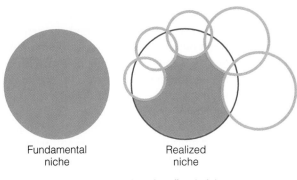

Figure 15.8 Fundamental and realized niches.

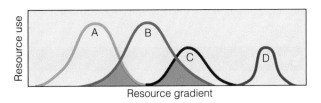

Figure 15.9 Types of niche overlap. Species A and B have overlapping niches of equal breadth. B and C have overlapping niches of unequal breadth. C and D have adjacent niches. D has a narrow niche, suggestive of a specialist species.

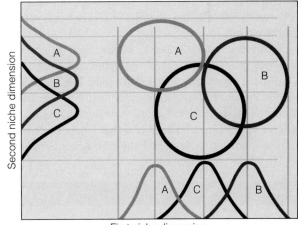

Figure 15.10 Niche relationships based on two gradients. Species may exhibit considerable overlap on one gradient and little or none on the other. When we sum niche dimensions (in the circles), we find niche overlap reduced considerably.

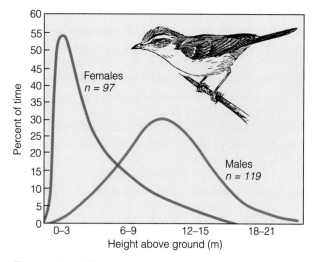

Figure 15.11 Niche separation of male and female red-eyed vireos *(Vireo olivaceus)* by height of foraging. Mean height for males is 11 m; mean height for females is 4 m.

Niche widths are usually described as narrow or broad. Generalist species, such as the fox with its broad food habits (see Figure 24.3), have wide niches. The wider the niche, the more generalized the species is considered to be. The narrower the niche, the more specialized is the species. For example the accipiter hawks specialize in preying on birds.

15.12 Organisms compress, expand, or shift their niches

As Figure 15.6 showed, intense competition may force species to restrict their use of space, range of foods, or other resource-oriented activities. This change is **niche compression.**

Conversely, if interspecific competition is reduced, a species may expand its realized niche, using space or a resource previously unavailable to it. Niche expansion in response to reduced interspecific competition is **competitive release.** Competitive release may occur when a species invades an island that is free of potential competitors, moves into habitats it never occupied

on a mainland, and increases its abundance. Such expansion may also follow when a competing species is removed from a community, allowing remaining species to move into microhabitats they previously could not occupy.

Instead of expanding or compressing their niches, two or more species may reduce competition by adjusting their use of the shared niche. This adjustment or **niche shift** may involve basic behavioral or morphological traits.

E. E. Werner and J. A. Hall demonstrated niche shift in three competing sunfish (Centrarchidae): the bluegill *(Lepomis macrochirus)*, the pumpkinseed *(L. gibbosus)*, and the green sunfish *(L. cynellus)*. When the biologists stocked the three species in separate experimental ponds, the food and habitat preferences of the fish were similar, and their average growth rate increased. All three preferred the emergent and submerged vegetation about the edges of ponds and the large food they found there.

When the three species were stocked together in equal densities, they all occupied the vegetated zone about the edges. As food resources declined, the bluegill and pumpkinseed left, leaving behind the green sunfish, a more efficient forager, in the vegetation. Bluegills and pumpkinseeds turned their foraging efforts to bottom invertebrates, mainly larval Chironomidae (midges). The pumpkinseed—with its widely spaced gill rakers that do not become fouled when the fish sorts through bottom sediments—was better able to exploit that food supply. The bluegill—with long, fine gill rakers that hold onto small prey—then shifted its foraging to open water.

The three fish also shifted their use of the pond habitat. When the biologists confined bluegill and green sunfish together in equal densities in the pond, the green sunfish eventually forced the bluegill to move to open water. In the absence of green sunfish, the bluegills moved back into the dense vegetation. Bluegill and pumpkinseed have a dietary overlap of 50 to 55 percent. When the two are together in a pond, the bluegill again is forced to move to the open water.

CHAPTER REVIEW

SUMMARY

Species Relationships (15.1) Relationships between species are positive (+), negative (−), or neutral (0). In mutualism (++) populations benefit each other; in competition (−−) both populations are affected adversely; in commensalism (0+) one population benefits and the other is unaffected; and in predation and parasitism (+−) one population benefits and the other is harmed.

Competition Models (15.2–15.3) In interspecific competition, individuals of two or more species seek a resource in short supply, reducing the fitness of both.

Exploitative competition depletes resources to a level of little value to either population. Interference competition involves aggressive interactions **(15.2)**.

The Lotka-Volterra equations describe four possible outcomes of interspecific competition. Species 1 may win over species 2; species 2 may win over species 1. Both of these outcomes represent competitive exclusion. The other two outcomes involve coexistence. One is unstable equilibrium, in which the potential winner often is the one most abundant at the outset. A final possible outcome is stable equilibrium, in which two species coexist, but at a lower population level than if each existed in the absence of the other **(15.3)**.

Tests of Competition (15.4–15.7)   Laboratory experiments with species interactions support the Lotka-Volterra model **(15.4)**. Results of the experiments led to formulation of the competitive exclusion principle—two species with exactly the same ecological requirements cannot coexist. This principle has stimulated critical examinations of competitive relationships outside the laboratory, especially how species coexist and how resources are partitioned **(15.5)**. Although some field experiments have suggested strong competitive interactions between two species, they fail to demonstrate interspecific competition explicitly **(15.6)**. Better examples of interspecific competition are observational. Some involve competition between invading exotic species and native species of flora and fauna **(15.7)**.

Coexistence (15.8–15.10)   Interspecific competition often results in an unstable equilibrium among species. Environmental variability may give each species a temporary advantage, promoting coexistence **(15.8)**. Sometimes species coexist by sharing resources, which limits the population size of each **(15.9)**. Many species that share the same habitat coexist by partitioning available resources. Groups of functionally similar species that share a spectrum of resources and interact strongly with each other are guilds **(15.10)**.

Niche (15.11–15.12)   Closely associated with interspecific competition is the concept of the niche. Basically, a niche is the functional role of an organism in the community. That role might be constrained by interspecific competition. In the absence of competition, an organism occupies its fundamental niche. In the presence of interspecific competition, the fundamental niche is reduced to a realized niche, the conditions under which an organism actually exists. When two different organisms use a portion of the same resource, such as food, their niches overlap. Overlap may or may not indicate competitive interaction. The range of resources used by an organism suggests its niche width. Species with broad niches are generalists, whereas those with narrow niches are specialists **(15.11)**.

A species compresses its niche when competition forces it to restrict its type of food or habitat. In the absence of competition, the species experiences competitive release and expands its niche. Organisms may also undergo niche shift, changing their behavior to reduce interspecific competition **(15.12)**.

STUDY QUESTIONS

1. Define commensalism, predation, parasitism, amensalism, mutualism, and competition. Use +, −, and 0.

2. What is interspecific competition, and what forms does it take?

3. Describe four outcomes of competition based on the Lotka-Volterra competition equations.

4. What is competitive exclusion? What is the problem with the concept?

5. What determines the outcome of unstable equilibrium in natural situations?

6. What permits coexistence among competing species?

7. What is allelopathy? Why is it a form of competition?

8. What is resource partitioning? What are some weaknesses of the concept?

9. Define niche. Distinguish between a fundamental niche and realized niche.

10. What is niche overlap? Niche width? How is niche width usually measured?

11. What is meant by niche compression, competitive release, and niche shift?

*12. Why are island flora and fauna so vulnerable to invasion by nonnative species? Discuss in terms of interspecific competition and competitive exclusion. Consider the Hawaiian islands as an example.

C H A P T E R   1 6

PREDATION

OBJECTIVES

On completion of this chapter, you should be able to:

- Define predation and distinguish among its forms.
- Sketch the Lotka-Volterra model of predation and discuss its weaknesses.
- Explain functional response, numerical response, and the three forms each can take.
- Define search image and switching and their importance to predation.
- Explain optimal foraging.
- Describe plant-herbivore and herbivore-carnivore systems.

Brown bears *(Ursus arctos)* and gulls prey on migrating salmon in an Alaskan river.

The poet who described nature as "red in tooth and claw" was thinking of predation. Most of us associate predation with a hawk taking a mouse or a wolf killing a deer, a form of predation called **carnivory.** That is a narrow view. A fly laying its eggs on a caterpillar to develop there at the expense of its victim is exhibiting a form of predation called **parasitoidism.** The parasitoid attacks the host (the prey) indirectly by laying its eggs on the host's body. When the eggs hatch, the larvae feed on the host, slowly killing it. In **parasitism,** too, a predator lives on or within the host (see Chapter 17). The parasite feeds on its host but seldom kills it outright (which, in effect, would destroy both its habitat and its food). A deer feeding on shrubs and grass and a mouse eating a seed are practicing a form of predation called **herbivory.** Seed consumption is outright predation because the embryonic plant is killed. Grazing on plants without killing them is a form of parasitism. When they kill the plant outright, grazers become true predators of plants. A special form of predation is **cannibalism,** in which the predator and the prey are the same species.

Predation is more than a transfer of energy. It is a direct and often complex interaction of two or more species, the eater and the eaten. The numbers of some predators may depend upon the abundance of their prey. The population of the prey may be controlled by its predator. Each can influence the population growth of the other and favor new adaptations.

16.1   Mathematical models describe the basics of predation

In the 1920s Lotka and Volterra turned their attention to the effects of predation on population growth. Independently, they proposed mathematical statements to express the relationship between predator and prey populations. They provided one equation for the prey population and another for the predator population.

The population growth equation for the prey population consists of two components: the maximum rate of increase per individual and the removal of the prey from the population by the predator:

$$\frac{dN_{prey}}{dt} = rN_{prey}\left(1 - \frac{N_{prey}}{K} - aN_{pred}N_{prey}\right)$$

where $aN_{pred}N_{prey}$ is the number of kills, which depends upon the number of predators (N_{pred}), the number of prey (N_{prey}), and the efficiency of predators (a).

The equation for the predator population is:

$$\frac{dN_{pred}}{dt} = b(aN_{pred}N_{prey}) - dN_{pred}$$

where b is the efficiency of converting kills ($aN_{pred}N_{prey}$) to new predators, and d is the death rate of predators in the absence of prey. The Lotka-Volterra equations assume that the prey population grows exponentially and that reproduction in the predator population depends upon the number of prey consumed.

These paired equations, when solved, show that as a single predator population increases, the single prey population decreases to a point at which the trend reverses. The prey increases, followed by an increase in the predator population. The two populations rise and fall in oscillations (Figure 16.1).

The reason for oscillation is that as the predator population increases, it consumes a progressively larger number of prey, until the prey population begins to decline. The declining prey population can no longer support the large predator population. The predators now face a food shortage, and many of them starve or fail to reproduce. The predator population declines

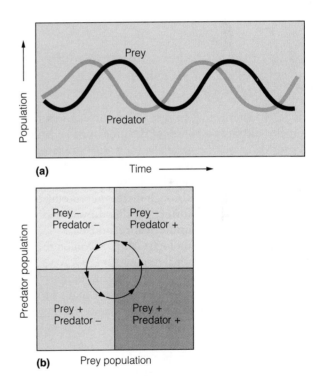
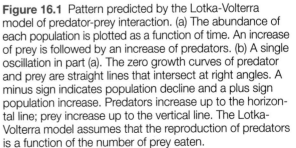

Figure 16.1 Pattern predicted by the Lotka-Volterra model of predator-prey interaction. (a) The abundance of each population is plotted as a function of time. An increase of prey is followed by an increase of predators. (b) A single oscillation in part (a). The zero growth curves of predator and prey are straight lines that intersect at right angles. A minus sign indicates population decline and a plus sign population increase. Predators increase up to the horizontal line; prey increase up to the vertical line. The Lotka-Volterra model assumes that the reproduction of predators is a function of the number of prey eaten.

sharply to a point where the reproduction of prey more than balances its losses through predation. The prey then increases, eventually followed by an increase in the population of predators. The cycle may continue indefinitely. The prey is never quite destroyed; the predator never completely dies out.

About a decade later A. H. Nicholson, an ecologist, and W. Bailey, an engineer and mathematician, developed a mathematical model for a host-parasitoid relationship. Their model predicted increasing oscillation in single-predator–single-prey populations living together in a limited area with all the external conditions remaining the same.

Both models are simplistic. They overemphasize the influence of predators on prey populations. They ignore genetic changes, stress, emigration, aggression, the availability of cover, the difficulty of locating prey as it becomes scarcer, and other aspects. The continuing appeal of the Lotka-Volterra equations to population ecologists lies in the straightforward mathematical descriptions.

16.2 Functional response models relate prey taken to prey density

Models of interactions between predator and prey suggest two distinct responses of the predator to changes in prey density. One is for the individual predator to eat more prey as the prey population increases or to take them sooner. The other is for predators to become more numerous through increased reproduction or immigration. The former is called **functional response** and the latter **numerical response.**

The English entomologist M. E. Solomon introduced the idea of functional response in 1949. A decade later, C. S. Holling explored the concept in more detail. The basis of a functional response is that a predator (or parasitoid) will take more prey as the density of the prey increases. Holling classified functional responses into three types (Figure 16.2).

In a Type I response the number of prey taken per predator increases in a linear fashion to a maximum as prey density increases (Figure 16.2a). Predators of any given abundance take a fixed number of prey during the time they are in contact, usually enough to satiate themselves. Trout feeding on an evening hatch of mayflies is an example. Type I produces density-independent mortality up to satiation.

In a Type II response the number of prey affected rises at a decreasing rate toward a maximum value (Fig-

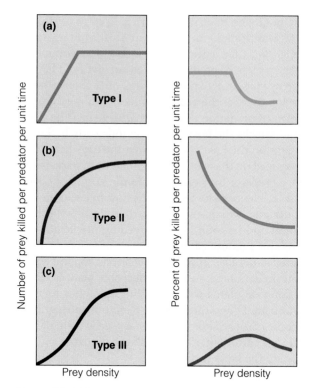

Figure 16.2 Three types of functional response curves, which relate predation to prey density. (a) Type I. The number of prey taken per predator increases linearly to a maximum as prey density increases. Graphed as a percentage, predation declines relative to the growth of the prey population. (b) Type II. The number of prey taken rises at a decreasing rate to a maximum level. The rate of predation declines as the prey population grows. Type II predation cannot stabilize a prey population. (c) Type III. The number of prey taken is low at first, then increases in a sigmoid fashion approaching an asymptote. Plotted as a percentage, the functional response retains some sigmoid features, but declines slowly as the prey population increases. The Type III functional response has the potential of stabilizing prey populations.

ure 16.2b). A dominant component of a Type II response is handling time for a predator. Handling time includes the time spent pursuing and subduing the prey and eating and digesting it. It both limits the amount of prey that a predator can process per unit time and reduces the time available to seek more prey. As a result the number of prey taken per unit of time slows to a plateau as the number of prey continues to increase (Figure 16.3a).

The plateau in Type II may be influenced by the number of predators attracted to patches of abundant prey, called aggregative response (Figure 16.4). When one or two members of a predator species discover and begin to feed on a particular prey, other members of

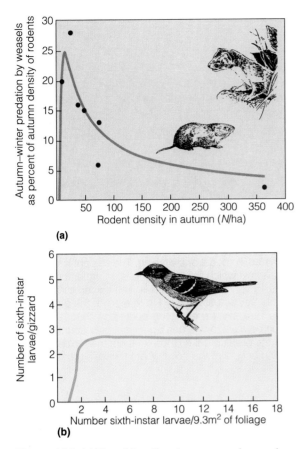

(a)

(b)

Figure 16.3 (a) Type II functional response of weasels preying on rodents in deciduous forests of Bialowieza National Park, Poland. Each point denotes one year. (b) Type III functional response of bay-breasted warblers feeding on spruce budworm *(Choristoneura fumiferana)* larvae.

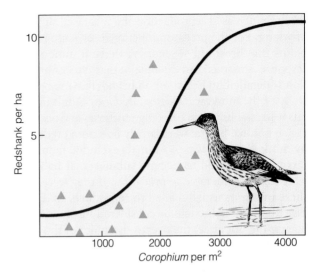

Figure 16.4 Aggregative response in the redshank *(Tringa totanus)*. The curve plots the density of the redshank in relation to the average density of its arthropod prey *(Corophium* spp.).

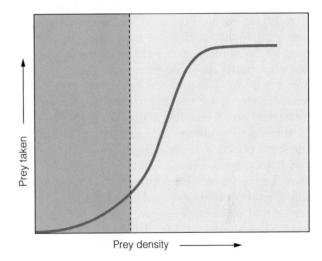

Figure 16.5 Compensatory predation. There is minimal response to the left of the vertical line, which represents the threshold of security for the prey.

the population see and join them. As the number of predators increases on the patch, they interfere with each other. That increases each predator's search time, reduces the efficiency of predation, and encourages many to leave the crowded area.

The Type III response is more complex than Type II and, as we will see, invariably involves two or more prey species. In a Type III response the number of prey taken is low at first, then increases in an S-shaped (sigmoid) fashion, approaching a plateau at which the attack rate remains constant (Figures 16.2c, 16.3b). A Type III functional response can potentially regulate a prey population, because the attack rate varies with prey density. At low densities the attack rate is negligible, but as prey density increases (as indicated by the upward sweep of the curve), predatory pressure increases in a density-dependent fashion.

In a Type III response predators take most or all of the individuals that are in excess of a minimum num-

ber, determined, perhaps, by the availability of escape cover and the prey's social behavior. The prey population level at which the predator no longer finds it profitable to hunt a particular prey species has been called the threshold of security by P. Errington.

Type III responses have been called compensatory (Figure 16.5). As prey numbers increase above the threshold, surplus animals are forced out of the best habitat into poorer, insecure habitat where they become vulnerable to predation. Below the threshold of security, the prey species compensates for its losses through increased reproduction and increased survival of young. For the predator, functional response is low below the threshold of security; above it, functional response is marked. An example of that type of predation is detailed by Errington in his notable long-term study of the muskrat *(Ondatra zibethicus)*. Adult muskrats well established on breeding territories and occupying an optimal habitat were largely free from predation by mink *(Mustela vison)*. Animals excluded from the territory or forced to occupy submarginal habitats were highly vulnerable to predation. If they were not, these muskrats usually failed to reproduce. In either case, these animals could be considered surplus. Their demise did not affect the overall dynamics of the prey population.

16.3 Predators develop a search image for prey

Another reason for the sigmoidal shape of the Type III functional response curve may be the **search image.** According to the search image hypothesis, first developed by the animal behaviorist L. Tinbergen, when a new prey species appears in the area, its risk of becoming prey is low. The predator has not yet acquired a search image—a way to recognize it. Once the predator has captured a desirable item of prey, it has an easier time locating others of the same kind. The more adept the predator becomes at securing a particular prey item, the more intensely it concentrates on it. In time the number of this particular prey species becomes so reduced or its population becomes so dispersed that encounters between it and the predator lessen. The search image for that prey item begins to wane, and the predator reacts to another prey species.

16.4 Predators may turn to alternate, more abundant prey

In a Type III response we assume there is a predator capable of choice and an alternate prey. Although a predator may have a strong preference for a certain prey, it can turn to another, more abundant prey species that provides more profitable hunting. If rodents, for example, are more abundant than rabbits and quail, foxes and hawks will concentrate on rodents.

Ecologists call the act of turning to more abundant alternate prey **switching** (Figure 16.6). In switching the predator concentrates a disproportionate amount

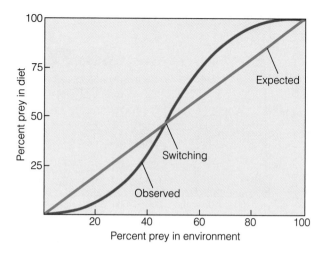

Figure 16.6 A model of switching. The straight line represents levels at which the predator has no preference for Species 1 over other prey species. The curved line represents the prey actually taken. Up to a point, the percent of Species 1 taken is lower than expected by chance when prey numbers are low. When Species 1 prey abundance is high, the predator takes more of them than expected. Switching occurs at the point where the lines cross.

of feeding on the more abundant species and pays little attention to the less abundant species. As the relative abundance of the second prey species increases, the predator turns its attention to it.

At what point in prey abundance a predator switches depends considerably on the threshold of security for the prey species involved. The threshold of security may be much lower for a highly desirable prey. A predator may hunt longer and harder for a palatable species before it turns to a more abundant, less palatable alternate prey. Conversely, the predator may turn from the less desirable species at a much higher level of abundance than it would from a more palatable species.

In spite of switching, some predators deliberately seek out certain prey, no matter how scarce. Predators in a California grassland showed a distinct preference for meadow voles *(Microtus californicus)* over harvest mice *(Reithrodontomys megalopis)* and other rodents even though alternate prey was more abundant. An abundance of alternate prey apparently enables carnivores to maintain a sufficiently high population and to obtain enough energy to allow continued predatory pressure on the preferred species. Among herbivores, deer show a pronounced preference for a certain species of woody plants. Meadow mice *(Microtus pennsylvanicus)* often concentrate on seeds of preferred grasses, even though the seeds of other species are more abundant.

16.5   Predators respond numerically to changing prey density

As the density of prey increases, the number of predators may also increase. Numerical response takes three basic forms (Figure 16.7): (1) direct response, in which the number of predators in a given area increases as the prey density increases; (2) no response, in which the predator population remains proportionately the same; and (3) inverse response, in which the predators cannot keep up with the increasing density of prey.

Most numerical responses involve an increase in reproductive effort. Because reproduction usually requires a certain minimal time, a lag exists between an increase of a prey population and a numerical response by a predator population. An example is the response of a weasel (*Mustela nivalis*) population to the density of two rodents, the bank vole (*Clethrionomys glareolus*) and the yellow-necked mouse (*Apodemus flavicollis*) in Biatowieza National Park in eastern Poland from 1985 through 1992.

During this time the rodents experienced five years of seasonal (noncyclic) fluctuations and a two-year irruption brought about by a heavy crop of oak, hornbeam, and maple seeds. The abundance of food stimulated the rodents to breed throughout the winter. Autumnal seasonal fluctuations of rodents ranged between 28 and 74 animals per hectare. During the irruption the rodent population reached 300 per hectare, and then declined precipitously to 8/ha (Figure 16.8).

The weasel population followed the fortunes of the rodent population. At moderate densities of rodents, the winter weasel density of 5 to 27 per 10 km² declined by early spring to 0 to 19. Following reproduction, the midsummer density rose to 42 to 47 per 10 km². No time lag exists between increased rodent reproduction and weasels' reproductive response. Weasels breed in the spring, and with an abundance of food they may have two litters or larger litter size. Young males and females breed during their birth year. During the irruption the number of weasels grew to 102 per 10 km², and during the crash it declined to 8 per 10 km². The increase and decline in weasels was positively correlated with the spring numbers of rodents. Neither during the regular seasonal fluctuations nor during the irruption and crash were the weasels able to keep up with the density of rodents. They showed an inverse numerical response to prey density (Figure 16.9).

16.6   Prey have evolved defenses against predators

The relationship between predator and prey is influenced considerably by prey defenses and the ability of predators to overcome them. The predator does not succeed in every encounter, as models of predation assume. Prey employ different means of defense.

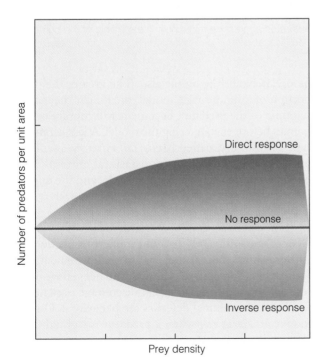

Figure 16.7 Basic forms of numerical response.

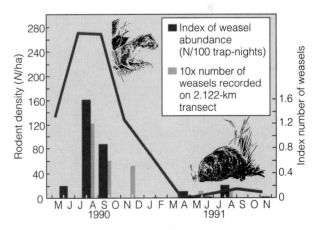

Figure 16.8 Numerical response of weasels to an irruption and crash of forest rodents. Brown bars represent weasel data from live trapping; orange bars represent data from captures, visual observations, and radio tracking.

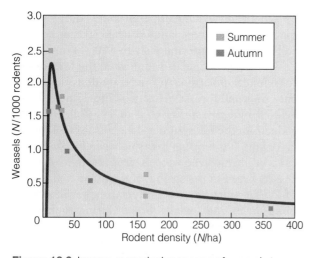

Figure 16.9 Inverse numerical response of weasels to rodent density in the reproductive seasons. Note that weasel response cannot keep up with rodent density.

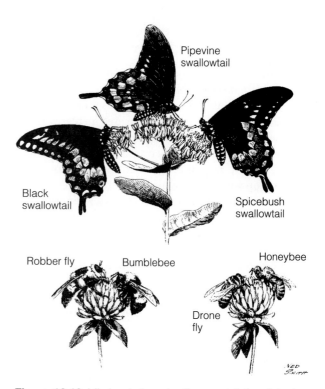

Figure 16.10 Mimicry in insects. One model, the distasteful pipevine swallowtail *(Battus philenor),* has as its mimics the black swallowtail *(Papilio polyxenes)* and the spicebush swallowtail *(Pterourus troilus).* All these butterflies occur in the same habitat. The hover, or drone fly *(Eristalis tenax),* deceives predatory birds by mimicking the honeybee *(Apis mellifera),* which has a stinger. The robber fly *(Laphria* spp.) illustrates aggressive mimicry. It is the mimic of the bumblebee *(Megabombus pennsylvanicus),* on which it preys.

Chemical defense is widespread among many groups of animals. Some species of fish release alarm pheromones, inducing flight reactions in members of the same and related species. Arthropods, amphibians, and snakes employ odorous secretions to repel predators. Many arthropods possess toxic secondary substances, which they have taken from plants and stored in their own bodies. Other arthropods and venomous snakes, frogs, and toads synthesize their own poisons.

Prey have evolved a number of other defense mechanisms. Cryptic coloration includes colors, patterns, shapes, and postures that allow prey to blend into the background. Movements and other behaviors also tend to hide the prey species from predators. Alternatively, some species use extremely visible color patches to distract and disorient predators. Some animals exhibit warning coloration to signal that they are highly toxic or possess other chemical defenses. The yellow and black coloration of many bees and wasps, for example, serves notice of danger to would-be predators. All predators, however, must have some experience with prey before they learn to associate the coloration with unpalatability or pain.

Some animals evolve a coloration that resembles or mimics the warning coloration of a toxic species (Figure 16.10). In this type of mimicry, called *Batesian mimicry* after the naturalist H. E. Bates, who described it, the mimic, an edible species, resembles the inedible species, called the model. Among North American butterflies, for example, the palatable viceroy butterfly mimics the distasteful monarch butterfly. Once a predator, usually a bird, has learned to avoid the distasteful

model, it avoids the mimic also. The greater the proportion of mimics to the model, the longer the learning time of the predator, because it will capture many mimics before encountering the model. A less common type of mimicry, called *Mullerian mimicry,* involves both distasteful models and distasteful mimics. The pooling of numbers between the model and the mimic reduces the losses of each. Predators learn quickly without having to handle both species.

Some animals employ physical means of defense. Clams, armadillos, turtles, and numerous beetles all withdraw into their armor coat or shell when danger approaches. Porcupines, echidnas, and hedgehogs possess quills (modified hairs) that discourage predators.

Still other animal defenses are behavioral. One is to give an alarm call when a predator is sighted. Because high-pitched alarm calls are not species-specific, a wide range of animals nearby recognize them. Alarm calls often bring in numbers of potential prey that mob the predator. Other behavioral defenses include distraction displays, most common among birds. They direct the attention of the predator away from the prey.

For some prey, living in groups is the simplest form of defense. Predators are less likely to attack a concentrated group of individuals. By maintaining a tight, cohesive group, prey make it difficult for any predator to obtain a victim.

A more subtle form of defense is the timing of reproduction so that most of the offspring are produced in a short period of time. Then prey is so abundant that the predator becomes satiated, allowing a percentage of the young to escape and grow to a less vulnerable size. Such is the strategy employed by ungulates such as the caribou.

16.7 Predators have evolved efficient hunting tactics

As prey have evolved ways of avoiding predators, predators have evolved better ways of hunting. Predators have three general methods of hunting: ambush, stalking, and pursuit. Ambush hunting means lying in wait for prey to come along. This method is typical among some frogs, alligators and crocodiles, lizards, and certain insects. Although ambush hunting has a low frequency of success, it requires minimal energy. Stalking, typical of herons and some cats, is a deliberate form of hunting with a quick attack. The predator's search time may be great, but pursuit time is minimal. Pursuit hunting, typical of many hawks, lions, and wolves, involves minimal search time, because the predator usually knows the location of the prey, but pursuit time is usually great. Stalkers spend more time and energy encountering prey. Pursuers spend more time capturing and handling prey.

Predators, like their prey, may use cryptic coloration to blend into the background or break up their outlines. Predators use deception by resembling the prey. Robber flies (*Mallophora bomboides*) mimic bumblebees, their prey. The female of certain species of fireflies imitates the mating flashes of other species, attracting males of those species, which she promptly kills and eats. Predators may also employ chemical poisons, as shrews and rattlesnakes do. They may form a group to attack large prey, as lions and wolves do.

16.8 Cannibalism is common to diverse animals

A special form of predation is cannibalism, euphemistically called intraspecific predation. **Cannibalism** is

killing and eating an individual of the same species. It is common to a wide range of animals, aquatic and terrestrial, from protozoans and rotifers through centipedes, mites, and insects to frogs, toads, fish, birds, and mammals, including humans. Interestingly, about 50 percent of terrestrial cannibals, mostly insects, are normally herbivorous species, the ones most apt to experience a shortage of protein. In freshwater habitats the bulk of cannibalistic species are normally carnivorous, as they all are in marine ecosystems.

Cannibalism has been associated with stressed populations, particularly those facing starvation. Although some animals do not become cannibalistic until other foods run out, others do so when alternative foods decline and individuals are malnourished. Other conditions that may promote cannibalism are (1) crowded conditions or dense populations, even when food is adequate; (2) stress, especially when individuals of low social rank are attacked by dominant individuals; and (3) the presence of vulnerable individuals—such as nestlings, eggs, or runty individuals that provide easy prey—even in the presence of food.

Whatever the cause, local conditions and the nature of local populations influence the intensity of cannibalism. Among some predatory fish, such as walleyes (*Stizostedium vitreum*), and insects such as freshwater backswimmers (*Notonecta hoffmanni*), cannibalism is most prevalent in summer. It coincides with a sudden decrease in normal prey and with low water levels, reducing hiding places. Those receiving the brunt of cannibalism are the small and young, preyed upon by older and larger individuals.

Even at low rates, cannibalism can produce demographic effects. Three percent cannibalism in the diet of walleyes could account for 88 percent of mortality among the young. Cannibalism causes 23 to 46 percent of the mortality among eggs and chicks of herring gulls, 8 percent of young Belding's ground squirrels, and 25 percent of lion cubs. Cannibalizing a large proportion of either a population or a vulnerable age class can result in dramatic fluctuations in recruitment.

Cannibalism can rapidly decrease the number of intraspecific competitors as food becomes scarce. It decreases as resources become more available to survivors and as vulnerable individuals become scarcer or harder to find. By reducing intraspecific competition at times of resource shortage, cannibalism may actually reduce the probability of local extinction of a population, improve conditions for survivors, and enhance their growth and fecundity. Conversely, cannibalism can be a selective disadvantage if certain survivors become too aggressive and destroy their own progeny.

16.9 A species may prey on its competitor

A three-way interaction complicates our understanding of predation. Suppose a predator, species B, eats an herbivore, species A. In turn, carnivores C and D feed on species B. What if species D also feeds on species C? That situation—called **intraguild predation**—is fairly common. It involves one species killing and eating another species that shares the same prey and thus is a potential competitor. By feeding on a competitor, the predator gains energy and reduces competition.

The simplest form of intraguild predation is a three-species system in which one predator preys on another predator that competes for a shared prey (Figure 16.11). For example, both the largemouth bass and the bluegill feed on aquatic insects. Small juvenile bluegills, however, are also major prey for the bass. The potential prey species or age class has a smaller body that matches normal prey size. Thus predation often falls most heavily on the young of the competitor.

Intraguild predation has important implications in resource management, especially fishery management. For example, fishery biologists introduced the opossum shrimp *(Mysis relicta)*, a voracious predator of zooplankton, especially cladocerans, into the Flathead Lake ecosystem in Montana. Their purpose was to provide a forage species for the kokanee salmon *(Oncorhynchus nerka)*, a landlocked species of sockeye salmon introduced into the lake in 1916. (These fish had eliminated the native population of cutthroat trout, *Salmo clarki.*) Shortly after the fishery managers introduced the shrimp population, they noticed that vari-

ous species of cladocerans declined sharply or went extinct. Instead of increasing with the supposedly additional food supply, the salmon population declined. The reason in part involved the behavior of the shrimp. They retreated to the bottom of the lake during the day and came up at night. By doing so, the shrimp escaped predation by the day-feeding salmon.

Rather than feeding on shrimp, the salmon preferred to feed on cladocerans that were also the major prey of both young salmon and shrimp. The shrimp, however, were superior competitors with the young salmon for cladocerans. The result was a collapse of the kokanee's population in the lake. The demise of the salmon reduced the autumnal congregation of bald eagles on the lake, as well as other wildlife attracted to the spawning run.

The moral of this story is two-fold. One, look beyond the single-predator–single-prey relationship, upon which much fishery and wildlife management is based. Two, never introduce a new or exotic species into an ecosystem in which you do not understand the feeding structure. To do so can upset evolved predator-prey relationships, eliminate native species, and reduce diversity.

16.10 Animals use a foraging strategy

Have you ever watched a robin hopping across the lawn? It flies in, lands, hops for a short distance as if sizing up the situation, and stops. Then it moves on deliberately in a series of irregular paths across the grass. The bird pauses every few feet, stares ahead as if in deep concentration, or cocks its head toward the ground. It either moves on or crouches low as if to brace itself. It pecks quickly in the ground and pulls out an earthworm. On occasion the earthworm pulls back, and in the tug-of-war the worm wins. The robin does not push the action. It lets go and hops to another spot to repeat the food-seeking activity.

The behavior of the robin has four parts. First the robin has to decide where to hunt for food. Once on the lawn, it has to search for palatable items. Having located some potential food, the robin has to decide whether to pursue it or not. If it begins pursuit by pecking, then the robin has to attempt a capture, in which it might or might not be successful. By capturing an earthworm, the robin earns some units of energy. The robin needs to gain more energy than its expends in its round of foraging.

The problem facing the robin and all other animals is securing sufficient energy to maintain itself, feed its

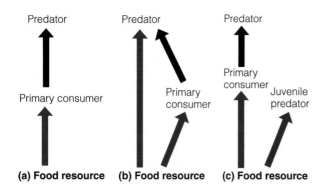

Figure 16.11 Intraguild predation. (a) A food relationship without intraguild predation. (b) A three-species feeding relationship with intraguild predation. (c) A feeding relationship with age-structured species whose juveniles compete with the consumer and whose adults eat the consumer. (Adapted from G. Polis and R. D. Holt, "Intraguild Predation: The Dynamic Complex Trophic Interactions," *TREE* (1992): 152. Reprinted by permission of Elsevier Science Ltd.)

young, and store fat for migration or winter, without spending too much time in doing so. To the robin, time is energy. To spend too much time in securing energy without a sufficient return for the effort results in a type of personal bankruptcy. If the robin fails to find suitable and sufficient food in the area of the lawn it is searching, it leaves to search elsewhere. If successful, the bird probably will return until the spot is no longer an economical place to feed.

The means animals employ to secure food are their **foraging strategy.** Of practical interest to ecologists is the optimal foraging strategy against which actual foraging strategy can be measured. An optimal foraging strategy is the one that provides the maximum net rate of energy gain. A foraging strategy involves two separate but related components. One is optimal diet; the other is optimal foraging efficiency.

When the robin searches the lawn for food, it should, by optimal foraging theory, select only those items that provide the greatest energy return for the energy expended. That places an upper and lower limit on the size of food items accepted. If a food item is too large, it requires too much time to handle; if too small, it does not deliver enough energy to cover the costs of capture. Our robin should reject or ignore less valuable items, such as small beetles and caterpillars, and give preference to small and medium-sized earthworms. These more valuable food items would be classified as preferred foods, ones taken out of proportion to their availability relative to all other food items.

Our robin, according to optimality theory, should take the most valuable food items first. When it depletes those items, the robin should turn its attention to the next most valuable food items. Eventually, the animal should expand its diet to include the poorer foods when the discovery of high-value foods falls below a certain rate.

Field studies provide some insight into how animals forage under natural conditions. N. Davies studied the feeding behavior of the pied wagtail (*Motacilla alba*) and the yellow wagtail (*Motacilla flava*) in a pasture near Oxford, England. The birds fed on various dung flies and beetles attracted to cattle droppings. Prey included large, medium, and small flies and small beetles. The wagtails showed a decided preference for medium-sized prey (Figure 16.12). The size of prey selected corresponded to the optimum-sized prey the birds could handle profitably (Figure 16.13). The birds ignored the small sizes. Easy-to-handle small prey did not return sufficient energy, and large sizes required too much time and effort.

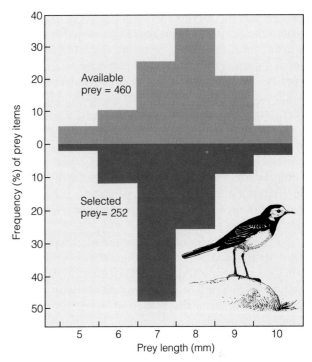

Figure 16.12 Pied wagtails show a definite preference for medium-sized prey, which are taken in disproportionate amounts compared to the frequencies of prey available.

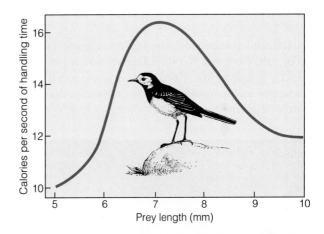

Figure 16.13 The prey size chosen by pied wagtails is the optimal size, providing maximum energy per handling time.

16.11 Foragers seek productive food patches

The robin lives in a patchy environment. It is made up of clumps of trees, shrubby thickets or plantings, and open lawns. The robin could forage in all these patches of vegetation, eating all food items it came across, but it does not. Rather, the bird concentrates its foraging activities in patches of open lawn. Moreover, the robin may restrict its searching to certain parts of the lawn, where the environment is more favorable for earthworms, until the earthworms are depleted or have moved deeper into the soil in response to a drying surface. Then the robin has to turn its attention to other patches and other food, such as ground beetles, flies, sow bugs, snails, and millipedes.

The robin has been following the rules of optimal foraging: (1) concentrate on the most productive patches; (2) stay with those patches until their profitability falls to a level equal to the average for the foraging area as a whole; (3) leave the patch once it has been reduced to the level of average productivity; (4) ignore patches of low productivity.

Compare the foraging pattern of the short-tailed fruit bat *(Carollia perspicillata)* (Figure 16.14). This bat inhabits the tropical dry forests of Central America and northern South America through northern Brazil. It feeds on highly nutritious fruits of understory shrubs, particularly members of the genus *Piper*. Three of the five species produce fruit in the wet season and four of the five in the dry season (two species fruit twice each year). Fruits are most abundant and evenly distributed during the wet season, and more highly concentrated or clumped during the dry season. The *Piper* species are "steady-staters." They produce flowers and ripe fruits daily in low densities over an extended period of time. By visiting these steady-state patches, the short-tailed fruit bat minimizes the time and energy it needs to expend for food. Nightly the bat visits a patch. It picks a ripe fruit and carries it to a roost to eat. Because new fruits ripen daily, the bat returns to the same patch or patches each night, as long as fruit is avail-

able. When it depletes the fruit in that patch, the bat moves to another.

Some social and environmental conditions force less than optimal foraging upon the short-tailed fruit bat. Because of their greater energy demands, female bats have to travel farther from day roosts to suitable patches of highly nutritious fruit than the less energy-demanding territorial males. The risk of predation inhibits the movements of both males and females. They avoid or reduce foraging during moonlit nights, especially during the dry season. At that time bright light makes them highly vulnerable to predators.

Risk of predation affects foraging activities and sites for many animals, notably birds and fish. This risk makes their foraging activities somewhat less than optimal. They may avoid food-rich patches and opt for poorer, safer patches.

Being opportunists, animals will take some less-than-optimal food items (by ecologists' standards) upon discovery. They may quit foraging in a patch before food items drop to some minimal level. They do not necessarily pass up profitable patches because the patches do not meet some theoretical expectation. Animals quickly learn where food is and where food is not. They will stay with a patch as long as the rate of replenishment exceeds the rate of depletion, as in the case of the short-tailed fruit bat. Some animals are highly restricted in their choices. Sedentary animals such as corals, barnacles, and blackfly larvae, filter feeders all, have to take what food flows past them. Others have severely restricted foraging patterns that limit their choices of food.

In spite of its weakness, the concept of optimal foraging is valuable. It provides a way to compare how animals actually forage, select, and harvest their food.

16.12 Herbivory affects plant growth and reproduction

If you measure the amount of biomass actually eaten by herbivores, it may be small, perhaps 6 to 10 percent of total plant biomass. In years of major insect outbreaks, however, or in the presence of overabundance of deer, consumption is considerably higher. Consumption, though, is not a good measure of the importance of herbivory. Grazing on plants has a subtle impact on both plants and herbivores.

Removal of plant tissue—leaf, bark, stems, roots, and sap—affects a plant's ability to survive, even though it may not be killed outright. Loss of foliage and subsequent loss of roots decrease plant biomass, reduce

Figure 16.14 The short-tailed fruit bat of Central America and northern South America is a nearly optimal forager.

the vigor of the plant, place it at a competitive disadvantage with surrounding vegetation, and lower its reproductive effort. That is especially true in the juvenile stage, when the plant is most vulnerable and least competitive with surrounding vegetation.

A plant may be able to compensate for the loss of leaves by increasing photosynthesis in the remaining leaves. However, it may be adversely affected by the loss of nutrients, depending on the age of the tissues removed. Young leaves are dependent structures, importers and consumers of nutrients drawn from reserves in roots and other plant tissues. As the leaf matures, it becomes a net exporter of nutrients, reaching its peak just before senescence. Grazing herbivores, such as sawfly and gypsy moth larvae, deer, and rabbits, concentrate on more palatable, more nutritious leaves. They tend to reject older leaves because they are tough and often contain secondary compounds (tannins, for example). If grazers concentrate on young leaves, they remove considerable quantities of nutrients.

Plants respond to defoliation with a flush of new growth that drains nutrients from reserves that otherwise would have gone into growth and reproduction. Defoliation also draws on the plant's chemical defenses, a costly response. Often the withdrawal of nutrients and phenols from roots exposes the roots to attack by root fungi while the plant marshals its defenses in the canopy. If defoliation of trees is complete, as often happens during an outbreak of gypsy moths *(Porthetria dispar)* or fall cankerworms *(Alsophila pometaria)*, replacement growth differs from the primary canopy removed. The leaves are smaller, and the total canopy may be reduced by as much as 30 to 60 percent. Defoliation in a subsequent year may further reduce leaf size and number. Some trees may end up with only 29 to 40 percent of the original leaf area with which to produce food in a shortened growing season.

Severe defoliation and subsequent regrowth alter the tree physiologically. Growth regulators controlling bud dormancy change with the removal of leaves. The plant uses reserve food to maintain living tissue until new leaves form. The drain on nutrient reserves adversely affects the tree over winter, because twigs and tissues are immature at the onset of cold weather. Such weakened trees are more vulnerable to insect and disease attack the next year. Defoliation kills coniferous species.

Some herbivores, such as aphids, do not consume tissue directly but tap plant juices instead, especially in new growth and young leaves. Sap-sucking insects can decrease growth rates and biomass of woody plants by 25 percent.

Damage to the cambium and the growing tip (apical meristem) may be more important in some plants. Deer, rabbits, mice, and bark-burrowing insects feed on those parts, often killing the plant or changing its growth form.

Moderate grazing, even in a forest canopy, may have a stimulating effect, increasing biomass production, but at some cost to vigor and to nutrients stored in roots. The degree of stimulation depends upon the nature of the plant, nutrient supply, and moisture. In general, grazing increases the biomass of grass up to a point. Then production of tissue declines. Adverse effects are greatest when new growth is developing. Defoliation, then, as in deciduous trees, results in loss of biomass, decreased growth, and delayed maturity.

Grasses, however, are well adapted to grazing, and, up to a point, most benefit from it (see Focus on Ecology 16.1: Grazing the West). Because the meristem, the source of new growth, is close to the ground, grazers first eat the older rather than the more expensive younger tissue. Grazing stimulates production by removing older tissue functioning at a lower rate of photosynthesis. It reduces the rate of leaf aging, thus prolonging active photosynthetic production, and it increases the light intensity on underlying young leaves, among other benefits. Some grasses can maintain their vigor only under the pressure of grazing, even though defoliation reduces sexual reproduction. In the absence of grazing, some dominant grasses disappear.

16.13 Plants defend themselves from herbivores

Herbivory is a two-way street. Plants, although seemingly passive in the process, have a pronounced effect on herbivores. For herbivores it is not the quantity of food that is critical—usually enough biomass is available—but the quality. Because of the complex digestive process needed to break down plant cellulose and convert plant tissue into animal flesh, high-quality forage rich in nitrogen is necessary. Without that richness, herbivores can starve to death on a full stomach. Low-quality foods are tough, woody, fibrous, and indigestible. High-quality foods are young, soft, and green, or they are storage organs such as roots, tubers, and seeds. Most food is low quality, and herbivores that have to live on such resources suffer high mortality or reproductive failure. Added to the problem of quality is the chore of overcoming the various defenses of plants.

The basis of chemical defense is the accumulation of an array of toxic proteins, terpenes, and tannins.

FOCUS ON ECOLOGY 16.1

GRAZING THE WEST

Grazing on public western rangeland is a highly debated topic in the popular press and among environmentalists, range managers, and grassland ecologists. Environmentalists are exerting pressure to eliminate grazing by sheep and cattle on rangelands. They say that grazing is destroying these ecosystems. On the other side are ranchers and livestock interests who want to maintain or increase grazing on public rangeland.

The issue revolves about the idea that grasses, when mowed or grazed, respond to the loss of leaves by increased growth, called overcompensation. (Lawn grasses overcompensate for mowing—that is why you have to mow often, especially early in the growing season.) Most experimental evidence for overcompensation comes from studies of how individual species of grass on highly fertile sites respond to defoliation. Other evidence comes from community-level studies of African grasslands grazed by a diversity of highly mobile native herbivores. Studies at both levels suggest that moderate grazing does promote the productivity of many species of grass above a level that prevails in the absence of grazing. Further, moderate grazing under productive conditions promotes the diversity of plants, whereas in the absence of grazing, certain species dominate and others disappear.

Some range managers and ranchers misinterpret or misunderstand the nature of the studies. They have turned the concept that grazing may benefit plants and increase grassland productivity into a range policy advocating heavy grazing of grassland.

Because evidence for overcompensation comes from highly productive monocultural grassland systems, many grassland ecologists and range managers question whether this concept can be broadly applied to arid and semiarid rangelands. Overcompensation may not occur in these less productive rangelands, where resources are limiting and crowding of plants does not occur. A number of studies of rangeland systems and semiarid grassland plants show that grazing often harms plants.

Some grassland ecologists point out that grazing affects several levels: individual plants, populations, and communities. Populations of various grassland species have direct and indirect responses to grazing. Direct responses to defoliation involve survival, fecundity, and growth. Indirect responses involve changes in competition between plants and in microclimatic conditions around the plant. Species that respond positively to grazing will disappear if grazing ceases. Plants less desirable to grazers can become dominant if palatable species are overgrazed. Types of herbivores also influence the response of plants. Sheep, which graze close to the ground, have a different effect than cattle, which remove leaves higher off the ground. At the ecosystem level, grazing can change the structure and plant composition of rangelands, redistribute nutrients through droppings, and cause erosion. Overgrazing of arid and semiarid rangeland can cause woody vegetation to replace grasses.

From this debate we learn caution. We must not lightly extrapolate findings gained from the study of one species or one community to other species and other communities. Results obtained from monocultural ecosystems seldom apply to natural ecosystems that are ecologically complex. The theory of overcompensation is most applicable to monocultural fertile domestic grasslands, not to arid and semiarid systems. It applies to systems that evolved under the influence of less abundant, migratory, large herbivores, not high-density, sedentary herds of cattle and sheep.

The debate also points to the need for better communication between ecologists and people who use the land. Ecologists should indicate explicitly the limitations of their findings and suggest how those findings might properly apply to management. We need better melding of agricultural research and ecological research in those areas where the findings of both impinge on the management of natural ecosystems.

These secondary products, poisonous to the plant itself, may be stored in vacuoles within cells and released only when the cells are broken. They also may be stored in and secreted by epidermal glands to function as a contact poison or volatile inhibitor. Production and storage of such poisons are expensive to a plant and require a trade-off between defense and reproductive effort.

The mode of defense varies with the nature of the plants, which fall into two broad groups. One group consists of large, conspicuous, long-lived woody plants available to herbivores. They possess the most expensive type of defense—**quantitative inhibitors** that reduce digestibility and thus potential energy gain from food. The chemicals are mostly tannins and resins concentrated near the surface tissues in leaves, bark, and seeds. They toughen plant tissues, reduce the ability of microorganisms to break them down in herbivore digestive systems, and lower palatability. The problem with such a defense is the slow response time. After defoliation by gypsy moths, oaks increase the phenol and tannin content and the toughness of leaves a year later.

The other group contains short-lived plants, mostly annuals and perennials. These plants employ **qualitative inhibitors**—toxic substances such as cyanogenic compounds and alkaloids such as nicotine, cocaine, morphine, and mescaline that interfere with metabolism. Plants can synthesize these substances quickly at little cost. They are effective at low concentrations and readily transported to the site of attack. They can be shuttled about the plant, from growing tips to leaves to stems, roots, and seeds. They can also be transferred from seed to seedling. These substances protect mostly against generalist herbivores.

A few herbivores have developed ways of breaching these chemical defenses. Some insects can absorb or metabolically detoxify these chemical substances. They even store the plant poisons to use them in their own defense, as the larvae of monarch butterflies do, or in the production of pheromones. Some beetles and certain caterpillars cut circular trenches in leaves before feeding, stopping the flow of chemical defenses to the leaf area on which they will feed. Others cut veins on milkweeds and other latex-producing plants, blocking the flow of latex to the intended feeding site.

Plants may also employ the least costly form of defense—hairy leaves, thorns, and spines, structures that evolved early, when they may have been subject to even greater predatory pressure.

16.14 Vegetation, herbivores, and carnivores interact

A common approach in ecology is to consider herbivory on plants and predation on animals as two separate entities. They are not separate. Predator-prey relationships at one level influence interactions at other levels. We cannot really understand a herbivore-carnivore system without understanding plants and their herbivores; nor can we understand plant-herbivore relations without understanding predator-herbivore relationships. All three—plants, herbivores, and carnivores—are related. Ecologists are beginning to understand these three-way relationships.

A classic case is the three-level interaction of vegetation, snowshoe hare (*Lepus americanus*), and lynx (*Felis lynx*) (Figure 16.15). The snowshoe hare inhabits the northern forest. In winter it feeds on the buds of conifers and twigs of aspen, alder, and willows, termed browse. Browse consists largely of stems about 3 mm in diameter, young growth rich in nutrients. The hare-vegetation interaction becomes critical when the amount of essential browse falls below that needed to support the population over winter, approximately 300 g per individual per day. Excessive browsing and girdling during periods when the hare population is high reduce future woody growth, bringing on a food shortage. Adding to the food crisis are the chemical defenses of alders, birches, and some willows. These defenses strongly influence the selection of winter forage among many browsing herbivores of the subarctic. Hares avoid these otherwise nutritious plants, especially juvenile growth and buds, because they contain more resin and phenolic glucosides than mature twigs.

The shortage and poor quality of food lead to malnutrition. Malnutrition and low winter temperatures weaken the hares, making them extremely vulnerable

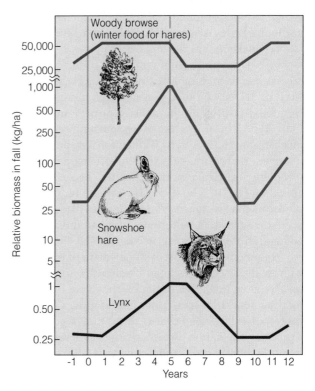

Figure 16.15 The three-way interaction of woody vegetation, snowshoe hare, and lynx. Note the time lag between the cycles of the three populations.

to predation. Intense predation by lynx and other predators causes a rapid decline in the number of hares and holds the population for several years at a low level. Now facing their own shortage of food, the predators, mostly lynx, fail to reproduce. Meanwhile,

released from the pressures of browsing by hares, vegetation growth rebounds. With a decline in predatory pressure and a growing abundance of winter food, the hare population begins to rise sharply, starting another cycle.

CHAPTER REVIEW

SUMMARY

Predation is the eating of one living organism by another. One organism benefits at the other's expense. However, there is a close interaction between predator and prey. Each influences the fitness of the other. Predation includes not only carnivory but parasitoidism, cannibalism, and herbivory.

Models of Predation (16.1–16.5) Interactions between predator and prey have been described in mathematical models by Lotka and Volterra, modified by others. All these models predict oscillations of predator and prey populations **(16.1)**. The interaction between predator and prey involves a functional response and a numerical response.

A functional response is one in which the number of prey taken increases with the density of prey. There are three types of functional responses. In Type I the number of prey affected increases in a linear fashion to a maximum as the number of prey continues to increase. In Type II the number of prey affected increases at a decreasing rate toward a maximum value. In Type III the number of prey taken increases in a sigmoidal fashion as the density of prey increases. Only Type III is important as a population regulating mechanism **(16.2)**. A Type III functional response may involve a search image. Predators develop a search image for a particular prey as that item becomes more abundant **(16.3)**. The Type III response implies that predators have a choice. Switching takes place when a predator turns to an alternate, more abundant prey, taking it in a disproportionate amount **(16.4)**.

A numerical response is the increase of predators with an increased food supply. Numerical response may involve an aggregative response, the influx of predators to a food-rich area. More importantly, a numerical response involves a change in the rate of growth of a predator population through changes in developmental time, survival rates, and fecundity **(16.5)**.

Prey Defense, Predator Offense (16.6–16.7) Chemical defense in animals usually takes the form of a distasteful or toxic secretion that repels, warns, or inhibits

a would-be attacker. Cryptic coloration and behavioral patterns enable the prey to escape detection. Warning coloration declares the prey is distasteful or disagreeable. Some palatable species mimic unpalatable species for protection. Armor and aggressive use of weapons defend some prey. Alarms and distraction displays help others. Another form of defense is predator satiation. Prey species reproduce so many young at once that predators can take only a fraction of them **(16.6)**.

Predators have evolved strategies of their own, with different methods of hunting, including ambush, stalking, and pursuit. Predators also employ cryptic coloration to hide and aggressive mimicry in which they mimic the appearance of the prey **(16.7)**.

Special Forms of Predation (16.8–16.9) Cannibalism is intraspecific predation. It often occurs when a population is under stress. Cannibalism is particularly common among spiders, insects, and fish. Predatory adults feed on grazing young **(16.8)**.

In interguild predation predators kill and eat species that use similar resources. By preying on a resource competitor, the predator gains energy and nutrients and reduces potential competition **(16.9)**.

Foraging Strategy (16.10–16.11) Central to the study of predation is the concept of optimal foraging. This strategy obtains for the predator a maximum rate of energy gain. There is a break-even point above which foraging is profitable and below which it is not. An optimal diet includes the most efficient size of prey for handling and net energy return **(16.10)**. Optimal foraging concentrates activity in the most profitable patches. When the predator depletes the food to the average profitability of the area as a whole, it abandons that patch for one more profitable. Because of risk of predators, social structure, and predator preference, foraging may be less than optimal **(16.11)**.

Herbivory (16.12–16.13) Herbivory affects plants by reducing the amount of photosynthate and the ability to produce more. Plants respond to defoliation with a flush of new growth, which draws down nutrient reserves. Such drawdown can weaken plants, especially woody ones, making them more vulnerable to insects and disease. Moderate grazing may stimulate leaf growth in grasses up to a point. By removing older leaves less active in photosynthesis, grazing stimulates the growth of new leaves **(16.12).** Plants affect herbivores by denying them palatable or digestible food or by producing toxic substances that interfere with growth and reproduction. Certain herbivore specialists are able to breach the chemical defenses. They detoxify the secretions, block their flow, or sequester them in their own tissues as a defense against predators **(16.13).**

Vegetation-Herbivore-Carnivore Systems (16.14)
Plant-herbivore and herbivore-carnivore systems are closely related. An example of a three-level feeding interaction is the cycle of vegetation, hare, and lynx. Malnourished hares fall quickly to predators. Recovery of hares follows recovery of vegetation and decline in predators.

STUDY QUESTIONS

1. Define predation, herbivory, cannibalism, and intraguild predation.
2. What are the basic features and assumptions of the Lotka-Volterra equations?
3. What is a functional response in predation?
4. Distinguish among Type I, Type II, and Type III functional responses. Which regulates population, and why?
5. What is a search image? Switching? Relate them to Type III functional responses.
6. Distinguish among three types of numerical response.
7. What defenses do prey use against predators?
8. Contrast ambush with pursuit and stalking. Evaluate costs and benefits.
*9. Relate the methods of hunting to several defensive and offensive strategies, such as cryptic coloration.
*10. Speculate on why cannibalism is widespread among animals. What triggers cannibalism, and what are its demographic effects?
11. What is an optimal foraging strategy? What is an optimal diet?
12. How is foraging influenced by a patchy environment?
13. How do plants defend themselves against herbivores? Include two approaches to chemical defense.
14. What effects do herbivores have on plants?
15. How do vegetation, herbivores, and carnivores relate in the scheme of predation?

PARASITISM AND MUTUALISM

OBJECTIVES

On completion of this chapter, you should be able to:

- Define parasitism and describe the types of parasites.
- Discuss the modes of transmission and the effects of parasitism on hosts.
- Describe the coevolutionary relationship between host and parasite.
- Discuss social parasitism.
- Explain mutualism and describe the types of mutualism.
- Discuss the role of mutualism in plant pollination and seed dispersal.
- Relate coevolution to obligate mutualism.

Mistletoe *(Viscum album)* with fruits on European birch *(Betula* sp.). The mistletoe is a parasite on the tree. Birds feeding on the berries act as facultative mutualists, carrying seeds from tree to tree.

We suggested in Chapter 16 that predator and prey have evolved in a reciprocal relationship. Prey evolved means of defense against predators, and predators evolved ways to breach those defenses. Such a relationship is even closer between parasite and host and between mutualists. We call a relationship in which two interacting populations appear to strongly influence the evolution of traits in each other **coevolution.** Any evolutionary change in one member changes the selective forces acting on the other. They play a game of adaptation and counteradaptation.

Most coevolutionary responses appear to be general. For example, plants have evolved chemical defenses against a diverse array of herbivorous insects. In turn, many insects have evolved the ability to detoxify a wide range of plant chemicals. Similarly, animals have evolved a generalized immune system in response to a wide range of parasites. However, paired interactions between certain parasites and their hosts or between certain mutualists suggest closer, more specific coevolution.

17.1 Parasites feed on a host and cause disease

Parasitism is a condition in which two organisms live together, one deriving its nourishment at the expense of the other. Parasites, strictly speaking, draw nourishment from the tissues of the larger host, a case of the weak attacking the strong. Typically parasites do not kill their hosts as predators do, although the host may die from secondary infection or suffer stunted growth, emaciation, or sterility.

Parasitic organisms belong to a wide range of taxonomic groups, including viruses, bacteria, protists, fungi, plants, and an array of invertebrates, among them arthropods. A heavy load of parasites is an infection, and the outcome of an infection is a **disease.**

Parasites are distinguished by size. Ecologically parasites may be classified as microparasites and macroparasites. **Microparasites** include viruses, bacteria, and protozoans. They are characterized by small size and a short generation time. They develop and multiply rapidly within the host and tend to induce immunity to reinfection in hosts that survive the initial infections. The duration of the infection is short relative to the expected life span of the host. Transmission from host to host is direct, although other species may serve as a carrier or **vector.**

Macroparasites are relatively large. Examples include flatworms, acanthocephalans, roundworms, flukes, lice, fleas, ticks, fungi, rusts, smuts, dodders, broomrape, and mistletoe. Macroparasites have a comparatively long generation time, and direct reproduction in the host is rare. Macroparasites stimulate a short immune response that depends on the number of parasites in the host. Macroparasites are persistent, causing continual reinfection. They may spread by direct transmission from host to host or by indirect transmission, involving intermediate hosts and carriers.

Although parasites and pathogens are extremely important in interspecific relations, only within the past decade or so have ecologists begun to study them intensely. Parasites have dramatic effects when they are introduced into host populations that have not evolved defenses against them. Then diseases sweep through and decimate the population. Examples are outbreaks of rinderpest in African ungulates, myxomatosis in European rabbits, blight in the American chestnut, the plague or black death in human populations in Eurasia in the 14th century, and recurrent outbreaks of flu, especially in 1918. In most natural populations, though, the effects of parasitism are elusive. Detritivores and scavengers quickly process the bodies of victims, eliminating traces of infection.

17.2 Hosts provide diverse habitats for parasites

Hosts are the habitats for parasites, which have exploited every conceivable habitat on and within these hosts. Parasites that live on the skin within the protective cover of feathers and hair are **ectoparasites.** Others, known as **endoparasites,** live within the host. Some burrow beneath the skin. They live in the bloodstream, in the heart, brain, digestive tract, liver, spleen, mucosal lining of the stomach, spinal cord and brain, in nasal tracts and lungs, in the gonads and in the bladder, in the pancreas, eyes, gills of fish, muscle tissue, and many other sites. Parasites of insects live on the legs, on the upper and lower body surfaces, and even on the mouthparts.

Parasites within the host colonize different sites in an organ system. Coccidian protozoans of the genus *Elmeria* occupy different regions of the digestive tract: one in the duodenum, another in the lower duodenum, a third in the upper small intestine, and a fourth in the caecum and rectum. Closely related parasites in many animals share a system in this manner.

Plant parasites, too, divide up the habitat. Some live on the roots and stems; others penetrate the roots and bark to live in the woody tissue beneath. Some live at the root collar, where the plants emerge from the soil. Others live within the leaves, on young leaves, on mature leaves, or on flowers, pollen, or fruits.

A major problem for parasites, especially parasites of animals, is gaining access to and escaping from the host. Parasites of the digestive tract enter the host through the mouth and escape through the rectum, a path used by other parasites as well. Parasites of the lungs enter the mouth or the skin and travel to the lungs through the pulmonary system. They escape mainly by being coughed up and swallowed into the alimentary tract. Liver parasites, exploiting one of the richest habitats in the animal body, arrive there by way of the circulatory system, bile duct, and hepatic portal system and escape through the same routes. Parasites of the urogenital system enter orally, travel through the gut to the site of infection, and exit by the urinary system. Blood parasites enter and escape through the skin, but always with the aid of some obliging vector such as mosquitoes and ticks. Parasites that end up in the muscle tissues, where they usually form capsules, reach a blind end. For them the only way out is for their host to be killed and eaten by a predator.

To survive and multiply, parasites have to escape from one host and locate another, something they cannot do at will. Parasites can escape only during an infective stage, when they must make contact with the next host. The infective stage is essential.

All parasites reach a stage in their life cycle when they develop no further. The definitive host is the one in which the parasite becomes an adult and reaches maturity. All others are intermediate hosts, which harbor some developmental phase. Parasites may require one, two, or even three intermediate hosts. Each infective stage can develop only when it is independent of the definitive host. It can continue its development only if it can find another intermediate host or its definitive host. For this reason, many parasites employ animals as intermediate hosts to aid in locating a definitive host or adapt to the host's habits. Thus the dynamics of a parasite population is closely tied to the population dynamics of the host.

17.3 Many parasites spread by direct contact

Direct transmission is the transfer of a parasite from one host to another by direct contact with a carrier. The parasite has no intermediate stages in a secondary host. Typically, microparasites are transmitted directly.

For example, an arthropod vector, the black-legged tick *Ixodes scapularis*, is responsible for transmitting Lyme disease, the major arthropod-borne disease in the United States. Named because the first noted occurrence was at Lyme, Connecticut, in 1975, the disease is caused by a bacterial spirochete, *Borrelia burgdorferi*.

It lives in the bloodstream of vertebrates, from birds and mice to deer and humans. The spirochete depends upon the tick for transmission from one host to another. The tick is small. The larva is the size of a grain of pepper, the nymph is the size of a poppy seed, and the adult the size of a sesame seed. White-footed mice, chipmunks, white-tailed deer, and ground-dwelling birds are reservoirs for the spirochete. However, the two main vertebrate hosts are the white-footed mouse and the white-tailed deer, both abundant in eastern North America. The bacterium depends upon the tick to gain entrance into new hosts.

Tick larvae hatch from eggs laid by the adult on the ground during the previous fall (Figure 17.1). They seek out a small host, preferably a mouse on which to feed once over a period of several days. If that host happens to harbor the spirochete, the larva picks it up and transfers it to the next host. The following spring the tick larva molts to become a nymph. The nymph seeks its blood meal from mice or chipmunks and carries its infection with it. Having fed once over a period of 3 to 4 days, the nymph molts to become an adult. Males and females move to a white-tailed deer, feed, mate, drop to the ground, lay eggs, and die.

Some adult ticks attach themselves to humans hiking or working in woods and fields. If the ticks carry the spirochete, they transmit it to humans. Unless treated, the infection, which may produce such visible symptoms as a bull's-eye rash, causes flu-like respiratory congestion, pain in joints and muscles, and permanent damage to the brain, central nervous system, and heart.

Microparasites that infect plants are transmitted by direct contact. In the presence of a suitable host, spores resting in the soil germinate and penetrate the

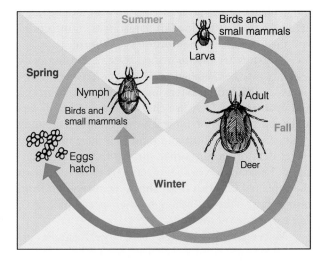

Figure 17.1 The two-year life cycle of the black-legged tick, the carrier of Lyme disease.

roots. Wind-carried spores come to rest on and infect leaves of the plant. The devastating Dutch elm disease *(Ceratocystis ulmi)* is carried from tree to tree by spore-bearing European and native elm bark beetles *(Scolytus mutisriatus* and *Hylurgopinys rufipes).*

Many important macroparasites of animals and plants move from infected to uninfected hosts by direct contact. Parasitic roundworms *(Ascaris)* live in the digestive tracts of mammals. Female roundworms lay thousands of eggs in the gut of the host, which are expelled with the feces. If they are swallowed by a host of the correct species, the eggs hatch in the intestines of the host, the larvae bore their way into the blood vessels, and come to rest in the lungs. From here they ascend to the mouth, usually by causing the host to cough, and are swallowed again to reach the stomach, where they mature and enter the intestines.

The most important external debilitating parasites of birds and mammals are spread by direct contact. They include lice, mites that cause mange, ticks, fleas, and botfly larvae. Many of these parasites lay their eggs directly on the host, but fleas lay their eggs and their larvae hatch in the nests and bedding of the host (even in rugs), from which they leap onto nearby hosts.

Some fungal parasites of plants spread through root grafts. For example, *Fomes annosus,* an important fungal infection of white pine *(Pinus strobus),* spreads

rapidly through pure stands of the tree when roots of one tree become grafted onto the roots of a neighbor.

A number of flowering plant macroparasites spread by direct transmission, too. One group is holoparasites, plants that lack chlorophyll and draw their water, nutrients, and carbon from the roots of host plants. Notable among them are members of the broomrape family (Orobanchaceae). Two examples are squawroot *(Conopholis americana),* which parasitizes the roots of oaks, and beechdrops *(Epifagus virginiana),* which parasitizes mostly the roots of beech trees.

Another group is the hemiparasites. They are photosynthetic, but they draw water and nutrients from their host plant. Among the hemiparasites are the mistletoes *(Phoradendron* spp.), whose sticky seeds attached to limbs send out rootlets that embrace the limb and enter the sapwood. Mistletoe can reduce the growth of its host.

17.4 Some parasites spread by indirect transmission

Many parasites, both plant and animal, use indirect transmission, spending different stages of the life cycle with different hosts. Figure 17.2 shows the brainworm *(Parelaphostrongylus tenuis)* of the white-tailed deer, which has as its intermediate host during its larval stage

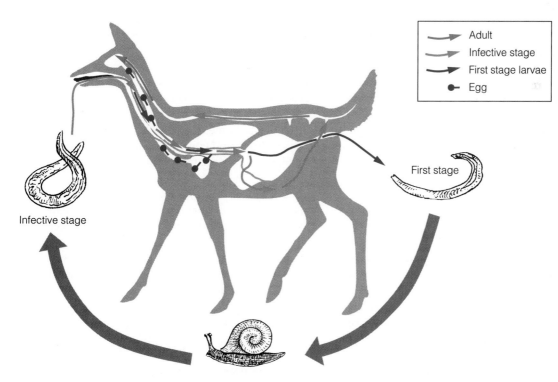

Figure 17.2 The life cycle of a macroparasite, the brainworm *Parelaphostrongylus tenuis,* with indirect transmission. Brainworms infect white-tailed deer, moose, and elk.

a snail or slug that lives in the grass. The deer picks up the infected snail while grazing. In the deer's stomach the larvae leave the snail, puncture the deer's stomach wall, enter the abdominal membranes, and travel via the spinal cord to reach spaces surrounding the brain. Here the worms mate and produce eggs. Eggs and larvae pass through the bloodstream to the lungs, where the larvae break into air sacs and are coughed up, swallowed, and passed out with the feces. The snail ingests the larvae, which continue to develop to the infective stage. There is little evidence that brainworms affect the health of white-tailed deer. Some deer may suffer lung damage, especially if they are infected with other species of lungworm. However, many other ungulates lacking a natural resistance to brainworms are adversely affected.

Indirect transmission among plant parasites is uncommon except among rusts. The wind carries infective stages from primary to intermediate hosts. Examples are white-pine blister rust and wheat rust, which have intermediate stages on shrubs of the genera *Ribes* and *Berberis*, respectively.

17.5 Transmission among hosts is essential to parasites

Transmission from host to host is the key to parasite survival. It can take place only with the dispersal of an infective stage independent of the definitive host. Parasites requiring more than one host can complete their life cycle only if they can infect the right series of hosts. For example, the brainworm of deer has to locate a snail as an intermediate host in which to continue its development to an infective stage that can be transmitted back to the deer. For this reason many animal parasites exploit the feeding habits of the definitive host and adapt to the habits of the intermediate host. The brainworm exploits the snail's habit of crawling up grass stems, where it risks being eaten along with the grass by grazing deer. Unless the deer swallows the snail, the parasite will perish.

Patchy or clumped distribution of hosts further complicates the transmission of parasites. A few members of the host population harbor major parasite loads. They act as reservoirs of infection. If uninfected hosts are widely dispersed or intermixed among populations of other species, the probability of a parasite or its carrier coming into contact with susceptible individuals is low. In other words, transmission depends upon the density of the host and the distance between the parasite and potential hosts.

Consider Lyme disease again. A direct correlation exists between the abundance of hosts, the abundance

Figure 17.3 As humans invade the habitat of deer, the animals are often spotted in suburbs such as this one in the Pocono Mountains of Pennsylvania. An abundance of highly nutritious forage—grass and ornamentals—compounds the problem.

of ticks, and the prevalence of the disease. The key lies with the abundance of vertebrate hosts, notably white-footed mice and chipmunks, for the larval and nymphal stages. There are more larval and nymphal ticks seeking mouse hosts than there are adult ticks seeking deer. Further, one white-tailed deer can support up to 500 adult ticks, whereas one mouse can support 40 juvenile ticks at most.

The abundance of hosts in many regions relates to acorn production. When the acorn crop is heavy, both mice and deer thrive. Deer concentrate in oak forests, and ticks feeding on deer drop off and lay their eggs on the forest litter. The abundance of acorns stimulates mice to breed in winter. By spring, when the larval ticks hatch and crawl about in the forest litter, they encounter an abundance of mice. Thus the abundance of mice, not of deer, determines the density of ticks.

Because of the explosive population growth of deer in the northeastern and midwestern United States and the intrusion of suburban developments into deer habitat, humans, deer, mice, and the spirochete causing Lyme disease are becoming close neighbors (Figure 17.3). This overlap of habitat accounts for the current prevalence of this disease in the northeast.

17.6 Hosts respond to parasitic invasions

Just as prey respond to predators, so hosts react to parasites. Some responses are defensive attempts to counter parasitic invasion. Other responses are negative reactions to parasitic infection.

The first line of defense involves immune and inflammatory responses. Inflammation, provoked by the death or destruction of host cells, stimulates the secretion of histamines and increased blood flow to the site. This reaction brings in white blood cells and associated cells to attack the infection. Scabs form on the skin, as in the case of heavy mange mite infestations on red fox and other canids. Internal reactions can produce calcareous cysts in muscle or skin that imprison the parasite. Examples are the cysts of the infective stage of flukes in their intermediate fish host and the cysts of the roundworm *Trichinella spiralis* (Nematoda), which causes trichinosis in humans, in the muscles of pigs and bears.

The host may also react to parasitic infections with abnormal growth. For example, plants respond to bacterial and fungal invasion by forming cysts in the roots and scabs in fruits and roots, cutting off contact of the fungus with healthy tissue. Plants react to attacks on leaf, stem, fruit, and seed by gall wasps, bees, and flies by forming abnormal growth structures unique to the particular gall insect (Figure 17.4). Gall formation exposes the larvae of some gall parasites to predation. The conspicuous, swollen knobs of the goldenrod ball gall, for example, attract the downy woodpecker (*Picoides pubescens*), which excavates and eats the larva within the gall. Malarial parasites in vertebrates stimulate the spleen into producing many red blood cells and antibodies, resulting in its enlargement. Larvae of warble flies (Oestridae) develop in boil-like swellings (warbles) just under the skin. These growths subside after the transformed larvae escape through a hole cut in the skin.

Infection of plants, notably grasses and sedges, by endophytic fungi that grow within the leaves and stems and by smuts and rusts that grow externally can prevent the plants from reproducing sexually, inhibiting or aborting flowers. This effect has been called "castration." Smuts and rusts, which comprise most of the parasitic fungi of plants, colonize new hosts by contagious spread of spores, as in white pine blister rust. Most castrating fungi are endophytes that spread by vegetative growth into tillers and tubers. Although inhibiting sexual reproduction, these endophytic fungi stimulate vegetative reproduction or cloning. By doing so, such fungi increase the availability of new hosts and prevent the evolution of resistant progeny through sexual reproduction.

Heavily parasitized animals often behave abnormally. Rabbits infected with the bacterial disease tu-

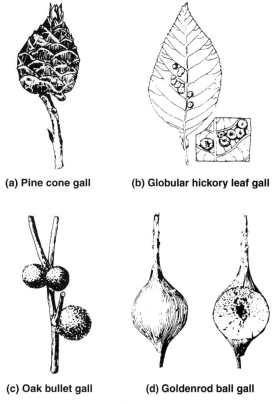

(a) Pine cone gall   **(b) Globular hickory leaf gall**

(c) Oak bullet gall   **(d) Goldenrod ball gall**

Figure 17.4 Galls are a growth response to an alien substance in plant tissues. In this case the presence of a parasitic egg stimulates a genetic transformation of the host's cells. (a) The pine cone gall, a bud gall on willows caused by the gall midge *Rhabdophaga strobiloides*. (b) The hickory leaf gall, induced by the gall aphid *Phylloxera canyae-globuli*. (c) The oak bullet gall, caused by the gall wasp *Disholcaspis globulus*. (d) The goldenrod ball gall, a stem gall induced by a gallfly, *Eurosta solidagin*.

laremia (*Francisella tularensis*), transmitted by the rabbit tick (*Haemaphysalis leporis-palustris*), are sluggish and difficult to move from cover. Rabid foxes and raccoons may be overly aggressive or overly tame and unafraid of human contact. Pacific killifish (*Fundulus parvipinnis*) parasitized by trematodes display abnormal behavior such as surfacing and jerking. This behavior calls attention to the trematodes' definitive host—fish-eating birds. By altering its intermediate host's behavior, making it more susceptible to predation, the trematode ensures the completion of its life cycle.

Some animals respond defensively to avoid parasitism. Birds and mammals rid themselves of ectoparasites by grooming. Among birds the major form of grooming is preening, which involves manipulating plumage with the bill and scratching with the foot. Both activities remove adults and nymphs of lice from the plumage. Deer seek dense, shaded places where they can avoid deerflies, common to open areas.

Birds may defend against parasites by mate selection. The hypothesis has been advanced that parasitic infections or resistance to such infections influence which mate a female chooses. Supposedly, males possessing the brightest plumage and other ornaments are the most free of parasites. By choosing males with the brightest colors, females are selecting the healthiest mates. This hypothesis is highly controversial. Evidence comes from studies correlating mating success with parasitic loads. The hypothesis has yet to be tested rigorously with experimental studies involving mate choice and controlled parasitic infection of males.

17.7   Parasites and hosts establish an uneasy truce

For parasites and host to coexist under a relationship that is hardly benign, the two have to come to an uneasy truce. The parasite gains no advantage if it kills its host. A dead host means dead parasites. The host gains no advantage if it wears itself out in resisting.

However, the parasite is invasive in the body of its host. This is particularly true among birds and mammals, who provide an ideal habitat for large endoparasites. The internal environment is stable, the food is rich, and the organism is long-lived. The host needs to resist this invasion by eliminating the parasites or at least minimizing their effects. A wide range of immune and inflammatory mechanisms may produce short-term or lifelong immunity to the infection.

When a foreign protein or **antigen** enters the bloodstream, it is taken up by white cells called lymphocytes (produced by lymph glands), which produce **antibodies.** The targets of these antibodies are the antigens present on the parasite's surface or released into the host. These antibodies cost a lot of energy to produce. They also are potentially damaging to the host's own tissues. Fortunately, the immune response does not have to kill the parasite to be effective. It only has to reduce the feeding, movements, and reproduction of the parasite to a tolerable level.

The immune response, however, can be breached. Some parasites vary their antigens more or less continuously. By doing so, they are able to keep one jump ahead of the host's response. The result is a chronic infection of the parasite in the host. Antibodies specific to an infection normally are composed of proteins. If the animal suffers from poor nutrition and its protein deficiency is severe, normal production of antibodies is inhibited. Depletion of energy reserves breaks down

the immune system and allows viruses or other parasites to become pathogenic. The ultimate breakdown in the immune system occurs in humans infected with the human immunodeficiency virus (HIV), the causal agent of AIDS, transmitted sexually, through the use of shared needles, or by infected donor blood. The virus attacks the immune system itself, exposing the host to a range of infections that prove fatal.

The parasite and its host have to strike a balance. The host has to reach a level of immune response between beneficial and harmful consequences. It has to direct enough of its metabolic resources to minimize the cost of parasitism, yet not unduly impair its own growth and reproduction. The parasite has to achieve optimal growth and reproduction without overwhelming its host. Not to do so would be highly detrimental to both. This mutual tolerance involves host and parasite adapting to each other to some degree.

An example of coevolutionary adaptation and counteradaptation is the parasite-host interaction of the European rabbit (*Oryctolagus cuniculus*) and the viral infection myxomatosis. To control the introduced rabbit, the Australian government introduced the rabbit's viral parasite into the population. The first epidemic of myxomatosis was fatal to 97 to 99 percent of the rabbits. The second resulted in a mortality of 85 to 95 percent; the third, 40 to 60 percent. The effect on the rabbit population was less severe with each succeeding outbreak, suggesting that the two populations had adjusted to each other.

In this adjustment, certain strains of the virus, intermediate in virulence, tended to replace highly virulent strains. Too high a virulence killed off the host; too low a virulence allowed the rabbits to recover before the virus could be transmitted to another host. Also involved was a passive immunity to myxomatosis conferred upon the young born to mothers that were immune. Finally, the rabbit population evolved a genetic strain that was resistant to the disease.

The transmission of myxomatosis depends upon *Aedes* and *Anopheles* mosquitoes, which feed only on the blood of living animals. Rabbits infected with a less virulent strain live longer than those infected with a more virulent strain. Because the mosquitoes have access to it for a longer time, the less virulent strain has a competitive advantage.

17.8   Parasites may regulate host populations

The impact of parasites on host populations depends on the mode of transmission and the density and dis-person of the host population. Microparasites, dependent for the most part on direct transmission, require a high host density to persist. For them, ideal hosts live in groups or herds. These parasites need a long-lived infective stage that does not ensure long-term immunity in the host population. Immunity reduces parasite populations, if it does not eliminate them. An example of a parasite in wild populations that does not confer long-term immunity is rabies. One that does confer immunity to animals that survive the disease is distemper.

Indirect transmission, typical of macroparasites, is more complex. Parasites with indirect transmission require a highly effective transmission stage. They do well and persist for a long time in low population densities of hosts. The longevity of each parasitic stage varies in different hosts. Longevity is high in the definitive host and much less in the intermediate host.

To prove that a parasite regulates a host population, we must show that it increases the host's death rate or decreases its reproductive capability. Such effects are most evident when parasites invade a population with no evolved defenses (see Focus on Ecology 17.1: Plagues upon Us). In such cases the disease may be density-independent, reducing populations, exterminating them locally, or restricting the distribution of the host. For example, the fungus *Endothia parasitica* spread rapidly through the American chestnut *(Castanea dentata)*. Once a major commercial timber species prominent in the eastern deciduous forest, the American chestnut has all but vanished.

Such extreme cases are hardly regulatory. What about directly transmitted endemic diseases maintained in the population by a small reservoir of infected carrier individuals? Outbreaks of these diseases appear to occur when the density of the host population is high, and they tend to reduce host populations sharply. Examples are distemper in raccoons and rabies in foxes, both significant in controlling their populations.

The distribution of macroparasites, especially those with indirect transmission, is highly clumped (Figure 17.5). Some individuals in the host population carry a higher load of parasites than others. These individuals are the ones that are most likely to succumb to parasite-induced mortality, suffer reduced reproductive rates, or both. Such deaths often are caused not directly by parasites but indirectly by secondary infection. Herds of bighorn sheep *(Ovis canadensis)* in western North America may be infected with up to seven different species of lungworms (Nematoda). Highest infections occur in the spring when the lambs are born. Heavy lungworm infections in the young bring about a secondary infection, pneumonia, that kills the lambs. Such infections tend to stabilize or sharply reduce moun-

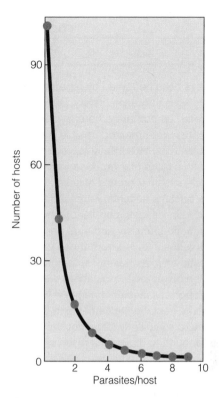

Figure 17.5 Clumped distribution of the tick *Ixodes trianguliceps* (Birula) on a population of the European field mouse, *Apodemus sylvaticus.* Most of the individuals in the host population carry no ticks. A few individuals carry most of the parasite load.

tain sheep populations by reducing reproductive success. Wildlife biologists successfully reduce the parasite load and increase lamb survival by treating pregnant ewes with an antiworm drug disguised in highly palatable fermented apple mash.

17.9   Social relations may be parasitic

Another form of parasitic relationship is social, in which one organism depends on the social structure of another. Social parasitism may be temporary or permanent, *facultative* (optional) or *obligatory* (necessary for species survival), within a species or between species.

There are four forms of social parasitism. First is a temporary facultative parasitism within a species. Intraspecific nest parasitism is an example. Parasitic females lay eggs in nests of host females of the same species. This practice is well developed among ants and wasps and among certain groups of birds, especially waterfowl (Anatidae). Among these waterfowl

PLAGUES UPON US

Humans have always been bedeviled by parasites, but more so in recent stages of human history. During our first two million years as hunter-gatherers, the most bothersome parasites were macroparasites such as roundworms, directly transmitted. Only microparasites with high transmission rates that produced no immunity could persist in such small groups of hosts.

Once humans became sedentary agriculturalists, however, aggregated in villages, populations became large enough to sustain bacterial and viral parasites. Many of these parasites evolved from those causing diseases in domestic animals. Measles, for instance, evolved from canine distemper. Populations were too small at first to support disease continuously, without reinfection from some neighboring settlement. Once a settlement grew into a sufficiently large city, the population was dense enough to maintain a reservoir of infection. As commerce developed between cities, people and goods began to move long distances. They introduced diseases from one part of the world to another where the populations lacked immunity. Periodic epidemics swept through the cities.

A classic example of the importation, direct transmission, and rapid spread of a disease is bubonic plague. Plague is caused by a rod-shaped bacillus, *Yersinia pestis (Pasteurella pestis)*. It is transmitted directly from host to host mostly by the bite of its vector (fleas from infected rodents) and from person to person by mucous droplets spread by coughing. Infected individuals become ill in a few hours to a few days, showing symptoms of high fever and swollen lymph nodes. Death often follows within several days. Plague was called the black death because of the dark color on the faces of many dead victims.

The reservoir of the bacillus is burrowing rodents. The bacillus is most closely associated with the black rat *(Rattus rattus),* a native of central India, also the original focal point of the disease. An agile climber, the black rat easily boarded cargo ships that carried them to port cities in Asia and the Mediterranean region. Hidden in the baggage of caravans, the black rat spread across the steppes of Asia, where undoubtedly it transferred the plague bacillus to burrowing rodents in the region. In 1331 an epidemic of the plague swept through China. Mongol armies carried the disease with them as they swept across Asia to the Mediterranean. At the siege of Caffa in 1346 on the Crimean peninsula, the Mongol army was devastated by the plague and withdrew after catapulting the victims' corpses into town. Trade resumed and ships carried infected black rats to ports of southern Europe.

Conditions were right for the spread of disease. Europe was experiencing huge population growth, climate was worsening, and crops were failing. By the end of December 1347 the disease that had decimated the Mongol army spread to Italy and southern France; by December 1348 it reached southern Germany and England; and by December 1350 it reached Scandinavia. Between 1348 and 1350, one-third of the European population, including entire villages, succumbed to the black death, upsetting the social, political, and economic stability of Europe. Later outbreaks occurred in 1630 in Milan, in 1665 in London, and in 1720–1721 in Marseilles. Sporadic local outbreaks occurred throughout the world until 1944, when antibiotics quickly cured the disease, if diagnosed early. The plague bacillus still thrives worldwide, including North America, harbored by burrowing rodents.

Other diseases introduced into populations lacking immunity mimic the spread and devastation of the black death. Smallpox, measles, and typhus, carried to the New World by Spanish and English explorers and settlers, spread rampant through indigenous populations of South, Central, and North America. Disease devastated the Aztecs in Peru and nearly exterminated the native American population of New England, allowing uncontested settlement of the region by the English. In more recent times a massive flu epidemic spread worldwide in 1918, killing 21 million people, including 500,000 people in the United States. Flu still remains a threat because of its high mutation rate. Strains evolve faster than resistance develops to the disease. As a result, flu returns in waves of different types.

Because many of the old plagues, like measles, have been contained by vaccinations and other health measures, many of us have become complacent about disease. Nevertheless new diseases—Ebola in Zaire and AIDS everywhere—warn us that plagues are still with us. New mutant forms of old diseases such as tuberculosis, once considered conquered, a massive increase in world population, a changing global climate, and rapid transcontinental movement of people and goods set the stage for future plagues.

are the black-bellied tree ducks (*Dendrocygna autumnalis*), goldeneye (*Bucephala clangula*), and wood ducks (*Aix sponsa*), hole-nesting species. The host female responds to parasitism by reducing the size of her own clutch by the number added. The earlier the parasitic female lays her eggs in her host's nest, the greater will be her share of the clutch. Such **brood parasitism** may have evolved among those waterfowl because suitable nest sites are scarce, nests are easy to locate, and ducks do not defend their nests.

A second type of social parasitism is temporary facultative parasitism between species. An example exists within the formicine ant genus *Lasius*. A newly mated queen of the species *L. regina* will enter the nest of a host species, *L. alienus*, and kill its queen. The *alienus* workers will care for the *regina* queen and her brood. In time the *alienus* workers, deprived of their own queen and thus replacements, die out, and the colony then consists of *regina* workers.

A third type of social parasitism is temporary obligatory parasitism between species. Although it is common in ants, the most conspicuous examples are in birds. Brood parasitism has been carried to the extreme by the cowbirds and old world cuckoos, both of which have lost the arts of building nests, incubating eggs, and caring for young. They pass off these duties by laying eggs in the nests of a host species. The brown-headed cowbird (*Molothrus ater*) of North America removes one egg from the nest of the intended host and lays one of her own as the replacement. Some host birds counter by ejecting the egg from the nest. Others hatch the egg and rear the young cowbird, usually to the detriment of their own offspring. The host's young may be pushed from the nest or starve because the young cowbird is more aggressive and larger.

A fourth type of social parasitism is permanent obligatory parasitism between species. The parasitic form spends its entire life cycle in the nest of the host. That type of social parasitism is common among ants and wasps. In most cases the species are workerless, and queens have lost the ability to build nests and care for young. The queen gains entrance to the nest of the host and either dominates the host queen or kills her outright and takes over the colony.

17.10 Parasitism can evolve into a positive relationship

Parasites and their hosts live together in one sort of symbiotic relationship. **Symbiosis** is a situation in which dissimilar organisms live together in close association. In the parasite-host relationship, the parasite draws its livelihood from the host. At some stage in the coevolution of the host and parasite, the relationship can become beneficial to both. For example, a host tolerant of parasitic infection may begin to exploit the relationship. In time, the two species may become totally dependent on each other. Such a relationship is called **mutualism.**

Mutualism is a positive reciprocal relationship at the individual or population level between two different species. Out of this relationship, both species enhance their survival, growth, or reproduction. Evidence suggests that the relationship is more a reciprocal exploitation than a cooperative effort between individuals.

Mutualism may be symbiotic or nonsymbiotic. In symbiotic mutualism individuals interact physically and their relationship is obligatory. At least one member of the pair becomes totally dependent on the other. At the extreme the two interacting organisms function as one individual, as algae and fungi do in lichens.

Among nonsymbiotic mutualists, obligate or facultative, the relationship may have begun with exploitation. In plant-pollinator relationships, birds and insects came to plants to feed on pollen. In the course of this exploitation, these animals happened to carry pollen to other plants of the same species. Such plants experienced improved reproductive success and began to exploit the visitors as a means of dispersing pollen. Selection favored the development of mechanisms to maintain the relationship, such as sugar-rich nectar.

17.11 Obligatory symbiotic mutualisms are permanent

Some forms of mutualism are so permanent and obligatory that the distinction between the two interacting populations becomes blurred. A good example is **mycorrhizae,** a mutualistic relationship between plant roots and fungi. The fungi assist the plant with the uptake of nutrients from the soil. In return the plant provides the fungi with carbon, a source of energy. So important is this mutualism to the growth of forest trees and the functioning of forest ecosystems that it is the object of expanding research.

Endomycorrhizae are common to many trees of temperate and tropical forests. Mycelia—masses of fine fungal filaments in the soil—infect the tree roots. They penetrate the cells of the host to form a finely bunched network called an arbuscle (Figure 17.6). The mycelia act as extended roots for the plant, but do not change the shape or structure of the roots. They draw in nitrogen and phosphorus at distances beyond those reached by the roots and root hairs. Another form,

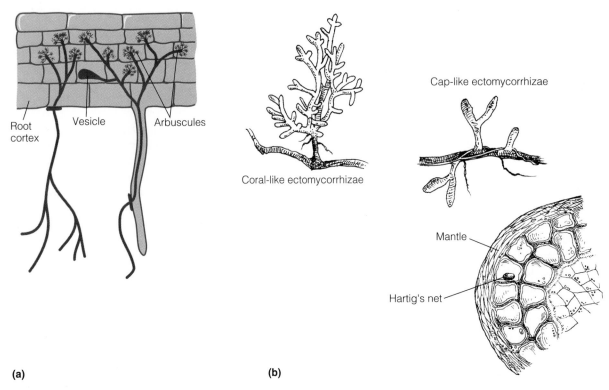

(a)   **(b)**

Figure 17.6 (a) Endomycorrhizae grow within tree roots, and fungal hyphae enter the cells. (b) Ectomycorrhizae form a mantle of fungi about the tips of rootlets. Hyphae invade the tissues of rootlets between the cells. The network is called Hartig's net.

ectomycorrhizae, produce shortened, thickened roots that look like coral (see Figure 9.4). The threads of the fungi work between the root cells. Outside the root they develop into a network that functions as extended roots.

Mycorrhizae are especially important in nutrient-poor soils (see Section 9.5). They aid in the decomposition of litter and the translocation of nutrients, especially nitrogen and phosphorus, from the soil into the root tissue. Mycorrhizae increase the ability of roots to absorb nutrients, providing selective ion accumulation and absorption. They produce growth regulators, mobilize nutrients in infertile soil, and make available certain nutrients bound up in silicate minerals. In addition, mycorrhizae reduce susceptibility of their hosts to disease by providing a physical barrier and stimulating the roots to produce chemical inhibitory substances. Without ectomycorrhizae, certain groups of trees, especially conifers, oaks, and birches, could not become established, survive, and grow. In fact, many mycorrhizae are specific to certain trees, such as pines and Douglas-fir.

The association between tree and fungus can be tenuous. Any alteration in the availability of light or nutrients for the host creates a deficiency of carbohydrates and thiamine for the fungus. Interruption of photosynthesis causes mycorrhizae to stop fruiting.

17.12 Mutualisms may be obligatory but nonsymbiotic

Many mutualistic relationships are obligatory without being symbiotic. An example involves mycorrhizae. Although some mycorrhizae have aboveground fruiting bodies, such as the familiar mushrooms that release their spores to the air, others have belowground fruiting bodies, called truffles. These mycorrhizae depend upon small mammals, especially voles, to disperse their spores. Small mammals are able to detect underground fruiting bodies by smell. They dig up the truffles, eat them, and defecate, spreading mycorrhizal spores across the forest floor. Mycorrhizal spores cannot germinate until they come into contact with tree roots. Thus, a three-way obligatory relationship exists (Figure 17.7). The tree depends upon mycorrhizae for nutrient uptake from the soil. The mycorrhizae depend upon the tree roots for energy and upon small mammals for dis-

17.13 Facultative mutualisms such as seed dispersal are diffuse

Most mutualisms are facultative, at least on one side. Such mutualisms are not confined to two species. A species may form a mutualistic relationship with any number of related species. Because the benefits of such mutualisms, usually associated with seed dispersal and pollination, spread over many plants, pollinators, and seed dispersers, these mutualisms are called **diffuse.**

Plants with heavy seeds depend upon animals such as jays, squirrels, and ants to carry the seeds some distance from the parent plant and deposit them in sites favorable for seedlings. Some seed-dispersing animals upon which the plant depends may be seed predators, eating the seeds for their own nutrition. Plants depending on such animals must produce a tremendous number of seeds over their reproductive lifetimes. They sacrifice most of them to ensure that a few will survive, come to rest on a suitable site, and germinate.

In the deserts of the southwestern United States, the sclerophyllous shrublands of Australia, and the deciduous forests of eastern North America, a number of herbaceous plants, including many violets (*Viola* spp.), depend upon ants to disperse their seeds. Such plants, called **myrmecochores,** have an ant-attracting food body on the seed coat called an **elaiosome.** Appearing as shiny tissue on the seed coat, the elaiosome contains certain chemical compounds essential for the ants. The ants carry seeds to their nests, where they sever the elaiosome and eat it or feed it to their larvae. The ants discard the intact seed within abandoned galleries of the nest. Ant nests, whose substrate is richer in nitrogen and phosphorus than the surrounding soil, provide a good substrate for seedlings. Further, by removing seeds far from the parent plant, the ants significantly reduce losses to seed-eating rodents.

Plants have an alternative approach for seed dispersal. They may enclose the seed in a nutritious fruit attractive to fruit-eating animals, the **frugivores.** Frugivores are not seed predators. They eat only the tissue surrounding the seed and, with some exceptions, do not impair the vitality of the seed. Most frugivores do not depend exclusively on fruits, which are deficient in proteins and are only seasonally available.

To use frugivorous animals as agents of dispersal, plants must attract them at the right time. Plants discourage the consumption of unripe fruit by cryptic coloration, such as green unripened fruit among green leaves, and by unpalatable texture, repellent substances, and hard outer coats. When seeds mature, plants attract fruit-eating animals by presenting attractive odors,

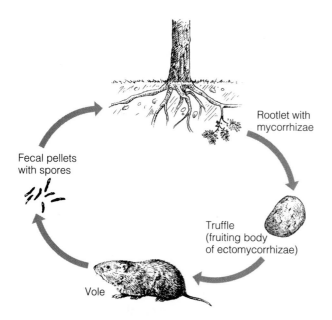

Figure 17.7 A nonsymbiotic mutualism. Voles eat truffles, the belowground fruiting bodies of some mycorrhizae, whose spores become concentrated in fecal pellets. The voles deposit some spores in nonforested areas. Without such dispersal forest trees cannot invade open areas.

persal of spores. Small mammals obtain a significant portion of their food from the fruiting bodies.

The relationship between the mycorrhizal spores and small mammals is nonsymbiotic obligatory mutualism. The mutualists lead separate lives yet depend on one another, a common relationship. The approximately 900 species of tropical figs (*Ficus*) have a complex obligatory relationship with pollinating aganoid fig wasps. The wasps lay their eggs in the developing seeds upon which the larvae feed. Figs have 44 to 77 percent seed mortality, a high cost for pollination.

Other such obligatory relationships involve shelter, protection against predators, and reproduction. Some of the most interesting exist between ants and plants. Attine ants culture a slow-growing fungus that cannot survive without them. A species of Central American ant lives in the swollen thorns of acacia (*Acacia* spp.), from which they derive shelter and a balanced and almost complete diet for all stages of development. In turn the ants protect the plants from herbivores. At the least disturbance the ants swarm out of their shelters, emitting repulsive odors and attacking the intruder until it is driven away. Neither the ants nor the acacias can survive in the absence of the other.

softening the texture, increasing sugar and oils, and "flagging" their fruits with colors.

Most plants have fruits that can be exploited by a large number of dispersal agents. Such plants opt for quantity dispersal, the scattering of a large number of seeds with the chance that a diversity of consumers will drop some seeds in a favorable site. Such a strategy is typical of, but not exclusive to, plants of the temperate regions. There fruit-eating birds and mammals rarely specialize in one kind of fruit and do not depend exclusively on fruit for sustenance. The fruits are usually succulent and rich in sugars and organic acids. They contain small seeds that pass through the digestive tract unharmed. To accomplish this passage, such plants have evolved seeds with hard coats resistant to digestive enzymes. Such seeds may not germinate unless they have been conditioned or scarified by passage through the digestive tract. Large numbers of small seeds may be so dispersed, but few are deposited on suitable sites. The length of time of such seeds remain within the digestive tracts of some small birds is no more than 30 minutes. The distance dispersed depends on how far the birds go after eating. Such dispersal is a lottery in the truest sense.

In temperate regions fruits ripen in early and late summer, when the young of the year are no longer dependent on highly proteinaceous food. Both adults and young turn their attention to a growing abundance of fruit. Such timing of ripening makes fruits available in early summer for resident animals, especially birds, and in late summer and fall for migrant birds.

In tropical forests 50 to 75 percent of the tree species produce fleshy fruits whose seeds are dispersed by animals. Rarely are these frugivores obligates of the fruits on which they feed. Exceptions include the oilbirds and a large number of tropical fruit-eating bats. The other species eat many different fruits. Dispersers of the seeds of one plant are also dispersers for others, for several reasons. Fruits vary widely in their nutritional value. By eating a variety of them, frugivores tend to balance their diets. Also, plants have few means available to restrict consumption of their fruits to a specific frugivore. However, because fruits have various sizes, shapes, colors, aromas, nutrients, and palatability, some appeal mostly to mammals and others to birds.

17.14 Mutualisms are often necessary for pollination

The goal of seed dispersal is to move seeds away from the parent plant to some site favorable for seedlings. The goal of pollination is much more specific and di-rect. The plant must transfer its pollen from the anthers of one plant to the stigma of a conspecific.

Some plants simply disperse their pollen in the wind. This method works well and costs little when plants grow in large homogeneous stands, as grasses and pine trees do. Wind dispersal is unreliable when individuals of the same species are scattered individually or in patches across a field or forest. Then pollen transfer depends upon insects, with some assistance from nectivorous birds and bats.

Like frugivores, nectivorous animals visit plants to exploit a source of food. While feeding, the nectivores inadvertently pick up pollen and carry it to the next plant they visit. With few exceptions, the nectivores are generalists. Like the frugivores, they find little advantage in specializing. Further, because each species flowers briefly, nectivores depend on a progression of flowering plants through the season. Nectivores cannot afford to commit themselves to a single species of flower, but they do concentrate on one species while its flowers are available.

Plants are the ones that have to specialize. They have to entice certain animals by color, fragrances, and odors, dusting them with pollen, and then rewarding them with a rich source of food: sugar-rich nectar, protein-rich pollen, and fat-rich oils. Providing such rewards is expensive for plants. Nectar and oils are of no value to the plant except as an attractant for potential pollinators. They represent energy that the plant otherwise might expend in growth.

Many species of plants, such as blackberries, elderberries, cherries, and goldenrods, are generalists themselves. They flower profusely and provide a glut of nectar that attracts a diversity of pollen-carrying insects, from bees and flies to beetles. Other plants are more selective, screening their visitors to ensure some efficiency in pollen transfer. These plants may have long corollas, allowing access only to insects and hummingbirds with long tongues and bills and keeping out small insects that eat nectar but do not carry pollen. Some plants have closed petals that only large bees can pry open. Orchids, whose individuals are scattered widely through their habitats, have evolved a variety of precise mechanisms for pollen transfer and reception so that pollen is not lost when the insect visits flowers of other species.

In addition to nectar, some plants provide oil as a floral reward. Families including Iridaceae, Orchidaceae, Scrophulariaceae, Concurbitaceae, Solanaceae, and Primulaceae, mostly in neotropical savannas and forest, and have specialized oil-secreting organs on the petals. The oil flowers are visited by highly specialized bees in four families that use the energy-rich floral oils in place of or along with pollen as provisions for developing larvae. These bees possess modified structures designed for mopping up, storing, and transporting oil to the nest.

17.15 Some mutualisms are defensive

A major problem for many livestock producers is the toxic effects of certain grasses, particularly perennial ryegrass and tall fescue. These grasses are infected by fungi that live inside plant tissues. The fungi (Clavicipitaceae, Ascomycetes) produce physiologically active alkaloids in the tissue of the host grasses. The alkaloids, which impart a bitter taste to the grass, are toxic to grazing mammals, particularly domestic animals, and a number of insect herbivores. In mammals the alkaloids constrict small blood vessels in the brain, causing convulsions, tremors, stupor, gangrene of the extremities, and death. At the same time these fungi seem to stimulate plant growth and seed production. This symbiotic relationship suggests a defensive mutualism between plant and fungi. The fungi defend the host plant against grazing.

There are costs to the plant. The fungal infection causes sterility in the host plant by inhibiting flowering or aborting seeds. Some plants have a few counter-adaptations that restore fertility. In most plants, however, the greater vegetative growth of infected plants and enhanced growth in the absence of herbivores balances the loss of sexual reproduction.

17.16 Population effects of mutualism may be complex

Mutualism is easy to appreciate at the individual level. We grasp the interaction between an ectomycorrhizal fungus and its oak or pine host; we count the acorns dispersed by squirrels and jays, and we measure the cost of dispersal to oaks in terms of seeds consumed. Mutualism improves the growth and reproduction of the fungus, the oak, and the seed predators. But what are the consequences at the level of the population?

Defining the population consequences of mutualism is considerably more difficult than defining those of predation and parasitism. The relationship is more difficult to model. Mutualism exists at the population level only if the growth rate of species A increases with the increasing density of species B, and vice versa.

For obligate symbiotic mutualists, the relationship is straightforward. Remove species A and the population of species B no longer exists. If ectoymycorrhizal spores fail to infect the rootlets of young pines, they will not develop. If the young pine invading a nutrient-poor field fails to acquire a mycorrhizal symbiont, it will not grow well. (Foresters and nursery owners innoculate nursery-grown pine seedlings with the appropriate mycorrhizae.)

For nonsymbionts, obligate or facultative, the effect on populations may be limited to that part of the life history cycle in which the mutualistic relationship occurs. A demographic analysis of an ant-seed mutualism by F. M. Hanzawa, A. J. Beattie, and D. C. Culver provides an example of the population consequences of mutualism. It involves a guild of ants and golden corydalis (*Corydalis aurea*), an annual or biennial widely distributed in open or disturbed sites in the northeastern and western continental United States, Canada, and Alaska. The three compared the survivorship of both seeds and plants, fecundity, reproduction, and growth rates of two seed cohorts of the plant. They relocated one cohort to ant nests occupied by undisturbed ant foragers. They hand-planted the other, the control cohort of equal numbers, near each nest. The ant-handled cohort had significantly higher survivorship than the control. The ant-handled cohort produced 90 percent more seeds than the control cohort, and its net reproductive rate R_0 was 8.0, whereas that of the control was 4.2. The finite rate of increase of the ant-handled cohort was 2.83 per year, compared to 2.05 per year for the control. The ant-handled cohort experienced greater reproductive success not because of any great difference in the fecundity of the plants but because of a significantly higher survival to reproductive age. This came about largely because of dispersal and the superior microsites of the ant nests.

This study makes the point that mutualistic relationships have complex effects on populations. They are essential to the integrity of ecosystems; yet most are unknown and unappreciated. This fact serves as a warning for us to go slow in determining what organisms we will tolerate and what organisms we will not. We do not know what we are destroying. Foresters consider the southern fox squirrel (*Sciurus niger*) in pine stands in the southern United States to be a seed-eating pest. Recently, however, studies have pointed out that the fox squirrel feeds heavily on truffles of the mycorrhizal fungus *Elaphomyces granulatus*, associated with longleaf pine (*Pinus palustris*), and serves as its dispersal agent. Therefore the squirrel, far from being a pest of longleaf pine, is essential for the pine's well-being. Most people consider ants and other insects to be pests. Given the choice, they would eradicate them, not appreciating their importance in the natural world.

SUMMARY

Coevolution Many species have some evolutionary effect on each other. Prey evolve defenses, and predators improve their hunting efficiency. A closer evolutionary relationship exists between parasites and their hosts, and between mutualists. We call their reciprocal evolutionary responses coevolution.

Characteristics of Parasites (17.1) Parasitism is a situation in which two organisms live together, one deriving its nourishment at the expense of the other. Parasitic infection can result in disease. Microparasites include the viruses, bacteria, and protozoa. They are small in size, have a short generation time, multiply rapidly in the host, tend to produce immunity, and spread by direct transmission. They are usually associated with dense populations of the host. Macroparasites, relatively large in size, include parasitic worms, lice, ticks, fleas, rusts, smuts, fungi, and other forms. They have a comparatively long generation time and rarely multiply directly in the host, persist with continual reinfection, and spread by both direct and indirect transmission.

Parasite-host Relationships (17.2–17.5) Parasites exploit every conceivable habitat in host organisms. Many are specialized to live at specific sites, such as the roots of a plant or the liver of an animal. The problem of parasites is to gain entrance into and escape from the host. The life cycles of parasites revolve about these two problems. The adult stages live in the definitive host. To complete their life cycle they need to escape and transfer to another host (17.2). Many species of parasites locate new hosts by direct contact with other hosts or by means of other organisms, called vectors. These carriers become intermediate hosts of some developmental or infective stage of the parasite (17.3). Other species of parasites require more than one type of host. Indirect transmission takes them from definitive to intermediate to definitive host. Indirect transmission often depends on the feeding habits of the host (17.4). Transmission of parasites, direct or indirect, is complicated by patchy or clumped distribution of the host. A few individuals in the population carry the greatest load of parasites; most remain free of infection (17.5).

Host Response to Parasitism (17.6–17.8) Hosts respond to parasitic infections by activating their immune systems, via inflammatory responses at the site of infection, and via behavioral changes. Abnormal growths

such as cysts in animals and galls in plants wall off parasites. Infection of plants by endophytic fungi can prevent sexual reproduction but stimulate vegetative reproduction (17.6). The death of a host does not benefit a parasite. Natural selection favors less virulent forms of the parasite that can live in the host without killing it. Hosts and parasites develop mutual tolerance with a low-grade, widespread infection (17.7). A heavy parasitic load can increase mortality and decrease fecundity of the host population. Under certain conditions parasitism can regulate a host population (17.8).

Social Parasitism (17.9) Another type of parasitism is social parasitism. It may be temporary, permanent, facultative, or obligatory. A common example is brood parasitism among some species of birds, ants, and wasps.

Types of Mutualism (17.10–17.15) Mutualism is a positive reciprocal relationship between two species that may have evolved from predator-prey, host-parasite, and commensal relationships. Mutualism may be symbiotic (living closely together) or nonsymbiotic, and obligatory or facultative (17.10). Obligate symbiotic mutualists are physically dependent on each other, one usually living within the tissues of the other. An example is mycorrhizal fungi. The fungi live symbiotically in the roots of plants. The fungi draw nutrients from the soil to the plants and protect plants from pathogens. The mycorrhizae gain support and nutrients from the host (17.11). Obligate nonsymbiotic mutualists depend upon each other, but they lead independent lives. Such relationships involve shelter, protection against predators, and pollination (17.12). Facultative mutualists include guilds of species involved in seed dispersal and pollination (17.13). In exchange for dispersal of pollen and seeds, plants reward animals with food—fruit, nectar, and oil. To reduce wastage of pollen some plants possess morphological structures that permit only certain animals to reach the nectar (17.14). Some mutualisms are defensive. Fungi growing within grass defend it against grazing and stimulate growth, at the cost of sterility in the host plant (17.15).

Population Consequences of Mutualism (17.16) Population consequences of mutualism are difficult to define. Benefits may be limited to one phase of the mutualist's life cycle. The natural world involves innumerable mutualistic relationships, few of which we understand. Disrupting them can have unforeseen adverse effects.

STUDY QUESTIONS

1. Define parasitism. Why is it more reasonable to classify parasites by size than by taxonomic groups?

2. Distinguish between a definitive host and intermediate hosts. Look up examples beyond those mentioned in the text.

3. Examine transmission of some common parasites such as tapeworms, roundworms, and the organisms responsible for malaria, Rocky Mountain spotted fever, and giardiasis. How might some of these parasitic diseases be controlled?

4. How might a patchy or clumped distribution of hosts affect the spread of parasites?

5. What is social parasitism? Speculate why brood parasitism evolved and why some individuals engage in intraspecific brood parasitism.

6. What is mutualism? Look up some examples of mutualism and examine them critically. Are they in fact mutualistic?

7. Distinguish among symbiosis, obligate symbiotic mutualism, and obligate nonsymbiotic mutualism.

8. Is fruit predation a chancy way of distributing seeds? Why?

*9. Is mutualism reciprocal exploitation, or two species acting together for mutual benefit?

*10. Explore examples of how parasitic infections have altered populations, ecosystems, even human history. (Hint: Consider the chestnut blight's effect on North American deciduous forests; the night mosquito, bird pox, and the fate of the Hawaiian honeycreepers; rats, fleas, and black death in Europe; rinderpest in African wild and domestic ungulates. There are many others.)

CHAPTER 18

HUMAN INTERACTIONS WITH NATURAL POPULATIONS

OBJECTIVES

On completion of this chapter, you should be able to:

- Discuss the concept and application of sustained yield management of exploited natural populations.
- List some of the ecological shortcomings of the sustained yield concept.
- Discuss why economics gets in the way of sustained yield management.
- Explain the relationship between human population growth, habitat destruction, and declining wildlife populations.
- Describe the efforts needed to restore and maintain endangered species.
- Explain the nature of pests and weeds.
- Discuss various methods and ecological aspects of pest and weed control.
- Describe integrated pest management.

Netting salmon off the Alaskan coast.

Humans have always had close relationships, positive and negative, with plants and animals. From the first we depended on them for food, clothing, shelter, and tools. In turn we were prey to large carnivores and hosts for an array of parasites. Early technology enabled hunters to exploit animals more successfully, driving some of them, such as the Pleistocene mammals, to extinction. Later we domesticated certain plants and animals, allowing us to form larger, more interdependent social units. Some animal species became competitors of or threats to domestic animals and plants. Some species of plants invaded the new habitats provided by agricultural fields and competed with crop plants.

Growing human populations and cultural changes demanded greater exploitation of resources. Forests were leveled to provide building materials for cities, ships, and armies. Large mammals were killed for food, and they decreased in ever widening circles about centers of human populations. Other animals, unable to survive in altered or diminished habitats, disappeared.

Out of this mixed dependency and enmity grew a cultural relationship to plants and animals. Mammals such as the cave bear, the wolf, and the cat and plants such as the oak and the lily became religious symbols and objects of worship. Even amid high technology this symbolic relationship persists in our use of animals as national and state symbols and as names and mascots of athletic teams. Plants and animals have influenced the development of drama, art, dance, and medicine.

In spite of this complex relationship, only recently has a segment of human society realized that the natural populations with which we share Earth are declining rapidly from overexploitation, habitat loss, and poisoned environments. For economic, aesthetic, or moral reasons, and for the sake of long-term human survival, these people have begun managing and conserving wild populations. They take three approaches: maintaining populations of exploited plants and animals, increasing populations of other species in danger of extinction, and reducing populations of species seen as detrimental to human interests.

18.1 Managing exploited populations began with fisheries

For thousands of years humans have been exploiting plant and animal populations without concern. Some species' populations have been able to absorb the effects of exploitation; others have been exterminated or driven to the point of extinction.

Not until the late 1800s did we make any effort to manage exploited populations to ensure their continuance. At that time wide fluctuations in the catch of fishes in the North Sea caused parallel fluctuations in commercial income. Debates raged over human impact on fisheries. Some people argued that the removal of fish had no effect on reproduction; others argued that it did. Not until a Danish fishery biologist, C. D. J. Petersen, developed a tagging and mark-recapture technique of estimating population size were biologists able to make some assessment of fish stocks. This technique was augmented by aging fish and by making egg surveys. All these studies suggested that overfishing indeed was the culprit.

The real answer surfaced after World War I. During the war, fishing in the North Sea had stopped. After the war, fishermen experienced sizable increases in their catches. Fishery biologists suggested that after fishing removed accumulated stock, population sizes and catches would stabilize. The size of the populations would be determined by the amount and size of fishes caught. In 1931 E. S. Russell of Great Britain presented this model of fishery yield:

$$S_2 = S_1 + (A + G) - (M + C)$$

where S_2 is preharvest population (time 2), S_1 is postharvest population (time 1), A is the weight of fish large enough to be caught, G is the increase in growth of these fish plus other fish already in the harvested population, M is mortality and decrease in weight, and C is fish caught. If $(A + G)$ was kept equal to $(M + C)$, then the population would remain stable.

The Norwegian fishery biologist Johannes Hjort added a new twist. He stated that the largest sustained yield could be obtained at the precise point where overfishing began. Fishing should cease at that point. This theory was not rational, argued the industry. "Rational fishing" was based on profit—an argument still with us today.

18.2 Continued exploitation depends on sustained yield

Harvesting at a level that will ensure a similar yield over and over without forcing the population into decline is called **sustained yield.** The crop removed per unit time is equal to production per unit time.

In a stable environment largely undisturbed, species populations tend to be dominated by large old individuals. When humans exploit such a population, those individuals are the first to go. To compensate, the population exhibits an increased growth rate, reduced age at sexual maturity, increased reproductive effort, and reduced mortality of small members of the population. As the harvest of the species declines, the exploiters are forced to increase the intensity of harvesting. If they overexploit the stock, the age classes become too young to carry on reproduction, and the population collapses. Unexploited, highly competitive, and closely sympatric or introduced species take over the exploited species' niche. The objective of sustained yield is to avoid such a collapse.

The rate of exploitation for sustained yield clearly depends on the rate of increase, r. Sustained yield does not imply holding a population at ecological carrying capacity (K), for at that level $dN/dt = 0$. A population stable in the absence of harvest can be managed under sustained yield only by manipulating the population to increase r. One way is to change the carrying capacity by increasing available resources; that method increases fecundity and survival. The usual way is to lower the density by removing a certain number and then to stabilize the population at some level below carrying capacity. Within limits, the lower the density of the population below the carrying capacity, the higher will be its rate of increase. The rate of harvest should equal the rate of increase to hold the reproductive population stable at lower density.

Sustained yield is not a particular value for a given population. There may be a number of sustained yield values corresponding to different population levels and different management techniques. The level of sustained yield above which the population declines is known as maximum sustained yield (MSY) (Figure 18.1). Maximum sustained yield implies the removal of all production above that needed to replace the amount harvested. Harvest takes the population down to a level at which the remaining stock can replace the amount removed before the next harvest period. In practice, maximum sustained yield is quite variable. If fluctuations in the environment and breeding stock are not taken into account, the margin between a level that stabilizes the population and one that drives it to extinction is dangerously thin.

An alternative to MSY is optimal sustained yield (OSY). It is more complex because it takes into account both biological and sociological factors. Much more conservative than MSY, it has built-in safety factors, taking yields less than MSY and not insisting on any particular proportion of the population to harvest. However, political and social pressure can raise the "optimum."

The higher the rate of increase of a population, the higher will be the rate of harvest that produces the max-

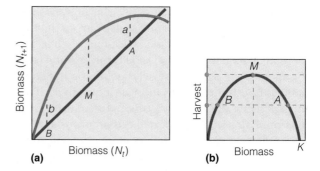

Figure 18.1 (a) A sustained yield model for K-strategists harvested under three regimes. The 45° line represents the replacement level of the population. Where it intercepts the curve, reproduction balances losses. In the first regime the population is harvested down from a steady state to size $N_t = A$. The dashed line a represents the number that could be harvested each year to hold the population stable at A. In the second regime the population is reduced to $N_t = B$. A number represented by dashed line b could be harvested each year to hold the population stable at B. The yield in this case is a large proportion of a small population. Maximum sustained yield is at M, where the diagonal line and the curve have maximum separation. (b) A parabolic recruitment curve illustrates the concept in a different fashion. Maximum sustained yield is approximately ½K, represented at M. At A the equilibrium is stable at high density, much above MSY. B is an unstable equilibrium point, much below MSY, vulnerable to chance extinction.

imum amount of harvestable biomass. Species characterized by scramble competition (r-strategists) lose much of their production to a high density-independent mortality, influenced by such variables as temperature and availability of nutrients. The management objective is to reduce wastage by taking all individuals that otherwise would be lost to natural mortality.

Such a population is difficult to manage. The stock can be depleted unless there is repeated reproduction. An example is the Pacific sardine (*Sardinops sagax*), a species for which there is little relationship between size of breeding stock and the number of progeny. Exploitation of the Pacific sardine population in the 1940s and 1950s shifted the age structure of the population to younger age classes. Before exploitation, 77 percent of the reproduction was distributed among the first five years. In the exploited population, 77 percent was associated with the first two years of life. The population approached the condition of single-stage reproduction under pronounced oscillations (Figure 18.2). Two consecutive years of environmentally induced reproductive failure caused a collapse of the population from which the species never recovered.

In populations characterized by density-dependent regulation (K-strategists), the maximum rate of har-

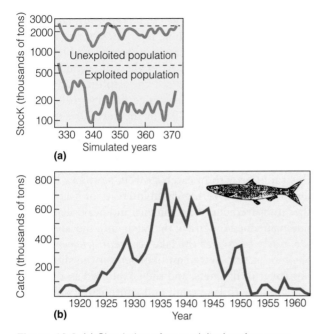

Figure 18.2 (a) Simulation of an exploited and an unexploited population of sardines, both subject to random environmental variation in reproductive success. The dashed lines indicate population size at *K*. Note how exploitation adds instability and how dangerously low the population can get. Compare this simulation with (b), the annual catch of the Pacific sardine along the Pacific coast of North America. Overfishing, environmental changes, and an increase in a competing fish, the anchovy, made the population collapse.

vest depends on age structure, frequency of harvest, numbers left behind after harvest, and fluctuations both in fecundity and in the environment. It also depends on the density of the population to be harvested and the rate of harvest needed to stabilize the density at that level.

Sustained yield is applied also to animals hunted for sport. Contest competition and density-dependent regulation characterize most game animals. Wildlife managers assume that animals taken by regulated hunting replace natural mortality. If the surplus were not harvested, they contend, the animals would succumb to disease, exposure, predators, and the like; therefore each individual removed by hunting reduces mortality by one. Supposedly, population stability and sustained yield can be maintained by harvesting some percentage of the number of young produced per year or of year-old animals in the population. This reasoning is sometimes faulty. Animals taken add to natural mortality in some species, such as certain waterfowl.

Several rules are applicable to exploitation. To obtain a croppable surplus, we first must reduce the population below a steady density. For each density to which

the population is reduced, there is an appropriate sustained yield. There are two levels of density from which a given sustained yield can be obtained, but maximum sustained yield can be harvested at one density only (see Figure 18.1). If the number taken exceeds the maximum sustained yield, the population declines to extinction. If a constant percentage of the standing crop of the year is removed, the population will decline and stabilize at a level of equilibrium with the rate of harvesting. This level may be above or below that generating maximum sustained yield.

Although sustained yield management, particularly maximum sustained yield, looks good on paper, it depends too heavily on the simplistic logistic equation. Maximum sustained yield (MSY) views management as a numbers game. Theoretically, an equal number of new recruits replaces the number or biomass removed. No proof of this theory exists. The usual approach to MSY fails to consider adequately: size and age classes, differential rates of growth among them, sex ratio, survival, reproduction, and environmental uncertainties, all data difficult to obtain. To add to the problem, the population may be common property, managed in several ways by different authorities. To attempt MSY without full information is to balance the population on the edge of catastrophe.

Several approaches to management of exploitable populations are in current use. One is the **fixed quota,** in which we remove a certain percentage, based on MSY estimates, each harvest period. Harvesting should match recruitment. Often used in fisheries, such an approach is risky. A fixed quota can easily drive a population to commercial if not actual extinction, if annual population fluctuations are not taken into account. Overharvest in spite of environmental changes has been responsible for the demise of the Pacific sardine, the Peruvian anchovy and the Atlantic halibut, flounder, and cod.

A second approach is **harvest effort,** often used in establishing seasons for sport hunting and fishing. We manipulate the number of animals killed by controlling the number of hunters in the field, the length of the hunting season, and the bag limit. Reducing the bag limit, shortening the season, or closing the season entirely reduces the kill. The reverse action increases the kill. This approach has been more successful than the fixed quota.

A third approach is the **dynamic pool model.** It assumes that natural mortality is concentrated in early life and that fishing mortality is not additive to natural mortality. In practice, the dynamic pool model controls fishing mortality by regulating equipment, such as size-selective gill nets that sort out age (size) classes. Few dynamic pool models have been developed, let alone put into practice. A general weakness of this model is the inability to estimate natural mortality accurately.

All three models have a major flaw. They fail to incorporate the most important component of population exploitation: economics (Figure 18.3). Once commercial exploitation begins, the pressure is on to increase it. Attempts to reduce the rate of exploitation meet strong opposition. People argue that reduction will mean unemployment and bankruptcy—that, in fact, the harvest effort should increase. This argument is short-sighted. An overused resource will fail, and the livelihoods it supports will collapse. That fact is written across the country in abandoned fish processing plants, rusting fishing fleets, and deserted logging towns. With conservative exploitation on lower economic and biological scales, the resource could be made to last.

18.3   Overexploited populations show danger signs

Overexploited populations exhibit symptoms of impending disaster. The catch per unit effort and the catch of related species decrease. There is a decreasing proportion of pregnant females, due both to a sparse population and to a high proportion of nonreproducing young. The species fails to increase its numbers rapidly after harvest. A change in productivity relative to age and age-specific survival shows that the ability of the population to replace harvested individuals has been impaired.

The history of the fishery in Lake Erie illustrates this point (Figure 18.4). Before the War of 1812, the lake, whose shores were lightly settled, held an abun-

dance of whitefish (*Coregonus clupeaformis*), lake trout (*Salvelinus namaycush*), blue pike (*Stizostedion vitreum glaucum*), sauger (*Stizostedion canadense*), and lake herring (*Coregonus artedii*). After the war settlement increased rapidly, and so did the exploitation of the fishery resource. A subsistence fishery grew into a thriving industry by 1820. For the next 70 years rapid improvement in transportation, fishing boats, gear, and techniques and an expanding market increased the average rate of the catch by 20 percent a year. By 1890 this rapid growth in catch slowed, as stocks became depleted. However, increased intensity of fishing, further improvement of equipment, and heavy capital input maintained the size of the catch until the late 1950s.

Early to go was the lake sturgeon (*Acipenser fulvescens*), at first netted and burned on the shore as a destroyer of fish nets, and then exploited as a market fish along with lake trout and whitefish. Fishing intensity grew with incentives to increase production during World War I and the introduction of a gill net that was extremely wasteful in its catch. In 1950 the development of the nylon gill net, which could be left in the water much longer, made fishing more intense than ever. Now walleye (*Stizostedion vitreum vitreum*), blue pike, and yellow perch (*Perca flavescens*) took the brunt of it. By 1960 the stocks of walleye and blue pike had been commercially depleted.

Adding to the stress was pollution in the lake. Industrial, agricultural, and urban wastes, toxic pollutants and biocides, and runoff from shore developments wiped out phytoplankton that supported fish life and stimulated the growth of cyanobacteria. Fallout of dead

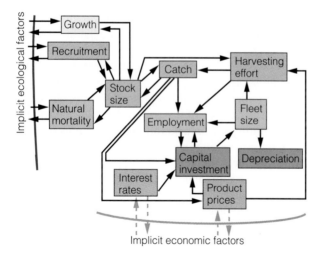

Figure 18.3 The relationship between sustained yield models and economic models of harvesting. Sustained yield models ignore powerful economic factors.

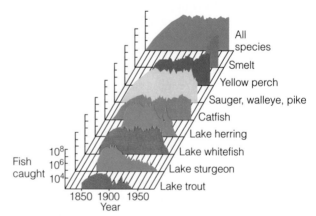

Figure 18.4 Annual catches of selected species of fish, and all species combined, by the Lake Erie commercial fishery since 1820. The vertical scale is logarithmic. (Reprinted with permission from H. A. Regier and W. L. Hartman, "Lake Erie's Fish Community: 150 Years of Cultural Stress," *Science* 180 (1973): 1248–1255. © 1973 American Association for the Advancement of Science.)

organic matter increased the oxygen demand of decomposers in the bottom sediments. Oxygen-demanding benthic life, such as certain mayflies on which fish fed, gave way to species with low oxygen tolerance, such as bloodworms and chironomids.

An invasion of rainbow smelt dealt the final blow. Young smelt feed on plankton and crustaceans, the older smelt on small fish. In turn, the smelt is valuable prey for adult lake trout, blue pike, walleyes, and sauger. With the decimation of those species, smelt increased rapidly and became the predators of young remnant stocks. Had the invasion of rainbow smelt (*Osmerus mordax*) been halted, and had fishing been regulated, the preferred commercial species would still be in harvestable numbers.

The history of this Great Lakes fishery is a microcosm of the history of marine fisheries. Certain stocks of fish, such as the Pacific sardine (*Sardinops sagax*), Atlantic halibut (*Hippoglossus hippoglossus*), and Peruvian anchovy, have been exploited to commercial extinction, causing ecological and economic damage. The sad plight of the whales is another example of decreased catches followed by increased hunting intensity, made possible by greater capital input and technological advances such as factory ships and fleets of hunters. Despite warnings of overharvesting, the marketplace and short-term profits dictated the take.

18.4 Economics interferes with sustained yield management

Although some international regulations, including a quota system for commercial species of fish, are in effect today, compliance is strictly voluntary. The system does not work. The problem is that traditional fisheries management considers stocks of individual species as single biological units rather than components of a larger ecosystem. Each stock is managed to bring in a maximum economic return, overlooking the need to leave behind a certain portion to continue its ecological role of predator or prey. This attitude encourages a tremendous discard problem, euphemistically called bypass.

Bypass includes all species of fish and other marine life pulled up in nets that have no economic value to the fishermen concerned. This bypass can include commercially valuable species. The discard includes more than nontarget fish. Taken in the nets in United States waters alone each year are nearly 1 million seabirds, more than 6000 sea lions and seals, over 10,000 small whales, and thousands of endangered sea turtles. Overall the world's marine fisheries annually catch and throw overboard 12 to 20 billion tons of sea life, about 10 to 20 percent of the total catch of fish and shellfish. Few of the discards survive. Shrimp fisheries remove billions of tons of live fish—at least 115 species—along with turtles. More than 21 pounds of fish are landed for every pound of shrimp. Commercially recognized groundfish species such as croaker, red snapper, and grouper make up 70 percent of the discard from shrimp fishing.

The ecological effects of such discard can be enormous. Because much of the bypass consists of juvenile or undersized fish of commercial species, the practice seriously affects the future of those fisheries. The biology and ecology of many species of fish are poorly understood. Their removal or reduction can interfere with predator-prey interactions and the dynamics of interspecific competition, modifying ecosystem structure and function. Tons of carcasses fall to the bottom to enrich the benthic detritus-based food chains, enhancing nutrient release and energy turnover. Such enrichment could change benthic community structure and function.

Economics also enters the management of some game species. In too many instances biologists have emphasized the increase of recreational opportunities for hunters rather than the welfare of the species involved. This emphasis is evident in the failure to reduce waterfowl bag limits when population levels seem to require it, in late-season hunting of grouse well after breeding males have established their spring territories, and in reluctance to restrict hunting of moose in Alaska. Hunting mortality is additive to predatory losses. Rather than restrict human exploitation of moose populations, we kill wolves to reduce natural predatory loss.

Reluctance to reduce seasons or to more tightly restrict bag limits relates in part to conservation funding. Most wildlife programs directly or indirectly depend upon hunting license revenues. Reducing such revenue lowers income to wildlife agencies. This loss has a two-edged effect. It may reduce pressure on some hunted species, but it also reduces money to support wildlife habitat restoration and acquisition (which benefits all species, including endangered ones) and other programs essential to wildlife welfare. Wildlife conservation should also be funded by other means.

18.5 "Sustained yield" in forestry may not be sustained yield

Forests have not fared much better than wildlife. They were destroyed rapidly over the centuries to clear land for agriculture and to supply building materials and firewood. Growing shortages of wood first stimulated some form of forest management in Great Britain in the 1600s. In Europe, forest management began in France, Germany, and Switzerland in the 1800s. Gifford Pinchot introduced the practices of forest management in 1892 on the Biltmore Estate in North Carolina. East of the Mississippi River, the seemingly endless forests, however, did not encourage management until exploitative logging, land clearing, and fires had decimated the forests. The same fate was happening to western forests until the federal government set much of the acreage aside, some in parks and most in national forests. In spite of federal ownership, western national forests still are being heavily clearcut.

The goal of sustained yield in forestry is to achieve a balance between net growth and harvest. To this end, the mature or unexploited forest must be cut in some fashion to stimulate regeneration, much as an unexploited fish or wildlife population must be reduced to increase r. This argument is used to justify the clearcutting of remaining old-growth forests. Management proceeds in a different time frame, because regenerating a forest requires decades. The time frame for the next harvest depends on the desired wood products. Pulp wood, posts, and poles requires only a short-term rotation of 30 to 40 years; sawlogs require 65 to 100 years. Sustained yield forestry works best on large acreages where blocks of timber of different age classes can be maintained.

Sustained yield in forestry is much more sophisticated than in wildlife and fisheries for the simple reason that it is easier to inventory, measure growth and potential yields of biomass, and manipulate populations or stands of trees. To achieve this end, foresters have an array of silvicultural and harvesting techniques, from clearcutting to selection cutting, with many variations in between. In recent years research foresters and forest ecologists have been giving considerable attention to nutrient cycling and nutrient conservation in managed forests.

The concept of sustained yield is well ingrained in forestry, and it is practiced to some degree by large timber companies and federal and state forestry agencies. However, industrial forestry's approach to sustained yield is to grow trees as a crop rather than to maintain a forest ecosystem. They take a cornfield approach: clearcut, spray herbicides, plant or seed the site to one species, and in the case of paper pulp, harvest in 30 to 40 years, clearcut, and plant again. Clearcutting practices in some national forests, especially in the Pacific Northwest and the Tongass National Forest in Alaska, hardly qualify as sustained yield management: below-cost timber sales are mandated by the govern-

ment to meet politically determined harvesting quotas. Even more extensive clearcutting of forests is taking place in the forests of northern Canada, especially British Columbia, and in large areas of Siberia. Such cuttings are forest mining, not harvesting. Sustained yield management has hardly filtered down to smaller parcels of private land. Poor cutting practices with little concern for regeneration and species composition has left behind forests of poor quality and quantity.

Forestry, like fisheries, focuses on the resource as an economic opportunity, not as a biological community. Once an old-growth forest is cut, that ecosystem will not return. A carefully managed stand of trees, reduced to one or two species, is not a forest in an ecological sense. Rarely will a naturally regenerated forest, and never will a planted one, become old growth. By the time the trees reach economic or financial maturity—based on the type of rotation—they are cut again. Economic maturity is not the same as ecological maturity.

18.6   Exploitation has decimated wildlife

When species arrive at the brink of extinction, we might undertake valiant efforts to bring them back; but we should never allow them to arrive at that condition in the first place. When human populations expand and colonize new regions, many species of wildlife suffer. Throughout the world the early abundance of wildlife was decimated by market hunting, wanton killing, and habitat destruction. In North America, species such as the ivory-billed woodpecker (*Campephilus principalis*), passenger pigeon (*Ectopistes migratorius*), and Carolina parakeet (*Conuropsis carolinensis*) have become extinct. In Africa settlers killed off the quagga (*Equus burchelli quagga*), a wild horse, and the bluebuck (*Hippotragus leucophaeus*), among others. In Australia it was marsupials, the Tasmanian wolf (*Thylacinus cynocephalus*) and eastern hare wallaby (*Lagorchestes leporides*).

In the United States in the early 1900s a number of individuals began to make strenuous efforts to halt the destruction of wildlife. The Lacey Act of 1900 prohibited interstate transportation of wildlife, dead or alive. Another milestone was the Migratory Bird Act of 1913, which gave international protection to migratory birds and ended market hunting for waterfowl. By the 1930s, when the United States was suffering from drought and severe soil erosion, wildlife was still in serious trouble. Then wildlife restoration efforts began in earnest with the passage of the Pittman-Robertson Act or Federal Aid in Wildlife Restoration Act, financed

FOCUS ON ECOLOGY 18.1

ENDANGERED SPECIES—ENDANGERED LEGISLATION

Endangered species are the subject of a hot political controversy. The root of the controversy is the Endangered Species Act of 1973. Administered by the National Biological Service (formerly the Fish and Wildlife Service), it covers vertebrate animals, invertebrates, and plants. The act has a number of provisions. First it requires the listing of endangered and threatened species. That list now includes over 1100 species. An endangered species is one that has so few individual survivors that the species could become extinct over its natural range. There are, unfortunately, numerous examples worldwide. In the United States the whooping crane, the black-footed ferret, and the California condor are among them. Threatened species, even more numerous, are still relatively abundant over their natural range, but are declining in numbers and are likely to become endangered. Among them are species of Pacific salmon, many grassland birds, neotropical warblers, and many species of cacti.

The act also mandates the designation of critical habitat to be protected and recovery programs to increase the abundance and distribution of endangered species. Some programs have been very successful, including the restoration of the bald eagle.

The act makes it illegal for the federal government to fund any projects that would harm listed species, including habitat destruction. It makes it illegal for United States citizens to harm, capture, or market animals that appear on the Federal Endangered and Threatened Species List. It also regulates the importation of endangered plants and animals from foreign lands and

provides for participation in international agreements regulating commerce in wild animals.

Up for reauthorization since 1992, the Endangered Species Act has vocal and powerful opponents. Because the act protects the individual organism as well as its habitat, land developers, loggers, and the mining and fishing industries see the act as an obstacle to their interests. They are exerting intense pressure to allow the use of economic factors to override the Endangered Species Act. With the Endangered Species Act out of the way, they could cut the remnant old-growth forest, drain wetlands, mine in parks, and ignore precautions to save threatened marine life, regardless of the consequences to a species. Succumbing to such pressures, congressional politicians withhold or reduce funds to slow down the listing of species and cut back protection of habitats. In spite of protests about the act, it has affected only one-tenth of one percent of all of the projects involving potential threats to listed species. The opponents of the act give little heed to the great contributions wild plants and animals make to the economy through natural products, recreation, and ecotourism, or to their financial value as a reservoir of genetic resources and a source of medicinal ingredients.

Unfortunately, the Endangered Species Act spotlights species rather than communities or ecosystems. The real issue is not protecting individual species such as the spotted owl, marbled murrelet, and red-cockaded woodpecker, but protecting the ecosystems of which they, associated species, and humans too are a part.

---

by an excise tax on shooting equipment that supported wildlife research and habitat restoration. An important recent piece of legislation for wildlife is the Endangered Species Act (see Focus on Ecology 18.1: Endangered Species—Endangered Legislation).

18.7 Wildlife restoration is a complex task

After the 1930s, a few species, such as the white-tailed deer, pronghorn antelope, and wild turkey, made a comeback. They received strict protection from hunt-

ing, followed by highly regulated hunting seasons once the populations were well on the way to recovery. States and the federal government set aside refuges and reserves to protect both animals and habitat. They reintroduced wild individuals taken from pockets of abundance to areas of scarcity and empty habitats.

What made the restoration efforts most successful, however, was the availability of large areas of empty habitat. During the depression of the 1930s, abandoned farmland and rangeland had reverted to natural vegetation. Devastated forest lands were growing back, providing outstanding food and cover for white-tailed deer and other species. So successful were restoration efforts that some of the species now are approaching a pest status. White-tailed deer are probably more abundant now than at the time of settlement. They cause accidents on highways, invade suburban gardens, threaten native plants, and destroy crops.

The wild turkey *(Meleagris gallopavo)* is a good example of the restoration of a species from the brink of extinction (Figure 18.5). Originally the turkey's range included all or part of 39 states and extended into Ontario, Mexico, and Central America. By the mid-1800s the species had been eliminated from the northeastern United States and by 1900 from the midwestern states. In 1949 only small populations of wild turkey survived on about 12 percent of their ancestral range. Most existed in remote areas of the Appalachian Mountains.

Intensive studies of the biology and ecology of the species, financed by Pittman-Robertson money, provided the informational base for restoration efforts. In the western United States small groups of live-trapped wild turkeys were transferred from areas of abundance into suitable vacant habitat. In South Dakota, 29 birds were reintroduced in 1948–1951. They had increased to 7000 birds by 1960.

It was easy to capture wild turkeys in the West with baited walk-in traps, but such traps did not work well in the eastern forest. Some states, notably Missouri and Pennsylvania, resorted to mating domestic turkey hens with wild gobblers. Such crosses were so maladapted genetically—nesting too early and lacking wildness—that the releases were failures. Worse, they hybridized with wild stock and introduced the fatal blackhead disease. Then the cannon-projected net trap, intended for capturing waterfowl, proved highly successful in capturing social groups of wild turkeys. Maturing forests, greater public support, continued intensive studies, and

the wild turkey's unforeseen ability to adapt to habitats previously thought unsuitable aided large-scale restoration efforts. Today the wild turkey population is about three million birds. Like the white-tailed deer, flocks have settled in wooded suburban areas.

Successful restoration, then, has a number of facets. Scientific studies of the bird's biology, ecology, and behavior provide the data necessary to manage growing populations. Legal protection, strict law enforcement, adequate financing, and public concern and cooperation also are critical. Protected reserves, large areas of suitable habitat, and the adaptability of the species further ensure recovery.

The task of wildlife restoration and preservation today is more difficult than in the past. Wildlife species are facing a dramatic rise in toxic chemicals in the environment, human populations, and loss of habitat (Chapter 21). Fragments of undeveloped land either hold small isolated populations of the species or are too small for occupancy.

Small isolated populations are subject to many problems and dangers. They are easy targets for poachers and predators. Isolated from others of the same species, the populations may not maintain the social cohesiveness necessary for successful mating and reproduction. They become vulnerable to inbreeding and lose heterozygosity (Chapter 19). If the population does expand, the surplus has no available habitat to colonize, a situation the opposite of early restoration efforts. Elephants restricted to parks and reserves in Asia and Africa and bison protected in Yellowstone National Park overpopulate the available habitat and move outside park boundaries, where they conflict with human interests. The reintroduction of the wolf to the Yellowstone ecosystem faced opposition by those who fear the wolf population may expand to areas outside the park.

18.8   Habitat restoration and protection is the key to saving wildlife

Simply restoring populations is not enough. The future of much of the world's wildlife depends upon the survival of their habitats. A current approach is to concentrate on the restoration, maintenance, and preservation of whole ecosystems, not on individual threatened species.

Habitat restoration involves planting, protecting, and managing an area. Eroded lands can be restored to grassland, cutover forest replanted to trees, and drained wetlands reflooded and replanted with aquatic vegeta-

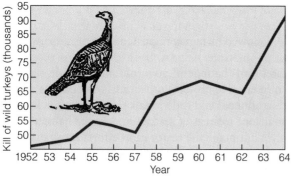

Figure 18.5 Growth of the wild turkey population in the United States following restoration (through 1964) as derived from hunting harvest data. Note the sharp increase in growth after 1962. Turkey populations are still increasing.

tion. Some species require early, ephemeral stages of succession (see Section 21.13), which must be maintained by such techniques as cutting and burning. The endangered Kirtland warbler *(Dendroica kirtlandii)*, for example, requires early-stage stands of jack pine. Such stands can be maintained only by periodic burning of blocks of overage pine to stimulate regeneration. In such cases the management objective is to keep blocks of vegetation in various stages of succession to ensure continuance of acceptable habitat.

Small populations cannot survive unless remaining available habitat is protected against such intrusions as development, agricultural clearing, and logging. Some of our most endangered species occupy natural reserves, many of which are too small to support a viable population of the species. Ideally we would increase the size of the reserve to provide habitat for an expanding population or to embrace the entire ecosystem. Most reserves do not. Human activities just outside park boundaries—logging, grazing, agriculture, housing, and recreational development—violate the integrity of many ecosystems.

For example, Yellowstone National Park is only part of an ecosystem that includes the surrounding national forests and private lands. The elk and bison in the park have expanded beyond the capability of the park itself to support them, especially in winter. For a species to survive, we have to preserve all components of its habitat. Because Yellowstone is mostly above 2100 m, their critical winter range lies outside the park. This area is under enormous conflicting human pressures, including grazing, logging, recreation, and development. The problem at Yellowstone repeats at other parks and reserves throughout the world. To protect the reserves, concentric buffer areas of land use should encircle the core area. Land use should be minimal about the immediate boundary and increase with distance.

Excessive human activity within their boundaries compromises many otherwise excellent parks and reserves. Intensive recreational development, trails, and human incursions into the backcountry that bring people into direct contact with large species of wildlife decrease the value of the reserve and actually endanger the species. The case of the grizzly bear in national parks is a good example. A certain segment of society would prefer to see the grizzly bear exterminated to make backcountry travel safe for humans.

Large reserves provide the greatest protection and support the largest populations, although even the largest reserves may be too small for certain large carnivores and herbivores. Associated with large reserves should be smaller reserves. Although isolated and surrounded by other types of land use, small reserves do hold populations that represent different samples of the gene pool. Subject to somewhat different selection pressures and to some degree of genetic drift, these smaller populations help maintain genetic diversity in the species' populations.

Corridors connecting one habitat patch with another (Chapter 21) can enhance the integrity and stability of fragmented populations. On a small scale hedgerows act as corridors between woodlands. On a much larger scale greenbelts in cities and suburbs, plantings along major highways, and vegetation along rivers and streams provide travel corridors for wild animals and dispersal routes for woody plants. Of particular value are corridors of woody vegetation along rivers and streams, because they provide water, food, and cover for a great diversity of species.

We too often view habitat protection only as a means of maintaining nesting habitat. Many wildlife species are migratory. Critical to their survival is winter habitat. It is impossible to maintain populations of neotropical migrants such as warblers if their tropical winter habitat is cleared for development. Other critical habitat lies along migration routes. Consider the problems of migrating warblers seeking forested resting and feeding areas across the megalopolis that extends along the Atlantic coast. Imagine the problems faced by migrating shorebirds that depend on sandy beaches uncompromised by development.

18.9   We may control wildlife populations to maintain habitat

If restoration is successful, a species may exceed the ability of the habitat to support it. Overpopulation causes deterioration of habitat, scarcity of food, and impairment of habitat for associated species. As the habitat is damaged, the species declines again.

Preventing such damage may require trapping and removing or cropping the excess individuals. These surplus individuals provide transplant stock for depleted habitats. For example, the highly productive white rhino *(Diceros simus)* population in the Umfolozi Game Reserve in South Africa has supplied a number of animals for reintroduction into other reserves.

Cropping requires knowledge of the relations and functions of species in the system. Are apparent overpopulation problems simply part of a natural plant-herbivore cycle, or does the situation require human intervention? For example, elephants, which feed on woody browse, are confined to parks and reserves, where they are converting tree savannas to grassland that cannot provide sufficient food for them. Eventually the elephants will begin to starve. When the elephant population declines, the vegetation will slowly return. The question is whether to allow such a cycle to proceed or to step in, reduce the elephant herd, and try to achieve some stability between the elephants and the trees. It is difficult to decide because of the longevity of elephants and the time trees take to regenerate. The decision affects not only the elephants but the entire ecosystem.

18.10 Reintroductions can return species to depleted habitats

If habitat within the natural range of a species is available but the species is locally extinct, then the habitat can be recolonized by introducing a core population into the habitat. The animals or plants for reintroduction may be trapped from wild populations, as was the case for the wild turkey, or captive-reared. Critical to introduction is the size of the initial or founding population (see Chapter 19), which must be large enough to form a cohesive social or reproductive unit.

Restoration through captive-bred individuals is a measure of last resort. However, it is the only way we can begin to increase some species. Problems with captive propagation include small population size, potential inbreeding, incompatibility of captive individuals as mates, and lack of social interaction. Captive propagation programs cannot be carried out indefinitely. After a number of generations, depending on population size, the captive stock will begin to show signs of inbreeding depression and domestication. For this reason it is important to consider reintroduction whenever feasible.

Introducing individuals from captive-bred populations to the wild (Figure 18.6) requires prerelease and postrelease conditioning. The animals must learn to acquire food, find shelter, interact with conspecifics, and avoid humans. Numerous attempts have failed. Some successful reintroductions include the whooping crane (*Grus americana*), masked bobwhite (*Colinus virginiana*), and Hawaiian goose or nene (*Branta sandvicensis*) among the birds and the American bison and its relative, the European wisent (*Bison bonasus*) among the mammals. More recent efforts involve the peregrine falcon (*Falco peregrinus*), the golden lion tamarin (*Leontopithecus rosalia*), and the Arabian oryx (*Oryx leucoryx*). The outcomes of these reintroductions are inconclusive. Problems have included high costs, logistical difficulties, and shortage of habitat, besides the question of adapting to the wild.

Figure 18.6 Peregrine chick *(Falco peregrinus) en route* to a cliff nest in Colorado.

Both wild and captive-bred individuals may be translocated to build up numbers and to introduce new genetic material into populations. Such translocations must be done carefully. Not only is there danger of introducing disease, but the new individuals must fit into the social and breeding structure of the native population.

Just as important is the genetic background of the transplanted animals. Unless they are adapted to their new environment, these individuals can weaken the resident stock. Attempts to increase depleted populations of the northern bobwhite quail in Pennsylvania and New England with birds of southern origin were disastrous. Both the introduced birds and their "hybrid" offspring died during cold winters to which they were not adapted. The northern populations went with them.

18.11 Protection and education must accompany restoration

Once any species is confined to an island of habitat, it becomes more vulnerable to poaching and predation. Waterfowl forced to nest in small wetlands surrounded by agricultural land are easy marks for such nest predators as skunks, raccoons, and opossum, highly adaptable to an agricultural landscape. Under such conditions protection from predators is necessary for successful nesting.

Larger mammals such as bear, bighorn sheep, rhinoceros, and elephant concentrated in a limited area become easy marks for poachers and unscrupulous trophy hunters. Protection of reintroductions needs the long-term support of governments and especially of communities in the vicinity of the parks and reserves. In most places it is necessary to demonstrate tangible economic benefits from parks and their wildlife before local people support the program. Protection must be accompanied by education about species and habitats. Education is absolutely necessary at the local level. Without local support reintroduction programs fail.

Ultimately the fate of wild creatures and the ecosystems of which they are a part rests with controlling the explosive growth of human populations. As the numbers of the dominant mammal on Earth increase, diversity of life decreases. Unless we teach this lesson, we will be left with only those plants and animals that are tolerant of or share the human-dominated landscape. The outstanding ecosystems—the tropical forests, old-growth forests, savannas, and grasslands, even marine environments and their unique forms of plants and animals—will exist only as memories preserved in books and films.

18.12   Many species clash with human interests

Humans have always had to contend with unwanted plants and animals that interfere with our well-being. What constitutes a pest or a weed depends upon your viewpoint and values.

Pests are organisms that humans consider undesirable, a classification that varies with time, place, circumstances, and individual attitudes. Mice, rats, cockroaches, fleas, mites, lice, and mosquitoes have long been unwelcome camp followers, adapting to human habitation and cultural practices. With the development of agriculture, large predators that threatened domestic animals, herbivorous mammals that invaded fields and gardens, and grain-feeding birds all became pests.

Plants also became competitors with humans. Wild plants in field and garden compete with crops for light, space, and nutrients, reducing yields and becoming, in human terms, weeds. A weed is a plant growing where it is not desired, a plant out of place. Violets growing in a lawn may be weeds (depending upon the lawn owner's value judgment). Electric companies regard woody plants and trees growing in power line rights-of-way weeds; to a naturalist, the same growth is wildlife habitat. Some foresters consider any trees that have no commercial value to be weed trees; to a naturalist those trees may be beautiful or important to wildlife.

We can look upon pests and weeds as occupying a continuum from r to K (see Focus on Ecology 12.1: Life History Strategies). Our most familiar and obnoxious pests, such as houseflies, cockroaches, fleas,

mites, and scale insects, possess certain r characteristics. They have a high rate of increase, small size, and a high rate of dispersal. They seek new and open habitats, adapt well to conditions provided by humans, and spread rapidly in homogeneous habitats where both shelter and food are abundant.

Common weeds are much the same. The most persistent and abundant weeds are easily dispersed, colonize highly disturbed sites, persist a long time in the soil as seeds, respond quickly to disturbed sites, and are tough and resilient. Good examples are ragweed and dandelions.

Other animals and plants become pests or weeds under different sets of conditions. They have low reproductive rates, occupy more specialized habitats, and need more specialized resources. Often they become pests or weeds because humans move into their habitats. The tsetse fly of Africa, a K species, became a pest only when humans moved their cattle into its natural range. Elephants and deer became pests when humans usurped their habitats for agricultural and settlements. Certain trees became weeds when humans changed the vegetation composition of forests and fields.

Most pests and weeds, however, fall somewhere along a continuum between the two. They show a mix of r and K characteristics. Normally they are regulated by natural predators and parasites. Native rats and mice fall into this category, as well as certain insect pests, such as the spruce budworm and the codling moth, whose larvae feed on apples.

Since the advent of agriculture, humans have been trying to control these pests. After centuries, it is evident that we cannot eradicate pests. The best we can do is control their numbers. Control should reduce pests to a level at which they will not cause economic injury (Figure 18.7). At this point costs of control are less than or equal to the net increase in value derived from such control. Even so, many people initiate control measures without a strong economic payoff. For example, they spray insecticides and herbicides on lawns. Appropriate control measures vary with the life history and the effect of the pest or the weed.

18.13   Widespread chemical control is harmful

One means of control is chemical. The ancient Sumerians used sulfur to combat crop pests, and the Chinese as long ago as 3000 B.C. used substances derived from plants as insecticides. By the early 1800s such chemicals as Paris green, Bordeaux mixture, and arsenic were commonly used to combat insect and fungal pests. The major chemical weapons, however, appeared

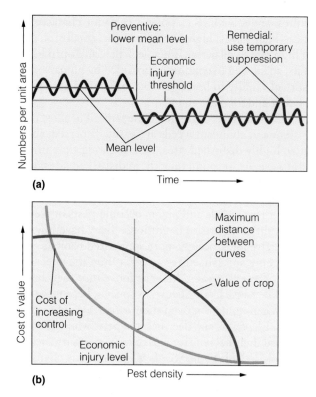

(a)

(b)

Figure 18.7 The economic injury threshold suggests when pest control measures should be taken. (a) When a pest exceeds the economic injury threshold, we should take strong measures to reduce the population. Below the threshold temporary suppression measures suffice. (b) The economic injury level is that point at which the value added to the crop exceeds the cost of increasing control.

after World War II with the development of organic insecticides (containing carbon) to combat insect vectors of human disease, especially in tropical areas. The success of these insecticides encouraged their rapid use in agriculture, for which the chemical industry provided an arsenal of over 500,000 biocides.

These organic chemicals, with varying degrees of toxicity and persistence, are either synthetic (of human manufacture) or botanical (derived from plants). Major groups of synthetic pesticides include chlorinated hydrocarbons, organophosphates, and carbamates (Table 18.1). All are broad spectrum insecticides. In one manner or another they disrupt the nervous system. Major groups of botanical insecticides include pyrethrum, nicotine, and rotenone, highly toxic to fish.

Ease of application, effectiveness in small doses, low cost, and toxicity gave chemical insecticides the appearance of a panacea. They became, as the entomologist Paul DeBach described them, ecological narcotics. Instead of solving the pest problem, the insecticides compounded it by killing natural predators as well as pests. Unchecked, the surviving pests resurged (Figure 18.8). Killing natural predators also released other insect pests that had been held in check. Now their populations increased dramatically, adding new pests to the situation.

There are numerous worldwide examples. One involves the boll weevil in the cotton fields of the Rio Grande Valley in Texas. For 15 years the insect was held in check with chlorinated hydrocarbons. Then in

TABLE 18.1

TYPES OF PESTICIDES

Pesticide	Characteristics	Examples
Chlorinated hydrocarbons	Fat-soluble; accumulate in fat tissues of animals; transferred through food chain; toxic to wide range of animals; long-term persistence	DDT, aldrin, lindane, chlordane, Mirex.
Organophosphates	Water-soluble; leach to groundwater; less persistent than chlorinated hydrocarbons; some systemic—taken up by plants, transferred to leaves and stems, where they then become available to leaf-eating and sap-sucking insects	Malathion, parathion
Carbamates	Derived from carbamate acids; kill a narrow spectrum of insects, but highly toxic to vertebrates; relatively low persistence	Sevin, carbaryl
Diflubenesuron	Interferes with formation of exoskeleton in molting insect larvae; used in gypsy moth control but nonselective, affecting all lepidopteran caterpillars developing at time of spraying	Dimelin
Botanicals	Affect nervous system; least persistent of pesticides; among safest to use; some in household sprays	Pyrethrins, nicotine-based sprays, rotenone
Insect pathogens	Only *Bacillus thuringiensis* (Bt) and its subspecies are used with frequency; applied to crops and forest pests, particularly gypsy moths; affect other caterpillars	Dispel, Foray, Thuricide

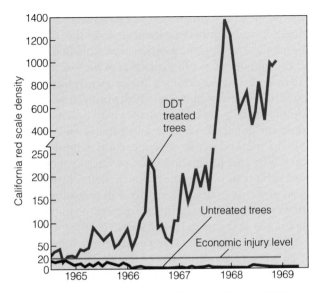

Figure 18.8 Experimental monthly applications of DDT on scale-infested lemon trees reduced not only the pest but its natural enemies. The scale resurged dramatically. Natural predators and parasites kept the scale in check on unsprayed trees.

the late 1950s, the weevil achieved resistance to the insecticides. When cotton growers switched to organophosphate insecticides, two other pests emerged, bollworm and tobacco budworm. By 1962 they, too, had become resistant to carbamates. By 1980 this second insecticide had lost its effectiveness. The original pest, the boll weevil, faded as the bollworm and especially the tobacco budworm became the major pests. Today the tobacco budworm is resistant to all registered pesticides, and cotton growing has collapsed in northeastern Mexico and southern Texas.

This example points out one of the major failings of chemical pesticides. Through natural selection (see Chapter 19), insects evolve a resistance to pesticides. As one pesticide replaces another, the pests acquire a resistance to them all. By 1988 over 1600 insect pests worldwide had developed resistance to one or more pesticides. Some insects, notably houseflies, certain mosquitoes, the Colorado potato beetle, and cotton bollworms, as well as cattle ticks and spider mites, have overcome the toxic effects of every pesticide to which they have been exposed. Insect pests need only about five years to evolve pesticide resistance; their predators do so much more slowly.

The chemicals for weed control are organic herbicides. They fall into three classes, based on their effects on plants. Contact herbicides, such as atrazine, kill foliage by interfering with photosynthesis. Systemic herbicides, such as 2,4-D and 2,4,5-T, are ab-

sorbed by plants and overstimulate growth hormones. The plants grow faster than they can obtain necessary nutrients and die. Soil sterilants kill microorganisms necessary for certain plants to grow. Although designed to kill plants, many herbicides are extremely toxic to humans, especially 2,4-D and 2,4,5-T, the two components of the notorious Agent Orange. These herbicides contain dioxin, which has been linked to human birth defects and cancers, including leukemia.

18.14 Biological control has had some success

Farmers observed long ago that natural enemies of pests act as controls. As early as 300 A.D. the Chinese were introducing predaceous ants into their citrus orchards to control leaf-cutting caterpillars and boring beetles. In the late Renaissance, naturalists made important observations on insect parasitism. In the 1800s entomologists made important advances in understanding the biological control of both insect pests and weeds.

Insect pests have their own array of enemies in their natural habitats. When an animal or a plant is introduced, intentionally or not, into new habitat outside of its natural range to which it can adapt, it leaves its enemies behind. Freed from predation and finding abundance, the species quickly becomes a pest or a weed. This fact has led to a search for suitable natural enemies to introduce.

The classic success story is the case of the introduced cottony-cushion scale moth *Icerya purchasi*, a native of Australia that sucks food from the leaves and twigs of citrus fruit trees. First seen in California in 1872, within 15 years the insect threatened the citrus fruit industry. In 1887 entomologists discovered that the ladybug beetle *Rodolia cardinalis* (*Vadalia cardinalis*) and a parasitic fly *Cryptochaetum iceryae* controlled the scale in its native habitat. The two parasites were reared and released in California citrus groves with amazing success. By the end of 1889, the scale was no longer a problem. The fly alone could have done the job, but the ladybug experienced such explosive population growth that it became the dominant predator on cottony-cushion scale. Then in 1946 and 1947 orchardists began using DDT, which killed the beetle, but not the scale. The scale rebounded, killing trees. Within three years, orchardists ceased or modified their spray program, and the beetles brought the scale under control again.

Biological control has also been used on weeds. One outstanding example is the reduction of prickly pear (*Opuntia* spp.) in Australia by the introduced moth *Cactoblastis cactorum*, a native of Argentina. Another example is the control of Klamath weed or St.-John's-wort (*Hypericum perforatum*). A native of Eurasia and northern Africa, this noxious weed was introduced in 1900 into California along the Klamath River. An aggressive colonizer of overgrazed pastures, it soon occupied large acreages in the northwestern United States. Among 600 species of insects that fed on the plant in its native habitat was the leaf-eating beetle *Chrysolina quadrigema*. Introduced in 1945, this beetle has greatly reduced Klamath weed in California.

18.15   Genetic approaches have advantages and drawbacks

Numerous wild plants and animals have evolved their own defenses against natural enemies (see Section 16.6). These defenses include allopathic effects against plant competitors and toxins in the tissues of plants and animals that inhibit or suppress predation. One approach to pest control, then, is to breed for genetic resistance. This tool has been used successfully in crop plants, such as corn, wheat, and rice. This process involves crossing cultivars with wild relatives to capture the gene into the gene pool of the cultivated plant. Such an approach requires considerable effort in locating, recognizing, and studying wild relatives. As our natural communities are destroyed—most crop plants are tropical in origin—we are losing the genetic storehouse upon which we can draw.

The latest technique to increase plant resistance to weeds and insect pests is genetic engineering, particularly the use of recombinant DNA material to introduce the gene for the desired trait into the plant. The transfer of a single gene may confer resistance to viruses or to herbicides, or encode for production of endotoxins, poisons that inhibit the feeding activity of insect enemies.

Genetic engineering for pest control has inherent dangers. Plants with engineered traits could evolve new genotypes with somewhat different life histories or physiological traits. Engineered crop plants might transfer genes over long distances by hybridizing with related plants that differ in their life history characteristics. Many crops, such as celery, asparagus, and carrot, have weedy relatives with high reproductive output and efficient seed dispersal. If these weedy relatives acquire the engineered gene, current herbicides would become ineffective against them. Insect-resistant plants could speed the evolution of even more resistant insect pests.

Another genetic approach to insect suppression is to release sterile males. Competitive sterile males reared in great numbers in the laboratory are introduced in sufficiently large numbers to ensure they will be involved in a high proportion of matings in the field. If the number of sterile males is kept high as the population of the pest declines, the proportion of sterile to fertile matings increases. If the population of the pest is initially high, insecticides may be used to reduce the population before sterile males are released. Such a method works only if certain conditions are met. The pest population must be fairly isolated and not subject to immigration of wild males or emigration of sterile males; genetically different subpopulations must not be present; and genetic changes must not occur in the reared population.

An example is the screwworm control program. The screwworm is the larva of the blowfly *Cochliomyia hominivorax*, which lays its eggs in open wounds of warm-blooded animals. The larvae enter the wound and feed on the flesh of the animal. The screwworm became a major livestock pest in the southeastern United States and Texas. A major eradication effort began in the 1950s, involving the release of factory-reared sterile males. The program was highly successful until 1972, when a major outbreak occurred. The cause was a major genetic difference between the sterile males released and the wild type. In 1977 the defective factory strain was replaced and control was regained. This experience points out the necessity of maintaining strict quality control to monitor genetic changes in the factory populations.

18.16   Mechanical and cultural control can be effective

Another approach to pest control is mechanical. For centuries agriculturalists have used fencing and other barriers not only to keep livestock in but to deter predators and to prevent herbivores such as deer and rabbits from feeding on crops. Sticky traps wrapped around trees prevent caterpillars from climbing to the crown. Traps baited with pheromones and sticky paper are effective in attracting and capturing males of specific insect pests such as gypsy moths. Light traps, although effective, indiscriminately attract and kill innumerable beneficial insects along with the insect pests. Various methods of trapping are used to catch certain mammalian and avian pests. Cultivation, hoeing, and hand weeding are typical mechanical methods of eliminating weeds from fields and gardens.

Homogeneous habitats provide the opportunity for many large outbreaks of pests. Enormous fields and extensive stands of forest trees of the same and closely related species, such as southern pines and balsam fir, provide huge areas of abundant food and cover. Spruce budworm and southern pine beetle, for example, have swept across hundreds of thousands of acres of forests. One deterrent to the rapid spread of pests is the creation of patchy environments in which homogeneous stands are broken up by other types of vegetation. For example, we can sow crops in alternate rows, intersperse hardwoods with conifers, and plant hedgerows. These patches not only scatter the food supply, checking the spread of the pest and breaking its population into smaller units more vulnerable to predation, but they also harbor enemies of the pests.

Another cultural approach, successful in controlling pests, is adjusting the timing of planting or harvest. Growers plant cotton after the peak emergence of the pink bollworm weevil and promptly harvest early-maturing varieties of cotton.

18.17 Integrated pest management combines approaches to control

No one method—chemical, biological, genetic, or mechanical—is always best. Therefore entomologists have developed a holistic approach to pest control, called **integrated pest management.** Integrated pest management (IPM) considers the biological, ecological, economic, social, and even aesthetic aspects of pest control and employs a variety of techniques. The objective of IPM is to meet the pest not at the point of a major outbreak but when the size of the population is easiest to control. Managers rely first on natural mortality caused by weather and natural enemies. They disrupt the natural system as little as possible as they work to hold the pest below the economic injury level (Figure 18.7).

Successful integrated pest management requires a knowledge of the population ecology of each pest and its associated species and the dynamics of the host species. It calls for considerable fieldwork. Monitoring the pest species and its natural enemies by such techniques as counting eggs and trapping adults, scientists determine the necessity, timing, and intensity of control measures. These measures sometimes include minimal chemical spraying at the appropriate time or thinning an affected forest. IPM makes minimal use of chemicals in order to reduce the development of genetic resistance to pesticides. The control methods must be adjusted from one location to another.

Based on the degree of damage sustainable, the costs of control, and the benefits, managers make a se-

ries of decisions (Figure 18.9). Integrated pest management has helped in the control of the spruce budworm in Canada and the southern pine beetle in the United States. It is used in apple orchards of the United States and Canada, in cotton fields in Texas, and in alfalfa-growing regions in the United States. It is the most sensitive approach to weed and pest control we have developed so far.

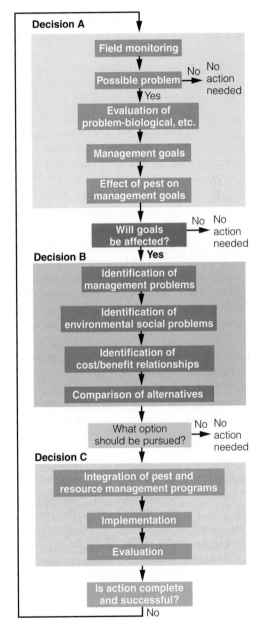

Figure 18.9 A decision-making model for integrated pest management. Note the inputs of biological, social, and environmental information.

CHAPTER REVIEW

SUMMARY

Humans interact in three ways with natural populations. They maintain exploited populations, support threatened or endangered species, and reduce pest populations.

Sustained Yield (18.1–18.5) First attempts at managing exploited populations involved fisheries (18.1). Out of these efforts came the concept of sustained yield. If plant and animal populations exploited for human use are to be continuously productive, the yield per unit time should equal production per unit time. The level of yield above which a population declines is maximum sustained yield. Management differs for *K*-selected and *r*-selected species. Based on the logistic equation, sustained yield models fail to consider all aspects of population dynamics, including natural mortality and environmental uncertainty. Sustained yield management considers the resource as a single biological unit, not as part of an ecosystem (18.2).

An overexploited commercial species shows warning signs, such as a sharp decrease in catch, dominance of young age classes, few pregnant females, and failure to replace losses (18.3). Too often economic considerations outweigh biological considerations, causing the commercial and even biological extinction of a species. Economics also encourages discarding marine life, from juvenile fish to dolphins caught in large nets along with commercial species. Euphemistically called bypass, this enormous tonnage of wasted marine life disturbs marine ecosystems and the overall fishery resource (18.4).

Sustained yield in forestry seeks to balance the volume of timber cut with the volume of timber growth. Economic considerations often override sustained yield, devastating forests (18.5).

Wildlife Restoration (18.6–18.11) Expanding human populations decimate wildlife through habitat destruction, wanton killing, and commercialization (18.6). Efforts at restoration of declining wildlife species in North America began in the 1930s. These efforts saved from extinction many of our most common species, such as wild turkey and deer. Current restoration efforts are difficult because of conflicting demands for land (18.7). Because loss of habitat is the major cause of the decline of most threatened and endangered species, the key to restoration is the protection, restoration, and management of habitat. Habitats are being frag-

mented, and isolated populations may be too small to remain viable. The future of threatened populations depends on protected parks and reserves. Surrounded by human development, small reserves should be connected by corridors (18.8). If protected populations do reproduce successfully, their numbers may be too great for the habitat fragment to support. In such situations we may need to control population growth (18.9). Sometimes surplus individuals can be used to recolonize empty habitats. In extreme cases captive breeding programs can save a species from extinction (18.10). All attempts at restoring threatened species must be accompanied by protection against poaching and by public education (18.11).

Pest and Weed Control (18.12–18.17) Pests and weeds are animals and plants growing anywhere humans do not want them. Most of them are *r*-strategists, dispersing easily and speedily colonizing disturbed sites. The aim of control is to hold the population below the level at which they cause economic injury (18.12).

The use of organic chemical pesticides and herbicides prompts the evolution of resistant strains and causes toxic pollution. It kills natural enemies, allowing a pest population to increase dramatically (18.13). Biological controls—natural enemies and parasites—are effective in keeping some pests below the economic injury level (18.14). Breeding for genetic resistance to pests and weeds is another important and effective tool. Transplanted genes produce plants that are resistant to viral diseases and insect attack. If engineered genes transfer to wild relatives, we risk creating highly resistant strains of weeds and pests (18.15). In some situations mechanical controls such as trapping pests and cultivating the ground to kill weeds are effective. A cultural approach creates habitat diversity or a patchy environment, which makes it difficult for pests to feed and disperse (18.16).

Integrated pest management is a holistic approach. More a philosophy than a specific strategy, it combines in appropriate ways chemical, biological, genetic, mechanical, and cultural control measures. Based on careful monitoring, IPM weighs environmental and social benefits and costs (18.17).

STUDY QUESTIONS

1. In what three major ways do we interact with natural populations?

2. Explain sustained yield. What is the difference between maximum sustained yield and optimal sustained yield?

*3. Report on the effects of modern commercial ocean fishing, especially with huge drift nets, on the fishery resource and on other ocean life. Can the ocean withstand such exploitation?

*4. What indirect ecological effect do you have on ocean life when you enjoy an "all you can eat" shrimp dinner?

*5. Why is there less outcry about overharvesting of the ocean than about destruction of tropical forests?

6. How do economic interests block sustained yield practices? Consider both commercial fishing and cutting of old-growth forests in the Pacific Northwest.

*7. Comment on the statement that a stand of trees is not necessarily a forest.

8. What approaches must we take to conserve and restore threatened species? Why is habitat a key element? Why should we emphasize habitat and not individual species?

*9. Why is captive breeding of endangered species controversial? All remaining California condors were captured to undertake a captive breeding program. What was the outcome?

10. Define a pest and a weed. Why are the terms value-laden?

11. Under what conditions do plants and animals become weeds and pests?

12. What are the major approaches to controlling pests and weeds? What are the ecological advantages and disadvantages of each?

13. What are the major synthetic organic pesticides, and what are their environmental impacts?

14. Why are most mechanical approaches to pest control ineffective?

15. What are the ecological dangers of plants genetically engineered to resist pests?

16. How can a patchy environment reduce pest outbreaks?

17. What is integrated pest management? How can it provide public input into a local or regional pest control program?

C H A P T E R 1 9

POPULATION GENETICS AND SPECIATION

OBJECTIVES

On completion of this chapter, you should be able to:

- Explain the differences among adaptation, natural selection, and evolution.
- Discuss the sources of genetic variation.
- Distinguish among directional, disruptive, and stabilizing selection.
- Discuss the special genetic problems of small populations.
- Explain what is meant by a viable population.
- Contrast the morphological, biological, and evolutionary view of species.
- Summarize how species vary across their geographical range.
- Compare allopatric and sympatric species.
- Define an isolating mechanism and describe several types.
- Define the types of speciation.
- Contrast adaptive radiation with convergent evolution.
- Discuss the rate of evolution.

A male large ground finch *(Geospiza magnarostris)*—
one of the Darwin finches—on Santa Cruz Island,
Galápagos Islands.

We have been treating the population as a demographic unit. Now we will look at it as a genetic unit, with each individual in the population containing a portion of the total genetic material that describes the species. The concepts of the population as a demographic and a genetic unit are closely tied.

Adaptation comes about through the interaction of organisms with their environment (see Chapter 2). If an organism can tolerate a given set of conditions well enough to leave mature, reproducing offspring in the population, it is adapted to its environment. If an organism leaves few or no mature reproducing progeny, it is poorly adapted.

Individuals in a population are not identical. This variation allows some of them to be better adapted than others. For a species to survive as a whole, enough of its members must be adapted to the changing environment. The problem is that many conditions are changing faster than organisms can adapt to them.

19.1 Individual variation is a characteristic of populations

A population consists of many different, locally defined groups of individuals similar in structure and behavior. Individuals within these groups interbreed, oak tree with oak tree, white-footed mouse with white-footed mouse, largemouth bass with largemouth bass. Interbreeding individuals within each group make up a genetic population or **deme.**

Individuals within the deme are not identical. Variations in features of biochemistry, physiology, or physical structure, such as size, coloration, or behavior, are common in most populations. Most of these features are to some degree genetically controlled. Some of the characteristics influence the probability that an individual will survive and reproduce. For example, if smaller individuals in a population were better able to avoid predators than larger individuals, the smaller individuals would have a higher rate of survival and would most likely leave behind a greater number of offspring.

Those individuals that leave behind the most viable offspring are considered the fittest. The **fitness** of an individual is measured by the number of its reproducing offspring, contributed over several generations. In time the differential reproductive success of individual organisms changes the genetic traits of a population. This process is **natural selection.**

19.2 Genetic variation is the raw material for natural selection

The heritable or genetic portion of the variations in individuals and in populations is the raw material of natural selection. Most of these genetic variations arise from the shuffling of genes and chromosomes in sexual reproduction.

Genetic information is carried on **chromosomes,** threadlike structures in the cell, consisting mostly of long coiled strands of a complex molecule, deoxyribonucleic acid or DNA. DNA is composed of smaller units, the **nucleotides,** which are arranged in a sequential pattern. Each species has a unique pattern.

Chromosomes come in matched pairs, called homologous chromosomes. One member of the pair is inherited from the mother through the egg, and the other from the father through the sperm. Each chromosome carries units of heredity called **genes,** the informational units of the DNA molecule. Genes existing in alternative forms are called **alleles.** Because chromosomes are paired in the body cells, alleles are also paired. The position an allele occupies on a chromosome is its **locus.** Members of the pair of alleles occupy the same locus on homologous chromosomes. If the alleles occupying the same loci on homologous chromosomes affect a given trait in the same manner, the individuals possessing them are called **homozygous.** If the alleles affect a trait differently, the individual is called **heterozygous.** In this case one allele is fully expressed and the other has no noticeable effect. The allele fully expressed is the dominant gene. The hidden, unexpressed gene is the recessive one.

When cells reproduce (a process called **mitosis**), each resulting cell nucleus receives the full complement of chromosomes, the double or diploid number. Organisms that reproduce sexually also produce germ cells or gametes (egg and sperm) by the process of **meiosis,** in which pairs of chromosomes are split. The germ cell nucleus receives only one-half the full complement, or the haploid number. When egg and sperm unite to form a zygote, the diploid number is restored.

When two gametes combine to form a zygote, the gene contents of the chromosomes of the parents recombine in the offspring. Because the possible number of recombinations is enormous, recombination is an immediate and major source of variation. Recombination does not change any genetic information, but it does provide different combinations of genes upon which selection can act. Because some combinations of genes are more adaptive than others, selection determines which combinations survive in the population.

The sum of hereditary information carried by the individual is the **genotype.** The genotype directs development and produces the individual's morphological, physiological, and behavioral makeup. The external, observable expression of the genotype is the **phenotype.**

Some of the conspicuous individual variants in the phenotype—such as shortened tails, missing appendages, enlarged muscles, or other features that show disease, injury, or constant use—are not inherited. They are acquired characteristics, which the early French evolutionist Jean-Baptiste Lamarck (1744–1829) hypothesized were passed down from one generation to another. He was mistaken, but an organism does inherit the ability to acquire environmentally induced characteristics.

The ability of a genotype to give rise to a range of phenotypic expressions under different environmental conditions is **phenotypic plasticity.** Some genotypes have a narrow range of reaction to environmental conditions and therefore give rise to fairly constant phenotypic expressions. Some of the best examples of phenotypic plasticity occur among plants.

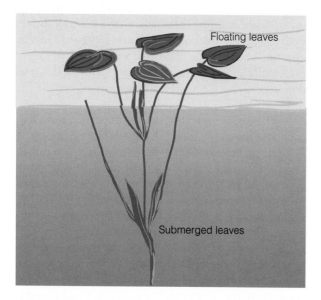

Figure 19.1 Plasticity of response to environmental conditions. Pondweed (*Potamogeton*) has lance-shaped underwater leaves, responsive to underwater movements, and broad, heart-shaped leaves that float on the surface.

The size of the plant, the ratio of reproductive tissue to vegetative tissue, and even the shape of the leaf may vary widely at different levels of nutrition, light, and moisture (Figure 19.1).

19.3　Alteration of genetic material results in inheritable change

Genetic material may be altered by an inheritable change in a gene or a chromosome. Such inheritable changes of genetic material are known as **mutations.**

Gene mutation is an alteration in the sequence of one or more nucleotides. During meiosis the gene at a given locus is usually copied exactly. On occasion the precision of this duplication breaks down, and the new gene is not an exact replica of the original. Most gene mutations have little or no apparent effect. Mutations of single genes that produce larger effects are usually harmful. Gene mutations are important because they add variation to the gene pool. Such mutations, however, do not direct evolutionary change.

Chromosomal mutations may result from a change in the structure of a chromosome or a change in the number of chromosomes. Structural changes involve a duplication, transposition, or deletion of part of a chromosome. Such changes result in abnormal phenotypic conditions. (One such condition is Down's syndrome in humans.) A change in chromosomal number can arise in two ways: (1) the complete or partial duplication of chromosomes, or (2) the deletion of one or more chromosomes.

Polyploidy is the duplication of entire sets of chromosomes. It can arise from an irregularity in meiosis or from the failure of the whole cell to divide at the end of the meiotic division of the nucleus. The individual body cell ordinarily is diploid ($2n$), or twice the haploid number. The union of a diploid gamete with a haploid gamete yields a *triploid,* usually sterile. Two diploid gametes ($2n$) may unite to form a *tetraploid,* also usually sterile. Naturally occurring tetraploids have arisen within a single species (*autotetraploid*). One example is *Epilobium,* the common fireweed of open fields and thickets.

The union of two diploid gametes from different species results in a hybrid tetraploid (*allotetraploid*). Such tetraploids are fertile because the chromosomes upon division behave as if they were diploids and produce balanced $2n$ chromosomes in the gametes (instead of the normal $1n$). Alloploidy is common in plants. The condition is rare in animals because an increase in sex chromosomes would interfere with the mechanism of sex determination and the animal would be sterile. Plants can self-fertilize and reproduce vegetatively, which gives polyploidy a selective advantage among many species. Polyploid plants differ in appearance from diploid individuals of the same species. They are generally larger, more vigorous, and more productive. Most of our agricultural plants are polyploids.

Mutations may be neutral, beneficial, or disadvantageous, depending upon the environmental circumstances and the genetic background in which they arise. If the effect on the phenotype is harmful, the mutant allele will be selected against. So will mutations that serve a function already filled by an established gene. If a mutant gene is advantageous or neutral, it may be retained, especially if it confers some selective advantage in a changing environment. Individual mutations, however, are rarely agents of recognizable change. Usually we notice an accumulation of mutations, each of which alters slightly the appearance or function of an organism. Most mutations just maintain variability in the gene pool of a population. Without mutations, a population would not respond any further to natural selection.

19.4   Variation is transmitted by sexual reproduction

Variations in a population are transmitted to the next generation through sexual reproduction when gametes join to form zygotes. If a gene occurs in two forms, *A* and *a*, then individuals can fall into three possible diploid classes: *AA*, *aa*, and *Aa*. Individuals

in which the alleles are the same, *AA* or *aa*, are homozygous. Haploid gametes produced by homozygous individuals are either all *A* or all *a*; those by heterozygous individuals, half *A* and half *a*. Figure 19.2 shows how two homozygous populations recombine in the first generation, or F_1 generation. The proportion of gametes carrying *A* and *a* is determined by the individual genotypes, the genes received from the parents.

If the mating pattern is random—that is, if the chance that an individual will mate with another individual having a certain genotype is equal to the frequency of that genotype in the population—then we can predict the genotypes in the next generation (Quantifying Ecology 19.1: Predicting Gene Frequencies).

Assume that individuals of a population homozygous for the dominant allele *AA* are mixed with an equal number from a population homozygous for the recessive *aa*. The allele frequency in the total population is 0.5 *A* and 0.5 *a*. The offspring of the first matings, the F_1 generation, will all be heterozygous, *Aa*, and the frequency of the alleles will still be 0.5 *A* and 0.5 *a*. The offspring of these heterozygotes, the F_2 generation, will consist of 0.25 *AA*, 0.50 *Aa*, and 0.25 *aa* (Figure 19.3). These proportions represent the **genotypic frequencies.** Notice that the allele frequencies still remain at 0.5 *A* and 0.5 *a*. Twenty-five percent of the population is homozygous *AA*, 25 percent is homozygous *aa*, and 50 percent of the population is heterozygous *Aa*. Dividing the frequency of the

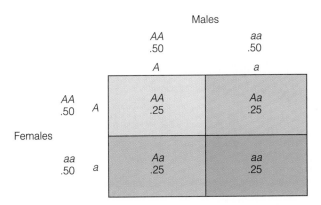

Figure 19.2 Mixing two homozygous populations to produce the F_1 generation.

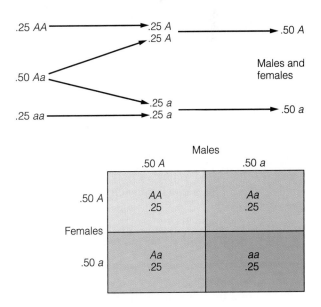

Figure 19.3 Proportions in the F_2 generation.

QUANTIFYING ECOLOGY 19.1

PREDICTING GENE FREQUENCIES

Gene frequencies are predicted by one of the important statements in population genetics, the Hardy-Weinberg law. This law holds that the various genotypes will remain the same in successive generations of a sexually reproducing population if certain criteria are met: (1) mating is random; (2) mutations do not occur; (3) the population is large, so that changes by chance in gene frequencies are insignificant; (4) natural selection does not occur; and (5) migrations do not occur.

We can express the frequency of the three genotypes as a binomial equation. Let p be the proportion of alleles that are A, and q the proportion of alleles that are a. Because all loci for that gene must be occupied either by A or a, $p + q = 1$. Start by assuming that A (or p) makes up 50 percent of the alleles, and a (or q) the other 50 percent, so $0.5 + 0.5 = 1.0$. In the F_1 generation, the genotypic frequency is $0.25\ AA + 0.50\ Aa + 0.25\ aa$. The frequency of the AA matings is $p \times p = p^2$; the frequency of aa matings is $q \times q = q^2$; and the frequency of Aa matings is $(p \times q) + (p \times q)$ or $2pq$. The frequency of the genotypes can be expressed as $p^2 + 2pq + q^2 = 1$ (see Figure 19.2).

The Hardy-Weinberg law predicts that the frequency of the alleles will not change. Consider the genotypic frequency of allele A. In the homozygote it is p^2, and in heterozygote it is $\frac{1}{2}(2pq)$. Combining them, we find the frequency of A in the F_2 generation is $p^2 + \frac{1}{2}(2pq) = p^2 + pq$. Since $q = (1 - p)$, the frequency of allele A equals:

$$p^2 + p(1 - p) = p^2 + p - p^2 = p$$

A similar situation exists for q. The frequency of the alleles does not change (see Figures 19.2, 19.3)

You can try another example, with different numbers. This time let the parental generation have the genotypic frequency of $0.36\ AA$, $0.48\ Aa$, and $0.16\ aa$. The allelic frequencies will be 0.6 for A and 0.4 for a; and the genotypic frequencies for the F_1 generation will still be $0.36\ AA$, $0.48\ Aa$, and $0.16\ aa$. We can conclude that all succeeding generations will carry the same proportions of the three genotypes, provided the criteria are met (Figure A).

In natural populations the criteria of the Hardy-Weinberg law are never fully met. Mutations do occur. Matings are not random. Individuals move between populations, and natural selection does take place. All of these circumstances change genotypic frequencies from one generation to another, and act as evolutionary forces in a population. For this reason the Hardy-Weinberg law must be considered theoretical—a distribution against which we compare observations.

The beauty of the Hardy-Weinberg law is that it allows us to determne allele frequencies in a population when we know the frequency of just one allele, q. Because q is the recessive or rarer allele, its expression in the homozygotic condition (frequency of q^2) provides all the information we need. Deviations in genotypic frequencies from the Hardy-Weinberg equilibrium will suggest the presence of selective pressures acting on the population. An important implication of this law is that in the absence of specific evolutionary or selective forces to change the frequencies, Mendelian inheritance alone is enough to maintain genetic variability in a population.

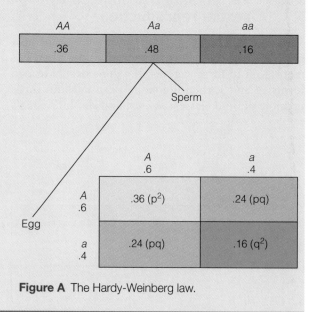

Figure A The Hardy-Weinberg law.

heterozygotes by 2, because half of the proportion of the population carries both A and a, we obtain 25 percent of A and 25 percent of a. Adding that 25 percent

A from the heterozygotes to the 25 percent from the homozygotes, we arrive at an allele frequency of 0.5 for A, and similarly a frequency of 0.5 for a.

19.5   Environmental forces may favor one phenotype

Reproduction is not random, because natural selection is at work. Environmental forces, abiotic or biotic, favor certain genotypes, as expressed by phenotypes, over others. Over time these selective pressures bring about a change in gene frequency in a population, or **evolution.** Biological systems have changed in response to selection pressures. Although natural selection is a major force in evolution, the two are not synonymous.

Natural selection can proceed without any major change in the phenotype. When it favors phenotypes near the population mean at the expense of the two extremes, selection is **stabilizing** (Figure 19.4a). It is characteristic of stable environments. In other situations natural selection favors one extreme phenotype over another (Figure 19.4b). In this case selection is **directional.** The mean phenotype shifts toward one extreme, provided that inheritable variations of an effective kind are present.

Consider the survival of Darwin's medium ground finch *(Geospiza fortis)* on the 40 ha islet of Daphne Major, one of the Galápagos Islands, from 1975 through

1978. During the early 1970s, the island received regular rainfall (127–137 mm), supporting an abundance of seed and a large finch population (1500 birds). In 1977, however, only 24 mm of rain fell. In the drought seed production declined drastically. Small seeds declined in abundance faster than large seeds, increasing the average size and hardness of seeds available. The finches, which normally fed on small seeds, had to turn to larger ones. Small birds had difficulty finding food. Large birds, especially males with large beaks, survived best because they were able to crack large, hard seeds. Females suffered heavy mortality. Overall, the population declined 85 percent from mortality and possibly emigration (Figure 19.5).

This example suggests that natural selection exerts its greatest influence during periods of environmental stress during a small portion of an organism's life history. It also suggests that a small, isolated, relatively contained, morphologically variable population of a species under the selective pressure of a variable environment can experience rapid evolution.

In some situations natural selection favors both extremes simultaneously, although not necessarily to the same degree (Figure 19.4c). Such selection, known

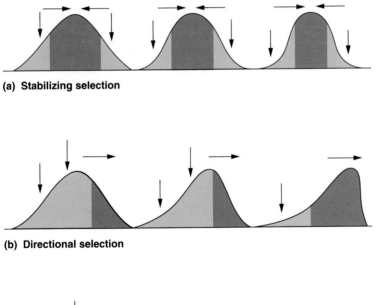

(a) Stabilizing selection

(b) Directional selection

(c) Disruptive selection

Figure 19.4 Three types of selection. (a) Stabilizing selection favors organisms with values close to the population mean. Little or no change takes place. (b) Directional selection moves the mean of the population toward one extreme. (c) Disruptive selection increases the frequencies of both extremes. Downward arrows represent selection pressures; horizontal arrows represent the direction of evolutionary change.

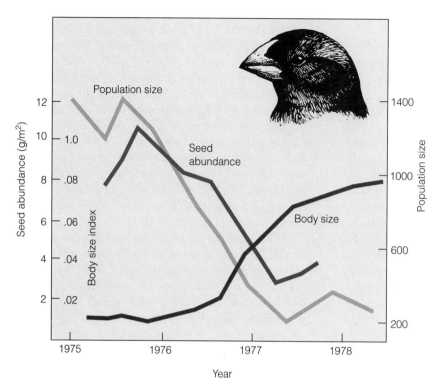

Figure 19.5 Evidence of directional selection in Darwin's medium ground finch, *Geospiza fortis*. The yellow line represents the population estimate on the island of Daphne Major based on the censuses of marked populations, and the green line estimates seed abundance, excluding two species of seeds never eaten by any Galápagos finches. Populations declined in the face of seed scarcity during a prolonged drought. The brown line plots changes in body size. Note how the body size of surviving birds increased during the drought period, suggesting that small-bodied birds were being selected against and large-bodied birds were being favored. Results suggest that the most intense selection in a species occurs under unfavorable environmental conditions. (Reprinted with permission from P. T. Boag and P. R. Grant, "Intense Natural Selection in a Natural Population of Darwin's Finches," *Science* 214 (1981):83. Copyright © 1981 American Association for the Advancement of Science.)

as **disruptive,** occurs when members of a population are subject to different selection pressures operating in different environments, such as microhabitats. It usually results in a population containing two or more genotypes.

Consider as an example the swallowtail butterfly *Papilio dardanus*, widely distributed across Africa (Figure 19.6). The females mimic an associated inedible species of butterfly that possesses warning coloration. (Warning coloration tells a potential predator that the bearer is inedible or dangerous; see Chapter 16.) The male is not a mimic. Instead, he retains a specific color pattern the females recognize, essential for mating and reproduction. Within any given part of the species range, the females mimic an inedible species common to that region. At least three different female mimics exist. Intermediate female forms that do not bear any resemblance to the inedible model are selected against. In regions where inedible models are absent, so are mimicking females. In this example predation promotes disruptive selection in the population.

Disruptive selection may also produce several distinct forms of a species in the same habitat at the same time. This phenomenon is called **polymorphism.** It may involve differences in morphological characters and in physiology. The important feature of polymorphism is that forms are distinct and discontinuous.

There are no intermediates. Such polymorphisms appear to be environmentally induced.

A classic example of genetic polymorphism is industrial melanism in the peppered moth *Biston betularia* in England. Before the middle of the nineteenth century, the moth, as far as is known, was always white with black speckling on wings and body. In 1850, near the manufacturing center of Manchester, a black form of the species was caught for the first time (Figure 19.7). The black form, *carbonaria*, increased steadily through the years until it became common, reaching a frequency of 95 percent or more in Manchester and other industrial areas. From these places *carbonaria* spread far into rural areas. The form came about through the spread of dominant and semidominant mutant genes under the selective pressure of predation by birds. The original light form of the peppered moth made it inconspicuous on lichen-covered tree trunks. Birds picked off the dark form, conspicuous on the pale background. After the rise of factories, the grime and soot of the industrial areas killed or reduced the lichens on the trees and turned the bark a nearly uniform black. Now the black form had the selective advantage. In the polluted woods the white form bore the brunt of predation.

In the case of the peppered moth, environmental changes converted a disadvantageous allele or mutant

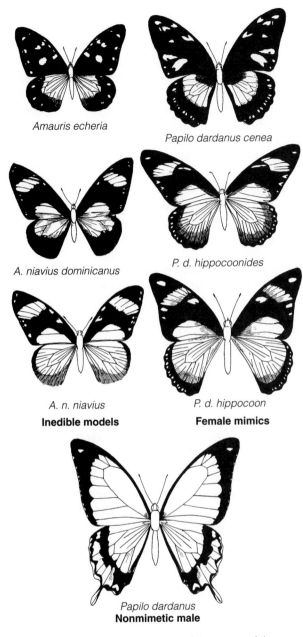

Figure 19.7 Melanistic and normal forms of the polymorphic moth *Biston betularia* at rest on a lichen-covered tree. On soot-darkened, lichen-free trunks, selection favors the dark form. Away from industrial areas, the light color is most frequent because black individuals resting on lichen-covered trunks are subject to heavy predation by birds.

Lichens returned to the tree trunks. In some areas since then, the dark form has decreased from 90 percent to 30 percent of the population.

19.6   Inbreeding reduces genetic variability

Populations, however widespread, consist of rather tight-knit local groups or demes. The white-tailed deer, common as it is throughout its range, consists of fairly independent groups. This lack of interchange of individuals between demes probably results from the matriarchal society of deer, the tendency of females to remain close to their birthplace, and short movements by males in search of mates. Although such behavior suggests that inbreeding should be common, apparently enough movement of males among demes takes place that genetic variability remains. In fact, the white-tailed deer exhibits high variability throughout its range. Collectively, demes across a region make up the population gene pool.

Now suppose that forest clearing, urban spread, industrial development, road construction, and the like fragment the population, eliminating some demes and isolating others. That is happening to many populations of plants and animals worldwide. Each isolated population represents only a sample of the total gene pool. The isolated population is now subject to inbreeding.

Inbreeding is mating between relatives. In small populations individuals may be forced into this situation. With inbreeding, mates on the average are more closely related than they would be if they had been chosen at random from the population. Inbreeding occurs because of small populations, close proximity

Amauris echeria

Papilo dardanus cenea

A. niavius dominicanus

P. d. hippocoonides

A. n. niavius
Inedible models

P. d. hippocoon
Female mimics

Papilo dardanus
Nonmimetic male

Figure 19.6 Disruptive selection in three races of the African swallowtail butterfly *Papilio dardanus.* The African swallowtail is widely distributed across the continent as a number of races, including *dardanus* in west and southwest Africa, *cenea* in southern Africa, and *polytropus* and *tribullus* in east Africa. Males of all races are nonmimetic. Females of each race (right) have wing patterns and coloration that mimic inedible species in their own region.

gene into an advantageous one, permitting its spread. Such a polymorphism will persist as long as environmental conditions favor the dark form. The passage of Britain's Clear Air Act in 1965 reduced air pollution.

of potential mates, ecological preferences, and the like. Inbreeding increases homozygosity and reduces heterozygosity.

The extreme form of inbreeding is self-fertilization, which occurs among some plants. This extreme provides the basis for comparisons with less intense forms of inbreeding. With each generation of inbreeding the homozygotes *AA* and *aa* breed true. Offspring from the heterozygotes *Aa* will be one-half heterozygous *Aa*, one-quarter homozygous *AA*, and one-quarter homozygous *aa*. The same situation will occur in the next generation. Add these new homozygotes to those already in the population, and you will discover that eventually self-fertilized populations will become exclusively homozygous, *AA* and *aa*. The more usual situation is close inbreeding, usually between brother and sister, parent and offspring, and first cousins, individuals who share a number of like genes.

In normally outbreeding populations, close inbreeding is detrimental. Rare, recessive, deleterious genes become expressed. They can cause decreased fertility, loss of vigor, reduced fitness, reduced pollen and seed fertility in plants, and even death. These consequences are **inbreeding depression.**

Of course, not all inbreeding is bad. Occasional inbreeding will fix rare alleles that otherwise might be lost. Animal and plant breeders use inbreeding to fix certain desirable genes that will breed true. These inbred lines are then outcrossed to produce hybrid vigor.

Except for self-fertilizing plants, close inbreeding in nature is rare. It amounts to less than 2 percent for natural populations for which there are data. Certain safeguards in nature reduce inbreeding. They include spatial separation or differences between sexes in the dispersal of young. One sex stays behind; the other leaves. Monogamous mating habits and the frequent loss of a mate during the reproductive season or between seasons reduces inbreeding in birds. Female mammals tend to stay near their birthplace, whereas young males leave or are driven away. Among prairie dogs (*Cynomys ludovicianus*) young males leave the family group before breeding, and adult males may move to new breeding groups if adult daughters are in the home place. (Although dispersal does reduce inbreeding, sex-biased dispersal may have evolved for other reasons, such as enhanced reproductive success.) Kin recognition may also reduce close inbreeding. Because of their close association in early life, siblings recognize one another over time. Females mate with unrelated males, leave the group if a related male returns, or fail to come into estrus if their father is in the group.

All these safeguards break down if the population is small and highly isolated.

The adverse effects of inbreeding can be mitigated by outcrossing or **outbreeding,** mating between unrelated individuals. Outbreeding can take place in demes large enough that many individuals are not closely related. It occurs among individuals of different populations when there are long-distance dispersers or when humans introduce individuals from distant populations. Although it adds genetic diversity, outbreeding can carry its own problems. When nonrelatives mate, descendants can experience reduced viability if each parent is adapted to its own local environment. The offspring, hybrids of two different local populations, may be poorly adapted to the local environment of either parent. For example, bobwhite quail from the warm southern United States were introduced to the northern United States to augment small populations of cold-adapted northern bobwhite quail. Poorly adapted to the cold weather, the offspring succumbed to winter, with the eventual loss of the affected quail population. Such maladaptation of the offspring is **outbreeding depression.**

19.7   Small populations may experience genetic drift

Small populations—whether small from habitat fragmentation, heavy mortality, or immigration to new or empty habitats—are more vulnerable to chance fluctuations in gene frequency. These populations may experience **founder effect.** Individuals carry only a small sample of the gene pool of the parent population. All genes in subsequent populations will stem from the limited genetic material carried by the founders. The small size of the population enforces inbreeding among related individuals. Recessive genes become more widely exposed, some reducing survival. A small sample of the parent population and a limited diversity of genetic material may have a powerful influence on the development of a new species. (See Section 19.14.)

Also altering the genetic composition of populations is **genetic drift,** a random change in gene frequency, especially pronounced in small populations. Random sampling of genes causes chance fluctuations in allele frequencies. In a small population each mating will include only a sample of the already small sample of the parent population's genes, because only a part of the population is breeding (Focus on Ecology 19.1: Genetic Drift in Small Populations). Through time

GENETIC DRIFT IN SMALL POPULATIONS

Measurements of genetic drift are based on an ideal population, characterized by a constant population size, equal sex ratios, equal probability of mating among all individuals, and a constant dispersal rate. Most populations, of course, are not ideal. They have age-related differences in reproduction, and particularly in polygamous populations, the ratio of breeding males to females is unequal. In such populations the number of males is more important than the number of females in determining the amount of random drift. For this reason the actual size of a small population or a subpopulation is of little meaning. Of greatest importance is the genetically effective population size, N_e. N_e is not the same as the actual number of breeding individuals. Unless the sexes are equal, N_e is less than N. The effective population size is defined as the size of an ideal deme (a randomly breeding one with a 1:1 sex ratio and with the number of progeny per family randomly distributed) that would undergo the same amount of random genetic drift as the actual population.

In a monogamous population in which one male mates with one female, all offspring are less related than in a polygamous population with, say, a ratio of one breeding male to four females. In the latter situation, the offspring would be one-half or full sibs. The chance of an allele being lost or fixed and therefore the amount of genetic drift is much greater in a polygamous population.

Population geneticists define the effective population size by the following formula:

$$N_e = \frac{4(N_m N_f)}{N_m + N_f}$$

where N_m and N_f are the numbers of breeding males and females. As the disparity in the ratio of males to females widens, the effective population size diminishes.

Consider a population of white-tailed deer consisting of 100 adult does and 40 adult bucks. The actual size of the population is 140, but because of the unequal sex ratio, the effective population size is $4(40 \times 100)/140 = 114$. In other words, the population of 140 deer will experience the same amount of genetic drift as an ideal population of 114 deer. Genetically, the population is smaller than the actual numbers imply.

---

some genes become fixed or homozygous for one allele, or other alleles are lost, and some continue to segregate, until there is a homozygous population. The smaller the population, the more rapid genetic drift becomes. If the genes involved are maladaptive, the small population will become extinct. If they confer a selective advantage in the new environment, they will enhance the fitness of the population. Thus chance may play an important part in evolutionary change.

Genetic drift is akin to inbreeding in its effect on genetic diversity. The major difference is that inbreeding involves nonrandom mating, whereas genetic drift involves random mating. Small populations with genetic drift may also undergo some inbreeding simply by chance. Both inbreeding and genetic drift increase homozygosity and decrease heterozygosity.

Populations are dynamic, fluctuating in numbers over time. Under adverse environmental conditions or a sudden loss of habitat, a population may decline sharply or "crash." The survivors of the crash, the pro-

genitors of future populations, possess only a sample of the original gene pool. This sharp reduction in numbers creates a **population bottleneck** (Figure 19.8). A bottleneck can severely reduce genetic diversity in the remaining population and future generations. An example is the northern elephant seal (*Mirouna angustirostris*). Fur hunters reduced the elephant seal population to about 20 animals in 1890. Now protected, the population has grown to over 30,000 individuals. Two biologists, M. C. Bonnell and R. K. Selander, examined the genetic variation among the seals by electrophoresis. They found no variation in a sample of 24 electrophoretic loci. In contrast, the population of southern elephant seals, never severely reduced, retained high genetic variability. The lack of genetic variation in the northern elephant seal increases its vulnerability to environmental change.

Unless they are separated by a wide expanse of inhospitable habitat, some interchange takes place among subpopulations. If these immigrants enter the

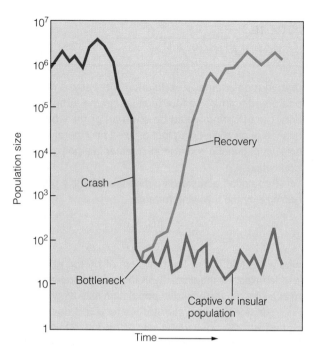

Figure 19.8 A population faced with an environmental catastrophe enters a bottleneck. Only a small sample of the gene pool survives. If the population makes a recovery, it may lack genetic diversity. If the population remains small, as in a captive or insular situation, it is subject to genetic drift.

breeding population, they introduce a different genetic sample that tends to reduce or slow genetic drift and helps maintain genetic diversity. Population geneticists believe that one successful immigrant per generation is the minimum number needed to slow genetic drift and five immigrants the maximum, depending upon whether the population is monogamous or polygamous. Just as inbreeding and drift take place more swiftly in a polygamous population, so does the spread of new genes, especially when the harem is large. New genes spread much more slowly in a monogamous population.

19.8 Species need a minimum number for survival

The concepts of effective population size, inbreeding, genetic drift, and dispersal are of paramount importance in the conservation of species and maintenance of biological diversity. Human population growth, fragmentation and elimination of habitats, and poaching are reducing populations of increasing numbers

of species. Many are precariously low, including the black rhinoceros (*Diceros bicorn*), Sumatran rhinoceros (*Dicerorhinus sumatrensis*), tiger (*Panthera tigris*), chimpanzee (*Pan troglodytes*), tamarins (*Leontopithecus* spp.), red-cockaded woodpecker (*Picoides borealis*), spotted owl (*Strix occidentalis*)—the list goes on and on. Already the gene pool has been depleted, and the question is what sizes of populations are needed to save the species.

A threshold number of individuals is necessary to ensure the persistence of a subpopulation in a viable state for a period of several hundred years. This threshold number is the **minimum viable population.** This population must be large enough to cope with chance variations in individual births and deaths, random series of environmental changes, genetic drift, and catastrophes. Genetic models suggest that populations with an effective population size of 100 or less and an actual size of less than 1000 are highly vulnerable to extinction. An accumulation of mildly harmful mutations that become fixed could drive such populations to extinction within 100 generations. To persist they need at least 1000 breeding individuals per generation.

When habitats are fragmented, local populations of various sizes combine to form a metapopulation (see Section 11.5). What is the minimum viable metapopulation? It takes both enough subpopulations and enough suitable habitat for them. We have to reserve unoccupied patches for potential colonization. It is difficult to estimate the size of a minimum viable metapopulation, but theoretically ten interacting viable local populations together with potential habitat patches are necessary for long-term persistence.

The inherent danger of a rule of thumb for minimum viable populations and metapopulations is that politicians, developers, and governments will try to get by with the theoretical minimum. A good example is the controversy over how much old growth forest in the Pacific Northwest of the United States is needed to save the spotted owl.

19.9 Defining a species is critical and difficult

Critical to the study of the genetics of populations and of ecological relationships and to the management and preservation of populations is the concept of the species. In this age of exploding human population, habitat destruction, and declining species populations, the survival of many species depends upon human intervention. The species then becomes the significant unit for conservation management.

Are separate populations of a species genetically the same? Are they different enough to warrant separate treatment? Is it in the best interests of species conservation to introduce individuals from remote populations into threatened local populations? These questions are crucial for some species. The orangutan (*Pongo pygmaeus*) populations on Sumatra and Borneo are chromosomally distinct, even though they look the same. When individuals of one population are introduced into another population, the hybrids between chromosomal types show outbreeding depression. A similar situation exists with spider monkeys (*Ateles* spp.), dik-diks (Madoquinae, small African antelopes), and other species. Not knowing such information, we do more harm than good. For this reason defining species has become a lively area of research.

Field guide in hand, we can distinguish a robin from a wood thrush or a white oak from a red oak. Each has morphological characteristics that set it apart. Each is an entity, a discrete unit to which a name has been given. That is the way Carl von Linne, who gave us our system of binomial classification (Focus on Ecology 19.2: Classifying Organisms) saw plants and animals. He, like others of his day, regarded the many different organisms as fixed and unchanging units, the so-called products of special creation. They differed in color, pattern, structure, proportion, and other characteristics. By these criteria biologists described, separated, and arranged species into groups. Each species was discrete. Some variation was permissible, but those variants were considered accidental. This classic **morphological species** concept is still alive, useful, and necessary for classifying the vast number of plants and animals. It is basic to field guide descriptions of organisms.

Later the studies of Charles Darwin on variation, of Alfred Wallace on geographical distribution, and of Gregor Mendel on genetics emphasized that variation within a species is the rule. Because of sexual dimorphism, male and female of the same species look like separate species. Many closely related forms replace each other geographically. Variants intergrade, so it is difficult to separate them precisely. A better definition of species was needed.

Ernst Mayr, an evolutionary biologist and taxonomist, advanced the idea of the **biological species,** a group of actually or potentially interbreeding populations that are reproductively isolated from other such groups. According to this concept, a species embodies a group of interbreeding individuals living together in a similar environment in a given region and under similar ecological relationships. The individuals recognize each other as potential mates. They are a genetic unit, in which each individual holds for a short period of time a portion of the contents of an intercommunicating gene pool.

This definition has limitations. It applies only to bisexual organisms, excluding asexual ones, and it overemphasizes between-population reproductive isolation. It views a species as a single biological entity, when in reality a species consists of numerous local populations, each somewhat different. These local populations interbreed enough to provide continuity and intergradation in characteristics from one population to another. Other distinguishable populations within a region that do not interbreed with that species are considered separate species. Evolutionary biologists suggest that we use the biological species for practical purposes to identify kinds of organisms, but that we think of the species as an evolutionary unit.

19.10 Species may be characterized by spatial relationships

Species may be sympatric or allopatric. **Sympatric species** occupy the same area at the same time, so they have the opportunity to interbreed. The fact that they do not marks them as "good" species, because they are reproductively isolated from others. **Allopatric species** occupy areas separated by time and space. Because they do not have the opportunity to meet similar species, there is no indication whether they are capable of interbreeding with other allopatric species. Only if the barriers are broken, allowing them to come together, can you test reproductive isolation. Often two allopatric species are not reproductively isolated, and without barriers the two species behave as one. Such is the case, for example, with the red-shafted flicker and the yellow-shafted flicker (*Colaptes auratus*) (Figure 19.9). The two, once allopatric, interbred when they became sympatric and are now considered one species.

Some sympatric species are **sibling species**—ones that are similar in appearance or morphology but do not interbreed. They may differ in behavior, ecology, or in physiology and chromosomal structure. To the human eye the species may be virtually indistinguishable, but not to the animal. Examples of sibling species are the three species of 17-year periodical cicadas,

(a)

(b)

Figure 19.9 Flickers: (a) the eastern form of the "yellow-shafted" flicker has yellowish wing linings and undertail surfaces, and a brown face with a black mustache. (b) The western form of the "red-shafted" flicker has pinkish-red wing linings and undertail surfaces, and a gray face with a red mustache. Once considered separate species, they are now regarded as one species because they intergrade and interbreed.

Magicicada septendecim, *M. cassini*, and *M. septendecula*. The members of each group emerge in synchrony. They maintain their separateness by differences in the type and timing of their mating choruses. The chorus of *M. septendecim* is most intense in the morning, an even, monotonous, nonsynchronized buzzing. The chorus of *M. cassini* is most intense in the afternoon, a shrill, sibilant, synchronized buzzing that rises and falls in intensity. The chorus of the small *M. septendecula* is most intense about midday, a more or less continuous, nonsynchronized repeating of short, separate buzzes with no regular fluctuations in pitch or intensity.

19.11 Species vary over a geographic range

There is widespread variation in many morphological, physiological, behavioral, and genetic characteristics in a widely distributed species. Significant differences often exist among populations in different regions. The greater the distance between populations, the more pronounced the differences may become. Geographic variants reflect the selective environmental forces acting on various phenotypes, each population adapting to the locality it inhabits. Geographic variation shows up as clines, ecotypes, and geographic isolates.

A **cline** is a measurable, gradual change over a geographic region in the average of some phenotypic character, such as size and coloration, or a gradient in genotypic frequency. Clines are usually associated with an ecological gradient such as temperature, moisture, light, or altitude. Continuous variation results from gene flow from one population to another along the gradient. Because environmental selection pressures vary along the gradient, any one population along the gradient will differ genetically to some degree from another, the difference increasing with the distance

CLASSIFYING ORGANISMS

Although humans have been classifying plants and animals for centuries, there was no uniform scheme. Then came a young Swedish naturalist and medical doctor, Carl von Linne, who Latinized his name to Carolus Linnaeus (1707–1778). Linnaeus had a passion for classifying. He developed a binomial or two-name system of naming plants and animals. This system finally brought order out of chaos. Linnaeus named all living organisms in Latin, then the universal language of science. Because it was a dead language, it would never change in grammar, syntax, or vocabulary. That gave it a great degree of stability.

Linnaeus explored Lapland and northern Europe; he had collectors send him plants from other parts of the world. Out of this research came a book in 1735, *Species Plantarum.* In it he listed and named all plants then known to science, giving each plant two Latin names—a genus name and a species name. In 1758, he published a tenth edition, *Systema Naturae,* in which he consistently applied the same idea to animals.

Before Linnaeus could assign names to each organism, he had to sort plants and animals into broad general groups. First, all living things, he reasoned, belonged to two great groups or kingdoms, plant and animal. The animal kingdom he subdivided into six parts—mammals, birds, reptiles, fishes, insects, and a catchall group, vermes (for worms).

Now each group divided again. Among the mammals, for example, he noted that some organisms were similar. It was not too difficult to identify a cat, although there are many different kinds. All have retractile claws, rounded heads, and well-developed but relatively small rounded ears. Some have long tails; others have short tails, ear tufts, and a ruff on the cheeks and throat. All the long-tailed cats, then, were placed in one genus, to which Linnaeus give the Latin name *Felis,* for cat. To the short-tailed cats he gave the name *Lynx.* Doglike mammals, too, are readily distinguishable, and he broke them down into groups similar in appearance: the sharp-nosed little foxes, the doglike wolves, and their descendants, the true dogs. To wolves and dogs, Linnaeus gave the name *Canis,* Latin for dog. Each genus then contained a number of different animals, with some broad structural characteristics in common, yet with differences in finer details—color, size, thickness of hair, and the like. Thus, the genus could be broken down into still finer groups, the individual species. Linnaeus gave a second part to the name of species. The wolf became *Canis lupus,* the mountain lion *Felis concolor,* and the African lion *Felis leo.*

Later all catlike animals were grouped together into a larger classification, the family, which takes its name from one of the genera it embraces. In the case of cats, the genus *Felis* was used. By adding the standardized suffix *idae* onto the root word *Felis,* the cat family became Felidae; and by a similar process the dog family became Canidae. Dogs and cats, skunks and weasels (Mustelidae), raccoons (Procyonidae), bears (Ursidae), seals (Phocidae), and so on were classified into families. All of them have several general features in common—five toes with claws, shearing teeth, well-developed canine teeth or fangs, and other skeletal structures. Therefore these animals were grouped together and placed in an even higher category, the order, in this case Carnivora. All animals giving birth to living young, possessing body hair, and nourishing young with milk were placed in an even higher category, the class Mammalia. All animals having a backbone were grouped into a phylum (in this instance Chordata) belonging to the animal Kingdom.

The final result was a hierarchy of classification in which similar species were grouped in a genus, similar genera were grouped in a family, and similar families were grouped into orders. The principal classification of the mountain lion *(Felis concolor)* looks like this:

> Kingdom: Animalia
> Phylum: Chordata
> Class: Mammalia
> Order: Carnivora
> Family: Felidae
> Genus: *Felis*
> Specific epithet: *Concolor*

As time went on, the number of known animals swelled, and the need arose to redefine this hierarchy. Plyla were divided into subphyla, classes into subclasses and infraclasses, orders into suborders, families into subfamilies (and even enlarged into superfamilies), and species into subspecies.

The principal categories used in plant classification are somewhat different. For example, here are the categories for the American beech *(Fagus grandifolia)*:

Kingdom: Plantae
Division: Tracheophyta
Subdivision: Pteropsida
Class: Angiospermae
Subclass: Dicotyledonae
Order: Fagales
Family: Fagaceae
Genus: *Fagus*
Specific epithet: *Grandifolia*

between the populations. For this reason, populations at the two extremes along the gradient may behave as different species. The length of the cline depends upon the gene flow and dispersal distance between populations.

Clinal differences exist in size, body proportions, coloration, and physiological adaptations among animals. For example, the white-tailed deer in North America exhibit a clinal variation in body weight. Deer in Canada and the northern United States are heaviest, weighing on the average more than 136 kg. Deer weigh 93 kg in Kansas, 60 kg in Louisiana, and 46 kg in Panama. The smallest, the Key deer in Florida, weighs less than 23 kg. Many species of plants exhibit clinal gradation, some in size and others in time of flowering, growth, or other physiological responses to the environment. A number of prairie grasses, among them blue grama *(Bouteloua gracilis)*, side-oats gamma *(B. curtipendula)*, big bluestem *(Andropogon gerardi)*, and switchgrass *(Panicum virgatum)*, flower earlier in the northern and western regions of their ranges and progressively later toward the south and east.

Clinal variations can show marked discontinuities. Such abrupt changes, or step clines, reflect abrupt changes in selection pressures in local environments. Such variants are called **ecotypes.** An ecotype is a genetic strain of a population adapted to its unique local environmental conditions. For example, a population inhabiting a mountaintop may differ from a population of the same species in the valley below. Often eco-

types will be scattered like a mosaic. That frequently is the situation when several habitats to which the species is adapted recur throughout the range of the species. Some ecotypes may have evolved independently from different local populations.

Yarrow, *Achillea millefolium*, blankets the temperate and subarctic Northern Hemisphere with an exceptional number of ecotypes. It exhibits considerable variation, an adaptive response to different climates at various latitudes. Populations at lower altitudes are tall and have high seed production. Montane populations have a distinctive small size and low seed production.

The southern Appalachian Mountains are noted for their diversity of salamanders, fostered in part by a rugged terrain, an array of environmental conditions, and the limited ability of salamanders to disperse (Figure 19.10). Populations become isolated from one another, preventing a free flow of genes. One species of salamander, *Plethodon jordani*, breaks into a number of semi-isolated populations, each characteristic of a particular part of the mountains. Groups of such populations make up **geographic isolates,** prevented by some extrinsic barrier—in the case of the salamanders, rivers and mountain ridges—from effecting a free flow of genes with other subpopulations. The degree of isolation depends upon the efficiency of the extrinsic barrier, but rarely is the isolation complete. These geographic isolates make up subspecies. Strong isolates with little gene flow between them may be in the first stages of speciation. Unlike clines, we can draw a line that will separate them all into subspecies.

19.12 Certain traits keep species apart

Each spring there is a rush of courtship and mating activity in woods and fields. Fish move to spawning areas; amphibians migrate to breeding pools; birds sing. During this frenzy of activity each species remains distinct. Song sparrows mate with song sparrows, brook trout with brook trout, wood frogs with wood frogs, and few mistakes occur, even between species similar in appearance.

The means by which diverse species remain distinct are **isolating mechanisms.** They include morphological characters, behavioral traits, ecological conditions, and genetic incompatibility. Isolating mechanisms

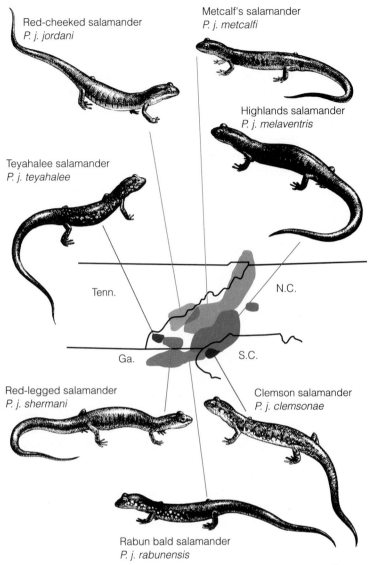

Red-cheeked salamander
P. j. jordani

Metcalf's salamander
P. j. metcalfi

Highlands salamander
P. j. melaventris

Teyahalee salamander
P. j. teyahalee

Tenn.

N.C.

Ga.

S.C.

Red-legged salamander
P. j. shermani

Clemson salamander
P. j. clemsonae

Rabun bald salamander
P. j. rabunensis

NED SMITH

Figure 19.10 Geographical isolates in *Plethodon jordani* of the Appalachian highlands. These salamanders originated when the population of the salamander *P. yonahlossee* became separated by the French Broad River valley. The eastern population developed into Metcalf's salamander, which spread northward, the only direction in which any group member could find suitable ecological conditions. South, southwest, and northwest the mountains end abruptly, limiting the remaining *jordani*. Metcalf's salamander is the most specialized and ecologically divergent and the least competitive. Next the red-cheeked salamander became isolated from the red-legged and the rest of the group by the deepening of the Little Tennessee River. Remaining members are still somewhat connected.

may be premating or postmating. Premating mechanisms—those that prevent interspecific crosses—include habitat selection, temporal isolation, behavior, and mechanical or structural incompatibility. Postmating mechanisms reduce the survival or reproductive success of interspecific crosses.

If two potential mates in breeding condition have little opportunity to meet, they are not likely to interbreed. Habitat selection reinforces this isolation among frogs and toads. Different calling and mating sites among concurrently breeding frogs and toads

tend to keep the species separated. The upland chorus frog *(Pseudoacris triseriata feriarum)* and the closely related southern chorus frog *(Pseudoacris nigrata)* breed in the same pools, but calling males tend to segregate themselves in different locations in the pond. The southern chorus frog calls from concealed positions at the base of grass clumps or among vegetation debris, whereas the upland chorus frog calls from more open locations.

Temporal isolation (differences in the timing of breeding and flowering seasons) separates some sympatric species. The American toad *(Bufo ameri-canus)* breeds early in the spring, whereas Flower's toad breeds a few weeks later. Fluctuations in environmental stimuli can time mating seasons. Among the narrow-mouthed toads, *Microhyla olivacea* breeds only after rains, whereas *M. carlinensis* is little influenced by rain. Besides this partial temporal isolation, call discrimination separates the toads; nevertheless, some hybridization does occur.

Ethological barriers (differences in courtship and mating behavior) are the most important isolating mechanisms. The males of animals have specific courtship displays, to which, in most instances, only females of the same species respond. These displays involve visual, auditory, and chemical stimuli. Some insects, such as certain species of butterflies and fruit flies, and some mammals possess species-specific scents. Birds, frogs and toads, some fish, and such singing insects as crickets, grasshoppers, and cicadas have specific calls that attract the "correct" species. Visual signals are highly developed in birds and some fish. Species-specific color patterns, structures, and displays, which give rise to a high degree of sexual dimorphism among such bird families as hummingbirds and ducks, have apparently evolved under sexual selection (see Section 12.4). Among insects, the flight patterns and flash patterns of fireflies on a summer night are most unusual visual stimuli. The light signals emitted by various species differ in timing, brightness, and color, which may range from white through blue, green, yellow, orange, and red.

Mechanical isolating mechanisms make copulation or pollination between closely related species impossible. Although evidence for such mechanical isolation among animals is scarce, differences in floral structure and intricate mechanisms for cross-pollination are common among plants. If hybrids should occur, they could possess such inharmonious combinations of floral structures that they would be unable to function together, either to attract insects or to permit insects to enter.

These four types of premating isolating mechanisms—ecological, temporal, behavioral, and mechanical—prevent the wastage of gametes, diminish the number of hybrids, and permit populations of incipient species to become wholly or partly sympatric.

The postmating isolating mechanism, the reduction of mating success, does not prevent the wastage of gametes, but it is highly effective in preventing cross-breeding. Hybrids that do not abort may be sterile or at a selective disadvantage.

19.13 Species arise through the evolutionary process

Speciation is the evolutionary multiplication of species. It comes about in most plants and animals by an interaction of heritable variation, natural selection, and barriers to gene flow. A single lineage of a species undergoes many changes over evolutionary time. If these changes affect reproductive compatibility, the descendants may become so distinctive they can be considered a new species.

The most common form of the speciation process is the divergence of different genetic lines from a common ancestral species. Over time the ancestral group splits into two or more species. Such speciation can occur only when part of the ancestral population becomes reproductively isolated.

The usual type of speciation is **allopatric speciation** or **geographic speciation.** The first step in geographic speciation is the splitting of a single interbreeding population into two spatially isolated populations, each taking its own evolutionary route (Figure 19.11a).

Imagine a piece of land, warm and dry, occupied by species A. At some point in geological time, mountains uplift, land sinks and floods with water, or some vegetational catastrophe occurs. That action splits the piece of land and separates a segment of species A from the rest of the population. The newly isolated population will become subpopulation A'. It now occupies an area with a cool, moist climate.

Because it represents only a random sample of species A, population A' will be slightly different genetically. The climatic conditions and the selective forces under which this population lives are different. Natural selection will favor individuals best adapted to a cool, moist climate. Similar selection for a warm, dry climate will continue in population A in the original land mass. With different selection forces acting on them, the two populations will diverge. Accompanying genetic divergence will be changes in physiology, morphology, color, and behavior, resulting in ever-increasing external differences. If the geological barrier breaks down before the species strongly diverge from parent stock, they simply rejoin, but if sufficient difference arises before the barrier breaks down, hybrids form. Eventually, however, the two populations diverge so much that inbreeding is impossible, even if they come together again.

A second form of geographic speciation arises from the establishment of a new colony by one founder

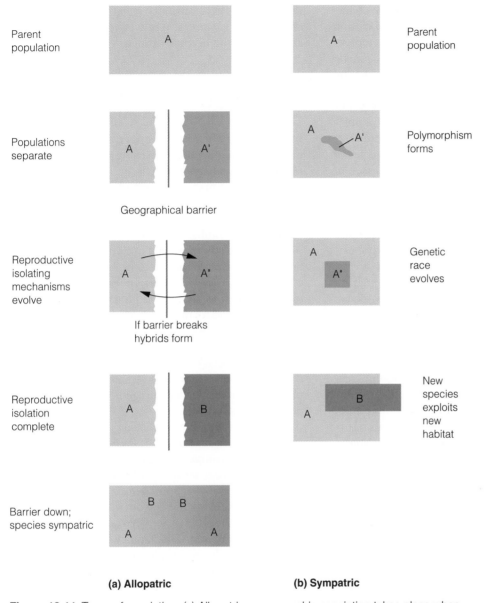

Parent population

Populations separate

Geographical barrier

Reproductive isolating mechanisms evolve

If barrier breaks hybrids form

Reproductive isolation complete

Barrier down; species sympatric

Parent population

Polymorphism forms

Genetic race evolves

New species exploits new habitat

(a) Allopatric   **(b) Sympatric**

Figure 19.11 Types of speciation. (a) Allopatric or geographic speciation takes place when two populations are isolated from each other over a long period of time. The first step in speciation comes about when species A becomes divided into two populations by a geographical barrier: A and A'. Over time under different selection pressures A' diverges even further into subspecies A". At this point if sufficient differences evolve before the barrier breaks down but isolating mechanisms are incomplete, hybrids form. When reproductive isolating mechanisms are complete, subspecies A" has evolved into a new species B. If the barrier should break down, species A and B can exist sympatrically. (b) Sympatric speciation takes place at the center of a population in a patchy environment. Premating isolation mechanisms arise in a small segment, A', of the population. A genetic race A" evolves, exploits a new habitat or niche, and evolves into a new species B.

(a gravid female) or a small number of founders (see Section 19.7). It may be a small population living on the periphery of the species's range that has become isolated from the parent populations; or it may be a small population that has found an unexploited area suitable for immigration. In all cases these small populations are isolated from genetically stable conditions in the center of the population, where variation in the gene pool is great and selection is for the heterozygous condition. Founder populations have less genetic variation and greater selection for homozygosity. Any adaptive mutation carried by the population becomes fixed rapidly. Such genetic changes may allow a population to exploit new habitats. Once a founder population is well established in a new habitat, reproductive isolation arises by chance. The numerous species of fruit flies on the Hawaiian Islands may have arisen in this fashion.

A new species may arise within a population occupying a single habitat or within the dispersal range of a population. Known as **sympatric speciation** (Figure 19.11b), this process occurs in the center of a population in a patchy environment. Premating reproductive isolating mechanisms arise before the population shifts to a new food source or habitat. The new species may exploit an underused or novel resource and invade an empty niche.

Sympatric speciation is most apt to occur among insects parasitic on plants and animals. Such insects are small and have short life spans. They possess high reproductive rates, readily adapt to new conditions, and experience some form of inbreeding. Among plants sympatric speciation shows up in orchids, noted for their specialized pollination systems. A single mutation in orchids can cause a shift in the insects that pollinate them.

Plants can undergo **abrupt speciation,** a form of sympatric speciation in which a new species arises spontaneously. The most common method of abrupt speciation is **polyploidy,** the doubling of the number of chromosomes (see Section 19.3). An organism that has a double number of chromosomes in its gametes cannot produce fertile offspring with a diploid member of its ancestral population, but it can do so by mating with another polyploid. Thus a polyploid has already achieved reproductive isolation. If the polyploid can spread into a new environment, a new species has formed.

Many of our common cultivated plants—potatoes, wheat, alfalfa, coffee, and grasses, to mention a few—are polyploids. Polyploidy is widespread among native plants, in which it produces a complex of species, such as blackberries (*Rubus*), willows (*Betula*), and birches (*Salix*). The common blue flag (*Iris versicolor*) of northern North America is a polyploid that probably originated from two other species, *Iris virginica* and *I. setosa*, when the two, once widely-distributed, met during the

retreat of the Wisconsin ice sheet. The sequoia (*Sequoiadendron giganta*) is a relict polyploid, its diploid ancestors having become extinct.

19.14 An outcome of speciation is adaptive radiation

In evolutionary history some species have spread outward from their center of distribution into new, unexploited environment. Before a species can enter a new mode of life, it first must have physical access to the new environment. Next the species must establish a beachhead. Then it must possess sufficient genetic variation to establish itself in the new environment. Adaptations that permit the organism to gain a foothold are only temporary; they must be altered, strengthened, and improved by selection. Finally, competition must be absent or slight enough to allow the new invader to survive its initial colonization. Such niches have been available to colonists of some remote islands, such as the Galápagos, the Hawaiian Islands, and the archipelagos of the South Pacific. The abundant empty niches in these diversified islands when the first colonists arrived encouraged the rapid evolution of species to fill them.

The Hawaiian honeycreepers (Drepanididae) are examples of colonization and diversification in an unexploited environment (Figure 19.12). The ancestor of the honeycreepers was probably a nectar-feeding insectivore somewhat similar to the present-day *Himatione*. After colonizing one or two islands, stragglers undoubtedly invaded other surrounding islands. Because each group was under somewhat different selective pressures, the geographically isolated populations gradually diverged. By colonizing one island after another, and after reaching the species level, recolonizing the islands from which they came (double invasion), the immigrants enriched the bird life, especially on the larger islands with more varied habitats. At the same time competition among sympatric species placed a selective premium on divergence. The honeycreepers evolved into finchlike, creeperlike, and woodpeckerlike forms. Such divergence of one species into a number of different forms—each adapted to fit a different ecological niche, exploit a new environment, or tap a new source of food—is called **adaptive radiation.**

The principle is nicely illustrated by the genus *Hemignathus* (see Figure 19.12e,f,g), all members of which are primarily insectivorous. The bill of the

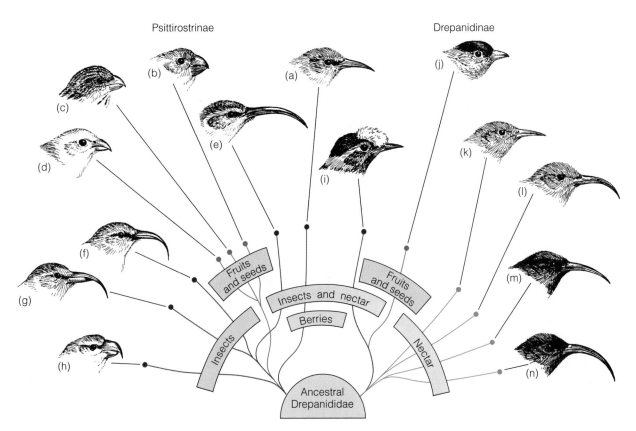

Figure 19.12 Adaptive radiation in the Hawaiian honeycreepers (Drepanididae). Selected representatives of the two sub-families, Drepanidinae and Psittirostrinae, illustrate how the family evolved through adaptive radiation from a common ancestral stock. Both subfamilies show a certain degree of parallel evolution based on diet. (a) *Loxops virens* probes for insects in the crevices of bark and folds of leaves and feeds on nectar and berries. (b) *Psittirostra kona,* a seed eater, is extinct. (c) *Psittirostra cantans* feeds on a wide diet of seeds and fruits. (d) *Psittrirostra psittacea,* a seed eater, feeds especially on the climbing screw pine *(Freycinetia arborea).* (e) *Hemignathus obscurus* is an insect-feeder and nectar-feeder. (f) *Hemignathus lucidus* is an insect-feeder. (g) *Hemignathus wilsoni* is an insect-feeder. (h) *Pseudonestor xanthophrys* feeds on larvae, pupae, and beetles of native Cerambycidae and grips branches with its curved upper beak. (i) *Palmeria dolei* feeds on insects and nectar of ohi'a *(Metrosideros).* (j) *Ciridops anna,* a fruit-eater and seed-eater, is extinct. (k) *Himatione sanguinea* feeds on the nectar of ohi'a. (l) *Vestiaria coccinea* feeds on nectar from a variety of flowers. (m) *Drepanis funerea,* a nectar-feeder, is extinct. (n) *Drepanis pacifica,* a nectar-feeder, is extinct.

genus *Hemignathus* was specialized at the start to feed on insects and nectar. *Hemignathus obscurus,* whose lower mandible is about the same length as its upper mandible, uses its decurved bill like forceps to pick insects from crevices as it hops along the trunks and limbs of trees. The bill of *H. lucidus* is also decurved, but the lower mandible is much shorter and thicker. The bird uses the lower bill to chip and pry loose bark as it seeks insects on the trunks of trees. *H. wilsoni* carries the modification even further. The lower mandible is straight and heavy. Holding its bill open to keep the slender upper mandible out of the way, the bird uses the lower mandible to pound, woodpeckerlike, into soft wood to expose insects.

The honeycreepers also exhibit parallel evolution, or adaptive changes in different organisms with a common evolutionary heritage in response to similar environmental demands. The long, thin decurved *Hemignathus obscurus* is adapted for a diet of insects and nectar; so, too, are the bills of several members of the subfamily Drepanidinae (see Figure 19.12 l–n).

19.15 Unrelated species may converge in form and function

In adaptive radiation differences between species become accentuated as they adapt to different niches.

Other wholly unrelated organisms living in similar environments and occupying similar niches come to resemble each other over evolutionary time. Figure 19.13 shows some striking examples among animals.

Similar convergence exists among plants. Within the chaparral vegetation of California, for example, the dominant plants belong to such diverse families as Ericaceae, Rhamnaceae, and Rosaceae; yet all are deep-rooted, evergreen, hard-leafed shrubs. In fact, throughout all areas with a Mediterranean-type climate, the vegetation has a similar appearance and character. Even though widely separated geographically and possessing different evolutionary histories, the vegetation has converged in both form and function. This convergence has been in response to similar selective forces, including fire, drought, high temperatures, and low rainfall.

19.16 Evolution of new species is a slow process

What are the rates of evolution? How fast do species arise? We have no definitive answers. If we keep in

mind that evolution is a change in gene frequency through time, involving some change in phenotype, and that speciation is the multiplication of species, we can keep the two related questions apart.

Evolution involves some adaptation to a changed environment. Such changes within a species can occur rapidly. (Consider how fast plant and animal breeders can change the phenotypes and growth efficiencies of domestic plants and animals through selective breeding. Such breeding is a form of goal-directed evolution brought about by human intervention.) In 50 years a single allele controlling melanism in the peppered moth changed in frequency from 0 to 98 percent in populations exposed to industrial pollution. The house sparrow (*Passer domesticus*), introduced into the United States about 100 years ago, has spread across North America and differentiated into a number of ecotypes. They differ in size, color, and other morphological characteristics—differences that represent adaptations to various North American environments, from the eastern deciduous forest to the desert Southwest. Such changes involve only some features; others do not change at all. Thus each species is a mo-

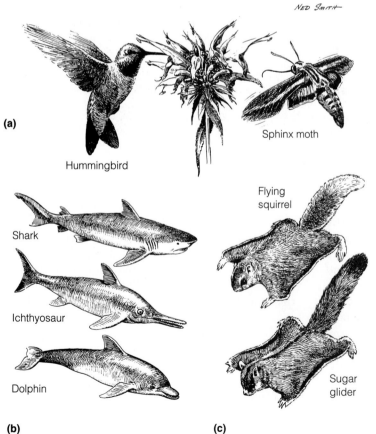

Figure 19.13 Convergent evolution of dissimilar species showing relationships between form and function. (a) The hummingbird and the sphinx moth both feed on nectar, the hummingbird by day and the sphinx moth by night. Both have bill and mouthparts adapted for probing flowers; both have a rapid, hovering mode of flight. (b) A prehistoric marine reptile, the ichthyosaur, the modern shark, and a marine mammal, the dolphin, all have the same streamlined shape for fast movement through water. (c) A North American rodent, the flying squirrel, and an Australian marsupial, the sugar glider, both have a flat, bushy tail and an extension of the skin between the foreleg and the hindleg that enable them to glide down from one tree limb to another.

saic of characteristics that have been passed on unchanged from ancestors and characteristics that have recently evolved.

The rates at which species evolve are difficult to determine. If species arise by bursts of speciation during periods of environmental change, as some evolutionary ecologists suggest, the determination of rates has little meaning. When species result from isolation of small populations from parent populations, then rates vary. The numerous species of Hawaiian fruit flies *(Drosophila)* diverged in a few thousand years. Five species of cichlid fishes in the African Lake Nabugabo apparently developed in less than 4000 years, since the lake was isolated from Lake Victoria. Banana-

feeding species of the Hawaiian moth genus *Hedylepta* diverged from a palm-feeding species in about 1000 years, after bananas were introduced to the islands by the Polynesians. Other species show little change. The American and Mediterranean species of the sycamore or plane tree (*Platanus orientalis* and *P. occidentalis*) have evolved little since their last separation at least 20 million years ago. When the two species were brought together in English gardens in the 1700s, they hybridized. The hybrid line, known as the London plane tree, is vigorous and fertile. Planted as an ornamental and as a timber tree, the London plane tree, able to grow in a variety of climates and latitudes, is becoming naturalized.

CHAPTER REVIEW

SUMMARY

To survive as a species, organisms must adapt to a changing environment. Adaptation comes about through the interaction of organisms with their environment. Differential survival and reproductive success, changing the frequency of genetic traits in a population, is natural selection.

Genetic Variation (19.1–19.5)   Local groups of interbreeding individuals of a species make up a deme. Individuals making up a deme vary. Those individuals that leave behind the greatest number of reproducing offspring are the fittest. The heritable portion of this variation is the raw material of natural selection **(19.1)**.

Most inherited variation arises from recombination of genes in sexual reproduction. The units of heredity are genes carried on chromosomes. The alternative forms of a gene are alleles. Individuals that possess alike pairs of alleles are homozygous. If the alleles are unlike, the individuals are heterozygous.

Germ cells, eggs, and sperm receive only one-half the chromosome complement (1*n*). When egg and sperm unite to form a zygote, the full complement (2*n*) is restored. Recombinations provide innumerable variations. The sum of heritable information carried by the individual is the genotype; its physical expression, on which natural selection acts, is the phenotype. The range of phenotypic expression under different environmental conditions is phenotypic plasticity **(19.2)**.

Some genetic material is altered by mutation. Gene mutations alter sequences of nucleotides. Chromosomal mutations change the structure or number of chromosomes. The duplication of entire sets of

chromosomes is polyploidy, most common in plants. Single gene mutations rarely are visible. Most gene mutations are neutral and maintain genetic variability in a population **(19.3)**.

Genetic variation in a population is transmitted from one generation to another through sexual reproduction. If genes occur in two forms, *A* and *a*, the gametes recombine as homozygous *AA*, heterozygous *Aa*, or homozygous *aa*. The proportion of the three combinations in a population is the genotypic frequency **(19.4)**.

Natural selection favors certain phenotypes, and thus genotypes, over time. The outcome of natural selection is evolution, a change in gene frequency through time, which in turn allows new change in biological systems in response to selection pressures. Stabilizing selection favors the population average in stable environments. Disruptive selection favors the extremes, when segments of a population are subjected to different selection pressures. Directional selection favors one extreme in a changing environment. In polymorphism, species possess several distinct forms in the same habitat at the same time. These differences appear to be environmentally induced, arising from disruptive selection **(19.5)**.

Small Populations (19.6–19.8)   Habitat fragmentation and human exploitation have reduced many species to isolated or semi-isolated populations. Small populations carry only a sample of the genetic variability of the total population. They are subject to inbreeding, mating between close relatives, which brings out hidden genetic variation and increases homozygous rare alleles. Inbreeding depression reduces fecundity, viability, and survival. However, occasional inbreeding can fix rare genes that otherwise might be lost. Dispersal of young, especially males, and kin recognition reduce inbreeding in natural populations.

Outbreeding, mating between unrelated individuals, can mitigate the effects of inbreeding by introducing new genetic diversity. However, offspring may be poorly adapted to the local environment of either parent. This effect is outbreeding depression (19.6).

A group that colonizes a new habitat area may experience a founder effect. The colonizers carry only a small sample of the parent gene pool. Survivors of populations that crash also possess a small sample of the original gene pool. This situation creates a population bottleneck, severely reducing genetic diversity in future generations. Genetic drift—chance fluctuations in allele frequency as a result of random sampling among gametes—affects small populations most strongly. Genetic drift mimics inbreeding, but it is the result of random mating. Exchange of a few individuals among populations in each generation will reduce random genetic drift (19.7).

To persist, a population must not fall below a threshold number, the minimum viable population. The population must be large enough to cope with chance variations in individual births and deaths, a random series of environmental changes, genetic drift, and catastrophes. Because populations are fragmented, it takes a minimum number of subpopulations to maintain a viable metapopulation (19.8).

Identifying Species (19.9–19.12) Taxonomists describe plants and animals as morphological species, discrete entities that exhibit little variation. Variation over space and time has given rise to the concept of the biological species, a group of interbreeding individuals living together in a similar environment in a given region and in similar ecological relationships. This concept, limited to bisexual organisms, does not explain where one species begins and another ends. Evolutionary biologists suggest that we regard the species as an evolutionary unit (19.9).

Species may be sympatric or allopatric. Sympatric species inhabit the same region at the same time and have the opportunity to interbreed. Allopatric species, separated by time and space, have no opportunity to interbreed. Thus it is difficult to determine whether allopatric species are "good" species. Some sympatric species are sibling species. They are similar in appearance but differ in behavior, ecological requirements, physiology, or chromosomal structure. They do not interbreed (19.10).

Widely distributed species may exhibit geographical variation, reflecting selective environmental forces. Variants show up as clines, ecotypes, or geographic isolates. A cline is a measurable, gradual change in some phenotypic or genotypic characteristic. Clines are associated with an ecological gradient such as temperature. Clinal variations that show marked discontinuities in a gradient are ecotypes. Ecotypes are genetic strains of a population adapted to unique local conditions. Isolates are populations prevented from interbreeding by some extrinsic barrier such as mountainous terrain. Strong isolates with little gene flow between them may become separate species (19.11).

Species maintain their identity by means of isolating mechanisms. Premating isolating mechanisms include distinctive behavioral patterns, habitat preferences, the timing of breeding, and physical differences that inhibit mating between species. Postmating isolating mechanisms reduce mating success, as when hybrids are sterile or at selective disadvantage (19.12).

Speciation (19.13–19.16) Speciation is the evolutionary multiplication of species. Species arise by the interaction of heritable variations, natural selection, and barriers to gene flow between populations. The most widely accepted mechanism of speciation is allopatric or geographic speciation. A single interbreeding population splits into spatially isolated populations, which diverge into distinct species. When reproductive isolation precedes differentiation and the process takes place within a population, we have sympatric speciation. The most common form of sympatric speciation is polyploidy among plants (19.13).

Adaptive radiation occurs when populations of a species subdivide and diversify under selective pressures of new situations and occupy new ecological niches (19.14). In convergent evolution, dissimilar species evolve a resemblance in response to selection pressures in widely separated but similar environments (19.15).

How rapidly evolution and speciation take place is a topic of debate. Changes within a population at the level of a single allele can take place over a few generations. Polyploidy allows instantaneous speciation among plants. Some species apparently never change for millions of years. Fossil records suggest that new species arise in bursts of speciation (19.16).

STUDY QUESTIONS

1. How does natural selection relate to evolution?
2. What is a deme?

3. Distinguish between a genotype and a phenotype. What is phenotypic plasticity?
4. Define allele, locus, homozygous, heterozygous, diploid, haploid, meiosis, and gene pool.
5. What are the major sources of variation in the gene pool?
6. What is a mutation? Distinguish between a gene mutation and a chromosomal mutation.
7. Contrast directional, stabilizing, and disruptive selection.
8. What is inbreeding? Outbreeding? What are inbreeding depression and outbreeding depression?
9. What is genetic drift? How does it relate to founder effect and a population bottleneck?
10. What is a species? Give three viewpoints.
11. Why is it difficult to define a species? Why is it important to try?
12. Distinguish between an allopatric species and a sympatric species.
13. What is a cline? An ecotype? A geographical isolate? How do they relate to the species problem?
14. What is polymorphism? What is its ecological and evolutionary significance?
15. What are isolating mechanisms? Describe four kinds.
16. What is speciation? Distinguish between allopatric and sympatric speciation.
17. What is polyploidy? What is its evolutionary significance? Why are plants more likely than animals to form hybrid polyploid species?
*18. A new male will be introduced into a small isolated population that suffers inbreeding depression. What are the potential problems?
*19. What problems, other than ecological ones, arise in maintaining or reestablishing populations of endangered species? Do a report on the plight of the spotted owl, the desert tortoise, or the sea turtle, on reintroduction of wolves to Yellowstone, or on maintaining elephants in African national parks.
*20. There appears to be little genetic difference between the Asiatic lion (*Panthera leo persica*) and the African lion (*Panthera leo leo*). The Asiatic species is highly endangered, restricted to the Gir Forest in India. The African lion is still widespread in Africa. Should we save the Asian subspecies? Why or why not?

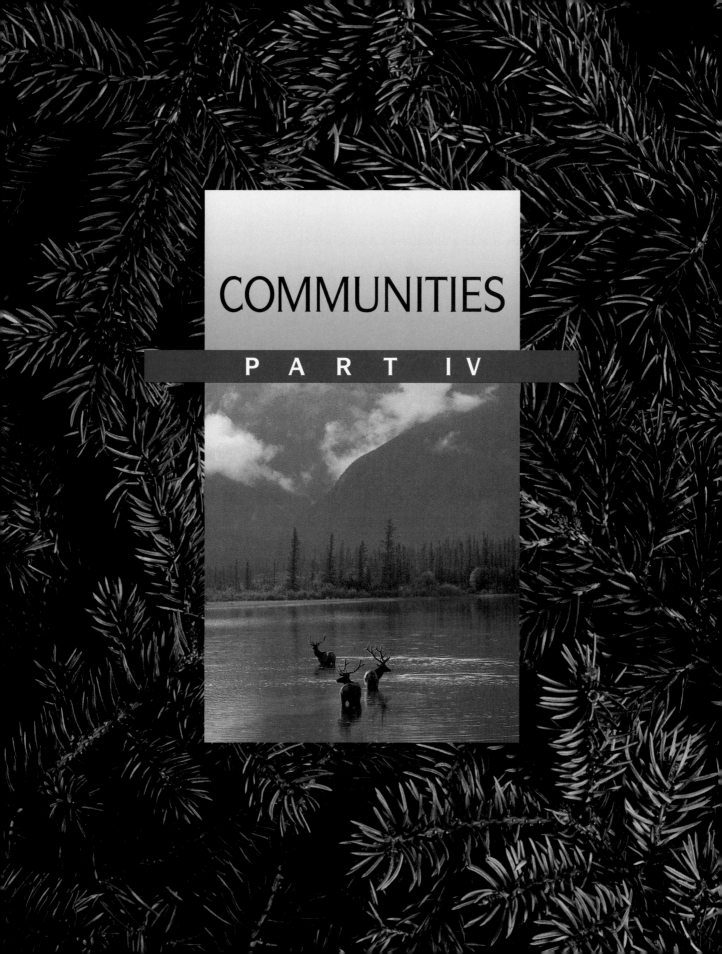

COMMUNITIES

PART IV

CHAPTER 20

COMMUNITY STRUCTURE

OBJECTIVES

On completion of this chapter, you should be able to:

- Explain how species dominance influences community structure.
- Define the concept of species diversity.
- Explain how vertical structure and horizontal patterns of vegetation influence the terrestrial community.
- Discuss how the physical environment influences the structure of aquatic communities.
- Show how physical and biological features of the environment influence community dispersion patterns.

Forest and pond—an illustration of the contrasts between terrestrial and aquatic communities

Populations of organisms do not live apart from one another as separate entities. Sharing environments and habitats, they interact in various ways. A collection of organisms interacting directly or indirectly is a **community.** Such a definition embraces the idea of community in its broadest sense. Within the community some species carry out similar functions or exploit the same resource. These groups are called **guilds.** For example, birds feeding mostly on insects make up an insect-feeding guild.

Other common definitions of community are more restrictive. Zoologists apply the term to related species, such as the bird community or mammal community of a particular forest or grassland. Botanists prefer the term **association** for a plant community possessing a definitive mix of species. Ecologists use the term to contrast basic characteristics. They speak of heterotrophic and autotrophic communities. Forests and grasslands are examples of **autotrophic** communities, which require only the energy of the sun to drive the process of photosynthesis. Groups of organisms in such habitats as a fallen log, a tiny pool of water in the hollow of a tree, or a cave are examples of **heterotrophic** communities, which depend on the autotrophic community for their energy.

Although ecologists classify communities in different ways, all communities have certain characteristics that define their biological and physical structure. These characteristics vary in both space and time.

20.1   Communities vary in biological structure

The mix of species, including both their number and relative abundance, defines the biological structure of a community. A community can be composed of a few common species; or it can have a wide variety of species, some common with high population density, but most rare with low population density. When a single or few species predominate within a community, these organisms are **dominants.**

It is not easy to determine the dominant species (see Quantifying Ecology 20.1: Some Measures of Dominance). The dominants in a community may be the most numerous, possess the highest biomass, preempt the most space, make the largest contribution to energy flow or nutrient cycling, or by some other means control or influence the rest of the community.

Some ecologists ascribe the dominant role to those organisms that are greatest in number, but abundance alone is not sufficient. A species of plant, for example,

---

QUANTIFYING ECOLOGY 20.1

SOME MEASURES OF DOMINANCE

1. Dominance =
$$\frac{\text{basal area or aerial coverage, species A}}{\text{area sampled}}$$

2. Relative dominance =
$$\frac{\text{basal area or coverage, species A}}{\text{total basal area or coverage, all species}}$$

3. Relative density =
$$\frac{\text{total individuals, species A}}{\text{total individuals, all species}}$$

4. Frequency =
$$\frac{\text{intervals or points where species A occurs}}{\text{total number of sample plots or points}}$$

5. Relative frequency =
$$\frac{\text{frequency value, species A}}{\text{total frequency value, all species}}$$

6. Importance value = relative frequency + relative dominance + relative density

All the above results may be multiplied by 100.

7. Simpson's index of dominance:
$$\text{dominance} = \frac{\Sigma n_i(n_i - 1)}{N(N - 1)}$$

where N is total number of individuals of all species and n_i is the total number of individuals of species A

---

can be widely distributed but exert little influence on the community as a whole. In the forest the small or understory trees can be numerically superior; yet the community is controlled by a few large trees that overshadow the smaller ones. In such a situation the dominant organisms are not those with the largest numbers but those that have the greatest biomass or that preempt most of the canopy space and thus control the distribution of light. Ecologists measure such dominants by biomass or basal area.

In other cases a scarce organism dominates by its activity. The predatory starfish *Pisaster,* for example, preys on several associated species and reduces competitive interactions among them, so they are able to coexist. If the predator is removed, a number of the prey species disappear and one becomes dominant. In effect, this predator controls the nature of the community and must be regarded as the dominant species.

The dominant species may not be the most essential species in the community from the standpoint of energy flow or nutrient cycling, although this is often the case. Dominant species achieve their status at the expense of other species in the community. For example, the American chestnut tree (*Castania dentata*) was a dominant component of the oak-chestnut forests in eastern North America until early in the twentieth century. At that time its populations were decimated by the chestnut blight imported from Asia. Since that time a variety of species, including oaks, hickories, and yellow-poplar, have taken over the position of the chestnut in the forest.

The concept of species dominance is context-dependent. A predator species that maintains one mix of species in the community may not be defined as dominant if the criterion is the cycling of nutrients or density. We must define our criteria before we determine dominance.

20.2 Number and relative abundance define species diversity

Among the array of species that make up the community, a few are abundant, but most are rare. You can discover this characteristic by counting all the individuals of each species in a number of sample plots within a community and determining what percentage each contributes to the whole. This measure is known as **relative abundance.**

As an example, samples representing the structure of two forest communities are presented in Tables 20.1 and 20.2. The sample from the first forest consists of 24 species of trees over 10 cm dbh (diameter measured at a defined height above the ground). Two species, yellow-poplar and white oak, make up nearly 44 percent of the stand. The four next most abundant trees—black oak, sugar maple, red maple, and American beech—each make up a little over 5 percent of the stand. Nine species range from 1.2 to 4.7 percent, while the nine remaining species as a group represent about 5 percent of the stand. The second forest presents a somewhat different picture. This community consists of ten species, of which two, yellow-poplar and sassafras, make up almost 84 percent of the total stand density. Both forest communities illustrate the pattern of a few common species associated with many rare ones.

These two forest communities differ in their **species diversity.** Species diversity relates to both the

TABLE 20.1

STRUCTURE OF ONE MATURE DECIDUOUS FOREST IN WEST VIRGINIA

Species	Number	Percentage of Stand
Yellow-poplar (*Liriodendron tulipifera*)	76	29.7
White oak (*Quercus alba*)	36	14.1
Black oak (*Quercus velutina*)	17	6.6
Sugar maple (*Acer saccharum*)	14	5.4
Red maple (*Acer rubrum*)	14	5.4
American beech (*Fagus grandifolia*)	13	5.1
Sassafras (*Sassafras albidum*)	12	4.7
Red oak (*Quercus rubra*)	12	4.7
Mockernut hickory (*Carya tomentosa*)	11	4.3
Black cherry (*Prunus serotina*)	11	4.3
Slippery elm (*Ulmus rubra*)	10	3.9
Shagbark hickory (*Carya ovata*)	7	2.7
Bitternut hickory (*Carya cordiformis*)	5	2.0
Pignut hickory (*Carya glabra*)	3	1.2
Flowering dogwood (*Cornus florida*)	3	1.2
White ash (*Fraxinus americana*)	2	.8
Hornbeam (*Carpinus caroliniana*)	2	.8
Cucumber magnolia (*Magnolia acuminata*)	2	.8
American elm (*Ulmus americana*)	1	.39
Black walnut (*Juglans nigra*)	1	.39
Black maple (*Acer nigra*)	1	.39
Black locust (*Robinia pseudoacacia*)	1	.39
Sourwood (*Oxydendrum arboreum*)	1	.39
Tree of heaven (*Ailanthus altissima*)	1	.39
	256	100.00

TABLE 20.2

STRUCTURE OF A SECOND DECIDUOUS FOREST IN WEST VIRGINIA

Species	Number	Percentage of Stand
Yellow-poplar (*Liriodendron tulipifera*)	122	44.5
Sassafras (*Sassafras albidum*)	107	39.0
Black cherry (*Prunus serotina*)	12	4.4
Cucumber magnolia (*Magnolia acuminata*)	11	4.0
Red maple (*Acer rubrum*)	10	3.6
Red oak (*Quercus rubra*)	8	2.9
Butternut (*Juglans cinerea*)	1	.4
Shagbark hickory (*Carya ovata*)	1	.4
American beech (*Fagus grandifolia*)	1	.4
Sugar maple (*Acer saccharum*)	1	.4
	274	100.0

number of species, **species richness,** and the relative abundance of individuals among the species, **species evenness.** The first forest community has both a greater species richness and a greater species evenness than the second.

The two components, species richness and species evenness, are useful in measuring species diversity. A community that contains a few individuals of many species is said to have a higher diversity than a community having the same number of individuals with most of them belonging to a few species. For example, a community with ten species of ten individuals each has a higher diversity than another community with ten species but with the 100 individuals apportioned 90, 1, 1, 1, 1, 1, 1, 1, 1, 1, 1.

To quantify species diversity, several indexes have been proposed (see Quantifying Ecology 20.2: Indexes of Diversity). The most widely used is the Shannon index, which ecologists have adapted from communication or information theory. The formula for the Shannon index is

$$H = -\sum_{i=1}^{s}(p_i)(\log_2 p_i)$$

QUANTIFYING ECOLOGY 20.2
INDEXES OF DIVERSITY

The Shannon index of diversity is one of a number of diversity indexes. Based on information theory, it measures the degree of uncertainty. If diversity is low, then the certainty of picking a particular species at random is high. If diversity is high, then it is difficult to predict the identity of a randomly picked individual. High diversity means high uncertainty.

Another common index is Simpson's. It takes a different approach—the number of times we would have to take pairs of individuals at random to find a pair of the same species. This index of diversity is the inverse of Simpson's dominance index (see Quantifying Ecology 20.1):

$$\text{diversity} = \frac{N(N-1)}{\Sigma n_i(n_i-1)}$$

or

$$1 - \frac{\Sigma n_i(n_i-1)}{N(N-1)}$$

In a collection of species high dominance means low diversity.

The Shannon and Simpson indexes take into consideration both the richness and evenness of species. A much simpler index of diversity that does not take evenness into account is Margalef's:

$$\text{diversity} = (s-1)/\log N$$

where s is the number of species and N is the number of individuals. Such an index does not express the differences among communities having the same s and N, so it is much less useful.

where H is diversity of species, s is the number of species, and p_i is the proportion of individuals in the total sample belonging to the ith species.

The index takes into consideration the number as well as the relative abundance of species. The woodland described in Table 20.1 has a diversity index of 3.59; the one described in Table 20.2 has a diversity index of 1.87. This index is obtained by using $\log_2$, usually used in the Shannon formula. Diversity calculated using $\log_n$ is 2.49 and 1.30, respectively.

The two components, species richness and species evenness, can be separated. The simplest determination of species richness is to count the number of species. In the first woodland that number is 24, and in the other, 10. To determine evenness, you first have to calculate H_{max}, what H would be if all the species in the community had an equal number of individuals:

$$H_{max} = l_n S$$

where l_n is the natural log and S is the number of species. For the first woodland H_{max} is 3.18, and for the second it is 2.30. Now we can calculate the evenness (J):

$$J = H/H_{max}$$

The evenness of the first woodland is 0.78, and that of the second is 0.57. The first woodland has a more even distribution than the second.

Up to this point we have considered measures of diversity within a community, or **alpha diversity.** Diversity between communities is **beta diversity.** We can find it by calculating coefficients of community, percent similarity in species composition, and other measures. Two examples of determining community similarity appear in Quantifying Ecology 20.3: Community Similarity. A third type is **gamma diversity,** which describes diversity on a regional basis, including species replacement over large geographical regions.

20.3 Communities have defining physical structure

Communities are characterized not only by the mix of species, the biological structure, but also by physical features. The physical structure of the community reflects abiotic factors, such as the depth and flow of water in aquatic environments. It also reflects biotic factors, such as the spatial configuration of organisms. In a forest, for example, the size and height of the trees and the density and dispersion of their populations define the physical attributes of the community.

The form and structure of terrestrial communities reflect the vegetation. The plants may be tall or short, evergreen or deciduous, herbaceous or woody. Such characteristics can describe growth forms. Thus we might speak of shrubs, trees, and herbs and further subdivide the categories into needle-leaf evergreens, broadleaf evergreens, broadleaf deciduous trees, thorn trees and shrubs, dwarf shrubs, ferns, grasses, forbs, mosses, and lichens.

A more useful system is the one designed in 1903 by the Danish botanist Christen Raunkiaer. Instead of considering the plants' growth form, he classified plant life by the relation of the embryonic or meristemic tissues that remain inactive over the winter or prolonged

---

QUANTIFYING ECOLOGY 20.3

COMMUNITY SIMILARITY

A number of methods are available for measuring similarity of communities. The one most often recommended is Morisita's index, based on Simpson's index of dominance. However, here are two simpler approaches: Sorensen's *coefficient of community* and *percent similarity.*

To find the **coefficient of community,** apply the equation

$$CC = \frac{2c}{s_1 + s_2}$$

where c is the number of species common to both communities and s_1 and s_2 are the number of species in communities 1 and 2.

For the woodland examples:

$$s_1 = 24 \text{ species}$$
$$s_2 = 10 \text{ species}$$
$$c = 9 \text{ species}$$
$$CC = \frac{2(9)}{24 + 10} = \frac{18}{34} = .529$$

Coefficient of community does not consider the relative abundance of species. It is most useful when the major interest is the presence or absence of species.

To calculate **percent similarity** (PS), first tabulate species abundance in each community as a percentage. Then add the lowest percentage for each species that the communites have in common. For the two woodlands, 16 species are exclusive to one community or the other. The lowest percentage for those 16 species is 0, so they need not be added in.

$$PS = 29.7 + 4.7 + 4.3 + 0.8 + 3.6 + 2.9 + 0.4 + 0.4 + 0.4 = 47.2$$

Percent similarity does consider relative abundance of species in each community.

---

dry periods—the **perennating tissues**—to their height above ground. Such perennating tissue includes buds, bulbs, tubers, roots, and seeds. Raunkiaer recognized six principal life forms, which are summarized in Figure 20.1. All of the species within a community can be grouped into these six classes. For example, trees and shrubs greater than 25 cm in height are classified as phanerophytes, because their leaf-producing buds are ele-

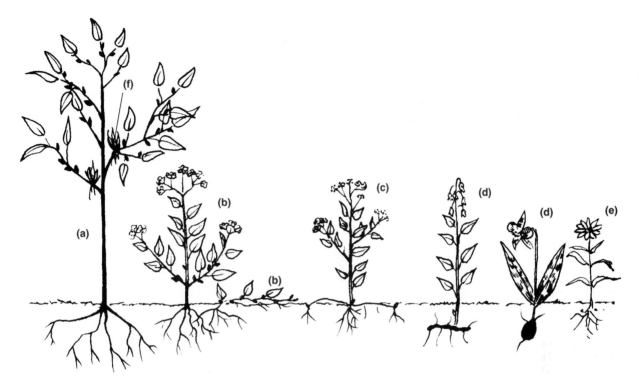

(a) Phanerophytes (Greek *phaneros*, "visible"). Perennial buds carried well up in the air and exposed to varying climatic conditions. Trees and shrubs over 25cm; typical of moist, warm environments.

(b) Chamaephytes (Greek *chamia*, "on the ground"). Perennial shoots or buds on the surface of the ground to about 25cm above the surface. Buds receive protein from fallen leaves and snow cover. Typical of cool, dry climates.

(c) Hemicryptophytes (Greek, *hemi*, "half" and *kryptos*, "hidden"). Perennial buds at the surface of the ground, where they are protected by soil and leaves. Many plants have rosette leaves. Characteristic of cold, moist climates.

(d) Cryptophytes (Greek, *kryptos*, "hidden"). Perennial buds buried in the ground on a bulb or rhizome, where they are protected from freezing and drying. Typical of cold, moist climates.

(e) Therophytes (Gr. *theros*, "summer"). Annuals, with complete life cycle from seed to seed in one season. Plants survive unfavorable periods as seeds. Typical of deserts and grasslands.

(f) Epiphytes (Gr. *epi*, "upon"). Plants growing on other plants; roots up in the air.

Figure 20.1 Raunkiaer's life forms. The parts of the plants that die back during winter or prolonged dry periods are unshaded; the persistent plant parts with buds (or seeds in the case of therophytes) are dark.

vated above the ground on stems. In contrast, grasses are classified as cryptophytes, because their aboveground tissues die back in winter or during prolonged dry periods. A community with a high percentage of perennating tissue well above ground is characteristic of warmer, wetter climates. A community where most of the plants are cryptophytes and hemicryptophytes is characteristic of colder or drier environments. The perennating tissue from which new leaves grow is underground.

When the species within a community are classified into life forms and each life form is expressed as a percentage, we get a life form spectrum that reflects the plants' adaptations to the environment, particularly the climate (Figure 20.2). Such a system of classification provides a standard means of describing the structure of a community for purposes of comparison.

20.4 Vertical layering is characteristic of all communities

Each community has a distinctive **vertical structure** (Figure 20.3). On land, vertical structure is determined largely by the life form of the plants—their size, branching, and leaves—which, in turn, influences and is influenced by the vertical gradient of light. The vertical structure of the plant community provides the physical framework in which many forms of animal life are adapted to live. A well-developed forest ecosystem, for example, has several layers of vegetation. From top to bottom, they are the *canopy*, the *understory*, the *shrub layer*, the *herb* or *ground layer*, and the *forest floor*. We could continue down into the root layer and soil strata.

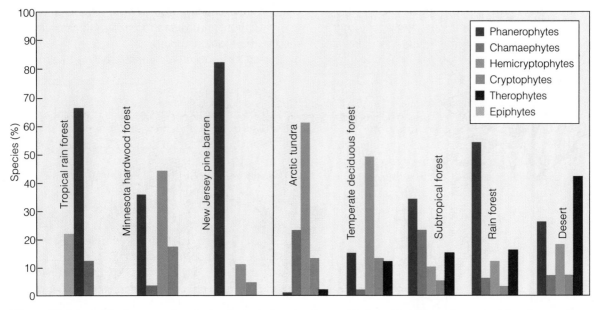

Figure 20.2 Left: Life form spectra of a tropical rain forest, a temperate forest in Minnesota, and a New Jersey pine barren. Note the absence of hemicryptophytes, chamaephytes, and therophytes in the tropical rain forest and the prominence of epiphytes. The pine barrens are dominated by phanerophytes. Right: Life form spectra of major ecosystems. Note the importance of hemicryptophytes in the arctic tundra and temperate deciduous forest, and the importance of phanerophytes and therophytes in the desert ecosystem.

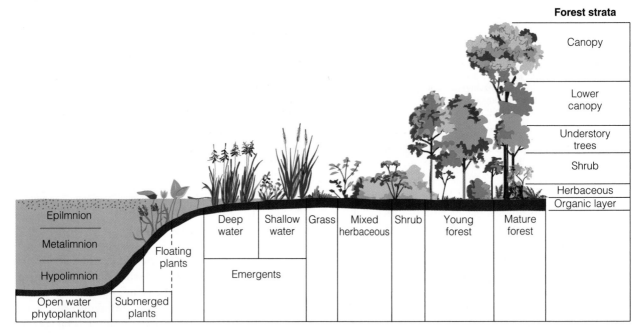

Figure 20.3 A vertical section view of communities from aquatic to terrestrial. In both, the zone of decomposition and regeneration is the bottom stratum and the zone of energy fixation is the upper stratum. From left to right, the stratification and complexity of the community become greater. Stratification in aquatic communities is largely physical, influenced by gradients of oxygen, temperature, and light. Stratification in terrestrial communities is largely biological. Dominant vegetation affects the physical structure of the community and the microclimatic conditions of temperature, moisture, and light. Because the forest has four or five strata, it can support a greater diversity of life than a grassland with two strata. Floating and emergent aquatic plant communities can support greater diversity than open water.

The **canopy,** which is the primary site of energy fixation through photosynthesis, has a major influence on the rest of the forest. If it is fairly open, considerable sunlight will reach the lower layers, which will have ample water and nutrients resulting in well-developed understory and shrub strata. If the canopy is dense and closed, light levels are low, and the understory and shrub layers will be poorly developed.

In the forests of the eastern United States the **understory** consists of tall shrubs such as witch hobble *(Viburnum alnifolium),* understory trees such as dogwood *(Cornus* spp.) and hornbeam *(Carpinus caroliniana),* and younger trees, some of which are the same species as those in the canopy. Species that are unable to tolerate shade will die. Survivors might eventually reach the canopy after older trees die or are harvested.

The nature of the **herb layer** will depend on the soil moisture and nutrient conditions, the slope position, the density of the canopy and understory, and the exposure of the slope, all of which vary from place to place throughout the forest.

The final layer, the forest floor, is the site where the important process of decomposition takes place and where decaying organic matter releases nutrients for reuse by the forest plants.

Aquatic ecosystems such as lakes and oceans have strata determined by light penetration. They have distinctive profiles of temperature and oxygen (see Chapter 32). In the summer well-stratified lakes have a layer of freely circulating water, the **epilimnion;** a second layer, the **metalimnion,** which is characterized by a **thermocline** (a steep and rapid decline in temperature); the **hypolimnion,** a deep, cold layer of dense water about 4° C (40° F), often low in oxygen; and a layer of bottom mud. In addition, two other structural layers are recognized, based on light penetration: an upper zone, the **trophogenic zone,** dominated by phytoplankton, which is the site of photosynthesis; and a lower zone, the **tropholytic zone,** in which decomposition is most active. The lower layer roughly corresponds to the hypolimnion and the bottom mud.

All communites, both terrestrial and aquatic, have a similar biological structure, related to these patterns of vertical layering. They possess an autotrophic layer concentrated where light is most available, which fixes the energy of the sun through photosynthesis, producing organic carbon compounds from CO_2. In forests this layer concentrates in the canopy; in grasslands, in the herbaceous layer; and in lakes and seas, in the upper layer of water. Communities also possess a heterotrophic layer that utilizes the carbon stored by autotrophs as a food source, transfers energy, and circulates matter by means of herbivory, predation in the broadest sense, and decomposition.

The degree of vertical layering has a pronounced influence on the diversity of animal life in the community. A strong relationship exists between foliage height diversity (the degree of vertical stratification) and bird species diversity (Figure 20.4). Increased vertical structure means more resources and living space and a greater diversity of habitats. Grasslands, with their two strata, support six or seven species of birds, all species that nest on the ground. A deciduous forest in eastern North America may support 30 or more species occupying different strata. The scarlet tanager *(Piranga olivacea)* and wood pewee *(Contopus virens)* occupy the canopy, the hooded warbler *(Wilsonia citrina)* is a forest shrub species, and the ovenbird *(Seiurus aurocapillus)* both forages and nests on the forest floor. Insects show similar stratification. Among the pine bark beetles inhabiting northeastern North America, the large turpentine beetle *(Buprestis apricans)* is restricted to the base of trees and the pine engraver beetle *(Ips oregoni)* to the upper trunk and large branches. A third species, the small and abundant *Pityogenes hopkinsi,* lives on smaller branches in the canopy.

20.5 Communities exhibit horizontal patterns

Walk across a typical eastern old field community. You move through patches of open grass, clumbs of goldenrods *(Solidago),* tangles of blackberry *(Rubus),* and small thickets of sumac *(Rhus)* and other tall shrubs. Continue into an adjacent woodland and you cross through open understory with patches of shade-tolerant undergrowth of laurel *(Kalmia)* and viburnum *(Viburnum).* You may come upon **gaps,** openings in the canopy caused by the death of a canopy tree, where dense thickets of new growth have claimed the sunlit openings. The vegetation patches form a quiltwork across the landscape (Figure 20.5). The horizontal patchiness adds to the physical complexity of the community.

This patchy distribution of plants shows influences of both the physical and biological environment. In terrestrial communities, soil structure, soil fertility, moisture conditions, and aspect influence the microdistribution of plants. Patterns of light and shade shape the development of the understory vegetation. Runoff and small variations in topography and microclimate

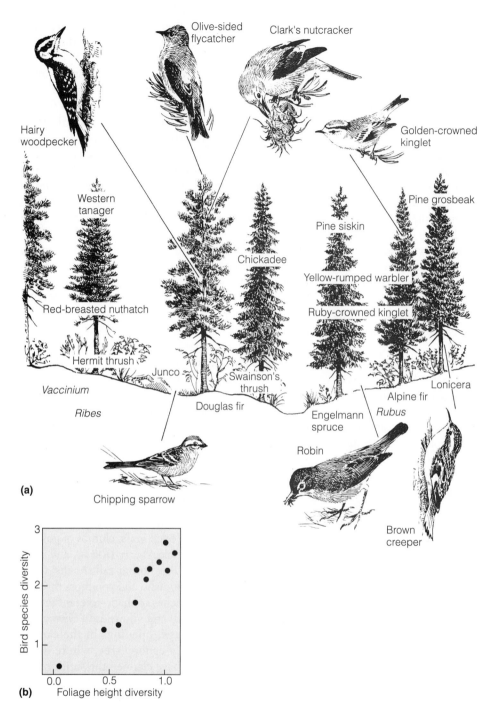

(a)

Chipping sparrow

Brown creeper

(b)

Figure 20.4 Bird species become more diverse as vertical stratification increases in forest ecosystems. (a) The vertical distribution of some bird species in a spruce-fir forest in Wyoming. (b) The relationship between bird species diversity and foliage height diversity for areas of deciduous forest in eastern North America. Foliage height diversity is a measure of the vertical structure of the forest. The greater the number of vertical layers of vegetation, the greater the diversity.

Figure 20.5 Horizontal patterns of vegetation patches in a Wisconsin countryside.

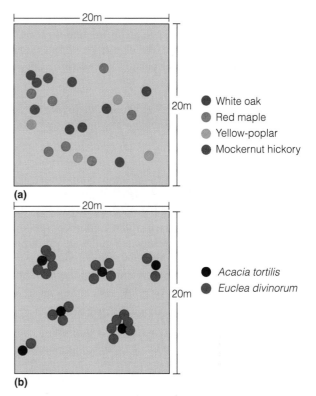

Figure 20.6 Horizontal positions of trees in two communities: (a) a forest in central Virginia, and (b) a savanna in Natal, South Africa.

produce well-defined patterns of plant growth. Additionally, grazing animals have subtle but important effects on the patterning of vegetation, as do abiotic disturbances such as wind and fire. Plants with airborne seeds may be distributed widely, whereas plants with heavy seeds or with pronounced vegetative reproduction will clump near the parent plant. Shading suppresses some plant species and encourages others. Like vertical structure, horizontal patchiness of plant life influences the distribution and diversity of animal life within the community.

20.6 Communities have characteristic patterns of dispersion

In Chapter 11 we saw that the spatial arrangement of individuals is characteristic for a population. Likewise, the dispersion of populations within a given area is char-

acteristic for a community. Figure 20.6 shows the horizontal position of trees within two communities, a forest in central Virginia and a savanna in Natal, South Africa. Note that in the forest, the individual populations as well as the total tree community (all individuals regardless of species) appear to be randomly dispersed. In contrast, the savanna community has a distinctive spatial pattern. If we examine the entire community (all trees regardless of species) the pattern appears to be aggregated or clumped. However, if we examine the dispersion of the two populations composing the community, the umbrella thorns (*Acacia tortilis*) appear to be regularly spaced, while the other species, diamond-leaved euclea (*Euclea divinorum*), exhibits a clumped dispersion. In fact, the *Euclea* trees appear to clump about the *Acacia* trees.

The dispersion patterns of the populations reflect the interactions among individuals within the community, both intraspecific and interspecific. In the case of the savanna, the dispersion of the two tree populations is a function of two different processes. The dispersion of *Acacia* is a result of competition for belowground resources (primarily water), whereas that of *Euclea* is a result of their means of dispersal. The seeds of *Euclea* are eaten and dispersed by birds that perch on *Acacia* trees and pass the seeds in their feces. Therefore, the distribution of *Euclea* is directly linked to the dispersion of *Acacia*.

CHAPTER REVIEW

SUMMARY

Community Defined (20.1) A biotic community is a naturally occurring assemblage of organisms sharing the same environment and habitats and interacting directly or indirectly with each other. Groups of species that carry out similar functions are guilds. Communities may be autotrophic, deriving their energy from the sun, or heterotrophic, deriving their energy from the autotrophic community.

Biological Structure (20.2–20.3) When a single or a few species predominate within a community, they are referred to as dominants. The dominants may be the most numerous, possess the greatest biomass, preempt the most space, or make the largest contribution to energy flow (20.2). Communities may be characterized by species diversity. Species diversity involves two components: species richness, the number of species in a community; and evenness, how individuals are apportioned among the species (20.3).

Physical Structure (20.4–20.6) Communities are also characterized by their physical structure. In terrestrial communities the structure is largely defined by the vegetation. Vertical structure on land reflects the life forms of plants. In aquatic communities it is largely defined by physical features such as light, temperature, and oxygen profiles. All communities possess an autotrophic and a heterotrophic layer. The autotrophic layer carries on photosynthesis. The heterotrophic layer uses carbon stored by the autotrophs as a food source. Vertical layering provides the physical structure in which many forms of animal life live. Communities that are most highly stratified offer the richest variety of animal life because they contain the greatest assortment of habitats. (20.4). Horizontal patchiness reflects physical changes in the environment such as soils, disturbance, light, moisture, and nutrients. The degree of horizontal patchiness influences the diversity of life in the community (20.5). The dispersion or spatial configuration of individuals and populations is a characteristic feature of communities related to intraspecific and interspecific interactions (20.6).

STUDY QUESTIONS

1. How should we measure species dominance?
2. Diversity is greatest in tropical rain forests. What does this fact tell you about dominance and species distribution there?
3. Contrast the stratification of an aquatic community with that of a terrestrial community.
4. Distinguish between species richness and species evenness.
5. How does vertical structure influence species diversity in a community?
6. What influence does the uppermost layer have on the rest of the community?
*7. If two communities have the same values for species diversity, do they necessarily have a high percent similarity value? Explain your answer.
*8. Consider a deciduous forest stand with a dense canopy but no understory (much like a city park) and a low diversity of animal life. How could you improve vertical structure of the forest and increase diversity? (You may want to refer to Chapter 5 for some ideas.) Would your action also improve horizontal structure?

CHAPTER 21

COMMUNITY DYNAMICS

OBJECTIVES

On completion of this chapter, you should be able to:

- Discuss spatial variation in community structure.
- Distinguish between edge and ecotone and discuss the edge effect.
- Discuss island biogeography theory.
- Relate the theories of island biogeography and metapopulations to habitat fragmentation.
- Define succession and describe the stages.
- Contrast primary and secondary succession.
- Evaluate the concept of the climax community.
- Define disturbance.
- Distinguish between small-scale and large-scale disturbances.
- Explain the relationship between plant and animal succession.
- Discuss the successional changes in vegetation following the retreat of the Pleistocene ice sheet.

Primary succession on sand dunes on Martha's Vineyard island off the coast of Massachusetts. Late-stage oak woods occupy the dunes in the foreground. Beach grasses colonize the distant dunes adjacent to the sea.

The physical and biological structure described in the preceding chapter is not a static characteristic of the community. Both change temporally and spatially. The vertical structure of the community changes with time as plants become established, grow, and die. The birth and death rates of species change in response to environmental conditions, shifting the pattern of species dominance and diversity. As environmental conditions change in time and space, the structure of the community, both physical and biological, likewise changes. The result is a dynamic mosaic of communities. It is this changing pattern of community structure that is the focus of community ecology.

21.1 Spatial variation in community structure is zonation

As you move across the landscape, the physical and biological structure of the community changes. Often these changes are small, subtle ones in the species composition or height of the vegetation. However, as you travel farther, these changes become more pronounced. For example, the region of central Virginia just east of the Blue Ridge mountains is a landscape of rolling hills. The area is a mosaic of forest and field. As you walk through a forest, the physical structure of the community—the canopy, understory, shrub layer, and forest floor—appear much the same. However, the biological structure, the mix of species that compose the community, may change quite dramatically. As you move from a hilltop to the bottomland along a stream, the mix of trees changes from oaks (*Quercus* spp.) and hickory (*Carya* spp.) to species associated with much wetter environments, such as sycamore (*Platanus occidentalis*), hornbeam (*Carpinus caroliniana*), and sweetgum (*Liquidambar styraciflua*) (Figure 21.1). These changes in the physical and biological structure of communities as you move across the landscape are referred to as **zonation.**

Patterns of spatial variation in community structure or zonation are common to all environments, aquatic and terrestrial. Figure 21.2 provides an example of zonation in a salt marsh. Notice the variations in the physical and biological structure of the communities as you move from the shore through the marsh to the upland. The dominant plant growth forms in the marsh are grasses and sedges. These growth forms give way to shrubs and trees as you move to dry land and the depth of the water table increases. Within the

zone dominated by grasses and sedges, the dominant species change as you move back from the tidal areas. These differences result from a variety of environmental changes across a spatial gradient, including microtopography, water depth, amount of tidal flooding, and salinity. The changes are marked by distinct plant communities that are defined by changes in dominant plants and in structural features such as height, density, and dispersion.

21.2 Defining boundaries between communities is often difficult

How different must two adjacent areas be before we call them separate communities? This is not a simple question. Consider the forest in Figure 21.1. Given the difference in species composition from hilltop to bottomland, most ecologists would define these two areas as different vegetation communities; but as you walk between them, the distinction may not seem so straightforward. If the transition between the two communities is abrupt, there may not be a problem in defining the community boundaries. However, the species composition and patterns of dominance may shift gradually. In this case, the boundary is not so clear.

Ecologists have a variety of sampling and statistical techniques to delineate and classify communities. Generally, all employ some measure of community similarity or difference (see Quantifying Ecology 20.3). Although it is easy to describe the similarities and differences between two areas in terms of species composition and structure, the actual classification of areas into distinct groups of communities involves a degree of subjectivity, and it often depends on the objectives of the particular study.

21.3 Transition zones are special environments

Closely associated with the transition zone between communities are edge and ecotone. Although the two terms are often used synonymously, they are different. An **edge** is where two or more different vegetation communities meet. An **ecotone** is where two or more communities not only meet but intergrade (Figure 21.3).

Edges may mark an abrupt change in environmental conditions such as soil type, topography, substrate,

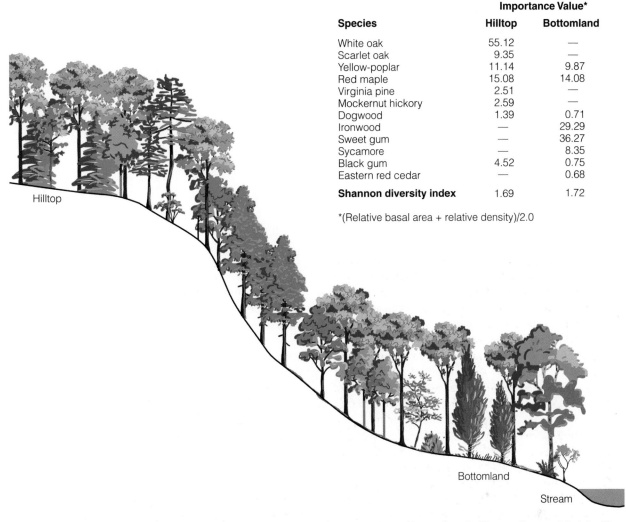

Species	Importance Value*	
	Hilltop	**Bottomland**
White oak	55.12	—
Scarlet oak	9.35	—
Yellow-poplar	11.14	9.87
Red maple	15.08	14.08
Virginia pine	2.51	—
Mockernut hickory	2.59	—
Dogwood	1.39	0.71
Ironwood	—	29.29
Sweet gum	—	36.27
Sycamore	—	8.35
Black gum	4.52	0.75
Eastern red cedar	—	0.68
Shannon diversity index	1.69	1.72

*(Relative basal area + relative density)/2.0

Hilltop

Bottomland

Stream

Figure 21.1 Changes in species composition of forest stands along a topographic gradient in Fluvana County, Virginia. The table summarizes stand composition as sampled at two points: on the hilltop and along the stream.

or microclimate. Where long-term natural features determine adjoining vegetation types, edges are usually stable and permanent. We call them **inherent.** Other edges result from natural disturbances such as fire, storms, and floods or from human-induced disturbances such as grazing, timber harvesting, land clearing, and agriculture. Then the adjoining vegetation types are possibly temporary. Such edges are **induced.** They can be maintained only by periodic disturbance. Induced edges, too, may be abrupt, or they may be transitional, producing an ecotone.

Ecotones arise in the transition zone between two communities that exhibit a shift in dominance. In some cases these areas may be composed of a mix of species that are found in the two adjacent communities, or they

may be characterized by a unique species or group of species not occurring in either.

Certain highly adaptable species colonize edge and ecotone environments. Edge species tend to be opportunistic. Plant species are often intolerant of shade and tolerant of a dry environment, which includes a high rate of evapotranspiration, reduced soil moisture, and fluctuating temperatures. Animal species of the edge are usually those that require two or more vegetation communities. For example, the ruffed grouse (*Bonasa umbellatus*) requires new forest openings with an abundance of herbaceous plants and low shrubs for feeding their young, a dense stand of sapling trees for nesting cover, and mature forests for winter food and cover. Because the ruffed grouse spends its life in an area of

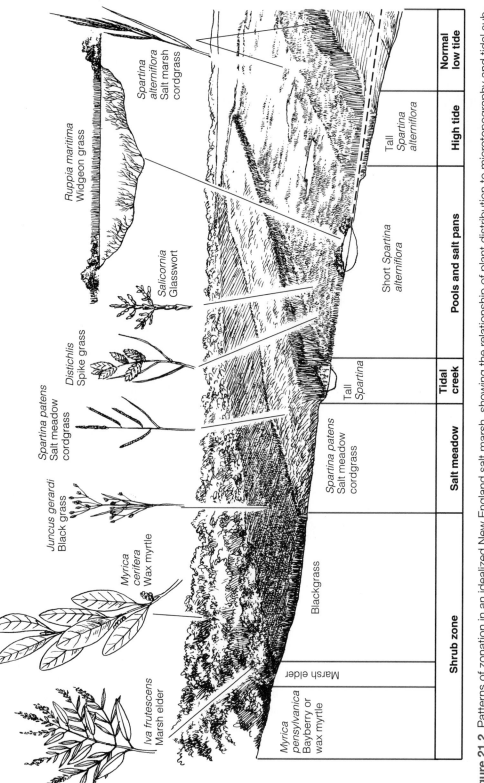

Figure 21.2 Patterns of zonation in an idealized New England salt marsh, showing the relationship of plant distribution to microtopography and tidal submergence.

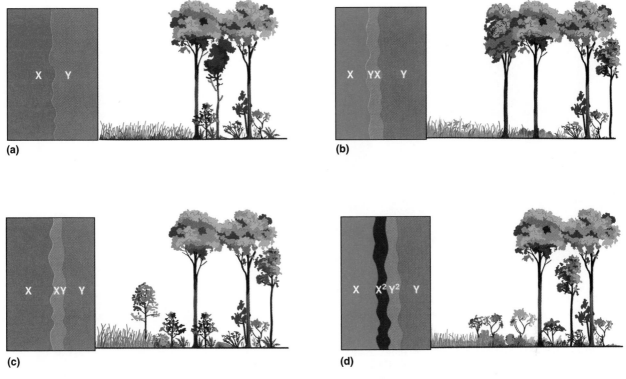

Figure 21.3 Edge between two adjacent communities and types of ecotones that might develop. (a) Abrupt, narrow edge with no development of an ecotone. (b) Narrow ecotone developed by the advance of community Y into community X. (c) Community X advances into community Y to produce ecotone XY. (d) Ideal ecotone development. Plants from both communities invade each other to create a wide ecotone, X^2Y^2. This type of ecotone will support the most edge species.

4 to 8 hectares, this amount of land must fulfill a whole range of requirements. Some species, such as the indigo bunting *(Passerina cyanea)*, are restricted exclusively to the edge environment (Figure 21.4).

The variety and density of life are often greatest in and about edges and ecotones. This phenomenon has been called the **edge effect.** Edge effect is influenced by the amount of edge available—its length and width, and the degree of contrast between adjoining vegetation communities. The greater the contrast between adjoining plant communities, the greater the species richness should be (Figure 21.5). An edge between forest and grassland should support more species than an edge between a young and a mature forest. The larger the adjoining communities, the more opportunity exists for flora and fauna of adjoining communities as well as species that favor edge habitat to occupy the area. If patches of vegetation are too small to support their characteristic species, the area becomes a homogeneous habitat occupied by edge species.

21.4 Fragmented habitats are like islands

Deforestation, suburban and urban expansion, road construction, land clearing, and other human activities are fragmenting large areas of forest and grassland, reducing them to edge communities. As large areas become fragmented, the total habitat is reduced. What is left is distributed in disjointed patches of varying size (Figure 21.6), set in a matrix of urban and suburban developments and agricultural land. The surrounding areas are also terrestrial habitats with their own sets of species. These include domestic animals such as cats, dogs, sheep, and cattle, and species of wildlife highly adaptable to human habitation. Acting as competitors or predators, these species have a significant impact on natural populations.

At one time, wildlife and conservation biologists thought of habitat fragments as islands. Early naturalist explorers had noted the relationship between area and species richness on oceanic islands. Smaller

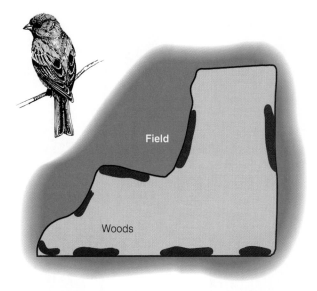

Figure 21.4 Map of territories of a true edge species, the indigo bunting *(Passerina cyanea),* which inhabits woodland edges, large gaps in forests creating edge conditions, hedgerows, and roadside thickets. The male requires tall, open song perches and the female a dense thicket in which to build a nest. (Adapted from J. Whitcomb et al., "Effects of the Forest Fragmentation on Avifauna of the Eastern Deciduous Forest," in R. L. Burger and D. M. Sharpe, eds., *Forest Island Dynamics in Man-Dominated Landscapes,* p. 183, Fig. 8.5c. New York: Springer-Verlag, 1981. Copyright © 1981 Springer-Verlag. Used by permission.)

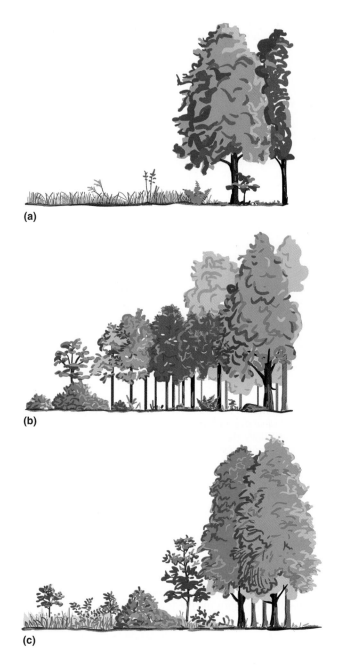

Figure 21.5 Contrast in edge is important in increasing species richness. A high-contrast edge (a) is more valuable to edge species than a low-contrast edge (b) because two quite different vegetation types adjoin. Low-contrast edges do not provide enough difference between vegetation communities to be of maximum value to edge species. Of greatest value is an advancing edge (c), such as woody vegetation invading an adjoining field. An advancing edge not only provides variation in height but, in effect, creates two edges on the site.

islands held fewer species than larger islands, and remote oceanic islands, large and small, held the fewest species. The zoogeographer P. Darlington suggested a rule of thumb: a tenfold increase in area leads to a doubling of the number of species.

Robert MacArthur and E. O. Wilson formally presented the theory of island biogeography in 1963. The theory is simple. The number of species of a given taxon (a taxonomic group of any rank) established on an island represents a dynamic equilibrium between the immigration of new colonizing species and the extinction of previously established ones (Figure 21.7). A new or uninhabited island would be colonized rapidly by species with the greatest ability to disperse to the island from the mainland. As the number of colonizing species increases, the number of immigrants arriving on the island decreases. As the number of colonizing species increases, fewer new immigrants are available from the mainland. Later immigrants may be unable to establish populations be-

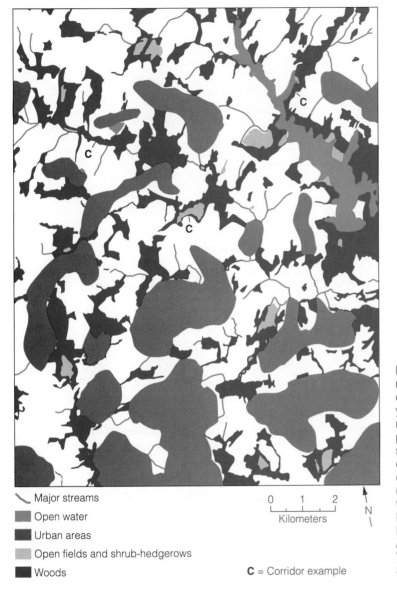

Figure 21.6 At the time of European settlement, this section of eastern Maryland was completely forested. Over several hundred years, land was cleared for agriculture, fragmenting the remaining forest land into isolated patches. In the past quarter century urban and suburban developments have broken up forests and farms even more. So far a few corridors remain, connecting various forest islands. (Adapted from J. Whitcomb et al., "Effects of the Forest Fragmentation on Avifauna of the Eastern Deciduous Forest," in R. L. Burger and D. M. Sharpe, eds., *Forest Island Dynamics in Man-Dominated Landscapes,* p. 127. New York: Springer-Verlag, 1981. Copyright © 1981 Springer-Verlag. Used by permission.)

Major streams
Open water
Urban areas
Open fields and shrub-hedgerows
Woods

0   1   2
Kilometers

N

C = Corridor example

cause available habitats are filled or resources are already utilized.

Not all species immigrating to the island are successful. Some go extinct. As more species arrive, the extinction rate increases. Interspecific competition and small population size make populations vulnerable to natural fluctuations in environmental conditions. An equilibrium is achieved when the rate of immigration is equal to the rate of extinction. At equilibrium the number of species remains stable, although the composition of the species may change. The rate at which one species is lost and a replacement gained is the **turnover rate.**

Influencing the equilibrium level are the size of the island and the distance of the island from the mainland, the pool of potential immigrants. Immigration varies with the distance of the island from the mainland; the farther off, the harder it is for immigrants to disperse. Extinction rates vary with area. They are influenced by the size of the population and the availability of resources. Extinction rates are lower on larger islands because they can support larger populations and

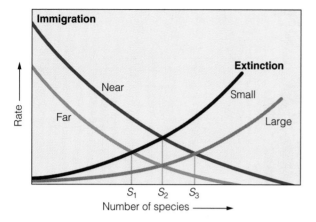

Figure 21.7 According to island equilibrium theory, immigration rates balance extinction rates. Immigration rates are distance-related. Islands near a mainland have a higher immigration rate than islands distant from a mainland. Extinction rates relate to area, being higher on small islands than on large ones.

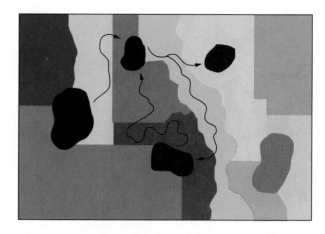

Figure 21.8 Habitat patches with their own subpopulations imbedded in a landscape mosaic. The subpopulations in the patches make up a metapopulation. Corridors influence the movement of individuals between patches and the probability that they will reach them.

contain a greater array of resources and habitats. Large islands can meet the needs of a wider array of species.

The number of species at equilibrium is greater on larger islands than on small ones, if both are the same distance from the mainland. The number of species expected on small islands close to the mainland is about the same as expected on large, distant islands. The turnover rate is higher on small, near islands. They have small populations, which are more likely to become extinct, but at the same time receive replacements easily from the nearby mainland. Large, far islands have a low turnover rate. Their large populations reduce the chance of random extinction as their distance from the mainland precludes a high immigration rate. Wildlife and conservation biologists have noted the potential application of this oceanic island theory to fragmented terrestrial and aquatic habitats.

21.5 Metapopulation theory mostly applies to habitat patches

Metapopulation theory has supplanted island biogeography to explain populations in fragmented habitats. As you will recall (Chapter 11), a **metapopulation** consists of a number of subpopulations that exchange individuals among them by immigration and emigration.

Each subpopulation has its own population fluctuations, its own mortality and natality rates, and its own probability of colonization and extinction. Maintenance of populations depends on movement among patches, not just between patches and one large "mainland."

Habitat patches, unlike oceanic islands, are imbedded in a varied landscape (Figure 21.8). The surrounding landscape influences the quality of the patch, the pathways for movement of species to other patches, and the fate of populations. If the landscape blocks movement, patches can shift in species composition. As habitats become fragmented, species requiring larger patches of habitat, called **interior species,** decline or disappear. Meanwhile other species, attracted by edge conditions, settle in the patch. The species composition of the habitat patch shifts, usually toward edge or generalized species. Interior species, however, may maintain their populations if large contiguous habitats exist as "mainland" sources of immigrants that could augment populations in fragments of similar habitat. The size and distance of fragments from mainland sources could influence the retention of interior species.

Interior species require conditions found in the interior of large habitat patches, away from the abrupt changes in environmental conditions associated with edge environments. Figure 21.9 shows the relationship between area of forest and the probability of occurrence for four different bird species of eastern

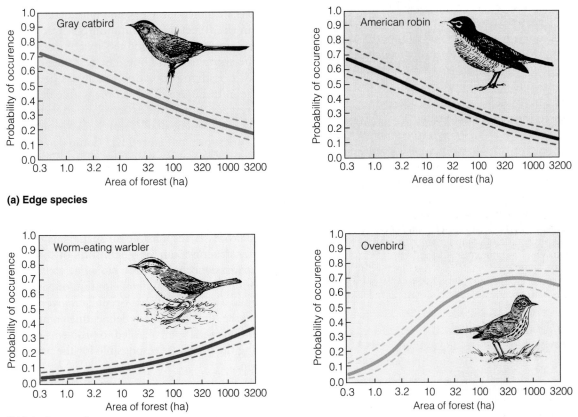

(a) Edge species

(b) Interior species

Figure 21.9 Difference in habitat responses between edge species and area-sensitive or interior species. The graphs indicate the probability of detecting these species from a random point in patches of various sizes. The dashed lines indicate the 95 percent confidence intervals for the predicted probabilities. (a) The catbird and the robin are familiar edge species. As patch size increases, the probability of finding them decreases. (b) The worm-eating warbler and the ovenbird are ground-nesting birds of the forest interior. The probability of finding them in small patches is low.

North America. Edge species, such as the gray catbird *(Dumetella carolinensis)* and American robin *(Turdus migratorius)*, have a high probability of occurrence in small forest patches dominated by edge environments. In contrast, species such as the ovenbird *(Seiurus auro-capillus)* and the worm-eating warbler *(Helmintheros vermivorus)* inhabit older forest stands with large trees and a sparse shrub cover and understory layer. These conditions are typically found in the interior of a forest rather than at the edges. Accordingly, the probability of occurrence for these species is low in small patches and rises with increased habitat size. There is a point in fragment size below which no interior species can exist. The size of a fragmented woodland, for example, may be so reduced that the edge merges with the interior. The fragment, for all purposes, is nothing but edge (Figure 21.10).

21.6 Classification of communities is scale-dependent

The forest in Figure 21.1 was small. As we consider ever larger areas, differences in community structure, both physical and biological, increase. An example is the pattern of forest zonation in the Great Smoky Mountains National Park (Figure 21.11). The zonation is a complex pattern related to elevation, slope position, and exposure. Note that the description of the forest communities in the park contains few species names. Names like hemlock forest are not meant to suggest a lack of species diversity; they are just a shorthand method of naming communities for the dominant species. Each community could be described by a complete list of species, their population sizes, and their contributions to the total biomass. However, such lengthy

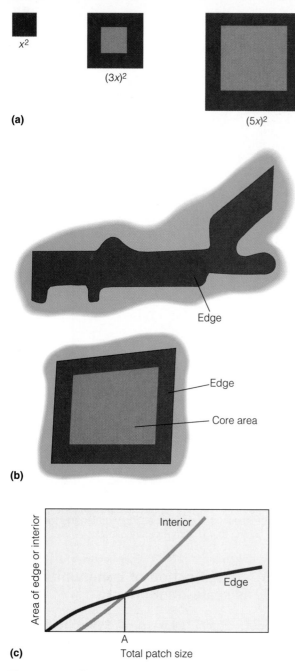

x^2

$(3x)^2$

(a)

$(5x)^2$

Edge

Edge

Core area

(b)

Interior

Edge

A

Area of edge or interior

Total patch size

(c)

Figure 21.10 Relationship of habitat patch size to edge and interior conditions. All habitat patches are surrounded by edge. (a) Assuming that the depth of the edge remains constant, the ratio of edge to interior decreases as the habitat size increases. When the patch is large enough to maintain shaded, moist conditions, an interior begins to develop. (b) This relationship of size to edge holds for a square or circular habitat patch. Long narrow woodland islands whose width does not exceed the depth of the edge would be edge communities, even though the area may be the same as that of square or circular ones. (c) The graph also shows this relationship. Below point A, the size at which interior species can exist, the habitat is all edge. As size increases, interior area increases and the ratio of edge to interior decreases.

descriptions are unnecessary to communicate the major changes in the structure of communities across the landscape. In fact, as we expand the area of interest to include the entire eastern United States, the nomenclature for classifying forest communities becomes even broader. In Figure 21.12, a broad-scale description of forest zonation in the eastern United States developed by Lucy Braun, all of Great Smoky Mountains National Park shown in Figure 21.11 (located in southeastern Tennessee and northwestern North Carolina) is described as a single forest community type—oak-chestnut, a type that extends from New York to Georgia.

These large-scale examples of zonation make an important point to which we shall return when we examine the processes responsible for spatial changes in community structure: our very definition of community is a spatial concept. Like the biological definition of population (Chapter 11), the definition of community refers to a spatial unit that occupies a given area. As we shall see, the difficulty in delineating communities as discrete spatial units has led to problems in understanding the processes responsible for the patterns of zonation.

21.7 Temporal variation in community structure is succession

Community structure varies not only in space, but in time. We have seen how community patterns, both physical and biological, change as you walk from hilltop to stream. Now suppose that you stand in one position, such as in a field adjacent to the forest, and observe the area as time passes. You return to the field each summer to stand in this same spot and look around. In the field that once was covered only in grasses and other herbaceous vegetation, you begin to see the arrival of woody plants, shrubs and small trees. The density of woody vegetation increases until the field becomes a forest (Figure 21.13). The forest initially is dominated by fast-growing, shade-intolerant species such as quaking aspen. Deciduous tree species such as red maple become established in the understory, eventually making their way to the canopy as the aspen trees die. Even these species eventually are replaced by slower-growing, shade-tolerant tree species such as oaks, hickories, and sugar maple.

The process that you have observed, the gradual and seemingly directional change in the structure of the community through time from field to forest, is **succession.** Succession in its most general definition

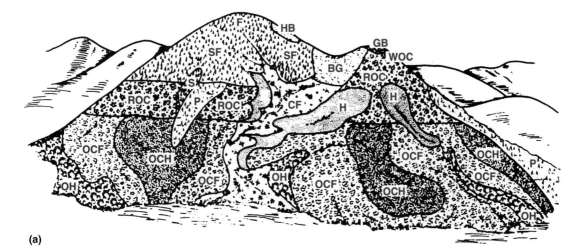

(a)

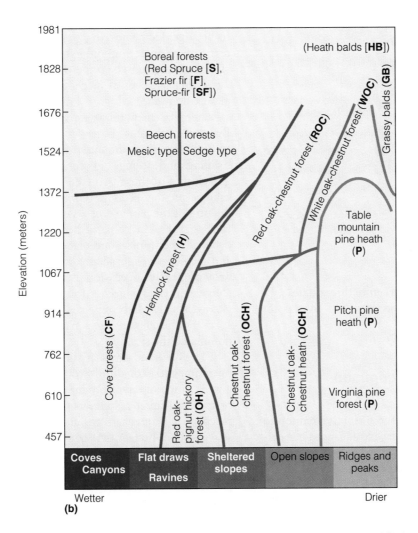

(b)

Figure 21.11 Two descriptions of forest communities in Great Smoky Mountains National Park.
(a) Topographic distribution of vegetation types on an idealized west-facing mountain and valley. (b) Idealized arrangement of community types according to elevation and aspect.

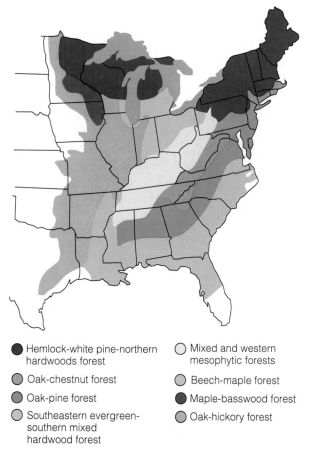

Hemlock-white pine-northern hardwoods forest

Mixed and western mesophytic forests

Oak-chestnut forest

Beech-maple forest

Oak-pine forest

Maple-basswood forest

Southeastern evergreen-southern mixed hardwood forest

Oak-hickory forest

Figure 21.12 Large-scale distribution of eastern deciduous forest communities in the eastern United States is defined by eight regions.

is the change in community structure through time. In contrast to zonation, succession refers to a given point in space—a single location.

21.8 Succession is common to all environments

Like zonation, the process of succession is common to all environments, both terrestrial and aquatic. The ecologist William Sousa carried out an interesting experiment to examine succession in a rocky intertidal algal community in southern California. A major form of nat-

ural disturbance in these communities is the overturning of rocks by the action of waves. Algae populations then recolonize these cleared surfaces. Sousa placed concrete blocks in the water to provide new surfaces for colonization. The results of his study showed a pattern of colonization and extinction. Populations that initially colonized the concrete blocks were displaced by other species as time progressed (Figure 21.14).

Initial or **early successional species** (often referred to as pioneer species) are usually characterized by a high growth rate, small size, wide dispersal, and fast population growth. In contrast, the **late successional species** generally have lower rates of dispersal and colonization, slower growth rates, larger size, and longer lives. As the terms *early* and *late successional species* imply, the pattern of species replacement with time is not random. In fact, if Sousa's experiment were to be repeated tomorrow, we would expect the patterns of colonization and extinction in the **successional sequence** to be somewhat similar.

Other patterns of succession occur in terrestrial plant communities. Figure 21.15 shows the pattern of species replacement following a clearcut at the Hubbard Brook Experimental Forest in New Hampshire. Prior to forest clearing, the understory was dominated by seedlings and saplings of beech and sugar maple. Large individuals of these two tree species dominated the canopy, and the seedlings represented successful reproduction of the parent trees. Following the removal of the larger trees in 1970, the beech and maple seedlings declined and were soon replaced by raspberry thickets and seedlings of sun-adapted (shade-intolerant), fast-growing, early successional tree species such as pin cherry and yellow birch. After many years, these species eventually will be replaced by the late successional species of beech and sugar maple that had previously dominated the site.

The observed similarity in patterns of species colonization and extinction through time for sites across a wide range of environmental conditions, both terrestrial and aquatic, suggests to ecologists a common mechanism or mechanisms influencing the process of succession. The fact that community structure does not vary randomly in either space (zonation) or time (succession), but exhibits repeatable, often predictable patterns, has been the motivating force for research

Figure 21.13 Successional changes over 55 years in a western Pennsylvania field. (a) In 1942 it was moderately grazed. (b) The same area 21 years later in 1963. (c) In 1972 quaking aspen has claimed some of the ground. (d) A view of the rail fence in the right background of (a). (e) In 1963 the rail fence has rotted and white pine and aspen grow in the area. (f) In 1997 the field has been completely claimed by a young stand of aspen and maple.

(a)

(b)

(c)

(d)

(e)

(f)

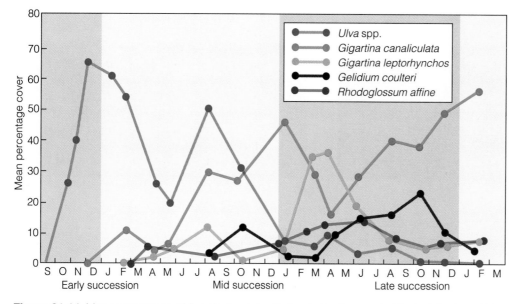

Figure 21.14 Mean percentage of five algal species that colonized concrete blocks introduced into the intertidal zone in September 1974. Note the change in species dominance over time.

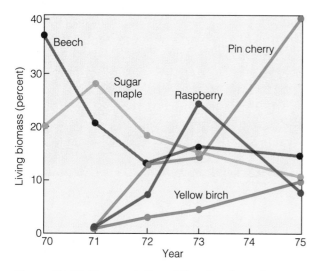

Figure 21.15 Changes in the relative abundance (percentage of living biomass) of woody species in the Hubbard Brook Experimental Forest after a clearcut in 1970. (From F. Bormann and G. E. Likens, *Pattern and Process in a Forested Ecosystem*, p. 116, Fig. 4.4b. New York: Springer-Verlag, 1979. Copyright © 1979 Springer-Verlag. Used by permission.)

throughout the history of ecology. As we will see in the next chapter, the search for causation has not been a simple task. It remains one of the major areas of ecological study.

21.9 Succession is either primary or secondary

The two studies above point out the similar nature of successional dynamics in very different environments. They also exemplify two different types of succession—primary and secondary. **Primary succession** occurs on a site previously unoccupied by a community—a newly exposed surface such as the cement blocks in the rocky intertidal environment. In contrast, **secondary succession** occurs on previously occupied sites following disturbance. Disturbance is any process that results in the removal (either partial or complete) of the existing vegetation community. In the Hubbard Brook example, the disturbance did not remove all individuals. In such cases, the amount (density and biomass) and composition of the surviving community have a major influence on successional dynamics.

Primary succession occurs on newly exposed land surfaces created by geological processes and in newly formed bodies of water. The island of Hawaii is known for its active volcanoes. Lava flows on Hawaii represent newly formed landscapes on which primary succession occurs (Figure 21.16). The previous plant communities in these areas have been destroyed by fire and heat, and the existing soils have been covered by a new layer of igneous rock.

Figure 21.16 Trees and ferns colonize a lava flow on the island of Hawaii.

Another example of primary succession is the colonization of newly exposed glacial till (soil) in the area of Glacier Bay National Park, Alaska. Over the past two hundred years, the glacier that once covered the entire region of the bay has been retreating (Figure 21.17a). As the glacier retreats, the newly exposed landscape is initially colonized by species such as alder (*Alnus* spp.) and cottonwood (*Populus* spp.) (Figure 21.17b). Eventually these early successional species are replaced by spruce (*Picea* spp.) and hemlock, and the resulting forest resembles the surrounding landscape.

21.10 Succession appears to move toward an end point

According to classic ecological theory, succession eventually stops when late successional species come to dominate the site. At this point the community is stable, and barring major disturbances it will persist indefinitely. This end point of succession is the **climax.** Many people misunderstand the concept and as-

sume that any stand of large trees represents a climax community.

The climax community, theoretically, has certain characteristics. The vegetation is tolerant of the environmental conditions it has imposed upon itself. There is equilibrium between gross primary production and total respiration, between energy captured from sunlight and energy released by decomposition, between the uptake of nutrients and the return of nutrients by litterfall. The community has a wide diversity of species, a well-developed spatial structure, and complex food chains. Every individual in the climax stage is supposedly replaced by another of the same kind; average species composition reaches an equilibrium.

A so-called climax forest, however, is not in a state of equilibrium. Self-destructive changes are continually taking place in it. Old trees die and more often than not new individuals and species replace them. Changes are constantly occurring in patches across the community. Thus the species abundance and composition changes, even though the forest's general appearance remains the same. The forest never achieves an equilibrium in growth and species composition. It consists of patches in various stages of successional development, from patches of young trees to patches of old and fallen trees. Rather than climax, such forests are now called old growth (see Chapter 30).

21.11 Disturbance initiates succession

A **disturbance** is a relatively discrete event in time that disrupts communities or populations, changes substrates and resource availability, and creates opportunities for new individuals or colonies. Disturbances have both spatial and temporal characteristics. These characteristics include the size of the area disturbed, the number of disturbances per unit time, the time between disturbances, intensity, and severity.

Disturbance is a matter of scale. There are small, frequent disturbances, such as the death of a single tree in a forest, and major disturbances that cover extensive areas swept by fire, buried under volcanic ash, torn by landslides, or denuded by human land-clearing. What constitutes a small-scale disturbance depends upon the scale of the landscape in which it occurs. The loss of a group of trees in a very small woodland would have more impact than the loss of scattered individual trees or small groups of trees in a large forested area.

Disturbances caused by the death of individuals or groups create a **gap** or opening in the canopy or in the substrate. Gaps in the forest are sites of increased availability of light, soil, temperature, and nutrients,

and decreased soil moisture and relative humidity. Suppressed growth is quickly stimulated by this sudden abundance of resources. Small gaps favor the growth of more shade-tolerant species. If a gap is large—the kind that results from clearcutting, insect damage, windstorms, and ice storms—then stump sprouts, suppressed seedlings, and invading opportunistic species may rapidly fill in the gap. The future composition of the gap will be determined in part by the competitive interactions of incoming species. Fast-growing, shade-intolerant species may outcompete more shade-tolerant species, which will remain in the understory, capable of filling in small gaps that might appear later in the established forest. Gaps create patches of different stages of succession in the community. Chronic small-scale disturbances are important in the maintenance of species richness and structural diversity within a mature forest ecosystem.

Some species depend on disturbances (see Focus on Ecology 21.1: Fire—The Rejuvenator). What species colonize the disturbed area depends upon seeds present on the site and sources outside the site, seedlings and saplings already present, soil conditions, amount of competition, and other ecological conditions. To take advantage of the disturbed conditions, opportunistic species should be nearby; their seeds

(a)

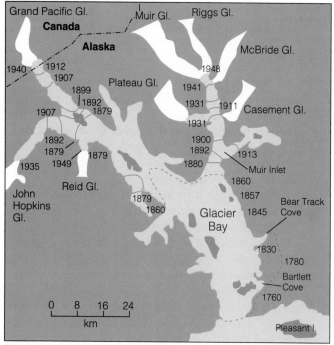
(b)

Figure 21.17 (a) Primary succession along riverine environments of Glacier Bay National Park, Alaska. (b) The Glacier Bay fjord complex in southeastern Alaska, showing the rate of ice recession since 1760. As the ice retreats, it leaves moraines along the edge of the bay in which primary succession occurs.

FIRE—THE REJUVENATOR

Fire, often regarded as a major disaster to forests and shrublands, is nevertheless a major force in the evolution of plants and the development of vegetation. In certain fire-prone regions of Earth—such as those with a mediterranean-type climate, grasslands, pinelands, and other coniferous forest—vegetation has evolved adaptations to periodic burns, usually set by lightning. Such plants can respond to fire in three general ways. They can survive as mature plants with little or no damage. In another response, the mature plants die, and a new generation of young plants of the same species springs up in their place. In a third response, fire destroys the aboveground parts of the plants and they regenerate from roots and associated structures. These plants require the rejuvenating effects of periodic fires.

Fire dependence in such vegetation appears to have evolved with flammability. The plants have fine branches, volatile resins and terpenes in foliage, and an accumulation of dead material, which increase their ability to burn.

Many fire-dependent plants and plant communities exist globally. They include chamise, one of the most abundant species in the California chaparral, and Australian eucalyptus. Longleaf pine (*Pinus palustris*), a fire-dependent species, reseeds on a fire-prepared seedbed. Giant sequoia (*Sequoiadendron giganteum*), with its serotinous cone, small seeds, and fire-resistant bark, requires fire for reestablishment and for the elimination of a shade-tolerant coniferous understory that increases the danger of fire in the crown. Fire-dependent grasslands, such as the tallgrass prairies and African grasslands, unless burned periodically, would become dominated by woody vegetation. Long-term exclusion of fires dooms fire-dependent communities.

Although fire over the short term may kill vertebrates and destroy habitat, many animal species survive the fire and recolonize the burned-over area. Many species depend upon periodic fires to maintain their habitat and to provide the proper mosaic of vegetation types. Among strongly fire-dependent species are Kirtland's warbler (*Dendroica kirtlandii*), restricted to the jack pine forests of the lower peninsula of Michigan, and the Dartford warbler (*Sylvia sarda*) and Marmora's warbler (*S. undata*) of the Mediterranean shrublands of Sardinia. Without periodic fires to maintain their restricted habitats, these birds face extinction.

should reach the site in large numbers to ensure some seed survival; and seed should settle on exposed mineral soil to favor rapid germination and growth.

A severe disturbance often changes the original community. It may be so greatly disturbed that it is unable to return to its original state and a different community takes its place. For example, when the forests of Scotch pine were cut in northern Britain, they were replaced by moorlands. Extensive land clearing in the Amazon basin has transformed tropical forests into scrub savannas.

21.12 Succession occurs in heterotrophic communities

So far we have focused on communities of photosynthetic organisms, the autotrophic communities (see Chapter 20). We also find succession in the heterotrophic communities involved in decomposition. Dead trees, animal carcasses, and droppings form substrates on which heterotrophic communities exist. Within these communities, groups of plants and animals succeed each other and disappear, becoming incorporated into the soil. In these instances succession is characterized by early dominance of fungi and invertebrates that feed on dead organic matter. Available energy and nutrients are most abundant in the early stages of succession, and decline steadily as succession proceeds.

When a windstorm uproots or breaks a tree, the fallen tree becomes the stage for succession of plant and animal colonists until the log becomes part of the forest soil (Figure 21.18). The newly fallen tree, its bark and wood intact, is a ready source of shelter and nutrients. The first to exploit this resource are bark beetles and wood-boring beetles that drill

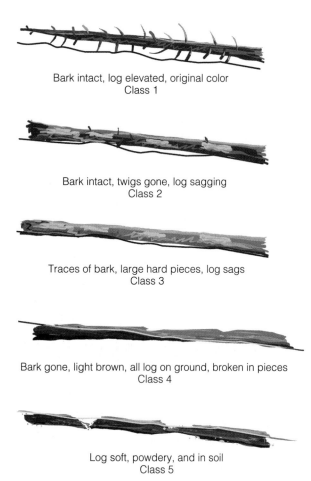

Bark intact, log elevated, original color
Class 1

Bark intact, twigs gone, log sagging
Class 2

Traces of bark, large hard pieces, log sags
Class 3

Bark gone, light brown, all log on ground, broken in pieces
Class 4

Log soft, powdery, and in soil
Class 5

Figure 21.18 Stages in the decomposition of a fallen log.

through the bark, feed on the inner bark and the cambium, reducing it to frass (feces) and fragments, and tunnel galleries in which to lay eggs. Both adults and larvae drill more tunnels as they feed. Ambrosia beetles tunnel into the sapwood, creating galleries in which fungi grow. The tunnels provide a passageway and the frass and softened wood form a substrate for bacteria. Loosened bark provides cover for predatory insects soon to follow: centipedes, mites, pseudoscorpions, and beetles.

As decay proceeds, the softened wood holds more moisture, but the accessible nutrients have been depleted, leaving behind more complex, decay-resistant compounds. The pioneering arthropods leave for other logs. Fungi possessing more sophisticated enzyme systems to break down cellulose and lignin remain. Moss and lichens find the softened wood an ideal habitat.

Plant seedlings, too, take root on the softened logs, and their roots penetrate the heartwood, providing a pathway for fungal growth into the depths of the log.

Eventually the log is broken into light brown, soft, blocky pieces, and the bark and sapwood are gone. At this advanced stage of decay the log provides the greatest array of microhabitats and the highest species diversity. Invertebrates of many kinds find shelter in the openings and passages; salamanders and mice dig tunnels and move into the rotten wood. Fungi and other microorganisms abound, and numerous species of mites feed on decomposed wood and fungi. At last the log crumbles into a red-brown mulchlike mound of lignin materials resistant to decay, its nutrients and energy largely depleted. The tree is incorporated into the soil.

21.13 Succession changes the distribution and abundance of animals

As plant succession advances, animal life changes too (Figure 21.19). Each successional stage has its own distinctive fauna. Because animal life is often influenced more by structural characteristics than by species composition, successional stages of animal life may not correspond to the stages identified by plant ecologists. For example, a stand of early successional trees that gives way to later successional species may represent two distinct stages of forest succession. However, if the stages do not differ in vertical and horizontal structure, they may have little effect on which bird species inhabit the canopy.

Animals can quickly lose their habitat by vegetation change. In eastern North America, grasslands and old fields support meadowlarks, meadow mice, and grasshoppers. When woody plants—both young trees and shrubs—invade, a new structural element appears. Grassland animals soon disappear, and shrubland animals take over. Towhees, catbirds, and goldfinches claim the thickets, and meadow mice give way to white-footed mice. When woody growth exceeds a height of 6 m and the canopy closes, species of the shrubland decline, replaced by birds and insects of the forest canopy. As succession proceeds to mature forest, the vertical structure becomes more complex. New species appear, such as tree squirrels, woodpeckers, and birds of the forest understory, such as the hooded warbler and ovenbird.

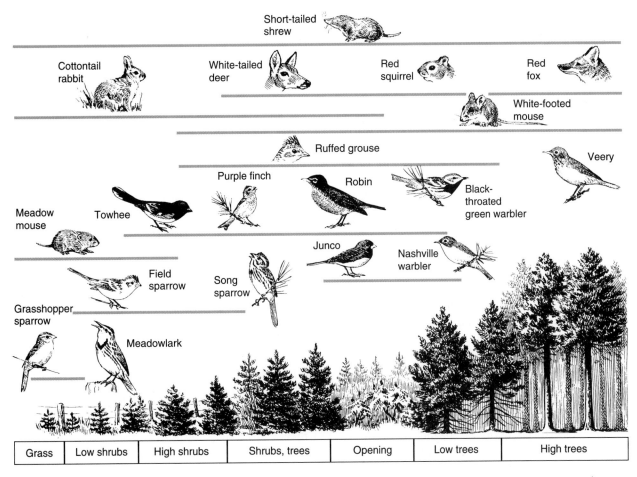

Figure 21.19 Wildlife succession in a conifer plantation in central New York. Species appear or disappear as vegetation density and height change.

The various stages of wetland succession support their own forms of animal life. Fish live in open water and waterfowl and herons feed there. Floating aquatic vegetation supports hydras, frogs, diving beetles, gill-breathing snails, and insects colonizing the undersides of floating leaves. Emergent vegetation is occupied by nesting waterfowl and bitterns, red-winged blackbirds, muskrats, amphibians, flies and mosquitoes, mayflies, and dragonflies. Incoming woody vegetation such as alder and willow creates habitats for swamp sparrows, yellowthroats, woodcocks, and white-footed mice.

In general, the diversity of animal life changes as succession proceeds (Figure 21.20). Herbaceous and shrubland stages support the greatest diversity of animal species. These species depend upon the early stages of plant succession for their habitat. As that stage passes away, so do the animals that occupy it.

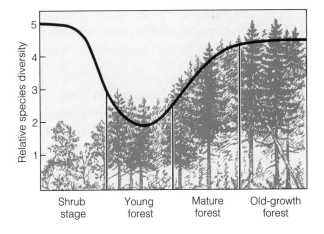

Figure 21.20 The relationship between successional stage and the number of mammal species present in Pacific Northwest forests.

Therefore certain species of animal life are highly dependent on disturbances that restore or maintain early stages of succession. Among such animals are bobwhite quail, cottontail rabbit, prairie warbler, and woodcock. Because of their dense canopy, lack of understory vegetation, and low degree of vertical stratification, young forest stands exhibit the lowest diversity. Diversity increases as the forest matures, with old-growth forests supporting a higher diversity than other forest stages. Old growth has more varied habitats, including dead and fallen trees and gaps in the canopy that stimulate new understory growth.

21.14   Community structure changes over geological time

Since its inception some 4.6 billion years ago, Earth has changed profoundly. Land masses emerged and broke into continents. Mountains formed and eroded, seas rose and fell, and ice sheets advanced to cover large expanses of the Northern and Southern Hemispheres and retreated. All these changes affected the climate and other environmental conditions from one region of Earth to another. Many species of plants and animals evolved, disappeared, and were replaced by others. As environmental conditions changed, so did the distribution and abundance of plant and animal species.

Records of plants and animals composing past communities lie buried as fossils: bones, insect exoskeletons, plant parts, and pollen grains. The study of the distribution and abundance of ancient organisms and their relationship to the environment is **paleoecology.** The key to present-day distributions of animals and plants can often be found in paleoecological studies. For example, paleoecologists have reconstructed the distribution of plants in eastern North America following the last glacial maximum of the Pleistocene.

The Pleistocene was an epoch of great climatic fluctuations throughout the world. At least four times in North America and three times in Europe an ice sheet advanced and retreated. With each movement the biota retreated and advanced again with a somewhat different mix of species.

Each glacial period was followed by an interglacial period. The climatic oscillations in each interglacial period had two major stages, cold and temperate. During the cold stage tundra-like vegetation dominated the landscape. As the glaciers retreated, light-demanding forest trees such as birch and pine advanced. Then as the soil improved and the climate warmed, these trees were replaced by more shade-tolerant species such as oak and ash. As the next glacial period began to develop, species such as firs and spruces took over the forest. Now both climate and soil began to deteriorate. Heaths began to dominate the vegetation and forest species disappeared.

The last great ice sheet, the Laurentian, reached its maximum advance about 20,000 BP to 18,000 BP during the Wisconsin glaciation stage in North America. Canada was under ice. A narrow belt of tundra about 60 to 100 km wide bordered the edge of the ice sheet and probably extended southward into the high Appalachians. A few relict examples of this tundra persist today. Boreal forest, dominated by spruce and jack pine, covered most of the eastern and central United States as far as western Kansas.

As the climate warmed and the ice sheet retreated northward, plant species invaded the glaciated areas. The maps in Figure 21.21 reflect the advances of four major tree genera in eastern North America following the retreat of the ice sheet. Margaret Davis developed these maps from patterns of pollen deposition in sediment cores taken from lakes in eastern North America. By examining the presence and quantity of pollen deposited in sediment layers and dating the sediments with radiocarbon, she was able to obtain a picture of the spatial and temporal dynamics of tree communities over the past 18,000 years.

These analyses identify plants at the level of genus rather than species, because in many cases we cannot identify species from pollen grains. Note that the rates at which different genera and probably species expanded their distribution northward with the retreat of the glacier are markedly different. The differences in the rates of range expansion are most likely due to the differences in the temperature responses of the species, the distances and rates at which seeds can disperse, and interactions among species. The implication is that over the past 18,000 years, the distribution and abundance of species and the subsequent structure of forest communities in eastern North America have changed dramatically. The next chapter will discuss the processes underlying these long-term changes in community structure and dynamics.

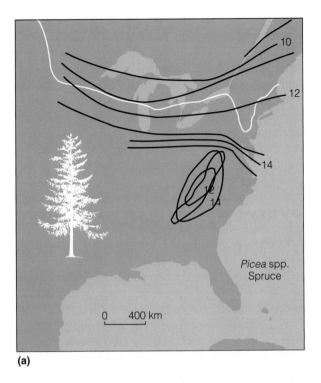

(a)

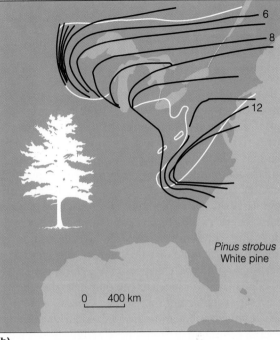

(b)

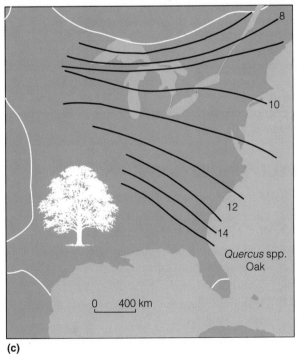

(c)

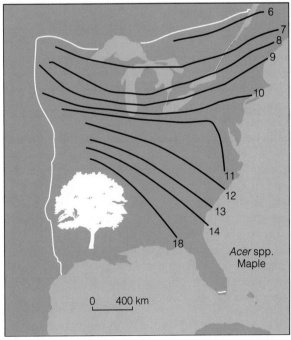

(d)

Figure 21.21 The postglacial migration of four tree genera: (a) spruce, (b) white pine, (c) oak, and (d) maple. The dark lines represent the leading edges of the northward expanding populations. The white lines indicate the boundaries of the present-day ranges. The numbers are thousands of years before the present.

SUMMARY

Spatial Variation in Community Structure (21.1–21.3)
Changes in the physical structure and biological communities across a landscape result in zonation. Zonation is common to all environments, aquatic and terrestrial. Zonation is most pronounced where sharp changes occur in the physical environment, as in aquatic communities (21.1). In most cases, transitions between communities are gradual, and defining the boundary between communities is difficult (21.2). The place where two different communities meet is an edge. The area of overlap between two communities is an ecotone. An edge may be inherent, produced by a sharp environmental change such as a shift in soil type. An edge may be induced, created by disturbance. Because they support not only selected species of adjoining communities but also a group of opportunistic edge species, edges and ecotones have a high species richness (21.3).

Habitat Fragmentation (21.4–21.6) Many edge communities result from the fragmentation of habitat. In habitat patches as in islands, large areas support more species than small areas. Island biogeography theory states that the number of species an island holds represents a balance between immigration and extinction. Immigration rates are influenced by the distance of the island from a mainland pool of potential immigrants. Extinction rates are influenced by the area of an island (21.4). Populations inhabiting fragmented habitats form metapopulations. Maintaining them depends upon movement of individuals among habitat patches. Diversity within patches is size-dependent. Small habitat patches are inhabited by edge species. Large habitat patches support interior species—those that require extensive habitats—as well as edge species. Extensive fragmentation of habitats eliminates interior species (21.5). The way we classify a community depends on the scale we use (21.6).

Succession (21.7–21.13) With time natural communities change. This gradual sequential change in the relative abundance of species in a community is succession. Opportunistic, early successional species yield to late successional species (21.7). Succession occurs in all environments. The similarity of succes-

sional patterns in different environments suggests a common set of processes (21.8). Primary succession begins on sites devoid of or unchanged by organisms. Secondary succession begins on sites where organisms are already present (21.9). Succession is often represented as having an end point, the climax community, although mortality and disturbance continue successional processes (21.10).

Disturbance, a discrete event that disrupts communities and populations, also initiates succession and creates diversity. Disturbances have spatial and temporal characteristics. Small-scale disturbances make gaps in the substrate or vegetation, creating patches of different composition or successional maturity. Large-scale disturbances favor opportunistic species. Severe disturbances can replace the community altogether (21.11).

Succession takes place in heterotrophic decomposer communities, such as fallen logs. Energy and nutrients dissipate as such succession proceeds (21.12). Successional changes in vegetation affect the nature and diversity of animal life. Certain sets of species are associated with each successional stage (21.13).

Long-term Changes (21.14) The present pattern of vegetational distribution reflects the glacial events of the Pleistocene. Plants retreated and advanced with the movements of the ice sheets. The rates and distances of their advances are reflected in the present-day ranges of species.

STUDY QUESTIONS

1. What are edge, ecotone, and the edge effect?
2. Contrast edge species with interior species. How do they relate to the size of habitat patches?
3. Contrast island biogeography theory with metapopulation theory.
4. What is succession?
5. Distinguish between primary succession and secondary succession.
6. How does succession work in heterotrophic communities?
7. What is the climax? Is it a valid concept?
8. What is disturbance? Contrast small-scale and large-scale disturbances.

9. What is the role of gaps in community dynamics?

10. What is the relationship of plant succession to animal habitats?

*11. Locate an area with which you were familiar years ago and compare the vegetation then and now. What changes have taken place? What brought them about?

*12. Neotropical birds—those that spend most of the year in the tropics, but nest in North America—are threatened by tropical deforestation. Might a greater threat be the fragmentation of nesting habitat, given that most of them are interior species? Discuss.

*13. Should we control fires in our national parks? Why or why not?

PROCESSES SHAPING COMMUNITIES

OBJECTIVES

On completion of this chapter, you should be able to:

- Contrast the organismal with the individualistic concept of the community.
- Discuss how the fundamental niche of a species constrains community structure.
- Explain autogenic and allogenic processes in succession.
- Show how organisms can influence succession by changing environmental conditions.
- Discuss why dominance changes along a temporal gradient in succession.

Douglas-fir and western hemlock with an abundance of dead wood and decomposing logs—a setting characteristic of old-growth forests.

The last two chapters examined the structure and dynamics of communities, exploring how patterns of species distribution and abundance vary in time and space. We described how dominance and diversity change as you move across the landscape or as time progresses. However, scientists want to do more than describe. They want to predict and understand. What processes shape these patterns of community structure? Why is the basic process of succession following disturbance consistent in different environments? Why are communities in some environments more or less diverse than others? The processes controlling community structure and dynamics have been the focus of research and debate for over a century, a debate that has extended to the definition of community itself.

22.1 Two contrasting views of the community exist

In Chapter 20 we defined the community as a collection of directly or indirectly interacting plant and animal populations. What does "interacting" mean? This question led to a major debate in ecology in the first half of the twentieth century, a debate that still influences our views of the community.

When you walk through most forests, you see a variety of plant, fungal, and animal species—a community. If you walk far enough, the dominant plant and animal species will change (see Figure 21.1). As you move from hilltop to valley, the structure of the community will differ. But what if you continue your walk over the next hilltop and into the adjacent valley? You will most likely notice that although the communities on the hilltop and valley are quite distinct, the communities on the two hilltops or valleys are quite similar. As a botanist might put it, they exhibit relatively consistent floristic composition. At the International Botanical Congress of 1910, botanists adopted the term **association** to describe this phenomenon. An association is a type of community with: (1) relatively consistent species composition, (2) a uniform general appearance (physiognomy), and (3) a distribution that is characteristic of a particular habitat, such as the hilltop or valley. Whenever the particular habitat or set of environmental conditions repeats itself in a given region, the same group of species occurs.

Some scientists of the time thought that association implied processes that might be responsible for structuring communities. The logic went that if clusters or groups of species repeatedly associate together, that is indirect evidence for either positive or neutral interactions among them. Such evidence favors a view

of communities as integrated units. One of the leading proponents of this thinking was the Nebraskan botanist Frederick Clements. Clements developed what has become known as the **organismal concept of communities.** Clements likened associations to organisms, with each species representing an interacting, integrated component of the whole.

As depicted in Figure 22.1a, in Clements's view the species in an association have similar distributional limits along the environmental gradient, and many of them rise to maximum abundance at the same point. The ecotones between adjacent associations are narrow, with few species in common. This view of the community suggests a common evolutionary history and similar fundamental responses and tolerances (see Chapter 2) for the component species. Mutualism and coevolution play an important role in the evolution of species making up the association. The community has evolved as an integrated whole.

In contrast to Clements's organismal view of communities was the botanist H. A. Gleason's view of community. It was almost exactly the opposite. Gleason stressed the individualistic nature of species distribution. His view became known as the **individualistic continuum concept.** The continuum concept states that the relationship among coexisting species (species within a community) is a result of similarities in their requirements and tolerances, not a result of strong interactions or common evolutionary history. In fact, Gleason concluded that changes in species abundance along environmental gradients occur so gradually that it is not practical to divide the vegetation (species) into associations. In contrast to Clements's view, species distributions along environmental gradients do not form clusters, but represent the independent responses of species. Ecotones are gradual and difficult to identify (Figure 22.1b). What is referred to as the community is merely the group of species found to coexist under any particular set of environmental conditions.

The major difference between these two views is the importance of interactions, evolutionary and current, in the structuring of communities. Although it is tempting to choose between these views, current thinking involves elements of both perspectives, as we will see.

22.2 The fundamental niche constrains community structure

Recall the concept of the fundamental niche from Chapter 15. All living organisms have a range of environmental conditions under which they can survive, grow,

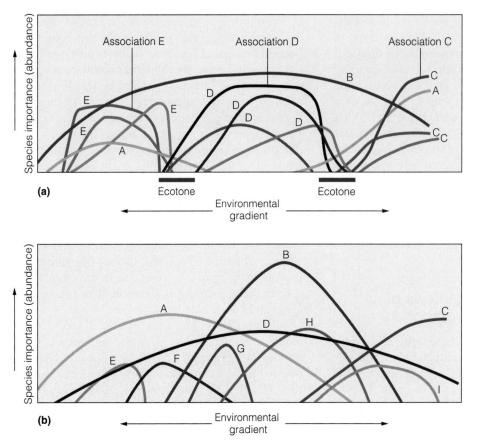

Figure 22.1 Two models of community. (a) The organismal or discrete view of communities proposed by Clements. Clusters of species (C's, D's, and E's) show similar distribution limits and peaks in abundance. Each cluster defines an association. A few species (for example, A) have sufficiently broad ranges of tolerance that they occur in adjacent associations, but in low numbers. A few other species (for example, B) are ubiquitous. (b) The individualistic or continuum view of communities proposed by Gleason. Clusters of species do not exist. Peaks of abundance of dominant species, such as A, B, and C, are merely arbitrary segments along a continuum. This view of community is similar to the groups of species associated with environments e_1 and e_2 in Figure 22.2.

and reproduce. Outside this range of conditions, they die. This range of environmental conditions is not the same for all organisms. The characteristics that allow an organism to prosper under one set of environmental conditions often limit its ability to do well under another set. We can represent the fundamental niches of a variety of species with bell-shaped curves along some environmental gradient, such as availability of water or light for plants (Figure 22.2). The response of each species is defined in terms of abundance. Although the fundamental niches overlap, each species has its limits, beyond which it cannot survive. Consider the fundamental niches of two species of cattail *(Typha)*, depicted in Figure 22.3a. Both species can survive in shallow waters, but only the narrow-leaved

cattail, *Typha angustifolia*, can grow in water deeper than 80 cm.

The distribution of fundamental niches along the environmental gradient shown in Figures 22.2 and 22.3a represents a primary constraint on the structure of communities. For any given range of environmental conditions, only a subset of species can survive, grow, and reproduce. As the environmental conditions change in either space or time, the possible distribution and abundance of species will change. For example, it is possible for both species of cattails to survive in the shallow water near shore and form components of the community. However, the broad-leaved cattail, *Typha latifolia*, will never be a part of the community in deeper waters, because it is unable to tolerate water deeper than 80 cm.

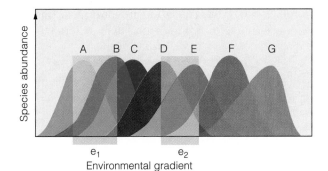

Figure 22.2 Fundamental niches of seven hypothetical species along an environmental gradient (e.g., moisture, temperature, or elevation) in the absence of competition from other species. The species all have bell-shaped responses to the gradient, but each has different tolerance limits defined by a minimum and maximum value along the gradient. As conditions change, for example from e_1 to e_2, the set of species that can potentially occur in the community changes. These changes can occur in either time or space.

We saw another example of the constraint on community structure imposed by species' fundamental niches in Figure 21.19. The diagram represents the changes in the structure of vegetation during secondary succession and the animal species that use these various plant communities as habitat. Species such as the meadowlark, meadow mouse, and field sparrow inhabit the herbaceous and shrub-dominated early stages of succession. Shrubs, grasses, and other herbaceous plants provide food, cover from predators, and nesting sites for these species. As the vegetation changes to forest, it no longer is suitable habitat. These species can no longer persist, and they are replaced by other species that require the forest for habitat. Thus vegetation is a primary constraint on the structure of the animal community. As the vegetation changes through time, so will the animal species that depend on it.

22.3 Organisms alter the environment for other species

Organisms can interact in either of two ways—directly, through interactions such as territoriality and predation; or indirectly, through modification of the physical environment. Direct interactions such as physical conflict are more noticeable. Indirect interactions through the modification of the physical environment, however, are by far the more common means by which populations and individuals interact and influence the

structure and dynamics of communities. We have seen numerous examples in the previous chapters. In plant communities, taller plants intercept solar radiation, reducing the availability of light to shorter plants, slowing their rates of photosynthesis and growth. The uptake of water and nutrients by the root system of one plant decreases the availability of those essential resources to neighboring plants. The consumption of seeds by one species of seed-eating bird reduces the availability of food to other species that share the same seed resource.

These are all examples of competition resulting from the sharing of a limiting resource. Individuals of one species modify the environment, in these cases the availability of an essential resource, for other species that inhabit the same area. These competitive interactions may limit the distribution and abundance of species by reducing rates of population growth or increasing rates of mortality. For an example of competition modifying the distribution and abundance of a species, we can return to the cattails. In Figure 22.3a each species grew along the water depth gradient in the absence of the other species. These responses therefore represent the potential response of the species, or the fundamental niche. When the two species grew together along the same gradient of water depth, their distributions, or realized niches, changed (Figure 22.3b). Even though *Typha angustifolia* can grow in shallow waters (0 to 20 cm depth) and above the shoreline (–20 to 0 cm depth), in the presence of *Typha latifolia* it is limited to depths of 20 cm or deeper. Individuals of *T. latifolia* outcompete individuals of *T. angustifolia* for the limited resources of nutrients, light, and space, limiting its distribution to the deeper waters. Note that the maximum abundance of *T. angustifolia* is limited to the deeper waters where *T. latifolia* is not able to survive.

In other situations one species modifies the environment in such a way as to make it more favorable for another species. An example is the mutualistic association between green plants and mycorrhizal fungi we discussed in Chapter 17. You will recall that the plant provides the fungi with carbon (food), while the fungi make nitrogen and phosphorus available to the plant. Therefore the fungi enable the plants to grow in environments where otherwise the low availability of these nutrients would limit growth and survival.

An example of how the fundamental niche of a tree species, the Brazilian sour orange (*Citrus aurantium*), is modified by its association with mycorrhizal fungi appears in Figure 22.4. In this experiment, the availability of phosphorus in the soil was controlled

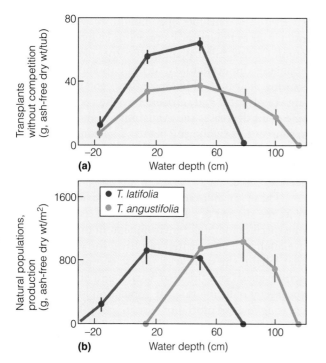

(a)

(b)

Figure 22.3 The distribution of two species of cattail (*Typha latifolia* and *Typha angustifolia*) along a gradient of water depth: (a) grown separately in an experiment; (b) grown together in natural populations. The response of the two species reflects their fundamental niche (physiological tolerances) in the absence of competition. The response of each species is altered by the presence of the other. They are forced to occupy only their realized niches.

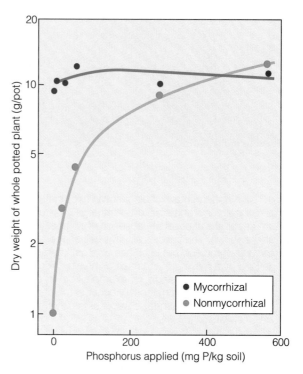

Figure 22.4 The effects of mycorrhizal fungi on the growth of Brazilian sour orange trees under different levels of phosphorus fertilization. Plants with mycorrhizae growing in soil with low phosphorus levels have a high growth rate because mycorrhizae increase phosphorus available to the plant. The mycorrhizal plants do well at all phosphorus levels in the soil, whereas nonmycorrhizal plants respond dramatically to phosphorus fertilization.

by the addition of fertilizer. When the amount of phosphorus added to the soil is high, plants with and without the fungi grow equally well. The uptake of phosphorus does not limit growth, and the fungi provide no advantage. However, when no phosphorus fertilizer is added to the soil, the plants with fungi grow ten times as large and fast as plants without them. The fungi enable the trees to grow in areas of low nutrient availability where they would otherwise be unable to survive.

In both of these examples, competition and mutualism, one species modifies the environmental conditions in such a way as to influence the distribution and abundance of the other species within the community. Competition is perhaps the most widely studied interaction among species that has been shown to influence community structure. If competition is such an important factor in community structure, then any theory or understanding of communities must address two questions: For a given set of environmental con-

ditions, why is one species a better competitor than another? Do the relative competitive abilities of species change as environmental conditions change?

22.4 Changing environments affect an organism's ability to compete

To address the question of why one species is a better competitor than another under a certain set of environmental conditions, we must return to the basic premise presented in Chapter 2. The set of characteristics that enable an organism to survive, grow, and reproduce under one set of environmental conditions restricts its ability to respond under different conditions. In Chapters 14 and 15, we saw how individuals of one species can "capture" some limited resource from individuals of other species that consume the same resource. Consider light as an example of an essential resource in plants. Taller plants intercept solar radiation, shading plants of smaller stature, in effect outcompeting them for the limited light resource (see Chapter 5). In the competition for light, the superior competitor is the plant that can grow quickly in height, overtopping other plants and maintaining access to the limited resource. Plants that grow fast in high light environments do so by maintaining high rates of photosynthesis and allocating much of the carbon that is fixed to growing new leaves and stem. The new leaves increase the amount of carbon that can be fixed, and the stem tissues enable the plant to increase in stature, maintaining access to the light resource.

However, this pattern of response has a cost. The allocation of carbon to produce leaves and stem reduces the amount the plant can allocate to produce roots. Because roots are responsible for the uptake of water and nutrients from the soil, the "strategy" of allocating most of the carbon to produce leaves and stems is effective only when these two resources are abundant. As water or nutrients become scarce, the limited root system may not be able to supply ample water to the leaves to maintain evapotranspiration, and the stomata may close. This action reduces photosynthesis or brings it to a halt. Under this set of circumstances the alternative strategy of allocating carbon to produce roots rather than stems and leaves may be more effective.

Although plants have the ability to change their patterns of carbon allocation, producing more or less leaf, stem, and root tissue in response to environmental conditions, this flexibility is limited. As a result, when environmental conditions change, either in space or time, the set of characteristics that allow a species to compete successfully will change. Plant species that are superior competitors for light under wet conditions (high rainfall) may be poor competitors under dry conditions. Plants that flourish under dry conditions may be shaded and outcompeted by faster-growing plants under wet conditions. As environmental conditions vary across the land or seascape, the competitive abilities of species shift, favoring one species over another. This process is a major factor influencing patterns of zonation and succession in communities.

22.5 Indirect interactions influence community structure

Competition, mutualism, predation, and parasitism are direct interactions between individuals of two popula-

tions. Individuals of one population have an impact on the second population, influencing its rates of mortality and birth. In the Lotka-Volterra equations of population growth in Chapter 15, these interactions were represented by terms (alpha and beta) that describe the per capita effect of an individual of one population on the growth rate of the second population.

Not all interactions among populations within a community are direct. Indirect interactions occur when one population does not interact with a second population directly, but influences a third population that does have a direct interaction with the second. An example is two species of grass that compete with each other, one of which is selectively eaten by an herbivore. The herbivore not only has a direct negative influence on the population of the species that it consumes, but an indirect positive effect on the population of the species on which it does not feed. Intensive grazing favors plants that are prostrate, have small leaves, or grow in rosettes close to the ground. Conversely, it selects against taller growth forms. When grazing pressure relaxes, tall-growing grasses become dominant. This positive effect on low-growing plants is a result of reducing the population of the competing grass species. As the saying goes, "The enemy of an enemy is a friend." These indirect interactions may well exert a major influence on community structure.

22.6 Changes in environmental conditions are allogenic or autogenic

The temporal and spatial changes in environmental conditions that influence community structure and dynamics are varied. We have seen how changes in temperature, rainfall, soil, nutrients, and light have a major influence on plant growth, competition, and community structure. Likewise, changes in the physical environment, the structure of vegetation, or the availability of food exert a major influence on animal populations.

The factors influencing these changes in environmental conditions are likewise varied. They can, however, be grouped into two general classes: allogenic and autogenic. **Autogenic** environmental change is a direct result of the organisms within the community. For example, the vertical profile of light in a forest is a direct result of the interception and reflection of solar radiation by the trees. In contrast, **allogenic** environmental change is not a direct function of the organisms, but a feature of the physical environment. Examples are the decline in average temperature with elevation in mountainous regions (Chapter 4) and the decrease in temperature with depth in a lake or ocean (Chapters 32 and 35).

Not all examples of environmental change fall nicely into these discrete classes. Patterns of continental and regional rainfall are large-scale features of the weather system (Chapter 4) or allogenic. However, in certain regions, such as tropical rain forests, a significant portion of the rainfall is water that originates locally from transpiration. In effect, the water is locally recycled. Computer experiments have shown that cutting the forest and reducing transpiration could reduce rainfall in the region, suggesting that in part the patterns of rainfall variation are autogenically controlled.

22.7 Dominance changes along a temporal, autogenic gradient

In Chapter 21, we defined succession as changes in community structure though time—specifically, changes in species dominance. One species initially colonizes an area, but as time progresses it declines, to be replaced by another species (see Figures 21.13 and 21.15). The fact that we find this general pattern of changing species dominance as time progresses in most environments suggests a common underlying mechanism.

The process of plant succession has been a major focus of study since the birth of ecology as a science. Although many hypotheses and models have been put forward to explain succession, no general consensus has been achieved. One obstacle is the diversity of environments and associated communities in which succession has been studied. No single cause fits all the examples. Despite this lack of consensus, a general model of plant succession has begun to emerge.

One feature common to all plant succession is autogenic environmental change. In both primary and secondary succession, colonization alters environmental conditions. One clear example is the alteration of the light environment. As we saw in Chapter 5, leaves reflecting and intercepting solar radiation create a vertical profile of light within a plant community. As you move from the canopy to ground level, the light available to drive the processes of photosynthesis declines. In the initial period of colonization, few if any plants are present. In the case of primary succession, the newly exposed site has never been occupied. In the case of secondary succession, plants have been killed or removed by some disturbance. Under these circumstances, the availability of light at the ground level is high, and seed-

lings are able to germinate. As plants grow, their leaves intercept sunlight, reducing the availability of light to shorter plants. This reduction in available light will most likely reduce rates of photosynthesis, slowing the growth of these shaded individuals. Assuming not all plant species photosynthesize and grow at the same rate, those species of plants that can grow tall the fastest will have access to the light resource. They reduce the availability of light to the slower-growing species. This reduction in light enables the fast-growing species to outcompete the other species and dominate the site. However, in changing the availability of light below the canopy, the dominant species create an environment that is more suitable for other species, species that will later displace them as dominants.

Recall from Chapter 5 that not all species of plants respond to variation in available light in the same manner. Sun-adapted, shade-intolerant plants exhibit a different response to light than do shade-adapted, shade-tolerant species. Shade-intolerant species exhibit high rates of photosynthesis and growth under high light environments. Under low light levels, they are not able to continue photosynthesis, growth, and survival. In contrast, shade-tolerant plant species exhibit much lower rates of photosynthesis and growth under high light conditions. However, they are able to continue photosynthesis, growth, and survival under lower light. There is a fundamental physiological tradeoff between the characteristics that enable high rates of growth under high light conditions and the ability to continue growth and survival under shaded conditions.

In the early stages of plant succession, shade-intolerant species come to dominate as a result of their high growth rates. Shade-intolerant species overtop and shade the slower-growing, shade-tolerant species. As time progresses and light levels decline below the canopy, seedlings of the shade-intolerant species cannot grow and survive in the shaded conditions. At this time, although the shade-intolerant species dominate the canopy, no new individuals are being recruited into their populations. In contrast, shade-tolerant species are able to germinate and grow under the canopy. As the shade-intolerant plants that make up the canopy die, shade-tolerant species in the understory replace them.

Figure 22.5 shows this pattern of changing population recruitment, mortality, and species composition through time for a forest community in the Piedmont region of North Carolina. Early in succession the forest is dominated by fast-growing, shade-intolerant pine species. As time progresses, the number of new pine

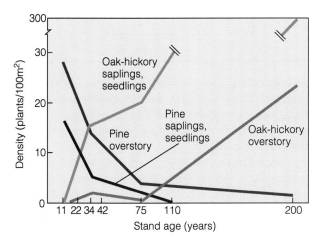

Figure 22.5 Dominance shift of overstory and understory (seedlings and saplings) pines and oaks and hickories during secondary succession in the Piedmont region of North Carolina. Early successional pine species initially dominate the site. Pine seedling regeneration declines as the light decreases in the understory. Shade-tolerant oak and hickory seedlings establish themselves under the reduced light conditions. As pine trees in the overstory die, oak and hickory replace them as the dominant species in the canopy.

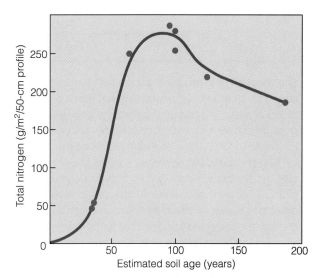

Figure 22.6 Changes in soil nitrogen during primary succession in Glacier Bay National Park since the retreating glacier exposed the surface for colonization by plants.

seedlings declines as the light decreases on the forest floor. Now the shade-tolerant oak and hickory species are able to establish seedlings in the shaded conditions of the understory. As the pine trees in the canopy die, the community shifts from a forest dominated by pine species to one dominated by oaks and hickories.

In this example, succession is the result of changing competitive ability under autogenically changing environmental conditions. The shade-intolerant species are able to dominate the early stages of succession because of their ability to grow quickly in the high light environment. However, as autogenic changes in the light environment occur, the ability to tolerate and grow under shaded conditions enables the shade-tolerant species to rise to dominance.

Light is not the only environmental factor that changes over the course of succession. Other autogenic changes in environmental conditions influence patterns of succession. Consider the example of primary succession on newly deposited glacial sediments and lava flows (Section 21.9). Because of the absence of a well-developed soil, little nitrogen is present in these newly exposed surfaces, restricting the establishment, growth, and survival of most plant species. However, those terrestrial plant species of the family Leguminosae that have the mutualistic association with nitrogen-fixing *Rhizobium* bacteria are able to grow and dominate

the site. These plants provide a source of carbon (food) to the bacteria that inhabit their root system. In return the plant has access to the atmospheric nitrogen fixed by the bacteria.

As leguminous plants shed their leaves or die, the nitrogen they contain is released to the soil by decomposition. Now other plant species can colonize the site. Figure 22.6 shows this pattern of increasing soil nitrogen during primary succession for Glacier Bay, Alaska (see Figure 21.17). As nitrogen becomes available in the soil, species that do not have the added cost of mutualistic association and exhibit faster rates of growth and recruitment come to dominate the site. As in the forest example, succession is a result of autogenic change in the environment and the relative competitive abilities of the species colonizing the site.

We have been focusing on competition as we explain succession. Other models of succession consider a wider range of species interactions and responses (Focus on Ecology 22.1: Other Views of Succession). However, in all cases, the role of temporal, autogenic changes in environmental conditions and the differential response of species to those changes explains community dynamics.

22.8 Herbivory influences patterns of community dynamics

Herbivores modify community dynamics both directly and indirectly. By selecting certain plant species,

OTHER VIEWS OF SUCCESSION

The question of what drives succession has stimulated a number of different models. One approach that considers a range of species interactions and responses through succession was proposed by J. Connell and R. Slatyer in 1977. They offered three models.

One is the *facilitation model.* Early successional species modify the environment so that it becomes more suitable for later successional species to invade and grow to maturity. In effect, early-stage species prepare the way for late-stage species, facilitating their success.

The *inhibition model* involves strong competitive interactions. No one species is completely superior to another. The site belongs to those species that, in the colorful words of a Civil War general describing how armies take and hold ground, are "the firstest with the mostest." The first species to come hold the site against all invaders. They make the site less suitable for both early and late successional species. As long as they live

and reproduce, they maintain their position. The species relinquish it only when they are damaged or die, releasing space to another species. Gradually, however, species composition shifts as short-lived species give way to long-lived ones.

A third model, the *tolerance model,* holds that later successional species are neither inhibited nor aided by species of earlier stages. Later-stage species can invade a newly exposed site, become established, and grow to maturity independent of species that precede or follow them. They can do so because they tolerate a lower level of some resources. Such interactions lead to communities composed of those species most efficient in exploiting available resources. An example might be a highly shade-tolerant species that could invade, persist, and grow beneath the canopy because it is able to exist at a lower level of one resource—light. Ultimately, through time, one species would prevail.

---

herbivores directly influence mortality and recruitment, favoring the population growth of one species over another. Moose on Isle Royale in Lake Superior selectively feed on the seedlings and saplings of deciduous hardwood tree species of aspen, birch, ash, and maple, ignoring the conifer species of spruce and fir. Long-term experiments using enclosures to exclude moose from certain areas have demonstrated that these selective patterns of herbivory have changed the community structure on the island. In the enclosures, the abundance of the deciduous hardwood species is much higher.

Herbivory also has indirect effects on community dynamics. Outbreaks of insect herbivores that feed on the forest canopy have a major influence on the forest environment. Defoliation of the canopy increases temperatures and available light on the forest floor, favoring the growth of seedlings and herbaceous plants.

Outbreaks of the gypsy moth larva *(Porthetria dispar)* in the northeastern United States have almost completely defoliated the canopy in oak forests. Not all of the leaf material is consumed. The larvae are sloppy eaters and many of the green leaves fall to the

forest floor. These leaves are high in nutrients and decompose rapidly, resulting in a large increase in the availability of nutrients in the soil. Because the defoliated trees are not growing, their nutrient uptake is dramatically reduced. Nutrients are leached from the soil into the streams and transported out of the forest. This loss of nutrients can have long-term effects on the growth and productivity of the forest.

22.9 Both views of the community reflect different perspectives

Our discussion of the processes influencing community structure and dynamics began by contrasting two views of the community. The organismal view stresses the community as an entity made up of interdependent species. In the individualistic or continuum view, the community is an arbitrary concept. Each species responds independently to the underlying features of the environment.

Research reveals that, as in most polarized debates, reality lies somewhere in the middle, and our viewpoints are colored by our perspective. The organismal community is a spatial concept. As you stand in the forest, you see a variety of plant and animal species, interacting and influencing the overall structure of the forest. The continuum view is a population concept, focusing on the response of the component species to the underlying features of the environment.

A simple example is presented in Figure 22.7, a transect up a mountain in an area with four plant species present. The distribution of the four plant species is presented in two ways. In one view, the species distribution is plotted as a function of altitude or elevation. Note that the four species exhibit a continuum of species regularly replacing each other in a sequence of A, B, C, and D with increasing altitude—a pattern in keeping with the individualistic view of communities. In the second view, species distribution is a function of distance along the mountainside. As you move up the mountainside, the distribution of the four species is not continuous. You might recognize

a number of species associations along the transect. These associations are identified by different symbols. These communities composed of coexisting species are a consequence of the spatial pattern of the landscape. This pattern fits the organismal view.

The two views are quite different, yet complementary. Each species shows a continuous response along an environmental gradient, elevation; yet it is the spatial distribution of such environmental variables across the landscape that determines the overlapping patterns of distribution—the composition of the community.

Figure 22.7 examines only one feature of the environment, elevation. The structure of communities is the product of complex patterns and processes. Species respond to a wide array of environmental factors that vary spatially and temporally across the landscape, and the interactions among organisms influence the nature of those responses.

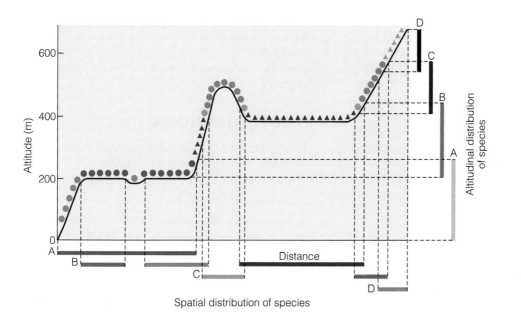

Figure 22.7 Patterns of co-occurrence for four species on a landscape along a gradient of altitude. The environmental distribution of the four species is presented in two ways: (1) the spatial distribution of the species along a transect of the mountainside, and (2) as a function of their response to altitude. Note that the species' responses to the altitudinal gradient are continuous, but their spatial distributions along the transect are discontinuous. The patterns of species composition (and community composition) along the mountain gradient are a result of the spatial pattern of environmental conditions (altitude) and the individual response of the species. The response of the species to the environmental gradient is consistent with the individualistic or continuum view of communities proposed by Gleason (see Figure 22.1b). However, consistent patterns of species co-occurrence across the landscape is a function of the spatial distribution of environmental conditions. Repeatable patterns of species co-occurrence in similar habitats is consistent with the idea of plant associations as supported by Clements.

CHAPTER REVIEW

SUMMARY

Concept of the Community (22.1)   Historically there have been two contrasting concepts of the community. The organismal concept views the community as a unit, an association of species, in which each species is a component of the integrated whole. The individualistic concept views the co-occurrence of species as a result of similarities in requirements and tolerances.

Fundamental Niche and Community Structure (22.2–22.5)   The range of environmental conditions tolerated by a species defines its fundamental niche. These constraints on species' ability to survive and flourish will limit their distribution and abundance to a certain range of environmental conditions. Species differ in the range of conditions they tolerate. As environmental conditions change in both time and space, the possible distribution and abundance of species will change **(22.2)**.

The growth and activities of organisms modify the environment both directly and indirectly. They make the environment more or less suitable for other species, even themselves **(22.3)**. As the environment changes, the ability of organisms to compete also changes for better or worse. If the changes exceed the tolerance limits for an organism, that species gives way to species more tolerant of those environmental conditions **(22.4)**. Interactions among populations in the community, especially competition, predation, and mutualism, can have a pronounced influence on community structure **(22.5)**.

Succession (22.6–22.7)   Environmental changes can be autogenic or allogenic. Autogenic changes are a direct result of the activities of organisms in the community. Changes in environmental conditions independent of organisms are allogenic. Weather is an example of allogenic environmental change **(22.6)**.

Succession is the progressive change in community composition through time in response to changing environmental conditions. Many of these environmental changes are autogenic. One example is the changing light environment and the shift in dominance from fast-growing shade-intolerant plants to slow-growing shade-tolerant plants observed in terrestrial plant succession. Autogenic changes in moisture and nutrient availability have a major influence on succession **(22.7)**.

Herbivory and Community Structure (22.8)   Herbivores influence community dynamics both directly and indirectly. By selecting certain plant species as food, herbivores favor the growth and abundance of certain species over others. Herbivory also influences the physical structure of the community, changing conditions that favor one species over another or influencing overall productivity.

Complementary Views (22.9)   The true nature of communities lies somewhere between the organismal concept and the individualistic concept. The community is a spatial concept; the individualistic continuum is a population concept. Each species has a continuous response along an environmental gradient. The patterns of spatial variation in the physical environment across the landscape interact with species responses to determine distribution and abundance.

STUDY QUESTIONS

1. Distinguish between the organismal and individualistic concepts of the community.
2. How do mutualism and herbivory influence succession?
3. How does competition for resources, especially light, influence the direction of succession?
4. In what ways can one species make the environment more favorable for another?
5. Distinguish between autogenic succession and allogenic succession processes.
6. How does the fundamental niche of species constrain community structure?
*7. Reconcile the organismal concept with the individualistic concept of community.

ECOSYSTEMS

PART V

CHAPTER 23

PRODUCTION IN ECOSYSTEMS

OBJECTIVES

On completion of this chapter, you should be able to:

- Describe the concept of the ecosystem.
- Relate the laws of thermodynamics to ecology.
- Define the types of ecological production.
- Discuss how plants allocate net primary production.
- Tell how net primary production varies among world ecosystems and why.
- Describe secondary production and its allocation.
- Compare assimilation and production efficiencies of poikilotherms and homeotherms.

Primary production halts during the dry season. The primary consumers, zebra *(Equus burchelli)*, depend upon the standing crop of dead grasses as an energy source until the rains return.

Up to this point we have looked at interactions of organisms with their physical environment, interactions in the same population, and interactions between populations. All of these interactions came together in the concept of the community, an assemblage of species in a given place interacting directly and indirectly with each other. The community involves the biota only. Now we add the last dimension: the interaction of the biotic community with the abiotic world to give us the ecosystem.

23.1 The ecosystem consists of the biotic and abiotic

The distribution and abundance of species and the biological structure of the community vary in response to environmental conditions. However, it is equally true that the organisms themselves, in part, define the physical environment. An example is the role of autogenic changes in light and nutrient availability in driving plant succession. It is this inseparable link between the biological environment (the community) and the physical environment that led A. G. Tansley to coin the term **ecosystem** in an article in the journal *Ecology* in 1935. Tansley wrote:

> The more fundamental conception is . . . the whole *system* (in the sense of physics) including not only the organism-complex, but also the whole complex of physical factors forming what we call the environment. . . . We cannot separate them [the organisms] from their special environment with which they form one physical system. . . . It is the systems so formed which . . . [are] the basic units of nature on the face of the earth. . . . These *ecosystems*, as we may call them, are of the most various kinds and sizes.

In the concept of the ecosystem, the biological and physical components of the environment are a single interactive system.

Like the community, the ecosystem is a spatial concept; it has defined boundaries. Like the community, these boundaries are often difficult to define. At first examination, a pond ecosystem is clearly separate and distinct from the surrounding terrestrial environment. A closer inspection, however, reveals a less distinct boundary between aquatic and adjacent terrestrial ecosystems. Some plants along the shoreline, such as cattails, may be either partially submerged or rooted in the surrounding land, able to tap the shallow water table with their roots. Amphibians move between the shoreline and the water. Surrounding trees drop leaves into the pond, adding to the dead organic matter that feeds the decomposer community on the pond bottom.

Regardless of these difficulties, ecosystems theoretically have spatial boundaries. Having defined the boundaries, we can view our ecosystem in the context of its surrounding environment.

The primary focus of ecosystem ecology is the exchange of energy and matter. Exchanges from the surrounding environment into the ecosystem are **inputs.** Exchanges from inside the ecosystem to the surrounding environment are **outputs.** An ecosystem with no inputs is called a **closed ecosystem;** one with inputs is an **open ecosystem.** Inputs and outputs, together with exchanges of energy and matter among components within the ecosystem, form the basis of our discussion in this and the following three chapters.

23.2 All ecosystems have three basic components

In the simplest terms, all ecosystems, both aquatic and terrestrial, consist of three basic components—the autotrophs, the consumers, and abiotic matter (see Figure 23.1). The producers, or autotrophs, are largely green plants. These organisms use the energy of the sun in photosynthesis (Chapter 3) to transform inorganic compounds into simple organic compounds.

The consumers, or heterotrophs, use the organic compounds produced by the autotrophs as a source of food. Through decomposition, heterotrophs eventually transform these complex organic compounds into simple inorganic compounds that are once again used by the producers. The heterotrophic component of the ecosystem is often subdivided into two subsystems, consumers and decomposers. The consumers feed largely on living tissue, and the decomposers break down dead matter into inorganic substances. No matter how we classify them, all heterotrophic organisms are consumers, and all in some way act as decomposers.

The third, or abiotic, component consists of the soil, sediments, particulate matter, dissolved organic matter in aquatic ecosystems, and litter in terrestrial ecosystems. All of the dead organic matter is derived from plant and consumer remains and is acted upon by the decomposers. Such dead organic matter is critical to the internal cycling of nutrients in the ecosystem.

The driving force of the system is the energy of the sun. This energy, harnessed by the producers,

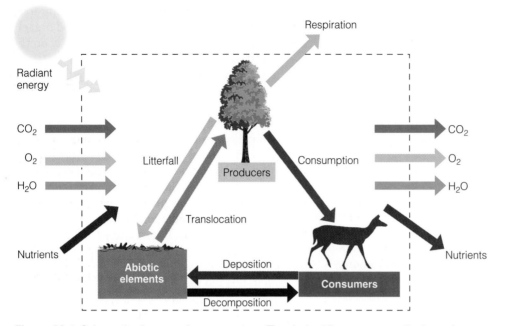

Figure 23.1 Schematic diagram of an ecosystem. The dashed lines represent the boundary of the system. The three major components are the producers, the consumers, and the abiotic elements: inactive or dead organic matter, the soil matrix, nutrients in solution in aquatic ecosystems, sediments, and so on. The arrows indicate interactions within the system and with the environment.

flows from producers to consumers to decomposers and eventually dissipates as heat.

23.3 The laws of thermodynamics govern energy flow

Production in ecosystems involves the fixation and transfer of energy from the sun. Green plants fix solar energy in the process of photosynthesis (Chapter 3). The products of photosynthesis, photosynthates, accumulate as plant biomass. Nonphotosynthetic organisms convert this stored energy into heterotrophic biomass. This fixation and transfer of energy through the ecosystem is governed by the laws of thermodynamics.

Energy exists in two forms, potential and kinetic. **Potential energy** is stored energy—that is, capable of and available for performing work. **Kinetic energy** is energy in motion. It performs work at the expense of potential energy. Work is of at least two kinds: the storage of energy and the arranging or ordering of matter.

The expenditure and storage of energy are governed by two laws of thermodynamics. The **first law**

of thermodynamics states that energy is neither created nor destroyed. It may change form, pass from one place to another, or act upon matter in various ways. Regardless of what transfers and transformations take place, however, no gain or loss in total energy occurs. Energy is simply transferred from one form or place to another. When wood burns, the potential energy lost from the molecular bonds of the wood equals the kinetic energy released as heat. When energy is lost from the system into the surrounding environment, the reaction is **exothermic.**

On the other hand, energy may be paid into a reaction. Here, too, the first law of thermodynamics holds true. In photosynthesis, for example, the molecules of the products (simple sugars) store more energy than the reactants that combined to form the products. The extra energy that is stored in the products is acquired from the sunlight that is harnessed by the chlorophyll within the leaf (see Section 3.2). Again there is no gain or loss in total energy. When energy from the outside is put into a system to raise it to a higher energy state, the reaction is **endothermic.**

Although the total amount of energy in any reaction, such as burning wood, does not increase or decrease, much of the potential energy degrades into a form incapable of doing further work. It ends up as heat, disorganized or randomly dispersed molecules in motion, useless for further transfer. The measure of this disorder is **entropy.**

The transfer of energy involves the **second law of thermodynamics.** It states that when energy is transferred or transformed, part of the energy assumes a form that cannot pass on any further. Entropy increases. When coal is burned in a boiler to produce steam, some of the energy creates steam, and part is dispersed as heat to the surrounding air. The same thing happens to energy in the ecosystem. As energy is transferred from one organism to another in the form of food, a large part of that energy is degraded as heat—no longer transferable. The remainder is stored as living tissue.

At first appearance, biological systems seemingly do not conform to the second law of thermodynamics. The tendency of life is to produce order out of disorder, to decrease rather than increase entropy. The second law theoretically applies to closed systems, in which no energy or matter is exchanged between the system and its surrounding environment. With the passage of time, closed systems tend toward maximum entropy; eventually no energy is available to do work. Living systems, however, are open systems with a constant input of energy.

23.4 Primary production fixes energy

The flow of energy through a terrestrial ecosystem starts with the harnessing of sunlight by green plants, a process that in itself demands the expenditure of energy. A plant gets its start by living on the food energy stored in the seed until it produces leaves. Energy accumulated by plants is called production, more specifically, **primary production,** because it is the first and basic form of energy storage. The rate at which energy accumulates is **primary productivity.** Total energy assimilated by the plant—the total photosynthesis—is **gross primary production.**

Plants, like all other organisms, must expend energy in production, maintenance, and reproduction. This energy is produced by the process of respiration (see Section 3.6), which liberates energy. Energy remaining after respiration and stored as organic matter is **net primary production.** Net primary production can be described by the following equation:

$$\text{net primary production (NPP)} =$$
$$\text{gross primary production (GPP)} -$$
$$\text{respiration by the autotroph (R)}$$

Production is usually expressed in units of energy per unit area: kilocalories per square meter (kcal/m^2). However, production may also be expressed as the mass of dry organic matter: grams per square meter (g/m^2).

Net primary production accumulates over time as plant biomass. The amount of accumulated organic matter found in an area at a given time is the **standing crop biomass.** Like production, biomass is usually expressed as grams of organic matter per square meter (g/m^2), or as calories per square meter (cal/m^2) or some other appropriate unit of area. Biomass differs from productivity. Biomass is the amount present at any given time. Productivity is the rate at which organic matter is created by photosynthesis.

23.5 Energy allocation varies in plants

Plants budget their production, distributing it in a systematic way to leaves, twigs, stems, bark, flowers, seeds, and other tissues. How much is allocated to each component is a function of both the life form of the plant (see Section 20.5) and environmental conditions. The patterns of allocation of the dominant plants within the ecosystem will largely define the physical structure of the ecosystem (see Section 20.1).

Annuals begin their life cycles in the spring with the germination of overwintering seeds. In regions with distinct dry and wet seasons, germination occurs with the onset of the rainy season. With only one growing season in which to complete its life cycle, an annual has to allocate its photosynthates first to leaves. Leaves, in turn, become involved in photosynthesis, which replenishes the supply of photosynthates and increases plant biomass. At the time of flowering, the plant decreases the amount of energy allocated to leaves, and diverts most of its photosynthate to reproduction. For example, in the sunflower, the biomass of leaves declines from approximately 60 percent of the total plant weight during the period of growth to 10 to 20 percent by the time the seeds are ripe. When in bloom, the sunflower allocates 90 percent of its photosynthate to the flower head and the remainder to the leaves, stem, and roots.

Perennial plants maintain a vegetative structure over several years. They begin their life cycles like an annual, but once established, they allocate their energy in a very different manner. Before perennials expend any energy on reproduction, they divert photosynthate to the roots. This allocation to roots is in excess of that required for the development of roots for the uptake of nutrients and water from the soil. In some species, such as the skunk cabbage *(Symplocarpus foetidus)*, the roots develop into large storage organs. Energy stored in the roots makes up a reserve upon which the plants draw when they begin growth the following growing season. When they are ready to flower, perennials divert energy from storage to the production of flowers and fruit. As the flowers fade and the fruits ripen, the plant once more sends photosynthate to the roots to build up the reserves they will need for the following spring.

Trees and woody shrubs live a long time, which greatly influences the manner in which they allocate energy. Early in life, leaves make up more than one-half of their biomass (dry weight); but as trees age, they accumulate more woody growth. Trunks and stems become thicker and heavier, and the ratio of leaves to woody tissue changes. Eventually leaves account for only 1 to 5 percent of the total mass of the tree. The production system (the leaf mass) that supplies the energy is considerably less than the biomass it supports. Thus, as the woody plant grows, much of the energy goes into support and maintenance (respiration), which increases as the plant ages.

When deciduous trees leaf out in the spring, they expend up to one-third of their reserve energy on the growth and expansion of leaves. This expenditure is repaid as the leaves carry out photosynthesis during the spring and summer. After leaves, trees give preference to flowers; then transport tissues, new buds, and deposits of starch in roots and bark; and finally, new flower buds.

Evergreen trees have a somewhat different approach. Because their photosynthetic tissues (leaves) can function year round when temperature and moisture conditions permit, they do not need to draw on root reserves for new growth in the spring. They can afford to wait until later in the growing season to produce new shoots. Then evergreens can draw upon energy built up earlier in the spring. For the same reason, new growth develops rapidly and matures within a few weeks.

Reproduction and vegetative growth compete for energy allotments. If photosynthesis is limited, vegetative growth gets first claim. Because the energy reproduction demands is high—up to 15 percent in pines, 20 percent in deciduous trees, and 35 percent or more in fruit trees—trees can afford an abundance of fruit only periodically, once every two to three years in deciduous trees and two to six years in conifers.

The proportionate allocation of net production to aboveground and belowground biomass, or shoot and root tissues, tells much about different ecosystems. In Chapters 5, 6, 7, and 9, we explored the influence of environmental conditions (light, water, and nutrients) on the allocation of biomass to the growth of shoot and root tissues. Low light conditions favor the allocation of energy to the production of leaves and stem (shoot) at the expense of roots. A reduction in water or nutrient availability results in the reverse, a greater allocation of energy to the production of roots at the expense of leaves and stem. As a result, ecosystems that are associated with low rainfall or low soil fertility have a lower shoot-to-root ratio than those associated with high belowground resources. For example, plants of tundra ecosystems (see Chapter 29), living in an environment with a long cold winter and a short growing season, have shoot-to-root ratios from 1:3 to 1:10. Midwest prairie grasses have a shoot-to-root ratio of 1:3, indicative of cold winters and limited moisture supply. Plants of arid regions have a low shoot-to-root ratio.

Within the forest ecosystem, different components of the plant community exhibit different patterns of aboveground and belowground allocation. For the Hubbard Brook Forest in New Hampshire, the shoot-to-root ratio for trees is 1:0.213; for shrubs, 1:0.5; and for herbaceous vegetation, 1:1.

23.6 Climate influences productivity of terrestrial ecosystems

Productivity of terrestrial ecosystems is influenced by climate. Such relationships reflect variations in production of different global ecosystems (Figure 23.2). Measured estimates of net primary production for a variety of terrestrial ecosystems are summarized in Table 23.1 and plotted in Figure 23.3 against the annual precipitation and mean daily temperature for each site. As the graphs show, net primary productivity increases with increasing temperature and rainfall. The reason is that temperature and precipitation influence the rate of photosynthesis, the amount of leaf

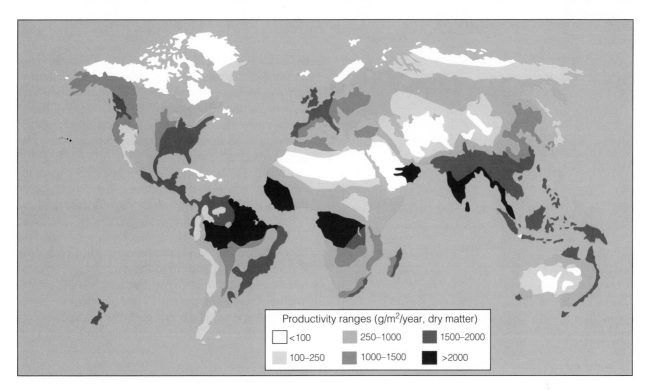

Figure 23.2 A map of primary production of terrestrial ecosystems.

TABLE 23.1

NET PRIMARY PRODUCTION AND PLANT BIOMASS OF WORLD ECOSYSTEMS

Ecosystems (in order of productivity)	Area (10^6 km^2)	Mean Net Primary Production per Unit Area (g/m^2/yr)	World Net Primary Production (10^9 mtn/yr)	Mean Biomass per Unit Area (kg/m^2)
Continental				
Tropical rain forest	17.0	2000.0	34.00	44.00
Tropical seasonal forest	7.5	1500.0	11.30	36.00
Temperate evergreen forest	5.0	1300.0	6.40	36.00
Temperate deciduous forest	7.0	1200.0	8.40	30.00
Boreal forest	12.0	800.0	9.50	20.00
Savanna	15.0	700.0	10.40	4.00
Cultivated land	14.0	644.0	9.10	1.10
Woodland and shrubland	8.0	600.0	4.90	6.80
Temperate grassland	9.0	500.0	4.40	1.60
Tundra and alpine meadow	8.0	144.0	1.10	0.67
Desert shrub	18.0	71.0	1.30	0.67
Rock, ice, sand	24.0	3.3	0.09	0.02
Swamp and marsh	2.0	2500.0	4.90	15.00
Lake and stream	2.5	500.0	1.30	0.02
Total continental	149.0	720.0	107.09	12.30
Marine				
Algal beds and reefs	0.6	2000.0	1.10	2.00
Estuaries	1.4	1800.0	2.40	1.00
Upwelling zones	0.4	500.0	0.22	0.02
Continental shelf	26.6	360.0	9.60	0.01
Open ocean	332.0	127.0	42.00	0.003
Total marine	361.0	153.0	55.32	0.01
World total	510.0	320.0	162.41	3.62

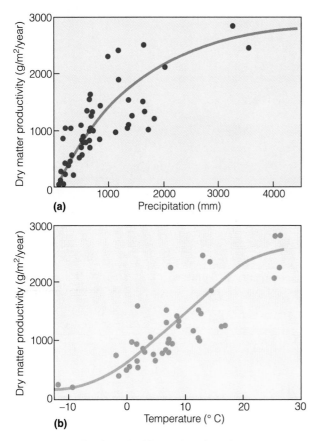

(a)

(b)

Figure 23.3 Productivity (a) as a function of mean annual precipitation, and (b) as a function of mean annual temperature.

area that can be supported, and the duration of the growing season.

Primary productivity is a function of the rate of photosynthesis (energy capture) and the total surface area of leaves that are photosynthesizing. Rates of photosynthesis are limited by extremely cold and hot temperatures. Within the range of temperatures that are tolerated, rates of photosynthesis rise with temperature (see Section 6.3). In addition, in the temperate zones, the length of the growing season is defined as the period when temperatures are sufficiently warm to support photosynthesis and net primary productivity. As a result, warmer temperatures typically support both higher rates of photosynthesis and a longer growing season.

As we discussed in Section 3.3, for photosynthesis and productivity to occur, the plant must open the stomata to take in CO_2. When the stomata are open, water is lost from the leaf to the surrounding air. For the plant to keep the stomata open, roots must replace

the lost water. The higher the rainfall, the more water is available to plants for transpiration. The amount of water available to the plant will therefore limit both the rate of photosynthesis and the amount of leaves (surface area that is transpiring) that can be supported. The combination of these factors determines the rate of primary productivity.

Although the two graphs in Figure 23.3 show independent effects of temperature and precipitation on primary productivity, in reality the influence of these two factors is closely related. Warm temperatures mean high water demand. If temperatures are warm but water availability is low, productivity will also be low. Conversely, if water availability is high but temperatures are low, productivity will be low. It is the combination of warm temperatures and an adequate water supply for transpiration that gives the highest primary productivity. This pattern is reflected in Figure 23.4,

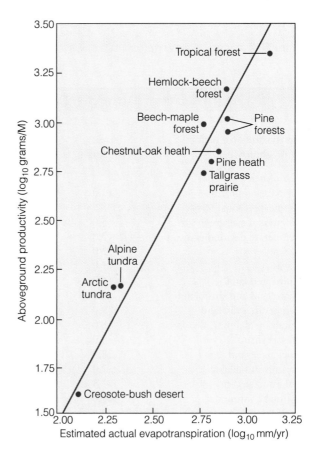

Figure 23.4 Range of productivities of world ecosystems based on evapotranspiration, which depends upon both precipitation and temperature. (From R. H. MacArthur and J. H. Connell, *The Biology of Populations*, p. 175, Fig. 7.3. New York: John Wiley, 1966. Used by permission.)

which relates the net primary productivity of various ecosystems to the measured value of evapotranspiration. Evapotranspiration is the combined value of surface evaporation and transpiration. It reflects both the demand and the supply of water to the ecosystem. The demand is a function of incoming radiation and temperature, whereas the supply is a function of precipitation.

Another way of expressing the relationship among net primary production, standing biomass, and climate is shown in Figure 23.5. A number of terrestrial ecosystems that we will discuss in Chapters 27 to 31 are plotted along the axes of mean annual temperature and precipitation. The patterns of production show the interaction between temperature and precipitation previously discussed. The highest terrestrial production and biomass are in the rain forests of the warm, wet tropical regions (but see Focus on Ecology 23.1: Productivity of a Cornfield). Deserts, where temperatures are high but rainfall is low, have very low production and biomass. So do tundra ecosystems, where precipitation is high but temperatures are low.

23.7 Nutrient availability limits productivity of oceans

In most ecosystems, there is a degree of vertical separation between the zone where primary productivity occurs and the zone in which decomposition and release of nutrients occurs. For terrestrial plants, this vertical separation does not present a problem, because the plants have height, allowing them to extend into both zones with their canopy and root system. In aquatic environments, particularly in the oceans, phyloplankton are near the surface. Nutrients in the deeper waters must be transported to the surface waters, where light is available to drive photosynthesis. As a result, nutrients, particularly nitrogen, phosphorus, and iron, are a major limitation on primary productivity in the oceans (see Chapter 35). The most productive waters of the oceans are coastal environments (Figure 23.6 and Table 23.1) and coral reefs, for two reasons. First, shallow waters mean less vertical separation, greater transport of nutrients to surface

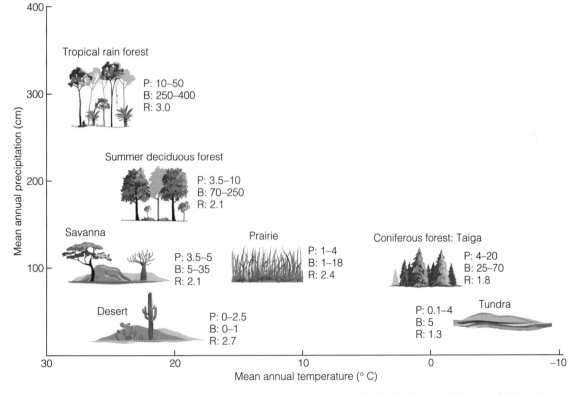

Figure 23.5 Distribution of primary production, standing biomass, and radiation input relative to rainfall and temperature. P = primary production (tn/ha); B = biomass (tn/ha); R = PAR solar radiation (kcal/m²/yr). (Adapted from J. R. Etherington, *Environmental and Plant Ecology,* 2nd ed., p. 355. New York, John Wiley, 1975. Used by permission.)

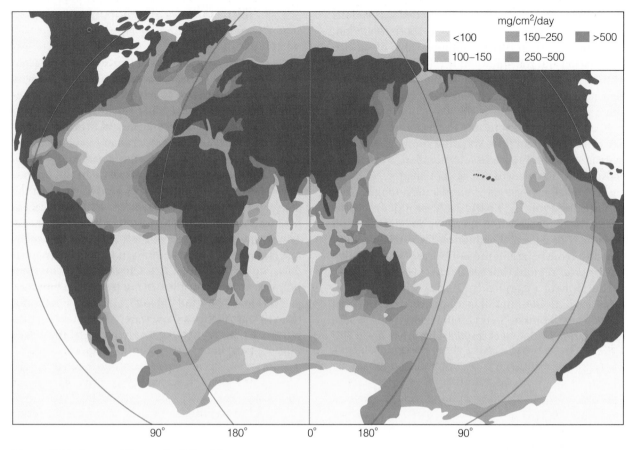

mg/cm²/day

<100	150–250	>500
100–150	250–500	

90°   180°   0°   180°   90°

Figure 23.6 A map of the productivity of the oceans.

waters, and changing tides. Second, estuaries receive a large input of nutrients carried from terrestrial ecosystems by rivers and streams (see Chapter 37).

23.8   Primary production varies with time

Primary production also varies within an ecosystem with time and age. Seasonal variations in environmental conditions directly influence photosynthesis and plant growth. Regions with cold winters or distinct wet and dry seasons have a period of plant dormancy when primary productivity ceases. In the wet regions of the tropics where conditions are favorable for plant growth year round, there is little seasonal variation in primary productivity.

Year to year variations also occur as a result of climatic variation and disturbances such as herbivory and fire. For example, grassland productivity of a site may vary by a factor of 8 between wet and dry years. Over-grazing of grasslands by cattle and sheep or defoliation of forests by such insects as the gypsy moth can seriously reduce net production. Fire in grasslands may increase productivity in wet years, but reduce it in dry years. An insufficient supply of nutrients, especially nitrogen and phosphorus, can limit net productivity, as can pollutants, both atmospheric and water-borne.

Net primary productivity also varies with age, particularly in ecosystems that are dominated by woody vegetation. As the age of a forest stand increases, more and more of the living biomass is in woody tissue, while the leaf area remains constant. As the stand ages, more of the gross production (photosynthates) goes for maintenance (respiration of woody tissues) and less remains for growth. The pattern is well illustrated by Douglas-fir in the Pacific Northwest of the United States. Seventy percent of the gross production of a 20- to 40-year-old forest stand accumulates as stored biomass (net production). In a 450-year-old stand, only 6 to 7 percent of gross photosynthesis is available for net production (see Chapter 30).

PRODUCTIVITY OF A CORNFIELD

About 30 percent of global land area has been converted to agroecosystems. These monocultural systems devoted to one crop species have replaced natural ecosystems with their diversity of species. Corn has replaced tallgrass prairie (see Chapter 27) in the midwestern United States. Corn is an annual that allocates much of its energy to seed production and a minimal amount to roots. The prairie grasses it replaced allocate much of their primary production to roots. For this reason the cornfield has a much higher aboveground annual production.

The first effort at estimating primary production took place not in a natural ecosystem but in an Illinois cornfield in the mid 1920s. In 1926 the agronomist Edgar Transeau published an estimated energy budget for a cornfield. It provided a foundation for future studies. He considered the amount of incoming solar radiation, the amount of energy fixed in gross production and lost through respiration and transpiration, and the amount of water transpired. He determined the allocation of net primary production to stalk, root, and grain, and estimated the efficiency of photosynthesis.

As a baseline Transeau used a 0.405 ha cornfield containing 10,000 plants that produced 100 bushels of corn with a dry weight of 2160 kg. At maturity the corn plant contains 80 percent water and 20 percent dry matter. Carbon fixed by photosynthesis makes up 45 percent of dry matter. The dry weight of the corn stalks was 600 g, of which 216 g was in grain, 200 g in stalk, 140 g in leaves, and 44 g in roots. Thus corn allocated 36 percent of its net primary production to grain.

Total carbon fixed by 10,000 plants was 2675 kg. The glucose equivalent of this carbon is 6687 kg. During the season the corn plants lost through respiration 2045 kg of glucose or 3000 kg of carbon, about one-fourth of the energy absorbed in photosynthesis. Over the growing season the corn plants manufactured 8732 kg of glucose. Transeau estimated that the amount of energy it took to produce one kg of glucose was 3760 kcal. The total solar energy available to the Illinois corn plant is 2043 million kcal. The total energy consumed by the corn plant in photosynthesis was 33 million kcal. Thus the corn plants used 1.6 percent of the available energy in photosynthesis. This amount may seem low, but consider that the corn plant uses only about 20 percent of the total light spectrum in photosynthesis. That raises the photosynthetic efficiency to about 8 percent.

Not all of the energy lost by the corn plant is through respiration. Transpiration is another source. The Illinois cornfield evaporated about 276 kg of water for every kilogram of its dry weight. Through the growing season the cornfield lost 1.5 million kg of water, or 408,000 gal., enough to cover a hectare to a depth of about 1m. The energy necessary to evaporate 1 kg of water at the average temperature during the growing season is 593 kcal. As a result 910 million kcal or 44.5 percent of available energy was expended in transpiration.

Transeau's final energy budget looked like this:

Total energy available	2043 million kcal
Used in photosynthesis	33 million kcal
Used in transpiration	910 million kcal
Total energy consumed	943 million kcal
Energy not used by plants	1100 million kcal
Energy released by respiration	8 million kcal
Efficiency of photosynthesis (assimilation efficiency)	1.6 percent
Efficiency at 20 percent of usable light spectrum	8.0 percent

Transeau's cornfield of the 1920s was exceptional in its productivity. Prior to 1950, corn allocated about 14 percent of its production to grain, and yields were more on the order of 100 bu/ha. Transeau's cornfield, growing on still highly fertile prairie soil, allocated 36 percent of its net primary production to grain. Since the introduction of hybrid corn in the 1950s, corn production has tripled. This increased production resulted from increased leaf area of corn plants, their ability to grow at densities of 50,000 to 70,000 plants per hectare, and heavy subsidies of fertilizers and pesticides. Although photosynthetic efficiency remains about the same, corn plants now allocate about 45 percent of their net production to grain.

Thus cornfields equal or exceed the productivity of the most highly productive natural ecosystems, the tropical rain forest and estuaries. All have one feature in common. They are heavily subsidized, the tropical forest by a complex interaction of temperature and high evaporation (see Chapter 31), estuaries by tidal flow and terrestrial inputs (see Chapter 37), the cornfield by fertilizer, protective chemicals, and soil manipulation.

This high cropland productivity, however, comes at a cost of low diversity, nutrient losses (see Chapter 25), and soil erosion (see Chapter 10).

23.9 Primary productivity limits secondary production

Net primary production is the energy available to the heterotrophic component of the ecosystem. All of it is eventually consumed, either by herbivores or decomposers, but often it is not all utilized within the same ecosystem. The net primary production of any given ecosystem may be dispersed to another food chain outside the ecosystem by humans or other agents, such as wind or water currents. For example, about 45 percent of the net production of a salt marsh is lost to estuarine water (see Chapter 37).

Some energy in the form of plant material, once consumed, passes from the body as feces and urine (Figure 23.7). Of the energy kept, part is utilized as heat for metabolism. The remainder is available for maintenance—capturing or harvesting food, perform-

ing muscular work, and keeping up with wear and tear on the animal's body. The energy used for maintenance is lost as heat.

The energy left over from maintenance and respiration goes into production, including both the growth of new tissues and the production of young. This net energy of production is called **secondary production.** Secondary production is greatest when the birthrate of the population and the growth rate of individuals are highest.

Secondary production depends on primary production for energy. Therefore, any environmental constraint on primary productivity, such as climate, will also constrain secondary productivity within the ecosystem. For example, Figure 23.8 shows the observed relationship between mean annual rainfall and the productivity of large herbivores in African ecosystems. The increase in large herbivore production with increasing rainfall is a direct result of the corresponding increase in net primary productivity, which provides the food source for the herbivore populations.

Herbivores are the energy source for carnivores. Just as when the herbivore eats a plant, the carnivore

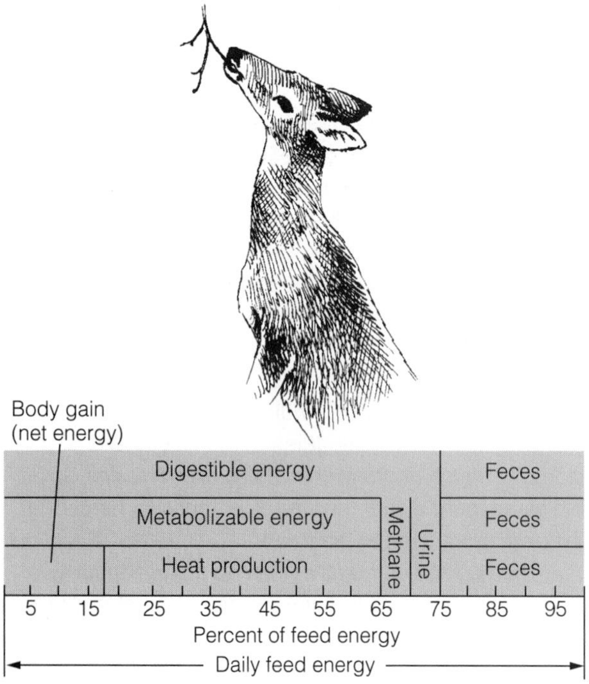

Body gain
(net energy)

Digestible energy		Feces
Metabolizable energy		Feces
Heat production		Feces

Methane | Urine

Percent of feed energy
5 15 25 35 45 55 65 75 85 95

Daily feed energy

Figure 23.7 End products of energy metabolism in the white-tailed deer. Note the small amount of net energy gained compared to that lost as heat, gas, urine, and feces.

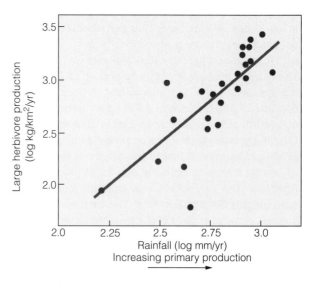

Figure 23.8 The relationship between rainfall (which affects primary productivity) and secondary productivity of large mammalian herbivores in Africa.

cannot use all of the energy in its food for production. Metabolic losses for heat production, maintenance, and respiration occur.

23.10 Consumers vary in efficiency of production

Even if they feed on the same tissues (either plant or animal) as a source of energy, not all consumer organisms have the same efficiency of transforming energy consumed into secondary production (see Quantifying Ecology: 23.1: Production Efficiencies). Of the food ingested by a consumer (I), a portion is assimilated across the gut wall (A) and the remainder is expelled from the body as waste products (W). Of the energy that is assimilated, some is utilized in respiration (R), while the remainder goes to production (P), which includes both the production of new tissues and reproduction. The ratio of assimilation to ingestion (A/I), the assimilation efficiency, is a measure of the efficiency with which the consumer extracts energy from food. The ratio of production to assimilation (P/A), the production efficiency, is a measure of the efficiency with which the consumer incorporates assimilated energy into secondary production.

The ability of a consumer to convert the energy it ingests varies with species and the type of consumer (Table 23.2). Insects that feed on plant tissues, such as grasshoppers, are more efficient producers than insects that feed on plant juices, such as aphids. Larval stages of insects are more efficient producers than the adult stage. Homeotherms have a high assimilation efficiency, but because they use about 98 per-

QUANTIFYING ECOLOGY 23.1
PRODUCTION EFFICIENCIES

Production, both primary and secondary, can be expressed in a number of ways. The following are the important ones, including those used in the text.

Terms
GPP Gross primary production
NPP Net primary production
R   Respiration
P   Secondary production, tissue growth, reproduction, biomass change
I   Ingestion or consumption
W   Egestion: feces, urine, gas, other products
A   Assimilation: food or energy absorbed

Equations
Photosynthetic efficiency = GPP/solar radiation
Assimilation efficiency, plants = GPP/light absorbed
Respiration = GPP − NPP
Effective primary production = NPP/GPP
Assimilation efficiency, animals = A/I
Ecological growth efficiency = P/I
Production efficiency = P/A

cent of that energy in metabolism, they have poor production efficiency. Poikilotherms use about 79 percent of their assimilation in metabolism. They convert a

TABLE 23.2
SECONDARY PRODUCTION OF SELECTED CONSUMERS (KCAL/M²/YR)

Species	Ingestion (I)	Assimilation (A)	Respiration (R)	Production (P)	A/I	P/I
Harvester ant (h)	34.50	31.00	30.90	0.10	0.90	0.003
Plant hopper (h)	41.30	27.50	20.50	7.00	0.67	0.169
Salt marsh grasshopper (h)	3.71	1.37	0.86	0.51	0.37	0.137
Spider, small < 1 mg (c)	12.60	11.90	10.00	0.91	0.94	0.072
Spider, large > 10 mg (c)	7.40	7.00	7.30	−3.00	0.95	—
Savanna sparrow (o)	4.00	3.60	3.60	0.00	0.90	0.000
Old-field mouse (h)	7.40	6.70	6.60	0.10	0.91	0.014
Ground squirrel (h)	5.60	3.80	3.69	0.11	0.68	0.020
Meadow mouse (h)	21.29	17.50	17.00	—	0.82	—
African elephant (h)	71.60	32.00	32.00	8.00	0.45	0.112
Weasel (c)	5.80	5.50	—	—	0.95	—

Note: h = herbivore; o = omnivore; c = carnivore.

greater portion of their assimilated energy into biomass than do homeotherms. The difference, however, is balanced by assimilation efficiency. Poikilotherms have an efficiency of around 30 percent in digesting food, whereas homeotherms have an efficiency of around 70 percent.

23.11   Productivity of decomposers is influenced by climate

The productivity of decomposers is limited by the amount of food energy available in dead organic matter. This food energy is linked to, and therefore limited by, primary productivity. Populations of microbial decomposers, bacteria and fungi, are also strongly influenced by climate. This limitation is reflected in the observed rates of decomposition in different environments. Low temperatures and water availability limit microbial populations and therefore the rates of decomposition.

We can see this limitation in Figure 23.9, which depicts the relationship between rate of decomposition and evapotranspiration for a number of ecosystems.

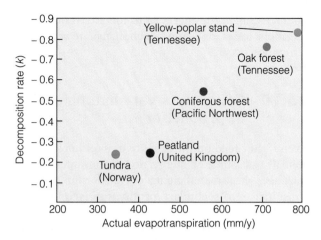

Figure 23.9 Decomposition rates are influenced by climate, as defined by actual evapotranspiration. The value k is the rate at which new litter accumulates as old litter disappears. The larger the negative number, the more rapidly decomposition takes place.

Just like primary productivity, the rate of decomposition increases with increasing evapotranspiration. Organic matter decomposes more quickly under warm, wet conditions as a direct result of greater microbial activity.

CHAPTER REVIEW

SUMMARY

Ecosystem Concept (23.1–23.2)   The biological community and the abiotic environment unite to form a single interactive system, the ecosystem. This system exchanges energy and matter with the surrounding environment **(23.1).** All ecosystems consist of three basic components—autotrophs, consumers, and abiotic matter—that exchange energy and material. The driving force is the energy of the sun. This energy, harnessed by producers, flows from producers to consumers and decomposers and eventually dissipates as heat **(23.2).**

Laws of Thermodynamics (23.3)   Energy flow in ecosystems supports life. Energy is governed by the laws of thermodynamics. The first law states that although energy can be transformed, it can neither be created nor destroyed. The second law states that as energy is transferred or transformed, a portion ceases to be usable. As energy moves through an ecosystem, much of it is lost as heat of respiration. Energy is degraded from a more organized to a less organized state, or entropy. However, a continuous flux of energy from the sun prevents ecosystems from running down.

Primary Production (23.4–23.8)   To carry on photosynthesis, plants must use part of the energy they fix. The total amount of energy fixed by plants is gross primary production. The amount of energy left after plants have met their respiratory need is net primary production, which shows up as plant biomass **(23.4).** Energy fixed by plants is allocated to different parts of the plant and to reproduction. Plants allocate energy first to leaves, then to flowers. Excess production goes to roots and other supporting tissue, where some of the reserves are available for growth the next year. The amount of energy allocated to aboveground stems and belowground roots tells much about different ecosystems. A low shoot-to-root ratio suggests a stressful environment for plants **(23.5).**

Productivity of terrestrial ecosystems is influenced by climate, especially temperature and precipitation. Temperature influences the rate of photosynthesis, and the amount of available water limits photosynthesis and the amount of leaves that can be supported.

Warm, wet conditions make tropical rain forest the most productive terrestrial ecosystem (23.6).

Nutrient availability is the most pervasive influence on the productivity of oceans. Most decomposition takes place at the bottom, so nutrients have to be transported back to the surface. The most productive ecosystems are shallow coastal waters, coral reefs, and estuaries, where nutrients are more available (23.7).

Primary production in an ecosystem varies with time. Seasonal and year to year variations in moisture and temperature directly influence primary production. In ecosystems dominated by woody vegetation, net primary production declines with age. As the ratio of woody biomass to foliage increases, more of gross production goes into maintenance (23.8).

Secondary Production (23.9–23.11)   Net primary production is available to consumers directly as plant tissue or indirectly through animal tissue. Once consumed and assimilated, energy is diverted to maintenance, growth, and reproduction and to feces, urine, and gas. Change in biomass, including both weight change and reproduction, is secondary production. Secondary production depends upon primary production. Any environmental constraint on primary production will constrain secondary production in the ecosystem (23.9). Efficiency of production varies. Homeotherms have high assimilation efficiency but low production efficiency, because they have to expend so much energy in maintenance. Poikilotherms have low assimilation efficiency but high production efficiency; they put more energy into growth (23.10). Productivity of decomposers is limited by the amount of food energy available in dead organic matter, and by temperature and moisture (23.11).

STUDY QUESTIONS

1. What is an ecosystem, and what are its basic components?
2. Distinguish between potential and kinetic energy.
3. How do the first and second laws of thermodynamics relate to ecology?
4. Define primary production, primary productivity, gross primary production, net primary production, and secondary production.
5. What is the significance of the shoot-to-root ratio?
6. How do plants allocate biomass and energy to growth, maintenance, and reproduction?
7. How does climate influence primary production?
8. What world ecosystems have high and low net productivity? Why?
9. How does secondary production differ from primary production?
*10. What is the basic reason for the low productivity of the oceans?
*11. What is the difference in energy allocation and energy efficiency between homeotherms and poikilotherms?

CHAPTER 24

TROPHIC STRUCTURE

OBJECTIVES

On completion of this chapter, you should be able to:

- Define a food chain and a food web.
- Distinguish between grazing and detrital food chains.
- Discuss the roles of microorganisms and macroorganisms in decomposition.
- Define trophic levels and ecological pyramids.
- Discuss the efficiency of energy transfer through the food chain.

Colorful decomposers such as these coral fungi
(*Clavulinopsis fusiformis* and *Russula emetica*) reside
in eastern deciduous forests.

Chapter 23 introduced the flow of energy between primary producers and secondary producers in its simplest forms. Now we will see how energy follows many different pathways through the ecosystem. These pathways influence the structure of the biotic component of the ecosystem.

24.1 Food chains describe the flow of energy through an ecosystem

Energy stored by plants moves through the ecosystem in a series of steps of eating and being eaten, the **food chain.** Food chains are descriptive diagrams—a series of arrows, each pointing from one species to another for which it is a source of food. Figure 24.1 contains several food chains. For example, grasshoppers eat grass, clay-colored sparrows eat the grasshoppers, and marsh hawks prey upon the sparrows. We write this relationship as follows:

grass → grasshopper → sparrow → marsh hawk

As Figure 24.1 indicates, the food chain is not linear. Resources are shared, especially at the beginning of the chain. The same plant is food for a variety of mammals and insects, and the same animal is food for several predators. Thus food chains link to form a **food web,** the complexity of which varies within and among ecosystems.

24.2 Species fit into trophic levels based on their food

Instead of drawing a web, we can group species into categories called **trophic levels** based on a common food source. For example, in Figure 24.1, all organisms that feed wholly on plants—the grasshopper, ground squirrel, pocket gopher, and prairie vole—occupy the same trophic level. The **first trophic level** belongs to the **producers** or plants. Their source of energy is the sun, and their nutrients come from soil, water, and atmosphere.

The second trophic level belongs to the plant eaters or herbivores. They make up the **first-level consumers.** Herbivores are capable of converting energy stored as plant tissue into animal tissue. Their role is essential to the ecosystem, for without them the higher trophic levels could not exist. Only herbivores can live on a high-cellulose diet. Modifications in the structure of the teeth, complicated stomachs, long in-

testines, a well-developed caecum, and symbiotic microbiota in the gut enable these animals to use plant tissue as a food source.

For example, ruminants such as deer have a four-compartment stomach (Figure 24.2). As they graze, these animals chew their food hurriedly. The material consumed descends to the first and second stomachs (the rumen and reticulum), where it is softened to a pulp by the addition of water, kneaded by muscular action, and fermented by bacteria. The bacteria convert part of the cellulose, starches, and sugars into fatty acids. These acids are absorbed into the bloodstream and oxidized to provide the mammal's chief form of energy. At leisure, the ruminants regurgitate the undigested portion (the cud), chew it more thoroughly, and swallow it again.

The lagomorphs—rabbits, hares, and pikas—have a simple stomach and a large caecum. They form two types of fecal pellets, hard and soft. Microorganisms attack part of the ingested material, which is expelled into the large intestine as moist, soft pellets surrounded by a membrane made of proteins. The soft pellets, much higher in protein and lower in crude fiber than the hard fecal pellets, are reingested. Eating feces is coprophagy. The amount of feces recycled by coprophagy may range from 5 percent to 80 percent. This reingestion is important, for it provides bacterially synthesized B vitamins and ensures more complete digestion of dry material and better utilization of protein.

Herbivores, in turn, are the energy source for **carnivores,** animals that feed on other animals. Organisms that feed directly on grazing herbivores are **first-level carnivores** and **second-level consumers.** First-level carnivores are an energy source for **second-level carnivores** and **third-level consumers.** In a popular sense, carnivores are large organisms that kill and eat smaller prey. In the broadest sense, any organism that feeds on another organism or on its tissue is a carnivore, functionally speaking. Some parasites are carnivores.

The typical carnivore is adapted for a diet of flesh. Hawks and owls have sharp talons for holding prey and hooked beaks for tearing flesh. Mammals have canine teeth for biting and piercing. Cheek teeth are reduced, but many species have sharp-crested shearing or carnassial teeth.

Not all consumers fit neatly into one trophic level. Many do not confine their feeding to one level alone. The red fox feeds on berries and small rodents; therefore it occupies herbivorous and carnivorous levels. Some fish feed on both plant and animal matter. The basically herbivorous white-footed mouse also feeds on insects, small birds, and bird eggs. The food habits

Figure 24.1 A food web for a prairie grassland community in the Midwest. Arrows flow from food to consumer.

of many animals vary with the seasons, with stages in the life cycle, and with size and growth. Consumers that feed on both plant and animal tissue are **omnivores** (Figure 24.3).

Scavengers are animals that eat dead plant and animal material. Among them are termites and various beetles that feed on dead and decaying wood, and clams and other marine invertebrates that feed on plant particles in the water. Botflies, dermestid beetles, vultures, gulls, and hyenas are a few of the animals that feed on animal remains. Scavengers are either herbivores or carnivores.

Saprophytes are the plant counterparts of scavengers. They draw their nourishment from dead plant and animal material, chiefly the former. Because they do not require sunlight as an energy source, they can

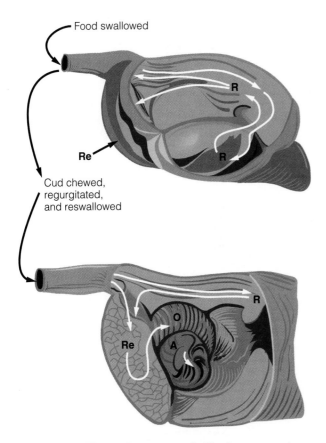

Food swallowed

Cud chewed,
regurgitated,
and reswallowed

Figure 24.2 The ruminant stomach. The four-compartment stomach consists of the rumen (R), reticulum (Re), omasum (O), and abomasum (A). Food enters the rumen and the reticulum. The ruminant regurgitates fermented material (cud) and rechews it. Finer material enters the reticulum, and then the omasum and abomasum. Coarser material reenters the rumen for further fermentation.

live in deep shade or dark caves. Fungi are examples of saprophytes. The majority are herbivores, but some do feed on animal matter.

24.3 Decomposers are a diverse group

Decomposers make up the final feeding group (see Sections 3.7 and 10.4). However, this category oversimplifies a complex group of organisms. All consumers, to some degree, function as decomposers. They either reduce food by digestion or fragment it into smaller pieces, making it more accessible to other consumers, including bacteria and fungi. What we typically refer to as true decomposers are organisms that feed on dead organic matter, called **detritus.**

Decomposers fall into two groups, microscopic and macroscopic. Together there may be over one million of these organisms in a square meter of the top 7 to 10 cm of a temperate forest soil. Approximately 40 percent of these organisms are bacteria; about 50 percent are microscopic fungi; 5 to 9 percent are protozoans; and 0.5 percent are fungi. Animals visible to the naked eye, the macroscopic decomposers, account for only 0.4 percent of the total decomposer community.

The dominant detritus-feeding organisms are bacteria and fungi. Bacteria are either aerobic, requiring oxygen, or anaerobic, carrying on their metabolic functions in the absence of free oxygen. Anaerobic bacteria commonly inhabit mud and sediments in aquatic environments. As a group bacteria are the major decomposers of animal matter. The major decomposers of plant material are the fungi. Both heterotrophic bacteria and fungi produce enzymes that catalyze specific chemical reactions. They secrete enzymes into the plant and animal matter and absorb the resulting products as food. Once a group of bacteria and fungi has exploited the material as much as possible, another group moves in to continue the process.

Macroorganisms include such small detritus-feeding animals as collembolas or springtails, mites, millipedes, earthworms, nematodes, and slugs in terrestrial ecosystems, and crabs, mollusks, and mayfly, stonefly, and caddisfly larvae in aquatic ecosystems. Larger detritus feeders, such as earthworms and caddisfly larvae, break organic matter into smaller pieces by both mechanical action and digestion, mixing it with soil in the case of earthworms, excreting it, and even adding substances that stimulate microbial growth. These same organisms also consume bacteria and fungi associated with the detritus, as well as small invertebrates and protozoans clinging to the material. Other invertebrates, the **microbivores,** feed directly on the microbes.

A complementary relationship exists between macrodecomposers and microdecomposers. Macroorganisms fragment the detrital material, making it available to smaller detritus feeders and to bacteria and fungi. Ultimately, the material becomes so small that even microbial activity cannot continue. At this point bacteria assimilate organic compounds, concentrating them into larger particles that in turn are made even larger by bacterial aggregation. This material is again available to the macrodecomposers. They, in turn, may produce fecal pellets larger than the material digested, providing surfaces for microbial colonization.

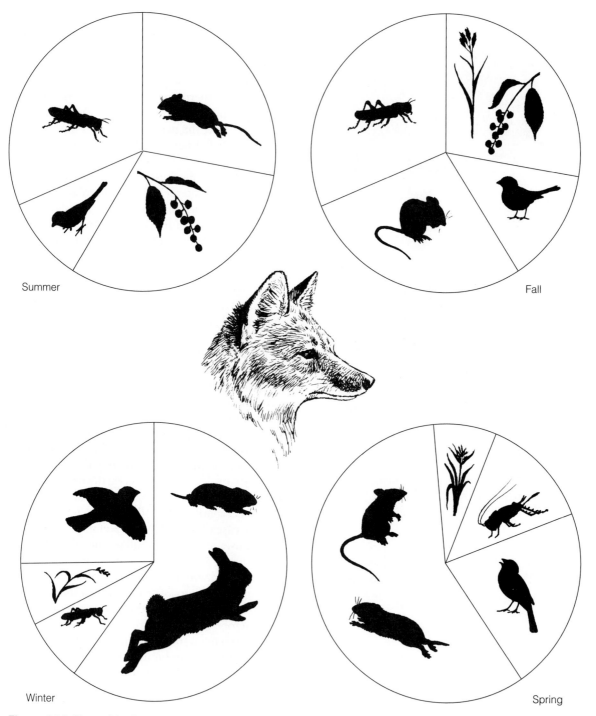

Summer

Fall

Winter

Spring

Figure 24.3 The red fox is an example of an omnivore. It feeds on both the herbivorous and the carnivorous levels. Its diet varies with the seasons.

In such a manner decomposing organic matter may be passed among feeding groups until it finally reaches an inorganic state.

In terrestrial environments bacteria are the main regenerators of nutrients and concentrators of energy that otherwise might be prematurely lost through respiration. In aquatic ecosystems phytoplankton and zooplankton play a major role in recycling nutrients and cycling energy. Phytoplankton, macroalgae, and zooplankton furnish dissolved organic matter, with algae being the main contributors. At certain stages of their lify cycle, particularly during rapid growth and reproduction, phytoplankton and other algae excrete large quantities of organic matter, which dissolves in the water. Bacteria concentrate these nutrients by incorporating them in their own biomass. Ciliates and zooplankton then eat these bacteria and excrete nutrients in the form of exudates and fecal pellets into the water. Zooplankton, in the presence of an abundance of food, consumes more than it needs and then excretes more than half as fecal pellets, which make up a significant fraction of suspended material. These pellets are attacked by bacteria that use the nutrients, growth substances, and energy they contain. Thus the cycle starts over again.

In the trophic structure of an ecosystem, we can call decomposer organisms herbivores or carnivores, depending on their source of food. Such an approach incorporates decomposers into general feeding groups at various levels of the food chain.

24.4 Ecosystems have two major food chains

A different approach puts decomposers into a separate chain. Within any ecosystem there are two major food chains, the grazing food chain and the detrital food chain (Figure 24.4). The distinction between these

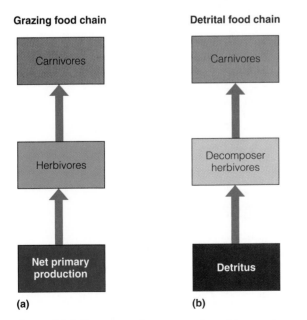

Figure 24.4 Two parts of any ecosystem: (a) a grazing food chain and (b) a detrital food chain.

two food chains is the source of energy for the first-level consumers, the herbivores. In the **grazing food chain** the source of energy is living plant biomass or net primary production (NPP). In the **detrital food chain** the source of energy is dead organic matter or detritus. In turn, the herbivores in each food chain are the source of energy for the carnivores, and so on. The grazing food chain is easier to see. Cattle grazing on a pastureland, deer browsing in the forest, rabbits feeding in old fields, and insects feeding on garden crops represent first-level consumers of the grazing food chain. In spite of its conspicuousness, the grazing food chain is not the major one in terrestrial and many aquatic ecosystems. Only in some open-water aquatic ecosystems do the grazing herbivores play a dominant role in energy flow.

In terrestrial ecosystems only a small portion of primary production goes into the grazing food chain. Over a three-year period, herbivores used only 2.6 percent of the net primary production of a temperate deciduous forest, although grazing insects made holes in the leaves, eating away 7.2 percent of the photosynthetic surface. R. D. Andrews and his associates studied energy flow through a shortgrass prairie ecosystem. By controlling the amount of cattle grazing, the scientists were able to examine ungrazed, lightly grazed, and heavily grazed plots. Even on heavily grazed grassland, cattle consumed only 30 to 50 percent of aboveground net primary production. Cattle return about 40 to 50 percent of energy they consume to the detrital food chain as feces.

Although aboveground herbivores are the conspicuous grazers, belowground herbivores can have a pronounced effect on primary production. Andrews and his associates found that the belowground herbivores—mainly nematodes (Nematoda), scarab beetles (Scarabaeidae), and adult ground beetles (Carabidae)—accounted for 81.7 percent of total herbivore assimilation on the ungrazed plots, 49.5 percent on the lightly grazed plots, and 29.1 percent on the heavily grazed plots.

The detrital food chain is common to all ecosystems. In terrestrial and littoral (shoreline) ecosystems, it is the major pathway of energy flow, because so little of the net primary production is used by grazing herbivores. Of the total amount of energy fixed by photosynthesis in a temperate deciduous forest, approximately 50 percent of gross production goes into maintenance and respiration, 13 percent becomes new tissue, 2 percent is consumed by herbivores, and 35 percent goes directly into the decomposer food chain. Two-thirds to three-fourths of gross production in a grassland ecosystem that is ungrazed by domestic animals returns to the soil as dead plant material, and less than one-fourth is consumed by herbivores. Of the quantity consumed by herbivores, about one-half returns to the soil as feces. In the salt marsh ecosystem, the dominant grazing herbivore, the grasshopper, consumes just 22 percent of net primary production.

Forest litter, the habitat of an array of detritus-feeding invertebrates, is a good place to seek an example of a detrital food web. One such food web (Figure 24.5) involves five groups of litter feeders, in this case macrodecomposers: millipedes (Diplopoda), orbatid mites (Cryptostigmata), springtails (Collembola), cave crickets (Orthoptera), and pulmonate snails (Pulmonata). Of these, the mites and springtails are the most important litter feeders. These herbivores are preyed upon by small spiders (Araneidae) and predatory mites (Mesostigmata). The spiders also feed on the predatory mites. Springtails, snails, small spiders, and cave crickets are fed upon by carabid beetles.

Medium-size spiders also eat the crickets and are eaten by the beetles. The beetles, snails, and spiders are consumed by birds and small mammals, members of the grazing food chain. In this manner, predation links the detrital food chain to the grazing food chain at higher consumer levels.

24.5   Energy flows through trophic levels

When ecologists trace the flow of energy through an ecosystem, they have to track the flow of energy between trophic levels. They also have to define the links between the two food chains, grazing and detrital, and measure the losses from the ecosystem through respiration (R).

Figure 24.6 combines the food chains to produce a generalized model of trophic structure and energy flow through an ecosystem. The two food chains are

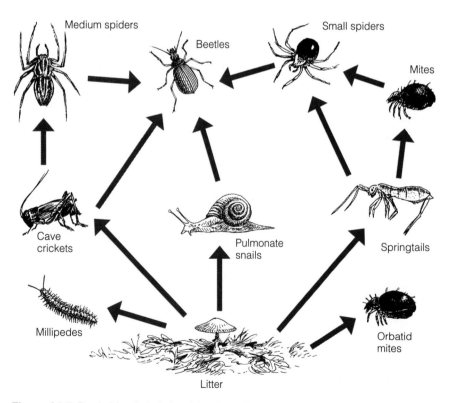

Figure 24.5 Detrital food chain involving forest litter-dwelling invertebrates in an Appalachian yellow-poplar forest.

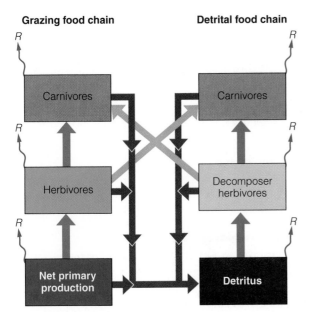

Grazing food chain   **Detrital food chain**

Figure 24.6 Grazing and detrital food chains from Figure 24.4 combined, showing their connections.

linked. The initial source of energy for the detrital food chain is the input of waste materials and dead organic matter from the grazing food chain. This linkage appears as a series of arrows from each of the trophic levels in the grazing food chain leading to the box designated as detritus or dead organic matter. There is one notable difference in the flow of energy between trophic levels in the grazing and decomposer food chains. In the grazing food chain the flow is unidirectional, with net primary production providing the energy source for herbivores, herbivores providing the energy for carnivores, and so on. In the decomposer food chain, the flow of energy is not unidirectional. The waste materials and dead organic matter (organisms) in each of the consumer trophic levels are "recycled," returning as an input to the dead organic matter box at the base of the detrital food chain. In addition, the higher trophic levels of the detrital food chain provide energy for higher trophic levels (through predation) of the grazing food chain.

To quantify the flux of energy through the ecosystem, we need to return to the processes involved in secondary production discussed in Sections 23.9 and 23.10: consumption, ingestion, assimilation, respiration, and production. We will diagram a single trophic compartment (Figure 24.7a). The energy available to a given trophic level (designated as n) is the production of the next lower level (n-1); for example, net primary production (P_{n-1}) is the available energy for grazing herbivores (trophic level n). Some proportion of that productivity is consumed or ingested (I); the remainder makes its way to the dead organic matter of the detrital food chain. Of the energy consumed, some portion is assimilated by the organisms (A) and the remainder is lost as waste materials (W) to the detritus food chain. Of the energy assimilated, some is lost to respiration, shown as the arrow labeled R that is leaving the upper left corner of the box, and the remainder goes to production (P_n).

We quantify this flow with the formulas in Quantifying Ecology 24.1: Ecological Efficiencies. Notice the **consumption efficiency,** the ratio of ingestion to production (I_n/P_{n-1}). The consumption efficiency defines the amount of available energy being consumed. The assimilation efficiency (A/I) defines the proportion of the energy ingested that is assimilated and the proportion that is lost as waste material. The production efficiency (P/A) defines the relative proportions of the assimilated energy that go to production and respiration. Sample values of these efficiencies for an invertebrate herbivore in the grazing food chain are provided in Figure 24.7. Using these efficiency values, we can track the fate of 1000 kcal of energy available to herbivores in the form of net primary productivity through the herbivore trophic level (Figure 24.7b).

If we apply efficiency values for each trophic level in the grazing and herbivore food chains, we will know the flow of energy through the whole ecosystem. The production from each trophic level provides the input to the next higher level, while unconsumed production, waste products, and dead individuals from each trophic level provide input into the dead organic matter compartment. The entire flow of energy through the ecosystem is a function of the initial transformation of solar energy into net primary productivity. All energy entering the ecosystem as net primary productivity eventually is lost through respiration.

24.6 Energy decreases in each successive trophic level

The quantity of energy flowing into a trophic level decreases with each increasing trophic level. This pattern occurs because not all energy is used for production. An ecological rule of thumb allows a magnitude of 10 reduction in energy as it passes from one trophic level to another. If herbivores eat 1000 kcal of plant energy, about 100 kcal will convert into herbivore tissue,

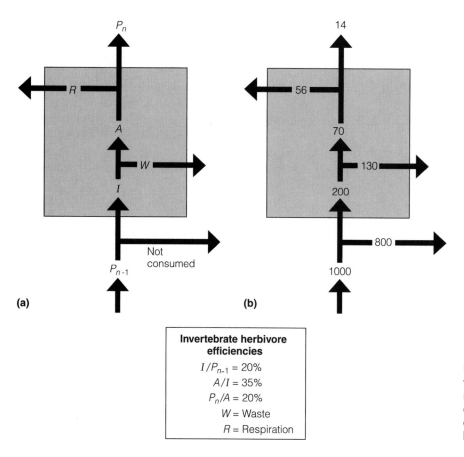

Invertebrate herbivore
efficiencies

$I/P_{n-1} = 20\%$
$A/I = 35\%$
$P_n/A = 20\%$
W = Waste
R = Respiration

Figure 24.7 (a) Energy flow within a single trophic compartment. (b) A quantified example of energy flow through that compartment for an invertebrate herbivore. Values are kcal.

10 kcal into first-level carnivore production, and 1 kcal into second-level carnivore production. However, data suggest that a 90 percent loss of energy from one trophic level to another may be too high.

Certainly, a wide range in the efficiency of conversion (assimilation and production efficiencies) exists among different feeding groups (see Table 23.2). Production efficiency in photosynthetic organisms (net production/solar radiation) is low, ranging from 0.34 percent in some phytoplankton to 0.8 to 0.9 percent in grassland vegetation. Plant production consumed by herbivores is used with varying efficiency. Herbi-

vores consuming green plants are wasteful feeders, but not nearly as wasteful as those feeding on plant sap.

Assimilation efficiencies vary widely among poikilotherms and homeotherms (Table 24.1). Homeotherms are much more efficient than poikilotherms. However, carnivorous animals, even poikilotherms, have high assimilation efficiency. Predatory spiders feeding on invertebrates have assimilation efficiencies of over 90 percent (see Table 23.2). Because of high maintenance and respiratory costs, homeotherms have low production efficiency compared to poikilotherms. Only about 2 to 10 percent of the energy consumed

TABLE 24.1

ASSIMILATION AND PRODUCTION EFFICIENCIES FOR HOMEOTHERMS AND POIKILOTHERMS

Efficiency	All Homeotherms	Grazing Arthropods	Sap-feeding Herbivores	Lepidoptera	All Poikilotherms
Assimilation					
A/I	77.5 ± 6.4	37.7 ± 3.5	48.9 ± 4.5	46.2 ± 4.0	41.9 ± 2.3
Production					
P/I	2.0 ± 0.46	16.6 ± 1.2	13.5 ± 1.8	22.8 ± 1.4	17.7 ± 1.0
P/A	2.46 ± 0.46	45.0 ± 1.9	29.2 ± 4.8	50.0 ± 3.9	44.6 ± 2.1

TABLE 24.2

		Producers		Herbivores		Carnivores	
Habitat	Growing Season (days)	Production (kcal/m²)	Efficiency (%)	Production (kcal/m²)	Efficiency (%)	Production (kcal/m²)	Efficiency (%)
Shortgrass plains	206	3767	0.8	53	11.9	6	13.2
Midgrass prairie	200	3591	0.9	127	16.5	37	23.7
Tallgrass prairie	275	5022	0.9	162	5.3	15	13.9

CONSUMPTION EFFICIENCY (SECONDARY PRODUCTION/SECONDARY CONSUMPTION)

by herbivore homeotherms goes into biomass production, less than the 10 percent average suggested by the rule of thumb. However, poikilotherms convert about 17 percent of their consumption to herbivore biomass. On midwestern grasslands average herbivore production efficiency, involving mostly poikilotherms, ranged from 5.3 to 16.5 percent (Table 24.2). Production efficiency on the carnivore level ranged from 13 to 24 percent.

Transfer of energy from one trophic level to another tells the real story, but such data are hard to collect. The ratio of phytoplankton to secondary zooplankton production in open freshwater ecosystems is about 7.1:1, and the ratio of herbivore zooplankton production to carnivore zooplankton production is 2.1:1. Efficiencies are lower in the benthic community—2.2 for herbivores and 0.3 for carnivores.

The proportion of production consumed by the next higher trophic level, the energy transfer efficiency, also varies greatly. Among invertebrate consumers on a shortgrass plain the proportion is about 9 percent for herbivores, 38 percent for aboveground predators, and 56 percent for belowground predators.

24.7   Ecological pyramids portray trophic levels

If we sum all of the biomass or energy contained in each trophic level, we can construct pyramids for the ecosystem (Figure 24.8). The pyramid of biomass indicates by weight, or other means of measuring living material, the total bulk of organisms or fixed energy present at any one time—the **standing crop.** Because some energy or material is lost at each successive trophic level, the total mass supported at each level is limited by the rate at which energy is being stored at the next lower level. In general, the biomass of producers must be greater than that of the herbivores they support, and the biomass of herbivores must be greater than that of carnivores. That circumstance results in a narrowing pyramid for most ecosystems.

This arrangement does not hold for all ecosystems. In such ecosystems as lakes and open seas, primary production is concentrated in the microscopic algae. These organisms have a short life cycle and rapid reproduction. They are heavily grazed by herbivorous zooplankton that are larger and longer-lived. As a result, despite the high productivity of algae, their biomass is low compared to that of zooplankton herbivores.

The energy pyramid indicates only the amount of energy flow at each level. The base on which it is constructed is the quantity of organisms produced per unit time. Stated differently, it is the rate at which food material passes through the food chain. Some organisms have a small biomass, but the total energy they assimilate and pass on may be considerably greater than that of organisms with a much larger biomass. On a pyramid of biomass these organisms would

QUANTIFYING ECOLOGY 24.1
ECOLOGICAL EFFICIENCIES

Assimilation efficiency (within a trophic level) =
$$\frac{\text{Assimilation}}{\text{Ingestion}} \quad \frac{A}{I}$$

Growth efficiency =
$$\frac{\text{Production}}{\text{Ingestion}} \quad \frac{P}{I}$$

Production efficiency =
$$\frac{\text{Production}}{\text{Assimilation}} \quad \frac{P}{A}$$

Consumption efficiency =
$$\frac{\text{Consumption at trophic level } n}{\text{Production at trophic level } n-1} \quad \frac{I_n}{P_n - 1}$$

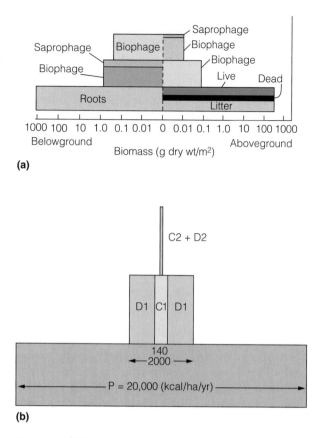

(a)

(b)

Figure 24.8 Examples of ecological pyramids. The detrital and grazing food chains have been collapsed into the same trophic levels. (a) A pyramid of biomass for a northern shortgrass prairie. The base of the pyramid represents biomass (g dry weight/m²) of producers; the second (middle) level, first-level consumers; and the top, second-level consumers. The dashed vertical line separates aboveground biomass (right) and belowground biomass (left). The trophic level magnitudes are plotted on a horizontal logarithmic scale. The compartments are divided into live, standing dead, and litter biomass; and into consumers of live biomass (biophages) and dead biomass (saprophages). (b) A pyramid of energy for the Lamto Savanna, Ivory Coast. P is primary production; C_1, first-level consumers; C_2, second-level consumers; D_1, decomposers of vegetable matter; D_2, decomposers of animal matter.

appear much less important in the ecosystem than they really are. Energy pyramids narrow because less energy is transferred from each level than was paid into it, in accordance with the second law of thermodynamics. In instances in which the producers have less bulk than consumers, as in open water, the energy they store and pass on must be greater than that of the next higher level. Otherwise the biomass that producers support could not be greater than that of the produc-

ers themselves. This high energy flow is maintained by a rapid turnover of individuals rather than an increase in total mass.

24.8   Food chains link to form food webs

Trophic relationships in nature are not simple, straight-line food chains. Numerous food chains link into a complex food web, with all links leading from producers through an array of primary and secondary consumers. When we unravel food webs, separating out the numerous food chains, certain patterns emerge. All chains are interconnected. No matter how productive an ecosystem, each food chain rarely exceeds four links, because the length is limited by the inefficiency of energy transfer. Highly productive ecosystems do not support longer food chains, but they may support more species and therefore more complex food webs.

Although omnivores at first glance may appear to be a significant component in food webs, they do not dominate food chains. When present, omnivores typically feed on species in adjacent trophic levels, involving two different types of food. Versatility is limited by the morphological and physiological makeup of the omnivore. Birds with beaks adapted to tearing flesh are not well adapted to feed on fruits or seeds. Carnivorous mammals with teeth adapted for shearing meat and a short digestive system designed to digest animal protein cannot cope with a diet of grass. At the best they are limited to highly digestible fruits. For this reason omnivory is not highly prevalent in food chains dominated by larger vertebrates and invertebrates. However, food webs dominated by insects and detritivores and their predators and parasitoids exhibit a more complex pattern of omnivory, which may involve feeding on nonadjacent trophic levels.

Predators may overlap in their exploitation of prey species. Foxes and kestrels, for example, feed on mice. Top predators feed on a number of species of primary and secondary consumers, or they are more or less restricted to prey species on the adjacent trophic level below them. In general, the more species of prey a species exploits, the fewer species of predators it faces. Consider the fox, which feeds on a wide range of prey species. Aside from humans, it has no major natural predator. A screech owl, which feeds principally on mice, can succumb to predation by the great horned owl.

The study of food webs raises interesting questions. What determines the size and complexity of

FOCUS ON ECOLOGY 24.1

ALTERING FOOD WEBS

What happens when a new top predator invades a food web or when one is removed? An example of the first situation took place when the Panamanian government and some businesspeople introduced peacock bass *(Cichla ocellaris),* a native of the Amazon River system, to Gatun Lake, Panama Canal. Their objective was to provide an outstanding sport fish and a highly edible market fish. As the population of the voracious peacock bass spread throughout the lake, it quickly reduced almost all secondary consumers. It especially decimated the major planktivorous fish, *Melaniris chagresi,* and several other planktivores on which the native top predators—tarpon, heron, and black tern—fed. These species consequently disappeared from most of the lake. The peacock bass became the top predator, and the planktivorous juvenile peacock bass replaced the native planktivores. The juvenile peacock bass and an algae-feeding fish became the major prey species of the peacock bass. Thus the invasion of the peacock bass highly simplified the food web.

A second example involves the removal of the large-mouth bass, a top predator, from a pond ecosystem. When present, the predatory bass reduces the biomass of vertebrate planktivores, allowing invertebrate planktivores and large zooplankton herbivores to increase, and causing small herbivorous zooplankton and primary producers to decrease. When the bass were removed, vertebrate planktivores increased, invertebrate planktivores and large herbivorous zooplankton decreased, and small herbivorous zooplankton and primary producers increased.

These two examples, among many, point out the dangers of deliberately or inadvertently introducing or removing predators from established food webs. Such introductions were once commonplace in the management of fisheries, with strongly adverse, unpredicted effects.

---

food webs and the number of trophic levels? How are food webs organized and structured? How are food webs affected by the successful invasion of a new species or the elimination of established species (see Focus on Ecology 24.1: Altering Food Webs)? Are complex food webs more stable than simple ones? What processes, including population dynamics and energy flow, affect the patterns of food webs?

Ecologists have analyzed real food webs, terrestrial, marine, and freshwater, and computer models of food webs in an effort to find some answers to these questions. Their analyses suggest that food webs in fluctuating environments—ones characterized by variations in temperature, salinity, acidity, moisture, and other conditions—tend to have shorter food chains with fewer trophic links than those in more constant environments. Food chains in constant environments, such as pelagic regions of the ocean, have greater species richness and more trophic links to the food chain. Environmental variability alone, however, does not appear to constrain the average or maximum length of a food chain. Highly stratified environments such as a forest or a pelagic water column have longer food chains than poorly stratified habitats such as grassland, tundra, and stream bottoms. The widest food webs, those with the greatest number of herbivores, are the shortest. In contrast, narrow food webs have the greatest fraction of top carnivores.

How food chains develop is another question. Are they a product of random assemblage or not? To seek the answer one ecologist, P. A. Yodzis, modeled the development of a food web based on natural food webs. Starting with a number of producer organisms, Yodzis added additional species, each of which had a certain ecological efficiency, making a fraction of its own consumption available to the next trophic level as a prey species. Each new species added had to obtain its energy from production available from other species. Each species had to choose a food source already used by another. Each new species introduced had a certain total production, a fraction of which had to be made available to a subsequently arriving species. Yodzis found an upper and lower limit to the number of species a predator could exploit. At some point total production was too low to allow a new species to enter. Thus, the limit of energy transformation (the second law) forced a pattern to a food web. Yodzis found a certain degree of similarity between his simulated food webs and real-world food webs.

Generalist species most easily invade simple food webs. Specialists, capable of exploiting a restricted source of energy, are best able to invade complex webs. Removal of a generalist prey species, a generalist predator species, or a member of a simple, straight-line food chain has little effect on a food web. The removal of a key predator species can have a pronounced effect on a food web. Its loss causes the greatest loss of species in the trophic level beneath when the predator has a controlling influence on the equilibrium density of the prey and the prey are generalists in their food habits. Removal has the least effect when the predator exerts a controlling influence on the equilibrium density of specialist prey species. In summary, field and computer simulation studies seem to support the hypotheses that complex food webs are less stable than simple ones and that food webs are not random assemblages but rather are regulated by properties of existing food webs and the nature of invading species.

Studies of published food webs and computer simulations have shortcomings. Published food webs are drawn to characterize certain environments and not to provide detailed trophic relationships. Most published food webs are rather general at the lower trophic levels, especially at the producer level, and more detailed at the upper trophic levels. Often the food relations between eater and eaten are based on general observations rather than detailed food habit studies of all species involved. Published food webs mostly cover grazing food chains and ignore detrital food chains, the most important of all. Further, food webs vary seasonally in any ecosystem. Food webs also change with the foraging behaviors of herbivores and carnivores and the heterogeneity or patchiness of the environment. Such spatial and temporal variations weaken any possibility of rigorous quantification.

CHAPTER REVIEW

SUMMARY

Food Chains and Webs (24.1–24.2) A basic function of the ecosystem is the flow of energy from the sun through various consumers to its final dissipation in a series of energy transfers known as the food chain. Food chains link to form food webs (24.1). The various members of a food web can be grouped into categories called trophic or feeding levels. Plants and phyloplankton, the autotrophic component of the ecosystem, occupy the first trophic level. Herbivores that feed on autotrophs make up the next trophic level. They convert autotrophs into animal tissue. Carnivores that feed on herbivores make up the third trophic level, and carnivores that feed on carnivores make up a fourth level. Omnivores do not confine their feeding to one level. Animals that feed on dead plant and animal matter are called scavengers. Plant counterparts of scavengers are saprophytes (24.2).

Role of Decomposers (24.3) Decomposers dissipate energy and return nutrients to the ecosystem for recycling. The true organisms of decay, responsible for the conversion of organic compounds to inorganic ions, are the heterotrophic bacteria and fungi. Consumers that feed on living organic matter excrete partially decomposed material. Macroscopic decomposers fragment detrital material into smaller particles more accessible to bacteria and fungi. Another group, the microbivores, feeds on detrital particles, mainly for the bacteria and fungi growing on them. In terrestrial

ecosystems, bacteria and fungi play the major role in decomposition. In aquatic ecosystems, phytoplankton and zooplankton play the major role.

Energy Flow (24.4–24.7) Energy flow in ecosystems takes two routes: one through the grazing food chain, the other through the detritus food chain. The bulk of production is used by organisms that feed on dead organic matter. The two food chains are linked at higher trophic levels of each through carnivory and omnivory (24.4). At each trophic level, efficiency is defined as consumption efficiency, the amount of available energy being consumed; assimilation efficiency, the portion of energy ingested that is assimilated and not lost as waste material; and production efficiency, the portion of assimilated energy that goes to growth and respiration (24.5). The efficiency of conversion varies among poikilotherms and homeotherms and between herbivores and carnivores. In general, poikilotherms have low assimilation efficiency and high production efficiency, whereas homeotherms have the reverse (24.6).

The loss of energy at each transfer limits the number of trophic levels or steps in the food chain to four or five. At each level, biomass usually declines. A plot of the total weight of individuals at each successive level produces a tapering pyramid. In aquatic ecosystems, however, where there is a rapid turnover of small aquatic producers, the pyramid of biomass becomes inverted. In either case, energy decreases from one trophic level to another, so trophic levels are always pyramidal (24.7).

Food Web Structure (24.8)   Food chains tangle in complex webs. Food web theory considers the size, organization, and structure of food webs, as influenced by environment, number of species, invasion and loss of species, and the relation of one trophic level to another. Food webs are important in understanding ecosystems.

STUDY QUESTIONS

1. What is a food chain? A food web?
2. What are the two major food chains, and how are they related?
3. Why is decomposition more than just the traditional end point in the food chain?
4. Relate the following to decomposition in terrestrial ecosystems: saprophytes, anaerobic bacteria, aerobic bacteria, fungi, microbial decomposers, macroorganisms, and microbial grazers.
5. What are the roles of bacteria, phytoplankton, and zooplankton in the decomposition process in the aquatic environment?
6. What is a trophic level? Relate the levels to ecological pyramids.
7. What is ecological efficiency?
8. What influences the length of food chains and patterns in food webs?
*9. Are natural food webs free of the influence of humans? How might food webs be affected by the use of pesticides? By the overexploitation of fish and shrimp? By land disturbance? By the introduction of exotic species?

BIOGEOCHEMICAL CYCLES

OBJECTIVES

On completion of this chapter, you should be able to:

- Define two types of biogeochemical cycles.
- Describe the oxygen, carbon, and nitrogen cycles.
- Outline the phosphorus cycle, both terrestrial and aquatic.
- Explain the sulfur cycle.
- Describe the roles of nitrogen and sulfur as environmental pollutants.
- Discuss the nature and effects of acid deposition.
- Discuss the effects of excess deposition of phosphorus and heavy metals on ecosystems.
- Describe how chlorinated hydrocarbons cycle through and affect ecosystems.

Impala *(Aepyceros melampus)* standing in the shade of acacia trees. Their urine and droppings make the impala important contributors to the internal nitrogen cycle of these trees.

The living world depends upon the flow of energy and the circulation of materials through the ecosystem. Both influence the abundance of organisms, the metabolic rate at which they live, and the complexity of the ecosystem. Energy and materials flow through the ecosystem together as organic matter; one cannot be separated from the other. The continuous round trip of materials, paid for by the one-way trip of energy, keeps ecosystems functioning.

25.1 All nutrients follow biogeochemical cycles

All nutrients flow from the nonliving to the living and back to the nonliving components of the ecosystem in a more less cyclic path known as a **biogeochemical cycle** (from *bio* for living, *geo* for the rocks and soil, and *chemical* for the processes involved). The important players in all nutrient cycles are the green plants, which organize the nutrients into biologically useful compounds; the decomposers, which return them to their simple elemental state; and the air and water, which transport nutrients between the abiotic and living components of the ecosystem. Without these factors no cyclic flow of nutrients would exist.

There are two basic types of biogeochemical cycles: gaseous and sedimentary. In **gaseous cycles** the main reservoirs of nutrients are the atmosphere and the oceans. Therefore gaseous cycles are pronouncedly global. The gases most important for life are nitrogen, oxygen, and carbon dioxide. These three gases in stable quantities of 78, 21, and 0.03 percent, respectively, are the dominant components of Earth's atmosphere. (See Focus on Ecology 25.1: The Gaia Hypothesis.)

In **sedimentary cycles** the main reservoir is the soil, rocks, and minerals. The mineral elements that living organisms require come initially from inorganic sources. Available forms occur as salts dissolved in soil water or in lakes, streams, and seas. The mineral cycle varies from one element to another, but essentially it consists of two phases: the salt solution phase and the rock phase. Mineral salts come directly from Earth's crust through weathering (see Section 10.3). The soluble salts then enter the water cycle. With water they move through the soil to streams and lakes and eventually reach the seas, where they remain indefinitely. Other salts return to Earth's crust through sedimentation. They become incorporated into salt beds, silts, and limestone. After weathering, they enter the cycle again.

There are many different kinds of sedimentary cycles. Cycles such as the sulfur cycle are a hybrid between the gaseous and the sedimentary, because they have reservoirs not only in Earth's crust but also in the atmosphere. Other cycles, such as the phosphorus cycle, are wholly sedimentary; the element is released from rock and deposited in both the shallow and deep sediments of the sea.

Both gaseous and sedimentary cycles involve biological and nonbiological agents, both are driven by the flow of energy through the ecosystem, and both are tied to the water cycle (see Section 7.4). Water is the medium by which elements and other materials move through the ecosystem. Without the cycling of water, biogeochemical cycles would cease.

25.2 Nutrients flow between reservoirs

We can represent the movement of nutrients (or matter in general) through ecosystems in an abstract form as a number of linked reservoirs. We also call these reservoirs, the major components of the ecosystems, pools or compartments. An exchange between two reservoirs is a **flux.**

Figure 25.1 is a generalized representation of a biogeochemical cycle in terrestrial ecosystem. In this example, the initial source of the nutrient is the atmosphere. In the initial flux the nutrient moves from the

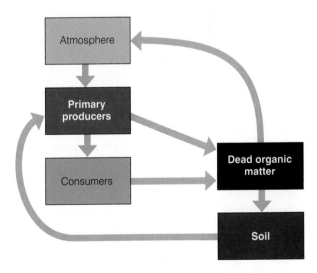

Figure 25.1 Nutrient flow between compartments in an ecosystem. Boxes represent compartments, reservoirs, or pools. Arrows indicate transfers or fluxes between compartments.

FOCUS ON ECOLOGY 25.1

THE GAIA HYPOTHESIS

Earth's atmosphere is extremely different from that predicted for a nonliving Earth and from that of other planets in the solar system. Those atmospheres are dominated by carbon dioxide and possess only a trace of oxygen. There are two views of the formation of Earth's atmosphere. One is that physical forces interacted to form life-sustaining conditions, and then life evolved to adapt to those conditions. The other is that organisms evolved in partnership with the physical environment. From the beginning these organisms helped to control geochemical cycles. For example, photosynthetic algae in the early oceans first released O_2 into the atmosphere. Ocean photosynthetic microbiota still provide 70 percent of our atmosphere's oxygen.

The constancy of Earth's atmosphere over 3.6 billion years, with its high O_2 and low CO_2 content and moderate temperatures, suggests some feedback system. It prompted James Lovelock, physical scientist, engineer, and inventor of instruments to measure the Martian environment, and microbiologist Lynn Margulis to postulate the Gaia hypothesis (*Gaia* is the Greek word for "Earth goddess") of global biogeochemical homeostasis. The theory says that Earth's biosphere, atmosphere, oceans, and soil together make up a feedback system that maintains an optimal physical and chemical environment for life on Earth. This feedback

system could not have developed nor be maintained without the critical buffering activity of early life forms and continued coordinated activity of plants and photosynthetic protists and monera. Together they damp the fluctuations of the physical environment that would occur in the absence of a well-organized living system.

No control mechanisms have been discovered, but microorganisms are the only life forms that could function like a chemostat, making Earth one large cybernetic system. For example, maintenance of 21 percent O_2 in the atmosphere—which maximizes aerobic metabolism just below the level that would make Earth's vegetation flammable—is possibly the outcome of microbial activity. Microbial production of CH_4 from the small amount of carbonaceous living matter buried each year might keep oxygen in check.

The Gaia hypothesis has not been accepted by all ecologists and atmospheric scientists, but it does help us to understand the behavior of ecosystems and the interactions of biogeochemical cycling. Evidence seems to indicate that organisms do play a dynamic role in determining the composition of many chemicals in the soil, water, and atmosphere. We need to look no further than the tremendous impact we humans have had on the physical aspects of Earth.

atmospheric reservoir into green plants, the primary producer reservoir. In the next flux herbivores eat the plants. The nutrient flows to the consumer reservoir. There are fluxes from both primary producers and consumers to the dead organic matter reservoir. Decomposers feeding on dead organic matter transform the nutrient from organic form to mineral form, transferring it to the soil reservoir. The next link is a flux from the soil to the primary producer reservoir. This flux represents the uptake of the nutrient by the root systems of green plants. There is also a flux from dead organic matter to the atmosphere. This flux represents a release of gases during decomposition.

In this nutrient cycle, there are two major fluxes to the primary producers, from the atmosphere and from the soil. The uptake of the nutrient from the atmosphere is an input to the ecosystem; the flux from

the dead organic matter to the atmosphere is an output from the ecosystem. Within the ecosystem, the nutrients cycle from primary producers to consumers to dead organic matter and decomposers, then to the soil and back to the plants (see Chapter 9). This movement or cycling of nutrients through the components of the ecosystem is **internal cycling.** The rate at which this cycle occurs depends on a variety of factors, including rates of uptake by primary producers and rates of release of nutrients from dead organic material. The latter is controlled by rates of decomposition and transformation of nutrients from organic to inorganic form.

The nutrient cycle in Figure 25.1 is a general representation. As we shall see, specific reservoirs and fluxes vary, depending on the nutrient and the type of cycle, gaseous or sedimentary.

25.3 The carbon cycle is closely tied to energy flow

Carbon is a basic constituent of all organic compounds and is involved in the fixation of energy by photosynthesis (see Section 3.2). Carbon is so closely tied to energy flow that the two are inseparable. In fact, we express ecosystem productivity in terms of grams of carbon fixed per square meter per year (see Chapter 23).

The source of all carbon in both living organisms and fossil deposits is carbon dioxide (CO_2) in the atmosphere and in the waters of Earth. Photosynthesis draws CO_2 from the air and water into the living component of the ecosystem (see Chapter 3). Just as energy flows through the grazing food chain, carbon passes to herbivores and then to carnivores. Primary producers and consumers release carbon back to the atmosphere in the form of CO_2 by respiration.

The carbon in plant and animal tissue eventually goes to the dead organic matter reservoir. Decomposers release it to the atmosphere through respiration.

In Figure 25.2 (a and b) we see the cycling of carbon through a terrestrial ecosystem. The difference between the carbon plants take up in photosynthesis and release by respiration is the net primary productivity (in units of carbon). The difference between net primary productivity and carbon lost through consumer and decomposer respiration is the net ecosystem productivity.

The rate at which carbon cycles through the ecosystem is determined by a number of processes, particularly the rates of primary productivity and decomposition. Both processes are strongly influenced by environmental conditions such as temperature and precipitation (see Sections 23.6 and 23.11). In warm, wet ecosystems such as a tropical rain forest, rates of productivity and decomposition are high, and carbon cycles through the ecosystem quickly. In cool, dry ecosystems, the process is slower. In ecosystems where temperatures are very low, decomposition is slow and dead organic matter accumulates (see Section 23.11 and Chapter 29). In swamps and marshes, where dead material falls into the water, organic material does not completely decompose. Stored as raw humus or peat (see Chapter 33), carbon circulates very slowly. Over geologic time, this buildup of partially decomposed

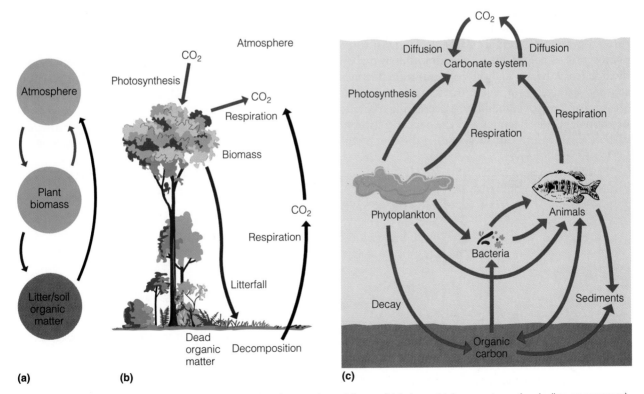

Figure 25.2 The carbon cycle (a) as a diagram showing pools and fluxes, (b) in terrestrial ecosystems (excluding consumers), and (c) in aquatic ecosystems.

organic matter in swamps and marshes has formed fossil fuels (oil, coal, and natural gas).

Similar cycling takes place in freshwater and marine environments (Figure 25.2c). Phytoplankton uses the carbon dioxide that diffuses into the upper layers of water or is present as carbonates and converts it into plant tissue. The carbon then passes from the primary producers through the aquatic food chain. The carbon dioxide produced through respiration is either reutilized or reintroduced to the atmosphere by diffusion from the water surface to the surrounding air.

Significant portions of carbon can be bound as carbonates in the bodies of mollusks and foraminifers. Some of these carbonates dissolve back into solution, while some become buried in the bottom mud at varying depths when the organisms die. Isolated from biotic activity, this carbon is removed from cycling. Incorporated into bottom sediments, over geologic time it may appear in coral reefs and limestone rocks.

25.4   The cycling of carbon varies daily and seasonally

If you were to measure the concentration of carbon dioxide in the atmosphere above and within a forest on a summer day, you would discover that it fluctuates throughout the day (Figure 25.3). At daylight, when photosynthesis begins, plants start to withdraw carbon dioxide from the air, and the concentration declines sharply. By afternoon, when the temperature is increasing and relative humidity is decreasing, the rate of photosynthesis declines and the concentration of carbon dioxide in the air surrounding the canopy in-

creases. By sunset, photosynthesis ceases, carbon dioxide no longer is being withdrawn from the atmosphere, respiration increases, and the atmospheric concentration of carbon dioxide increases sharply. A similar diurnal fluctuation takes place in aquatic ecosystems.

Likewise, there is a seasonal fluctuation in the production and utilization of carbon dioxide that relates both to temperature and to the timing of the growing and dormant seasons (Figure 25.4). With the onset of the growing season, when the landscape is greening, the atmospheric concentration begins to drop as plants withdraw carbon dioxide through photosynthesis. As the growing season reaches its end, photosynthesis declines or ceases, respiration is the dominant process, and atmospheric concentrations of carbon dioxide rise. Although these patterns of seasonal rise and decline occur in both aquatic and terrestrial ecosystems, the fluctuations are much greater in terrestrial environments. As a result, these fluctuations are more pronounced in the Northern Hemisphere with its much larger land area.

25.5   The oxygen cycle is largely under biological control

The major source of free oxygen (O_2) that supports life is the atmosphere. There are two significant sources of atmospheric oxygen. One is the breakup of water vapor through a process driven by sunlight. In this reaction, the water molecules (H_2O) are disassociated to produce hydrogen and oxygen. Most of the hydrogen escapes into space. If the hydrogen did not escape, it would recombine with the oxygen to form water vapor again.

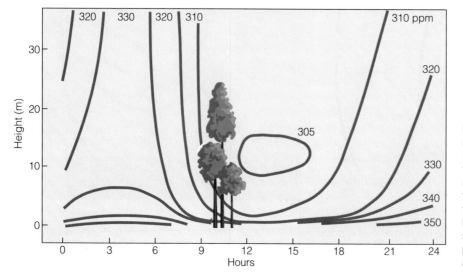

Figure 25.3 Daily flux of CO_2 in a forest. Note the consistently high level of CO_2 on the forest floor, the site of microbial respiration. Atmospheric CO_2 in the forest is lowest from midmorning to late afternoon. CO_2 levels are highest at night, when photosynthesis shuts down and respiration pumps CO_2 into the atmosphere.

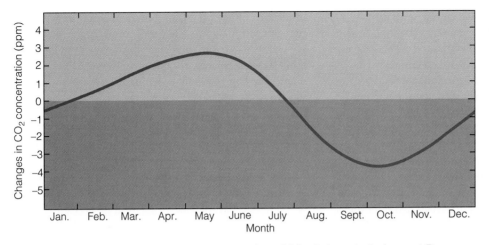

Figure 25.4 Variation in atmospheric concentration of CO_2 during a typical year at Barrow, Alaska.

The other oxygenic source is photosynthesis, active only since life began on Earth. Oxygen is produced by green plants (see Section 3.2) and consumed by both plants and animals (see Section 3.6). Because oxygenic photosynthesis and aerobic respiration involve the alternate release and utilization of oxygen, one would seem to balance the other, so no significant quantity of oxygen would accumulate in the atmosphere. Nevertheless, at some time in Earth's history, the amount of oxygen introduced into the atmosphere had to exceed the amount taken up in respiration (including the decay of organic matter) and geological processes, such as the oxidation of sedimentary rocks. Part of the oxygen present in the atmosphere is from the past imbalance between photosynthesis and respiration in plants. Undecomposed organic matter in the form of fossil fuels and carbon in sedimentary rocks represent a net positive flux of oxygen to the atmosphere. The amount of stored carbon suggests that 150×10^{20} grams of oxygen has been available to the atmosphere, fifteen times as much as is now present (10×10^{20} grams).

The other main reservoirs of oxygen are water and carbon dioxide. All the reservoirs are linked through photosynthesis. Oxygen is also biologically exchangeable in such compounds as nitrates and sulfates, which organisms transform to ammonia and hydrogen sulfide.

Because oxygen is so reactive, its cycling in the ecosystem is complex. As a constituent of carbon dioxide, it circulates throughout the ecosystem. Some carbon dioxide combines with calcium to form carbonates. Oxygen combines with nitrogen compounds to form nitrates, with iron to form ferric oxides, and with other minerals to form various oxides. In these states, oxygen is temporarily withdrawn from circula-

tion. In photosynthesis, the oxygen freed is split from the water molecule. The oxygen is then reconstituted into water during plant and animal respiration. Part of the atmospheric oxygen is reduced to ozone (O_3) by high-energy ultraviolet radiation.

25.6 Pollutants influence the dynamics of ozone

Ozone (O_3) is an ambivalent atmospheric gas. In the stratosphere, 10 to 40 km above Earth, it shields the planet from biologically harmful ultraviolet radiation. Close to the ground ozone is a damaging pollutant, cutting visibility, irritating eyes and respiratory systems, and injuring or killing plant life. In the stratosphere, ozone is diminished by its reaction with human-caused pollutants. In the troposphere, ozone is born from the union of nitrogen oxides with oxygen in the presence of sunlight.

A cycling reaction requiring sunlight maintains ozone in the stratosphere. Solar radiation breaks the O—O bond in O_2. Freed oxygen atoms rapidly combine with O_2 to form O_3. At the same time a reverse reaction consumes ozone to form O and O_2. Under natural conditions in the stratosphere, a balance exists between the rates of ozone formation and destruction. In recent times, however, a number of human-caused and some biologically derived catalysts injected into the stratosphere have been reactive enough to reduce stratospheric ozone. Among them are chlorofluorocarbons (CFCs), methane (CH_4), both natural and human-caused, and nitrous oxide from denitrification and synthetic nitrogen fertilizer. Of particular concern is chlorine monoxide (ClO) derived from chlorofluorocarbons in aerosol spray propellants (banned in the United States), refrigerants, solvents, and other sources. This form of chlorine can break down ozone.

In 1985 atmospheric scientists discovered a pronounced springtime thinning in the ozone layer over Antarctica. They questioned whether the thinning is due to pollution or is a natural effect of upper-level winds. Detection of chlorine monoxide in the hole points to pollution. Reduction of the ozone layer has adverse ecological effects on Earth. It alters DNA and increases skin cancer (see Section 5.6). In addition, changes in the ozone layer could increase the temperature of the lower atmosphere, change air circulation patterns, and contribute to the greenhouse effect (see Chapter 26).

Down in the troposphere, nitrogen oxides and volatile hydrocarbons react with O_2 in the presence of sunlight to form ozone and a number of secondary pollutants, such as formaldehydes, aldehydes, and peroxacetylnitrates, known as PAN. All of these substances collectively form photochemical smog.

Ozone, relatively insoluble in water, readily diffuses through stomatal cavities, where it reacts rapidly. The degree of reaction varies among crop plants and forest trees. Highly sensitive tobacco becomes flecked with white lesions; bean leaves show stippling and bleached areas; leaves of woody plants have reddish-brown lesions. Plants especially sensitive to ozone include white pine and ponderosa pine, red spruce, alfalfa, oats, spinach, and tomato. Foliar damage reduces the photosynthetic capacity of plants, annual radial growth in trees, and nutrient retention in foliage, predisposes trees to insect and fungal infections, and is associated with forest decline.

Other pollutants, especially PAN, are extremely toxic. By destroying some of the lower epidermal and internal cells of leaves, they interfere with plants' metabolic processes and reduce growth. Damage to crop plants in the United States alone is an estimated 2 billion dollars a year.

25.7 The nitrogen cycle begins with fixing atmospheric nitrogen

Nitrogen is an essential constituent of protein, which is a building block of all living tissue. It is also the major constituent (79 percent) of the atmosphere. The paradox is that in its gaseous state (N_2), nitrogen, abundant though it is, is unavailable to most life. Before it can be utilized, nitrogen must be converted into some

other form. Getting it into that form comprises a major part of the nitrogen cycle.

To be used, free molecular nitrogen has to be fixed. This **fixation** comes about in two ways. One is high-energy fixation. Cosmic radiation, meteorite trails, and lightning provide the high energy needed to combine nitrogen with the oxygen and hydrogen of water. The resulting ammonia and nitrates are carried to Earth's surface in rain water. Estimates suggest that less than 8.9 kg N/ha comes to Earth annually in this manner. About two-thirds of this amount comes as ammonia and one-third as nitric acid.

The second method of fixation is biological (Figure 25.5). This method produces 100 to 200 kg N/ha, or roughly 90 percent of the fixed nitrogen contributed to Earth each year. This fixation is accomplished by symbiotic bacteria living in association with leguminous and root-noduled nonleguminous plants (see Section 17.13), by free-living aerobic bacteria, and by cyanobacteria (blue-green algae). Fixation splits molecular nitrogen (N_2) into two atoms of free N. The free N atoms then combine with hydrogen to form two molecules of ammonia (NH_3). This process requires considerable energy. To fix 1 g of nitrogen, nitrogen-fixing bacteria associated with the root system of a leguminous plant must expend about 10 g of glucose, a simple sugar produced in photosynthesis.

In agricultural ecosystems legumes of approximately 200 species are the preeminent nitrogen fixers. In nonagricultural systems some 12,000 species, from cyanobacteria to nodule-bearing plants, are responsible for nitrogen fixation. Also contributing to the fixation of nitrogen are free-living soil bacteria. The most prominent of the 15 known genera are the aerobic

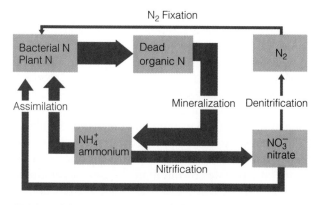

Figure 25.5 The bacterial processes involved in the nitrogen cycle. The widths of arrows approximately represent the relative rates of the processes.

Azotobacter and the anaerobic *Clostridium.* Cyanobacteria (blue-green algae) are another important group of largely nonsymbiotic nitrogen fixers. Of some 40 known species, the most common are in the genera *Nostoc* and *Calothrix,* which are found both in soil and in aquatic habitats. Certain lichens *(Collema tunaeforme* and *Peltigera rufescens)* are also implicated in nitrogen fixation. Lichens with nitrogen-fixing ability possess nitrogen-fixing cyanobacteria as their algal component.

Another source of nitrogen is organic matter. Dead organic matter broken down by decomposition releases nitrogen into the ecosystem in the form of nitrates and ammonia. All of these nitrogen compounds are involved in another phase of the nitrogen cycle: the processes of ammonification, nitrification, and denitrification.

In **ammonification** decomposers break down the amino acids in dead organic material to release energy. It is a one-way reaction. Ammonia is then directly absorbed by plant roots and incorporated into amino acids, which pass through the food chain.

Nitrification is a biological process in which ammonia is oxidized to nitrite and nitrate, yielding energy. Two groups of microorganisms are involved. *Nitrosomonas* bacteria utilize the ammonia in the soil as their sole source of energy. They promote its transformation into nitrite and water. Nitrite is then further transformed into nitrate by another group of bacteria, *Nitrobacter.*

Nitrogen in the form of nitrate is transformed through the process of **denitrification** into gaseous nitrogen by denitrifiers, represented by fungi and *Pseudomonas* bacteria. Like nitrification, denitrification takes place under certain conditions: a sufficient supply of oxygen, a pH range from 6 to 7, and an optimum temperature of 60° C.

Combining these basic and necessary processes, we can construct the nitrogen cycle (Figure 25.6). The sources of nitrogen under natural conditions are the biological fixation of atmospheric nitrogen; additions of inorganic nitrogen in rainfall from such sources as lightning and volcanic activity; ammonia absorption from the atmosphere by plants and soil; and nitrogen accretion from windblown aerosols, which contain both organic and inorganic forms of nitrogen.

In terrestrial ecosystem nitrogen, largely in the form of ammonia or nitrates, is taken up by plants, which convert it to amino acids. The amino acids are transferred to consumers, which convert them to different types of amino acids. Eventually, animal wastes and dead plant and animal tissue are broken down by

bacteria and fungi into ammonia. Ammonia may be lost as gas to the atmosphere, acted upon by nitrifying bacteria, or taken up directly by plants. Nitrates may be utilized by plants, immobilized by microbes (see Section 9.4), stored in decomposing organic matter, or leached away. Leached material runs off to streams, lakes, and eventually the sea, where it is available for use in aquatic ecosystems.

In aquatic ecosystems, nitrogen cycles in a similar manner, except that the large reservoir in the soil is lacking. Life in the water contributes organic matter and dead organisms that undergo decomposition, releasing ammonia and nitrates (see Chapters 32–34).

Under natural conditions nitrogen lost from ecosystems by denitrification, volatilization, leaching, erosion, and windblown aerosols is balanced by biological fixation and other sources.

25.8 Human-made emissions of nitrogen act as pollutants

Nitrogen is one of the most essential nutrients for life, and it increases the fertility of soil and water. However, it is an important catalyst for the formation of ozone, it contributes to acid precipitation, and it damages human health. Nitrogen has become a pollutant because of human intrusion into the natural cycle and balance of nitrogen. Humans pollute the atmosphere with various oxides of nitrogen; they pollute the water with nitrates leached from the soil.

Natural leaching of nitrates and natural input of nitrogen oxides into the atmosphere have always occurred. During nitrification and denitrification microorganisms in marine, freshwater aquatic, and terrestrial ecosystems release nitrous oxide (N_2O) to the atmosphere. Tropical forests and woodlands alone release three-fourths of the natural global flux of nitrous oxide. However, for several decades nitrous oxides have been increasing at the rate of 0.2 to 0.3 percent a year. Most of this increase comes from human activity.

Major human sources of nitrogen pollution are agriculture, industry, and automobiles. The first major intrusion probably came from agriculture, when people began burning forests and clearing land for crops and pasture. Conversion of natural grasslands into grain fields has caused a steady decline in the nitrogen content of their soils. Breaking up and mixing soil increases the rate of decomposition of deep organic matter, releasing nitrates and nitrous oxides. Harvesting crops and logging result in a heavy loss of nitrogen from agricultural and forest ecosystems, not only

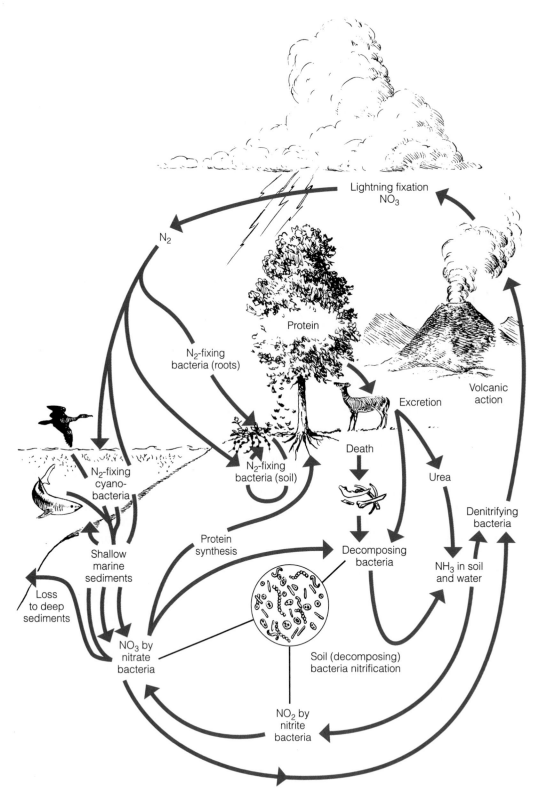

Figure 25.6 The nitrogen cycle.

in the material removed but in nitrate losses from the soil. Heavy application of chemical fertilizers to croplands disturbs the natural balance between denitrification and nitrogen fixation. A considerable portion of nitrogen fertilizers is lost as nitrates to groundwater and runoff, which find a way into aquatic ecosystems, reducing water quality and harming aquatic life (see Chapter 32). Excess nitrogen not leached out as nitrates is removed by microbial denitrification, increasing atmospheric levels of nitrous oxide. About 10 percent of applied nitrogen fertilizers in time evaporates into the atmosphere. Added to these inputs are nitrogenous inputs from animal wastes at concentrated livestock feeding yards, from municipal sewage treatment plants, and from chemical fertilizer plants.

Automobile exhaust and industrial high-temperature combustion add nitrous oxide (N_2O), and nitric oxide (NO), and nitrogen dioxide (NO_2). These oxides are relatively unreactive. They can reside in the atmosphere for 20 years, drifting slowly up to the strato-sphere. There ultraviolet light reduces nitrous oxide to nitric oxide and atomic oxygen (O). Atomic oxygen reacts with oxygen (O_2) to form ozone (O_3).

One outcome of this atmospheric pollution is an increased deposition of nitrogen, which benefits ecosystems that are traditionally nitrogen-limited, especially northern and high-altitude forests. Such ecosystems, however, can suffer from too much of a good thing. Because nitrogen is limiting, these forests are efficient at retaining and recycling nitrogen from precipitation and organic matter. Only a few lose a significant amount of nitrates to streams. Now many of these forests are receiving more nitrogen in the form of ammonium and nitrates than the trees and their associated microbial populations can handle and accumulate.

The first response to increased availability of nitrogen in a nitrogen-limited ecosystem is increased growth. Evidence suggests, however, that mounting levels of nitrogen lead to the decline and dieback of coniferous forests at high elevations (Figure 25.7). If

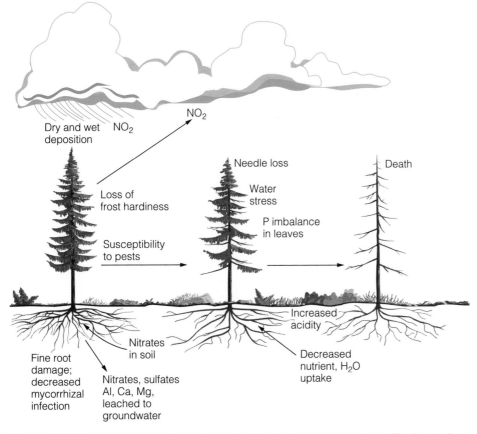

Figure 25.7 Effects of excess nitrogen in nutrient-limited forest ecosystems. The forest dies back, and the ecosystem changes from a nitrogen sink to a nitrogen source.

the increased growth in foliage continues into the summer, the late new growth may not have time to become frost-hardened, and it is killed during the winter. Overstimulated by nitrogen, tree growth exceeds the availability of other necessary nutrients in the soil, particularly phosphorus, and the tree begins to experience nutrient deficiencies. Experimental evidence suggests that the production of fine roots and mycorrhizae, which take up nutrients from the soil, is lower on sites rich in nutrients, especially nitrogen, than in nutrient-poor soils, and root turnover is higher. Trees on nutrient-poor soils have a longer-lived and higher density root system and a lower turnover of root biomass, a condition that helps them to scavenge nutrients from poor soils. Thus as nitrogen levels increase, root biomass decreases, further inhibiting the uptake of nutrients other than nitrogen and impairing the ability of trees to pull water from the soil during periods of drought.

25.9   The phosphorus cycle has no atmospheric reservoir

Phosphorus is unknown in the atmosphere, and none of its known compounds has an appreciable vapor pressure. It can follow the hydrological cycle only part of the way, from land to sea (Figure 25.8). Because phosphorus lost from the ecosystem in this fashion is not returned via the biogeochemical cycle, under undisturbed natural conditions phosphorus is in short supply. Phosphorus's natural scarcity in aquatic ecosystems is emphasized by the explosive growth of algae in water receiving heavy discharges of phosphorus-rich wastes.

The main reservoirs of phosphorus are rock and natural phosphate deposits, from which the elements are released by weathering, leaching, erosion, and mining for agricultural use. Some of it passes through terrestrial and aquatic ecosystems in the grazing food chain—plants, grazers, predators, and parasites—and is cycled back to soil and water by excretion, death, and decomposition. In terrestrial ecosystems organic phosphates unavailable to plants are transformed by bacteria to inorganic phosphates. Some inorganic phosphates are recycled to plants, some become transformed by various chemical processes into unavailable compounds, and some are immobilized in the bodies of microorganisms. Some of the phosphorus of terrestrial ecosystems escapes to lakes and seas.

In marine and freshwater ecosystems, the phosphorus cycle moves through three states: particulate organic phosphorus, dissolved organic phosphates, and inorganic phosphates. Organic phosphates are taken up quickly by all forms of phytoplankton and eaten in turn by zooplankton and detritus-feeding organisms

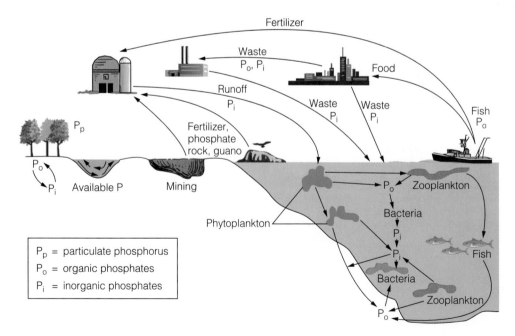

Figure 25.8 The phosphorus cycle in aquatic and terrestrial ecosystems.

(see also Chapter 35). Zooplankton may excrete as much phosphorus daily as it stores in its biomass, returning it to the cycle. More than half of the phosphorus zooplankton excretes is inorganic phosphate, which is taken up by phytoplankton. The remainder of the phosphorus in aquatic ecosystems is in organic compounds that may be utilized by bacteria, which fail to regenerate much dissolved inorganic phosphate. Bacteria are consumed by the microbial grazers, which then excrete the phosphate they ingest. Part of the phosphate is deposited in shallow sediments and part in deep water. In the ocean upwelling, the movement of deep waters to the surface, brings some phosphates from the dark depths to shallow waters where light is available to drive the process of photosynthesis. These phosphates are taken up by phytoplankton. Part of the phosphorus contained in the bodies of plants and animals sinks to the bottom and is deposited in the sediments. As a result, surface waters may become depleted of phosphorus and the deep waters saturated. Much of this phosphorus becomes locked up for long periods of time in the bottom sediments, while some is returned to the surface waters by upwelling.

Humans have altered the phosphorus cycle through the disposal of waste products and the application of phosphate to cropland. Some of the phosphate applied as fertilizer reacts with calcium, iron, and aluminum in the soil and becomes immobilized as insoluble salts. Another part goes along with the harvested crop. Transported far from the point of origin, the phosphorus in vegetables and grains is released as waste when food is processed or consumed. Concentrations of phosphorus in the wastes of food-processing plants and feedlots add excessive phosphates to natural waters. Greater quantities are released in urban areas; elsewhere phosphates concentrate in sewage systems. Primary sewage treatment in only 30 percent effective in removing phosphorus, so 70 percent remains in the effluent and is added to waterways. In aquatic ecosystems vegetation takes up phosphorus rapidly, causing a sudden increase in algal biomass. Eventually all the phosphorus mobilized by humans becomes immobilized in the soil or in the bottom sediments of ponds, lakes, and seas.

25.10   The sulfur cycle is both sedimentary and gaseous

The sulfur cycle has both sedimentary and gaseous phases (Figure 25.9). In the long-term sedimentary phase sulfur is tied up in organic and inorganic deposits, released by weathering and decomposition, and carried to terrestrial ecosystems in salt solution. The

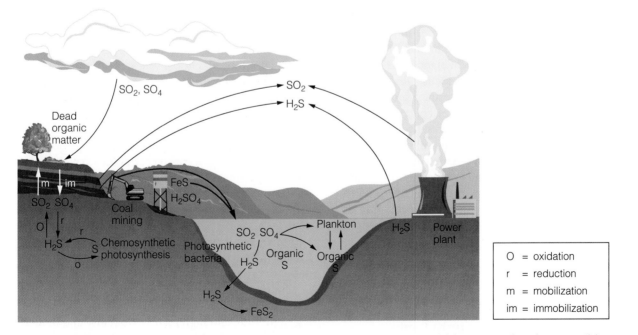

Figure 25.9 The sulfur cycle. Note the two components, sedimentary and gaseous. Major sources from human activity are the burning of fossil fuels and acidic drainage from coal mines.

gaseous phase of the cycle permits the circulation of sulfur on a global scale.

Sulfur enters the atmosphere from several sources: the combustion of fossil fuels, volcanic eruptions, exchange at the surface of the oceans, and gases released by decomposition. It enters the atmosphere initially as hydrogen sulfide, H_2S, which quickly interacts with oxygen to form sulfur dioxide, SO_2. Atmospheric sulfur dioxide, soluble in water, is carried back to the surface in rain water as weak sulfuric acid, H_2SO_4. Whatever the source, sulfur in a soluble form is taken up by plants and incorporated through a series of metabolic processes, starting with photosynthesis, into sulfur-bearing amino acids. From the producers, sulfur in amino acid is transferred to consumers.

Excretion and death carry sulfur from living material back to the soil and to the bottom of ponds, lakes, and seas, where bacteria release it as hydrogen sulfide or sulfate. One group, the colorless sulfur bacteria, both reduces hydrogen sulfide to elemental sulfur and oxidizes it to sulfuric acid. Green and purple bacteria, in the presence of light, utilize hydrogen sulfide in the process of photosynthesis. Best known are the purple bacteria found in salt marshes and in the mud flats of estuaries. These organisms are able to transform hydrogen sulfide into sulfate, which is then recirculated and taken up by producers or used by bacteria that further transform the sulfates. Green bacteria can transform hydrogen sulfide into elemental sulfur.

Sulfur, in the presence of iron and under anaerobic conditions, will precipitate as ferrous sulfide, FeS_2. This compound is highly insoluble in neutral and low pH conditions, and it is firmly held in mud and wet soil. Sedimentary rocks containing ferrous sulfide, called pyritic rocks, may overlie coal deposits. Exposed to air during deep and surface mining for coal, the ferrous sulfide reacts with oxygen. In the presence of water it produces ferrous sulfate ($FeSO_4$) and sulfuric acid.

In this manner, sulfur in pyritic rocks, suddenly exposed to weathering by human activities, discharges sulfuric acid, ferrous sulfate, and other sulfur compounds into aquatic ecosystems. These compounds destroy aquatic life. They have converted hundreds of kilometers of streams in the eastern United States to lifeless, highly acidic water.

25.11   Human-made sulfur dioxide is a major air pollutant

Of all the atmospheric gases, sulfur dioxide (SO_2) is most strongly implicated in air pollution. Major sources fall into two categories, natural and human-made. Natural sources include microbial activity, volcanoes, sea spray, and weathering. These sources make up about 60 percent of sulfur emissions to the atmosphere. Human-made emissions comprise the remaining 40 percent.

Of the human-made emissions, 68 percent comes from the burning of fossil fuels, 40 percent of that from the burning of coal. Natural emissions are widely distributed about the globe; human-made emissions are concentrated regionally. Ninety percent of those inputs come from the urban and industrialized areas of Europe, North America, India, and the Far East. Europe and North America release between 110 and 125 million metric tons of sulfur into the atmosphere every year. In these regions industrial input far exceeds natural inputs. Because of the imposition of emission controls, new technologies, and changes in patterns of fuel consumption, sulfur dioxide emissions have been declining. However, if coal consumption increases as predicted, sulfur dioxide levels could rise above former levels.

Sulfur dioxide produces acute toxicity and major damage to vegetation in areas surrounding the source of emission. Examples are the destruction of forest cover in the vicinity of iron smelters at Sudbury and Wana, Ontario, and the copper smelters at Duckberry and Copperhill, Tennessee. Damage away from point sources is more long-term and insidious. Tall stacks at power plants and industrial complexes have greatly reduced local damage, but they eject into the atmosphere gases that are carried by upper-level winds to points far removed. On the way some of the gases and particulate matter drop out close to the source. In the moving plume, sulfur dioxide combines with atmospheric moisture to form sulfuric acid, which falls on land and water and forms a significant part of acid rain.

Sulfur dioxide in the atmosphere affects both humans and plants. In a concentration of a few parts per million, it irritates the respiratory tract. In a fine mist or adsorbed onto small particles, sulfur dioxide moves into the lungs and attacks sensitive tissues. High concentrations (over 1000 millimicrons/m³) have caused a number of air pollution disasters characterized by higher than expected death rates and increased incidence of bronchial asthma.

Sulfur dioxide injures exposed plants or kills them outright. Acidic aerosols present during periods of fog, light rain, and high relative humidity and moderate temperatures do most of the injury. External surfaces of leaves absorb the aerosols. When dry, leaves and needles take up sulfur dioxide through the stomata. In the leaf it rapidly reacts with moisture to form sulfuric acid. Symptoms of sulfur damage are a bleached look to deciduous leaves and red-brown needles on conifers, partial defoliation, and reduced growth.

25.12   The sulfur and nitrogen cycles produce acid deposition

The gaseous sulfur dioxide component of the sulfur cycle and the nitrogen oxides of the nitrogen cycle mix in the atmosphere. Some of this mixture returns to Earth as particulate matter and airborne gases, known as **dry deposition.** A major portion is transported away from the source. The direction it takes is strongly influenced by the general atmospheric circulation. During their atmospheric transport SO_2 and NO_2 and their oxidative products participate in complex reactions involving hydrogen chloride and other compounds, oxygen, and water vapor. These reactions dilute solutions of strong acids, notably nitric acid and sulfuric acid (Figure 25.10). Eventually they come to Earth in acidic rain, snow, and fog, known as **wet deposition.**

Rain water is naturally acidic. Unpolluted rain water, considered pure water, has a pH of 5.6, but rarely is rain water pure. Even in regions not subject to industrial pollution, atmospheric moisture is exposed to varying amounts of acids of natural origin, so precipitation has a pH of about 5. In regions extending hundreds of kilometers about centers of human activity, however, the pH of precipitation is much lower, 3.5 to 4.5, or occasionally lower still.

Acid rain has been with us for well over a century. In 1872, when the Industrial Revolution was in full swing, Robert Angus Smith introduced the term to describe the rainfall around Manchester, England. In some areas acid rain was probably worse than now, but it was more localized. Only when we constructed large stacks in industrial and power plants, sending emissions much higher into atmospheric circulation, did acid rain become a regional problem. The industrial midwestern United States and Ohio River valley send acidic pollutants to eastern Canada and the northeastern United States; eastern Canada sends them to the northeastern United States; and the industrialized regions of central Europe and the United Kingdom send them to Scandinavia.

Little evidence exists to show that acid rain has a direct effect on most plants. Acid rain, intercepted by vegetation, leaches nutrients, particularly calcium,

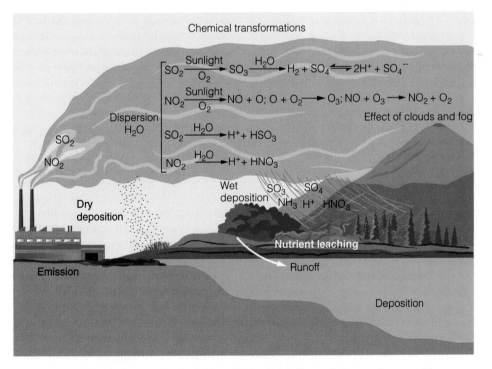

Figure 25.10 Formation of acid deposition. Excessive sulfur and nitrogen in several forms are poured into the atmosphere. They are then converted to sulfites (SO_3), sulfates (SO_4), sulfuric acid (H_2SO_4), and nitric acid (HNO_3) and carried to Earth.

magnesium, and potassium, from the leaves and needles. Such leaching, a normal process in nutrient cycling, has little effect on trees' health, provided the trees can replace the lost nutrients by uptake from the soil. Forests at high elevations, however, are frequently enveloped by mists and fog. Cloud droplets are more acidic and hold higher concentrations of other pollutants than rainfall. When immersed in fog, needle-leaved conifers comb moisture out of the air, and their wet surfaces permit the uptake of the pollutants it contains. As the water evaporates, it leaves behind high concentrations of pollutants, some of which wash off during the next rain and fall on the soil. Added to this wet deposition is a dusting of dry deposition that leaches nutrients from the leaves.

25.13   Acid deposition impacts soils and aquatic systems

Acid precipitation has its greatest impact on soils that are low in cations and poorly buffered. Such soils, mostly podzolic (see Chapter 10) and derived largely from granitic bedrock, are characteristic of the eastern and upper north-central United States, the southeastern United States, Canada, and northern Europe. Although not yet affected, parts of Asia, Africa, and South America are also acid-sensitive. In all these regions, terrestrial ecosystems are nutrient-poor and their soils acidic. Over a period of time acid precipitation can have adverse effects. It increases the leaching of calcium, magnesium, and potassium from the receiving soil and replaces these cations with hydrogen ions, further increasing soil acidity. Acid precipitation can reduce the solubility and availability of phosphorus and the rate of nitrogen fixation. If the rate of leaching outstrips the replacement of these nutrients by weathering, acid precipitation upsets the nutrient balance of trees and other vegetation. Further, acid rain can inhibit the activity of fungi and bacteria in the soil. By so doing it reduces the rates of humus production, mineralization, and fixation of nutrients. All of these interactions cause nutrient-deficient soils.

The most serious effect, however, is the mobilization of toxic elements in the soil, particularly aluminum and manganese, as they are replaced by hydrogen ions on soil particles (see Sections 9.9, 10.7). Aluminum affects the structure and function of fine roots and interferes with the uptake of calcium from the soil. It also suppresses growth of the cambium in trees, which in turn reduces the formation of new sapwood. As sapwood growth in conifers declines, the ratio between living sapwood and dead heartwood declines. When sapwood forms less than 25 percent of the cross section of a tree, that tree succumbs (see Focus on Ecology 25.2: Air Pollution and Forest Decline).

Acid deposition has its most pronounced effects on aquatic ecosystems in acid-sensitive regions. Acidic inputs to aquatic ecosystems come directly from rainfall and snowmelt and indirectly from the soils of the surrounding watershed. Acidic water leached from soils increases the nutrient levels of streams and lakes in a watershed, provided the soil has appreciable reserves of calcium (see Chapter 32). As acid rain percolates through the soil, it is neutralized while releasing basic ions and carrying them to streams and lakes. Such enrichment, however, is often canceled by snowmelt and spring rain water flowing over the surface of frozen ground, following old root channels and animal burrows into receiving waters. Such inflow discharges much of the winter precipitation in a slug of acidic waters. The receiving water then can become acidic in spite of the buffering effects of the soil.

In the water, sulfate and nitrogen ions replace bicarbonate ions, pH declines, and the concentration of metallic ions increases. When the pH of the groundwater and surface water in the surrounding watershed is 5 or lower, high concentrations of aluminum ions are carried to lakes and streams. Aluminum then tends to precipitate the dark humics, increasing the transparency of the water. Increased light penetration into the water can stimulate phytoplankton production and the growth of benthic algae and bryophytes, but the number of species and biomass of zooplankton decrease.

Although adult fish and some other aquatic organisms can tolerate high acidity, a combination of high acidity and a high level of aluminum, a typical situation during snowmelt, can kill them. At the level of 0.1 to 0.3 mg/l, aluminum retards growth and gonadal development of fish and increases their mortality.

Filling depressions and temporary ponds with snowmelt and surface runoff, even in regions without acidified lakes and streams, inhibits the reproduction of frogs and salamanders, whose eggs and larvae are sensitive to acidic water. This effect may account in part for the rapid decline of amphibians. Acidic waters are toxic to invertebrates also, either killing them directly or interfering with calcium metabolism, causing crustaceans to lose the ability to recalcify their shells after molting. As recruitment fails and food declines, fish life disappears from the affected waters.

Acidification already has damaged many lakes and streams in the northeastern United States, Canada, Norway, Sweden, and the United Kingdom. Unassisted

FOCUS ON ECOLOGY 25.2

AIR POLLUTION AND FOREST DECLINE

Over much of Europe, forests, especially coniferous ones, are declining and dying. In Germany 80 percent of the fir, 54 percent of both spruce and pine, and 60 percent of beech and oak show symptoms of damage—defoliation, thinned crowns, yellowed and discolored foliage, premature leaf fall, and death. Over central Europe, Norway spruce, pine, beech, and oak suffer from chlorosis and defoliation. Fourteen percent of all Swiss forests, 22 percent of Austria's forests, and 29 percent of Holland's forests show symptoms of decline.

The story is the same in North America. Pine forests in southern California northeast of Los Angeles and San Diego suffer from chlorosis, decrease in radial growth, and death. In the high Appalachians from North Carolina to Vermont, red spruce has been declining in growth and dying since the 1960s. In the Great Smoky Mountains one-half of red spruce and Fraser fir are dead. In the southern United States, loblolly pine and slash pine are experiencing a 30 to 50 percent decline in radial growth and increasing mortality. Sugar maples in the northeastern United States and eastern Canada are showing crown dieback and bark peeling on larger branches. In the province of Quebec dieback affects 52 percent of the sugar maples. Throughout northeastern North America white pine shows needle discoloration and decreased growth in height, diameter, and needle length, all symptoms of ozone damage.

Forest decline is not a new phenomenon. During the past two centuries our forests have experienced several declines, with different species affected. What sets the current decline apart from all others is differences among the symptoms in the past and the similarity of symptoms among species today. Past declines could be attributed to natural stresses, such as drought and disease. What causes forest decline and dieback today is not established, but the widespread similarity of symptoms suggests a common cause, air pollution. All of the affected forests are in the path of pollutants from industrial and urban sources.

Forests close to the point of origin of pollutants experience the most direct effects of air pollution, and their decline and death can be directly attributed to it. Little evidence exists that acid precipitation alone is the cause of forest decline and death at distant points. The effects of acid deposition, however, can so weaken trees that they succumb to other stresses such as drought and insect attack. The stressed stands of Fraser fir in the Great Smoky Mountains are succumbing to the attacks of the introduced balsam woolly adelgid *(Adelgid picea).* The once deep, fragrant stands of Fraser fir, especially on the windward side and peaks of the Great Smoky Mountains, are now stands of skeleton trees (Figure A).

Air pollution and acid rain also are altering succession by changing the species composition of forests. Just as the chestnut blight shifted dominance in the central hardwoods forest from chestnut to oaks, so air pollution is shifting dominance from pines and other conifers to deciduous trees more tolerant of air pollution.

Figure A Forest decline in the Great Smoky Mountains. Air pollution makes these trees vulnerable to the attacks of the balsam woolly adelgid.

biological recovery probably will never happen. Restoration will depend upon restocking components of the original lake community. Even with that help, restoration of the original food chains and community structure will be difficult, if not impossible.

Added to the damage to forests, crops, and aquatic ecosystems is the enormous cost of $75–80 billion per year attributed to the corrosion of bridges, highways, buildings, and monuments exposed to the atmosphere. Iron and its alloys are rusted and weakened by chlorides and sulfides, aluminum by chlorides, copper and its alloys by sulfides, and masonry and marble by sulfides and other atmospheric aerosols, whose corrosive effects appear to be enhanced by ozone and solar radiation.

25.14 Heavy metals also cycle through ecosystems

Heavy metals, such as mercury, cadmium, chromium, and lead, toxic to life in varying amounts, have their own biogeochemical cycles. Metals such as mercury and cadmium are associated with industrial pollution, runoff from agricultural fields, toxic dumps, and landfills. Discharged into rivers or lakes or seeping into groundwater, these elements contaminate water supplies and build up in food chains. The serious problems they create, such as the contamination of fish and adverse effects on reproduction of some birds, are local; but we feel their effects over a much wider area.

Other heavy metals, notably lead, join sulfur and nitrogen in atmospheric circulation, moving great distances from the point of origin. Automobiles burning leaded gasoline poured most of the lead into the air until the use of unleaded fuel was mandated in the United States, Japan, Brazil, and the European Economic Community. Mining, smelting, and refining of lead, lead-consuming industries, coal combustion, burning of refuse and sewage sludge, and the burning and decay of lead-painted surfaces add additional quantities to the atmosphere.

Emitted into the air as very small particles, less than 0.5 μm, lead is widely distributed to all parts of Earth. Areas near point sources of pollution, particularly roadsides, receive the heaviest deposition. Urban areas close to industry and heavy automobile traffic may have a flux rate greater than 3000 g/ha/yr, whereas remote areas may have a flux rate of less than 20 g/ha/yr.

Lead particles settle on the surface of the soil and on vegetation. Forest canopies are particularly efficient at collecting lead from the atmosphere as dry deposition. Lead accumulates in the canopies during the summer and is carried to the ground by rain as throughfall and stemflow and by leaf fall in autumn. In the forest soil, lead becomes bound to organic matter in the litter layer and reacts with sulfate, phosphate, and carbonate anions in the soil. In such an insoluble form, lead moves slowly, if at all, into the lower horizons. It persists in the upper soil layer for about 5000 years.

Once in the soil and on plants, lead enters the food chain. Plant roots take up lead from the soil, and leaves pick it up from contaminated air or from particulate matter on the leaf. Lead is then taken up by herbivorous insects and grazing mammals, who pass it on to higher consumers. This uptake through the food chain is most pronounced along roadsides. Microbial systems also pick up lead and immobilize substantial quantities of it.

The long-term increase in concentrations of atmospheric lead in the industrialized areas of Earth has resulted in significant increases of lead in humans. The average concentration of lead in bodies of adults and children in the United States is 100 times greater than the natural burden, and existing rates of lead absorption are 30 times the level in preindustrial society. An intake of lead can cause mental retardation, palsy, partial paralysis, loss of hearing, and death.

25.15 Chlorinated hydrocarbons establish cycles

Biogeochemical cycles are nearly as old as Earth. Now substances that humans manufacture and release to the environment have established their own biogeochemical cycles that mimic natural ones. These substances include chlorinated hydrocarbons—pesticides (see Section 18.13), dioxin-containing herbicides, and the industrial chemical PCB.

The substance that first received attention was DDT, the dangers of which were widely exposed by Rachel Carson in her book *Silent Spring* (1962). The detection of DDT in the tissues of animals in the Antarctic, far removed from any applied source of the insecticide, emphasized the fact that chlorinated hydrocarbons do indeed establish their own global biogeochemical cycles that disperse them around the Earth. DDT has been banned in the United States and most industrialized nations, but it still is used extensively in other parts of the world, notably Central and South America and Asia. We can use it to illustrate the cycling of chlorinated hydrocarbons.

Chlorinated hydrocarbons have certain characteristics that make global circulation possible (Figure 25.11). They are highly soluble in fats or lipids and poorly soluble in water. Therefore they tend to accumulate in plants and animals. They are persistent and stable, but undergo some degradation. DDT, for example, degrades to DDE and has a half-life of approximately 20 years. It has a vapor pressure high enough to ensure direct losses from plants. It can become ad-

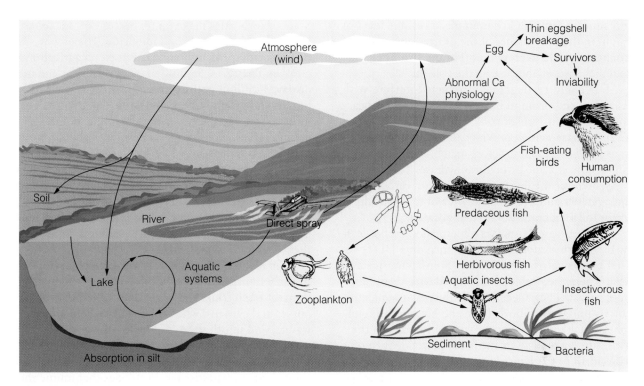

Figure 25.11 The movement of chlorinated hydrocarbons in terrestrial and aquatic systems. The initial input comes from spraying on vegetation. A large portion fails to reach the ground and is carried on water droplets and particulate matter through the atmosphere.

sorbed to particles or remain as a vapor and in either state be transported by atmospheric circulation. It can return to land or sea in rain water.

Insecticides are applied on a large scale by aerial spraying. One-half or more of the toxicant applied in this manner is dispersed to the atmosphere and never reaches the ground. If the vegetative cover is dense, only about 20 percent reaches the ground. On the ground some of the pesticide is lost through volatilization, chemical degradation, bacterial decomposition, and runoff.

DDT sprayed on forests and croplands enters streams and lakes, where it is subject to further distribution and dilution as it moves downstream. Insecticides released in oil solutions penetrate to the bottom and kill fish and aquatic invertebrates. Trapped in the bottom rubble and mud, the insecticide may continue to circulate and kill for some days. DDT in oil solutions floats on the surface and is moved about by the wind. Eventually it reaches the ocean, where it may concentrate in surface slicks. These slicks, which attract plankton, are carried across the seas by ocean currents. In the ocean, part of the DDT residue may circulate in the mixed layer. Some of it may be trans-

ferred to the deep waters, and more may be lost through the sedimentation of organic matter.

The major movement of pesticide residues takes place in the atmosphere. Not only does the atmosphere receive the bulk of the pesticide spray, but it also picks up the fractions volatized from soil, water, and vegetation. The capacity of the atmosphere to hold DDT is increased greatly by the adsorption of residues on fine particulate matter. Thus the atmosphere becomes a large circulating reservoir of DDT and other chlorinated hydrocarbons. Residues are removed from the atmosphere by chemical degradation, diffusion across the air-sea interface, and mostly by rainfall and dry deposition.

The high solubility of DDT in lipids leads to its concentration through the food chain. Most of the DDT contained in ingested food is retained in the fatty tissue of consumers. Because it breaks down slowly, DDT accumulates to high and even toxic levels. DDT so concentrated is passed on to the next trophic level. The carnivores on the top level of the food chain receive massive amounts of pesticides. With high concentrations of DDT in their tissues, consumers die or experience impaired reproduction and genetic damage. For example, a residue level of 5 ppm in the fatty tissues of the ovaries of freshwater trout causes 100 percent dieoff of the fry. They pick up lethal doses as they use up the yolk sac.

DDT and its degraded products interfere with calcium metabolism in birds. Chlorinated hydrocarbons block ion transport by inhibiting the enzyme ATP synthase, which provides the energy needed to transport ionic calcium across membranes. DDT also inhibits the enzyme carbonic anhydrase. This enzyme is essential for the deposition of calcium carbonate in the eggshell and the maintenance of a pH gradient across the membrane of the shell gland.

The ecological problem of pesticides has not lessened. Chlorinated hydrocarbons, including DDT, are used extensively in Central and South America and Asia. There the problems associated with these pesticides persist. They affect not only the year-round fauna but also migratory birds from the Northern Hemisphere that come in contact with pesticides on their wintering grounds. Quantities of pesticides are sent back north to the United States, Canada, and Europe on fruits and vegetables grown for the winter market.

The use of biocides (pesticides and herbicides) other than DDT continues unabated in the United States. Herbicides make up 60 percent of this usage, insecticides 24 percent, and fungicides 16 percent. Of these biocides, 341 million kg are used on agricultural crops and pastures, 55 million by government and industry, 4 million on forests, and a surprising 55 million in and around urban and suburban homes. The most concentrated use is about the home, not in agricultural fields. The dosage of pesticides about homes is 14 kg/ha compared to 3 kg/ha on agricultural crops. As much as one-third of these household pesticides is never used and is thrown into the trash and ultimately into the environment. These excess pesticides, together with losses from croplands and roadsides, expose pests to widespread selective pressures, increasing their resistance to pesticides.

Finally, we must consider the effects on human health. Humans are widely exposed to pesticides when they eat fresh fruits and vegetables, when they spray homes and gardens, when they swim in contaminated waters, and in other ways. Those who work directly with pesticides are particularly at risk. Annually in the United States some 45,000 humans are poisoned to some degree by pesticides. Pesticides that mimic the effects of estrogen may increase human infertility and birth defects.

CHAPTER REVIEW

SUMMARY

Nutrient Cycles (25.1–25.2) Nutrients flow from the living to the nonliving and back to the living parts of the ecosystem in a perpetual cycle. By means of these cycles, plants and animals obtain necessary nutrients. There are two kinds of biogeochemical cycle: the gaseous cycle (represented by oxygen, carbon, and nitrogen) and the sedimentary cycle (represented by phosphorus). The sulfur cycle is a combination of the two. The gaseous cycle involves two main reservoirs, the atmosphere and the oceans, and is pronouncedly global. The sedimentary cycle involves two phases, salt solution and rock. Minerals become available through the weathering of Earth's crust, enter the water cycle as salt solutions, take diverse pathways through the ecosystem, and return to the sea or Earth's crust through sedimentation (25.1).

Nutrients move through the ecosystem by means of exchanges called fluxes between major components or reservoirs. These fluxes in terrestrial ecosystems move nutrients from atmosphere and soil to producers, then to consumers, and then to dead organic matter. Decomposers feeding on dead organic matter transform nutrients from organic to mineral form, transferring them to the soil reservoir. Then they flow back to the primary producers. The movement of nutrients through the ecosystem is internal cycling. The rate of recycling is influenced by the rate of uptake of nutrients by primary producers and the rate of nutrient release by decomposers (25.2).

The Carbon Cycle (25.3–25.4) The carbon cycle is inseparable from energy flow. Carbon is assimilated as carbon dioxide by plants, consumed in the form of plant and animal tissue by heterotrophs, released through respiration, mineralized by decomposers, accumulated in standing biomass, and withdrawn into long-term reserves. The rate at which carbon cycles through the ecosystem depends on the rates of primary productivity and decomposition. Both processes are faster in warm, wet ecosystems. In swamps and marshes, organic material stored as raw humus or peat circulates slowly, forming oil, coal, and natural gas. Similar cycling takes place in freshwater and marine environments (25.3).

Cycling of carbon exhibits daily and seasonal fluctuations. Carbon dioxide builds up at night, when respiration increases. During the day plants withdraw carbon dioxide from the air and its concentration drops sharply. During the growing season atmospheric concentration drops (25.4).

The Oxygen Cycle (25.5–25.6) The oxygen cycle is complex. The major sources of oxygen are the photodissociation of water vapor and photosynthesis. Oxygen also circulates freely as a constituent of carbon dioxide. Active chemically, it combines with a wide range of inorganic chemicals, organic substances, and reduced compounds, and it is involved in the oxidation of carbohydrates, releasing energy, carbon dioxide, and water (25.5).

Ozone (O_3), related to the oxygen cycle, is produced by photochemical reactions in the atmosphere. In the stratosphere ozone is essential to reduce the influx of harmful ultraviolet radiation to Earth. However, human-produced chlorofluorocarbons and other pollutants rise to the stratosphere and destroy this ozone. In the troposphere ozone is created from nitrogen oxides by photochemical reactions. With related pollutants it forms smog, toxic to vegetation (25.6).

The Nitrogen Cycle (25.7–25.8) Atmospheric nitrogen is fixed by physical processes in the atmosphere and by nitrogen-fixing bacteria found in symbiosis with plants, mostly legumes, and cyanobacteria. The nitrogen cycle includes ammonification, nitrification, and denitrification (25.7).

Humans pour nitrogen oxides into the atmosphere and nitrates into aquatic ecosystems. The major source of nitrogen oxides are the automobile and the burning of fossil fuels. Large quantities of nitrates are added to aquatic systems by fertilizers, animal wastes, and sewage effluents. Increased deposition of nitrogen on high-elevation coniferous forests overstimulates growth and produces nutrient deficiencies (25.8).

The Phosphorus Cycle (25.9) The phosphorus cycle is wholly sedimentary. The main reservoirs of phosphorus are rock and natural phosphate deposits. The terrestrial phosphorus cycle follows the typical biogeochemical pathways. In marine and freshwater ecosystems, however, the phosphorus cycle moves through three states: particulate organic phosphorus, dissolved organic phosphates, and inorganic phosphates. Involved in the cycling are phytoplankton, zooplankton, bacteria, and microbial grazers. Humans overload the phosphorus cycle by applying phosphate fertilizers and disposing of wastes. In aquatic systems, this excess contributes to abundant algal growth and water pollution until it is immobilized in bottom sediments.

The Sulfur Cycle (25.10–25.11) Sulfur has both gaseous and sedimentary phases. Sedimentary sulfur comes from the weathering of rocks, erosional runoff, and decomposition of organic matter. Sources of gaseous sulfur are decomposition of organic matter, evaporation of oceans, and volcanic eruptions. A significant portion of the sulfur released to the atmosphere is a by-product of the burning of fossil fuels (25.10). Sulfur enters the atmosphere mostly as hydrogen sulfide, which quickly oxidizes to sulfur dioxide, SO_2. Sulfur dioxide reacts with moisture in the atmosphere to form sulfuric acid, carried to Earth in precipitation. Plants incorporate it into sulfur-bearing amino acids. Consumption, excretion, and death carry sulfur back to soil and aquatic sediments, where bacteria release it in inorganic form. Sulfur is an essential nutrient, but sulfur dioxide is a major atmospheric pollutant, damaging and killing plants, causing respiratory afflictions, and adding to acid deposition (25.11).

Acid Deposition (25.12, 25.13) Acid deposition develops when sulfur dioxide and nitrogen oxides combine with water and hydrogen in the atmosphere to produce sulfuric and nitric acids. These acids reach Earth as wet deposition in the form of acidic rain, snow, and fog and as dry deposition in the form of particulate matter and gases. The acidification of lakes and streams kills fish, crustaceans, and insects. Acid precipitation promotes forest decline by increasing the acidity of poorly buffered soils, nutrient depletion, and aluminum toxicity in the soil, and by inhibiting the activity of soil fungi and bacteria. Acid deposition can so weaken trees that they succumb to other stresses. Acid pollution also causes crop losses and ruins roads, bridges, buildings, and monuments.

Cycling of Heavy Metals (25.14) Heavy metals, toxic to life, follow their own biogeochemical cycles. Some cycles are local; others, such as lead, can become regional and even global. Lead is one of the most pervasive heavy metals. Mining, smelting, lead-consuming industries, combustion of coal, burning of leaded fuel, and decay of lead-painted surfaces release small particles of lead into the atmosphere. These particles settle on soil and vegetation and pass through the food chain. Lead may remain in upper levels of soil for millennia.

Chlorinated Hydrocarbons (25.15) Of serious consequence globally is the insecticidal use of chlorinated hydrocarbons. Following biogeochemical cycles, these pesticides have contaminated global ecosystems.

Because they become concentrated at higher trophic levels, chlorinated hydrocarbons affect predaceous animals most, interfering with their reproduction. Such insecticides kill more than the targeted pest species, reducing species diversity. They endanger human health.

STUDY QUESTIONS

1. What are biogeochemical cycles? Describe the types.
2. Explain the cycling of nutrients among pools.
3. Trace the carbon cycle. Why is it closely tied to energy flow?
4. Why does carbon dioxide exhibit fluctuations?
5. What are the features of the oxygen cycle?
6. Explain the paradox of ozone: its beneficial role in the stratosphere and its harmful effects in the lower atmosphere.
7. What four steps are involved in making atmospheric nitrogen available to life?
8. What are the major sources of nitrogen pollution? Discuss the relationship between excessive nitrogen and forest decline.
9. Why is excess nitrogen a pollutant?
10. What are the sources of phosphorus? How does it circulate in aquatic ecosystems?
11. Under what conditions does phosphorus become a pollutant?
12. What are the ecological effects of sulfur pollution?
13. What is acid deposition? What effects does it have on soils and aquatic ecosystems?
14. How do heavy metals and chlorinated hydrocarbons circulate in the biosphere?
15. What is the impact of chlorinated hydrocarbons on ecosystems?
*16. In natural ecosystems nutrients are recycled in place; however, few ecosystems are natural and nutrients are often removed from ecosystems by human actions such as timber harvest and food production. What role do we as individuals play in nutrient cycling? What was the source of the nutrients you consumed yesterday? Make a list of the foods you ate, their point of origin, and their nutrient content.
*17. Discuss methods of garbage disposal in relation to nutrient cycling and in relation to input and output. Are we impoverishing the very ecosystems upon which we depend for food?
*18. What are the major shortcomings of chemical fertilizers in relation to nutrient cycling?

CHAPTER 26

GLOBAL ENVIRONMENTAL CHANGE

OBJECTIVES

On completion of this chapter, you should be able to:

- Identify the major causes of rising atmospheric CO_2 concentration.
- Describe how increasing atmospheric concentrations of CO_2 influence the oceans.
- Tell how terrestrial plants respond to elevated CO_2.
- Describe how CO_2 and other greenhouse gases influence global climate patterns.
- Discuss how changing global climate will influence the distribution and abundance of ecosystems.
- Explain how global warming will influence sea level and coastal ecosystems.

Earth hangs like a lonely blue sphere in space.

The term *environmental change* is almost redundant. Change is an inherent characteristic of Earth's environment. Paleoecology has recorded the responses of populations, communities, and ecosystems to changes in climate during periods of glacial expansion and retreat over the past 100,000 years. On an even longer time scale, the geological and fossil records recount environmental and evolutionary change since the origin of the planet.

Humans, like other species, are part of Earth's environment. Anthropologists tell us that the first humans were hunters and gatherers with low populations. At that time humans functioned as natural predators and herbivores. With the adoption of agriculture, believed to have begun with the cultivation of grasses in the Middle East some 10,000 years ago, human populations and associated conversion of land to agricultural production began to increase exponentially. In the mid 1800s, the Industrial Revolution once again changed the nature of human interaction with the global environment (Figure 26.1). The demand for energy to fuel industrialization and the concentration of populations, or urbanization, brought environmental problems of unprecedented magnitude.

In this chapter we will examine a small subset of the environmental changes that are global in scale brought about by human activities. The nitrogen and sulfur compounds released in industrial emissions (see Sections 25.8, 25.11, and 25.12), although extensive in their distribution, are largely a regional concern. In contrast, other gaseous emissions, such as carbon dioxide, methane, and chlorofluorocarbons (CFCs), mix readily in the air and circulate globally in the atmosphere. To understand and predict the consequences of these emissions on the local and regional environments, science must study them on a global scale.

26.1 Atmospheric concentration of carbon dioxide is rising

The atmospheric concentration of carbon dioxide has increased by more than 25 percent over the past 100 years. The evidence for this rise comes primarily from continuous observations of atmospheric CO_2 started in 1958 at Mauna Loa, Hawaii, by Charles Keeling (Figure 26.2), and from parallel records around the world. Evidence before the direct observations of 1958 comes from a variety of sources, including the analysis of air bubbles trapped in the ice of glaciers in Greenland and Antarctica.

Reconstructing atmospheric CO_2 concentrations over the past 300 years, we see values that fluctuate from 280 to 290 ppm until the mid 1800s (Figure 26.3). After 1860, the onset of the Industrial Revolution, the value soars exponentially. The change reflects the combus-

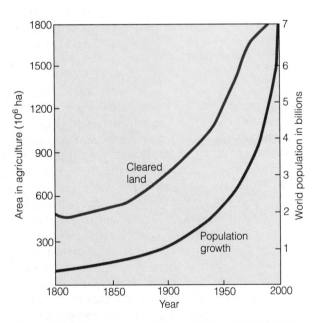

Figure 26.1 Changes in global population and land area cleared for agriculture over the past 200 years.

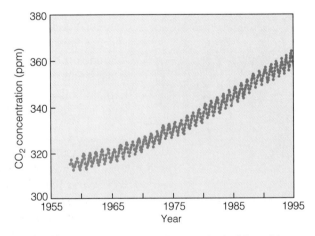

Figure 26.2 Concentrations of atmospheric CO_2 at Mauna Loa Observatory, Hawaii. The dots depict monthly averages. Data prior to direct observation (1958–present) are estimated from various techniques, including analysis of air trapped in Antarctic ice sheets.

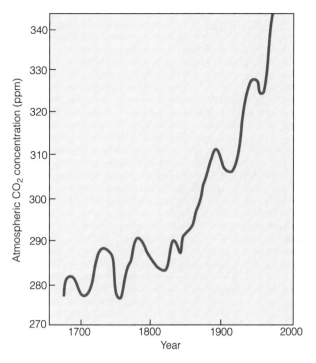

Figure 26.3 The rise of atmospheric CO_2 over the past 300 years.

tion of fossil fuels (coal, oil, and gas) as an energy source for industrialized nations (Figure 26.4).

The burning of fossil fuels is not the only cause of rising atmospheric CO_2 concentration. Deforestation is also a major cause (Figure 26.5). Forested lands are typically cleared and burned to prepare for farming. Although the trees may be harvested for timber or wood pulp, a large part of the biomass, litter layer, and soil organic matter is burned, releasing the carbon to the atmosphere as CO_2.

Calculations of the contribution of land clearing to atmospheric CO_2 are complex. Following timber harvest on lands managed for forest production, or on lands that have been cultivated and then abandoned, vegetation and soil organic matter become reestablished. We calculate the net contribution to the atmosphere as the difference between CO_2 released by clearing and burning and CO_2 taken up by photosynthesis and the accumulation of biomass during reestablishment. At one time scientists used regional estimates of population growth and land use (forestry and agriculture) together with simple models of vegetation and soil succession to estimate the contribution due to changing land use. More recent estimates

use satellite images to quantify these changes (see Focus on Ecology 26.1: Monitoring Vegetation Change from Space on page 368).

26.2 Higher atmospheric CO_2 concentrations increase uptake by oceans

Scientists estimate that the amount of carbon released to the atmosphere from 1958 to 1980 by the combustion of fossil fuels is approximately 85.5 gigatons (a gigaton is 10^9 metric tons). To put this number into perspective, if the average weight of a human is 70 kg, 1 Gt would be the weight of over 14 billion people, or almost three times the world's population. This number also puts the per capita use of fossil fuels into perspective.

Direct measurements of atmospheric CO_2 over this same period show only a 49.9 Gt increase in the amount of carbon in the atmosphere. The difference, 35.6 Gt, must have flowed from the atmosphere into the other main pools in the global carbon cycle (Figure 26.6), the oceans and terrestrial ecosystems. Carbon dioxide diffuses from the atmosphere into the surface waters of the ocean, where it dissolves and undergoes a number of chemical reactions, including transformation to carbonates and bicarbonates (Figure 26.7). If the concentration of CO_2 in the water is greater than that in the air, the process is reversed, and CO_2 diffuses from the surface waters of the ocean to the atmosphere.

As the concentration of CO_2 in the atmosphere rises, the diffusion of CO_2 into the surface waters of the oceans increases. Once the concentrations in the atmosphere and the surface waters are equal, the flow would stop. However, there are two main reasons why it does not occur. First, the concentration of CO_2 in the atmosphere continues to rise as a result of fossil fuel combustion. Second, CO_2 in the surface waters transforms into bicarbonates and other carbon compounds, and plants take up CO_2 and bicarbonates in photosynthesis. Both of these processes lower CO_2 concentrations in the surface waters, maintaining the diffusion gradient. The ocean, then, acts as a sink, drawing CO_2 from the atmosphere as its concentration rises.

Given their volume, the oceans have the potential to absorb most of the carbon that is being transferred to the atmosphere by fossil fuel combustion and land clearing. The oceans, however, do not act as a

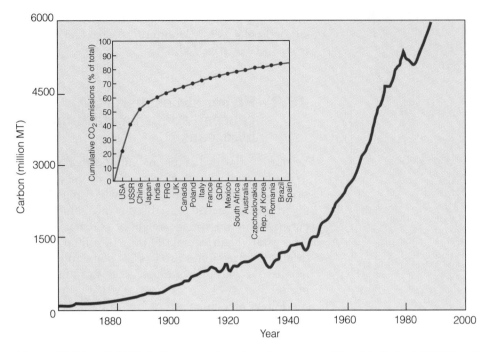

Figure 26.4 Input of CO_2 to the atmosphere from the burning of fossil fuels since 1860. Inset graph shows cumulative amounts of CO_2 emmissions from the 20 largest CO_2-emitting nations in 1989, plotted as a fraction of emissions from all countries.

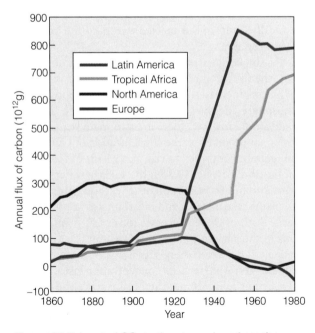

Figure 26.5 Input of CO_2 to the atmosphere from the clearing and burning of forests since 1860.

homogeneous sponge, absorbing CO_2 equally into the entire volume of water.

The oceans consist of two basic layers, the surface waters and deep waters (see Figure 26.7). The average depth of the ocean is 2000 m. Intercepted solar radiation warms the surface waters. Depending on the amount of radiation reaching the surface, the zone of warm water ranges from 75 m to 200 m in depth. The average temperature of this surface layer is 18° C. The remainder of the vertical profile (200 m to 3000 m depth) is deep waters, whose average temperature is 3° C. The transition between these two zones is abrupt; we call it the thermocline (see Section 20.4 and Chapter 32). In effect, the ocean can be viewed as a thin layer of warm waters floating on a much deeper layer of cold waters. The temperature difference between these two layers leads to a separation of many processes. Turbulence caused by winds mixes the surface waters, transferring CO_2 absorbed at the surface to the waters below. Due to the thermocline, however, this mixing does not extend into the deep waters. Mixing between surface and deep waters depends on deep

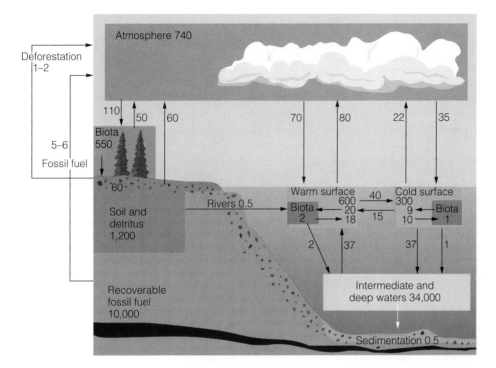

Figure 26.6 Global compartments of CO_2 and their connected subsystems. Arrows indicate the major fluxes (in Gt/yr) and pool sizes (in Gt) of carbon.

ocean currents caused by the sinking of surface waters as they move toward the poles (Chapter 4). This process occurs on a time scale of hundreds of years, limiting the short-term uptake of CO_2 by the deep waters. The result is that the amount of CO_2 that the oceans can absorb over the short term is very limited despite the large volume of water.

26.3   Plants respond to increased atmospheric carbon dioxide

Carbon dioxide flows from the atmosphere into terrestrial ecosystems via photosynthesis (Chapter 3). To understand how rising atmospheric CO_2 concentrations will influence the productivity of terrestrial ecosystems, we must understand how photosynthesis will change in an enriched CO_2 environment.

Recall that CO_2 diffuses from the outside air into the leaf through the stomatal openings (see Section 3.3). The higher the CO_2 concentration in the outside air, the greater the rate of diffusion into the leaf. A higher rate of diffusion increases the availability of CO_2 for photosynthesis in the mesophyll cells of the leaf, so it generally results in a higher rate of photosynthesis. The higher rates of diffusion and photosynthesis

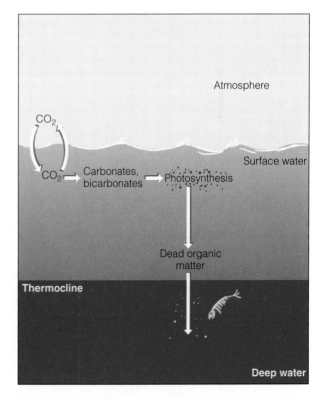

Figure 26.7 Processes involved in the uptake of CO_2 from the atmosphere into the oceans.

MONITORING VEGETATION CHANGE FROM SPACE

Before the mid 1980s, scientists had to estimate vegetation cover and land use from aerial photographs and ground surveys. In the absence of these types of data, they relied on regional statistics for population or agricultural production. Now satellite remote sensing provides images of land cover on a regional and global scale.

Satellite systems sample the energy returning to space from Earth's surface in different ways. Two satellite

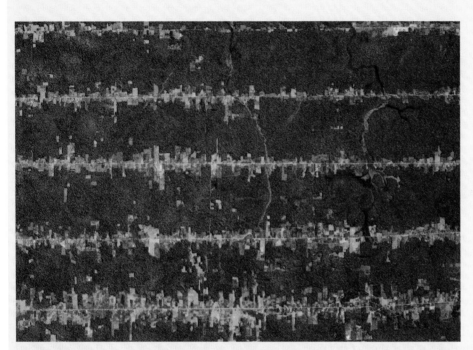

Figure A AVHRR image of Rondonia, Brazil.

Figure B LANDSAT image of Rondonia, Brazil.

remote sensing systems widely used for terrestrial research are LANDSAT and AVHRR (Advanced Very High Resolution Radiometer). These two systems measure different regions of the energy spectrum reflected by Earth's surface (see Section 5.1, Figure 5.2), but both use the fact that plants absorb a specific range of wavelengths (PAR, photosynthetically active radiation—see Section 5.1). The systems differ in their spatial resolution and in how often they return to a given region to record an image. LANDSAT has a spatial resolution of 30 m × 30 m and returns to a given point on the Earth once every 16 days. AVHRR has a resolution of 1000 m and provides daily global coverage.

Satellite images of forest clearing in the Brazilian province of Rondonia are shown in Figures A and B. To stimulate settlement in the Amazon basin, the government constructs roads in remote areas of the rain forest. These two images show both the pattern of clearing (light-colored areas) and the difference in resolution of information between the two systems.

under elevated atmospheric concentrations of CO_2 have been termed the "CO_2 fertilization effect." Plant species vary in their response to elevated CO_2, although most species show an increase in photosynthetic rates under a doubled concentration of CO_2. A review of over 200 experiments examining the photosynthetic response of tree species to a doubling of CO_2 concentration showed an average increase of 44 percent (Figure 26.8).

In addition to increased rates of photosynthesis, plants exposed to a doubled CO_2 concentration exhibit a partial closure of the stomata (Chapter 3). This increased stomatal resistance reduces water loss. Thus under elevated CO_2, plants increase their water-use efficiency (carbon uptake/water loss).

The effects of longer-term exposure to elevated CO_2 on plant growth, however, may be more complicated. In some studies the enhanced effects of elevated CO_2 on plant photosynthesis have been short-lived. Some plants reduce the production of the photosynthetic enzyme rubisco under elevated CO_2, reducing photosynthesis to rates comparable to those measured at lower concentrations. Other studies reveal that plants grown under increased CO_2 allocate less carbon to the production of leaves and more to the production of roots. In addition, plants grown under elevated concentrations of CO_2 appear to produce fewer stomata on the leaf surface. The smaller leaf area and lower stomatal density reduce both water loss and growth.

Whether the results observed for leaves or single plants translate into changes in the net primary productivity of terrestrial ecosystems is largely unknown. The availability of water or nutrients in many ecosystems may limit potential increases in plant productivity under elevated CO_2. A number of large-scale experiments seek to examine the effects of elevated CO_2 on whole ecosystems (see Focus on Ecology 26.2: Raising CO_2 in a Forest). By exposing whole areas of forest and grassland to elevated CO_2, scientists hope to examine the variety of processes influencing primary productivity, decomposition, and nutrient cycling in terrestrial ecosystems. Only then will we understand

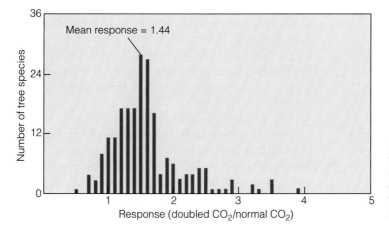

Figure 26.8 The change in photosynthesis for tree species grown under a doubled CO_2 concentration. Response is expressed as the ratio of observed photosynthesis under doubled CO_2 to that under normal CO_2. A value of 1 represents no detected change.

RAISING CO₂ IN A FOREST: WHAT HAPPENS?

Most of our understanding of how terrestrial ecosystems respond to elevated CO_2 comes from experiments on single plants or small groups of plants grown either in the greenhouse or under controlled conditions in the field. Although important, these experiments provide limited insight into ecosystem dynamics.

A new experimental approach modifies CO_2 concentrations in whole ecosystems. The new experimental approach is called FACE—**F**ree **A**ir **C**arbon Dioxide **E**nrichment. Figure A shows the FACE facility at the Duke University Forest in North Carolina. The tower reaching above the forest canopy at the center of the ring is 35 m tall. It contains instruments to measure features of the climate (including temperature, water vapor, solar radiation, and wind speed and direction) and the exchange of CO_2 between the atmosphere and the forest (net ecosystem productivity). The vertical pipes that form a ring around the site have a series of nozzles along their length that release CO_2 into the air. A computer-controlled system continuously monitors CO_2 concentrations in the forest (within the circle of pipes); it releases CO_2 from the nozzles as a function of CO_2 concentration, wind direction, and speed. This system provides a fairly uniform concentration of doubled CO_2 in the forest.

Scientists are conducting a variety of experiments at the site. They measure photosynthesis, transpiration, productivity, decomposition, and nutrient cycling and compare the experimental site with the surrounding forest.

Figure A The Free Air Carbon Dioxide Enrichment facility at Duke University.

the influence of elevated atmospheric CO_2 concentrations on the transfer of carbon from the atmosphere to terrestrial ecosystems.

26.4 Greenhouse gases may change the global climate

CO_2 in the atmosphere does not readily absorb shortwave radiation from the sun. However, it does absorb longwave or thermal radiation (see Figure 4.1). Earth's surface emits absorbed solar radiation to the atmosphere as longwave or thermal radiation. CO_2 and water vapor trap this thermal energy, warming the atmosphere. Because it traps heat, we call carbon dioxide a **greenhouse gas.** Were it not for the warming effect of carbon dioxide, Earth would be several degrees cooler.

As human activities increase the atmospheric concentration of CO_2, will they influence the global

climate? Scientists estimate that at current rates of emission, the preindustrial level of 280 ppm of CO_2 in the atmosphere will double by 2020. Moreover, CO_2 is not the only greenhouse gas increasing in concentration as a result of human activities. Other greenhouse gases include methane (CH_4), chlorofluorocarbons (CFCs), hydrogenated chlorofluorocarbons (HCFCs), nitrous oxide (N_2O), ozone (O_3), and sulfur dioxide (SO_2). Although much lower in concentration, some of these gases are much more effective at trapping heat than CO_2. They are significant components of the total greenhouse effect.

Although the role of greenhouse gases in warming Earth's surface is well established, the specific influence that doubling the CO_2 concentration of the atmosphere will exert on the global climate system is much more uncertain. Atmospheric scientists have developed complex computer models of Earth's climate systems, called General Circulation Models, or GCMs for short. Using these computer models, scientists conduct experiments to examine how increasing concentrations of greenhouse gases may influence large-scale patterns of global climate. GCMs at numerous research institutions all use the same underlying physics. However, they differ in spatial resolution and in how they describe certain features of Earth's surface and atmosphere. As a result, the models differ in their predictions.

Despite these differences, certain consistent patterns emerge. All of the models predict an increase in the average global temperature as well as a corresponding increase in global precipitation. Findings published in 1996 by the Intergovernmental Panel on Climate Change (IPCC) suggest an increase in the average global temperature between 1° C and 3.5° C by 2100. These changes would not be evenly distributed over Earth's surface. Warming would be greatest during the winter months and in the northern latitudes.

You can see the spatial variation in changes in mean annual temperature and precipitation for the 48 contiguous states of the United States predicted by one such GCM in Figure 26.9. This model predicts regional temperature increases ranging from 0.6° C

to 7.0° C, with the highest values in the polar regions and the lowest values in the equatorial region.

Although in popular speech "the greenhouse effect" is synonymous with global warming, the models predict more than just hotter days. One of the most notable predictions is an increased variability of climate. More storms and hurricanes, deeper snow, and uneven rainfall are also among the predictions for different regions.

The development of General Circulation Models continues. As these models improve, there will no doubt be changes in the patterns and the severity of change they predict. However, our understanding of the physics of greenhouse gases and the consistent qualitative predictions of the GCMs lead scientists to believe that rising concentrations of atmospheric CO_2 will have a significant impact on global climate.

26.5 Changing climates affect ecosystems

Climate influences almost every aspect of the ecosystem: the physiological and behavioral responses of organisms (Chapters 5–7); the birth, death, and growth

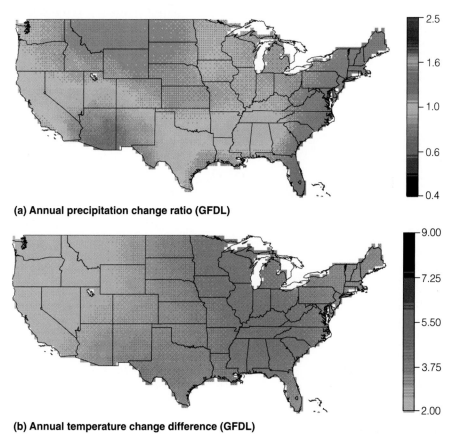

(a) Annual precipitation change ratio (GFDL)

(b) Annual temperature change difference (GFDL)

Figure 26.9 Changes in (a) annual precipitation and (b) temperature for doubled CO_2, estimated by the general circulation model developed by the Geophysical Fluid Dynamics Laboratory (GFDL) at Princeton University. Changes in temperature are expressed as absolute increases in ° C. Changes in precipitation are expressed as the ratio of current to predicted annual precipitation. A value of 1.0 represents no change, whereas a value of 1.5 represents a 50 percent increase, and a value of 0.8 a 20 percent decrease.

rates of populations (Chapters 11–13); the competitive abilities of species (Chapter 15); community structure (Chapter 20); productivity (Chapter 23); and the cycling of nutrients (Chapters 9 and 25). As they change Earth's climate, greenhouse gases will have a major influence on these processes and patterns.

Ecologists have learned a great deal about the response of organisms to changing climate conditions from the study of past climate change. Pollen samples from sediment cores taken in lake beds have allowed paleobotanists to reconstruct the vegetation of many regions during the last 20,000 years. The work of Margaret Davis in reconstructing the distribution of tree species in eastern North America since the last glacial maximum (see Section 21.14 and Figure 21.21) is a good example. Tree genera migrated northward at different rates following the retreat of the glaciers. The migration rates depended on how well a species's physiology, dispersal ability, and competitive interactions let it respond to changes in climate. Such studies show us that the existing forest communities in eastern North America are a recent result of different responses of tree species to changing climate. As Earth's climate has changed in the past, the distribution and abundance of organisms and the communities and ecosystems they compose have changed.

Current research on the potential impact of greenhouse warming focuses on the responses of organisms at all levels of organization: individuals, populations, communities, and ecosystems. Changes in temperature and water availability will have a direct effect on the distribution and abundance of individual species. For example, the northern limit of the winter range of the Eastern phoebe is associated with average minimum January temperatures of −4° C. The phoebe is not found in areas where temperature drops below this value. Two lines, or isotherms, defining the region of eastern North America where average minimum January temperatures of −4° C occur are plotted in Figure 26.10. Minimum temperatures drop below −4° C in areas to the north and west of the lines. The two isotherms show the current −4° C average minimum January temperature isotherm, and the −4° C isotherm predicted by the Geophysical Fluid Dynamics Laboratory General Circulation Model (GCM) for a doubled atmospheric concentration of CO_2. A change in the isotherm would be expected to result in a northern expansion of the Eastern phoebe's winter range.

Changes in climate will have a direct influence on the growth and reproductive rates of species. These changes will influence their relative competitive abilities, altering patterns of zonation and succession (Chapters 20 and 21). Given the difficulty of experimentally changing climate conditions in the field, few studies have examined the effects of changing climate conditions on community dynamics. One such experiment was conducted in a meadow community in the Rocky Mountains of Colorado. Using electric heaters suspended 2.6 m above experimental plots (Figure 26.11), scientists were able to raise soil temperature and influence soil moisture and the timing of snowmelt. In heated plots, the density of shrubs increased at the ex-

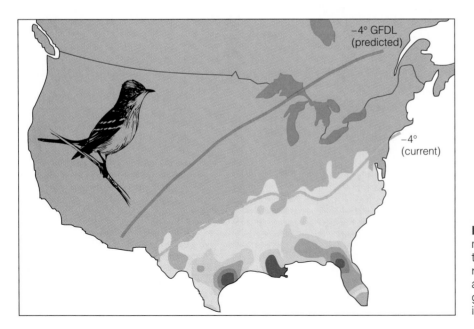

Figure 26.10 Isotherms demarcating the current range of the Eastern phoebe and the range predicted for doubled atmospheric CO_2 by the GFDL general circulation model shown in Figure 26.9a.

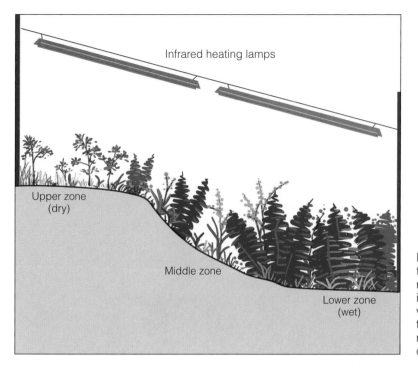

Infrared heating lamps

Upper zone
(dry)

Middle zone

Lower zone
(wet)

Figure 26.11 Experimental setup used to elevate temperatures in a Colorado meadow community to examine possible impacts of global warming. Changes in vegetation composition were monitored through time to track how the community responded to elevated temperatures and drier conditions.

pense of grass and forb species. Results suggest that the increased warming expected under an atmosphere with a doubled concentration of CO_2 would shift the dominant vegetation of the widespread mountain meadow habitat. Shrubs would compete better in the altered environment. Such shifts have a major impact not only on plant communities, but on associated animal species as well.

Changes in climate will also affect vegetation indirectly, through decomposition and nutrient cycling. In terrestrial ecosystems these processes depend on temperature and available moisture (Chapters 7, 6, 25). Decomposition proceeds faster under warmer, wetter conditions. An ongoing experiment at Harvard Forest in Massachusetts is examining the impact of elevated soil temperatures on rates of decomposition and nutrient cycling in a forested ecosystem. Buried heating cables raise the soil temperature by 5° C. Initial results show a 60 percent increase in soil respiration (CO_2 emissions), a direct result of increased microbial and root respiration. The former is associated with an increased rate of decomposition. The results are consistent with patterns of soil respiration observed in other forests in warmer regions around the world. They indicate that greenhouse warming will increase rates of decomposition and microbial respiration. There will be a significant rise in emissions of CO_2 from the soil to the atmosphere.

It is virtually impossible to develop experiments in the field to examine the longer-term response of terrestrial ecosystems to climate change. This limitation on direct observation means that scientists must base predictions on computer models of ecosystems. The simplest, but most telling, of these ecosystem models are the biogeographical models that relate the distribution of ecosystems to climate. From the days of the early naturalists, plant ecologists have recognized the link between climate and plant distribution. For example, tropical rain forests are found in the wet tropical regions of Central and South America, Africa, Asia, and Australia. According to the biogeographical model developed by L. R. Holdridge, within these regions tropical forest distribution is limited to areas where mean annual temperatures are at or above 24° C and annual precipitation is above 2000 mm. The regions of the tropics that meet these climate restrictions are shown in Figure 26.12a. Under the temperature and rainfall patterns predicted by the United Kingdom Meteorological Office GCM for a doubled atmospheric CO_2 concentration, this distribution changes dramatically (Figure 26.12b). The region that can support wet rain forest under this scenario of greenhouse warming shrinks by 25 percent. This decline is a direct result of drying due to decreased precipitation and higher temperatures. Together with the demands of agriculture and forestry (see Section 26.1),

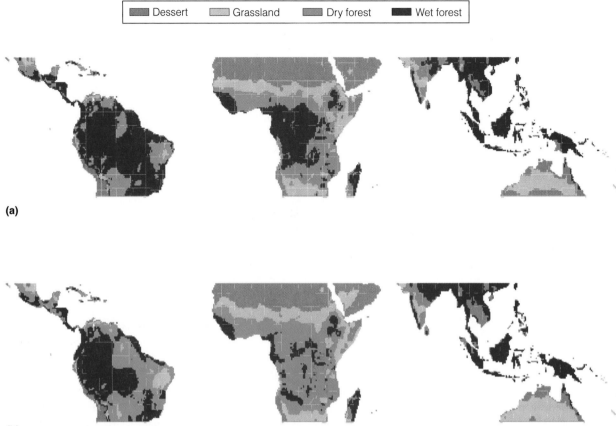

Dessert ☐ Grassland ☐ Dry forest ■ Wet forest

(a)

(b)

Figure 26.12 Maps of the areas in the tropical zone that could support wet forest ecosystems as predicted by the Holdridge biogeographical model of ecosystem distribution. (a) Vegetation zones under current climate conditions. (b) Vegetation zones under changed climate conditions predicted by the United Kingdom Meteorological Office General Circulation Model. Note changes in the distribution of wet forest.

this scenario would devastate the tropical rain forest ecosystems of the world. It would cause large-scale extinction of rain forest flora and fauna.

Changes in the global pattern of temperature will also affect the distribution of aquatic ecosystems. The global distribution of coral reefs is limited to tropical waters where mean surface temperatures are at or above 20° C. Reef development is not possible where the mean minimum temperature is below 18° C. Optimal reef development occurs in waters where the mean annual temperatures are 23–25° C. Some corals can tolerate temperatures up to 40° C. Warming of the world's oceans will alter the range of waters in which reef development is possible, allowing reefs to form farther up the eastern coast of North America.

Ecologists are far from a complete analysis of the potential impact of global climate change. There is little question, however, that changes in temperature and precipitation of the magnitude predicted by cli-

mate models will have a significant influence on the distribution and functioning of both terrestrial and aquatic ecosystems.

26.6 Global warming would raise sea level

During the last glacial maximum, some 18,000 years ago, sea level was 100 m lower than the current level. Today's highly productive, shallow coastal waters, such as the continental shelf of eastern North America, were above sea level and covered by terrestrial ecosystems. As the climate warmed and the glaciers melted, sea level rose. Over the last century, sea level has risen at a rate of 1.8 (±0.1) mm per year. This rise is a result of the general pattern of global warming over this period and the associated expansion of ocean waters and melting of glaciers. The 1996 report of the Intergovernmental Panel on Climate Change (IPCC) estimates that under the scenarios of global warming outlined in Section 26.5, sea level will rise from 0.3 to 1.1 m above the current level by the year 2100. The rise is primarily a function of melting polar ice caps and sea ice. A 1 m rise in sea level will have serious effects on coastal environments, eroding beaches and flooding low-lying islands.

Estuarine ecosystems (see Chapter 37) will be adversely affected by a sea level rise of this magnitude. Current patterns of water depth, temperature, salinity, and turbidity and the inflow and outflow of tidal waters are critical to maintaining estuarine ecosystems. Rising sea levels will submerge coastal salt marshes as they now exist. Invasion of salt water farther up the estuary would be disastrous to estuarine organisms. It could cause salinization of adjacent lands and in extreme cases the salinization of freshwater aquifers.

26.7   Understanding global change requires the study of ecology on a global scale

The increasing atmospheric concentrations of CO_2 and other greenhouse gases, and the potential changes in global climate patterns that may result, present a new class of ecological problems. To understand the impact of rising CO_2 emissions from fossil fuel burning and land clearing, we have to examine the carbon cycle (see Figure 26.6) on a global scale, linking the atmosphere, hydrosphere, biosphere, and lithosphere. Although the discussion in the previous sections focuses on the impacts of rising CO_2 concentrations and changes in climate on populations, communities, and

ecosystems, the possible impacts are not in one direction. Ecosystems also influence atmospheric CO_2 and regional climate patterns. For example, if climate changes as shown in Figure 26.12, the global distribution and abundance of tropical rain forests will decline dramatically. Tropical rain forests are the most productive terrestrial ecosystems on the planet. A significant decline in these ecosystems will reduce global primary productivity, the uptake of CO_2 from the atmosphere and the storage of organic carbon in biomass. In fact, as tropical rain forests shrink, atmospheric CO_2 will increase. The drying of these regions will kill trees, increase fires, and transfer carbon stored in living biomass to the atmosphere as CO_2 in much the same way forest clearing does in these regions (Section 26.1). The rise in atmospheric CO_2 will increase the greenhouse effect, further exacerbating the problem. In this case, the changes in the terrestrial surface act as a positive feedback loop to rising atmospheric concentrations of CO_2.

On the other hand, if rising CO_2 and changing climate increase the productivity of the world's ecosystems, they will take up more CO_2 from the atmosphere. Increased productivity will function as negative feedback, drawing down atmospheric CO_2 concentrations.

These are not simple connections. To understand the interactions among the atmosphere, oceans, and terrestrial ecosystems, ecologists must study Earth as a single, integrated system. It is only through the development of a global ecology that ecologists, working with oceanographers and atmospheric scientists, will come to understand the potential consequences of doubling the concentration of CO_2 in the atmosphere over the next century.

CHAPTER REVIEW

SUMMARY

Rising Atmospheric Concentrations of CO_2 (26.1–26.3) Direct observations beginning in 1958 reveal an exponential increase in the atmospheric concentration of CO_2. The rise is a direct result of fossil fuel combustion and the clearing of land for agriculture **(26.1)**.

Of the CO_2 released from fossil fuel combustion and land clearing, only about 60 percent has remained in the atmosphere. The remainder is taken up by the oceans and terrestrial ecosystems. The ocean has two layers, surface waters (0–200 m) and deep waters. Over

85 percent of the ocean volume is deep water. The thermocline prevents mixing. Carbon dioxide diffuses from the atmosphere into the surface waters of the oceans. The rise in atmospheric concentrations causes increased uptake of CO_2 into surface waters. Transfer of dissolved carbon dioxide from the surface waters into the deep waters of the ocean takes hundreds of years. This fact limits the short-term uptake of CO_2 by the oceans **(26.2)**.

In general, plants respond to increased atmospheric CO_2 with higher rates of photosynthesis and partial closure of stomata. These responses increase water-use efficiency. Responses to long-term exposure vary, including diversion of carbon from leaves to roots and reduction in stomatal density. Scientists are studying long-term effects on net primary productivity **(26.3)**.

Global Climate Change (26.4–26.7) Carbon dioxide is a greenhouse gas. It traps long-wave radiation emitted from Earth's surface, warming the atmosphere. Rising atmospheric concentrations of CO_2 and other greenhouse gases could raise the global mean temperature by 1° C to 3.5° C by the year 2100. Warming will not be uniform over Earth. The greatest warming is predicted during the winter months and at northern latitudes. Increased variability in climate is predicted, including changes in precipitation and the frequency of storms **(26.4)**.

The distribution and abundance of species will shift as temperature and precipitation change. Changes in climate will influence the competitive ability of species and thus change patterns of community zonation and succession. Ecosystem processes such as decomposition and nutrient cycling are sensitive to temperature and moisture, and changing climate will affect them. Changes in climate also will shift the distribution and abundance of both terrestrial and aquatic ecosystems **(26.5)**.

Global warming will cause sea level to rise by 0.3 m to 1.1 m by 2100, as the polar ice caps melt and warmer ocean waters expand. A sea level rise of this magnitude will have major impacts on beaches, estuaries, and other coastal environments **(26.6)**.

To grasp global warming, we have to study the whole Earth as a single, complex system **(26.7)**.

STUDY QUESTIONS

1. What are the major sources of greenhouse gases, especially CO_2?
2. Not all of the CO_2 released to the atmosphere by human activities remains there. What happens to the rest?
*3. How do greenhouse gases maintain Earth's temperature? If greenhouse gases benefit life, why should increasing concentrations of CO_2 be a concern?
4. How do greenhouse gases contribute to global warming?
5. How do forest burning and land clearing affect the global climate?
6. In what ways does elevated CO_2 affect plants?
7. What limits the transfer of CO_2 from the surface waters of the ocean to the deep waters?
*8. What impact would changing climate have on ecosystems in general, and coastal environments in particular?
*9. Why does understanding global climate change require research on a bigger scale than ever?

A GUIDE TO ECOSYSTEMS

PART VI

ECOSYSTEM PATTERNS

A view from the window of a plane on a transcontinental flight from Boston to California is revealing to an ecology-minded passenger. Below, the pattern of vegetation changes from the mixed coniferous-hardwood forests of the northeast to the oak forests of the central Appalachians with patches of high-elevation spruce forests. Then the forest cover merges with midwestern croplands of corn, soybean, and wheat, land that once was the domain of tallgrass prairie. Wheat fields yield to high-elevation shortgrass plains, and then the plains give way to the coniferous forest of the Rocky Mountains, capped by tundra and snowfields. Beyond the mountains to the southwest lie the tan-colored desert regions.

On a trip of less than eight hours the airborne ecologist can observe a wide range of vegetation that took years to discover. Botanists were the first to note that the world could be divided into great blocks of vegetation—deserts, grasslands, and coniferous, temperate, and tropical forests. They called the divisions *formations*. In time plant geographers attempted to correlate vegetation formations with climatic differences and found that blocks of climate reflected blocks of vegetation with their own life forms.

Zoogeographers lagged behind plant geographers in their study of animal distribution. Complicating their studies were the great number of animal species and the lack of a clear relationship between animal distribution and climate. Ultimately, zoogeographers did accumulate basic information on the global distribution of animals. Alfred Wallace, also known for having developed the same general theory of evolution as Darwin, provided the major synthesis of animal distribution. Wallace's realms, with some modification, still stand today.

BIOGEOGRAPHICAL REALMS

There are six biogeographical **realms,** each more or less embracing a major continental land mass and separated by oceans, mountain ranges, or desert (see Figure VI.1). They are the Palearctic, the Nearctic, the Neotropical, the Ethiopian, the Oriental, and the Australian. Because some zoogeographers consider the Neotropical and the Aus-

tralian realms to be so different from the rest of the world, these two are often considered as realms equal to the other four combined. Then there are just three realms: Neogea (Neotropical), Notogea (Australian), and Metagea (everything else). Each region possesses a certain distinction and uniformity in the taxonomic units it contains, although each shares some of the families of animals with other regions. Each has at some time in Earth's history had some land connection with another across which animals and plants could pass.

Two realms, the Palearctic and Nearctic, are quite closely related. In fact, the two are often considered as one, the Holarctic. Both are much alike in their faunal composition, and together they share, particularly in the north, such animals as the wolf, hare, moose (called elk in Europe), stag (called elk in North America), caribou, wolverine, and bison.

Below the coniferous forest belt, the two regions become more distinct. The Palearctic is not rich in vertebrate fauna, of which few are endemic. Palearctic reptiles are few and are usually related to those of the African and Oriental tropics. The Nearctic, in contrast, is the home of many reptiles and has more endemic families of vertebrates. The Nearctic fauna is a complex of New World tropical and Old World temperate families. The Palearctic is a complex of Old World tropical and New World temperate families.

Isolated until 15 million years ago, the fauna of the Neotropical is most distinctive and varied. In fact, about half of the South American mammals, such as the tapir and llama, are descendants of North American invaders, whereas the only South American mammals to survive in North America are the opossum and the porcupine. Lacking in the Neotropical is a well-developed ungulate fauna of the plains, characteristic of North America and Africa. However, the Neotropical is rich in endemic families of vertebrates. Of the 32 families of mammals, excluding bats, 16 are restricted to the Neotropical. In addition, 5 families of bats, including the vampire, are endemic.

The Ethiopian realm embraces tropical forests in central Africa, and savanna, grasslands, and desert in the mountains of east Africa. During the Miocene and Pli-

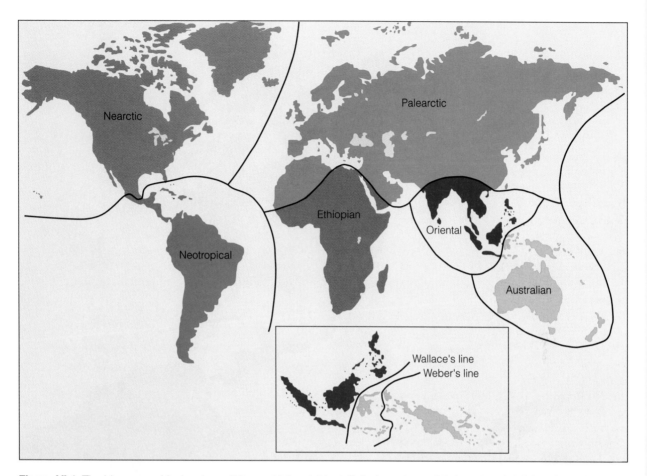

Figure VI.1 The biogeographical realms of the world. Inset: No definite boundary exists between the Oriental and Australian region, where the islands of the Malay Archipelago stretch toward Australia. Two lines have been proposed to separate the two regions. One, Wallace's line, runs from the Philippines through Borneo and the Celebes. The other, Weber's line, lies east of Wallace's line. It separates the islands with the majority of Oriental animals from those with a majority of Australian animals. Because the islands between these two lines are a transitional zone, some zoogeographers call the area Wallacea.

ocene epochs, Africa, Arabia, and India shared a moist climate and a continuous land bridge, which allowed animals to move freely among them. That connection accounts for some similarity in the fauna between the Ethiopian and Oriental regions. Of all the regions, the Ethiopian contains the most varied vertebrate fauna, and in endemic families it is second only to the Neotropical.

Of the tropical realms, the Oriental, once covered with lush forests, possesses the fewest endemic species and lacks a variety of widespread families. It is rich in primate species, including two families confined to the region, the tree shrews and tarsiers.

Perhaps the most interesting and the strangest region, and certainly the most impoverished in vertebrate species, is the Australian. Partly tropical and partly south temperate, this region is noted for its lack of a land connection with other regions; the few freshwater fish,

amphibians, and reptiles; native placental mammals restricted to bats, mice, and rats; and the dominance of marsupials. Included are the monotremes, with two egg-laying families, the duckbilled platypus and the spiny anteaters. The marsupials have become diverse and have evolved ways of life similar to those of the placental mammals of other regions.

BIOMES

Another approach, pioneered by V. E. Shelford, was simply to accept plant formations as biotic units and to associate animals with plants. Such an approach works fairly well, because animal life does depend upon a plant base. These natural broad biotic units are called **biomes** (Figure VI.2). Each biome consists of a distinctive combination of plants and animals in the fully developed climax

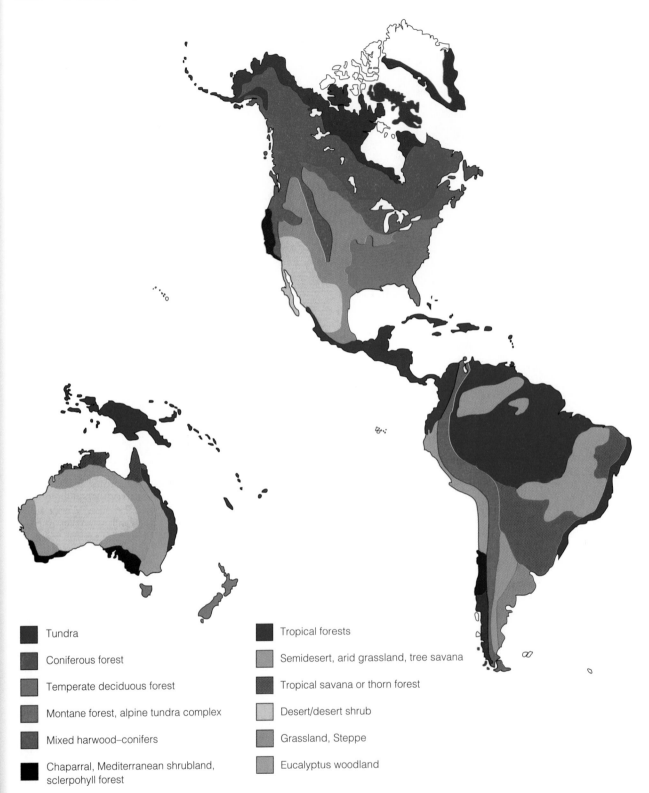

Tundra

Coniferous forest

Temperate deciduous forest

Montane forest, alpine tundra complex

Mixed harwood–conifers

Chaparral, Mediterranean shrubland, sclerpohyll forest

Tropical forests

Semidesert, arid grassland, tree savana

Tropical savana or thorn forest

Desert/desert shrub

Grassland, Steppe

Eucalyptus woodland

Figure VI.2 The major biomes of the world. The arctic and boreal forest are circum-polar and hold many taxonomically related and functionally similar species. Other biomes possess different genera and species that function as ecological equivalents.

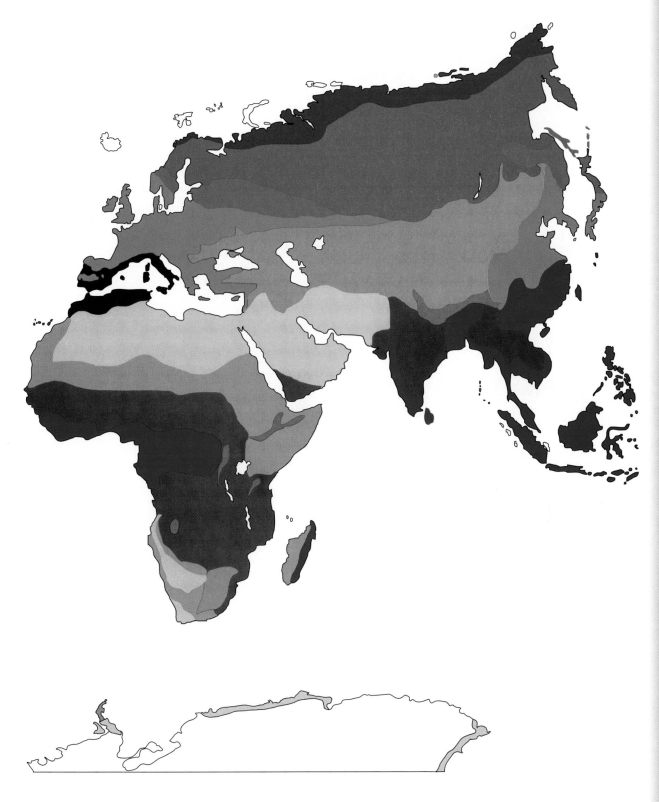

community, and each is characterized by a uniform life form of vegetation, such as grass or coniferous trees. The biome also includes developmental stages, which may be dominated by other life forms. Because the species that dominate the seral stages are more widely distributed than those of the climax, they are of little value in defining the limits of the biome.

On a local and regional scale, communities lie along gradients, in which the combination of species varies as the individual species respond to environmental gradients. On a larger, continental scale, we can consider the terrestrial and even some of the aquatic ecosystems as forming gradients. We call such gradients of ecosystems **ecoclines** (Figure VI.3).

In addition to gradual changes in vegetation, there are gradual changes in other ecosystem characteristics. From highly mesic situations and warm temperatures to xeric situations and cold temperatures, productivity, species diversity, and the amount of organic matter decrease. There is a corresponding decline in the complexity and organization of ecosystems, in the size of plants, and in the number of strata to vegetation. Growth form

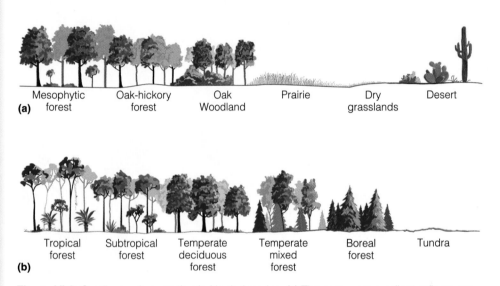

(a)
Mesophytic forest   Oak-hickory forest   Oak Woodland   Prairie   Dry grasslands   Desert

(b)
Tropical forest   Subtropical forest   Temperate deciduous forest   Temperate mixed forest   Boreal forest   Tundra

Figure VI.3 Gradients of vegetation in North America. (a) The east-west gradient reflects precipitation. This transect runs from the mixed mesophytic forest of the Appalachians through the oak hickory forests of the central states, the ecotone of bur oak and grassland, the prairie, the short-grass plains, and the desert. It does not cross the Rocky Mountains. (b) The north-south gradient reflects temperature. The transect cuts across the tundra of Canada, the boreal coniferous forest, the mixed northern hardwood forest, the mixed mesophytic forests of the Appalachians, the subtropical forests of Florida, and the tropical forests of Mexico.

changes. The tropical rain forest is dominated by pha-nerophytes and epiphytes, the arctic tundra by hemicryp-tophytes, geophytes, and therophytes. Wherever similar environments exist on Earth, the same growth forms exist, even though species differences may be great. Thus, different continents tend to have communities of similar physiognomy.

There are six major terrestrial biomes: forest, grass-land, woodland, shrubland, semidesert shrub, and des-ert. These six further divide into biome types, depending upon climatic conditions and elevation. These biomes of the world fall into a distinctive pattern when plotted on a gradient of mean annual temperature and mean annual precipitation (Figure VI.4). The plots obviously are rough. Many types intergrade, and adaptations of various growth forms may differ among continents.

Climate alone is not responsible for biome types. Soil and fire also influence which one of several biomes will occupy a region. Structure of biomes is further influ-enced by whether the climate is marine or continental. The same amount of rain, for example, can support ei-ther shrubland or grassland.

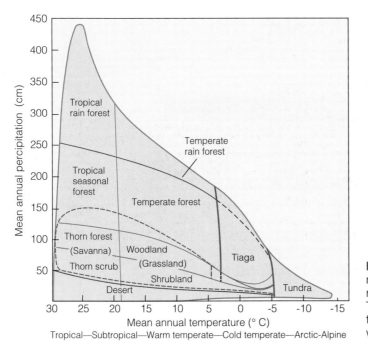

Figure VI.4 The pattern of world plant biomes in relation to temperature and moisture. Where the cli-mate varies, soil can shift the balance between types. The dashed line encloses environments in which ei-ther grassland or one of the types dominated by woody plants may prevail.

GRASSLANDS AND SAVANNAS

OBJECTIVES

On completion of this chapter, you should be able to:

- Describe the types and general features of grasslands.
- Describe the forms of animal life unique to grassland ecosystems.
- Discuss energy flow and nutrient cycling in grassland ecosystems, contrasting belowground and aboveground production and consumption.
- Describe the distribution and characteristics of savannas.
- Point out the major structural and functional features of savanna ecosystems.
- Discuss human impacts on grasslands and savannas.

The midgrass (often called mixed-grass) prairie stands between the tallgrass prairie and the shortgrass plains.

27.1 DOMESTIC GRASSLANDS

At one time grasslands covered about 42 percent of the land surface of Earth. In the Northern Hemisphere great expanses of grassland covered the mid-continent of North America and extended across the central part of Eurasia. In the Southern Hemisphere grasses covered much of the southern tip of South America and the high plateau of southern Africa. Today grasslands probably occupy less than 12 percent, most of them plowed under for cropland and degraded by overgrazing.

All grasslands have in common a climate characterized by rainfall between 250 and 800 mm (too light to support a heavy forest and too great to result in a desert), a high rate of evaporation, and periodic severe droughts. They share a rolling to flat terrain. Grazing and burrowing species are the dominant animals. Most grasslands require periodic fires for maintenance, renewal, and elimination of woody growth.

Grasses have a mode of growth that adapts them to grazing and fire. The grass plant consists of leafy shoots called tillers. Each shoot has a leaflike blade or lamina, the base of which has a tubelike sheath. These tillers grow from short, underground stems, which grow upward only when the plant begins flowering. Tillers that group closely about a central stem and buds make up bunch or tussock grasses. Species that spread lateral buds on underground stems, producing a sod, are sod or turf grasses (Figure 27.1). Associated with grasses are a variety of legumes and composite plants.

Types of Grasslands

Domestic Grasslands

The grasslands most familiar to a majority of us are hayfields and pasturelands, created and maintained by human efforts. Mostly they occupy forested land cleared for settlements and agriculture. Some grasslands, especially in Britain, Switzerland, and Scandinavia, have existed for centuries, becoming a climax community supporting distinctive vegetation. In other areas, such as eastern North America, abandoned agricultural grasslands revert to forest.

Domestic grasslands are permanent, rotational (plowed every few years for other crops), or rough. The last are marginal, unimproved, semiwild lands used principally for grazing. Many successional grasslands fit this category.

North American Grasslands

In North America, grasslands once covered much of the interior between the Rocky Mountains and the

Figure 27.1 Growth forms and root penetration (maximum depth about 2.5 m) of a sod grass (right) and a bunchgrass (left).

eastern deciduous forest. There were three main types, distinguished by the height of the dominant species, influenced by climate and rainfall.

Tallgrass prairie occupied a narrow belt running north and south next to the deciduous forest of eastern North America. It was well developed in a region that could support forests. Oak-hickory forests did extend into the grassland along streams and rivers, on well-drained soils, sandy areas, and hills. Fires, often set by Native Americans in the fall, stimulated a vigorous growth of grass and eliminated the encroaching forest. Little tallgrass prairie remains.

Big bluestem (*Andropogon gerardi*), growing 1 m tall with flowering stalks 1 to 3½ m tall, was the dominant grass of moist soils and occupied the valleys of rivers and streams and lower slopes of hills. The drier uplands were dominated by bunch-forming needlegrass (*Stipa*), side-oats grama (*Bouteloua curtipendula*), and dropseed (*Sporobolus* spp.). The drier uplands grew such a diversity of composites that they were nicknamed "daisy land."

West of the tallgrass prairie was the **mixed-grass prairie.** Typical of the Great Plains, it embraced largely the needlegrass-grama grass (*Bouteloua stipa*) community. Extremes in precipitation changed its makeup from year to year. The grasses were largely bunch and cool-season species that begin their growth in early April, flower in June, and mature in late July and August.

South and west of the mixed prairie and grading into the desert are the **shortgrass plains,** one grassland that has remained somewhat intact (Figure 27.2). The shortgrass plains reflect a climate in which rainfall is infrequent and light (up to 400 mm in the west and 500 mm in the east), humidity low, winds high, and evaporation rapid. The shallow-rooted grasses utilize moisture in the upper soil layer, beneath which the roots do not penetrate. Sod-forming blue grama *(Bouteloua gracilis)* and buffalo grass *(Buchloe dactyloides)* dominate the shortgrass plains. Because of the dense sod, few forbs grow on the plains, but prominent among them are lupines *(Lupinus* spp.).

From southeastern Texas to southern Arizona and south into Mexico lies the **desert grassland,** similar in many respects to the shortgrass plains, except that three-awn grass *(Aristida* spp.) replaces buffalo grass. Composed largely of bunchgrasses, desert grasslands are widely interspersed with other vegetation types, such as oak savanna and mesquite. The climate is hot and dry. Rain falls only during two seasons, summer (July and August) and winter (December to February), in amounts that vary from 300 mm to 400 mm in the west and 500 mm in the east; but evaporation is rapid, up to 2000 mm per year. Vegetation puts on most of its annual growth in August.

Confined largely to the Central Valley of California is **annual grassland.** It is associated with Mediterranean-type climate, characterized by rainy winters and hot, dry summers. Growth occurs during early spring, and most plants are dormant in summer, turning the hills a dry tan color accented by the deep green foliage of scattered California oaks. The original vegetation was perennial grasses dominated by purple needlegrass *(Stipa pulchra),* but since settlement, native grasses have been replaced by vigorous annual species well-adapted to a Mediterranean-type climate. Dominant species are wild oats *(Avena fatua)* and slender oats *(Avena barbata).*

Eurasian Steppes

At one time the great grasslands of the Eurasian continent extended from eastern Europe to western Siberia south to Kazakhstan. These **steppes,** treeless except for ribbons and patches of forest, are divided into four belts of latitude, from the mesic meadow steppes in the north to semiarid grasslands in the south. The meadow steppes occupy a region in which the rainfall is 500–600 mm, extending south from the taiga. Dominated by bunch-forming fescues *(Festuca)* and feather grass *(Stipa)* along with many species of daisy (Compositae), the meadow steppes were once outstandingly beautiful in spring and early summer. Little remains of meadow steppes, turned under the plow for cereal grains. Further south where rainfall is 400 to 500 mm, tussock-forming species of *Stipa* dominate and flowering herbs are fewer. In the central Asian steppes with their cold dry spring, no ephemeral plants exist, and grasses give way to woody and herbaceous species of drought-resistant *Artemisia.* About the Black Sea and in Kazakhstan, where the humidity is higher, steppe vegetation is dominated by large feather grasses and sheep's fescue *(Festuca ovina)* and by ephemeral spring plants such as tulips *(Tulipa).*

South American Pampas

In the Southern Hemisphere the major grasslands exist in southern Africa and southern South America. Known as **pampas,** the South American grasslands extend westward in a large semicircle from Buenos Aires to cover about 15 percent of Argentina. In the eastern part of the pampas rainfall exceeds 900 mm, well distributed throughout the year. In this humid east the tallgrasses dominate pampas. South and west, where rainfall is about 450 mm, semidesert vegetation becomes prominent. South into Patagonia, where the rainfall averages about 250 mm, the pampas change to open steppe grasses dominated by *Stipa* and *Festuca* and xerophytic cushion plants. These pampas have been modified by the introduction of European forage grasses and alfalfa *(Medicago sativa),* and the eastern tallgrass pampas have been converted to wheat and corn.

Figure 27.2 Shortgrass plains give way in places to forest vegetation in the Black Hills of South Dakota.

South African Veld

The pampas of Argentina occupy the lowlands; by contrast, the **velds** of southern Africa (not to be confused with the savanna) occupy the eastern part of a high plateau 1500 to 2000 m above sea level in the Transvaal and the Orange Free State. Most of the rainfall comes in the summer, brought in by the moist air masses from the Indian Ocean. The heaviest rainfall is in the east, the lowest in the west where the grasslands grade into the semiarid shrubland known as the karoo.

Australian Grasslands

Australia has four types of grasslands: arid tussock grassland in the northern part of the continent, where the rainfall averages between 200 and 500 mm, mostly in the summer; arid hummock grasslands dominated by *Triodia* and *Plectrachne* in areas with less than 200 mm rainfall; coastal grasslands dominated by *Sporobolus* in the tropical summer rainfall region; and subhumid grasslands dominated by such grasses as *Poa* and kangaroo grass *(Themeda)* along coastal areas where rainfall is between 500 and 1000 mm. Most of these grasslands have been changed by fertilization, introduced grasses and legumes, and sheep grazing.

Structure

Vegetation

The most visible feature of a grassland is the tall, green, ephemeral herbaceous growth that develops in spring and dies back in autumn. One of the three strata in the grassland, it arises from the crowns, nodes, and rosettes of plants hugging the soil. The ground layer and the belowground root layer are the other two major strata of grasslands.

The herbaceous layer, consisting of both grasses and forbs, has three or more sublayers, more or less variable in height, according to the grassland type (Figure 27.3). Low-growing and ground-hugging plants such as dandelion, strawberry, and mosses make up the first layer. As the growing season progresses, these plants become hidden beneath the middle and upper layers. The middle layer consists of shorter grasses and such forbs as wild mustard and daisy. The upper layer consists of leaves and flowering stems of tall-grasses and the leafy stalks and flowers of forbs.

The ground layer is easy to see in late winter and early spring. Exposed to high light, the plants respond to warmth and moisture. As the grasses and forbs grow taller and shade the ground, light intensity reaching

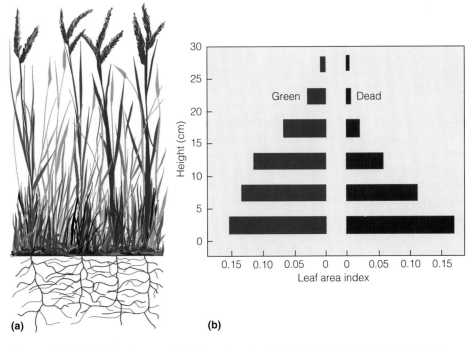

Figure 27.3 (a) Profile of a grassland showing physical stratification during the summer, energy flow, structure, and stratification of the physical environment. (b) Leaf area indexes, as indicated by bar graphs, at different levels in a grassland for both green and dead plant structures.

the ground layer decreases. Temperature declines, relative humidity increases, and wind flow decreases, creating a region of calm near the ground. Conditions on grazed lands are much different. Because the grass cover is closely cropped, the ground layer continues to receive much higher solar radiation, higher temperatures, and greater wind velocity.

Grasslands that are unmowed, unburned, and ungrazed accumulate a thick layer of mulch (on your lawn it is called thatch). The oldest bottom layer, humic mulch, consists of decayed and fragmented remains of fresh mulch; the top layer consists of fresh herbage, leafy and largely undecayed, that is deposited throughout the growing season. As the mat increases in depth, it retains more moisture, creating favorable conditions for microbial activity. Three or four years must pass before natural grassland mulch decomposes completely.

Grazing reduces the mulch layer, as do fire and mowing. Light grazing tends to increase the weight of decayed humic mulch at the expense of fresh mulch; moderate grazing increases compaction, which favors microbial activity and a subsequent reduction in both fresh and humic mulch. Heavy grazing greatly reduces mulch accumulation. Burning reduces both fresh and humic mulch, but the mulch structure returns after a fire on lightly grazed and ungrazed lands. Mowing greatly reduces both fresh mulch and humic mulch. Haylands have a minimal amount of mulch on the ground.

The amount of mulch is ecologically important. Mulch increases soil moisture through its effects on infiltration and evaporation; it decreases runoff and erosion, stabilizes soil temperatures, and improves conditions for seed germination. How much mulch is necessary is the question. Where mulch can accumulate to the proper degree, grassland maintains itself. Heavy mulches can suppress growth of grasses and allow the invasion of forbs and woody vegetation. In areas of no accumulation, grassland regresses to weedy plants. Deep litter provides habitat for meadow mice and certain ground-nesting birds such as the bobolink, but inhibits the presence of others.

The root layer is more highly developed in grasslands than in any other major community. Half or more of the plant is hidden beneath the soil; in winter, roots represent almost the total grass plant. Most roots are fibrous and occupy rather uniformly the upper 15 cm or so of the soil profile; they decrease in abundance with depth. The depth to which roots extend is considerable. Little bluestem, for example, reaches 1 to 2 m and forms a dense mat as deep as 0.8 m. In addition, many grasses possess underground stems or rhizomes that serve both to propagate the plant and to store food. Constantly dying, roots add finely divided organic matter to the mineral soil.

Roots develop in three or more zones. Some plants are shallow-rooted and seldom extend much below 0.5 m. Others go well below the shallow-rooted species but seldom more than 1.5 m. Deep-rooted plants extend even farther into the soil and absorb little moisture from the surface soil. Thus plant roots absorb nutrients from different depths in the soil at different times, depending on moisture.

Animal Life

Natural or domestic, grasslands support similar forms of life, vertebrate and invertebrate. The invertebrate life includes an incredible number and variety of species and occupies all strata during some time of the year. During winter in temperate grasslands, insect life is confined largely to soil, litter, and grass crowns where they exist as eggs or pupae. In spring, soil occupants are chiefly earthworms and ants, the latter being the most prevalent if not the most conspicuous. The ground and mulch layers harbor scavenger carabid beetles and predaceous spiders, of which the majority are hunters rather than web builders. Life in the herbaceous layer varies as the strata become more pronounced from spring to fall. Here invertebrate life is most abundant and varied. Homoptera, Coleoptera, Diptera, Hymenoptera, and Hemiptera are all represented. Insect life reaches two highs during the year, a major peak in summer and a less defined one in the fall.

Large grazing ungulates and burrowing mammals are the most conspicuous vertebrates. All of the world's native grasslands support similar forms. The North American grasslands once were dominated by huge migratory herds of bison, numbering in the millions (Figure 27.4), and the forb-consuming pronghorned antelope (*Antilocarpa americana*). The most common burrowing rodent was the prairie dog (*Cynomys* spp.), which along with gophers (*Thomomys* and *Geomys*) and the mound-building harvester ants appeared to be instrumental in the development and maintenance of the ecological structure of the shortgrass prairie.

The Eurasian steppes lack herds of large ungulates. The western steppes are home to the small migratory goat antelope, the saiga (*Saiga tartarica*), characterized by a large proboscis-like nose, which increases in size in the male during rut. Nearly extinct in 1917, it now numbers over one million animals. Farther east lives the Mongolian gazelle (*Procarpa gutterosa*) and several

Figure 27.4 Bison, which once roamed the shortgrass plains in countless numbers, epitomize the North American grasslands.

species of rare wild horses. The dominant burrowing animals are the bobak marmot *(Marmota bobak)*, which looks like an oversized prairie dog, the sousliks or ground squirrels *(Citellus)*, and the common hamster *(Cricetus cricetus)*.

The Argentine pampas also lack a large ungulate fauna. The two major large herbivores are the pampas deer *(Ozotoceras bezoarticus)* and farther south the guanaco *(Lama guanaco)*, small relatives of the camel, greatly reduced in number compared to historical times. Major burrowing rodents are the viscacha *(Lagostomus maximus)* and the Patagonian hare or mara *(Dolichotis patagonium)*, a monogamous, cavylike rodent, with the long ears of a hare and the body and long legs of an antelope.

The African grassveld once supported great migratory herds of antelope and zebra along with their associated carnivores, the lion, leopard, and hyena. Burrowing rodents include the kangaroo-like springhare *(Pedetes capensis)* and the gerbil *(Tatera brantsii)*, and a most interesting carnivore, the meerkat *(Cynictis penicillata)*, whose burrowing habits suggest those of the prairie dog. The rodents remain, but the great ungulate herds have been destroyed and replaced with sheep, cattle, and horses.

The Australian marsupial mammals evolved many forms that are the ecological equivalents of placental grassland mammals. The dominant grazing animals are a number of species of kangaroos, especially the red kangaroo *(Macropus rufus)* and the gray kangaroo *(M. giganteus)*. The wombats *(Vombatus)* occupy the ecological niche of the viscachas of the pampas and the gophers of the prairies.

Three of the world's grasslands evolved unique unrelated birds with a poor ability to fly, large size, and high running speed. Australia has the emu *(Dromiceius casuarius)*, the pampas the rhea *(Rhea americana)*, and Africa the ostrich *(Struthio)*. New Zealand, whose grasslands (not discussed) lacked herbivorous mammals, had flocks of the now extinct grass-consuming moa *(Dinorus)*. Although the grasslands of the Northern Hemisphere lack such large birds, the European steppes do have the large great bustard *(Otis tarda)*, weighing up to 16 kg. Its numbers have been reduced by loss of habitat to agriculture.

Vertebrate life in seral and tame grasslands is strongly affected by human management. Mowing hayfields destroys habitat at a critical nesting time. Losses to birds, rabbits, and mice from mechanical injury and predation on exposed nests are often heavy, but most species will remain on the area to complete or reattempt nesting. Early mowing at the very start of the nesting season eliminates nesting cover and forces the animals elsewhere. Early mowing is one of the reasons for the sharp decline in grassland birds. Pasturelands more often than not are so badly overgrazed they support little vertebrate life. The two most common inhabitants in eastern North America are the killdeer *(Charadrius vociferus)* and the horned lark *(Eremophila alpestris)*.

Function

Grasslands are adapted to periods of drought and survive under low rainfall, but grasses grow best under optimal moisture and temperature. Grasslands do the poorest where precipitation is the lowest and temperatures are high; they do best where the mean annual precipitation is greater than 800 mm and mean annual temperature is above 15° C. Production, however, is most directly related to precipitation (Figure 27.5). The greater the mean annual precipitation, the greater is aboveground production. This increased production comes about because increased moisture reduces water stress and enhances the uptake of nutrients.

We associate grassland production with the aboveground growth of grass, but much of the net production is below ground (Figure 27.6). Except for tropical grasslands, which put most of their net production into aboveground biomass, seminatural and temperate grasslands send most of their production into the roots. Except for a short period of maximum aboveground biomass during the growing season, belowground biomass is two to three times that above ground.

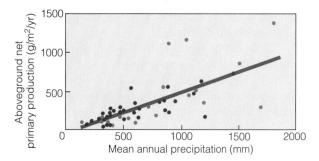

Figure 27.5 Relationship between aboveground primary production and mean annual precipitation for 52 grassland sites around the world. North American grasslands are indicated by dark green dots.

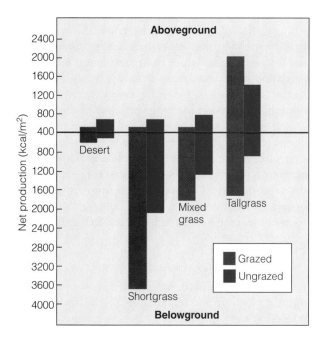

Figure 27.6 Aboveground and belowground net primary production for grazed and ungrazed North American grassland types.

Seventy-five to 85 percent of grassland photosynthate is translocated to the roots for storage.

Grasslands have evolved under grazing pressures of ungulates since the Cenozoic. Their structure and growth habits reflect this selective pressure. Critical growth tissues are at or below ground surface, protected from grazing and fire. As the grazers clip and eat the leaves, grasses respond by increasing the photosynthetic rate in remaining tissue, stimulating new growth, and reallocating nutrients and photosynthates from one part of the plant to another, especially from roots to stems (see Chapter 16). In addition, grazers recycle nutrients in grass through dung and urine.

Grassland ecosystems respond to grazing in still another way by changing species composition. Some grasses and forbs tend to disappear, while other species increase. On desert grasslands of North America black grama is replaced by weedy species; on shortgrass plains, which are the most stable under grazing pressure, blue grama and prickly pear increase; on mixed-grass prairies midgrasses decrease and shortgrasses and sedges increase. On tallgrass sites tallgrasses disappear and little bluestem and tall dropseed increase; if grazing pressure is heavy, the site may be invaded by the weedy Japanese chess. In domestic pasturelands, heavy grazing increases the amount of unpalatable forbs, such as thistles and ironweed.

Because they are so conspicuous, we equate grazing pressure with aboveground herbivores such as cattle and rabbits and invertebrate grazers such as grasshoppers. In reality, much more intense grazing takes place underground. The aboveground biomass of invertebrates—including plant consumers, saprovores, and predators—ranges from 1 to 50 g/m^2, whereas grazing mammals amount to about 2 to 5 g/m^2. Belowground invertebrates exceed 135 g/m^2, most of them nematodes. They account for 90, 95, and 93 percent of all belowground herbivory, carnivory, and saprophagous activity, respectively.

Not only is a large proportion of primary production eaten below ground, but also a greater proportion is used at each trophic level there. Some invertebrate aboveground consumers, particularly grasshoppers, are wasteful. The amount of aboveground vegetation they detach or otherwise kill about equals that consumed by vertebrate grazing herbivores.

Invertebrate consumers are also highly inefficient in assimilating ingested material and deposit much of their intake as highly soluble feces, or frass. The nutrients feces contains return rapidly to the system. Large grazing herbivores return a portion of their intake as dung, which is fed upon by a well-developed coprophagous fauna that speeds the decay of manure and accelerates the activity of bacteria in feces.

Most of the primary production, however, goes to the decomposers, dominated by fungi whose biomass is two to seven times that of bacteria. Overall the decomposer biomass exceeds that of invertebrates.

Central to the cycling of nutrients in grasslands is mulch or detritus, a large standing crop of which can have detrimental effects on nitrogen cycling, particularly in tallgrass ecosystems. Detritus intercepts rainfall, from which microbes can assimilate inorganic nitrogen directly before it reaches plant roots, while the mulch itself inhibits nitrogen fixation by free-living nitrogen-fixing microbes. By insulating the soil surface from solar radiation, mulch reduces production of new roots and inhibits the activity of soil microbes and invertebrates. Periodic grassland fires clear away the mulch layer and release nutrients in detritus to the soil, but nitrogen equal to about two years of nitrogen inputs to the system through rainfall is lost to the atmosphere. Fires, however, stimulate the growth of nitrogen-fixing leguminous forbs and improve conditions for earthworms.

Human Impact

Humans have broken up grasslands with the plow and converted the most productive of them into the breadbaskets of the world, dominated by a monoculture of cereal grains. Conversion of the shortgrass plains of North America to wheat resulted in the Dust Bowl, when a seven-year drought hit the region in the 1930s.

On less productive grasslands we have replaced native plant species with highly productive forage plants accompanied by fertilization, pest control, irrigation, and other intensive practices. We have removed great herds of wild ungulates from natural grasslands and replaced them with domestic stock. Natural grasslands had experienced grazing pressure from free-ranging and often migratory populations of mammals. When we replaced these wild ungulates with domestic ones, we confined them with fences and overstocked the ranges, causing serious deterioration of grassland systems (Figure 27.7). Overgrazing desert grasslands in North America has increased the spread of mesquite because of lessened competition from grass and dispersal of seed by livestock. On other overgrazed grasslands mulch deteriorates and disappears because only a small amount of litter is added to the ground. Water flows over the surface, taking topsoil with it. Lacking moisture and nutrients, the original species cannot maintain themselves, and the vegetation cover continues to decrease until only an erosion pavement remains. In the semiarid regions of Africa, overgrazing has obliterated vegetation, converting those regions to deserts and dust bowls.

In forested regions, we have created new grasslands by clearing the forests and planting forage grasses. At first such activity permitted the eastward spread of such grassland species as bobolinks, meadowlarks, and rabbits as their habitat in the plains was being destroyed. In Old World countries such as Britain and Scandinavia, human-created grasslands have existed for centuries, creating a seminatural climax vegetation. However, even these grasslands have been destroyed by plowing in recent times, threatening the

Figure 27.7 Cattle stand in this severely overgrazed range on Bureau of Land Management land in Utah.

extinction of such plants as bird's-eye primrose *(Primula farinosa)* and pasqueflower *(Anemone pulsatilla)*, which cannot tolerate disturbance.

27.2 TROPICAL SAVANNAS

The one ecosystem that defies any general description is the tropical **savanna.** The problem is an old one, involving even its name. The word in its several origins, largely Spanish, referred to grasslands or plains; but over time the word was applied to an array of vegetation types representing a continuum of increasing cover of woody vegetation, from open grassland to widely spaced shrubs or trees to closed woodland (Figure 27.8). Moisture appears to control the density of woody vegetation, a function of both rainfall (amount and distribution) and soil—its texture, structure, and water-holding capacity.

Savannas cover much of central and southern Africa, western India, northern Australia, large areas of northwestern Brazil where they are known as cerrados, Colombia and Venezuela where they are called llanos, and to a more limited extent Malaysia. Some savannas are natural. Others are seminatural, brought about and maintained by centuries of human interference. In the African savannas, in particular, it is difficult to separate the effects of humans from the effects of climate. The savannas of central India, however, are the result of human degradation of original forest land.

Savannas, in spite of their vegetational differences, exhibit a certain set of characteristics. Savannas

(a)

(b)

(c)

(d)

Figure 27.8 Savanna ecosystems in Africa. (a) Grass savanna with giraffe on the Maasai Mara reserve in Kenya. (b) Shrub savanna in southern Africa. (c) Tree savanna with a well-developed growth of *Acacia* in the background and a bull elephant at its edge. (d) Savanna thorn woodland cloaks a hillside in South Africa.

occur on land surfaces of little relief, often old alluvial plains. The soils are low in nutrients, due in part to infertile parent material and a long period of weathering. Savanna regions are associated with a warm continental climate with precipitation ranging between 500 mm and 2000 mm. Precipitation exhibits extreme seasonal fluctuations; in South American savannas, in particular, the soil-water regime may fluctuate from excessively wet to extremely dry, often below the permanent wilting point. Savannas are subject to recurrent fires, and the dominant vegetation is fire-adapted. Grass cover with or without woody vegetation is always present. When present the woody component is short-lived, with individuals seldom surviving for more than several decades (except for the African baobab trees). Detrital-processing termites are a conspicuous component of savanna animal life, especially in Africa.

Structure

Vegetation

The major and most essential stratum of the savanna ecosystem is grass, mostly bunch or tussock, with no vertical structure; its biomass decreases with height. A woody component adds one or two more vertical layers, ranging from about 50 to 80 cm when small woody shrubs are present to about 8 m in the tree savannas. Highly developed root systems make up the larger part of the living herbaceous biomass. The root system is concentrated in the upper 10 cm but extends down to about 30 cm. Savanna trees have extensive horizontal roots that go below the layer of grass roots. Competition may exist between grass and woody vegetation for soil moisture, but more intense competition takes place among trees, accounting for the spacing patterns of woody vegetation.

In contrast to the poorly developed vertical structure is a well-developed, although often unapparent, horizontal structure. The tussock grasses form an array of clumps set in a matrix of open ground, creating patches of low vegetation with frequent changes in microclimatic conditions. The addition of woody growth, the widely spaced shrubs and trees, increases horizontal structure extending to the soil. Trees add some organic matter and nutrients to the soil beneath them, reduce evapotranspiration, resulting in increased herbaceous and woody shrub growth, and provide patches of shade. On the African savanna in particular, large grazing herbivores rest in the shade during the heat of the day and concentrate nutrients from dung and urine beneath them.

Breaking up the monotony of the savannas are numerous marshy depressions that support wetland wildlife. Large ribbons of riverine or gallery forests weave through the savannas. The gallery forests support a diversity of wildlife and provide forage in the dry season for buffalo, waterbuck, and other large ungulates.

Animal Life

Savannas are capable of supporting a large and varied assemblage of herbivores, invertebrate and vertebrate, grazing and browsing. Dominant herbivores are the invertebrates, including acrid mites, acridid grasshoppers, seed-eating ants, and detrital-feeding dung beetles and termites. Savanna vegetation supports an incredible number of insects: flies, grasshoppers, locusts, crickets, carabid beetles, and especially termites and ants, which dominate insect life. Insect abundance is seasonal and is strongly affected by burning, which can reduce populations by more than 60 percent.

In South American savannas, there is a strong element of grazing ungulates represented by pampas deer and the capybara *(Hydrochaeris hydrochaeris)*. Granivorous, insectivorous, and frugivorous birds become an important component of the consumer community.

The African savanna, visually at least, is dominated by a large and diverse ungulate fauna of at least 60 species that partition the vegetative resource among them. Some, such as the wildebeest and zebra, are migratory during the dry season. Others, such as the impala, partially disperse during the dry season. Still others, such as the giraffe and Grant's gazelle, have little or no seasonal dispersal. Among the ungulates zebras and wildebeest are generalist grazers. Zebras, especially during the migratory period, feed on upper grass leaves, low in protein. Wildebeest feed on the more nutritious grasses, and the small gazelles, being more refined feeders, live on the lower grasses left behind, especially the new short growth at the beginning of the rainy season. Other ungulates, such as giraffe (Figure 27.9), Thompson's gazelle, kudu, and black rhino, are woody browsers. A close interaction exists among the grazing herbivores, and intensive grazing pressure by one species can affect the populations of others. In spite of their visual dominance, large ungulates consume only about 10 percent of primary production.

Putting the level of consumption aside, herbivores have short-term and long-term impacts on the savanna. Over the short term, the grazing ungulates affect vegetation structure. Elephants can convert woodland to grassland, and large concentrations of grazers can turn grassland to eroded, bare ground. The species composition and structure of the African savanna vegetation would be different if it were not subject to heavy grazing, which alters competitive interactions among plants. Heavy grazing that reduces grass cover can cause competitive release of woody growth.

Figure 27.9 Two major woody browsers of the African woodland savanna. The giraffe feeds on the tall woody growth; the endangered black rhino feeds on low shrubs.

Over evolutionary time the selective pressures of grazing have resulted in the development of structural and chemical defenses against grazing, such as concentration of silica in leaves, and in the alteration of growth processes to respond favorably to grazing. Highly palatable plants retain a high carbohydrate concentration in their crowns and roots and respond quickly to defoliation. Some acacia trees respond to browsing by increasing growth. Others have their growth form altered and size reduced by browsing, especially by giraffes.

Living on the ungulate fauna is an array of carnivores, including the lion, leopard, cheetah, hyena, and wild dog. Subsisting on leftover prey are scavengers, including vultures and jackals.

Function

Because of the wide diversity of savanna types and limited studies, it is difficult to make any strong generalizations about primary production. Probably a wide range of production exists between grass savanna on one end of the gradient and tree savanna and woodlands on the other.

At the beginning of the wet season moisture releases nutrients from materials accumulated in the dry season and stimulates nutrient translocation from the roots. This action is followed by a quick flush of growth into grass and woody plants. Nutrient movement between soil and vegetation is generally higher under the trees than in the open, because of greater organic matter accumulation and reduced evaporation, which keeps the soil moist.

Savanna trees, especially the African acacias, exhibit tight internal cycling. Nitrogen concentration in the leaves, for example, decreases as the dry season approaches, with maximum withdrawal before leaf fall. The trees transfer some of the nitrogen into new woody growth, but much of it goes to the root reserve, where it is available to stimulate the flush of new season growth. A similar tight circulation exists in neotropical savannas. Most of the nitrogen in the dry aboveground biomass is lost to the atmosphere by volatilization if fire sweeps the savanna; otherwise a fraction will be transferred to the soil through leaching effects of rain water.

The influence of large herbivores on nutrient dynamics over the long term is debatable, but the impact of ants and especially termites cannot be questioned. Ants and termites consume and break down plant litter and modify the soil. Mound-building termites excavate and move tons of soil, mixing mineral soil with organic matter. Some species construct extensive subterranean galleries, and others accumulate organic matter. Comprising over 50 percent of soil biomass, termites have a considerable impact on the physical and chemical properties of savanna soil.

Human Impact

Ever since their early evolution, humans have had such a close association with savanna vegetation, especially in Africa, that it is difficult to separate human influences from natural influences on the shaping of savanna ecosystems. In South America, savanna vegetation still is relatively free of human impact. In Africa and Australia, however, the arrival of early humans increased the importance of fire.

In modern times humans have had a severe and often adverse impact on the savanna ecosystem. Introduction of crops, grazing animals, and settlements has accentuated the dry season and increased the desiccation of the drier savannas. This desiccation has allowed the desert to encroach; such encroachment is most evident in the Sahelian zone south of the Sahara Desert in North Africa and in the Great Indian Desert. Cutting and burning of trees for fuel wood, destruction by domestic grazing animals, and the loss of grass cover help to expose the soil to wind and water erosion.

In some areas of Africa, savanna vegetation is being converted to pine and eucalyptus forests for wood and paper pulp. Widespread slaughter of the large grazing herbivores has also changed the character of the savanna vegetation. Many thousands of hectares have been converted to croplands of corn, pineapple, and sisal, or to pastures by addition of fertilizers and the introduction of exotic forage grasses and legumes.

In Africa, savanna vegetation can support five times as much standing biomass of wild ungulates as it can domestic livestock, because wild ungulates are better at foraging, withstanding heat stress, and resisting disease than are domestic cattle.

STUDY QUESTIONS

1. What is the difference between a bunchgrass and a sod grass?
2. What characteristics do all grasslands have in common?
3. Why does the root system assume such importance in the grassland ecosystem?
4. What is the role of mulch in grassland ecosystems?
5. How have grasses adapted to grazing?

6. What characterizes grassland animal life?

7. Contrast the production and function of the aboveground and belowground components of grassland ecosystems. Why does so much production and nutrient cycling take place below ground?

8. What distinguishes savannas from grasslands in structure and function?

9. Where are the world's savannas located? Under what climatic conditions have they developed?

10. What is the relationship of humans to savannas?

*11. Speculate on why there are so few grassland reserves in North America.

*12. Prior to 1930, Frederick Clements, the plant ecologist, considered plowing the shortgrass plains for wheat short-sighted and urged their preservation by restocking them with commercial herds of bison. Comment on his rejected suggestion.

CHAPTER 28

SHRUBLANDS AND DESERTS

OBJECTIVES

On completion of this chapter, you should be able to:

- Describe the major characteristics of shrubs and shrublands.
- Characterize the major types of world shrublands.
- Discuss the relationship between growth forms of shrubs and nutrient cycling.
- Describe the major features of deserts.
- Discuss the unique aspects of nutrient cycling in desert ecosystems.
- Discuss human impacts on both shrublands and deserts.

The Namib Desert—a narrow strip of land along the Atlantic coast of Namibia in southern Africa.

28.1 CLIMAX SHRUBLANDS

Climax shrubby vegetation covers large portions of the arid and semiarid world. In addition, climax shrubland covers parts of temperate regions because historical disturbances of landscapes have seriously affected their potential to support forest. Among such shrub-dominated human-induced climax communities are the moors of Scotland and the macchia of South America. Outside these regions, shrublands are seral, a stage in the land's progress back to forest. There they are second-class citizens of the plant world, given little attention by botanists, who tend to emphasize dominant plants. As a result, not enough work has been done on seral shrub communities.

Characteristics

Shrubs are difficult to characterize. They have, as W. G. McGinnes points out, a "problem in establishing their identity." They constitute neither a taxonomic nor an evolutionary category. A rough definition is that a shrub is a plant with multiple woody, persistent stems but no central trunk and a height from 4.5 to 8 m. However, under severe environmental conditions many trees will not exceed that size. Some trees, particularly coppice stands, are multistemmed, and some shrubs have large, single stems. Shrubs may have evolved either from trees or from herbs.

The success of shrubs depends on their ability to compete for nutrients, energy, and space. In certain environments shrubs have many advantages. They invest less energy and nutrients in aboveground parts than trees. Their structural modifications improve light interception, heat dissipation, and evaporation. The more arid the site, the more common is drought deciduousness and the less common is evergreenness. The multistemmed forms influence interception of moisture and stemflow, increasing or decreasing infiltration into the soil. Because most shrubs can get their roots down quickly and form extensive root systems, they use moisture deep in the soil. This feature gives them a competitive advantage over trees and grasses in regions where soil recharges between growing seasons. Because they do not have a high root-shoot ratio, shrubs draw fewer nutrients into aboveground biomass and more into roots. Their perennial nature allows immobilization of limiting nutrients and slows nutrient recycling, favoring further shrub invasion of grasslands.

Subject to strong competition from herbs, some shrubs, such as chamise (*Adenostoma fasciculatum*), inhibit the growth of herbs by means of allelopathy (see Section 15.10). Only when fire destroys mature shrubs and degrades the toxins do herbs appear in great numbers. As the shrubs recover, herbs decline. The herb species affected have apparently evolved the ability to let their seeds lie dormant in the soil until they are released from suppression by fire.

Types of Shrubland

Mediterranean-Type Shrublands

In five regions of the world, lying for the most part between 32° and 40° north and south of the equator, are areas with a mediterranean climate: the semiarid regions of western North America, the regions bordering the Mediterranean Sea, central Chile, the Cape region of South Africa, and southwestern and southern Australia. The mediterranean climate has hot, dry summers with at least one month of protracted drought and cool, moist winters. About 65 percent of the annual precipitation falls during the winter months and for at least one month the temperature remains below 15° C.

All five areas support similar-looking communities of xeric broadleaf evergreen shrubs and dwarf trees known as sclerophyll (*scleros*, "hard"; *phyll*, "leaf") vegetation with an herbaceous understory. Sclerophyllous vegetation possesses small leaves, thickened cuticles, glandular hairs, and sunken stomata. In the Northern Hemisphere this vegetation evolved from tropical floras in dry summer climates that began during the Pleistocene.

Vegetation in each of the mediterranean systems also shares adaptations to fire and to low nutrient levels in the soil.

Despite convergence, each area has distinct flora and fauna. In the Mediterranean region, shrub vegetation falls into three major types. The **garigue,** resulting from degradation of pine forests, includes several types of dwarf-shrub communities less than 0.5 m high, dominated by aromatic evergreen shrublets on well-drained to dry, calcareous soil. The **maquis,** replacing cork forests, is a dense evergreen sclerophyllous shrub community where the climate is moist. The **mattoral** appears to be equivalent to the North American chaparral. The Chilean mediterranean system, also called mattoral, varies from the coast to the foothills of the Andean cordillera.

In North America the sclerophyllous shrub community is known as **chaparral,** a word of Spanish origin meaning a thicket of shrubby evergreen oaks. California chaparral is dominated by scrub oak (*Quercus* *berberidifolia)* and chamise *(Adenostoma fasciculatum).* Another shrub type, also designated as chaparral, is associated with the Rocky Mountain foothills. It differs from California chaparral in two ways. It is dominated by Gambel oak *(Q. gambelii)* and other species and lacks chamise; it is summer-active and winter-deciduous, whereas California chapparal is evergreen, winter-active, and summer-dormant.

In their original presettlement state, both the Mediterranean and California mediterranean plant communities were dominated by oaks, both shrub and tree. Natural oak forests remain now only in scattered patches. In Spain, cork oak is a plantation tree, and many oak forests throughout the region have been converted to olive plantations. In California, four major oak forest communities included two endemics, the evergreen blue oak *(Q. douglassii)* and valley oak *(Q. lobata).* The ranges of both have been greatly reduced by settlement.

Much of the South African mediterranean shrubland is heathland, discussed later. The rest, dominated by a broad-sclerophyll woody shrub, goes by the names of *strandveld, coastal renosterveld,* and *inland renosterveld.*

In southwest Australia the mediterranean shrub country, known as **mallee,** is dominated by low-growing *Eucalyptus,* 5 to 8 m high with broad sclerophyllous leaves. There are six types of mallee ecosystems, which intergrade. Three of them fall into mediterranean-type ecosystems, with a grassy and herbaceous understory (Figure 28.1). The other three types occur on nutrient-poor soils and fall under the category of heathland shrubs. Razed by fire at irregular intervals, the mallee retains a summer growth rhythm evolved in the subtropical Tertiary, which is out of phase with the mediterranean climate of the area. Growth takes place during the summer, the driest part of the year, using extensive root systems to draw on water conserved in the soil during the wet winter and spring.

For the most part, mediterranean-type shrublands lack an understory and ground litter, are highly inflammable, and are heavy seeders. Many species require the heat and scarring action of fire to induce germination. Others sprout vigorously after a fire.

For centuries periodic fires have roared through mediterranean-type vegetation, clearing away the old growth, making way for the new, and recycling nutrients through the ecosystem. When humans intruded on this type of vegetation, they changed the fire regime, either by attempting to exclude fire completely or by overburning. In the absence of fire, chaparral grows tall and dense and yearly adds more leaves and twigs

Figure 28.1 Tall shrub mallee in Victoria, Australia, is an example of a mediterranean-type shrubland dominated by *Eucalyptus.* Note the canopy structure and the open understory of grass at the beginning of spring rains. This type of vegetation supports a rich diversity of bird life.

to those already on the ground. During the dry season the shrubs, even though alive, nearly explode when ignited. Once set on fire by lightning or humans, an inferno follows.

After fire the land returns either to lush green sprouts coming up from buried root crowns or to grass, if a seed source is nearby. New grass and vigorous young sprouts are excellent food for deer, sheep, and cattle. As the sprout growth matures, chaparral becomes dense, the canopy closes, the litter accumulates, and the stage is set for another fire.

Northern Desert Scrub

In the Great Basin of North America, the northern, cool, arid region lying west of the Rocky Mountains, is the northern desert scrub. The climate is continental, with warm summers and prolonged cold winters. Although this region is perhaps more appropriately considered a desert, it is one of the most important shrublands in North America (Figure 28.2). Its physiognomy differs greatly from the southern hot desert and the dominant vegetation is shrub. The vegetation falls into two main associations: One is sagebrush, dominated by *Artemisia tridentata,* which often forms pure stands; the other is shadscale, *Atriplex conifertifolia,* a C_4 species, and other chenopods, halophytes tolerant of saline soils. Inhabiting this shrubland are pocket and kangaroo mice, lizards, and sage grouse, sage thrasher, sage sparrow, and Brewer's sparrow, four birds that depend on sagebrush.

Figure 28.2 The northern desert shrubland in Wyoming is dominated by sagebrush. Although classified as cold desert, sagebrush forms one of the most important shrub types in North America.

Figure 28.3 Saltbrush shrubland in Victoria, Australia, is dominated by *Atriplex*. It is an ecological equivalent of the shrublands of the Great Basin in North America.

A similar type of shrubland exists in the semiarid inland of southwestern Australia. Numerous chenopod species, particularly the saltbushes of the genera *Atriplex* and *Maireana*, form extensive low shrublands on low riverine plains (Figure 28.3).

Heathlands

Typically, heathlands have been associated with cool to cold temperate climatic regions of northwestern Europe. It was probably coincidental that the original name came to refer to land dominated by Ericaceae (the heath family). The word *heath* comes from the German *heide*, meaning "an uncultivated stretch of land," regardless of the vegetation.

Heathlands are found in all parts of the world, from the tropics to polar regions and from lowland to alpine altitudes. Heathland flora probably evolved in the Mesozoic in the eastern to central portion of Gondwanaland and retained most of its characteristics. It expanded from Africa into western Europe and the northern part of Eurasia and North America, from India into southeastern Asia, and from Australasia into the Malay Archipelago.

Heathland vegetation is an assemblage of dense to mid-dense growth of ancient or primitive genera adapted to fire (Figure 28.4). Heathland shrubs have leaves with thick cuticles, sunken stomata, thick-walled cells, and hard and waxy upper surfaces. Many species have leaves with small surface area—less than 25 mm² —and others roll their edges in toward the midrib. Although

Figure 28.4 Heathlands, dominated by ericaceous shrubs, have a similar physiognomy around the world. Typical is this heathland in the Brindabella Range, Australia. In the background is a stand of snow gum.

mostly associated with heathlands, many heathland shrubs are usually present as shrubby understory in other ecosystems, such as the deciduous forest.

Heathlands invariably occur on nutrient-poor soils especially deficient in phosphorus and nitrogen. Although heathlands are most extensive in the arctic regions, they are also prominent in the mediterranean-type regions of South Africa, where they are known as *fynbos*, and in southeastern and western Australia. In subtropical to tropical climates true heathlands are confined to alpine areas and to lowland, poor soils subject to seasonal waterlogging. Some heathlands, such as the heather-dominated moors of Scotland, are human-induced and maintained only by periodic fires.

There are two distinct heathland ecosystems: dry heathlands and wet heathlands. Dry heathlands are on well-drained soils subject to seasonal drought, and wet heathlands are subject to seasonal waterlogging. In wet heathlands, grasses and sedges may become codominant with heathland shrubs; and in extreme wet heathlands the grass component is suppressed by *Sphagnum* moss (see Chapter 33). Because both foliage cover and height vary considerably with the habitat, heathlands are divided according to height of the uppermost stratum: shrubs taller than 2 m, scrub; shrubs 1–2 m, tall heathland; shrubs 25–100 cm, heathland; and shrubs less than 25 cm, dwarf heathland.

Successional Shrublands

On drier uplands, shrubs rarely exert complete dominance over herbs and grass. Instead the plants are scattered or clumped in grassy fields, the open areas between filled with the seedlings of forest trees, which in the sapling stage of growth occupy the same ecological position as tall shrubs (Figure 28.5). Typical are thicket-forming hazel (*Corylus* spp.), sumacs (*Rhus* spp.), and shrub dogwoods (*Cornus* spp.).

On wet ground the plant community often is dominated by tall shrubs and contains an understory intermediate between those of a meadow and a forest. In northern regions the common tall shrub communities found along streams and lake shores are thickets composed of alder or alder and a mixture of other species such as willow (*Salix* spp.) and red osier dogwood (*Cornus stolonifera*). Alder thickets are relatively stable and remain for some time before being replaced by forests. Outside of alder country, a shrub or *carr* community occupies the low places. Dogwoods are some of the most important species in the carr. Growing with them are a number of willows, which usually dominate.

Shrub thickets provide excellent food and cover for wildlife. Many shrubs, such as blackberry, hawthorn, greenbriar, and dogwoods, rank high as wildlife food. However, the overall value of different types of shrub cover, its composition, quality, and minimum amounts needed, have never been assessed. Seral shrub communities are ephemeral, lasting only about 15 to 20 years. There is some evidence, however, that even where forest is the normal end of succession, shrubs can form a stable community that will persist for many years. If incoming tree growth is removed either by selective herbicidal spraying or by cutting, shrubs eventually form a closed community resistant to further invasion by trees. This response might be more widely used in vegetation management of power line rights-of-way.

Structure

Shrub ecosystems, seral or climax, are characterized by woody structure, increased stratification over grasslands, dense branching on a fine scale, and low height, up to 8 m. Typically there are three layers—a broken upper canopy, an irregular low shrub canopy, and a grass/herbaceous layer—but the presence of these layers varies. Dense shrubland may have only a canopy layer, and stratification often decreases as the shrubs reach maximum height. This condition is particularly true in seral shrublands. Horizontal patterns vary with the vegetation type. Heathlands may exhibit little patchiness across the landscape. Mediterranean-type shrublands, notably the mattoral and the mallee, may be very patchy, with woody growth well interspersed with open areas of grass. Greatest patchiness probably occurs in seral shrublands, with scattered clumps of invading shrubs and trees.

Because of this structure, shrub communities have distinctive animal life. Seral shrub communities support not only species common to shrubby edges of forest and shrubby borders of fields but a number of species dependent on them, such as bobwhite quail, cottontail rabbit, prairie warbler (*Dendroica discolor*), and yellow-breasted chat (*Ictera virens*). In Great Britain some shrub communities, especially hedgerows, have

Figure 28.5 In eastern North America and in northern and western Europe, shrublands are usually successional communities. This old field in southern New York State is in the early stages of shrub invasion.

been stable for centuries, and many forms of animal life, invertebrate and vertebrate, have become adapted to or dependent on them. Among these species are the whitethroat *(Sylvia communis)*, linnet *(Acanthis cannabina)*, blackbird *(Turdus merula)*, and yellowhammer *(Emberiza citrinella)*.

Climax shrub communities have a complex of animal life that varies with the region. Within the Mediterranean-type shrublands and heathlands, similarity in habitat structure and in the nature and number of niches has resulted in pronounced parallel and convergent evolution among bird species and some lizard species, especially between the Chilean mattoral and the California chaparral. In North America chaparral and sagebrush communities support mule deer *(Odocoileus hemionus)*, coyotes *(Canis latrans)*, a variety of rodents, jackrabbits *(Lepus* spp.*)*, and sage grouse *(Centrocercus urophasianus)*. The Australian mallee is rich in birds, including the endemic mallee-fowl *(Leipoa ocellata)*, which incubates its eggs in a large mound. Among the mammalian life are the gray kangaroo *(Macropus giganteus)* and various species of wallaby. Rash clearance of mallee vegetation is endangering mallee wildlife as well as affecting millions of migrant, nectar-feeding birds supported by the mallee in spring.

Function

Precipitation, temperature, soil moisture, and nutrients are major influences on the function of mediterranean-type systems. Precipitation falls mostly in the cool winter months, and most of the plant growth and flowering is concentrated in spring, much of it at the end of the rainy season. Then the plants have to respond to the dry season, which imposes a great deal of environmental stress. How they respond is reflected in plant growth forms. Aridity, cold temperatures, a short growing season, long periods of drought, and a low nutrient supply favor shrubs. Where drought periods are shorter and neither temperature nor nutrients strongly limit growth, trees grow. If the physiological costs of growing new leaves each year are less than maintaining the same leaves, the plants possess a deciduous habit. If the costs are greater, then the plants are evergreen.

Soils of mediterranean-type ecosystems are low in nutrients and are especially deficient in nitrogen and phosphorus. During the dry period, nitrogen and other nutrients accumulate beneath the woody plants and remain fixed as the topsoil dries out. Wetting of the soil during the winter stimulates a flush of microbial activity involving decomposition of humus and mineralization of nitrogen and carbon. The concentration of nutrients stimulates a flush of growth. If heavy rains suddenly enter dry topsoil, quantities of nutrients may be lost by leaching and erosion.

Some plants of the mediterranean systems conserve nutrients. *Ceanothus*, an early successional species of California chaparral, is a nonleguminous nitrogen-fixer (see Sections 17.13, 25.7). In the Australian mallee *Atriplex vesticana*, a dominant plant, lowers the nitrogen content of the surrounding soil during the growing season and concentrates nitrogen directly beneath it through litterfall. It transfers nitrogen and phosphorus from its leaves to its stems before leaf fall.

Human Impact

In the lands around the Mediterranean Sea, human civilization dating back 10,000 years has had a profound influence on shrubland vegetation. Cycles of logging for oak and pine to build ships and to construct settlements, wood cutting for fuel, clearing for agricultural lands, terracing hillsides for crops, vineyards, and olive groves, irrigation, land abandonment followed by destruction of terraced walls and decay of terraced hill land, overgrazing by cattle, sheep, and goats brought about by pastoral nomadism, followed by soil erosion, have destroyed shrublands. More recently, spreading urbanization, overcrowded settlements and recreational areas, and pollution threaten the future of these landscapes.

Mediterranean-type vegetation of California has undergone similar degradation, over a much shorter period of time. Although native peoples exploited the plant communities for food, the major impact of humans began with Spanish colonization in 1779. Many California oak communities have disappeared because of stock raising, charcoal burning, agricultural clearing, and rapid development. What oak communities remain occur as scattered patches in inaccessible areas and in managed grazing land. Shrubby plant communities are extensive in rugged terrain, but they have been invaded by housing developments.

In Chile the mattoral has been degraded by the grazing of sheep and goats, or cleared and replaced by cropland and tree plantations. In Australia grazing not only by sheep but by the introduced rabbit threatens to cause the disappearance of numerous species of plants. Because of rough terrain, only the South African fynbos remains in its natural state.

Seral shrub communities for the most part result from human disturbances such as land clearing and abandonment and logging. Once established, they are valuable as wildlife habitat and sources of food, such as fruits and nuts. Nevertheless these shrublands are subject to destruction. Regarded as worthless waste-land, they are cleared for grazing land, converted into housing developments, and destroyed by land reclamation projects. Wetland shrub communities are drowned by flood control and hydroelectric projects or drained for other uses. Hedgerows, once a familiar part of the landscape in the United States and western Europe, have been torn out to enlarge agricultural fields to sizes required by mechanization. Because seral shrub-lands are ephemeral, lasting only 15 to 20 years before entering the forest stage of succession, they must be maintained by cutting, burning, or other techniques if they are to provide stable habitats for shrubland wildlife.

28.2   DESERTS

Geographers define deserts as land where evaporation exceeds rainfall. No specific amount of rainfall serves as a criterion; deserts range from extremely arid regions to those with sufficient moisture to support a variety of life. Deserts have been classified according to rainfall into *semideserts*, ones that have precipitation between 150 and 300 to 400 mm per year; *true deserts*, regions with rainfall below 150 mm per year; and *extreme deserts*, areas with rainfall below 70 mm per year. Deserts, which occupy about 26 percent of the continental area, occur in two distinct belts between 15° and 35° latitude in both the Northern and Southern Hemispheres—the Tropic of Cancer and the Tropic of Capricorn.

Deserts are the result of several forces. One force that leads to the formation of deserts is the movement of air masses (see Chapter 4). High-pressure areas alter the course of rain. The high-pressure cell off the coast of California and Mexico deflects rainstorms moving south from Alaska to the east and prevents moisture from reaching the Southwest. In winter high-pressure areas move southward, allowing winter rains to reach southern California and parts of the North American desert. Winds blowing over cold waters become cold also. They carry little moisture and produce little rain. Thus the west coast of California and Baja California, the Namid Desert on coastal southwest Africa, and the coastal edge of the Atacama in Chile may be shrouded in mist, yet remain extremely dry.

Mountain ranges also play a role in desert formation by causing a rain shadow on their lee side. The high Sierras and Cascade Mountains intercept rain from the Pacific and help maintain the arid conditions of the North American desert. The low eastern highlands of Australia block the southeast trade winds from the interior. Other deserts, such as the Gobi and the interior of the Sahara, are so remote from the ocean that all of the water has been wrung from the winds by the time they reach those regions.

Characteristics

All deserts have in common low rainfall, high evaporation (from 7 to 50 times as much as precipitation), and a wide daily range in temperature from hot by day to cool by night. Low humidity allows up to 90 percent of solar radiation to penetrate the atmosphere and heat the ground. At night the desert yields the accumulated heat of the day back to the atmosphere. Rain, when it falls, is often heavy and, unable to soak into the dry earth, rushes off in torrents to basins below.

Deserts are not the same everywhere. Differences in moisture, temperature, soil drainage, topography, alkalinity, and salinity create variations in vegetation cover, dominant plants, and groups of associated species. There are hot deserts and cool deserts, extreme deserts and semideserts, ones with sufficient moisture to verge on being grasslands or shrublands, and gradations between those extremes within continental deserts.

There is a certain degree of similarity among hot deserts and cold deserts of the world. The cold deserts, including the Great Basin of North America, the Gobi, Takla Makan, and Turkestan deserts of Asia, and high elevations of hot deserts are dominated by *Artemisia* and chenopod shrubs, and may be considered shrub steppes. The northern part of the North American Great Basin is dominated by nearly pure stands of big sagebrush (see Figure 28.2) and the southern part by shadscale and bud sage (*Artemisia*). The hot deserts range from no or scattered vegetation to ones with some combination of chenopods, dwarf shrubs, and succulents. The deserts of southwestern North America—the Mojave, the Sonoran, and the Chihuahuan—are dominated by creosote bush (*Larrea divaricata*) and bur sage (*Franseria* spp.). Areas of favorable moisture support tall growths of *Acacia* spp., saguaro (*Cereus giganteus*), palo verde (*Cercidium* spp.), and ocotillo (*Fouquieria* spp.).

Structure

The topography of the desert, unobscured by vegetation, is stark and, paradoxically, partially shaped by water. The unprotected soil erodes easily during violent storms and is further cut away by the wind. Alluvial fans stretch away from eroded, angular peaks of more resistant rocks. They join to form deep expanses of debris, the *bajadas*. Eventually, the slopes level off to low basins, or **playas,** which receive waters that rush down from the hills and water-cut canyons, or

arroyos. These basins hold temporary lakes after the rains, but water soon evaporates and leaves behind a dry bed of glistening salt.

Woody-stemmed and soft brittle-stemmed shrubs are characteristic desert plants (Figure 28.6). In a matrix of shrubs grows a wide assortment of other plants, the yucca, cacti, small trees, and ephemerals. In the Sonoran desert, in the Peru-Chilean, and South African Karoo and southern Namib deserts, large succulents rise above the shrub level and change the appearance of the desert far out of proportion to their numbers. The giant saguaro, the most massive of all cacti, grows on the bajadas of the Sonoran desert (Figure 28.7). Ironwood, smoketree, and palo verde grow best along the banks of intermittent streams, not so much because they require the moisture, but because their hard-coated seeds must be scraped and bruised by the grinding action of sand and gravel during flash floods before they can germinate.

Both plants and animals are adapted to the scarcity of water either by drought evasion or by drought resistance (see Chapter 1). Plant drought-evaders flower only in the presence of moisture. They persist as seeds during drought periods, ready to sprout, flower, and produce seeds when moisture and temperature are favorable. There are two periods of flowering in the North American deserts: after winter rains come in from the Pacific Northwest, and after summer rains move up from the south out of the Gulf of Mexico. Some species flower only after winter rains, others only after summer rains, but a few bloom during both

seasons. If no rains come, these ephemeral species do not grow. Drought-evading animals, like their plant counterparts, adopt an annual lifestyle or go into estivation or some other stage of dormancy during the dry season. For example, the spadefoot toad *(Scaphiopus)* remains underground in a gelatinous-lined underground cell, making brief reproductive appearances during periods of winter and summer rains. If extreme drought develops during the breeding season, birds fail to nest and lizards do not reproduce.

Belowground biomass in the desert can be as patchy as the aboveground biomass. Desert plants may be deep-rooted woody shrubs, such as mesquite *(Prosopis* spp.) and *Tamarix*, whose taproots reach the water

(a)

(b)

(c)

Figure 28.6 Three hot deserts. (a) The Chihuahuan Desert in Nuevo Leon, Mexico. The substrate of this desert is sand-sized particles of gypsum. (b) The edge of the Great Victorian Desert in Australia. This desert and the Chihuahuan Desert are dominated by woody, brittle-stemmed shrubs. (c) Dunes in the Saudi Arabian desert near Riyadh. Note the extreme sparseness of vegetation.

Figure 28.7 Organpipe cactus and saguaro dominate this part of the Sonoran Desert in the southwestern United States.

table, rendering them independent of water supplied by rainfall. Some, such as *Larrea* and *Atriplex*, are deep-rooted perennials with superficial laterals that extend as far as 15 to 30 m from the stems. Other perennials, such as the various species of cactus, have shallow roots, often extending no more than a few centimeters below the surface. Ephemerals have shallow and poorly branched roots reaching a depth of about 30 cm, where they pick up moisture quickly from light rains.

The desert floor is stark, a raw mineral substrate of various types devoid of a continuous litter layer. Dead leaves, bud scales, and dead twigs, mostly associated with drought-resistant species that shed them to reduce transpiring surfaces, accumulate in wind-protected areas beneath the plants and in depressions in the soil.

Function

Primary production in the desert depends on the proportion of available water used and the efficiency of its use. Data from various deserts in the world suggest that annual primary production of aboveground vegetation varies from 30 to 200 g/m². Belowground production is also low but greater than aboveground production. It ranges from 100 to 400 g/m² in arid regions and from 250 to 1000 g/m² in semiarid regions.

The amount of biomass that accumulates and the ratio of annual production to biomass depend on the dominant type of vegetation. In those deserts in which trees, shrubs, and cactuslike plants dominate, annual production is about 10 to 20 percent of the total aboveground standing crop biomass. Annual or ephemeral

communities have a 100 percent turnover of both roots and aboveground foliage, and their annual production is the same as peak biomass. In general, desert plants do not have a high root biomass relative to aboveground shoot biomass.

Adding to primary production in the desert are lichens, green algae, and cyanobacteria, abundant as soil crusts. Cyanobacteria have unusually high rates of nitrogen fixation, but less than one-half of their total nitrogen input becomes part of higher plants. Approximately 70 percent of the nitrogen is short-circuited back to the atmosphere as volatilized ammonia and as N_2 from denitrification, speeded by dry alkaline soils.

Nutrient cycling in arid ecosystems is tight. Two major nutrients, phosphorus and nitrogen, are in short supply; much of them is tied up in plant biomass, living and dead. Desert plants tend to retain in stems and roots certain elements, particularly nitrogen, phosophorus, and with some, potassium, before shedding any parts. The nutrients remaining in the shed parts collect and decompose beneath the plants, where microclimate conditions created by the plants favor biological activity. The soil is further enriched by animals attracted to the shade. The plants, in effect, create islands of fertility beneath themselves.

In spite of their aridity, desert ecosystems support a surprising diversity of animal life, including a wide assortment of beetles, ants, locusts, lizards, snakes, birds, and mammals. The mammals are notably herbivorous species (Figure 28.8). Grazing herbivores of the desert tend to be generalists and opportunists in their mode of feeding. They consume a wide range of species, plant types, and parts. Desert sheep feed on succulents and ephemerals when available and then switch to woody browse during the dry period. As a last resort, herbivores consume dead litter and lichens. Small herbivores—the desert rodents, particularly the family Heteromyidae, and ants—tend to be granivores, feeding largely on seeds, and are important in the dynamics of desert ecosystems.

Herbivores can have a pronounced impact on desert vegetation, especially if they are more abundant than the range's capacity to support them. Once grazers have eaten the annual production, they eat plant reserves, especially during long dry periods. Over-browsing can so weaken the plant that the vegetation is destroyed or irreparably damaged. Areas protected from grazing, especially grazing by goats and sheep, have a higher biomass and a greater percentage of palatable species than grazed areas.

Native plant-eating herbivores in a shrubby desert, under most conditions, consume only a small part of the aboveground primary production, but seed-

Figure 28.8 Typical herbivores of the desert. (a) A five-toed jerboa (*Alactaga* sp.), a Middle Eastern rodent whose long ears radiate heat. (b) The desert oryx *(Oryx gazella)* copes with the desert environment by reducing unnecessary energy expenditure.

eating herbivores can eat most of the seed production, close to 90 percent of it. This consumption can have a pronounced effect on plant composition and plant populations.

Desert carnivores, like the herbivores, are opportunistic feeders, with few specialists. Most desert carnivores, such as foxes and coyotes, have mixed diets that include leaves and fruits; even insectivorous birds and rodents eat some plant material. Omnivory, rather than carnivory and complex food webs, seems to be the rule in desert ecosystems.

The detrital food chain seems to be less important in the desert than in other ecosystems. Fungi and actinomycetes are the most prominent. Microbial decomposition, like the blooming of ephemerals, is limited to short periods when moisture is available. For this reason dry litter tends to accumulate until the detrital biomass may be greater than the aboveground living biomass. Most of the ephemeral biomass disappears through grazing, weathering, and erosion. Decomposition proceeds mostly through detritus-feeding arthropods such as termites that ingest and break down woody tissue in their guts. In some deserts, considerable amounts of nutrients are locked up in termite structures, to be released when the structure is destroyed. Other important detritivores are acarids and isopods.

Human Impact

In spite of their aridity, deserts have not been spared from human impact. Only extremely arid deserts have escaped significant disturbance. Ancient human intrusions were limited to food-gathering and hunting for-

ays by aborigines or to grazing by nomadic pastoralists. In recent times most aborigines have vanished and many of the pastoralists have settled into agricultural communities. Today desert regions, particularly in the Middle East, have been invaded by the oil industry, radically changing and polluting the desert environment. In parts of the Middle East and in North America, urban development has expanded into the desert, complete with lawns and swimming pools (Figure 28.9). Irrigation agriculture also has turned some areas of desert green. These developments are depleting the fossil water supply. Deserts are also suffering massive degradation by unrestricted recreational use of all-terrain vehicles. Widespread collection of cacti of many species and sizes for the world plant trade in the deserts of the United States, Mexico, Peru, Chile, and Brazil is destroying the integrity of desert ecosystems and threatening some species with extinction in the wild. In the United States, the desert southwest has the greatest number of endangered species, many of which are endemic to the desert region.

The greatest impact occurs on the semiarid edges of the natural deserts of the world, which support some agriculture and grazing. There mismanagement of land has created new deserts. Virgin lands, even in dry climates, are able to support some vegetation. The roots of trees, shrubs, and grasses tap the deeper water supply and bind the soil. However, expanding populations and the periods of adequate rainfall encourage encroachment on marginal lands. Overcultivated and overgrazed, the land is exposed to wind and water erosion. Because these regions are subject to unpredictable droughts, the human population finds itself faced with

Figure 28.9 Water-demanding suburban developments in Phoenix, Arizona, encroach upon the surrounding desert.

devastating famines and increasing land degradation. Eventually the destruction is total. Vegetation and top-soil are gone, dust storms become frequent, and sand dunes advance across the land. The land has reached a point of no return. The result is **desertification**—the creation of new deserts on the periphery of natural deserts in northern Africa, India, China, Argentina, Chile, Mexico, and the southwestern United States. Formation of new deserts has doubled over the past 100 years.

Supplied with water and managed well, many desert areas can be converted into productive agricultural land, but poor irrigation practices that allow the seepage of water from canals and overwatering of soil cause the water table to rise. As the moisture evaporates from the surface, it leaves behind a glistening surface layer of salt toxic to plants, and the land is abandoned to the wind. This salinization process has affected irrigated lands in India, Syria, Iraq, central Asia, California's San Joaquin Valley, and the Colorado River Basin. Because irrigation depends on "mined" fossil water beneath the desert and water drawn from rivers, irrigation severely affects the hydrology of deserts, further threatening the future of these regions.

STUDY QUESTIONS

1. What characteristics do seral and climax shrub-lands share?
2. Describe mediterranean-type shrublands.
3. Argue that successional shrublands are not wastelands.
4. Contrast the benefits of evergreen and deciduous leaves in a mediterranean-type shrubland.
5. What climatic forces lead to deserts? Human influences?
6. In what general ways are plants and animals of the desert adapted to aridity?
7. What is unique about nutrient cycling in the desert ecosystem?
8. Human relationships to the desert are two-way. On the one hand, through the mismanagement of desert ecosystems, humans cause the spread of deserts in semiarid lands (desertification). On the other hand, humans attempt to bring the desert into greater productivity by irrigation. Discuss this two-way relationship and how both lead to the spread of deserts over time. (*Hint:* What is the relationship of desert reclamation to the mining of fossil water and increasing salinity of the soil?)
*9. Project future problems in the southwestern United States if development continues in the hot desert regions.
*10. How have mismanagement and poor understanding of desert ecology led to massive economic and social problems across several north African nations?
*11. What would the Middle East and other Mediterranean lands be like today had the people given more consideration to good land management? Were good land management techniques known and available in ancient times? Why are we making the same disastrous mistakes today?

CHAPTER 29

TUNDRA AND TAIGA

OBJECTIVES

On completion of this chapter, you should be able to:

- Describe the characteristics of the tundra ecosystem.
- Compare the arctic, alpine, and tropical alpine tundras.
- Explain the role of permafrost in the arctic environment.
- Describe the effects of human disturbances on arctic and alpine tundras.
- Describe the characteristics of the taiga and its vegetation.
- Explain the roles of mosses and lichens in the taiga.
- Discuss the effects of human disturbance on the taiga.

A Rocky Mountain alpine tundra carpeted with cotton grass *(Eriophorum angustifolium)* in full bloom.

29.1 TUNDRA

Encircling the top of the Northern Hemisphere is a frozen plain, clothed in sedges, heaths, and willows. Called the **tundra,** its name comes from the Finnish *tunturi,* meaning "a treeless plain." At lower latitudes similar landscapes, the alpine tundra, occur in the mountains of the world. Arctic or alpine, the tundra is characterized by low temperatures, a short growing season, and low precipitation (cold air carries little water vapor).

The arctic tundra is a land dotted with lakes and crossed by streams. Where the ground is low and moist, extensive bogs exist. On high drier areas and places exposed to the wind, vegetation is scant and scattered, and the ground is bare and rock-covered. These regions are the fell-fields, an anglicization of the Danish *fjoeldmark,* or rock deserts. Lichen-covered, the fell-fields are most characteristic of highly exposed alpine tundra. The arctic tundra falls into two broad types: *tundra* with 100 percent cover and wet to moist soil, and *polar desert* with less than 5 percent cover and dry soil.

Conditions unique to the arctic tundra are a product of at least three interacting forces: permafrost, vegetation, and the transfer of heat. **Permafrost** is the perennially frozen subsurface that may be hundreds of meters deep. It develops where the ground temperatures remain below 0° C for many years. Its upper layers thaw in summer and refreeze in winter. Because the permafrost is impervious to water, it forces all water to remain and move above it. Thus the ground stays soggy on the tundra even though precipitation is low, enabling plants to exist in the driest parts of the Arctic.

Vegetation and its accumulated organic matter protect the permafrost by shading and insulation, which reduce the warming and retard thawing of the soil in summer. Any natural or human disturbance, however slight, can cause the permafrost to melt. If the vegetation is removed, the depth of the thaw is 1.5 to 3 times that of the area still retaining vegetation. Thus vegetation and its organic debris impede the thawing of the permafrost and act to conserve it.

In turn, permafrost chills the soil, retarding the general growth of both aboveground and belowground parts of plants, limiting the activity of soil microorganisms, and diminishing the aeration and nutrient content of the soil. The effect becomes more pronounced the closer the permafrost is to the surface of the soil, where it contributes to the formation of shallow root systems.

Alternate freezing and thawing of the upper layer of soil create the unique, symmetrically patterned landforms typical of the tundra. This action of frost pushes stones and other material upward and outward from the mass to form a patterned surface. Frost hummocks, frost boils, and earth stripes, all typical nonsorted, or irregular patterns, are associated with seasonally high water tables (Figure 29.1). Sorted, or regular patterns appear on better-drained sites. Best known are stone

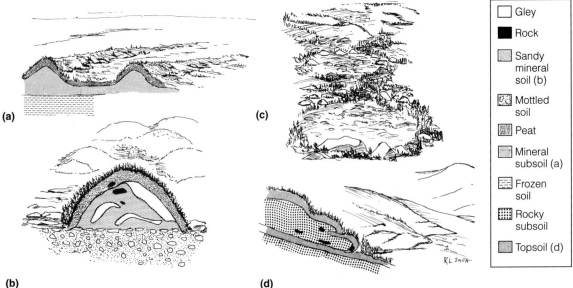

Figure 29.1 Patterned landforms typical of the tundra region: (a) unsorted earth stripes; (b) frost hummocks; (c) sorted stone nets and polygons; and (d) a solifluction terrace.

polygons, whose size is related to frost intensity and size of the material. On sloping ground, creep, frost thrusting, and downward flow of soil changes polygons into sorted stripes running downhill. Mass movement of supersaturated soil over the permafrost forms *solifluction terraces*, or "flowing soil." This gradual downward creep of soils and rocks eventually rounds off ridges and other irregularities in topography. This molding of the landscape by frost action, called *cryoplanation*, is far more important than erosion in wearing down the arctic landscape.

Alpine tundras have little permafrost, confined mostly to very high elevations; but frost-induced processes, such as small solifluction terraces and stone polygons, are present nevertheless. Lacking permafrost, soils are drier; only in alpine wet meadows and bogs do soil moisture conditions compare with those of the Arctic. Precipitation, especially snowfall and humidity, is higher in the alpine regions than in the arctic tundra, but steep topography induces a rapid runoff of water.

Structure

Vegetation

In spite of its distinctive climate and many endemic species, the tundra does not possess a vegetation type unique to itself. Structurally, the vegetation of the tundra is simple. The number of species tends to be few; the growth is slow; and most of the biomass and functional activity are confined to a few groups. In the Arctic only those species able to withstand constant disturbance of the soil, buffeting by the wind, and abrasion from wind-carried particles of soil and ice can survive. In the alpine tundra, the environment is even more severe for plants. It is a land of strong winds, snow, cold, and widely fluctuating temperatures. During the summer the temperature on the surface of the soil ranges from 40° to 0° C. The atmosphere is thin, so light intensity, especially ultraviolet, is high on clear days.

Although it appears homogeneous, the pattern of vegetation is patchy. A combination of microrelief, snowmelt, frost heaving, and aspect, among other conditions, produces an endless change in plant associations from spot to spot. In the arctic tundra (Figure 29.2), low ground is covered with a cotton grass, sedge, dwarf heath, sphagnum moss complex. Well-drained sites support heath shrubs, dwarf willows and birches, dryland sedges and rushes, herbs, mosses, and lichens. The driest and most exposed sites—the flat-topped domes, rolling hills, and low-lying terraces, all usually covered

Figure 29.2 The wide expanse of the Arctic tundra in the Northwest Territories of Canada.

with coarse, rocky material and subject to extreme action by frost—support only sparse vegetation, often confined to small depressions. Plant cover consists of scattered heaths and mats of dryads, as well as crustose and foliose lichens growing on the rocks.

The alpine tundra is a land of rock-strewn slopes, bogs, alpine meadows, and shrubby thickets (Figure 29.3). Cushion and mat-forming plants, rare in the Arctic, are important in the alpine tundra. Low and ground-hugging, they are able to withstand the buffeting of the wind, and their cushionlike blanket traps heat. The interior of the cushion may be 20° C warmer than the surrounding air, a microclimate that insects use.

In spite of similar conditions, only about 20 percent of the plant species of the arctic and Rocky Mountain alpine tundra are the same, and they are of different ecotypes. Lacking in the Rocky Mountain alpine tundra are heaths and heavy growth of lichens and mosses between other plants; lichens are confined mostly to rocks, and the ground is bare between plants. The alpine tundra of the Appalachian Mountains, however, is dominated by heaths and sedge meadows, as is the alpine tundra of Europe, and mosses are common. The Australian alpine region supports a growth of heaths on rocky sites, and wet areas are covered with sphagnum bogs, cushion heaths, and sod tussock grasslands.

Alpine plant communities are not restricted to the northern and southern temperate regions of Earth. They also exist above the tree line of the high mountains in tropical regions: Central America, South America, Africa, Borneo, New Guinea, Java, Sumatra, and

(a) **(b)**

Figure 29.3 Alpine tundra (a) in the Rocky Mountains and (b) in the Australian Alps. Although the species of vegetation are different, the growth forms are convergent and the physiognomy is similar.

Hawaii. Tropical alpine vegetation and its environment contrast with those of temperate alpine regions. Tropical alpine regions undergo great seasonal variation in rainfall and cloud cover but little seasonal variation in mean daily temperature. Instead there is a strong daily fluctuation in temperature, from below freezing conditions at night to hot, summerlike temperatures during the day. Such a diurnal freeze-thaw cycle is unique to tropical alpine regions.

Tropical alpine tundras support tussock grasses, small-leaved shrubs, and heaths; but the one feature that sets tropical alpine vegetation apart from the temperate alpine is the presence of unbranched or little-branched, giant, treelike rosette plants (Figure 29.4), many of which belong to the family Compositae. Although the genera and species differ among the tropical alpine areas (for example, *Senecio* species in Africa and *Espeletia* species in the Andes), their growth forms and physiology are strikingly similar, suggesting convergent evolution. These species are the antithesis of the usual low-growing tundra species. The higher the elevation, the taller these plants grow, reaching up to 6 meters. Many of these giant rosettes have well-developed water-storing pith in the xylem; retain dead rosette leaves about the stem, apparently as insulation against the cold; secrete a mucilaginous fluid about the bases of leaves that seems to function as a heat-storage device (water has a high heat-storing capacity); and possess dense pubescent hairs that reduce convective loss of heat.

Arctic plants propagate themselves almost entirely by vegetative means, although viable seeds many hun-

Figure 29.4 Tropical alpine tundra on Mt. Kenya, East Africa, with giant rosettes of *Lobelia*. Plant height tends to increase with elevation.

dreds of years old exist in the soil. Alpine plants, including those of the tropics, propagate mostly by seeds. The short-lived adventitious roots of arctic plants are short and parallel to the rhizomes. Adventitious roots of alpine plants are long-lived, long, and unimpeded by permafrost, able to penetrate to considerable depths.

In both arctic and alpine tundras, topographic location and snow cover delimit a number of plant communities. On the arctic tundra, steep, south-facing slopes and river bottoms support the most luxuriant and tallest shrubs, grasses, and legumes, whereas cotton grass dominates the gentle north-facing and south-facing slopes, reflecting higher air and soil temperatures and greater snow depth. Pockets of heavy snow create two types of plant habitat in both arctic and alpine tundra, the snow patch and the snow bed. Snowpatch communities occur where wind-driven snow collects in shallow depressions and protects the plants beneath. Snow beds are found where large masses of snow accumulate because of topographic peculiarities. Not only does the deep snow protect the plants beneath, but the meltwater from the slowly retreating snowbank provides a continuous supply of water throughout the growing season. Snow bed plants have a short growing season, but they break into leaf and flower quickly because of the advanced stage of growth beneath the snow.

Microbiota, the bacteria and fungi, live near the surface and tolerate cold temperatures. Bacteria are active at $-7.5°$ C, and fungal growth continues at $0°$ C. Fungi, however, can break down plant structural carbohydrates at temperatures below $0°$ C, whereas aerobic bacteria at $0°$ C are restricted to nonstructural carbohydrates and products of fungal decomposition. Fungi and aerobic bacteria live mostly in the upper 7 to 10 cm of soil. Below that depth, only anaerobic bacteria exist.

Animal Life

The tundra world holds fascinating animal life even though the diversity of species is low. Invertebrate fauna are concentrated near the surface, where there are abundant populations of segmented whiteworms (Enchytraeidae), collembolas, and flies (Diptera), chiefly craneflies. Summer in the arctic tundra brings hordes of blackflies, deerflies, and mosquitoes. In alpine regions flies and mosquitoes are scarce, but collembolas, beetles, grasshoppers, and butterflies are common. Because of ever-present winds, butterflies keep close to the ground; other insects have short wings or no wings at all. Insect development is slow; some butterflies may take two years to mature, and grasshoppers three.

Dominant vertebrates on the arctic tundra are herbivores, including lemmings, arctic hare, caribou, and musk-ox. Although caribou have the greatest herbivore biomass, lemmings, which breed through the year, undergo three- to four-year cycles. At their peak they may reach densities as great as 125 to 250 per hectare, consuming three to six times as much forage as caribou. Arctic hares that feed on willows disperse over the range in winter and congregate in more restricted areas in summer. Caribou are extensive grazers, spreading out over the tundra in summer to feed on sedges. Musk-ox are more intensive grazers, restricted to more localized areas where they feed on sedges, grasses, and dwarf willow. Herbivorous birds are few, dominated by ptarmigan and migratory geese.

The major arctic carnivore is the wolf, which preys on musk-ox, caribou, and when they are abundant, lemmings. Medium-sized to small predators include the arctic fox (*Alopex lagopus*) that preys on arctic hare and several species of weasel that prey on lemmings. Also feeding on lemmings are snowy owls (*Nyctea scandiaca*) and the hawklike jaegers (*Stercorarius* spp.). Sandpipers, plovers, longspurs, and waterfowl, which nest on the wide expanse of ponds and boggy ground, feed heavily on insects.

The alpine tundra, which extends upward like islands in mountain ranges, is small in area and contains few characteristic species. The alpine regions of western North America are inhabited by the hay-cutting pika, marmots, mountain woodchucks that hibernate over winter, mountain goats (not goats at all but related to the alpine-dwelling chamois of South America), mountain sheep, elk, voles, and pocket gophers. Eurasian alpine mammals include marmots and wild goats. The African alpine tundra is the home of the rock hydrax (*Procavia capensis*).

Function

Primary production on the tundra is low. Low temperature, a short growing season ranging from 50 to 60 days in the high arctic to 160 days in the low-latitude alpine tundra, and the low availability of nutrients help to keep it that way.

Plants are photosynthetically active on the arctic tundra about three months out of the year. As quickly as snow cover disappears, plants start photosynthetic activity, but it is limited initially because plant leaves are poorly developed. Alpine species, however, undergo a rapid burst of growth following snowmelt, at the expense of belowground root and rhizome carbohydrate reserves.

Arctic plants make maximum use of the growing season and light by carrying on photosynthesis during the 24-hour daylight period, even at midnight when light is one-tenth that of noon. They rarely become light-saturated, and they possess a high leaf area index (0.5 to 1.0). The nearly erect leaves of some arctic plants permit the almost complete interception of the slanting rays of arctic sun.

Much of the photosynthate goes into the production of new growth, but about one month before the growing season ends, plants cease to allocate photosynthate to aboveground biomass. They withdraw nutrients from the leaves and move them to roots and belowground biomass, sequestering ten times the amount stored by temperate grasslands. In general, alpine tundras are more productive than arctic tundras.

Structurally, most of the tundra vegetation is underground. Root-to-shoot ratios of vascular plants range from 3:1 to 10:1. Roots are concentrated in the upper soil that thaws during the summer, and aboveground parts seldom grow taller than 30 cm. It is not surprising, then, that the net annual production of aboveground vegetation ranges from 40 to 110 g/m^2, whereas net annual belowground production ranges from 130 to 360 g/m^2. Efficiency of primary production ranges from 0.20 to 0.5 for the growing season. That rate of production is comparable to temperate ecosystems, but the growing season is so short—50 to 75 days—that overall annual production is greatly reduced.

Arctic and alpine tundras are short on nutrients because the short growing season, cold temperatures, and low precipitation slow weathering and restrict decomposition. Nutrient cycling has to be conservative and tight, with minimal loss outside the system. Of all the terrestrial ecosystems, the tundra has the smallest proportion of its nutrient capital in live biomass. Dead organic matter functions as the nutrient pool, but most of it is not directly available to plants. To conserve nutrients, vascular plants retain and reincorporate nutrients, especially nitrogen, phosphorus, potassium, and calcium, in their tissues rather than release them to decomposers.

Because the tundra soil does not store available nutrients in any great quantity, plants depend on the release of nutrients from decomposition, the uptake of which is often aided by mycorrhizae. Decomposition is stimulated by rising temperature in spring. Because 60 percent of the active roots are in the upper 5 cm of soil, the root mass, once thawed, takes up most of the nutrients, except early in the season. Then plants have to compete with microbes and mosses, whose uptake of nutrients is greater. Once plants establish aboveground biomass, they have the competitive advantage over microbes.

Leaching or removal of nutrients is minimal, occurring mostly at the beginning of the growing season. Melting snow releases nutrients frozen over winter in the litter, excreta of animals, and microbes. Spring rains leach nutrients from dead plant material remaining from the previous growing season. In summer vascular plants leak nutrients, particularly phosphorus and potassium, from the cuticle of the leaves to the surface, where they are washed off by summer rains. Mosses often capture these nutrients before they reach the soil and release them slowly, thus functioning as a temporary nutrient sink.

A rapid upward movement of nutrients early in the season at the expense of belowground biomass supports fast shoot growth. Although more nutrients become available as the depth of thaw increases, they usually are in short supply. Tundra plants respond to the shortage, especially of nitrogen and phosphorus, by producing only a small biomass of leaves and stems that is well supplied with nutrients. Six weeks into the growing season, plants start to send nutrients belowground. As the cold approaches, the aboveground tissues die, and their dead parts add to the accumulation of organic matter. Nutrients leached from the dead leaves are accumulated by mosses or are frozen into place until the following summer's snowmelt.

The two nutrients most limiting are nitrogen and phosphorus. The major sources of nitrogen are precipitation and biological fixation. Nitrogen-fixers are anaerobic and free-living aerobic bacteria; cyanobacteria in soil, water, and in the foliage of mosses where they live epiphytically; and lichens. Precipitation adds nearly as much nitrogen. Phosphorus comes from decomposition of organic matter and animal feces.

Animals contribute to the release of nutrients by either stimulating or short-circuiting decomposition. Soil invertebrates consume nearly all of the microbial populations near the surface of the soil. Grazing herbivores, especially lemmings, eat large amounts of aboveground production and distribute nutrients across the tundra through their droppings. During a cyclic high, lemmings may consume over 25 percent of the aboveground primary production, or 10 percent of total plant production. They return about 70 percent as feces. Musk-ox, being selective grazers, restrict their feeding activity to certain areas, from which they may remove up to 85 percent of the herbage available. Overall this removal amounts to only about 15 percent of production. Decomposition of their dung is slow, taking 5 to 12 years. Grazing herbivores also fell standing litter and live plant biomass, improving conditions for decomposers. Preying on the herbivores, mostly lemmings, are a number of carnivores. Efficient assimilators, the carnivores rarely pass off as feces more than 5 percent of total energy assimilated.

The activity of consumers stimulates a continuous overturn in both soluble and exchangeable pools. Nitrogen on the average is recycled 10 or more times and phosphorus 200 times during the growing season. Such efficient short-season cycling allows tundra ecosystems to make maximum use of a limited supply of nutrients.

Human Impact

You might expect that the arctic tundra—remote, cold, and largely empty—would be immune to serious disturbance. Once native Eskimos lived along the coast in close ecological harmony with the arctic environment. The arrival of Western culture has broken down that culture and weakened that strong ecological relationship. Winter igloos and summer tents have given way to permanent settlements with wooden houses, dog-drawn sleds have yielded to snowmobiles, and rifles have replaced harpoons and spears as hunting instruments. All these social and cultural changes have affected the environment.

The discovery of oil has opened up the tundra for exploitation (Figure 29.5). Movement of heavy equipment, construction of all-season roads, airstrips, supply depots, oil pipelines, and oil spills have destroyed cover of mosses and grass, allowing the permafrost to melt and causing soil subsidence and gully erosion. Roads now obstruct wildlife movement and expose them to increased hunting pressure. Solid wastes and sewage, a disposal problem in the frozen landscape, pollute streams and surface waters, and toxic chemicals and heavy metals drain into arctic wetlands. Burning of vented gas from oil installations and sulfur dioxide originating from coal mining and smelting facilities, mostly in Russia and Ukraine, pollute the arctic air. Be-

Figure 29.5 A part of the Trans-Alaskan pipeline, carrying oil from Prudhoe Bay across the arctic tundra.

cause of the cold environment, tundra vegetation recovers slowly, if ever, from major disturbances.

The alpine tundra has not fared much better. Many alpine tundras, both temperate and tropical, have been overgrazed by domestic animals. Roads have opened the alpine tundra to recreational developments, ski trails, off-road vehicle tracks, and hiking trails. Such increased human activity damages thin soil and vegetation. Restricted in its distribution to mountaintop islands, tundra wildlife is vulnerable to human disturbance and faces a threatened future.

Because the arctic tundra is circumpolar, and because its exploitation is massive and worldwide, protecting the arctic ecosystem is an international problem. Saving and managing alpine tundras depends on the action of individual countries.

29.2 KRUMMHOLZ

At high altitudes where the winds are too steady and strong for all but low ground-hugging plants, the forest is reduced to pockets of stunted, wind-shaped trees. This area where forest gives way to tundra is the **Krummholz,** or "crooked wood" (Figure 29.6). The Krummholz in the North American alpine region is best developed in the Appalachian and Adirondack mountains. On the high ridges, trees begin to show signs of stunting far below the timber line. As you climb upward, stunting increases until spruces and birches, deformed and semiprostrate, form carpets 0.6–1 m high, impossible to walk through but often dense enough to walk upon. Where strong winds come in from a constant direction, the trees are sheared until the tops resemble close-cropped heads, although the trees on the lee of the clumps grow taller than those on the windward side.

In the Rocky Mountains the Krummholz is much less marked, for there the timber line ends abruptly with little lessening of height. Most of the trees are flagged; that is, the branches remain only on the lee side. In the alpine regions of Europe, the Krummholz is characterized by dwarf mountain pine (*Pinus mugo*), which forms dense thickets 1–2 m high on calcareous soils. On acid soils it is replaced by dwarf juniper (*Juniperus communis* subsp. *nana*). In the Australian Brindabella Range, the Krummholz is marked by pockets of low, twisted snow gum (*Eucalyptus pauciflora*).

Though wind, cold, and winter desiccation are regarded as the cause of the dwarfed and misshapen condition of the trees, the ability of some tree species to show a Krummholz effect is genetically determined,

(a)　　　　　　　　　　　　　　　　　　　　　**(b)**

Figure 29.6 The Krummholz. (a) In the Rocky Mountains the tree line is sharply defined. Note the narrow pockets of stunted trees. (b) The tree line in the Australian Alps in the Brindabella Range is marked by low, twisted snow gum *(Eucalyptus pauciflora)* in protected pockets that give way to low-growing plants.

with species such as mountain (mugo) pine adapted to high alpine conditions. Eventually conditions become too severe even for the prostrate forms, and trees drop out entirely, except for those that have taken root behind the protection of high rocks. Tundra vegetation then takes over completely.

29.3 TAIGA

The largest vegetation formation on Earth is the boreal forest or taiga. This belt of coniferous forest encompassing the high latitudes of the Northern Hemisphere covers about 11 percent of Earth's terrestrial surface. Its northern limit is roughly along the July 13 isotherm, the southern extent of the arctic front in summer, which also marks the beginning of the northward-stretching tundra. Its southern limit, much less abrupt, is more or less marked by the winter position of the arctic front, roughly just north of 58° N latitude. In North America the boreal forest covers much of Alaska and Canada and spills into northern New England with a finger extending down the high Appalachians. In Eurasia the boreal forest begins in Scotland and Scandinavia and extends across the continent, covering much of Siberia, to northern Japan.

Four major vegetation zones make up the taiga: the forest-tundra ecotone with open stands of stunted spruce, lichens, and moss; the open boreal woodland with stands of lichens and black spruce; the main bo-

Figure 29.7 Scots pine is dominant in the Scandinavian taiga.

real forest with continuous stands of spruce and pine broken by poplar and birch on disturbed areas; and the boreal-mixed forest ecotone where the boreal forest grades into the mixed forest of southern Canada and the northern United States. Occupying, for the most part, glaciated land, the taiga is also a region of cold lakes, bogs, rivers, and alder thickets.

In Europe the forest is dominated by Norway spruce *(Picea abies)*, Scots pine *(Pinus sylvestris)* (Figure 29.7), and downy birch *(Betula pubescens)*; in Siberia by Siberian spruce *(P. obovata)*, Siberian stone

Figure 29.8 Black spruce is a dominant conifer in the North American taiga.

pine *(Pinus sibirica)*, and larch *(Larix sibirica)*; and in the Far East by Yeddo spruce *(Picea jezoensis)*. The North American taiga, richer in species, has four genera of conifers, *Picea*, *Abies*, *Pinus*, and *Larix*, and two genera of deciduous trees, *Populus* and *Betula*. Dominant tree species include black spruce *(Picea mariana)* (Figure 29.8) and jack pine *(Pinus banksiana)*.

A cold continental climate with strong seasonal variation dominates the taiga. The summers are short, cool, and moist, and the winters are prolonged, harsh, and dry, with a long-lasting snowfall. The driest winters and the most extreme seasonal fluctuations are in interior Alaska and central Siberia, which experience as much as 100° C seasonal temperature extremes. Like the tundra, much of the taiga is under the controlling influence of permafrost. As in the tundra, permafrost impedes infiltration and maintains high soil moisture.

Structure

Vegetation

Like the tundra, the taiga is an endless sweep of sameness—a blanket of spire-shaped evergreens over the landscape. The appearance of the land can be deceptive, because variations in slope, aspect, topography, drainage, and permafrost add variety to the vegetation. In the North American taiga, black spruce with its low nutrient requirements and ability to tolerate wet soils occupies cold, wet, north-facing slopes and bottomlands. White spruce *(Picea glauca)* and birch grow on permafrost-free south-facing slopes, and jack pine

grows on the high, drier, warmer sites. Areas swept by fires come back in early successional hardwoods— quaking aspen, balsam poplar *(Populus balsamifera)*, paper birch *(Betula papyrifera)*—and jack pine.

The boreal forest conifers fall into three growth forms: (1) the spire-shaped spruces and fir, with an open, narrow, upper canopy and a dense lower canopy that casts a deep shade on the forest floor; (2) the open, thin, light-penetrating upper canopy of pines; and (3) the deciduous larch. Only a thick carpet of mosses grows in the dense shade of spruce, whereas under pine, light-tolerant lichens replace the shade-loving mosses.

The conifers are well suited to the cold taiga environment. The narrow, needlelike leaves with their thickened cuticles and sunken stomata reduce transpiration and assist in moisture conservation during periods of summer drought and winter freeze. Because they retain several years' growth of foliage at any one time, conifers can start photosynthesis quickly when environmental conditions are favorable.

Permafrost imposes its own authority on patterns of vegetation, which paradoxically encourages its formation. Permafrost impedes soil drainage, chills the soil, reduces its depth, slows decomposition, and reduces the availability of nutrients. Trees grow best where permafrost lies deep beneath the soil or is absent altogether; but the taiga trees worsen the situation for themselves by encouraging and maintaining the permafrost. Stands of spruce shade the ground and encourage a heavy growth of moss. Moss and an accumulation of the fine litter of undecomposed needles insulate the soil, immobilize nutrients, and increase soil moisture. The colder the soil becomes, the closer to the surface moves the permafrost, and the more shallow becomes the soil. With little space in which to anchor their roots, the trees are subjected to frost heaving. During early periods of warm weather, the roots encased in frozen soil are unable to replace the moisture lost through the crowns. The result is winter kill.

In stands of pine, larch, and scattered spruce, conditions do not improve. Heavy mats of lichens replace mosses on the dry, nutrient-poor, highly acidic soil. Lichens retain soil moisture through the growing season, encouraging growth of trees on sites that otherwise would be too dry, but the thick lichen mat also insulates the soil, chilling it and inhibiting the decomposition of organic matter. In spite of these effects, lichens do appear to improve tree growth.

Fires are recurring events in the taiga. During periods of drought, fires can sweep over hundreds of thousands of hectares. All of the boreal species, both broadleaf trees and conifers, are well adapted to fire. Unless too severe, fire provides a seedbed for regeneration of trees. Light surface burns favor successional hardwoods. More severe fires eliminate hardwood competition and favor spruce and jack pine regeneration.

Animal Life

As on the tundra, caribou are the major herbivores. Inhabiting open spruce-lichen woodlands, caribou are wide-ranging and feed on grasses, sedges, and especially lichens. Joining the caribou is the moose, the largest of all deer. Called elk in Eurasia, the somewhat solitary moose is a lowland mammal feeding on aquatic and emergent vegetation as well as alder and willow. Competing with moose for browse is the cyclic snowshoe hare. The arboreal red squirrel inhabits the conifers and feeds on young pollen-bearing cones and seeds of spruce and fir; and the quill-bearing porcupine (*Erethizon dorsatum*) feeds on leaves, twigs, and the inner bark of trees. Major mammalian predators are the wolf, which feeds on caribou and moose; the lynx, which preys on snowshoe hare and other small mammals; the pine marten (*Martes americana*), the major predator on red squirrels; and other species of weasels. Ruffed grouse and spruce grouse (*Dendragapus canadensis*) are conspicuous birds of the North American boreal forest, and capercaillie (*Tetrao* spp.) and hazel grouse (*Tetrastes bonasia*) in the Eurasian taiga. Crossbills (*Loxia* spp.) and siskins (*Carduelis* spp.) extract conifer seeds from cones and occasionally move south in winter when the food supply fails. Importantly, the taiga is the nesting ground of many species of both neotropical and tropical migrant warblers. Major avian predators include numerous species of owls, such as the great gray owl (*Strix nebulosa*), and goshawks (*Accipiter* spp.).

Of great ecological and economic importance are major herbivorous insects, the larch sawfly (*Pristiphora erichsonii*), the pine sawfly (*Neodiprion sertifer*), and the spruce budworm (*Choristoneura fumiferana*). Although major food items for the insectivorous summer birds, these insects experience periodic outbreaks and defoliate and kill large expanses of forest.

Function

A cold environment, a short growing season, continuous or discontinuous permafrost, slow decomposition, low availability of nutrients, and infertile, acidic soils impede productivity in the taiga. Particularly important is the relationship between ground cover, above-ground biomass, and nutrient availability. Growing in the already cool dark shade of spruces, mosses lower soil temperature and increase soil moisture, both of which reduce decomposition and the release of nutrients. Increasing the nutrient shortage to trees is the sequestering of nutrients by mosses, which depend upon precipitation and tree wash for their mineral nutrition. Nutrients become available only when mosses, which have a slow turnover rate, die. Associated with drier, highly acidic soil, lichens, too, immobilize nutrients. Because lichens hold moisture, however, they do lessen the effects of summer drought.

Particularly acute is the shortage of nitrogen. In addition to being immobilized in mosses and lichens, nitrogen is tied up in poorly decomposed soil organic matter. The major source of nitrogen is precipitation and biological fixation by lichens, by alders, bog myrtle, soapberry, and other woody plants possessing root nodules, and by some free-living bacteria. Uptake of nitrogen by trees, however, exceeds the annual input by their litter. Such a slow turnover in nitrogen can have an adverse effect on the growth and vigor of trees and associated vegetation. Depending on its severity, fire releases various accumulated nutrients, and more important, improves the conditions for decomposition by warming the soil and exposing organic matter to sunlight.

Because of the wide range of environmental conditions across the taiga, primary production is highly variable. In a general way, productivity there is much less than in deciduous forest, and biomass accumulation is slow. Herbivores influence primary production not only by consuming but also by trampling and selectively browsing certain species, especially willow, birch, aspen, and alder. Moose can reduce young plant growth by as much as 50 percent. A study of plant-moose-wolf relationships at Isle Royale National Park showed that the moose ate about 3000 tons of vegetation annually. In turn their major predator, the wolf, consumed about 45 tons of moose. Just as the nutrient and energy turnover among the plant components in the boreal forest is slow, so too is the turnover of biomass slow in the long-lived moose, which in turn controls the biomass of wolves.

Human Impact

Exploitation and exploration of the taiga began in the late 1600s with the fur trade, over which wars were fought. Exploitation of the taiga's wildlife resources was followed by logging. The taiga is the world's richest source of softwood timber and pulpwood. One-half of the total exploitable reserve is in Russia (including Siberia); the other half is in North America and Europe. Much of the logging is little more than timber mining, with no overall effort at regeneration and restoration.

In Russia up to 3 million hectares are cut annually, and only about 35 percent is restored. Much of the boreal forest has a slow rate of regeneration; and in many areas removal of the forest results in the formation of bogs. Removal of trees for firewood and excessive grazing by reindeer along the taiga-tundra ecotone is causing a southward expansion of the tundra at the expense of the taiga, a situation somewhat analogous to the spread of deserts into the African savanna.

The taiga is rich in metal ores, supplying a large portion of the world's minerals, including iron ore and gold, as well as in coal, gas, and oil. The extraction of these products began in the late nineteenth century with marked effects on the environment. Development gave rise to settlements, roads, and industrial complexes, including pulp and paper mills, mines, and ore smelting plants, which have contributed to massive air pollution in both Canada and Russia. Huge freshwater systems of lakes and rivers have encouraged the development of massive hydroelectric schemes such as the ones at Angara in Siberia and on the Peace River in Canada, the installation of which lowered water in the Mackenzie Basin, damaging trapping grounds of northern native tribes and destroying waterfowl breeding habitat. Such hydroelectric developments drown hundreds of thousands of hectares of boreal forest, the breeding grounds of waterfowl and other northern wildlife, and destroy the cultural and social fabric of northern native American and Lapp communities.

Further ecological damage is caused by mining peat for fuel and moss litter and organic soils for the horticultural and gardening trade, and by draining land for gardens, crops, and pasture. Drainage of peatlands is extensive in Europe, especially in Russia.

Opening the taiga by logging and industrial development makes the region accessible for expanded recreational use, including summer fishing camps and winter hunting lodges. What effects these and future developments hold for taiga wildlife and indeed the entire taiga ecosystem is a troublesome and unanswered question.

STUDY QUESTIONS

1. What physical and biological features characterize the tundra?
2. How does the alpine tundra differ from the arctic tundra?
3. What are the major contrasts between alpine tundras of temperate and tropical regions?
4. What is the relationship among permafrost, plant life, and nutrient cycling in the tundra?
5. Where does the bulk of net primary production of the arctic tundra accumulate and why?
6. What is the role of herbivores in the tundra ecosystem?
7. Differentiate the major vegetation zones that make up the taiga.
8. What is the relationship between tree cover and moss and lichen growth on the ground? How do they relate to permafrost?
9. How does the homogeneous forest cover relate to forest insect outbreaks and fire?
*10. Speculate why the taiga holds such a large number and diversity of important fur-bearing mammals. What effect did this abundance of valuable wildlife have on the history and settlement of the taiga?
*11. How does economic development clash with maintaining the integrity of the tundra and taiga ecosystems? Consider the ecological impacts of oil drilling in the Arctic National Wildlife Refuge, the James Bay hydroelectric development scheme in Quebec, and the proposed diversion of water southward from the taiga to the semiarid regions of Russia.

CHAPTER 30

TEMPERATE FORESTS

OBJECTIVES

On completion of this chapter, you should be able to:

- Describe the types and general features of coniferous and broadleaf forest ecosystems.
- Describe the effects and role of environmental and biological stratification in forest ecosystems.
- Compare the structure and function of coniferous and temperate deciduous forests.
- Discuss the impact humans have had on temperate forest ecosystems.

Spectacular fall color is a hallmark of the eastern deciduous mixed hardwood forest.

Temperate forests, in spite of their name, do not have a temperate environment. They occupy topographic positions that range from low-lying lands to mountaintops, and environmental conditions that range from warm and semiarid to cold and wet. They face great fluctuations in daily and seasonal temperatures, which place physiological stress on plants and animals. Deciduous trees are leafless during the winter and in northern regions remain so for the greater part of the year. They are exposed to droughts and in places to flooding. In spite of their intemperate environment, temperate forest ecosystems maintain high productivity.

Temperate forests embrace coniferous, deciduous, and mixed stands. Regardless of type, all forests possess large aboveground biomass. This biomass creates several layers or strata of vegetation, which influence both the vertical structure and environmental conditions within the stand: light, moisture, temperature, wind, and carbon dioxide. These factors differ among forest types.

30.1 CONIFEROUS FORESTS

Types

Montane Forests

Many temperate coniferous forests are in the mountains. In Central Europe extensive coniferous forests, dominated by Norway spruce *(Picea abies)*, cover the slopes up to the subalpine zone in the Carpathian Mountains and the Alps (Figure 30.1a). In North America several coniferous forest associations blanket the Rocky, Wasatch, Sierra Nevada, and Cascade mountains. In the southwestern United States such forests occur between 2500 and 4200 m elevation and in the northern United States and Canada between 1700 and 3500 m elevation. At high elevations in the Rocky Mountains, where winters are long and snowfall is heavy, grows a subalpine forest dominated by Engelmann spruce *(Picea engelmannii)* and subalpine fir *(Abies lasiocarpa)* (Figure 30.1b). Mid-elevations have stands of Douglas-fir, and lower elevations are dominated by open stands of ponderosa pine *(Pinus ponderosa)* (Figure 30.1c) and thick stands of the early successional pioneering conifer, lodgepole pine *(P. contorta)*.

Similar forests grow in the Sierra and Cascades. There high-elevation forests consist largely of mountain hemlock *(Tsuga mertensiana)*, red fir *(Abies magnifica)*, and lodgepole pine. We also find sugar pine *(Pinus lambertiana)*, incense cedar *(Libocedrus decurrens)*, and the largest tree of all, the giant sequoia *(Se-quoiadendron giganteum)*, which grows only in scattered groves on the west slopes of the California Sierra.

A deciduous seral, but occasionally permanent, species common both to the montane and the boreal forest is quaking aspen *(Populus tremuloides)*. It is the most widespread tree of North America (Figure 30.1d).

Pine Forests

Pines form extensive stands in both Eurasia and North America. A major component of the Eurasian boreal forest (see Chapter 29), Scots pine is also widespread through central Europe, where it grows from the lowlands to the tree line in the mountains. It occurs largely as planted or seminatural stands in southern England and western France.

Unlike the Scots pine forests of Eurasia, the pine forests of the coastal plains of the South Atlantic and Gulf states are considered a seral stage of the temperate deciduous forest, because without disturbance from fire and logging they give way to it. These pines maintain their presence by possessing a competitive advantage over hardwoods on nutrient-poor, dry, sandy soil and by their adaptation to a fire regime. At the northern end of the coastal pine forest in New Jersey, pitch pine *(Pinus rigida)* is the dominant species. Farther south, loblolly *(P. taeda)*, longleaf *(P. australis)* (Figure 30.2), and slash *(P. elliottii)* pine are most abundant.

Temperate Rain Forests

South of Alaska the coniferous forest differs from the northern boreal forest, both floristically and ecologically. The reasons for the change are both climatic and topographic. Moisture-laden winds move in from the Pacific, meet the barrier of the Coast Range, and rise abruptly. Suddenly cooled, the moisture in the air is released as rain and snow in amounts up to 635 cm/yr. During the summer, when winds shift to the northwest, the air is cooled over chilly northern seas. Although rainfall is low, cool air brings in heavy fog, which collects on the forest foliage and drips to the ground to add 127 cm or more of moisture.

This land of superabundant moisture, high humidity, and warm temperatures supports the temperate rain forest, a community of luxuriant vegetation dominated by conifers adapted to wet, mild winters, dry warm summers, and nutrient-poor soils. The forests are dominated by western hemlock *(T. heterophylla)*, mountain hemlock, Pacific silver fir *(Abies amabilis)*, and Douglas-fir, all trees with high foliage and stem biomass (Figure 30.3). Farther south, where precipitation is lower, grows the redwood *(Sequoia sempervirens)* forest, occupying a strip of land about 724 km wide.

Figure 30.1 Some coniferous forest types. (a) A Norway spruce forest in the Carpathian Mountains of central Europe. (b) Rocky Mountain subalpine forest dominated by subalpine fir *(Abies lasiocarpa).* This tree grows with Engelmann spruce and mountain hemlock. (c) A montane coniferous forest in the Rocky Mountains. The dry lower slopes support ponderosa pine; the upper slopes are cloaked with Douglas-fir. (d) Quaking aspen *(Populus tremuloides)* is the dominant deciduous tree in the western montane forest of North America.

Structure

Coniferous forests fall into three broad classes of growth form and growth behavior: (1) pines with straight, cylindrical trunks, whorled spreading branches, and a crown density that varies with the species from the dense crowns of red and white pine to the open, thin crowns of Virginia, jack, Scots, and lodgepole pine; (2) spire-shaped evergreens, including spruce, fir, Douglas-fir, and (with some exceptions) the cedars, with more or less tall pyramidal crowns, gradually tapering trunks, and whorled, horizontal branches; and (3) deciduous conifers such as larch (*Larix* spp.) and bald cypress, with pyramidal, open crowns that shed their needles annually. Growth form and behavior influence animal life and other aspects of coniferous ecosystems.

Vertical stratification in coniferous forests is not well developed. Because of a high crown density, the lower strata are poorly developed in spruce and fir forests, and the ground layer consists largely of ferns and mosses with few herbs. The maximum canopy development in spire-shaped conifers is about one-third down from the open crown, a profile different from that of pines. Pine forests with a well developed

Figure 30.2 A longleaf pine *(Pinus palustris)* forest in Florida. Regular burning, which simulates natural ground fires, eliminates the buildup of unburned fuel that could cause a devastating fire if ignited.

Figure 30.3 Old-growth forest in the Pacific Northwest. Typical is the Douglas-fir stand with an abundant western hemlock understory. Such forests are the home of the spotted owl and murrelet.

high canopy lack lower strata. The litter layer in coniferous forests is usually deep and poorly decomposed, resting on top of instead of mixing in with the mineral soil.

This poor vertical stratification influences the environmental stratification within the stand. When you walk into a spruce or fir forest, you are struck by the sharp diminution in light. Light intensity is progressively reduced through the canopy to only a fraction of full sunlight. The upper crown of spruce and firs, a zone of widely spaced narrow spires, is open and well lighted, whereas the lower crown is dense and intercepts most of the solar radiation. Most pines form a dense upper canopy that excludes so much sunlight that lower strata cannot develop. Open-crowned pines allow more light to reach the forest floor, stimulating a grassy or shrubby understory. Because conifers retain their foliage through the year, the degree of light interceptions in coniferous forests is about the same throughout the year. Illumination is greater during midsummer, when the sun's rays are most direct, and lower in winter, when the intensity of incident sunlight is the lowest.

The temperature profile of a coniferous forest also varies with the growth form. For example, in forests of sprucelike trees, temperatures tend to be coolest in the upper canopy, perhaps because of greater air circulation, and hottest in the lower canopy.

Animal life in the coniferous forest varies widely, depending upon the nature of the stand. Soil invertebrate litter fauna is dominated by mites. Earthworm species are few and their numbers low. Insect populations, although not diverse, are high in numbers and, encouraged by the homogeneity of the stands, are often destructive. Sawflies *(Neodiprion)*, for example, attack a wide variety of pines, including pitch, Virginia, shortleaf, and loblolly, and the southern pine beetle *(Dendroctonus frontalis)* can reach outbreak proportions in southern pinelands.

A number of bird species are closely associated with coniferous forests. In North America they include chickadees, kinglets, pine siskins, crossbills, purple finches, and hermit thrushes. Related species, the tits and grosbeaks, are common to European coniferous forests.

Except for strictly boreal species, such as the pine marten and lynx, mammals have much less affinity for coniferous forests. Most are associated with both coniferous and deciduous forest; the white-tailed deer, moose, black bear, and mountain lion are examples. Their north-south distribution seems to be limited more by climate, especially temperature, than by vegetation. The red squirrel, commonly associated with coniferous forests, is quite common in deciduous woodlands in the southern part of its range.

Function

The great economic importance of forests and the increasing intensity of forest management practices demand considerable understanding of forest ecosystems. For these reasons forests have received the most intensive studies of ecosystem function. One coniferous forest ecosystem studied in detail is Douglas-fir, an important forest type in the western United States. Forest ecologists have researched nutrient budgets and cycling for both a 36-year-old young stand and a 450-year-old old-growth stand. The old-growth stand had a considerably larger total biomass, both living and dead, than the younger stand, and both had similar foliage biomass. The younger forest had the higher percentage of its living biomass in foliage and roots—21 percent compared to 14 percent for the old-growth stand. Both age classes had most of their living biomass in branches and trunk.

Decaying logs and litter accounted for the most detrital matter (55 percent) in the old-growth stand; in the young stand, soil held most of the detrital organic matter (83 percent). Litter organic matter exceeded soil organic matter in the old-growth stand because of the accumulation and slow decomposition of needles and the long-term decomposition, over hundreds of years, of large fallen limbs and trunks.

Both old-growth and young stands cycle and store nutrients internally, particularly nitrogen. The young stand accumulated most of its nitrogen in foliage and bole, whereas the old stand stored most of it in bole and root. The 450-year-old stand accumulated considerably more nitrogen, but invested a lower percentage of it in the soil. The young forest banked 98 percent of its detrital nitrogen in the soil.

Both systems had a low input of nitrogen, about 2 kg/ha/yr. Of this input the 450-year-old stand returned somewhat more nitrogen to the forest floor than the young stand, mostly because of greater throughfall. The young growing stand needed considerably more nitrogen. Uptake of nitrogen by the young stand amounted to 55 percent of its needs; uptake met 44 percent of the old-growth's requirement. To make up the deficiency, both stands recycled nitrogen within the biomass. This internal cycling is also typical for other nutrients, with the exception of calcium and magnesium. Uptake of those nutrients far exceeds requirements, so internal cycling is not necessary.

Some coniferous ecosystems scavenge nutrients directly from rainfall through microcommunities of algae and lichens that colonize canopy leaves in balsam fir and western coniferous forests. Of particular interest is the complex community, including primary producers, consumers, and decomposers, supported by the canopy of old-growth Douglas-fir (Figure 30.4). Cyanophycophilous lichens (those whose algae component consists of cyanobacteria) fix atmospheric nitrogen. Organic nitrogen leached from these lichens combines with canopy moisture to form a dilute organic solution that in turn is taken up by microorganisms and other canopy epiphytes. Part of this microbial production is consumed by canopy arthropods. These nutrient cycles in the canopy tend to influence and even restrict the amount of nutrients that reach the forest floor by throughfall and stemflow.

Roots, fungi, and mycorrhizae (see Sections 9.5, 17.11) have a role in nutrient cycling in coniferous stands. Fungal hyphae or rhizomorphs concentrate nutrients in their tissues, especially the fruiting body, and act as a living sink of nutrients. Resistant to leaching, the fungal rhizomorphs hold biologically important elements in the litter, which they may slowly release to the soil through exudates.

Dead wood is often overlooked as a critical component of the forest ecosystem. The mass of dead wood waxes and wanes with tree mortality and disturbances, never achieving equilibrium within a stand. Woody litterfall increases as the forest ages and becomes most conspicuous in old temperate forests, both deciduous and coniferous, where logs and large limbs decay slowly. The volume of woody debris is much greater in old-growth coniferous forest of the Pacific Northwest than in eastern deciduous forests.

Large standing dead trees or snags (Figure 30.5) and downed trunks and limbs provide critical habitat for many animals of the forest. Standing dead trees provide essential nesting and den sites for cavity-nesting birds and mammals, food, and foraging areas. Fallen trees, which may make up 10 to 20 percent of the ground surface in forests, provide food, protection, and pathways for small mammals, and reproductive sites for certain woody plants. The elimination of standing and fallen dead trees greatly impoverishes animal life in the forest.

Human Impact

Because of their successional nature and dependence on disturbance for regeneration, temperate coniferous forest ecosystems can reoccupy a site that has been logged, although their return may follow a deciduous forest interlude. Historically, however, logging often has been followed by fire, which fed on the large amount of debris left behind. These human-caused fires have covered hundreds of thousands of hectares. The Cloquet-Moose Lake fire in Minnesota in fall of 1918 burned 4000 hectares in one day on logged-over white pine forest. The forest has never regenerated. In

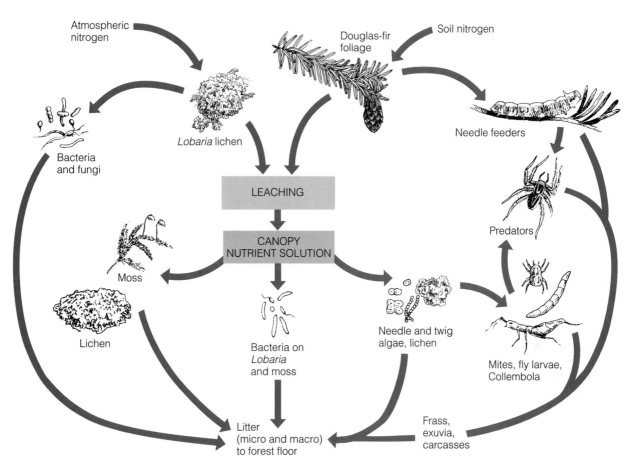

Figure 30.4 Nitrogen cycle in the canopy of old-growth Douglas-fir. Lichens (primary producers) support biophage and saprophage consumers. Such forest microecosystems conserve and recycle nutrients and influence nutrient return to the forest floor.

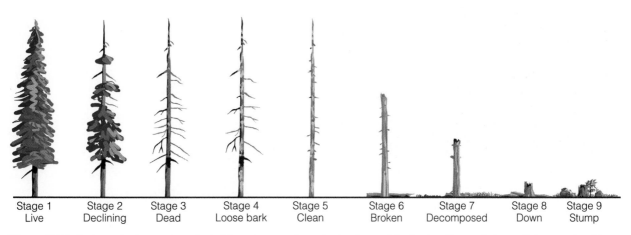

Figure 30.5 An important component of all forest ecosystems is dead wood, both standing and down. Snags or dead trees go through a process of decay, each stage of which supports its own group of animal life.

the spring of 1987 the Black Dragon fire, possibly the largest forest fire ever, burned over huge areas of large forest in China and the Soviet Union that still fail to show strong signs of recovering. The devastation of such fires is much greater than fires in uncut forest.

Montane coniferous forests and pinelands are a major source of lumber. Commercial foresters try to regenerate cutover forest for future cuts. One of the preferred methods is to clearcut blocks of forest, then burn the debris on the ground to eliminate the fire hazard and to expose a seedbed. Then foresters allow the area to regenerate naturally, or they replant it to leapfrog an incoming deciduous forest stage. The latter practice is common on southern pinelands and in Douglas-fir and Engelmann spruce forests. Most European coniferous forests are artificially planted and maintained. An artificially regenerated stand, even of the species harvested, is not a forest (Figure 30.6). Although some foresters argue that this method is an improvement on nature, the simplified, monocultural stand lacks all the complexity of a natural forest.

Even a naturally regenerated forest will not replace the original old-growth forest. In the Pacific Northwest the old-growth forest is an extremely complex ecosystem that evolved over hundreds of years. When such forests are cut, the system cannot be replaced, unless large tracts of old-growth forests are allowed to remain among younger stands, and the young stands are allowed to reach old age. Because commercial foresters consider old-growth forests senescent, unproductive stands, they are replacing them with younger, faster-growing stands that will not be allowed to reach an ancient age before they are cut. They will be harvested in cycles of 100 to 150 years, and the original ecosystem will have no time to return. The only way to maintain true old-growth ecosystems is to set aside sufficiently large tracts to sustain the ecosystem for centuries. Foresters regard retaining stands for that length of time as highly uneconomical and unproductive.

Serious damage is done to many montane coniferous forests by logging on steep slopes. Severe soil erosion degrades the site to a point where forest regeneration is impossible (Figure 30.7).

Montane coniferous forests are popular places for recreational use. Many forests have been carved up for ski slopes, trails, and access roads. These developments compact the soil, kill trees, and destroy natural values.

Coniferous forests have their share of diseases and insects. Humans often introduce a disease or insect or encourage them by logging and fire suppression. Examples are outbreaks of mountain pine beetle (*Dendroctonus monticolae*), which have wiped out thousands of hectares of lodgepole pine, and the western spruce budworm (*Choristoneura occidentalis*), which has become a problem because of the absence of light ground fires and selective cutting practices that encourage understory development in fir and spruce forests.

Figure 30.6 A planting of trees does not make a forest. This monocultural fast-growing southern pine plantation lacks the complexity of a natural pine stand. Such plantations make up industrial forests. Usually managed on short rotations, they provide the pulp, wood, poles, and treated lumber used in construction.

Figure 30.7 This clearcut mountainside on the Olympic Peninsula records the fate of old-growth forests in the Pacific Northwest.

30.2   BROADLEAF FORESTS

Types

Deciduous Forests

The deciduous forest once covered large areas of Europe and China, parts of South America and the middle American highlands, and even North America. The deciduous forests of Europe and Asia have largely disappeared, cleared over the centuries for agriculture. What remains is seminatural, except for pockets in the more mountainous regions of central Europe. The two major forest types there are beech-oak-hornbeam and oak-hornbeam. Beech *(Fagus sylvatica)* is the most uniformly distributed tree throughout central Europe. Beech forests, which grow from lowlands into the mountains, are characterized by dense canopy and poorly developed understory. Occupying damper, more acid soils are oak-hornbeam forests dominated by pedunculate oak *(Quercus robur)* and sessile oak *(Q. petraea)*. The Atlantic deciduous forest, originally dominated by beech, oaks, ash *(Fraxinus* spp.), and birch *(Betula* spp.), exists only in a seminatural state. Because of glacial history, the species diversity of the European deciduous forests does not compare with that of North America or China.

The Asiatic broadleaf forest, found in eastern China, Japan, Taiwan, and Korea, is similar to the North American deciduous forest and contains a number of plant species of the same genera as those found in North America and western Europe.

In North America the temperate deciduous forest reaches its greatest development in the mixed mesophytic forest of the central Appalachians, where the number of temperate tree species is unsurpassed by any other area in the world (Figure 30.8). In eastern North America the deciduous forest consists of a number of associations, including the mixed mesophytic forest of the unglaciated Appalachian plateau; the beech-maple and northern hardwood forests (with pine and hemlock) in northern regions that eventually grade into the boreal forest; the maple-basswood forests of the Lake states; the oak-chestnut (now oak since the die-off of the American chestnut) or central hardwood forests, which cover most of the Appalachian Mountains; the magnolia-oak forests of the Gulf Coast states; and the oak-hickory forests of the Ozarks. Fingers of forest extend along the rivers of the prairies, plains, and semiarid southwestern United States and Mexico. These forests, growing on the floodplains and banks of rivers and streams, are known as riparian woodlands. Growing on rich, moist, alluvial soil, riparian ecosystems are highly productive, add diversity to the landscape, and provide habitat for a number of wildlife species disproportionate to their area.

Temperate Woodlands

In western parts of North America where the climate is too dry for montane coniferous forest, we find the temperate woodlands. These forests are characterized by open-growth small trees with a well-developed understory of grass or shrubs. They contain needle-leaved trees, deciduous broadleaf trees, sclerophylls, or any combination. An outstanding example is the piñon-juniper woodland (Figure 30.9). This ecosystem occurs in the southwestern United States, primarily Utah, Airzona, New Mexico, and Colorado. In southern Arizona, New Mexico, and northern Mexico grow oak-juniper and oak woodlands, and in the Rocky Mountains, in particular, there are oak-sagebrush woodlands. In the Central Valley of California grows still another type—evergreen oak woodlands with grassy undergrowth.

Temperate Evergreen Forests

Extensive mixed forests of both broadleaf evergreen and coniferous trees occur in several subtropical areas of the world. Such forests include the eucalyptus forests of Australia (Figure 30.10), paramo forests and anacardia gallery forests of South America and New Caledonia, and false beech *(Nothofagus* spp.) forests of Patagonia. Representatives of temperate evergreen

Figure 30.8 An Appalachian hardwood forest in spring, dominated by oaks and yellow-poplar, with an understory of redbud in bloom.

Figure 30.9 A piñon *(Pinus edulis)* and juniper *(Juniperus osteosperma)* woodland in Utah, typical of temperate woodland in the southwestern United States. Seeds of piñon and fruits of juniper were staples in the diet of southwestern Native Americans.

Figure 30.10 A eucalyptus forest in the Grampian Mountains of Australia.

forests also occur in the Caribbean and on the North American continent along the Gulf Coast, in the hummocks of the Florida Everglades, and in the Florida Keys. Depending on location, these forests are characterized by oaks, magnolias, gumbo-limbo *(Bursera simaruba)*, and royal and cabbage palms.

Structure

Highly developed, uneven-aged deciduous forests usually have four strata (Figure 30.11). The upper canopy

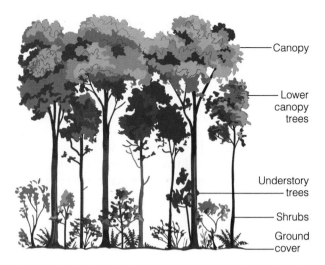

Figure 30.11 Stratification in an eastern deciduous forest.

consists of dominant and codominant trees, below which is the lower tree canopy and then the shrub layer. The ground layer has herbs, ferns, and mosses.

Even-aged stands, the results of fire, clearcut logging, and other large-scale disturbances, often have poorly developed strata beneath the canopy because of dense shade. The low tree and shrub strata are thin and the ground layer is poorly developed, except in small, open areas.

The physical stratification of the forest influences the microclimate within the forest. The highest temperatures are in the upper canopy, because this stratum intercepts solar radiation. Temperatures tend to decrease through the lower strata. The most rapid decline takes place from the leaf litter down through the soil.

Humidity in the forest interior is high in summer because of plant transpiration and poor air circulation. During the day, when the air warms and its water-holding capacity increases, relative humidity is lowest. At night, when temperature and moisture-holding capacities are low, relative humidity rises. The lowest humidity in the forest is a few feet above the canopy, where air circulation is best. The highest humidity is near the forest floor, kept that way by the evaporation of moisture from the ground and settling of cold air from the strata above.

Variation of humidity within the forest is influenced in part by the degree to which the lower strata are developed. Leaves add moisture to the immediate surrounding air; well-developed strata with more leaves have higher humidity. Thus layers of increasing and decreasing humidity may exist from the floor to the canopy.

Bathed in full sunlight, the uppermost layer of the canopy is the brightest part of the forest. Down through the forest strata, light intensity dims. In an oak forest only about 6 percent of the total midday sunlight reaches the forest floor; the forest floor is about 0.4 percent as bright as the upper canopy.

Light intensity within the forest varies seasonally (see Chapter 5). The forest floor receives its maximum illumination during early spring before the leaves appear; a second, lower peak of maximum illumination during the growing season occurs in the fall. The darkest period is midsummer. Light intensity during summer is highly variable from point to point and time to time as sun shines through gaps in the canopy. Sun flecks can influence the distribution of herbaceous vegetation on the forest floor.

In general, the diversity of animal life is associated with stratification and the growth forms of plants (see Chapter 20). Some animals, particularly forest arthropods, are associated with or spend the major part of their lives in a single stratum; others range over two or more strata. The greatest concentration and diversity of life in the forest occurs on and just below the ground layer. Many animals, the soil and litter invertebrates in particular, remain in the subterranean stratum. Others, such as mice, shrews, ground squirrels, and forest salamanders, burrow into the soil or litter for shelter and food. Larger mammals live on the ground layer and feed on herbs, shrubs, and low trees. Birds move rather freely among several strata, but favor one layer over another. Some occupy the ground layer but move into the upper strata to feed, roost, or advertise territory.

Other species occupy the upper strata—the shrub, low tree, and canopy layers. The red-eyed vireo, the most abundant bird of the eastern deciduous forests of North America, inhabits the lower tree stratum and the wood pewee the lower canopy. The black-throated green warbler and scarlet tanager live in the upper canopy. Squirrels are mammalian inhabitants of the canopy, and woodpeckers, nuthatches, and creepers live amid the tree trunks between shrubs and the canopy.

Function

Compared to other ecosystems, an abundance of data exists on energy flow and nutrient cycling in deciduous forests. One of the most intensively studied was a mesic yellow-poplar (*Liriodendron tulipifera*) forest on Walker Branch, Oak Ridge, Tennessee, with an understory of redbud, dogwood, Virginia creeper (*Parthenocissus quinquefolis*), and Christmas fern (*Polystichum acrostichoides*).

The forest expended 66 percent of its gross primary production of 2162 g $C/m^2/yr$ in plant respiration. This expenditure left a balance of 726 g $C/m^2/yr$ as net primary production.

Forty-two percent of the forest's total organic matter (329,100 kg/ha) was in living biomass. Ninety percent of the living biomass was in branches, bole, and root. Ninety-nine percent of total detrital biomass (190,500 kg/ha) was in the soil rooting zone. Only 1 percent was in the thin layer of forest litter. Thus at Walker Branch, total detrital mass exceeded living organic biomass.

Nutrient cycling involves a balance between inputs to the biological system and outputs or losses through streamflow. The difference represents the amount of recharge to the system from the soil pool. In the yellow-poplar forest on Walker Branch, total input through precipitation for five elements—N, P, K, Ca, and Mg—amounted to 20.44 kg/ha, consisting mostly of nitrogen and calcium; losses amounted to 212.78 kg/ha, mostly calcium. Nitrogen and phosphorus were cycled more tightly than the other three elements. More of these two elements came into the system through precipitation than was lost through leaching. Calcium and magnesium sustained the greatest losses, probably from the weathering of dolomitic limestone substrate.

Ecologists drew up an annual nutrient balance sheet for the forest. They found that the yellow-poplar forest at Oak Ridge (Table 30.1) could not meet its nutrient requirements by uptake from the soil, except for calcium, which it took up in excess of need (probably because of the limestone soil). Its uptake of nitrogen, for example, met only 66 percent of its requirements. The forest had to make up the deficiency by recycling the nutrient within its tree biomass.

The litter layer is the most important short-term nutrient pool, because it quickly decomposes (average turnover time is four years) and recycles, although the bulk of the nutrient pool is in mineral soil. Nutrients

TABLE 30.1

ANNUAL ELEMENT BALANCE OF A 30- TO 80-YEAR-OLD YELLOW-POPLAR–OAK FOREST (KG/HA/YR)					
	N	**P**	**K**	**Ca**	**Mg**
Requirement	87.9	6.3	47.5	82.6	21.7
Uptake	58.1	3.4	40.0	87.6	12.4
Internal recycling	29.8	2.9	7.5	(–5.0)	9.3

stored in living biomass, especially in roots, are translocated and recycled through the living biomass, particularly the foliage. The foliage, in turn, translocates a considerable portion of its nutrients back to the roots before leaf fall. However, the forest ecosystem can maintain adequate mineral cycling only if nutrients are pumped from soil reserves, maintained in part by weathering of parent material (see Section 10.3).

Long-term studies of the nitrogen cycle (see Section 25.7) in deciduous forests indicate that natural forest ecosystems tend to accumulate and cycle large amounts of nitrogen. Nearly 90 percent of the nitrogen is incorporated in the mineral soil. The remaining nitrogen is in vegetation and the forest floor. The most important mechanism for cycling nitrogen from vegetation to soil is death of small lateral roots, replaced by seasonal growth. The second most important mechanism is litterfall, 90 percent of which is leaves and reproductive parts. The third is leaching of nutrients from the leaves. Added to this input is nitrogen released through the decomposition of forest litter, input from precipitation, nitrogen fixation, and finally uptake from mineral soil.

Such studies of nitrogen cycling in the deciduous forest ecosystem emphasize important aspects of nutrient economics of the forest. Of the nitrogen added to long-term storage, about 54 percent is deposited in living matter and 46 percent is stored in organic matter in the forest soil. Of the amount of nitrogen coming into the system each year, the forest retains about 81 percent. Of the nitrogen used in plant growth, the forest withdraws about one-third from storage in the living plants. To replace that withdrawal, the forest takes a like amount from leaves before leaf fall and deposits it in the stems. So conservative are forest ecosystems in nitrogen cycling that only a small fraction of the nitrogen they add to the inorganic pool within the ecosystem is lost in streamflow. Thus an internal source of nitrogen normally not subject to loss from the system, annual uptake by living vegetation, and annual additions of nitrogen to woody biomass are important in promoting a tight cycling of nitrogen.

Human Impact

Although humans originally evolved in grassland and savanna ecosystems, we achieved heavy cultural and economic development in temperate broadleaf deciduous forest regions. Destruction of the deciduous forest began early in Europe and was largely complete by the Middle Ages. Over most of Europe the broadleaf deciduous forests that do remain are seminatural, highly modified by humans.

In North America great expanses of natural deciduous forest still exist in spite of mismanagement, logging, fires, and land clearing. However, logging wiped out magnificent virgin stands in a wave of deforestation unmatched for speed in any other part of the world. Most of the eastern forest was cleared for agriculture so intensely that New England had a wood shortage in colonial days. As farming moved west into the prairie region, forest returned to abandoned land. Much of the east now appears to be heavily forested. This appearance is illusionary, however. Much of the forest has been and is being fragmented by highways and industrial, urban, suburban, and recreational development (see Chapter 21). Lands once held under one ownership are being divided into smaller and smaller tracts for suburban, recreational, and retirement housing, forever changing the nature and integrity of large forested tracts. This fragmentation of ownership has serious implications for both wildlife and wood production. Large reserves of strippable coal coupled with energy demands threaten the rich mesic forest of the Appalachians with its great diversity of tree species and endemic salamanders.

Many of the second, third, and fourth growth forests we see bear little resemblance in species to the original forest cover. A good proportion of these forests became reestablished by invading abandoned agricultural land, so the species composition was influenced by the available seed source and the response of colonizing species to the site conditions.

After a deciduous forest is logged, the species composition of the regenerating forest may differ from the original. The nature of the incoming forest on cutover land depends on the effects and type of logging, the amount of disturbance to the residual stand and the forest floor, availability of seeds, suitable seedbed, and environmental conditions at the time of disturbance. For example, foresters have a difficult time regenerating Appalachian oak forests after harvesting. They grow back in red maple and black cherry, which may mean the demise of the great oak forests.

Introduced diseases and insects have not helped. Chestnut blight accidentally introduced into the United States from China by way of Europe in 1904 eliminated the American chestnut as a major component of the eastern forest. It holds on precariously by means of persistent, short-lived root sprouts. The gypsy moth, which escaped from a silk-moth breeder in Massachusetts in 1869, threatens the oaks of the central hardwood forest. Dutch elm disease, spread by a bark beetle, has devastated the American elm in North America.

Adding to the problem is a rise in tree mortality in the temperate deciduous forest. Termed forest decline, the cause for this change is not fully understood. An inordinate number of sugar maple (*Acer saccarum*) are in decline, with some stands experiencing 59 percent mortality. Other trees in decline are beech, red maple, and white ash. Forest decline presents a serious threat to deciduous forest, especially in northeastern North America (see Focus on Ecology 25.2).

Especially impacted by human activity are the riparian forests of the prairie regions and particularly of the southwestern United States and Mexico. Occupying rich alluvial soils and drawing moisture from the rivers and streams, these areas appealed to settlers. They cut forests for wood, cleared land for urban development, crops, and grazing, and destroyed other forests by stream channelization and flood-control works. The loss of riparian woodlands has contributed to flooding and seriously reduced wildlife habitat, both nesting and wintering grounds and protective pathways for movement and migration.

As available forests dwindle, we see efforts to improve the situation. People have begun to study and protect the remaining riparian ecosystems in the prairie regions and the southwest. Forest ecologists have been studying nutrient cycling, biomass accumulation, successional trends, and forest development. Silviculturists have been investigating ways to improve harvesting and regeneration methods to reduce ecological damage to the forest. They have amassed a greater amount of information; but the problem is to get commercial and government foresters to apply those findings.

STUDY QUESTIONS

1. Discuss the types of coniferous and deciduous forests.
2. How does structural stratification of a forest affect its environmental stratification?
3. What are the major life forms among conifers and what effect do they have on the structure of coniferous forests?
4. Contrast structural stratification in coniferous and deciduous forests.
5. Contrast nutrient cycling in coniferous forests with that in deciduous forests.
6. What is the role of dead wood on the forest floor? Why should dead trees be allowed to stand?
*7. The effort to save the endangered spotted owl in the Pacific Northwest is an effort not just to save a species but to save an ecosystem. Why is it easier to gain support for an endangered species than an endangered ecosystem? Aside from the spotted owl, why should old-growth forests be saved?
*8. Trace human relationships to the forest through the centuries, especially in northern Europe, in North America, and in Australia. What are the similarities, and why?

TROPICAL FORESTS

OBJECTIVES

On completion of this chapter, you should be able to:

- Describe the types of tropical forests.
- Explain the high diversity of life in tropical rain forests.
- Describe the structure of tropical rain forests.
- Discuss the role of mutualism in functioning of tropical forests.
- Describe the ways in which humans have affected tropical forests.

A tropical rain forest in Costa Rica.

The term "tropical forest" brings to mind images of the lush green jungles of the Amazon. Tropical forests, however, embrace more than the rain forest. Forty percent of the tropical land mass is open or closed forest; but of this forest only 25 percent is wet or rain forest, 32 percent is moist, and 43 percent is dry.

31.1 TYPES OF TROPICAL FORESTS

Plant geographers divide tropical forests into a number of types and subtypes. One type grades into another, with no sharp boundary.

Rain Forests

The prototype of tropical forests is the tropical **rain forest,** restricted mostly to the equatorial climatic zone between latitudes 10° N and 10° S. The rain forest occupies those regions of the world where the temperatures are warm through the year and rainfall, measured in meters, occurs almost daily. Although there is little annual variation in temperature and precipitation, daily variations in heat, rainfall, and humidity are great.

Tropical forests do not form a continuous belt around the equator. They are discontinuous, broken up by differences in precipitation (governed by the direction of the winds) and in the land masses. Most tropical forests, for example, grow below 1000 m altitude, and they are absent from the eastern part of equatorial Africa, the northwest part of South America, and the southern tip of India, all places that experience long annual droughts.

Tropical rain forests fall into three main groups. The largest and most continuous is in the Amazon basin of South America. The second is in the Indo-Malaysian area from the west coast of India and southeast China through Malaysia and Java to New Guinea. This forest has the greatest diversity of plant species. The third forest is in West Africa around the Gulf of Guinea, into the Congo basin. Smaller rain forests occur on the eastern coast of Australia, the windward side of the Hawaiian Islands, the South Sea Islands, and the east coast of Madagascar. There the trees are mainly evergreen with little or no bud protection.

Within the rain forest are many subtypes. The most luxuriant is the multilayered lowland rain forest (Figure 31.1). At higher elevations the lowland forest grades into mountain forest, with abundant undergrowth, tree ferns, and small palms. The mountain forest gives way to cloud forest, with understory thickets and trees burdened with epiphytes. Although it

Figure 31.1 An interior view of an Amazonian rain forest. Note the vigorous undergrowth and vines.

receives less rainfall, the cloud forest is continually wrapped in clouds and mists. Fingers of rain forest, called gallery forests, follow river courses into savannas. Swamp forests occupy perennially wet soils, and peat forests grow on nutrient-poor ones.

Seasonal Forests

The rain forest flows into tropical and subtropical seasonal forests, also called semi-evergreen and semi-deciduous forest. Although they retain many characteristics of the rain forest, these forests are subject to droughts of two to four months. Some 30 percent of the trees of the upper canopy lose their leaves during the dry period, but the lower canopy trees and understory retain their leaves through the year. In the Indo-Malaysian forest about a month after the coming of the monsoon rains, rising temperature triggers the emergence of new leaves. In other regions leaves emerge about a month before the rainy season begins, often coinciding with flowering. Fruits develop at the beginning of the dry season. Like the rain forest, the seasonal forest has lowland and mountain types.

Dry Forests

Overlooked in our concern over tropical forests are the dry tropical forests (Figure 31.2). The largest proportion of dry tropical forest is in Africa and on tropical islands, where it comprises about 80 percent of the forested area. About 22 percent of South American and 55 percent of Central American forested areas are dry tropical forest. Much of the original forest is gone,

Figure 31.2 A dry tropical forest in Costa Rica. Most of the unique dry tropical forests in Central America have disappeared from land-clearing schemes.

especially in Central America and India. It has been converted to agricultural and grazing land, or it has regressed through disturbance to thorn woodland, savanna, and grassland.

Dry tropical forests undergo a dry period, the length of which is based on latitude. The more distant the forest is from the equator, the longer is the dry season, up to eight months (see Chapter 4). During the dry period the trees and shrubs drop their leaves. Before the start of the rainy season, which may be much wetter than the wettest time in the rain forest, the trees begin to leaf; during the rainy season the landscape becomes uniformly green.

31.2   TROPICAL RAIN FORESTS

Tropical rain forests are noted for their diversity of plant and animal life. Tree species number in the thousands. A 10-square-kilometer area of tropical rain forest may contain 1500 species of flowering plants and up to 750 species of trees. The richest area is the lowland tropical forest of peninsular Malaysia, which contains some 7900 species. There one of the major groups, the Dipterocarpaceae, contains 9 genera and 155 species, of which 27 are endemic. (The Asian dipterocarps have 12 genera and 470 species.) Tropical rain forests also account for several million species of flora and fauna, one-half of all known plant and animal species, and 20 to 25 percent of all known arthropods.

Structure
Vegetation
The tropical rain forest divides into five general layers (Figure 31.3), most apparent in the undisturbed for-

Figure 31.3 Vertical stratification of a tropical rain forest.

est. Stratification, however, is often poorly defined, because the growth plan of many tree species is the same, differing only in size. The uppermost layer consists of emergent trees over 40 to 80 m high, whose deep crowns billow above the rest of the forest to form a discontinuous canopy. The second layer, consisting of mop-crowned trees, forms another, lower, discontinuous canopy. Not clearly separate from one another, these two layers form an almost complete canopy. The third layer, the lowest tree stratum, is made up of trees with conical crowns. It is continuous, often the deepest layer, and well defined. The fourth layer, usually poorly developed in deep shade, consists of shrubs, young trees, tall herbs, and ferns. Many of these plants have elongated, downward-curving leaf blades, called drip tips. These blades apparently enable the leaves to rid themselves of excess water in their permanently wet environment, increase transpiration, and reduce nutrient leaching. The fifth stratum is the ground layer of tree seedlings and low herbaceous plants and ferns.

A conspicuous part of the rain forest is plant life dependent on trees for support. Such plants include epiphytes, climbers, and stranglers. Climbers, the lianas, are vines with fine stringlike to massive cable-like stems that reach the tops of trees and expand into the form and size of a tree crown. They may loop to the ground and ascend again. Climbers grow prolifically in openings, giving rise to the image of the impenetrable jungle. This popular image applies more to secondary forest, second growth that develops where primary forest has been disturbed.

Stranglers and epiphytes share some characteristics. Stranglers start life as epiphytes. As they grow, they send roots to the ground and increase in number and girth until they eventually encompass the host tree and claim the crown and limbs as support for their own leafy growth. Epiphytes inhabit niches on the trunks, limbs, and branches, and even leaves of trees, shrubs, and climbers. Their roots are aerial. Epiphytes come in various types. One group, the microepiphytes, consists of mosses, lichens, and algae. Macroepiphytes, such as orchids, bromeliads, and members of the Ericaceae, are vascular plants. Their roots never reach the ground. They attach themselves to a tree and take up nutrients from the air, rain water, and organic debris on their supports. Some of the epiphytes are important in recycling minerals leached from the canopy.

The floor of the tropical rain forest is thickly laced with roots, both large and small, forming a dense mat on the ground. Except for a few that reach down to weathered parent rock, rain forest roots are shallowly concentrated in the upper 0.3 m of soil, where inorganic nutrients are available. Associated with the fine roots are mycorrhizae (see Chapter 17) that aid in the uptake of nutrients from the decaying organic matter.

Reaching upward as much as 5 m from the wide-spreading roots of many species of large trees are thin, strong, planklike outgrowths called buttresses (Figure 31.4). These buttresses function as prop roots, providing support for trees rooted in soil that offers poor anchorage.

The mature tropical forest, like the mature temperate forest, is a mosaic of continually changing vegetation. Death of tall trees, brought about by senescence, lightning, wind storms, hurricanes, defoliation by caterpillars, and other causes, creates gaps (see Chapter 21), which shade-intolerant pioneer species quickly fill. These trees are replaced eventually by shade-tolerant late successional species, perhaps over a period of 100 years; but continuous random disturbances across the forest ensure persistence of the species in the mature forest. A high frequency of tree fall may account for the low density of large trees (1 m dbh and larger) in mature rain forests. Enhancing this diversity are local changes in soil, topography, and drainage that support varying arrays of species.

Most tropical rain forest trees reach full height when they have achieved only about one-third to one-half of their final bole diameter. Layering comes about when a group of species of similar mature height dominates a stand. Layering is also influenced by crown

Figure 31.4 Planklike buttresses help to support tall rain forest trees.

shape, which in turn correlates with tree growth. Young trees still growing in height have a single stem and a tall narrow crown; they are *monopodial*. As many species mature, large limbs diverge from the upper stem or trunk; they become *sympodial*. This change happens when the bud of the main stem axis ceases to grow and the lateral buds take over their role. The process repeats itself, adding to crown growth and producing a pattern that suggests the spokes of an umbrella. Looking up into the canopy, the observer gains the impression that crowns of trees fit together like a jigsaw puzzle with the pieces about a meter apart. This growth pattern is called **crown shyness** (Figure 31.5). Among the dipterocarps of the Far East, this reiteration of the growth pattern within the crown to produce many dense subcrowns gives a cauliflowerlike appearance to the canopy. Within the crown there is minimal overlap among the leaves, each positioned to receive the maximum amount of light available.

Layering of vegetation influences the internal microclimate of the forest. Crowns of emergent trees experience conditions similar to open land. The level of CO_2 and amount of humidity increase going down through the canopy, and temperature and evaporation decrease. From the ground up to about 1 m the levels of CO_2 are high, and humidity at over 90 percent is oppressive. Temperature on the average is 6° C cooler inside than outside the forest, with a strong nocturnal inversion. Although light penetrates the upper canopy, the lianas, epiphytes, and lower tree layers block out

most of it. The amount of light that reaches the floor of a Malaysian rain forest is about 2 to 3 percent of incident radiation, and half of that comes from sunflecks; about 6 percent comes from breaks in the canopy, and 44 percent from reflected and transmitted light.

Animal Life

Stratification of animal life in the tropical rain forest is pronounced. J. L. Harrison in 1962 described six distinct feeding communities of birds and mammals in a lowland tropical forest of the Far East. (1) A group feeding above the canopy is made up mostly of insectivorous and some carnivorous birds and bats. (2) A canopy group consists of a large variety of birds, fruit bats, and other species of mammals that eat leaves, fruit, and nectar. A few are insectivorous and mixed feeders. (3) Below the canopy, in a zone of tree trunks, is a world of flying animals—birds and insectivorous bats. (4) Also in the middle canopy are scansorial mammals, such as squirrels, that range up and down the trunks, entering the canopy, and the ground zone to feed on the fruits of epiphytes, on insects, and on other animals. (5) The forest floor is occupied by large herbivores, such as the gaur, tapir, and elephant, which feed on ground vegetation and low-hanging leaves, and their attendant carnivores, such as leopards and tigers, all of which range over a large area. (6) The final feeding stratum includes the small ground and undergrowth animals, birds and small mammals capable of some climbing, that search the ground litter and lower parts of tree trunks for food. This stratum includes insectivorous, herbivorous, carnivorous, and mixed feeders.

The enormous diversity, but low species population density, of animal life in the tropical rain forest mirrors the great diversity of microhabitats and niches. Insect fauna number in the millions, with many species still to be discovered. Not only is invertebrate life distributed vertically, but horizontally as well. Many species are restricted to certain forests, and within them inhabit only certain plants. Numerous species live in the epiphytes or in small pools of water caught in epiphytic plants, where they may be joined by small canopy-dwelling frogs.

Nearly 90 percent of all non-human primates live in the tropical rain forests of the world. Sixty-four species of New World primates, small with prehensile tails, live in the trees. The Indo-Malaysian forests are inhabited by a number of primates, many of which are restricted to certain regions. The orangutan, an arboreal ape, is confined to the island of Borneo. Peninsular Malaysia has seven species of non-human primates,

Figure 31.5 A view through the canopy of a secondary forest reveals crown shyness. The crowns are separated from one another, giving the impression of a jigsaw puzzle.

including three gibbons, two langurs, and two macaques. The long-tailed macaque is common to disturbed or secondary forests, and the pig-tailed macaque is a terrestrial species, adaptable to human settlements. The tropical forest of Africa is home to mountain gorillas and chimpanzees. The diminished rain forest of Madagascar holds 39 species of lemurs.

Function

Tropical forests are among the most productive ecosystems in the world. Because sites, soils, location, precipitation, and other conditions vary among tropical forests, so does their productivity. In general, tropical forests use about 70 to 80 percent of their assimilated energy in maintenance and 20 to 30 percent for net production. Mean annual net production is about 22 mtn/ha. This amount exceeds the net production of temperate forests (about 13 mtn/ha/yr) by a factor of 1.7 and that of boreal forests (an average of 8 mtn/ha/yr) by a factor of 2.7. The rate of wood production by tropical forests is about the same as that of temperate hardwood forests, but tropical forests turn more of their net production into foliage and fruit.

Because high year-round temperatures and abundant rainfall accelerate geological cycling, biological cycles function as retainers of nutrients in the living portion of the system. Tropical forests may store nutrients in living biomass, where they are protected from leaching, or they may reduce to a minimum the time nutrient elements remain in the soil. This mechanism of nutrient cycling contrasts with that of the temperate forest, in which the soil plays a significant role in nutrient cycling.

Tropical forests have a large standing crop biomass, averaging about 300 tn/ha, two times that of the temperate hardwood forest. They tend to concentrate more calcium, silica, sulfur, iron, magnesium, and sodium, and less potassium and phosphorus in their biomass than do temperate forests. Because of the great variation within and among tropical rain forests, large differences in mineral storage and nutrient cycling exist among them.

Tropical rain forests maintain their nutrient balance by internal cycling (see Chapter 25). Nutrients leached from foliage and epiphytes by throughfall are rapidly taken up by the roots and returned to the vegetation. Nutrients not taken up are captured and held by fungal rhizomorphs and mycorrhizal fungi. From their studies, ecologists learned that a tropical forest on the Ivory Coast of Africa recycled more than 60 per-

cent of its potassium and 15 to 56 percent of other nutrients such as calcium, magnesium, and nitrogen gained by throughfall. At the end of the dry season, when leaching is the greatest, the soil-root system picked up the nutrients leached from dead leaves.

The soil-root system is highly dependent on mycorrhizal fungi (see Chapter 17). The mycorrhizae transfer nutrients directly from dead organic matter to the living roots with a minimum of leakage to the soil, so minerals remain tied up in living and dead organic matter. This concentration and rapid recycling of nutrients in the uppermost layers of the soil explain why many tropical soils cleared of their forest vegetation are nutrient-poor and can support agricultural systems for only a few years.

Other important mutualistic relationships keep the tropical forest running. Ninety-eight to 99 percent of all flowering species of the tropical lowland rain forest are pollinated by animals. The major exception is the emergent dipterocarps, mainly the genus *Shorea*, whose pollen is carried by the wind. Animal pollinators range from the tiny fig wasps, only 1 to 2 mm in length, to flying fox bats with a wingspread of 2 m. Bats, birds, bees, moths, beetles, large flies, and wasps comprise the major pollinators. Bats in both the neotropical and paleotropical rain forests are important pollinators in the high canopy. Flowers pollinated by bats open at dusk. They have a sour scent, a sticky nectar, and pale yellow or white flowers. Among the birds, hummingbirds are the dominant pollinators in the neotropical forests, whereas in paleotropical forests sunbirds, white-eyes, and honey eaters are the major pollinators. Bird-pollinated flowers are usually scentless with bright colors—reds, oranges, yellows, and greens—and with abundant, watery nectar. In all rain forests, bees are the major insect pollinators, aided by moths and beetles. Bee flowers are rich in nectar, weakly scented, and often colored yellow, blue-green, or blue. Beetle flowers are heavily scented with a rich sweet smell or a carrion odor. Pollination systems are so specialized that a given plant species is pollinated by only one species or a few belonging to the same taxonomic order. A classic example is the 600 species of fig trees (*Ficus* spp.), each pollinated by its own unique species of fig wasp (Family agaonidae).

Tropical plants also use animals to disperse their seeds. Across all the tropical forests, 50 to 90 percent of trees and shrubs depend on animal dispersers. To attract them, plants bear fruit easy to obtain and digest. Seeds are discarded or pass through the digestive tract (see Chapter 17). The problem for fruit eaters is that fruits can be scarce at the end of the rainy season and the beginning of the dry season. Fruits may be spatially or temporally patchy, depending on the species, the sequence of fruiting, and the distribution of plants through the forest. The wide-ranging fruit-eating bats generally eat the larger fruits of widely dispersed tropical trees. In the neotropical forest little dietary overlap exists among birds, bats, and primates. In the tropical forests of the Far East, however, which hold a lower diversity of fruit-eating species, considerable dietary overlap exists, especially among the primates.

Human Impact

For at least 20 centuries, humans have been associated with tropical forests. They entered the rain forest first as hunter-gatherers. Nomadic, they lived off the forest, hunting its wildlife and gathering its abundance of fruits, roots, and tubers for food, and leaves, bark, and wood for clothing and shelter. Because they were a part of the forest ecosystem and their populations were small—about one person per 4 km²—they had little impact on the forest.

Other less nomadic groups, the hunter-gardeners, established small semipermanent settlements and cultivated garden plots of native plants to supplement food obtained by hunting and gathering. Still others became shifting cultivators, developing the technique of slash-and-burn agriculture. These people cut down and burned the vegetation on small plots. The ashes fertilized the soil in which the cultivators interplanted the seeds of a variety of crops. For one to two years the plots remained productive; after that the people abandoned them to vigorous native weeds and cleared new ones. Meanwhile the old plots, now supporting pioneering tropical vegetation, remained fallow for 8 to 20 years, at which time they were cleared again for crops. As long as the populations were not large, the tropical forest was able to absorb shifting cultivation; but as populations grew and pressure on the land increased, long fallow periods were no longer possible. Land degradation and erosion ate away at the forest.

The real onslaught to the tropical forest came with the Europeans. They exploited the forests for hardwood timber (Figure 31.6a), cleared it for pasture and cropland, and in Southeast Asia converted it to rubber and palm plantations (Figure 31.6b). Since World War II the rate of exploitation has accelerated. Worldwide demand for tropical hardwood timber has greatly reduced some species, such as various species of *Shorea*, commercially sold as lavan and Philippine mahogany,

and has driven others, including African and Central American mahogany (*Khaya* spp.) and Brazilian rosewood (*Dalbergia nigra*), to the brink of extinction.

Logging is not the direct cause of loss of tropical forests. Secondary forest can grow back. The problems with logging are that felling large-crowned trees damages residual vegetation and that road building and heavy machinery cause erosion. More important, roads open up inaccessible areas to settlers, who swarm in and clear off the remaining forest for cultivation. In a short time, the land grows to unproductive shrub and grassland with little hope of reverting to tropical forest.

In addition to logging, millions of hectares of lowland tropical rain forest have been and are being converted to other uses. The hillsides of Malaysia are green not because they are covered with tropical forests but because they are blanketed with rubber and palm plantations (Figure 31.6c, d). In the Amazon and Central America, large areas of tropical forest have been cleared, burned, and converted to pastureland, with devastating results. The response of pasture grasses, stimulated by the fertilizing effects of the ashes, is vigorous at first, but the fertility quickly declines because of leaching, binding of some essential nutrients to the soil, and removal in crops and cattle. Within five to eight years the pasture degenerates into weedy fields. Because the cleared areas are so large, the soil so infertile, and the distance from any primary forest so great, the tropical forest cannot regenerate as it did in smaller plots of slash-and-burn agriculture.

As logging and clearing eat deeper into the tropical forests, these activities come in direct conflict with native peoples of the forests. Contact with the western world has tragic effects: loss of home and livelihood because of a fragmented environment, outright murder, introduction of disease, and destruction of native culture. Unable to exist in the forest any longer, tribes are forced to move to settlements and take up a foreign way of life or resist removal, slowing the inevitable. For example, the Bateq in peninsular Malaysia cling to their fragmented tropical forest home, too small to support them through the year. During the fruit-poor monsoon season, they either move to nearby villages and leave the forest by day to earn money for food or stay in settlement camps. At the end of the monsoon season, they disappear into the forest.

The greatest tragedy of tropical deforestation is the loss of biological diversity. Although occupying only 7 percent of Earth's surface and disappearing at the rate of an estimated 56 × 10⁶ ha per year, the trop-

Figure 31.6 The tropical forest faces destruction. (a) A logging truck delivers a load of *Shorea* logs to a large sawmill on the Malaysian peninsula. (b) Hundreds of hectares of tropical rain forest were cleared to establish this new palm oil plantation. (c) Oil palm plantations cover thousands of hectares of former tropical forest. Note the lone *Shorea* dipterocarp tree among the palm trees. (d) An old rubber plantation has been clearcut prior to replanting a new one.

ical rain forest region holds 50 to 80 percent of the world's plant species; and most of the animal species are endemic, resident, and nonmigratory. As tropical rain forests are reduced and fragmented, species lose their habitat, and mutualistic relationships involving pollination and seed dispersal necessary for the survival of both plants and animals are broken. Lost, too, are major and potential sources of medicine and wild genotypes needed for improving agricultural crops and livestock. As native peoples disappear or lose their culture, we will lose their knowledge of the ecology

and usefulness of the biotic resources of the tropical rain forest.

There is hope that the destruction may slow or be contained. The costs of developing pastures and rangeland in the Amazonian tropical forests do not balance short-term yields, and investments in fertilizer and weed control to maintain productivity are prohibitive. Agriculturalists are developing tree crops and new agricultural methods for tropical soils to slow slash-and-burn agriculture. Interest in managing rain forests on a sustained yield basis, especially in the Far

East, is gaining momentum. The Forestry Research Institute of Malaysia is developing methods and markets for utilizing tree species that loggers now cut and discard. Tropical forest silviculturists have developed management plans to ensure continued production from forest reserves and are establishing plantations of fast-growing tropical species. Many species of wildlife, fortunately, do well in secondary forests, including gibbons, deer, gaur, and elephants.

Because of burgeoning world populations and demand for wood, ceasing all cutting of tropical forests is unrealistic; but we must ensure that we use the tropical rain forests without destroying them.

STUDY QUESTIONS

1. Name and characterize the types of tropical forests.
2. What are the major strata in the tropical forest?
3. What are emergents, lianas, and epiphytes? What are their positions in the rain forest?
4. What are buttresses?
5. What is crown shyness?
6. If tropical forest soils are so nutrient-poor, how can they support such a high plant biomass and diversity?
7. How are mutualistic interactions involved in the functioning of the tropical rain forest?
8. How do logging, shifting cultivation, cattle ranching, and other agricultural developments affect tropical rain forests?
9. What are the long-term effects of tropical deforestation?
*10. The media have focused on deforestation in the Amazon. Investigate the rate and consequences of deforestation in Central America, Sarawak, Borneo, or Thailand.
*11. Select a native people of South or Central America or the Ibans of Sarawak, and report on the impact of deforestation on them.

C H A P T E R   3 2

LAKES AND PONDS

Moraine Lake in Banff National Park, Alberta, Canada.

No feature in the landscape attracts more interest than ponds and lakes. We are drawn toward them as if by a magnet. We seek them out for beauty and for recreation. Yet how many of us give any thought to the life they hold and how it functions beneath the surface, or how lakes came about in the first place, or how our intrusion affects them?

Lakes and ponds are inland depressions containing standing water (Figure 32.1). They vary in depth from 1 m to over 2000 m. They range in size from small ponds of less than a hectare to large lakes covering thousands of square kilometers. Ponds are small bodies of water so shallow that rooted plants can grow over much of the bottom. Some lakes are so large that they mimic marine environments. Most ponds and lakes have outlet streams; and both may be more or less temporary features on the landscape, geologically speaking.

Some lakes have formed by glacial erosion and deposition. Abrading slopes in high mountain valleys, glaciers carved basins that filled with water from rain and melting snow to form tarns. Retreating valley glaciers left behind crescent-shaped ridges of rock debris that dammed up water behind them. Numerous shallow kettle lakes and potholes were left behind by the glaciers that covered much of northern North America and northern Eurasia.

Lakes also form when silt, driftwood, and other debris deposited in beds of slow-moving streams dam up water behind them. Loops of streams that meander over flat valleys and floodplains often become cut off, forming crescent-shaped oxbow lakes.

Shifts in Earth's crust, uplifting mountains or displacing rock strata, sometimes develop water-filled depressions. Craters of some extinct volcanoes have also become lakes. Landslides block off streams and valleys to form new lakes and ponds. In any given area all natural ponds and lakes have the same geological origin and similar characteristics; but because of varying depths at time of origin, they may represent several stages of development.

Many lakes and ponds form through nongeological activity. Beavers dam streams to make shallow but

(b)

(a)

(c)

Figure 32.1 Lakes and ponds fill basins or depressions in the land. (a) A swampy tundra in Siberia is dotted with numerous ponds and lakes. (b) A beaver dam forms a pond in this Colorado Rocky Mountain meadow. (c) A human-constructed old New England mill pond. Note the floating vegetation.

often extensive ponds. Humans create huge lakes by damming rivers and streams for power, irrigation, or water storage (see Chapter 34) and construct smaller ponds and marshes for recreation, fishing, and wildlife. Quarries and strip mines form other ponds.

32.1 PHYSICAL CHARACTERISTICS

Unlike most terrestrial ecosystems, lakes and ponds have well-defined boundaries—the shoreline, the sides of the basin, the surface of the water, and the bottom sediments. Within these boundaries, environmental conditions vary from one pond or lake to another. However, all still-water ecosystems share certain characteristics. Life in still-water ecosystems depends on light. The amount of light penetrating the water is influenced not only by natural attenuation (see Chapter 5), but also by silt and other material carried into the lake and by the growth of phytoplankton. Temperatures vary seasonally and with depth. Oxygen can be limiting, especially in summer, because only a small proportion of the water is in direct contact with air, and decomposition on the bottom consumes it. These variations in oxygen, temperature, and light strongly influence the distribution and adaptations of life in lakes and ponds.

Life in lakes and larger ponds experiences seasonal shifts. The heating and cooling of surface waters changes temperature and oxygen levels throughout the basin. In late spring and early summer both increasingly direct solar radiation and warming air temperatures heat surface water faster than deep water. Because water reaches its maximum density at 4° C (see Chapter 7), the surface water becomes lighter as its temperature increases. Soon a layer of lighter, warm water, called the **epilimnion,** rests on top of a heavier mixed layer of cooler water (Figure 32.2). This layer, known as the **metalimnion,** becomes cooler with depth. For approximately every 1 m downward, the temperature declines 1° C, a drop called the **thermocline.** (If you dive into deep water, you become suddenly aware of the thermocline.) When the temperature of the water reaches 4° C and its greatest density, it lies as a layer of cold water on the bottom called the **hypolimnion.** The thermocline acts as a barrier between the epilimnion and hypolimnion. The lake basin is much like a sandwich, with the epilimnion and hypolimnion forming the top and bottom of the roll and the metalimnion the filling in between. The filling is thick enough to prevent any contact between the top and bottom water, and little circulation takes place.

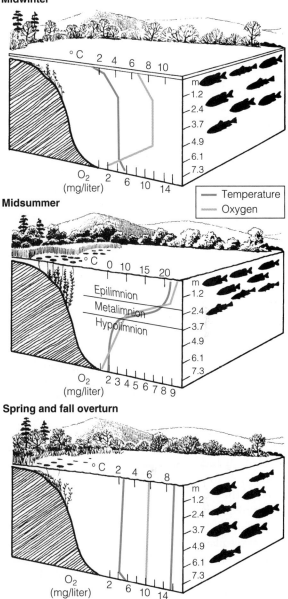

Figure 32.2 The distribution of oxygen and temperature in a lake during different seasons affects the distribution of fish. The narrow fish silhouettes represent trout, or cold-water species. The wider silhouettes are bass, or warm-water species. Note the pronounced horizontal stratification in midsummer and the nearly vertical oxygen and temperature curves during the spring and fall overturn.

Oxygen produced by the action of wind and phytoplankton keep the upper layer of water aerated. The waters below, however, may be deficient in oxygen, consumed by decomposers (Figure 32.3). Because

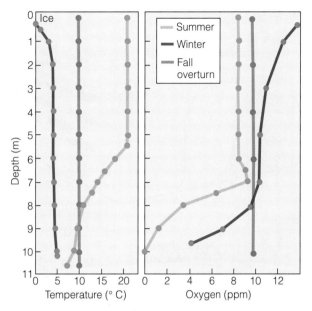

Figure 32.3 Oxygen stratification in Mirror Lake, New Hampshire, in winter, summer, and late fall. The late fall overturn results in both a uniform temperature and uniform distribution of oxygen throughout the lake basin. In summer a pronounced stratification of both temperature and oxygen exists. Oxygen declines sharply in the thermocline and is nonexistent on the bottom because of decomposition taking place in the sediments. In winter oxygen is also stratified, but it is present at a low concentration in deep water.

sediments accumulate on the bottom, the deep waters are relatively high in nutrients, but those nutrients are not available to phytoplankton in the upper layer. Deprived of the riches below, phytoplankton may suffer nutrient depletion late in summer.

By fall, conditions begin to change, and a turnabout takes place. Air temperatures and sunlight decrease, and the surface water starts to cool. As it does, the water becomes denser and sinks, displacing the warmer water below to the surface, where it cools in turn. This cooling continues until the temperature is uniform throughout the basin (see Figure 32.3). Now pond and lake water circulate throughout the basin and the nutrients denied the phytoplankton in late summer are at the surface. This circulation, which recharges oxygen and nutrients through the basin, is called the fall **overturn.** Stirred by wind, the overturn may last until ice forms.

Then comes winter, and the surface water cools to below 4° C. It becomes lighter again and remains on the surface. (Remember, water becomes lighter above and below that temperature.) If the winter is cold enough, surface water freezes; otherwise it remains close to 0°. Now the warmest place in the pond

or lake is on the bottom. A slight temperature inversion develops, in which the water becomes warmer, up to 4° C, with depth. Water beneath the ice may be warmed by solar radiation through the ice. Because that increases its density, this water flows to the bottom, where it mixes with water warmed by heat conducted from the bottom mud. The result is a higher temperature on the bottom, although the overall stability of the water is undisturbed.

With the spring breakup of ice and the heating of surface water up to 4° C, another overturn occurs. Again oxygen and nutrients are recharged throughout the basin. The surface waters are now both nutrient and oxygen rich, ready for the spring growth of phytoplankton. As the season wears on, the lake water again becomes stratified into the three familiar layers.

Not all lakes experience such seasonal changes in stratification, and you should not consider this phenomenon as characteristic of all deep bodies of water. In shallow lakes and ponds, temporary stratification of short duration may occur; in others stratification may exist, but no thermocline develops. In some very deep lakes, the thermocline may simply descend during periods of overturn and not disappear at all. In such lakes the bottom water never becomes mixed with the top layer. However, some form of thermal stratification occurs in all very deep lakes, including those of the tropics.

32.2   STRUCTURE

Ponds and lakes may be divided into both vertical and horizontal strata based on penetration of light and photosynthetic activity (Figure 32.4). The horizontal zones are obvious to the eye; the vertical ones, influenced by depth of light penetration, are not. Surrounding most lakes and ponds and engulfing some ponds completely is the **littoral zone** or shallow-water zone, in which light reaches the bottom, stimulating the growth of rooted plants. Beyond the littoral is open water, the **limnetic zone,** which extends to the depth of light penetration. It is inhabited by plant and animal plankton and **nekton,** free-swimming organisms such as fish that can move about freely. Beyond the depth of effective light penetration is the **profundal zone.** Its beginning is marked by the **compensation level** of light, the point at which respiration balances photosynthesis. The profundal zone depends on a rain of organic material from the limnetic zone for energy. Common to both the littoral and profundal zones is the third vertical stratum, the **benthic zone** or bottom region, which is the place of decomposition. Although these zones are named and often described separately, all are closely dependent on one another in the dynamics of lake ecosystems.

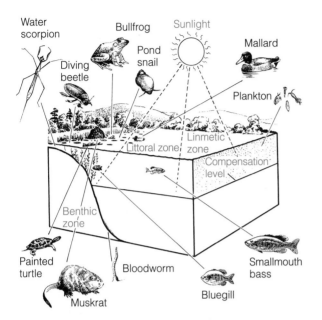

Figure 32.4 The major zones of a lake in midsummer: littoral, limnetic, profundal, and benthic. The compensation level is the depth at which light becomes too low for photosynthesis. The surrounding organisms are typical of a lake community.

Littoral Zone

Aquatic life is richest and most abundant in the shallow water about the edges and in other places within lakes and ponds where sediments have accumulated on the bottom, decreasing water depth. Dominating these areas is emergent vegetation, plants whose roots are anchored in the bottom mud, whose lower stems are immersed in water, and whose upper stems and leaves stand above water (Figure 32.5). The distribution and variety of plants vary with water depth and fluctuation of water levels. Very shallow depths support spike rushes and small sedges; deeper water is occupied by plants with narrow tubular or linear leaves, such as bulrushes, reeds, and cattails. With them are associated such broadleaf emergents as pickerelweed (*Pontederia* spp.) and arrowhead (*Sagittaria* spp.). Beyond the emergents and occupying even deeper water is a zone of floating plants such as pondweed (*Potamogeton*) and pond lily (*Nuphar* spp.). Many of these floating plants have poorly developed root systems but highly developed aerating systems. In depths too great for floating plants live submerged plants, such as certain species of pondweed. Lacking cuticles, these plants absorb nutrients and gases directly from the water through thin and finely dissected or ribbonlike leaves.

Associated with the emergents and floating plants is a rich community of organisms, among them hydras, snails, protozoans, and sponges. Insects include

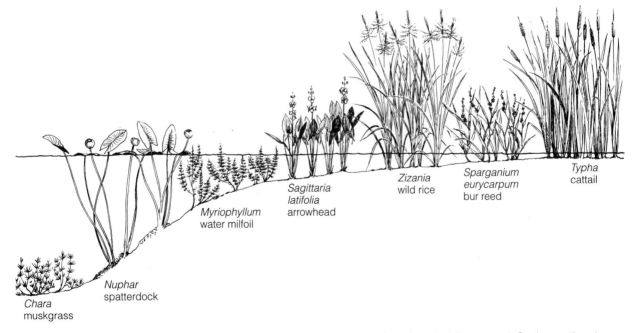

Figure 32.5 Zonation of emergent, floating, and submerged vegetation at the edge of a lake or pond. Such zonation does not necessarily reflect successional stages, but rather response to water depth.

dragonflies and diving insects such as water boatmen and diving beetles that carry a bubble of air with them when they go underwater in search of prey. Fish such as pickerel and sunfish find shelter, food, and protection among the emergent and floating plants. Fish of lakes and ponds lack strong lateral muscles characteristic of fish living in swift water, and some, such as sunfish, have compressed bodies that permit them to move with ease through the masses of aquatic plants. The littoral zone contributes heavily to the large input of organic matter into the system.

Limnetic Zone

People usually associate the open water of a lake with fish; but the main forms of life in the limnetic zone are minute protistan and animal organisms—phytoplankton and zooplankton. Because the tiny protists that make up the phytoplankton, including desmids, diatoms, and filamentous algae, carry on photosynthesis in open water, they are the base on which the rest of life in open water depends. Their presence is noticeable during bloom, the season when the populations of these protists are most abundant. Suspended with the phytoplankton are small animals, mostly tiny crustaceans, that graze on minute protists. These animals form an important link in energy flow in the limnetic zone.

Light sets the lower limit at which phytoplankton can survive, so populations of these protists concentrate in the epilimnion. Because the zooplankton feeds on these minute protists, it too is concentrated in the limnetic zone. By its own growth phytoplankton limits light penetration into the water. As summer progresses, growth lessens the depth at which phytoplankton can live. As the zone becomes shallower, phytoplankton can absorb more light, increasing organic production.

Within the limits of light penetration, the depth at which various species of phytoplankton can live depends on the optimum conditions for their development. Some phytoplankton species live just below the surface; others are more abundant a few feet beneath; and those requiring colder temperatures live deeper still. Cold-water plankton, in fact, is restricted to lakes in which phytoplankton growth is scarce in the epilimnion and in which the oxygen content of the deep water is not depleted by decomposition of organic matter.

Animal plankton may be seasonally stratified, because it is capable of independent movement. In winter some animal plankton species distribute themselves evenly to considerable depths; in summer they concentrate in layers most favorable to them and to their stages of development. At that season zooplankton

species undertake a vertical migration during some part of the 24-hour period. Depending on the species, they rest in the deep water or on the bottom and move up to the surface during the alternate period to feed on phytoplankton.

During the spring and fall overturns, plankton is carried downward, but at the same time nutrients released by decomposition on the bottom are carried upward to the impoverished surface layers. In spring when surface waters warm and stratification develops, phytoplankton has access to both nutrients and light. A spring bloom develops, followed by a rapid depletion of nutrients and a reduction in planktonic populations, especially in shallow water.

Fish make up most of the nekton in the limnetic zone. Their distribution is influenced mostly by food supply, oxygen, and temperature. During the summer largemouth bass, pike, and muskellunge inhabit the warmer epilimnion waters, where food is abundant. In winter they retreat to deeper water. Lake trout, on the other hand, move to greater depths as summer advances. During the spring and fall overturn, when oxygen and temperature are fairly uniform throughout, both warm-water and cold-water species occupy all levels.

Profundal Zone

Life in the profundal zone depends not only on the supply of energy and nutrients from the limnetic zone above but also on the temperature and availability of oxygen. In highly productive waters, oxygen may be limiting, because the decomposer organisms so deplete it that little aerobic life can survive. The profundal zone of a deep lake is much larger in proportion to total volume, so production of the epilimnion is relatively low, and decomposition does not deplete the oxygen. In these lakes the profundal zone supports some life, particularly fish, some plankton, and such organisms as certain cladocerans that live in the bottom ooze. Some zooplankton may occupy this zone during some part of the day, but migrate up to the surface to feed. Only during spring and fall overturns, when organisms from the upper layers enter this zone, is life abundant in profundal waters.

Easily decomposed substances drifting down through the profundal zone are partly mineralized while sinking. The remaining organic debris—dead bodies of plants and animals of the open water, and decomposing plant matter from shallow-water areas—settles on the bottom. Together with quantities of material washed in, they make up the bottom sediments, the habitat of benthic organisms.

Benthic Zone

The bottom ooze is a region of great biological activity, so great, in fact, that the oxygen curves for lakes and ponds show a sharp drop in the profundal water just above the bottom (see Figure 32.3). Because the organic muck is so low in oxygen, the dominant organisms there are anaerobic bacteria. Under anaerobic conditions, however, decomposition cannot proceed to inorganic end products. When the amounts of organic matter reaching the bottom are greater than can be utilized by bottom fauna, they form an odoriferous muck rich in hydrogen sulfide and methane. Thus lakes and ponds with highly productive limnetic and littoral zones have an impoverished fauna on the profundal bottom. Life in the bottom ooze is most abundant in the lakes with a deep hypolimnion in which some oxygen is still available.

As the water becomes shallower, the benthos changes. The action of water, plant growth, drift materials, and recent organic deposits modifies the bottom material—stones, rubble, gravel, marl, and clay. Increased oxygen, light, and food encourage a richness of life not found on the profundal bottom.

Closely associated with the benthic community are organisms collectively called **periphyton** or **aufwuchs.** They are attached to or move on a submerged substrate, but do not penetrate it. Small aufwuchs communities colonize the leaves of submerged aquatic plants, sticks, rocks, and debris. Periphyton, mostly algae and diatoms, living on plants is fast-growing and lightly attached. Because the substrate is so short-lived, the associated periphyton rarely lives for more than one summer. Aufwuchs on stones, wood, and debris form a more crustlike growth of cyanobacteria, diatoms, water moss, and sponges.

32.3 FUNCTION

Lakes and ponds appear to be self-contained aquatic ecosystems, surrounded by totally different terrestrial ecosystems. Nevertheless they are strongly influenced by inputs of materials from surrounding terrestrial ecosystems and other sources outside the basin (Figure 32.6). Nutrients and other substances move across the boundaries along biological, geological, meteorological, and hydrological pathways.

Wind-borne particulate matter, dissolved substances in rain and snow, and atmospheric gases make up meteorological inputs. Outputs along the same pathway are small, mainly spray aerosols and gases such as carbon dioxide and methane. Geological inputs include nutrients dissolved in groundwater and inflowing streams and particulate matter flowing into the basin from the surrounding watershed. Geological outputs include dissolved and particulate matter carried out of the lake by outflowing waters and nutrients buried in deep sediments that are removed from circulation for a long time. Biological inputs and outputs are relatively small, mostly animals such as fish that move in and out of the lake. Hydrological inputs involve precipitation and drainage of surface waters. Outputs include seepage through the walls of the lake basin, subsurface flows, and evaporation. Nutrients and energy move through lakes and ponds by the way of grazing and detrital food chains.

Although studies of lake metabolism have emphasized the phytoplankton-zooplankton grazing food chain, in reality lakes, like terrestrial communities, are dominated by the detrital food chain. Most of this detritus (which is all dead organic carbon) comes from the littoral zone. In total, however, detritus includes particulate and dissolved organic carbon (POC and DOC) that comes from external sources and cycles within the system, and organic matter lost to a particular trophic level through egestion, excretion, and secretion.

Lake ecosystems function mostly within a framework of organic carbon transfer. The central pool comes from both internal and external sources: (1) imports from outside; (2) the littoral zone; (3) the limnetic zone. Most detrital metabolism takes place in the open-water zone during sedimentation and in the benthic zone, where particulate matter is decomposed.

Phytoplankton contributes the primary production in the limnetic zone, and macrophytes provide the same in the littoral zone. The contribution each makes varies among systems. Availability of nutrients in the water influences phytoplankton production. If nutrients are not limiting and the only losses are respiratory, the rate of net photosynthesis and biomass accumulation is high. In fact, a linear relationship exists between phytoplankton production and phytoplankton biomass. However, as phytoplankton biomass increases, shading also increases, which reduces net photosynthesis and increases respiration. As a result production declines. When nutrients are low, respiration and mortality increase, reducing net photosynthesis and biomass. However, if zooplankton grazing and bacterial decomposition are high, nutrients recycle rapidly, resulting in a high rate of net photosynthesis even though the concentrations of nutrients and biomass accumulation are low.

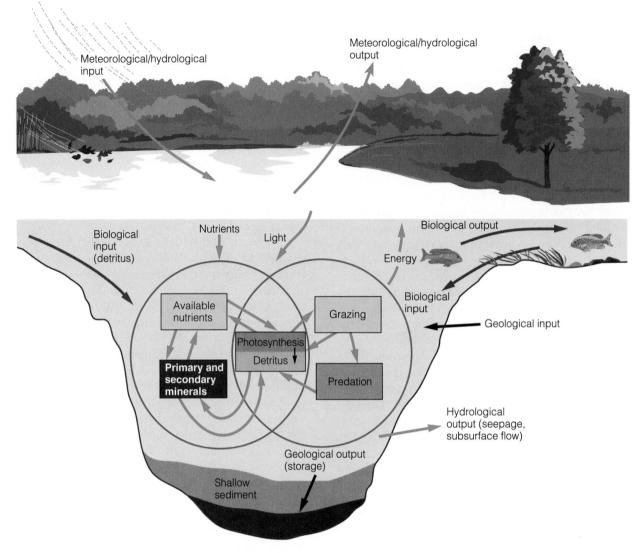

Figure 32.6 Model of nutrient cycling and energy flow in a lake ecosystem. Meteorological, geological, and biological inputs enter the lentic system from the watershed. Nutrients and energy move through a number of pathways. Part of the nutrients and energy accumulate in bottom sediments.

Macrophytes also contribute heavily to lake production. The ratio of macrophytic production in the littoral to microphytic production in the limnetic is influenced by the fertility of the lake. Highly fertile lakes support a heavy growth of phytoplankton that shades out macrophytes and reduces their contribution. In less fertile lakes where phytoplankton production is low, light penetrates much deeper into the water and rooted aquatics grow. Macrophytes are little affected by nutrient availability in the open water because they draw their nutrients from the bottom sediments.

Nutrient transfers within lake ecosystems take place largely between the water column and sediments.

Phytoplankton, zooplankton, bacteria, and other consumers take up nutrients that settle in the water column and benthic muds. In spring when phytoplankton bloom is at its height, nitrogen and phosphorus become depleted in the limnetic zone, because of the high rates of photosynthesis, sinking of dead phytoplankton, and sedimentation. At the same time decomposition decreases particulate N and P. This action increases dissolved P, but the dissolved N is lost through denitrification.

In summer conditions change. Because of a decline in phytoplankton in the limnetic zone and a slower sinking rate, as much N and P enter solution as phytoplankton takes up in photosynthesis. N and P increase in the dissolved and particulate pools and in bottom sediments. Phosphorus, in particular, becomes trapped in the hypolimnion, unavailable to phytoplankton until the fall overturn.

Macrophytes, however, can change this situation by moving phosphorus from sediments to the water column and on to phytoplankton, at the same time building up more bottom sediments. Steven Carpenter found that in Lake Wingra, Wisconsin, macrophytes significantly increased the amount of phosphorus available to phytoplankton that it otherwise would have to obtain directly from sediment. Macrophytes obtained 73 percent of the phosphorus incorporated in shoot tissue from sediments. Eventually some 550 kg of P from this source became available to phytoplankton, compared to 470 kg available from sediments through overturns and disturbance and resuspension of sediments. This uptake stimulates the production of both macrophyte and phytoplankton biomass, adding to sediment accumulation on the bottom. This buildup of sediment creates new areas for colonization and provides additional phosphorus for phytoplankton. Thus rooted aquatic plants enhance the recycling of phosphorus by mobilizing it from the sediments. Such activities accelerate the nutrient enrichment of lakes.

Although the productivity of a lake relates to the nutrient richness of its waters, other internal forces influence it. Any two lakes with a similar nutrient load may differ in productivity. Various species of phytoplankton differ in their rates of metabolic activity and nutrient recycling, both size-dependent. Small phytoplankters, for example, have higher maximum growth rates overall, higher maximum growth at low levels of nutrients, and a lower sinking rate than large phytoplankton species.

Feeding on phytoplankters are the zooplankton, essential to the recycling of nutrients, particularly N and P (see Chapter 25). Various species and sizes of zooplankton graze on different sizes of phytoplankton. The size of the dominant zooplankters influences the species composition and size structure of the phytoplankton community. Zooplankters, in turn, are eaten by invertebrate planktivores (insect larvae and crustaceans) and vertebrate planktivores (minnows and small spiny fish). These predators, too, are size-dependent in their food selection (see Chapter 16). The vertebrate planktivores may eat their associated invertebrate planktivores as well. The vertebrate planktivores in turn become prey for fish-eating predators (piscivores).

Interactions among these feeding groups flow down through the food web, influencing productivity at each trophic level. A rise in the biomass of predatory fish (bass, pike, trout) can change the density, species composition, and behavior of zooplanktivorous fish. That relationship, in turn, can affect invertebrate planktivores. Vertebrate planktivores, taking the largest available prey, reduce the density of large zooplankters, forcing invertebrate planktivores to select smaller species. Any changes in the relative densities of these two groups of planktivores influence the structure and density of zooplankton, thus affecting grazing intensity and rates of nutrient recycling. With fewer herbivorous grazers, phytoplankton increases. Each change in biomass at one trophic level has the opposite response at the next trophic level. Throughout the food web, however, maximum production is achieved at intermediate levels of density. Such trophic interactions influence and regulate the productivity of lake ecosystems.

32.4   NUTRIENT STATUS

A close relationship exists between land and water ecosystems. Primarily through the hydrological cycle, one feeds on the other. The water that falls on land runs from the surface or moves through the soil to enter streams, springs, and eventually lakes. The water carries with it silt and nutrients in solution. Human activities, including road construction, logging, mining, construction, and agriculture, add another heavy load of silt and nutrients, especially nitrogen, phosphorus, and organic matter. These inputs enrich aquatic systems, a process we call **eutrophication.**

The term **eutrophy** (from the Greek *eutrophos,* "well nourished") means a condition of being nutrient-rich. The opposite of eutrophy is **oligotrophy,** the condition of being nutrient-poor. The German limnologist C. A. Weber coined these terms in 1907 for the development of peat bogs. E. Naumann later associated the terms with phytoplankton production in lakes. He said that eutrophic lakes, found in fertile lowland regions, hold high populations of phytoplankton; oligotrophic lakes, common to regions of primary rocks, contain little plankton. This concept of oligotrophy and eutrophy ignores the input of highly productive littoral zones.

Eutrophic Systems

A typical eutrophic lake (Figure 32.7) has a high surface-to-volume ratio; that is, the surface area is large relative to depth. Nutrient-rich deciduous forest and farmland surround it. An abundance of nutrients, especially nitrogen and phosphorus, stimulates a heavy

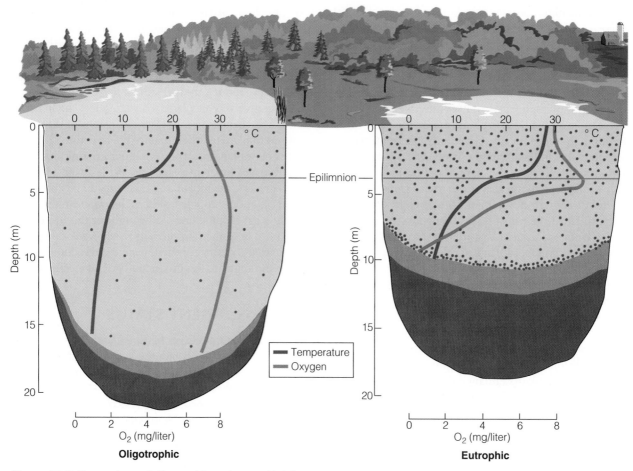

Figure 32.7 Comparison of oligotrophic and eutrophic lakes.

growth of algae and other aquatic plants. Increased photosynthetic production leads to an increased regeneration of nutrients and organic compounds, stimulating even further growth.

Phytoplankton concentrates in the warm upper layer of the water, giving it a murky green cast (Figure 32.8a). The turbidity reduces light penetration and restricts biological productivity to a narrow layer of surface water. Algae, inflowing organic debris and sediment, and remains of rooted plants drift to the bottom, adding to the highly organic sediments. Bacteria partially convert this dead organic matter into inorganic substances. The activities of these decomposers deplete the oxygen supply of the bottom sediments and deep water to the point at which this region of the lake cannot support aerobic life. The number of bottom species declines, although the biomass and numbers of organisms remain high.

As the basin continues to fill, the volume decreases, and shallowness speeds the cycling of available nutri-

ents and plant production. Positive feedback carries the lake or pond to extinction. It becomes a marsh or swamp, then a terrestrial community.

Oligotrophic Systems

Oligotrophic lakes have a low surface-to-volume ratio. The water is clear and appears blue to blue-green in the sunlight. The epilimnion is cool, the hypolimnion is high in oxygen, and the bottom sediments are largely inorganic. The nutrient content of the water, however, is low; and although nitrogen may be abundant, phosphorus is highly limited. A low input of nutrients from surrounding terrestrial ecosystems and other external sources is mostly responsible for this condition. Typically, coniferous forests on thin, acid soil dominate the watershed (Figure 32.8b).

Low availability of nutrients causes low production of organic matter, particularly phytoplankton. Low organic matter production leaves little for de-

(a)   **(b)**

Figure 32.8 (a) A eutrophic pond. Note the floating algal mats on the water. (b) An oligotrophic lake in Maine.

composers, so oxygen concentration remains high in the hypolimnion. These oxidizing conditions are responsible for the low release of nutrients from the sediment. The lack of decomposable organic matter means low bacterial populations and slow rates of microbial metabolism.

Although the numbers of organisms in oligotrophic lakes and ponds may be low, species diversity is often high. Members of the salmon family predominate.

Dystrophic Systems

Lakes that receive large amounts of organic matter from surrounding land, particularly in the form of humic materials that stain the water brown, are called **dystrophic** (from *dystrophos*, "ill-nourished"). Although the productivity of dystrophic lakes is considered low, they are low only in planktonic production. Dystrophic lakes generally have highly productive littoral zones, particularly those that develop bog flora. This littoral vegetation dominates the metabolism of the lake, providing a source of both dissolved and particulate organic matter.

32.5   HUMAN IMPACT

A pristine lake is one of the ecosystems highly vulnerable to human changes. Its ruin begins when the first house appears on its shores or when it is opened to recreational use, however limited. Any activity and settlement begins to unravel the fabric of the lake ecosystem. As more people are attracted to the lake,

developers move in. The shore is parceled into lots, cottages and residences spring up, boats stir up the placid waters, nutrients seep from septic tanks, and pesticides find their way into the lake. Before long the entire structure of life changes.

Consider an oligotrophic lake whose shoreline is newly developed. Drainage from the development moves into the lake. When nutrients in moderate amounts are added to this oligotrophic lake, they are taken up rapidly and circulated. It so happens that tissues of aquatic algae growing there in low populations contain phosphorus, nitrogen, and carbon in the ratios of 1 P : 7 N : 40 C per 500 g wet weight. If nitrogen and carbon are in excess and phosphorus is limiting, the addition of phosphorus will stimulate algal growth; if nitrogen is limiting, the addition of that nutrient will do the same. In most oligotrophic lakes, phosphorus rather than nitrogen is limiting. Because of its low ratio, 1 P per 500 g, an addition of even a moderate amount of phosphorus generates considerable growth of algae. As increasing quantities of nutrients are added to a lake or pond, it begins to change from oligotrophic to metatrophic (having a moderate amount of nutrients) to eutrophic. This change has been happening to clear oligotrophic lakes around the world at an increasing rate.

In fact, this galloping eutrophication has been changing naturally eutrophic lakes into **hypertrophic** ones. A heavy influx of wastes, raw sewage, drainage from agricultural lands, river basin development, runoff from urban areas, and burning of fossil fuels overloads the hypertrophic lake with nutrients. This accelerated enrichment causes chemical and environmental changes and major shifts in plant and animal life. We call this process **cultural eutrophication.**

People destroy floating and emergent vegetation around lakes as homes and marinas proliferate along the shores. Wakes created by motorboating disturb littoral vegetation and birds that nest within it, notably loons on northern lakes. Motorboats discharge an oily mixture with gas exhausts beneath the surface of the water, where it escapes immediate detection. One gallon of oil per million of gallons of water imparts an odor to lake water; eight gallons per million taint fish. These oily discharges can lower oxygen levels and adversely affect the growth and longevity of fish.

There are worse problems than excessive quantities of nutrients flowing into the lake. Life-threatening, organism-deforming pesticides feed in from surrounding farmland, suburban lawns, and golf courses. Toxic wastes enter from lakeshore industries. Some areas of the Great Lakes have become so contaminated with these wastes that fish taken from them are unfit for human consumption. Silt from construction, road building, logging, and other sources fills in the shores and destroys littoral vegetation. Many lakes are suffering from acid deposition (see Chapter 25). Already over 10 percent of northern lakes are so acidified they cannot support fish and their associated invertebrate fauna.

Pollution, overfishing, and introduction of exotic species, accidentally or on purpose, into lakes and larger ponds have upset the original assemblage of species, changing and destroying food webs. So badly have many lakes been modified that even with the best of pollution controls and other restoration efforts, they cannot be brought back to their original condition, although they can be greatly improved.

Because of the great amount of water they contain and the volume of water flowing into them, some lakes have been exploited for urban consumption and irrigation. Because water has been diverted for irrigation, the Aral Sea in the former Soviet Union has dropped 9 m and is expected to drop another 8 to 10 m, which would reduce its volume by one-half. The surrounding shoreline and exposed lake bottom are nearly desert, and a thriving fishing industry has been destroyed. On a smaller scale, diversion of water from Mono Lake in California has shrunk its surface area by one-third, increasing its salinity and threatening the future of resident and migratory wildlife that depend on it.

STUDY QUESTIONS

1. Explain why seasonal stratification of temperature and oxygen takes place in lakes and deep ponds.
2. What characterizes the epilimnion, the hypolimnion, and the metalimnion?
3. What distinguishes the littoral zone from the limnetic, and the limnetic from the profundal?
4. What conditions distinguish the benthic zone from the other strata, and what is its role in the lake ecosystem?
5. What are the main sources of nutrients and energy in the lake ecosystem?
6. What is the relationship between nutrient availability and phytoplankton production?
7. Describe the mechanisms of nutrient transfer in the lake ecosystem.
8. What are the major feeding groups in a lake ecosystem, and how do they interact?
9. Distinguish among oligotrophy, eutrophy, and dystrophy.
10. What is cultural eutrophication?
11. How has human intrusion affected lakes' function and structure?
*12. Consider a residential town built on the shores of a small natural lake. The town has no central sewage system and relies on residential septic tanks. The lake, once known for its recreation and fishing, is now badly polluted. Fish life has declined, algal growth has increased, and in summer bacterial counts become so high that swimming is restricted. The town's solution to the problem is to dump chlorine into the swimming areas in summer. The town is not high above the level of the lake, and the soils are sandy. What is the reason for the problem in the lake, what ecological events are taking place, and what is the long-term solution to the problem?
*13. Investigate the ecological health of a pond or lake in your area. How have the water quality and life in the lake changed over the years? What current human pressures impinge on it?

FRESHWATER WETLANDS

OBJECTIVES

On completion of this chapter, you should be able to:

- Define a wetland.
- Describe the various types of wetlands.
- Explain the role of hydrology and hydroperiod in wetlands.
- Compare production and nutrient cycling among wetlands.
- Explain the ecological and economic value of wetlands.
- Discuss effects of human activities on wetlands.

Pond pine *(Pinus serotina)* and a diversity of floating and emergent aquatic plants dominate this piece of Okefenokee Swamp in Georgia.

What is a wetland? This question seems to require only a simple answer: an area covered with water and supporting aquatic plants. Although this answer is not wrong, it is not correct either, not in this day when the life or death of a wetland rests on a precise definition.

Some wetlands are easy to distinguish. A water area supporting submerged plants such as pondweed, floating plants such as pond lily, and emergents such as cattails and sedges is unquestionably a wetland. But what about a piece of ground where the soil is more or less permanently wet and supports some ferns and such trees as maple that also grow on the uplands? Where do you draw the line between wetlands and uplands on this gradient of soil wetness?

Vegetation alone does not define a wetland. First we must consider the hydrological conditions; then we may use vegetation as an indicator.

Wetlands range along a gradient from permanently flooded to periodically saturated soil (Figure 33.1) and support hydrophytic (water-loving) vegetation at some time during the growing season. Hydrophytic plants are adapted to grow in water or on soil that is periodically anaerobic (deficient in oxygen) be-

cause of excess water (see Chapter 7). Hydrophytic plants include several groups: (1) obligate wetland plants, such as the submerged pondweeds, floating pond lily, and emergent cattails and bulrushes, and trees such as baldcypress *(Taxodium distichum)*; (2) facultative wetland or amphibious plants that can grow in standing water or saturated soil and rarely grow elsewhere, such as certain sedges and alders; (3) facultative species such as red maple *(Acer rubrum)*, which have about a 50:50 probability of growing in either wetland or nonwetland situations; and (4) facultative upland species such as beech *(Fagus grandifolia)*, which have a 1 to 30 percent probability of growing in a wetland. It is the last group of plants that is critical in determining the upper limit of a wetland on the soil moisture gradient, when species designation alone is insufficient.

For example, some species of trees usually associated with uplands adapt and grow quite well in wetland environments. An example from eastern North America is red maple. It thrives in the drier uplands, but is also a conspicuous species in forested wetlands. In the uplands red maple has a deep taproot; in wetland situations the tree has a shallow root system that

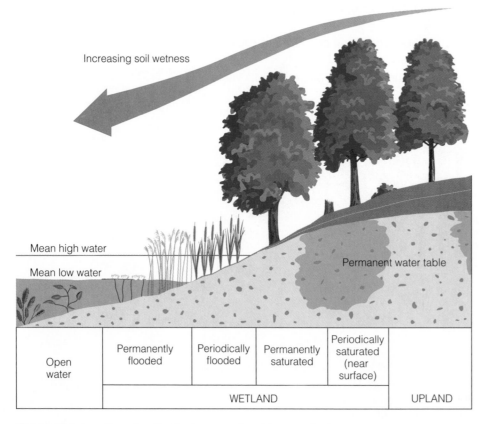

Figure 33.1 Location of wetlands along a soil moisture gradient.

enables it to avoid anaerobic stress. The red maple is a facultative species that has evolved ecotypes adapted to different soil moisture conditions. Black gum *(Nyssa sylvatica)*, too, grows in both upland and wetland situations. Pitch pine *(Pinus rigida)*, associated with the drier ridgetops of the southern Appalachians, has ecotypes that grow in poorly drained soils and the muck of swamps. In fact, pitch pine is a dominant species in the wetlands of the extensive New Jersey pine barrens. Hemlock *(Tsuga canadensis)*, a shallow-rooted species, is at home in wetland situations.

The point is that species of vegetation alone do not define a wetland. They are important indicators, especially in the wettest situations, but ecotypes of upland species confuse the situation. It is essential to consider hydrological conditions and soil properties along with the vegetation.

33.1   TYPES OF WETLANDS

A wide variety of wetlands exists, and classifying them for management and conservation has presented problems. An old, short, but still useful classification appears in Table 33.1. A much more comprehensive classification is *Classification of Wetlands and Deepwater Habitats of the United States.*

Wetlands most commonly occur in three topographic situations (Figure 33.2). Many develop in shallow basins, ranging from upland depressions to

TABLE 33.1

TYPES OF WETLANDS		
Type	**Site Characteristics**	**Plant and Animal Populations**
Inland Fresh Areas		
Seasonally flooded basins or flats	Soil covered with water or waterlogged during variable periods, but well drained during much of the growing season; in upland depressions and bottomlands	Bottomland hardwoods to herbaceous growth
Fresh meadows	Without standing water during growing season; waterlogged to within a few inches of surface	Grasses, sedges, rushes, broadleaf plants
Shallow fresh marshes	Soil waterlogged during growing season; often covered with 15 cm or more of water	Grasses, bulrushes, spike rushes, cattails, arrowhead, smartweed, pickerelweed; major waterfowl production areas
Deep fresh marshes	Soil covered with 15 cm to 1 m of water	Cattails, reeds, bulrushes, spike rushes, wild rice; principal duck breeding areas
Open fresh water	Water less than 3 m deep	Bordered by emergent vegetation such as pondweed, naiads, wild celery, water lily; brooding, feeding, nesting areas for ducks
Shrub swamps	Soil waterlogged; often covered with 15 cm or more of water	Alder, willow, buttonbush, dogwoods; nesting and feeding areas for ducks to limited extent
Wooded swamps	Soil waterlogged; often covered with 0.3 m of water; along sluggish streams, flat uplands, shallow lake basins	North: tamarack, arborvitae, spruce, red maple, silver maple; South: water oak, overcup oak, tupelo swamp, black gum, cypress
Bogs	Soil waterlogged; spongy covering of mosses	Heath shrubs, *Sphagnum*, sedges
Coastal Fresh Areas		
Shallow fresh marshes	Soil waterlogged during growing season; at high tide as much as 15 cm of water; on landward side, deep marshes along tidal rivers, sounds, deltas	Grasses and sedges; important waterfowl areas
Deep fresh marshes	At high tide covered with 15 cm to 1 m of water; along tidal rivers and bays	Cattails, wild rice, giant cutgrass
Open fresh water	Shallow portions of open water along fresh tidal rivers and sounds	Vegetation scarce or absent; important waterfowl areas

(continued)

TABLE 33.1 *(continued)*

TYPES OF WETLANDS

Type	Site Characteristics	Plant and Animal Populations
Inland Saline Areas		
Saline flats	Flooded after periods of heavy precipitation; waterlogged within few inches of surface during the growing season	Sea blite, salt grass, saltbush; fall waterfowl feeding areas
Saline marshes	Soil waterlogged during growing season; often covered with 0.61 to 1 m of water; shallow lake basins	Alkali hard-stemmed bulrush, widgeon grass, sago pondweed; valuable waterfowl areas
Open saline water	Permanent areas of shallow saline water; depth variable	Sago pondweed, muskgrasses; important waterfowl feeding areas
Coastal Saline Areas		
Salt flats	Soil waterlogged during growing season; sites occasionally to fairly regularly covered by high tide; landward sides or islands within salt meadows and marshes	Salt grass, sea blite, saltwort
Salt meadows	Soil waterlogged during growing season; rarely covered with tide water; landward side of salt marshes	Cordgrass, salt grass, black rush; waterfowl feeding areas
Irregularly flooded salt marshes	Covered by wind tides at irregular intervals during the growing season; along shores of nearly enclosed bays, sounds, etc.	Needlerush; waterfowl cover areas
Regularly flooded salt marshes	Covered at average high tide with 15 cm or more of water; along open ocean and along sounds	Atlantic: salt marsh cordgrass; Pacific: alkali bulrush, glassworts; feeding area for ducks and geese
Sounds and bays	Portions of saltwater sounds and bays shallow enough to be diked and filled; all water landward from average low-tide line	Wintering areas for waterfowl
Mangrove swamps	Soil covered at average high tide with 15 cm to 1 m of water; along coast of southern Florida	Red and black mangroves

filled-in lakes and ponds. They are basin wetlands. Other wetlands develop along shallow and periodically flooded banks of rivers and streams. They are riverine wetlands. A third type occurs along the coastal areas of large lakes and seas. They are fringe wetlands. Some of the best-developed fringe wetlands are mangrove swamps associated with a marine environment, considered in Chapter 37.

What separates the three types is the direction of water flow (Figure 33.2). Water flow in the basin wetlands is vertical, involving precipitation and capillary flow. In riverine wetlands water flow is unidirectional. In fringe wetlands flow is in two directions, because it involves rising lake levels or tidal action. The flows bring in and carry away nutrients and sediments. They may stress systems by exporting or importing too much.

Wetlands dominated by emergent herbaceous vegetation are **marshes** (Figure 33.3). Growing to reeds, sedges, grasses, and cattails, marshes are essentially wet prairies. Forested wetlands are commonly called **swamps.** They may be deep-water swamps dominated by cypress, tupelo, and swamp oaks; or they may be shrub swamps dominated by alder and willows. Along many large river systems are extensive tracts of **riparian woodlands** (Figure 33.4), which are occasionally or seasonally flooded by river waters but are dry for most of the growing season.

Wetlands in which considerable amounts of water are retained by an accumulation of partially decayed organic matter are **peatlands** or **mires** (Figure 33.5). Mires fed by water moving through mineral soil, from which they obtain most of their nutrients, and dominated by sedges are known as **fens.** Mires dependent largely on precipitation for their water supply and nutrients and dominated by *Sphagnum* moss are **bogs.** Mires that develop on upland situations where decomposed, compressed peat forms a barrier to the downward movement of water, resulting in a perched water

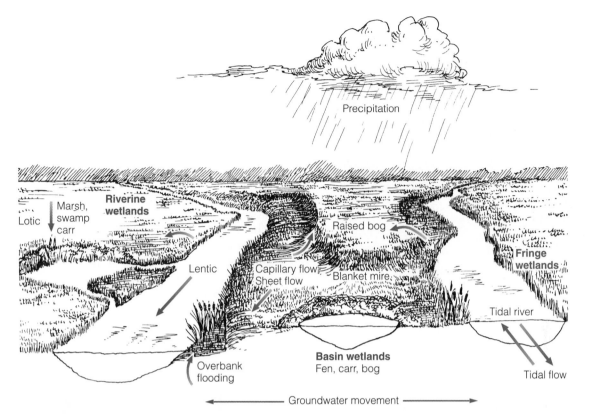

Figure 33.2 Water flow in various types of freshwater wetlands.

Figure 33.3 The Horicon marsh in Wisconsin is an outstanding example of a northern marsh with well developed emergent vegetation and patches of open water—an ideal environment for wildlife.

Figure 33.4 A riparian forest in Alabama.

table above mineral soil, are **blanket mires** and **raised bogs** (see Figure 33.2). Raised bogs are popularly known as **moors.** Because bogs depend on precipitation for nutrient inputs, they are highly deficient in mineral salts and low in pH. Bogs also develop when a lake basin fills with sediments and organic matter carried by inflowing water. These sediments divert water around the lake basin and raise the surface of the mire

Figure 33.5 An upland black spruce-tamarack bog in the Adirondack Mountains of New York.

above the influence of groundwater. Other bogs form when a lake basin fills in from above rather than from below, creating a floating mat of peat over open water. Such bogs are often termed **quaking** (Figure 33.6).

33.2 STRUCTURE

The structure of a wetland is influenced by the phenomenon that creates it—its hydrology. Hydrology has two components. One is the physical aspects of water and its movement: precipitation, surface and subsurface flow, the direction and kinetic energy of water, and the chemistry of the water. The other is **hydroperiod,** which includes the duration, frequency, depth, and season of flooding. The length of the hydroperiod varies among types of wetlands. Basin wetlands have a longer hydroperiod. They usually flood during peri-

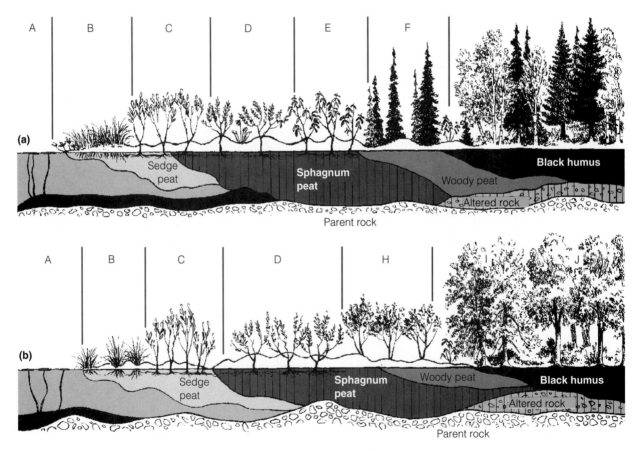

Figure 33.6 (a) Transect through a quaking bog, showing zones of vegetation, sphagnum mounds, peat deposits, and floating mats. A, pond lily in open water; B, buckbean *(Menyanthes trifoliata)* and sedge; C, sweetgale *(Myrica gale);* D, leatherleaf *(Chamaedaphne calyculata);* E, Labrador tea *(Ledum groenlandicum);* F, black spruce; G, birch-black spruce-balsam fir forest. (b) An alternative vegetational sequence. H, alder; I, aspen, red maple; J, mixed deciduous forest.

ods of high rainfall and draw down during dry periods. Both phenomena appear to be essential to the long-term existence of wetlands (see Chapter 26). Riverine wetlands have a short period of flooding associated with peak stream flow. The hydroperiod of fringe wetlands, influenced by wind and lake waves, may be short and regular, and does not undergo the seasonal fluctuation characteristic of many basin marshes.

Hydroperiod influences plant composition, for it affects germination, survival, and mortality at various stages of the plants' life cycles. The effect of hydroperiod is most pronounced in basin wetlands, especially those of the prairie regions of North America. In basins (called potholes in the prairie region) deep enough to have standing water throughout periods of drought, the dominant plants will be submergents. If the wetland goes dry annually or during a period of drought, tall or midheight emergent species such as cattails will dominate the marsh. If the pothole is shallow and flooded only briefly in the spring, then grasses, sedges, and forbs will make up a wet-meadow community.

If the basin is sufficiently deep toward its center and large enough, then zones of vegetation may develop (see Chapter 32), ranging from submerged plants to deep-water emergents such as cattails and bulrushes, shallow-water emergents, and wet-ground species such as spike rush. Zonation reflects the response of plants to hydroperiod. Those areas of wetland subjected to a long hydroperiod will support submerged and deep-water emergents; those with a short hydroperiod and shallow water are occupied by shallow-water emergents and wet-ground plants.

Periods of drought and wetness can induce vegetation cycles associated with changes in water levels. Periods of above-normal precipitation can raise the water level and drown the emergents to create a lake marsh dominated by submergents. During a drought the marsh bottom is exposed by receding water, stimulating the germination of seeds of emergents and mudflat annuals. When water levels rise again the mudflat species drown, and the emergents survive and spread vegetatively.

Peatlands differ from other freshwater wetlands in that their rate of organic production exceeds the rate of decomposition, and much of the production accumulates as peat. In northern regions acid-forming, water-holding sphagnum mosses add new growth on top of the accumulating remains of past moss generations; and their spongelike ability to hold water increases water retention on the site. As the peat blanket thickens, the water-saturated mat of moss and associ-

ated vegetation is raised above and insulated from mineral soil. The peat mat then becomes its own reservoir of water, creating a perched water table.

Peat bogs and mires generally form under oligotrophic and dystrophic conditions (see Chapter 32). Although usually associated with and most abundant in boreal regions of the Northern Hemisphere, peatlands also exist in tropical and subtropical regions. They develop in mountainous regions or in lowland or estuarine regions where hydrological situations encourage an accumulation of partly decayed organic matter. Examples are the Everglades in Florida and the pocosins of the southeastern United States coastal plains.

Biologically, wetlands are among the richest and most interesting ecosystems. They support a diverse community of benthic, limnetic, and littoral invertebrates, especially crustaceans and insects. These invertebrates, along with small fishes, provide a food base for waterfowl, herons, gulls, and other birds, and supply the fat-rich nutrients ducks need for egg production and the growth of young. Wetlands support a diversity of amphibians and reptiles, notably frogs, toads, and turtles.

Herbivores make up a conspicuous component of animal life. Microcrustaceans filter algae from the water column. Snails eat algae growing on the leaves and litter; geese graze on new emergent growth; coots and mallards and other surface-feeding ducks feed on algal mats. The dominant herbivore in the prairie marshes is the muskrat *(Ondatra zibethicus)*. During population highs, muskrats can eliminate emergent vegetation, creating "eat-outs" and transforming an emergent-dominated marsh into an open-water one. Introduced into Eurasia, the muskrat has become the major herbivore in many marshes on that continent. Muskrats are the major prey for mink, the dominant carnivore on the marshes. Other predators include raccoon, fox, weasel, and skunk, which can seriously reduce the reproductive success of waterfowl on small marshes surrounded by agricultural land.

33.3 FUNCTION

Freshwater wetlands are highly productive ecosystems, but their complexity and differences make generalizations about their functions difficult. In fact, we know much less about wetland functions than about forests and grasslands.

Wetlands are sedimentary or detrital systems. They accumulate carbon, nitrogen, phosphorus, and other materials and exchange them among the wetland, the atmosphere, and the landscape. Productivity of fresh-water marshes is influenced by hydrological regimes: groundwater, surface runoff, precipitation, drought cycles, flooding in riverine wetlands, and the like. Wetlands, whatever their type, are closely associated with the total landscape in which they reside—its size, soils, land use, nutrient availability, and types of vegetation. Basin wetlands accumulate muck and peat. Riverine wetlands experience a throughflow of water, importing and exporting materials. All of these conditions influence the nature of each wetland's vegetation. In turn, the life-history pattern of the species involved further influences the productivity of the wetlands.

Aboveground biomass varies with the proportionate abundance of annual and perennial species, whose dominance changes through the growing season. Annual emergents increase their biomass through the growing season, reaching a maximum in late summer; perennials increase their biomass during the first part of the growing season, then see it decline or level off as they become senescent (Figure 33.7). In general, however, the average maximum standing crop of biomass matches annual aboveground productivity.

Belowground production, much more difficult to estimate, appears for some species to be highest in summer, at the same time the peak aboveground biomass is achieved. Others, such as cattails *(Typha)* and sedges *(Scirpus)*, reach peak production in the fall, when nutrients are stored in the roots. Such species

may have minimal root biomass in the summer because of nutrient transfer to aboveground biomass.

To balance losses, cattails and other emergent plants must draw on the phosphorus supply in the soil, which is derived from decomposition of the litter of previous years. By doing so, the plants act as a nutrient pump, drawing nutrients from the soil, translocating them into the shoots, and then releasing them to the surface soil by leaching and death of shoots during the growing and postgrowing season. In this way marsh plants make nutrients sequestered in the soil available for growth.

Nitrogen, possessing a gaseous form, undergoes considerable exchange between the wetland and atmosphere, involving nitrogen fixation, volatilization of NH_3, denitrification, and possibly nitrification. Denitrification may be the major source of loss of nitrogen from wetlands. Wetland plants mobilize nitrogen from the soil, much as they do phosphorus, and concentrate it in their tissues. Under eutrophic conditions, the accumulation of nitrogen may be high. Nevertheless, wetland plants undertake considerable internal cycling, through which they meet 40 percent of their requirements for both nitrogen and phosphorus.

Wetlands contribute a great amount of litter or detrital material to the system. This material at first decomposes rapidly; leaching in a watery environment removes soluble compounds as dissolved organic matter (Figure 33.8). After initial leaching, decomposition

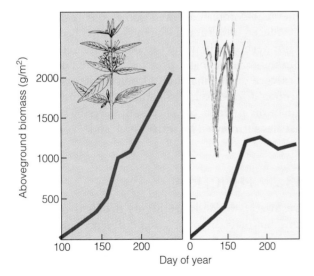

Figure 33.7 Pattern of aboveground biomass accumulation through the growing season for a freshwater annual, loosestrife *(Lythrum),* and a perennial, cattail *(Typha).* Note the linear increase in biomass in the annual and the sigmoid growth of the perennial.

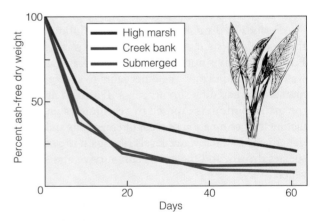

Figure 33.8 Decomposition of leaves of arrow arum *(Peltandra virginica)* as measured by the percentage of original ash-free dry weight remaining in litter bags under three conditions: irregular flooding in a high marsh exposed to alternate wetting; flooding twice a day on a creek bank; and permanent submersion. Note that the detrital material that was consistently wet showed the highest rate of decomposition, although the overall patterns of decomposition are similar.

proceeds more slowly. Permanently submerged leaves decompose more rapidly than those on the marsh surface, because they are more accessible to aquatic detritivores such as crayfish, and because the constantly moist environment is more favorable for microbial decomposition. Microbes that use the litter as a carbon source add to it nitrogen and phosphorus, which they obtain from the surrounding water and sediments. This accumulation in the decomposed material helps to retain N and P in the wetland, where it is available to emergents and other plants. How long these nutrients remain in the depository depends on how rapidly the litter decomposes. Much of the finer material becomes incorporated into anaerobic bottom muds, where decomposition produces methane and hydrogen sulfide.

Peatlands present a different situation. Because bog vegetation is not in contact with mineral soil and because inflowing groundwater is blocked, bogs depend mostly on precipitation for their nutrients. In addition, both cyanobacteria living in close association with bog mosses and bog myrtle fix nitrogen; and carnivorous plants, such as sundews, extract nitrogen from captured and digested insects. Bogs, however, face a scarcity of nutrients, a shortage compounded by the plants themselves. Most of the nutrients they fix in their tissues remains in the accumulating peat.

However, bog plants do possess some means of conserving nutrients. Consider how the trailing plant cloudberry *(Rubus chamaemorus)* manages its phosphorus budget. This plant increases its uptake of phosphorus through the roots prior to budbreak. After budbreak, the cloudberry increases the amount of phosphorus in stem, leaf, and root. In summer after the plant has completed its shoot growth, it moves phosphorus from its shoots to developing fruits and to roots and rhizomes. As senescence sets in, cloudberry moves most of the phosphorus remaining in its shoots to its winter buds, where it accumulates and is available for the next year's growth.

Energy flow in peatlands differs from that in other wetlands because the detrital food chain is impaired. In most ecosystems, material that enters the detrital food web is eventually recycled, and the energy that enters the system is liberated or stored in living material. In bogs, material from primary production accumulates in a partially decomposed state, and energy is locked up in peat until environmental conditions change to favor decomposition or until the material burns.

Because of low temperatures, acidity, and nutrient immobilization, primary production in peatlands is low, as little as 300 g/m²/year in sphagnum bogs. Likewise, decomposition of that primary production is slow.

Shrub litter decomposes slowly, and sphagnum hardly decomposes at all. Ecologists have estimated that in English mires, turnover of 95 percent of the organic matter for the system as a whole takes 3000 years. For the top 20 cm of material, it takes 70 years. This slow rate of decomposition accounts for the accumulation of peat.

33.4 VALUE OF WETLANDS

Just as we like to dam rivers, so we are motivated to drain wetlands and convert them into dry land. The Romans drained the great marshes about the Tiber to make room for the city of Rome. In spite of the enormous amount of vacant dry land about him, George Washington proposed draining the Great Dismal Swamp (most of which has been done since his time). Many of us consider wetlands to be wastelands, areas to be drained for more productive uses by human standards: agricultural land, solid waste dumps, housing, industrial developments, and roads. We also look on wetlands as forbidding mysterious places, sources of pestilence, the home of dangerous and pestiferous insects, the abode of slimy sinister creatures that rise out of swamp waters. We are blind to the ecological, hydrological, and economic values of wetlands.

Wetlands assume an importance, ecologically and economically, out of proportion to their size. Their major contribution is to the hydrology of a region. Basin wetlands, in particular, are groundwater recharge points. They hold rainwater, snowmelt, and surface runoff in their basins and discharge the water slowly into the aquifers. These same basins also function as natural flood-control reservoirs. As little as 5 percent of the watershed or catchment area in wetlands can reduce flood flows by as much as 50 percent (Figure 33.9).

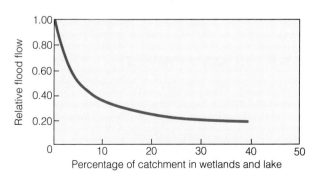

Figure 33.9 The influence of different percentages of wetlands in a watershed on relative flood flows in Wisconsin.

Wetlands act as water-filtration systems. Wetland vegetation takes up excessive nitrogen, phosphorus, sulfates, copper, iron, and other heavy metals brought by surface runoff and inflow, incorporates them into plant biomass, and deposits much of them in anaerobic bottom muds. Because of their ability to filter out heavy metals and to reduce pH, we are beginning to treat urban wastewater and drainage from surface mines by diverting these flows into natural or specially created wetlands.

Wetlands contribute to the human economy in other ways. They provide places of recreation, sources of horticultural peat and timber—notably baldcypress and bottomland hardwoods in the southern United States—and sites for growing cranberries in the northeastern United States. These uses, however, tend to interfere with the natural function and integrity of the wetland ecosystems.

Wetlands are vitally important as wildlife nesting and wintering habitat. Many species of wildlife, some of them endangered, are dependent on wetlands. Worldwide, wetlands are home to numerous species of amphibians and reptiles, including alligators and crocodiles. Many species of fish of the Amazon and other tropical rivers depend on seasonal flooding of riverine swamps and floodplains to forage for terrestrial foods and to spawn. Waterfowl, wading birds, gulls and terns, herons, and storks depend on marshes and wooded swamps for nesting and foraging. In fact, the prairie pothole region of the north central United States and Canada is used by two-thirds of the continent's 10 to 12 million waterfowl as a nesting area. Waterfowl in concentrated numbers use southern marshes and swamps as wintering habitat. Moose, hippopotamus, waterbuck, otters, and muskrat are mammalian inhabitants of wetlands.

In addition to wetland dwellers, wetlands support animal life in other ecosystems. A mosaic of wetlands in an upland terrestrial environment increases the abundance and diversity of wildlife populations.

33.5 HUMAN IMPACT

That we have little regard for wetlands and their values is underscored by the destruction we have imposed on them. Wetlands, both forested and nonforested, once made up about 3 percent of Earth's surface, but much of that area, especially in the Northern Hemisphere, has been converted to other land uses. In colonial times the area now embraced by the 50 United States contained some 392 million acres of wetlands. Of these, 221 million acres were in the lower 48 states, 170 mil-

lion acres in Alaska, and 59,000 acres in Hawaii. Now, over 200 years later, Alaska has lost a fraction under 1 percent, Hawaii 12 percent, and the lower 48 states well over 50 percent of their wetlands. Among the states California has lost 91 percent of its wetlands. Wetlands, which once made up 5 percent of that state's total land area, have shrunk to one-half of 1 percent. Over the continental United States the 392 million acres of wetland have decreased to 274 million acres (Figure 33.10), and many of these remnants are degraded.

The rationales for drainage are many. The most persuasive relates to agriculture. Drainage of wetlands opens many hectares of rich organic soil for crop production. In the prairie country the innumerable potholes are viewed as a nuisance to efficient agriculture.

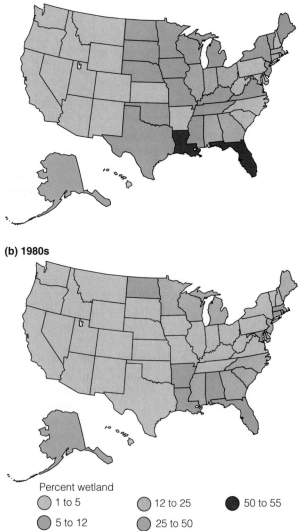

(a) 1780s

(b) 1980s

Percent wetland
1 to 5   12 to 25   50 to 55
5 to 12   25 to 50

Figure 33.10 The loss of wetland in the United States over 200 years.

Draining them tidies up fields and allows unhindered use of large agricultural machinery. There are other reasons, too. Wetlands are viewed as an economic liability by landowners and by local governments. They produce no economic return and they provide little tax revenue. Many regard the wildlife wetlands support as threats to grain crops. Elsewhere wetlands are considered valueless lands, at best filled in and used for development. Some major wetlands have been in the way of dams. For example, the large Pymatuning Lake in the states of Pennsylvania and Ohio covers a 4200 ha sphagnum-tamarack bog. Peat bogs in the northern United States, Canada, Ireland, and northern Europe are excavated for fuel, horticultural peat, and organic soil. In some areas such exploitation threatens to wipe out peatland ecosystems.

Many remaining wetlands, especially in the north central and southwestern United States, are contaminated and degraded by pesticides and heavy metals carried into them by surface and subsurface drainage and sediments from surrounding croplands (Figure 33.11). Although inputs of nitrogen and phosphorus increase the productivity of wetlands, a concentration of herbicides, pesticides, and heavy metals poisons the water, destroys invertebrate life, and has debilitating effects on wildlife, including deformities, lowered reproduction, and death. Waterfowl in wetlands scattered throughout agricultural lands are also more subject to predation, and without access to natural upland vegetation they breed less successfully.

Wetlands can suffer even from the best intentions. Management attempts to maintain stability reduce or eliminate the fluctuations in water level so necessary to the health of the wetland. Lack of fluctuations increases anaerobic decomposition, tying up nutrients in the bottom muck, and changes plant composition, eliminating those species that require a cycle of drawdown for germination and flooding for growth.

The loss of wetlands has reached a point where both environmental and socioeconomic values—including waterfowl habitat, groundwater supply and quality, floodwater storage, and sediment trapping—are in jeopardy. Although we have made some progress in the United States toward preserving the remaining wetlands through legislative action and land purchase, the future of freshwater wetlands is not secure. Apathy, hostility toward wetland preservation, political maneuvering, court decisions, and arguments over what constitutes a wetland allow the continued destruction of wetlands at a rate of over 200,000 ha per year.

STUDY QUESTIONS

1. What is a wetland? A hydrophyte?
2. How does the definition of a wetland relate to the gradient of soil wetness?
3. What are the three major types of wetlands in terms of hydrology?
4. Characterize the types of wetlands by their vegetation.
5. What is the hydroperiod, and how does it relate to the structure of wetlands?
6. How do wetlands relate to the landscapes in which they occur?
7. Contrast nutrient cycling in marshes and peatlands.
8. Why are wetlands worth saving?
9. What major impacts have humans made on wetlands, and why?
10. What is the paradox of draining wetlands, then building large flood-control dams?
*11. What has been the impact of draining prairie potholes, swamps, and bottomland hardwood forests on waterfowl populations?
*12. What has been the fate of wetlands in your region? How much has been lost? To what use has the drained land been put?
*13. Trace the history of the attitudes toward wetlands from Roman times to the present. Why are we fearful of them?
*14. How could you argue for preservation of wetlands on an economic basis to developers and local governments? Can we put monetary values on wetlands?

Figure 33.11 Drainage, such as this project in Florida, has destroyed millions of acres of wetlands, vital as wildlife habitat and as regulators of water flow and water tables.

C H A P T E R   3 4

STREAMS AND RIVERS

OBJECTIVES

On completion of this chapter, you should be able to:

- Describe the physical characteristics of flowing-water ecosystems.
- Compare fast streams with slow streams and rivers.
- Describe nutrient cycling in flowing water.
- Discuss the role of various feeding groups in streams and rivers.
- Explain the role of detritus in flowing-water ecosystems.
- Point out the effects of pollution on streams and rivers.
- Discuss the problems channelization causes.
- Explain the meaning of regulated rivers, and discuss the impact of dams on flowing-water ecosystems.

A fast-flowing mountain stream hurries over a rocky bed in Finke Gorge National Park, Northern Territory, Australia.

Even the largest rivers begin somewhere back in the hinterlands as springs or seepage areas, becoming headwater brooks and streams; or they arise as outlets of ponds or lakes. A very few emerge full-blown from glaciers. As a brook drains away from its source, it flows in a direction and manner dictated by the lay of the land and underlying rock formations. Its course may be determined by the original slope; or water, seeking the least resistant route to lower land, may follow joints and fissures in bedrock near the surface and shallow depressions in the ground. Whatever its direction, water concentrates in rills that erode small furrows, which soon grow into gullies. Moving downstream, especially where the gradient is steep, the moving water carries with it a load of debris collected from its surroundings that cuts the channel wider and deeper. Sooner or later, the stream deposits this material on its bed or along its banks. In mountainous areas, erosion continues to eat away at the head of the gully, cutting backward into the slope and increasing the drainage area. Joining the new stream are other small streams, spring seeps, and surface water.

Just below its source the stream may be small, straight, and swift, with waterfalls and rapids. Farther downstream, where the gradient is less, velocity decreases, meanders become common, and the stream deposits its load of sediment as silt, sand, or mud. At flood time, a stream drops its load of sediment on surrounding level land, over which floodwaters spread to form floodplain deposits. These floodplains are a part of a stream or river channel used at the time of high water—a fact few people recognize.

Where a stream flows into a lake or a river into the sea, the velocity of water is suddenly checked. The river then is forced to deposit its load of sediment in a fan-shaped area about its mouth to form a delta. Here its course is carved into a number of channels, which are blocked or opened with subsequent deposits. As a result, the delta becomes an area of small lakes, swamps, and marshy islands. Material the river fails to deposit in the delta is carried out to open water and deposited on the bottom.

Because streams become larger on their course to rivers and are joined along the way by many others, we can classify them according to order. A small headwater stream without any tributaries is a first-order stream. When two streams of the same order join, the stream becomes one of higher order. If two first-order streams unite, the resulting stream becomes a second-order one; and when two second-order streams unite, the stream becomes a third-order one. The order of a stream can increase only when a stream of the same order joins it. It cannot increase with the entry of a lower-order stream. In general, headwater streams are orders 1 to 3; medium-sized streams, 4 to 6; and rivers, greater than 6.

The area of land a stream or river drains is its **watershed.** Each watershed is different, characterized by vegetative cover, geology, soils, topography, and land use. Streams and rivers provide the drainage pathways. Ponds, lakes, and wetlands act as the catch basins. Thus a watershed includes **lotic** or flowing water systems and **lentic** or still-water systems.

34.1 STRUCTURE

Physical Structure

The velocity of a current molds the character and the structure of a stream. The shape and steepness of the stream channel, its width, depth, and roughness of the bottom, and the intensity of rainfall and rapidity of snowmelt affect velocity. Fast streams (Figure 34.1) are those whose velocity of flow is 50 cm per second or higher. At this velocity, the current will remove all particles less than 5 mm in diameter and will leave behind a stony bottom. High water increases the velocity; it moves bottom stones and rubble, scours the streambed, and cuts new banks and channels. As the gradient decreases and the width, depth, and volume of water increase, silt and decaying organic matter accumulate on the bottom. The character of the stream changes from fast water to slow, with an associated change in species composition (Figure 34.2).

Figure 34.1 A fast mountain stream. The gradient is steep and the bottom is largely bedrock.

Figure 34.2 A slow stream is deeper and has a lower slope gradient.

Figure 34.3 Two different but related habitats in a stream: a riffle (background) and a pool (foreground).

Flowing-water ecosystems often alternate two different but related habitats, the turbulent riffle and the quiet pool (Figure 34.3). The waters of the pool are influenced by processes occurring in the rapids above, and the waters of the rapids are influenced by events in the pool.

Riffles are the sites of primary production in the stream. Here the **periphyton** or aufwuchs, organisms that are attached to or move on submerged rocks and logs, assume dominance. Periphyton, which occupies a position of the same importance as phytoplankton in lakes and ponds, consists chiefly of diatoms, cyanobacteria, and water moss.

Above and below the riffles are the pools. Here the environment differs in chemistry, intensity of current, and depth. Just as the riffles are the sites of organic production, so the pools are the sites of decomposition. They are catch basins of organic materials, for here the velocity of the current is reduced enough to allow part of the load to settle. Pools are the major sites of carbon dioxide production during the summer and fall. That work is necessary for the maintenance of a constant supply of bicarbonate in solution. Without pools, photosynthesis in the riffles would deplete the bicarbonates and result in smaller and smaller quantities of available carbon dioxide downstream.

Free carbon dioxide in rapid water is in equilibrium with that in the atmosphere. The amount of bound carbon dioxide is influenced by the nature of the surrounding terrain and decomposition taking place in pools of still water. Most of the carbon dioxide in flowing water occurs as carbonate and bicarbonate salts.

Streams fed by groundwater from limestone springs receive the greatest amount of carbonates in solution.

The degree of acidity or alkalinity, or pH, of the water reflects the CO_2 content as well as the presence of organic acids and pollution. The higher the pH of stream water, the richer natural waters generally are in carbonates, bicarbonates, and associated salts. Such streams support more abundant aquatic life and larger fish populations than streams with acid waters, generally low in nutrients.

The constant churning and swirling of stream water over riffles and falls give greater contact with the atmosphere; therefore the oxygen content of the water is high, often near the saturation point for existing temperatures. Only in deep holes or in polluted waters does dissolved oxygen show any significant decline.

The temperature of a stream is variable. Small, shallow streams tend to follow, but lag behind, air temperatures, warming and cooling with the seasons but rarely falling below freezing in winter. Streams with large areas exposed to sunlight are warmer than those shaded by trees, shrubs, and high banks. That fact is ecologically important because temperature affects the stream community, influencing the presence or absence of cool-water and warm-water organisms.

Adaptations

Living in moving water, inhabitants of streams and rivers have a major problem remaining in place and not being swept downstream. They have evolved unique adaptations for dealing with life in the current (Figure 34.4). A streamlined form, which offers less resis-

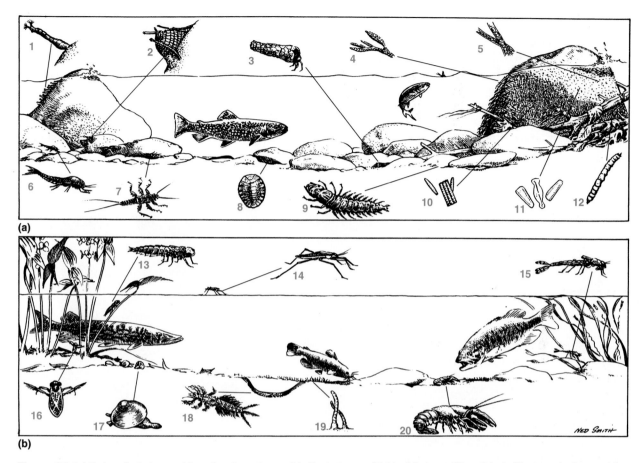

Figure 34.4 Life in a fast stream (a) and a slow stream (b). Fast stream: (1) blackfly larva (Simuliidae); (2) net-spinning caddisfly (*Hydropsyche* spp.); (3) stone case of caddisfly; (4) water moss *(Fontinalis);* (5) algae *(Ulothrix);* (6) mayfly nymph *(Isonychia);* (7) stonefly nymph *(Perla* spp.); (8) water penny *(Psephenus);* (9) hellgrammite (dobsonfly larva, *Corydalis cornuta);* (10) diatoms *(Diatoma);* (11) diatoms *(Gomphonema);* (12) cranefly larva (Tipulidae). Slow stream: (13) dragonfly nymph (Odonata, Anisoptera); (14) water strider *(Gerris);* (15) damselfly larva (Odonata, Zygoptera); (16) water boatman (Corixidae); (17) fingernail clam *(Sphaerium);* (18) burrowing mayfly nymph *(Hexegenia);* (19) bloodworm (Oligochaeta, *Tubifex* spp.); (20) crayfish *(Cambarus* spp.). The fish in the fast stream is a brook trout *(Salvelinas fontinalis).* The fish in the slow stream are, from left to right: northern pike *(Esox lucius),* bullhead *(Ameiurus melas),* and smallmouth bass *(Micropterus dolommieu).*

tance to water current, is typical of many animals of fast water, such as the dace and the brook trout. The larval forms of many species of insects cling to the undersurfaces of stones, where the current is weak. They possess extremely flattened bodies and broad, flat limbs that allow the current to flow over them. Typical are many species of mayflies and stoneflies. Other forms, such as the blackfly (Simuliidae) larvae, attach themselves in one way or another to the substrate, and they obtain food by straining particles carried to them by the current. The larvae of certain species of caddisflies construct protective cases of sand or small pebbles and cement them to the bottoms of stones. Larvae of net-spinning caddisflies *(Hydropsyche)* firmly attach to stones funnel-shaped, food-collecting nets whose open ends

face upstream. Sticky undersurfaces help snails and planarians cling tightly and move about on stones and rubble in the current.

Among the plants, water moss *(Fontinalis)* and heavily branched filamentous algae cling to rocks by strong holdfasts. Other algae grow in cushionlike colonies or closely appressed sheets that are covered with a slippery, gelatinous coating and follow the contours of stones and rocks.

All animal inhabitants of fast-water streams require high, near-saturation concentrations of oxygen and moving water to keep their absorbing and respiratory surfaces in continuous contact with oxygenated water. Otherwise a closely adhering film of liquid impoverished of oxygen forms a cloak about their bodies.

In slow-flowing streams where current is at a minimum, streamlined forms of fish give way to species such as smallmouth bass (Figure 34.4b), shiners, and darters. They trade strong lateral muscles needed in fast current for compressed bodies that enable them to move through beds of aquatic vegetation. Pulmonate snails and burrowing mayflies replace rubble-dwelling insect larvae. Bottom-feeding fish, such as catfish, feed on life in the silty bottom, and back swimmers and water striders inhabit sluggish stretches and still backwaters.

34.2 FUNCTION

Flowing-water or lotic systems are open and largely heterotrophic (Figure 34.5). A major energy source

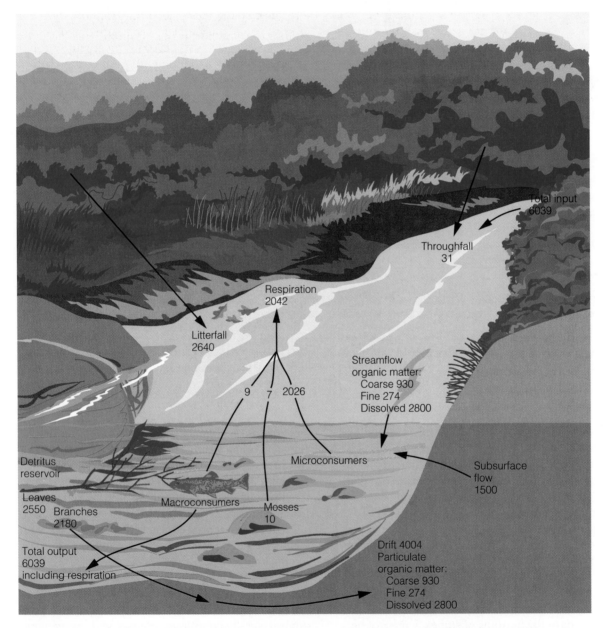

Figure 34.5 Energy flow in a stream ecosystem. Note the great dependence on materials from terrestrial sources and inflow from upstream, and the role of coarse particulate organic matter (CPOM), fine particulate organic matter (FPOM), and dissolved organic matter (DOM). Primary production contributes little to energy flow. Energy values in kcal/m²/yr are based on Bear Brook, Hubbard Forest, New Hampshire.

is detrital material carried in from the outside. Much of this organic matter input comes as coarse particulate organic matter (CPOM), leaves and woody debris dropped from streamside vegetation, with particles larger than 1 mm in size. Another type of organic input is fine particulate organic matter (FPOM), material less than 1 mm in size, including leaf fragments, invertebrate feces, and precipitated dissolved organic matter. A third input is dissolved organic matter (DOM), material less than 0.5 micron in solution.

One source of DOM is rainwater dripping through overhanging leaves, dissolving the nutrient-rich exudates on them. Other DOM input comes by a geological pathway. Subsurface seepage brings nutrients leached from adjoining forest, agricultural, and residential lands. Many streams receive inputs from mechanical pathways through the dumping of industrial and residential effluents. Supplementing this detrital

input is autotrophic production in streams by diatomaceous algae growing on rocks and by rooted aquatics such as water moss. Energy is lost through two pathways: geological (streamflow feeding downstream systems) and biological (respiration).

Food Webs

The processing of this organic matter involves both physical and biological mechanisms (Figure 34.6). In fall, leaves drift down from overhanging trees, settle on the water, float downstream, and lodge against banks, debris, and stones. Soaked with water, the leaves sink to the bottom, where they quickly lose 5 to 30 percent of their dry matter as water leaches soluble organic matter from their tissues. Much of this DOM is either incorporated onto detrital particles or precipitated to become part of the FPOM. Another part is incorporated into microbial biomass.

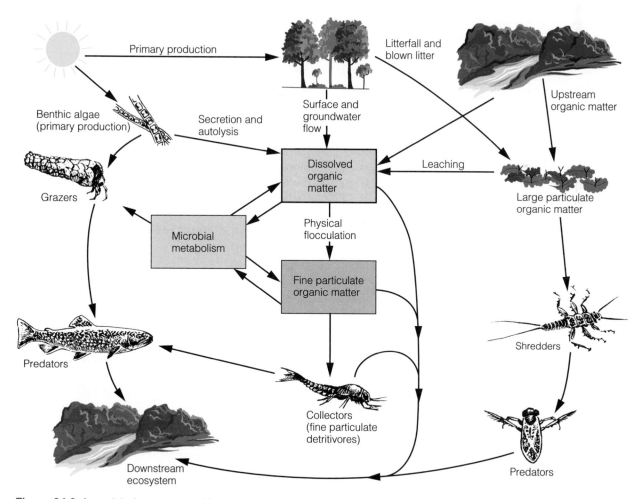

Figure 34.6 A model of structure and function in a lotic system, showing the processing of leaves and other particulate matter and dissolved organic matter.

Within a week or two, depending on the temperature, the surface of the leaves is colonized by bacteria and fungi, largely aquatic hyphomycetes. Fungi are more important on CPOM, because large particles offer more surface for mycelial development. Bacteria are associated more with FPOM. Microorganisms degrade cellulose and metabolize lignin. Their populations form a layer on the surface of leaves and detrital particles that is much richer nutritionally than the detrital particles themselves.

Soon the leaves and other detrital particles are attacked by a major feeding group, the **shredders,** insect larvae that feed on coarse particulate organic matter. Among these shredders are craneflies (Tipulidae), caddisflies (Trichoptera), and stoneflies (Plecoptera). They break down the CPOM, feeding on the material not so much for the energy it contains but for the bacteria and fungi growing on it. Shredders assimilate about 40 percent of the material they ingest and pass off 60 percent as feces.

Broken up by the shredders and partially decomposed by microbes, the leaves, along with invertebrate feces, become part of the FPOM, which also includes some precipitated DOM. Drifting downstream and settling on the stream bottom, FPOM is picked up by another feeding group of stream invertebrates, the **filtering** and **gathering collectors.** The filtering collectors include, among others, the larvae of blackflies (Simuliidae), with filtering fans, and net-spinning caddisflies, including *Hydropsyche.* Gathering collectors, such as the larvae of midges, pick up particles from stream bottom sediments. Collectors obtain much of their nutrition from bacteria associated with the fine detrital particles.

While shredders and collectors feed on detrital material, another group feeds on the algal coating of stones and rubble. These are the **grazers,** which include the beetle larvae, water penny (*Psephenus* spp.), and a number of mobile caddisfly larvae. Much of the material they scrape loose enters the drift as FPOM. Another group, associated with woody debris, are the **gougers,** invertebrates that burrow into water-logged limbs and trunks of fallen trees.

Feeding on the detrital feeders and grazers are predaceous insect larvae such as the powerful dobsonfly larvae *(Corydalus cornutus)* and fish such as the sculpin *(Cottus)* and trout. Even these predators do not depend solely on aquatic insects; they also feed heavily on terrestrial invertebrates that fall or wash into the stream.

Because of current, quantities of CPOM, FPOM, and invertebrates tend to drift downstream to form a traveling benthos. This drift is a normal process in streams, even in the absence of high water and abnormal currents. Drift is so characteristic of streams that a mean rate of drift can serve as an index of the production rate of a stream.

Energy Flow and Nutrient Cycling

Energy flow in lotic ecosystems has been documented for only a few streams. One energy budget is for the well-studied, small, forested Bear Brook in Hubbard Forest of northern New Hampshire (see Figure 34.5). Over 90 percent of the energy input came from the surrounding forested watershed or from upstream. Primary production by mosses accounted for less than 1 percent of the total energy supply. Algae were absent from the brook. Inputs from litter and throughfall accounted for 44 percent of the energy supply, and geological inputs from subsurface flows accounted for 56 percent. Energy was introduced in three forms: CPOM represented by leaves and other debris; FPOM represented by drift and small particles; and DOM. In Bear Brook 83 percent of input from surface and subsurface flow and 47 percent of the total energy input was in the form of DOM. Sixty-six percent of the organic input was exported downstream, leaving 34 percent to be utilized locally.

Although nutrient cycling is downhill in all ecosystems, the problem in flowing water is how to keep nutrients upstream and reduce losses to downstream. Nutrients in terrestrial and lentic systems are recycled more or less in place. An atom of nutrients passes from the soil or water column to plants and consumers and back to soil or water in the form of detrital material or exudates. Then it is recycled within the same segment of the system, although losses do occur. Cycling essentially involves time. Flowing water has an added element, a spatial cycle. Nutrients in the form of DOM and POM are constantly being carried downstream. How quickly these materials are carried downstream depends on how fast the water moves and what physical and biological means hold nutrients in place. Physical retention involves storage in wood detritus such as logs and snags, in debris caught in pools behind logs and boulders, in leaf sediments, and in beds of macrophytes. Biological retention is uptake and storage in animal and plant tissue.

The processes of recycling, retention, and downstream displacement may be pictured as a spiral lying longitudinally in a stream (Figure 34.7). **Spiraling** combines nutrient cycling and downstream transport. One cycle in the spiral is the uptake of an atom or nutrient from DOM, its passage through the food chain, and its return to water, where it is available for reuse.

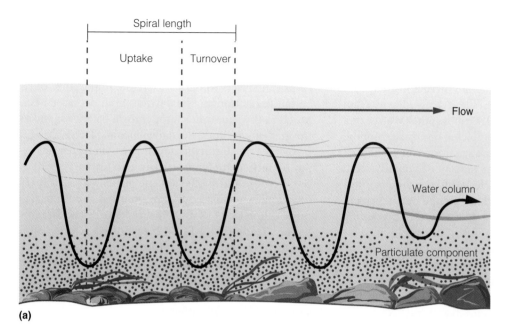

(a)

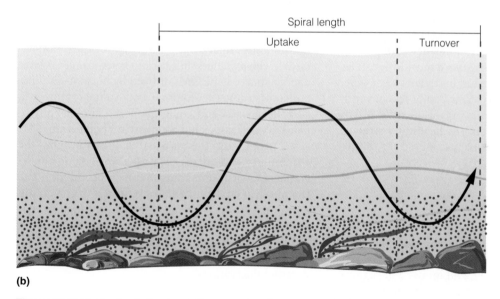

(b)

Figure 34.7 Nutrient spiraling in a lotic ecosystem between particulate organic matter and the water column. Uptake and turnover take place as nutrients flow downstream. The tighter the spiraling, the longer the nutrients remain in place. (a) Tight spiraling. (b) Open spiraling.

Spiraling is measured as the distance needed for completion of one cycle. The longer the distance required, the more open the spiral; the shorter the distance, the tighter the spiral. If leafy detritus is physically held in place long enough to allow the biological component of the stream, especially the shredders and microbes, to process the organic matter, then the spiral is tight. Retention is especially important in fast headwater streams, which can rapidly lose unprocessed particulate organic matter downstream.

Shredders open the spiral by fragmenting CPOM and excreting fecal material as FPOM, which joins with invertebrate drift and algal growth torn from the stream bottom. By trapping some of this FPOM and consuming it in place, collector organisms, especially net-spinning filter feeders, tighten the spiral.

Ecologists at Oak Ridge, Tennessee, experimentally determined how quickly one nutrient, exchangeable phosphorus in the form of $^{32}PO_4$, moved downstream in a small woodland brook, Walker Branch. The tagged P moved downstream at the rate of 10.4 m a day and cycled once every 18.4 days. The average downstream distance of one spiral was 190 m. In other words, one atom of P on the average completed one cycle from the water compartment and back again for every 190 m of downstream travel. Only 2.8 percent of P uptake from particulate matter was transferred to consumers; most of the P was transferred back to water. About 30 percent of the consumer uptake was transferred to predators.

The bottom and width influence overall production of a stream. Pools with sandy bottoms are the least productive because they offer little substrate for the aufwuchs. Bedrock, although a solid substrate, is so exposed to currents that only the most tenacious organisms can maintain themselves. Gravel and rubble bottoms support the most abundant life because they provide the greatest surface area for aufwuchs, offer many crannies and protected places for insect larvae, and are the most stable. Food production decreases as the particles become larger or smaller than rubble.

Bottom production in streams 6 m wide decreases by one-half from sides to center; in streams 30 m wide, it decreases by one-third. Streams 2 m or less in width are four times as rich in bottom organisms as those 6 to 7 m wide. That is one reason why headwater streams make such excellent trout nurseries.

34.3 THE RIVER CONTINUUM

From its headwaters to its mouth, the flowing-water ecosystem is a continuum of changing environmental conditions (Figure 34.8). Headwater streams (orders 1 to 3) are usually swift, cold, and in forested regions shaded. They are strongly heterotrophic, heavily dependent on the input of detritus from terrestrial streamside vegetation, which contributes more than 90 percent of the organic input. Even when headwater streams are exposed to sunlight and autotrophic production exceeds heterotrophic inputs, organic matter produced invariably enters detrital food chains. Dominant organisms are shredders, processing large-sized litter and feeding on CPOM, and collectors, processors of FPOM. Populations of grazers are minimal, reflecting the small amount of autotrophic production, and predators are mostly small fish—sculpins, darters, and trout. Headwater streams, then, are accumulators, processors, and transporters of particulate organic matter of terrestrial origin. As a result, the ratio of gross primary production to community respiration is less than 1.

As streams increase in width to medium-sized creeks and rivers (orders 4 to 6), the importance of riparian vegetation and its detrital input decreases. Exposed to the sun, water temperature increases; and as the gradient declines, the current slows. These changes bring about a shift from dependence on terrestrial input of particulate organic matter to primary production by algae and rooted aquatic plants. Gross primary production now exceeds community respiration. Because of the lack of CPOM, shredders disappear; collectors, feeding on FPOM transported downstream, and grazers, feeding on autotrophic production, become the dominant consumers. Predators show little increase in biomass but shift from cold-water species to warm-water species, including bottom-feeding fish such as suckers and catfish.

As the stream order increases from 6 through 10 and higher, riverine conditions develop. The channel is wider and deeper. The volume of flow increases, and the current becomes slower. Sediments accumulate on the bottom. Both riparian and autotrophic production decrease, with a gradual shift back to heterotrophy. A basic energy source is FPOM, utilized by bottom-dwelling collectors, now the dominant consumers. However, slow, deep water and DOM support a minimal phytoplankton and associated zooplankton population.

Throughout the downstream continuum, the community capitalizes on upstream feeding inefficiency. Downstream adjustments in production and the physical environment are reflected in changes in consumer groups. Through the continuum the ecosystem achieves some balance between the forces of stability, such as natural obstructions in flow, and the forces of instability, such as flooding, drought, and temperature fluctuations.

34.4 HUMAN IMPACT

Pollution

For centuries humans have used streams and rivers as depositories of human, industrial, and solid wastes with the idea that these materials would be diluted and carried downstream. So pervasive has that idea been that few streams and rivers have escaped pollution. The magnitude of ecological changes brought about depends on the type of pollutant, its quantity, and how long it has been dumped.

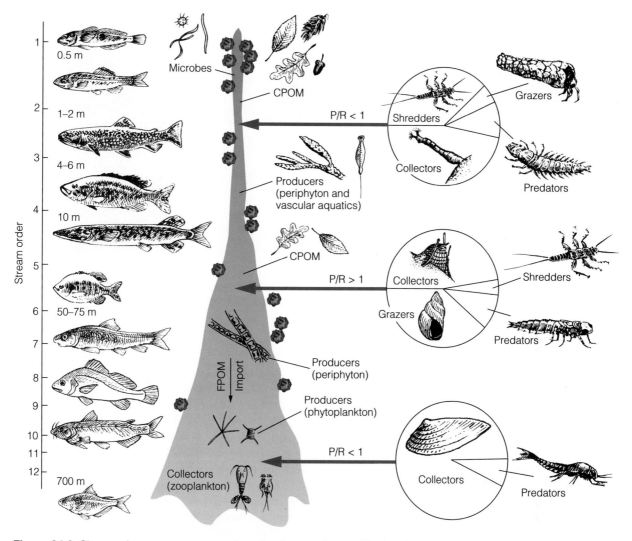

Figure 34.8 Changes in consumer groups along the river continuum. The headwater stream is strongly heterotrophic, dependent on terrestrial input of detritus. The dominant consumers are shredders and collectors. As stream size increases, the input of organic matter shifts to primary production by algae and rooted vascular plants. The major consumers are now collectors and grazers. As the stream grows into a river, the lotic system shifts back to heterotrophy. A phytoplankton population may develop. The consumers are mostly bottom-dwelling collectors.

Industrial pollution of large streams and rivers is the most serious contaminant, because of its concentration and chemical complexity. Water withdrawn from rivers for cooling in power plants and certain industrial processes, heated, and then returned raises river temperatures and lowers dissolved oxygen. Water used for flushing and chemical treatment and returned to the river imparts bad tastes and odors and introduces toxic substances that affect downstream use. Discharges from chemical plants and sulfurous wastes from pulp and paper mills are highly poisonous to aquatic life. Many chemical wastes, perhaps harmless alone, react with other chemicals to produce highly toxic conditions. Sudden influxes of such chemicals have frequently caused spectacular and tragic kills of fish and other aquatic life. Acid water from both deep and surface coal mines has destroyed stream life in coal country and has so reduced bacterial activity that biological purification of sewage and other organic wastes in water is impeded.

Radioactive wastes from uranium mills and nuclear power plants that find their way into streams do not break down naturally but do become increasingly diluted as they move downstream from the source.

Some radioactive materials are deposited on the bottom, whereas others are taken up by aquatic organisms, both plant and animal. Even minute amounts of radioactive substances may become concentrated many thousands of times in living tissue and be passed along in the food chain (Chapter 26).

Raw and inadequately treated sewage poured into streams and rivers sharply changes biological conditions. Even effluents from sewage treatment plants can upset ecological stability. As sewage enters a stream, it is dispersed and the solids settle to the bottom, where they are attacked by aerobic bacteria. This bacterial activity depletes oxygen, but the loss is offset by the absorption of more oxygen from the air into the stream. Streams can purify themselves by natural, bacterial breakdown of organic matter. The time required depends on the degree of pollution and the character of the stream. A fast-flowing stream constantly saturated with oxygen can purify itself much faster than a slow stream, which does not have the luxury of rapid oxygenation.

The carbon dioxide and hydrogen content of the water is high at the point of discharge. This condition eliminates normal stream life, particularly vertebrates and mollusks, and replaces them with a new group of dominant organisms, including protozoans, mosquito larvae, and tubifex worms. Below this zone of active decomposition, flowing waters dilute the pollutants. Although conditions improve, the stream still is far from normal. Green algae are present, but reduced in numbers; bacteria are abundant, and oxygen is low. Downstream the pollutants are diluted further, dissolved oxygen is higher, and organisms tolerant of such conditions, such as carp, catfish, chironomid larvae, and protozoans, inhabit the area. Eventually the water becomes clean and fresh again, and normal populations of fish and invertebrates reappear.

In far too many streams, however, conditions become worse downstream. No sooner has a river somewhat recovered from its polluted condition than another town, city, or industry dumps its sewage. As a result the river carries a load greater than it can handle and aerobic conditions no longer exist. Aerobic bacteria are replaced by anaerobic ones, and normal stream life is destroyed. Putrefying bacteria alone remain, and the stream becomes a foul-smelling, open sewer. Thus pollution by upstream industry and communities severely affects downstream communities. It decreases the quality of water for commercial and domestic use and increases the costs of water filtration and purification.

Figure 34.9 Banks of pasture streams broken down by livestock are an important, but often overlooked, source of siltation in smaller streams.

Siltation, caused by the erosion of farmlands, road construction, surface mining, logging, and other forms of soil disturbance (Figure 34.9), is the most insidious form of pollution, for it is widespread, it often goes unnoticed, and the damage it does is often permanent. Clay soils suspended in water block out light and prevent the growth of aquatic plants. Silt settles on the stream bottom, covering substrates for insect larvae and smothering larvae, mussels, and other bottom organisms. It blankets sewage and other organic material and retains them in place, reducing the oxygen supply. Silt clogs the opercular cavities and gill filaments of fish and the mantles and gills of mollusks, killing them both. Silty water flowing through the gravel nests of trout and salmon causes heavy mortality of eggs. Thousands of miles of trout and salmon streams have been destroyed by siltation, which, more than any other cause, limits the natural reproduction of these fish.

Regulated Rivers

The effects of damming are still worse. Small to massive schemes abound to straighten streams and rivers and throw dams across them. We can trace such damming activities as far back as 5000 years, but the greatest outburst of dam construction began after World War II. It has continued until over 60 percent of the world's streams and rivers have been dammed. Dams' flows are regulated (Figure 34.10), which profoundly

Figure 34.10 The Grand Coulee Dam on the Columbia River, one of the world's most highly regulated rivers. Built for hydroelectric power, the dam regulates both upstream and downstream flow, interfering with the natural movement of aquatic life, especially salmon.

affects the river's hydrology, ecology, and biology. Dams change the environment in which lotic organisms live, more often than not to their detriment.

Under normal conditions free-flowing streams and rivers experience seasonal fluctuation in flow. Snowmelt and early spring rains bring scouring high water; summer brings low water levels that expose some of the streambed and speed decomposition of organic matter along the edges. Life has adapted to these seasonal changes. Damming a river or stream interrupts both nutrient spiraling and the river continuum.

Downstream flow is greatly reduced as a pool of water fills behind the dam, developing characteristics similar to those of a natural lake, yet retaining some features of the lotic system, such as a constant inflow of water. Heavily fertilized by decaying material on the newly flooded land, the lake develops a heavy bloom of phytoplankton and in tropical regions dense growths of floating plants. Species of fish, often introduced exotics, adapted to lakelike conditions replace fish of flowing water.

The type of pool allowed to develop depends on the purpose of the dam and has a strong effect on downstream conditions. Single-purpose dams serve only for flood control or water storage; multipurpose dams provide hydroelectric power, irrigation water, and recreation, among other uses. Flood control dams have a minimum pool; the dam fills only during a flood, at which time inflow exceeds outflow. Engineers release

the water slowly to minimize downstream flooding. In time the water behind the dam recedes to the original pool depth. During flood and postflood periods the river below carries a strong flow for some time, scouring the riverbed. During normal times, flow below the dam is stabilized. If the dam is for water storage, the reservoir holds its maximum pool; but during periods of water shortage and drought, drawdown of the pool can be considerable, exposing large expanses of shoreline for a long time and stressing or killing littoral life. Only a minimal quantity of water is released downstream, usually an amount required by law, if such exists. Hydroelectric and multipurpose dams hold a variable amount of water, determined by consumer needs. During periods of power production pulsed releases are strong enough to wipe out or dislodge benthic life downstream, which under the best of conditions has a difficult time becoming established.

Reservoirs with a large pool of water become stratified, with a well-developed epilimnion, metalimnion, and hypolimnion (see Chapter 32). If water is discharged from the upper layer of the reservoir, the effect of the flow downstream is similar to that of a natural lake. Warm, nutrient-rich, well-oxygenated water creates highly favorable conditions for some species of fish below the spillway and on downstream. If the discharge is from the cold hypolimnion, downstream receives cold, oxygen-poor water carrying an accumulation of iron and other minerals and a concentration of soluble organic materials. Such conditions inimical to stream life may persist for hundreds of kilometers downstream before the river reaches anything near normal conditions. Gated selective withdrawal structures or induced artificial circulation to increase oxygen concentration reduce such problems at some dams.

The impact of dams is compounded when a series of multipurpose dams is built on a river. The amount of water released and moving downstream becomes less with each dam until eventually all available water is consumed and the river simply dries up. That is the situation on the Colorado River, the most regulated river in the world. The river is nearly dry by the time it reaches Mexico.

The effects of dams go beyond changing the nature of the ecosystem. Large dams on such rivers as the Columbia in North America interfere with the migratory patterns of fish such as salmon, which swim upstream to spawn. Although fish ladders are of assistance, many local populations have been excluded from their traditional spawning streams. Even if spawning is successful, unnatural timing of high and low flows in the rivers may induce premature seaward migration of the young or lengthen their downstream passage, exposing them to high temperatures. Because of these environmental conditions and losses caused by passage through hydroelectric dams, 90 percent of juveniles perish on the journey from their home stream to the estuary.

Dams also have an impact on the economic, social, and cultural patterns of human society. Although dams provide economic benefits—at great ecological cost—such as power, water, industrial and agricultural developments, and some fisheries, they have exacted some great economic and human costs. Communities have been displaced from lands to be flooded, and historic sites and productive agricultural lands have been covered with water. Sediments that might have built up floodplains or seasonally fertilized croplands, as the annual flooding of the Nile once did, remain behind the dam, lowering agricultural production. Reduction of freshwater inflow with its rich supply of nutrients impoverishes estuaries and allows salt water to intrude upriver, destroying or greatly reducing the catch of fish. In tropical regions dams, with their hundreds of miles of shoreline, provide excellent conditions for vectors of such parasitic diseases as malaria, arboviruses, and schistosomiasis. Schistosomiasis, the intermediate host of which is an aquatic snail, is picked up by humans, the definitive host, when they play, wash, bathe, walk, or work in infested waters. The problem is especially prevalent and widespread in tropical Africa. Because damming affects human health and welfare, and because of the increasing rareness of free-flowing streams with their valuable ecological and economic contributions, a moratorium on dam-building may be a wise decision.

Associated with dam-building is stream and river channelization, the dredging and straightening of streams and rivers for flood control, navigation, and agricultural development. Thousands of kilometers of meandering, productive, fish-filled streams and rivers have been channeled into sterile, unattractive drainage ditches. Such channelization has eliminated streamside vegetation and important wildlife habitat, and has destroyed associated wetlands (see Chapter 33). Newly cut channels support little bottom fauna, and fish lack food, shelter, and breeding sites. Paradoxically, channelization, designed to speed water from the uplands to the rivers and sea, actually intensifies downstream flooding because it increases the volume and rapidity of flow.

STUDY QUESTIONS

1. What physical characteristics are unique to flowing-water ecosystems?
2. In what way are organisms adapted to flowing water? How do adaptations change as fast streams become slow?
3. What is the basic energy source of headwater streams?
4. Characterize the major functional groups of stream invertebrates and their roles in the food web.
5. What is spiraling, and how does it function in nutrient cycling in streams?
6. How do downstream systems relate to upstream systems?
7. How does channelization—the straightening and deepening of a streambed—affect the structure and function of a stream?
*8. Determine the extent of stream channelization in your area. What was the reason for channelization? Use an old aerial photograph to assess the amount of change. What steps could be taken to minimize the effects of stream channelization? (See, for example, Gore and Petts 1989.)
*9. What types of dams exist in your area? What have been their ecological and economic benefits and costs? Describe the downstream conditions.
*10. Select one major dam—for example, the dams on the Columbia River, the Glen Canyon Dam, the Aswan High Dam in Egypt, the Ord River Dam of tropical Australia, the Kariba Dam on the Zambezi River, or the large dams on the River Volga. How is it harmful? How is it helpful?

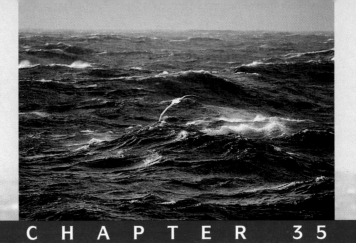

OCEANS

OBJECTIVES

On completion of this chapter, you should be able to:

- Discuss the salinity of the oceans.
- Describe the formation and types of waves and currents in the ocean.
- Explain the cause of tides.
- Describe the major zones in the sea and their relationship to temperature stratification.
- Discuss the roles of phytoplankton, zooplankton, and nekton in nutrient cycling in the open sea.
- Discuss the effect of human activities on the structure and function of the open sea.

An albatross flying close to the water epitomizes the open sea.

Freshwater rivers eventually empty into the oceans, and terrestrial ecosystems end at the edge of the sea. For some distance there is a region of transition. Rivers enter the saline waters of the ocean, creating a gradient of salinity. That gradient provides a habitat for organisms uniquely adapted to the half-world between salt water and fresh. The coastal regions exposed to the open sea are inhabited by other organisms, able to live in the often severe environments dominated by tides. Beyond lies the open ocean—the shallow seas overlying continental shelves and the deep oceans.

35.1 SPECIAL FEATURES

The marine environment is marked by a number of differences from the freshwater world. It is large, occupying 70 percent of Earth's surface, and it is deep, in places nearly 10 km. The surface area lighted by the sun is small compared to the total volume of water. This small volume of sunlit water and the dilute solution of nutrients limit primary production. All of the seas are interconnected by currents, dominated by waves, influenced by tides, and characterized by salinity, restricting life (see Chapter 4).

Salinity

The salinity of the open sea is fairly constant, averaging about 35 practical salinity units, psu (‰). Two elements, sodium and chlorine, make up some 86 percent of sea salt. These, along with other major elements such as sulfur, magnesium, potassium, and calcium, whose relative proportions vary little, comprise 99 percent of sea salts (Table 35.1). Determination of the most abundant element, chlorine, is used as an index of salinity of a given volume of seawater. Salinity is expressed in ‰ as the amount of chlorine in grams in a kilogram of seawater.

The salinity of parts of the ocean is variable because of physical processes. Salinity is affected by evaporation and precipitation, most pronounced at the interface of sea and air; by the movement of water masses; by the mixing of water masses of different salinities, especially near coastal areas; by the formation of insoluble precipitates that sink to the ocean floor; and by the diffusion of one water mass into another.

The elements most affected by these physical processes are the conservative ones not involved in biological processes. The most variable elements in the sea are the nonconservative ones, such as phosphorus and nitrogen, because their concentrations relate to

TABLE 35.1

COMPOSITION OF SEAWATER OF 35‰ SALINITY			
Elements	g/kg	Milli-moles/kg	Milli-equiva-lents/kg
Cations			
Sodium	10.752	467.56	467.56
Potassium	0.395	10.10	10.10
Magnesium	1.295	53.25	106.50
Calcium	0.416	10.38	20.76
Strontium	0.008	0.09	0.18
			605.10
Anions			
Chlorine	19.345	545.59	545.59
Bromine	0.066	0.83	0.83
Fluorine	0.0013	0.07	0.07
Sulphate	2.701	28.12	56.23
Bicarbonate	0.145	2.38	—
Boric acid	0.027	0.44	—
			602.72

biological activity. Taken up by organisms, these elements are usually depleted near the surface and enriched at lower depths. In parts of the ocean, some of these nutrients are returned by upwelling.

Temperature and Pressure

What you have learned about temperature in fresh water (Chapter 32) also applies to the sea. The range of temperature is far less than that on land, although it is considerable—from $-2°$ C in antarctic waters to warmer than $27°$ C in parts of the western Pacific. In general, seawater is never more than $2°$ to $3°$ below the freezing point of fresh water nor warmer than $27°$ C. At any given place the temperature of deep water is almost constant and cold. Seawater has no definite freezing point, although there is a temperature for seawater of any given salinity at which ice crystals form. Thus, pure water freezes out, leaving even more saline water behind. Eventually, it becomes a frozen block of mixed ice and salt crystals. With rising temperatures the process is reversed.

Unlike fresh water, seawater (with a salinity of 24.7‰ or higher) becomes heavier as it cools and does not reach its greatest density at $4°$ C; therefore the limitation of $4°$ C as the temperature of bottom water does not apply to the sea. The temperature of the sea bottom generally averages around $2°$ C even in the tropics if the water is deep enough. The temperature of the ocean floor over 1 km deep is 3° C.

Another aspect of the marine environment is pressure. Pressure in the ocean varies from 1 atmosphere at the surface to 1000 atmospheres at the greatest depth. Pressure changes are many times greater in the sea than in terrestrial environments, and pressure has a pronounced effect on the distribution of life. Certain organisms are restricted to surface waters, where the pressure is not great, whereas others are adapted to life at great depths. Some marine organisms, such as the sperm whale and certain seals, can dive to great depths and return to the surface without difficulty.

Waves and Currents

Wind generates waves on the open sea. The frictional drag of the wind on the surface of smooth water ripples the water. As the wind continues to blow, it applies more pressure to the steep side of the ripple, and wave size begins to grow. As the wind becomes stronger, short, choppy waves of all sizes appear; and as they absorb more energy, they continue to grow. When the waves reach a point at which the energy supplied by the wind is equal to the energy lost by the breaking waves, they become whitecaps. Up to a certain point, the stronger the wind, the higher the waves.

The waves that break on a beach are not composed of water driven in from distant seas. Each particle of water remains largely in the same place and follows an elliptical orbit with the passage of the wave form. As a wave moves forward, it loses energy to the waves behind and disappears, its place taken by another. The swells that break on a beach are distant descendants of waves generated far out at sea.

As the waves approach land, they advance into increasingly shallow water. The height of each wave rises until the wave front grows too steep and topples over. As the waves break on shore, they dissipate their energy, pounding rocky shores or tearing away sandy beaches at one point and building up new beaches elsewhere.

Surface waves command attention, but in the ocean there are also internal waves. Similar to surface waves, internal waves appear at the interface of layers of waters of different densities. In addition, there are stationary waves or seiches.

Just as there are internal waves, so there are internal currents in the sea. Surface currents are produced by wind, heat budgets, salinity, and the rotation of Earth (see Chapter 4). Water moving in surface currents must be replaced by a corresponding inflow from elsewhere. Because the surface waters are cooled and salinity changes, high-density water formed on the surface, largely at high latitudes, sinks and flows toward low latitudes. These currents are subject to the Coriolis effect (see Chapter 4) and are deflected or obstructed by submarine ridges and modified by the presence of other water masses. The result is three main systems of subsurface water movements: the bottom, the deep, and the intermediate ocean currents, each of which runs counter to the others.

In coastal regions, winds blowing parallel to the coast cause surface waters to be blown offshore. This water is replaced by water moving upward from the deep, a process known as **upwelling.** Although cold, upwelling water is rich in nutrients that support an abundant growth of phytoplankton. For this reason, regions of upwellings are highly productive, teeming with fish and bird life.

Tides

The gravitational pulls of the sun and the moon each cause two bulges in the waters of the oceans. The two caused by the moon occur at the same time on opposite sides of Earth on an imaginary line extending from the moon through the center of Earth. The tidal bulge on the moon side is due to gravitational attraction; the bulge on the opposite side occurs because the gravitational force there is less than at the center of Earth. As Earth rotates eastward on its axis, the tides advance westward. Thus, any given place on Earth will in the course of one daily rotation pass through two of the lunar tidal bulges, or high tides, and two of the lows, or low tides, at right angles to the high tides. Since the moon revolves in a 29½-day orbit around Earth, the average period between successive high tides is approximately 12 hours 25 minutes.

The sun also causes two tides on opposite sides of Earth, and these tides have a relation to the sun like that of the lunar tides to the moon. Because the gravitational pull of the sun is less than that of the moon, solar tides are partially masked by lunar tides except for two times during the month—when the moon is full and when it is new. At these times, Earth, moon, and sun are nearly in line, and the gravitational pulls of the sun and the moon are additive. This combination causes the high tides of those periods to be exceptionally large, with maximum rise and fall. These are the fortnightly **spring tides,** a name derived from the Saxon *sprungen*, which refers to the brimming fullness and active movement of the water. When the moon is at either quarter, its pull is at right angles to the pull of the sun, and the two forces interfere with each other. At this time the differences between high and low tide are exceptionally small. These are the **neap tides,** from an old Scandinavian word meaning "barely enough."

Tides are not entirely regular, nor are they the same all over Earth. They vary from day to day in the same place, following the waxing and waning of the moon. They may act differently in several localities within the same general area. In the Atlantic, semidaily tides are the rule. In the Gulf of Mexico, the alternate highs and lows more or less efface each other, and flood and ebb follow one another at about 24-hour intervals to produce one daily tide. Mixed tides are common in the Pacific and Indian oceans. These tides are combinations of daily and semidaily tides in which one partially cancels out the other.

Local tides around the world are inconsistent for many reasons. These reasons include variations in the gravitational pull of the moon and the sun due to the elliptical orbit of Earth, the angle of the moon in relation to the axis of Earth, onshore and offshore winds, the depth of water, the contour of the shore, and internal waves.

35.2 STRUCTURE

Zonation and Stratification

Just as lakes exhibit stratification and zonation, so do the seas. The ocean itself has two main divisions: the **pelagic,** or whole body of water, and the **benthic zone,** or bottom region (Figure 35.1). The pelagic is further divided into two provinces: the **neritic province,** water that overlies the continental shelf, and the **oceanic**

province. Because conditions change with depth, the pelagic is divided into three vertical layers or zones. From the surface to about 200 m is the **photic zone,** in which there are sharp gradients in illumination, temperature, and salinity. From 200 to 1000 m is the **mesopelagic zone,** where little light penetrates and the temperature gradient is more even and gradual, without much seasonal variation. It contains an oxygen-minimum layer and often the maximum concentration of nitrate and phosphate. Below the mesopelagic is the **bathypelagic zone,** where darkness is virtually complete, except for bioluminescence; temperature is low, and the pressure is great.

The upper layers of ocean water are thermally stratified. Depths below 200 m are usually thermally stable. In high and low latitudes, temperatures remain fairly constant throughout the year. Polar seas, covered with ice most of the year, exhibit no thermocline in winter, spring, and fall. The waters are well mixed and nutrients are not limiting. Slight stratification takes place in the polar summer (July and August). At that time the ice melts, the water warms enough, and light is sufficient to support a bloom of phytoplankton.

In tropical seas, the upper waters are well lighted and the continuous input of energy maintains a high temperature throughout the year. Light and temperature are optimal for phytoplankton production, but the waters are permanently stratified. That prevents mixing and upward circulation of nutrients. The result is low productivity.

In temperate seas, thermal structure changes seasonally, reflecting the amount of light and solar thermal energy entering the water. Water in the summer is thermally stratified with no mixing. In spring and

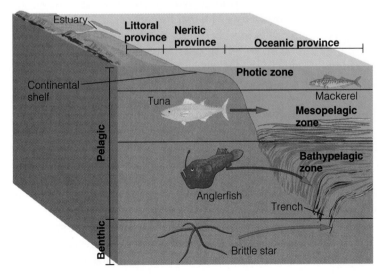

Figure 35.1 Major regions of the ocean.

fall, when the surface water warms and cools, respectively, thermal stratification decreases, and the waters mix to varying degrees, recharging nutrients in the surface waters.

Structural Groups

Viewed from the deck of a ship or from an airplane, the open sea appears to be monotonously the same. Nowhere can you detect any strong pattern of life or well-defined communities, as you can over land. The reason is that pelagic ecosystems lack the supporting structures and framework of large dominant plant life, and their major herbivores are not large, conspicuous mammals like elephants and deer, but tiny zooplankton.

There is a reason for the smallness of sea plants. Surrounded by a chemical medium that contains in varying quantities the nutrients necessary for life, they absorb their food directly from the water. The smaller the organism, the greater the surface-to-volume ratio. More surface is exposed for the absorption of nutrients and solar energy. Seawater is so dense that there is little need for supporting structures.

Nevertheless, differences based on physical characteristics and life forms do allow a division of the sea into ecological regions. The Arctic Ocean lies north of the land masses in the Northern Hemisphere and is open to the Atlantic and Pacific oceans via the Bering Strait. It holds unique forms of life, as does the Southern or Antarctic Ocean, which lies about the continent of Antarctica and is open to three oceans, the Atlantic, Pacific, and Indian. Warm oceanic waters making up the Atlantic and Pacific have their own distinctive communities. The deep-sea benthic ecosystems are quite different from the lighted waters. Other marine ecosystems include the coral reefs (Chapter 36) and upwelling systems off the coasts of California, Peru, northwest Africa, southwest Africa, India, and Pakistan. Important and distinctive are the shelf-sea ecosystems, such as the Georges and Grand Banks of the North Atlantic. Shallow, productive, and nutrient-rich, they support a diversity of fish and invertebrate life. All these regions vary, from oligotrophic to eutrophic and from cold to tropical, but they share structural groups.

Phytoplankton

Requiring light, phytoplankton is restricted to the upper surface waters. Light penetration varies from tens to hundreds of meters. Because of seasonal, annual, and geographic variations in light, temperature, and nutrients, as well as grazing by zooplankton, the distribution and species composition of phytoplankton vary from ocean to ocean and place to place within them.

Each ocean or region within an ocean appears to have its own dominant forms. Littoral and neritic waters and regions of upwelling are richer in plankton than mid-oceans. In regions of downwelling, the dinoflagellates, a large, diverse group characterized by two whiplike flagellae, concentrate near the surface in areas of low turbulence. They attain their greatest abundance in warmer waters. In summer they may concentrate in the surface waters in such numbers that they color it red or brown. Often toxic to vertebrates, such concentrations of dinoflagellates are responsible for red tides. In regions of upwelling, the dominant forms of phytoplankton are diatoms. Enclosed in a silica case, diatoms are particularly abundant in arctic waters. Smaller than diatoms, the **nanoplankton** make up the largest biomass in temperate and tropical waters. Most abundant are the tiny prochlorophytes. Distributed in all waters except the polar seas, the coccolithophores are a major source of primary production. Because they have calcareous plates and a threadlike appendage, coccolithophores can swim. Droplets of oil aid in buoyancy and serve as a means of storing food. In equatorial currents in shallow seas, the concentration of phytoplankton is variable. Where both lateral and vertical circulation of water is rapid, the composition reflects in part the ability of the species to grow, reproduce, and survive under local conditions.

Zooplankton

Grazing on the phytoplankton is the herbivorous zooplankton. Except for inshore water and upwelling areas, ciliate protozoans feed on nanoplankton. Copepods (planktonic arthropods and the most numerous animals of the sea) and the shrimp-like euhausiids (commonly known as krill) eat larger phytoplankton, the diatoms and dinoflagellates. Other planktonic forms are the larval stages of gastropods, oysters, and cephalopods. Feeding on the herbivorous zooplankton is the carnivorous zooplankton, which includes such organisms as the larval forms of comb jellies (Ctenophora) and arrowworms (Chaetognatha).

The composition of zooplankton varies. Zooplankton falls into two main groups based on size. These are the larger net zooplankton (so called because they cannot pass through plankton nets) and the smaller microzooplankton. Zooplankton of the continental shelf contains a large portion of the larvae of fish and benthic organisms. It includes a greater diversity of species, reflecting a greater diversity of environmental and chemical conditions. The open ocean, more homogeneous and nutrient-poor, supports a less diverse zooplankton. Zooplanktonic species of polar and temperate waters, having spent the winter in a dormant state in the deep water, rise to the surface during short periods of diatom blooms to reproduce. In temperate regions, distribution and abundance depend on the temperature of the water. In tropical regions, where temperature is nearly uniform, zooplankton is not so restricted, and reproduction occurs throughout the year.

Like phytoplankton, zooplankton lives mainly at the mercy of the currents; but possessing sufficient swimming power, many forms of zooplankton exercise some control. Some species migrate vertically each day to arrive at a preferred level of light intensity. As darkness falls, zooplankton rapidly rises to the surface to feed on phytoplankton. At dawn, it moves back down to preferred depths.

Nekton

Feeding on zooplankton and passing energy along to higher trophic levels is the nekton, swimming organisms that can move at will in the water column. They range in size from small fish to large predatory sharks and whales, seals, and marine birds such as penguins. Some of the predatory fish, such as tuna, are more or less restricted to the photic zone. Others are found in the deeper mesopelagic and bathypelagic zones, or move between them as the sperm whale does. Although the ratio in size of predator to prey falls within limits, some of the largest nekton organisms in the sea, the baleen whales, feed on disproportionately small prey, euphausiids or krill (Figure 35.2). By contrast, the sperm whale attacks very large prey, the giant squid.

Living in a world that lacks refuges against predation or sites for ambush, inhabitants of the pelagic zone have evolved various means of defense and of securing prey. Among them are the stinging cells of the jellyfish, streamlined shapes that allow speed both for escape and for pursuit, unusual coloration, advanced sonar, a highly developed sense of smell, and social organization involving schools or packs. Some animals, such as baleen whales, have specialized structures that permit them to strain krill and other plankton from the water. Others, such as the sperm whale and certain seals, dive to great depths to secure food. Phytoplankton lights up darkened seas, and fish take advantage of that bioluminescence to detect their prey.

Residents of the deep also have special adaptations for securing food. Some, like the zooplankton, swim to the surface to feed by night; others remain in the dimly lit or dark waters. Darkly pigmented and weak-bodied, many of the deep-sea fish depend on luminescent lures, mimicry of prey, extensible jaws, and expandable abdomens (which enable them to consume large items of food). Although most of the fish are small (usually 15 cm or less in length), the region is inhabited by rarely seen large species such as the giant squid. In the mesopelagic region bioluminescence reaches its greatest development—two-thirds of the species produce light. Fish have rows of luminous organs along their sides and lighted lures that enable them to bait prey and recognize other individuals of the same species. Bioluminescence is not restricted to fish. Squid and euphausiids possess searchlightlike structures complete with lens and iris; and squid and shrimp discharge luminous clouds to escape predators.

The Benthos

The term *benthic* refers to the floor of the sea, and **benthos** refers to plants and animals that live there. There is a gradual transition of life from the benthos on the rocky and sandy shores to that in the ocean's depths. From the tide line to the abyss, organisms that colonize the bottom are influenced by the substrate. Where the bottom is rocky or hard, the populations consist largely of organisms that live on the surface of the substrate, the **epifauna** and the **epiflora.** Where the bottom is largely covered with sediment, most of the inhabitants, chiefly animals, live within the deposits and are known collectively as the **infauna.** The kind of organism that burrows into the substrate is influenced by particle size. The mode of burrowing is often specialized for a certain type of substrate.

The substrate varies with the depth of the ocean and with the relationship of the benthic region to land areas and continental shelves. Near the coast, bottom sediments derive from the weathering and erosion of land areas along with organic matter from marine life. The sediments of deep water are fine-textured ma-

Figure 35.2 Antarctic krill are essential to the food chain.

terial that varies with depth and with the types of organisms in overlying waters. Although we call these sediments organic, they contain little decomposable carbon, consisting largely of skeletal fragments of planktonic organisms. In general, with regional variations, organic deposits down to 4000 m are rich in calcareous matter. Below 4000 m, hydrostatic pressure causes some forms of calcium carbonate to dissolve. At 6000 m and lower, sediments contain even less organic matter and consist largely of red clays rich in aluminum oxides and silica.

Within the sediments are layers that relate to oxidation-reduction reactions. The surface, or oxidized layer, yellowish in color, is relatively rich in oxygen, ferric oxides, nitrates, and nitrites. It supports the bulk of benthic animals, such as polychaete worms, bivalves, and copepods, and a rich growth of aerobic bacteria. Below this surface is a grayish transition zone to the black layer, characterized by a lack of oxygen, iron in the ferrous state, nitrogen in the form of ammonia, and hydrogen sulfide. This layer is inhabited by anaerobic bacteria, chiefly reducers of sulfates.

In a world of darkness, no photosynthesis takes place, so the bottom community is strictly heterotrophic (except in vent areas), depending entirely on what organic matter finally reaches the bottom as a source of energy. In spite of the darkness and depth, the benthic communities support a high diversity of species. In the shallow benthic regions the recorded number of polychaete worms is over 250 species and of pericarid crustaceans (shrimplike mysidaceans, cumaceans, the small tanaidaceans, and isopods) well over 100. But the deep-sea benthos supports a surprisingly higher diversity. The number of species collected in over 500 samples of which the total surface area sampled was only 50 m² was 707 species of polychaetes and 426 species of pericarid crustaceans. Most of the species are small, but among the deep-sea amphipods, isopods, and copepods, a few are large by crustacean standards, reaching 15 to 42 cm in length.

There are several hypotheses about this deep-sea diversity. One attributes it to the lack of widespread disturbance or environmental extremes. The temperature is nearly constant and the bottom is not stirred by storms, as the shallow bottoms are. Small local disturbances are created as the crustaceans and other bottom dwellers move, creating mounds and pits on the surface, interrupting the smoothness and adding diversity. Increasing the variety of living conditions is the patchy distribution of food. The benthos depends on the rain of organic matter drifting to the bottom, which is small and scattered. Patches of dead phytoplankton, the bodies of dead whales, seals, birds, fish,

and invertebrates, all provide a diversity of foods for different feeding groups and species. The low input and patchy distribution of food probably mean less interference among species.

Bottom organisms have four feeding strategies. They may filter suspended material from the water, as the stalked cnidarian do; they may collect food particles that settle on the surface of the sediment, as sea cucumbers do; they may be selective or unselective deposit feeders, as the polychaete worms are; or they may be predatory, like the brittle stars and the spiderlike pycnogonids.

Important in the benthic food chain are the bacteria of the sediments. Common where large quantities of organic matter are present, bacteria may reach several tenths of a gram per square meter in the topmost layer of silt. Bacteria synthesize protein from dissolved nutrients, and in turn become a source of protein, fat, and oils for deposit feeders.

Hydrothermal Vents

In 1977 oceanographers first discovered high-temperature deep-sea springs along volcanic ridges in the ocean floor of the Pacific near the Galápagos Islands. These springs vent jets of hydrothermal fluids that heat the surrounding water to 8° to 16° C, considerably higher than the 2° ambient water. Since then oceanographers have discovered similar vents on other volcanic ridges along fast-spreading centers of ocean floor, particularly in the Atlantic and eastern Pacific.

Vents form when cold seawater flows down through fissures and cracks in the basaltic lava floor deep into the underlying crust. The waters react chemically with the hot basalt, giving up some minerals but becoming enriched with others such as copper, iron, sulfur, and zinc. Heated to a high temperature, the water reemerges through mineralized chimneys rising up to 13 m above the sea floor. Among the chimneys are white smokers and black smokers (Figure 35.3). White-smoker chimneys rich in zinc sulfides issue a milky fluid under 300° C. Black smokers, narrower chimneys rich in copper sulfides, issue jets of clear water from 300° to over 450° C that are soon blackened by precipitation of fine-grained sulfur-mineral particles.

Associated with these vents is a rich diversity of unique deep-sea life confined within a few meters of the vent system. They include giant clams, mussels, polychaete worms that encrust the white smokers, crabs, and vestimentiferan worms lacking a digestive system.

The primary producers are chemosynthetic bacteria, which oxidize reduced sulfur compounds such as H_2S to release energy, which they use to form organic matter from carbon dioxide. Primary consumers,

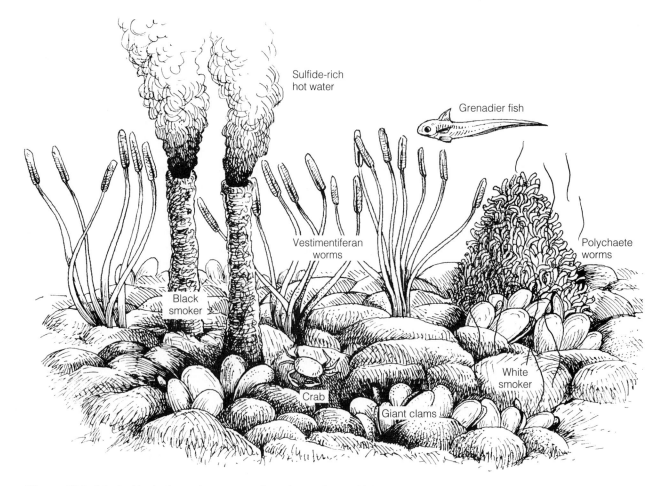

Figure 35.3 A typical hydrothermal vent mound, resting on flows of black basaltic lava.

the clams, mussels, and worms, filter bacteria from water and graze on bacterial film on rocks. The giant clam *Calyptogena magnifica* and the large vestimentiferan worm *Riftia pachyptila* posess symbiotic chemosynthetic bacteria. These bacteria need reduced sulfide, which the blood of these animals carries. *Riftia* has in its blood a sulfide-binding protein that concentrates sulfide from the environment and transports it to the bacteria. Such concentrations would poison other animals, but the sulfide-bearing protein of the worm and apparently the clam has a high affinity for free sulfides, preventing them from accumulating in the blood and entering the cells.

35.3   FUNCTION

Although the oceans dominate Earth's surface, they contribute much less to Earth's primary production than do terrestrial ecosystems. Oceans are less pro-

ductive because only a superficial illuminated area can support plant life; and most of the open sea is nutrient-poor, with an almost nonexistent nutrient reserve. Phytoplankton, zooplankton, and other organisms' remains sink below the lit zone into the dark benthic water. Although this sinking supplies nutrients to the deep, it robs the upper layers.

This depletion of nutrients is most pronounced in tropical waters. There, a permanent thermal stratification—a layer of warmer, less dense water lying on top of a colder, denser layer of deep water—prevents an exchange of nutrients between the surface and the deep. Therefore in spite of high light intensity and warm temperatures, tropical seas have the lowest productivity, an estimated 18–50 g C/m^2/yr.

The temperate oceans are more productive, largely because a permanent thermocline does not exist. During the spring and to a limited extent during the fall, temperate seas, like temperate lakes, undergo nutrient overturn (see Chapter 32). Recirculation of phosphorus and nitrogen from the deep stimulates a surge of spring phytoplankton growth. As spring wears on, the temperature of the water becomes stratified and a thermocline develops, preventing a nutrient exchange. The phytoplankton growth depletes the nutrients, and the phytoplankton population suddenly declines. In the fall a similar overturn takes place, but the rise in phytoplankton production is slight because of decreasing light intensity and falling temperatures. Reduced production in winter holds down the annual productivity of temperate seas to a level a little above that of tropical seas, 70–110 g C/m²/yr.

Most productive are coastal waters and regions of upwelling, whose annual productivity may amount to 1000 g C/m²/yr. Major areas of upwelling are largely on the western sides of continents: off the southern California coast, Peru, northern and southwestern Africa, and the Antarctic. Upwellings result from the differential heating of polar and equatorial regions that produces the equatorial currents and the winds (see Chapter 4). For example, as water is pushed northward to the equator by winds blowing out of the south, it is deflected from the coast by the Coriolis effect. As the deflected surface waters move away, they are replaced by an upwelling of colder, deeper water that brings a supply of nutrients into the warm, sunlit portions of the sea. As a result, regions of upwelling are highly productive, supporting an abundance of life. Because of their high productivity, upwellings support important commercial fisheries such as tuna fisheries, the anchovy fishery off Peru, and the sardine fishery off Portugal. Other zones of high production are coastal waters and estuaries, where productivity may run as high as 380 g C/m²/yr. Turbid, nutrient-rich waters are major areas of fish production.

Measurements show that a great deal of the productivity of the coastal fringes comes from the benthic as well as the surface waters. Benthic production deeper in the open sea remains unknown. An estimate of the total production for marine plankton is 50 Gt of dry matter per year; if we included benthic production, total production might be 55 Gt of dry matter per year.

Carbohydrate production by photosynthetic nanoplankton is the base on which the life of the seas rests (Figure 35.4). Conversion of primary production into animal tissue is the work of zooplankton, the most important of which are the copepods. To feed on the minute phytoplankton, most of the grazing herbivores must also be small, between 0.5 and 5.0 mm. In the oceans most of the grazing herbivores are ciliate pro-

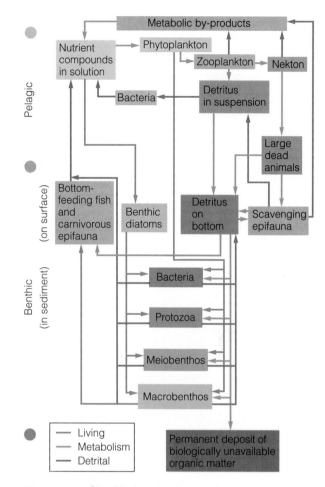

Figure 35.4 Simplified marine food web.

tozoans and members of the genera *Calanus, Acartia, Temora,* and *Metridia,* probably the most abundant animals in the world. The single most abundant copepods are *Calanus finmarchicus* and its close relative *C. helgolandicus.* In the Antarctic the shrimplike euphausiids, or krill, fed on by the baleen whales and penguins, are the dominant herbivores. The herbivorous copepods, then, are the link in the food chain between the phytoplankton and the second-level consumers, as illustrated in the North Sea food web (Figure 35.5).

However, part of the food chain begins not with the phytoplankton, but with organisms even smaller. Bacteria and protists, both heterotrophic and photosynthetic, make up one-half of the biomass of the sea and are responsible for the largest part of energy flow in pelagic systems. Photosynthetic nanoflagellates (2–20 μm) and cyanobacteria (1–2 μm) are responsible for a large part of photosynthesis in the sea. These cells excrete a substantial fraction of their photosynthate

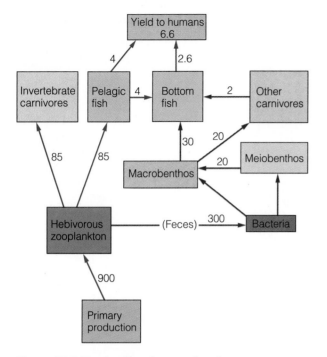

Figure 35.5 Food web and energy flow from an area of the North Sea. Note the low-energy yield to humans, 6.6 kcal/m²/yr from a primary production of 900 kcal/m²/yr.

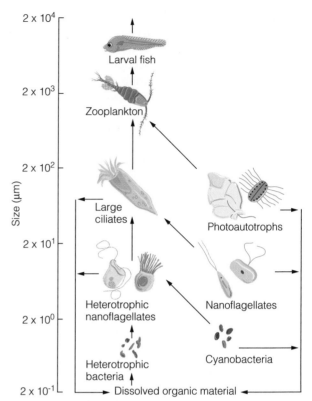

Figure 35.6 Microbial loops such as this feed into the classic marine food web.

in the form of dissolved organic material that heterotrophic bacteria use. Populations of such bacteria are dense, around 1 million cells per milliliter of seawater. These heterotrophic bacteria account for about 20 percent of primary production. Bacterial growth efficiency does not exceed 50 percent, so one-half of phytoplankton primary production in the form of dissolved organic material is consumed by bacteria. Bacterial numbers in the sea remain relatively stable, suggesting predation, but filter-feeding zooplankton cannot retain particles of bacterial size, so the consumption of bacteria by heterotrophic nanoflagellates, experimentally demonstrated, accounts for the disappearance of bacterial production. This interaction introduces a feeding loop, the **microbial loop** (Figure 35.6), and adds several trophic levels to the plankton food chain.

35.4 HUMAN IMPACT

The ocean is polluted, Thor Heyerdahl reported after completing his trip across the southern Atlantic in 1968 on the papyrus raft *Ra II*. "For weeks along the route," wrote Norman Baker, navigator on the *Ra II*, "We saw no land, no ship, no light, no man; all we saw

was his garbage, and we saw that all the time." Among this garbage were floating gobs of oil. ". . . As far as I could see," Baker continued, "below as well as to the sides, the entire visible layer of the ocean was infested with hanging bits of hardened oil." Conditions have not improved since then. They have only gotten worse.

Despite natural leakages from deposits on the ocean floor, only since oil became the major energy source, exploited, transported, and used about the world, has it become a major pollutant of the seas. Leakage from rusting oceanic pipelines, errors in handling oil cargo, accidents involving tankers and barges, sinking of ships during wars, illegal washing of tanker bilges, and spills and leakage from off-shore drilling rigs all contribute to a growing oil pollution problem.

Oil is a collection of hundreds of substances that react with the environment. When released in water, it spreads in a film over the surface. The lighter fractions evaporate or are absorbed by particulate matter and sink to the bottom. Some dissolves in seawater. Much is oxidized by bacteria, yeast, and molds that attack different fractions of crude oil. Bacteria can digest straight-chain hydrocarbons. They have more problems with branch-chained hydrocarbons, which persist for a long time as tarry chunks bobbing on the surface and lying on the bottom of the sea.

The aspect of oil pollution that captures our attention is the damage to coastal ecosystems (Chapter 36). We deplore massive oil spills from tanker wrecks along the coast, or the disastrous releases of oil from the Kuwait oil fields. Much less obvious but in the long run much more important are the insidious effects of oil and other pollutants on the world's oceans. We know little about what effect oil has on the benthos, where it comes in direct contact with life in the deep. We do know that oil reaching the bottom of bays and harbors kills bottom life, and that certain fractions of oil soluble in water are highly toxic to many forms of marine life.

The problem is aggravated by the presence of many other toxic materials in the seas. The oceans have become the great dumping grounds for the world. Toxic substances of all sorts—pesticides, herbicides, and heavy metals such as mercury, cadmium, lead, zinc, and copper—enter the oceans from many sources and add to the dissolved hydrocarbons from oil. These pollutants, although in low concentration, are foreign to the marine ecosystem, and some, such as mercury, become magnified in the food web. Particularly sensitive to this chronic pollution are the phytoplankton and many forms of zooplankton and crustaceans. Such pollution inhibits photosynthesis, growth, and cell division of marine phytoplankton. It affects growth and development of microzooplankton filter feeders, and the early developmental stages of other forms of life.

The impact of marine pollution has barely been studied, but we know that pollution in the North Atlantic has decreased the number of species of phytoplankton and zooplankton. The long-term effect will be a decrease in primary production in the ocean, accompanied by a disruption of trophic interactions, beginning with phytoplankton and zooplankton and working up the food web.

The vastness and depth of the ocean, far removed from direct human contact, make it a seemingly ideal place for disposing of the wastes of human activity: sewage, sewage sludge, industrial wastes, garbage, and radioactive wastes. Discharge of sewage from southern California cities out to the mainland shelf has contaminated a 3640 km² area of ocean bottom. This pollution has degraded benthic invertebrates, killed beds of kelp, and caused disease in fish. The bottom of a 105 km² area of the New York Bight is covered with black toxic sludge. The ocean has long been a favorite place to dump solid wastes, including urban garbage, industrial and hazardous wastes, construction materials, and junk, from cars to military shells and old ships. Although some of the metallic junk has provided sheltering reefs for fish, other material entangles or poisons marine birds and mammals.

Already the recipient of low-level radioactive wastes, the deep benthic regions, because of the relative stability of their sediments, are viewed as possible safe places for the disposal of accumulating highly radioactive wastes from nuclear power plants. Given our general ignorance of the structure and function of the deep-sea benthos and deep-sea food webs, we have no idea of the long-term effects and dangers of such activity.

STUDY QUESTIONS

1. Why is the ocean salty?
2. What causes tides? What are spring tides and neap tides?
3. What currents other than surface currents exist in oceans? What is their significance?
4. What is upwelling, and what is its ecological importance?
5. Characterize the major regions of the ocean, both vertical and horizontal.
6. How does temperature stratification in the tropical seas differ from that in the polar seas? What sea develops the most pronounced thermocline, and why?
7. What might account for the high diversity of the deep-sea benthos?
8. What are hydrothermal vents, and what makes life about them unique?
9. What is the role of phytoplankton, zooplankton, and the nekton in the marine food web?
10. What is the significance of the microbial loop in the marine food web?
11. Why is primary production in the seas so low? What effect does that have on the overall productivity of the oceans?
12. What characteristics of oil make it a chronic pollutant of the open ocean?
13. What are some other major sources of ocean pollution and their potential effects?
*14. Find out the extent of the use of oceans for sewage and solid waste disposal. Does any of your garbage end up there?
*15. Report on the cause and effects of a major oil spill.
*16. How can interior agricultural regions pollute the ocean?
*17. Debate the pros and cons of using the deep-sea bottom for disposal of radioactive wastes.

C H A P T E R   3 6

INTERTIDAL ZONES AND CORAL REEFS

OBJECTIVES

On completion of this chapter, you should be able to:

- Describe the major features and zonation of rocky, sandy, and muddy shores and coral reefs.
- Discuss the adaptations of life to an intertidal environment.
- Explain the role of disturbance in intertidal and coral reef ecosystems.
- Describe the formation of coral reefs.
- Explain the symbiotic relationship between coral anthozoans and algae and its ecological significance.
- Discuss the impact of human activity on the intertidal zones and coral reefs.

Low tide exposes highly predatory starfish (*Pisaster* spp.) resting on algal covered rocks in Olympic National Park, Washington.

Where the land meets the sea we find the fascinating and complex world of the seashore. Rocky, sandy, or muddy, protected or pounded by incoming swells, all shores have one feature in common: they are alternately exposed and submerged by the tides. Roughly, the region of the seashore is bounded on one side by the height of extreme high tide and on the other by the height of extreme low tide. Within these confines conditions change from hour to hour with the ebb and flow of the tides. At flood tide the seashore is a water world; at ebb tide it belongs to the terrestrial environment, with its extremes in temperature, moisture, and solar radiation. In spite of all this change, seashore inhabitants are essentially marine, adapted to withstand some degree of exposure to the air for varying periods of time.

36.1   ROCKY SHORES

Structure

As the sea recedes at ebb tide, rocks, glistening and dripping with water, begin to appear. Life hidden by tidal water emerges into the open air layer by layer. The uppermost layers of life are exposed to air, wide temperature fluctuations, intense solar radiation, and desiccation for a considerable period, while the lowest fringes on the intertidal shore may be exposed only briefly before the flood tide submerges them again. These varying conditions result in one of the most striking features of the rocky shore, the zonation of

life (Figure 36.1). Although this zonation differs from place to place as a result of local variations in aspect, substrate, wave action, light intensity, shore profile, exposure to prevailing winds, climatic differences, and the like, the same general features are always present. All rocky shores have three basic zones, each characterized by dominant organisms (Figure 36.2).

Where the land ends and the seashore begins is hard to determine. The approach to a rocky shore from

Figure 36.1 The broad zones of life exposed at low tide on the rocky shore of the Bay of Fundy in Canada. Note the heavy growth of knotted wrack on the lower portion and the white zone of barnacles above.

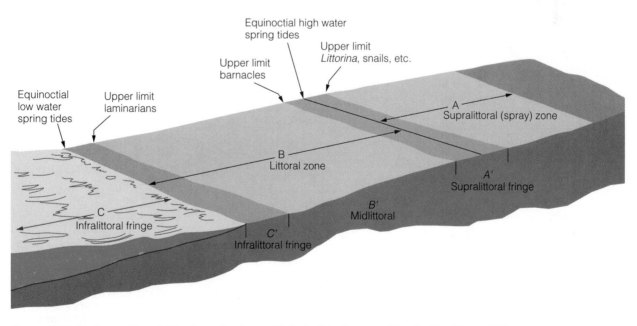

Figure 36.2 Basic zonation of Atlantic rocky shores. Refer to this diagram while studying Figure 36.3.

the landward side is marked by a gradual transition from lichens and other land plants to marine life dependent at least partly on the tidal waters (Figure 36.3). The first major change from land shows up on the **supralittoral fringe,** where salt water comes only once every fortnight on the spring tides. It is marked by the black zone, a patchlike or beltlike encrustation of Verrucaria lichens, *Calothrix*, a cyanobacteria, and the green alga *Entophysalis*. Living under conditions that few other plants could survive, these organisms, enclosed in slimy, gelatinous sheaths, and their associated lichens represent an essentially nonmarine community. Common to this black zone are periwinkles, basically marine animals, that graze on wet algae covering the rocks. On European shores lives a similarly adapted species, the rock periwinkle, the most highly resistant to desiccation of all the shore animals.

Below the black zone lies the **littoral zone,** covered and uncovered daily by the tides. The littoral tends to be divided into subzones. In the upper reaches bar-

nacles are most abundant. The oyster, the blue mussel, and the limpets appear in the middle and lower portions of the littoral, as does the common periwinkle.

Occupying the lower half of the littoral zone (midlittoral) of colder climates and in places overlying the barnacles is an ancient group of plants, the brown algae, commonly known as rockweeds (*Fucus* spp.) and wrack (*Ascophyllum nodosum*). Rockweeds attain their finest growth on protected shores, where they may grow 2 m long; on wave-whipped shores they are considerably shorter.

The lower reaches of the littoral zone may be occupied by blue mussels instead of rockweeds, particularly on shores where hard surfaces have been covered in part by sand and mud. No other shore animals grow in such abundance; the blue-black shells packed closely together may blanket the area.

Near the lower reaches of the littoral zone, mussels may grow in association with red algae, *Gigartina*, a low-growing, carpetlike plant. Algae and mussels to-

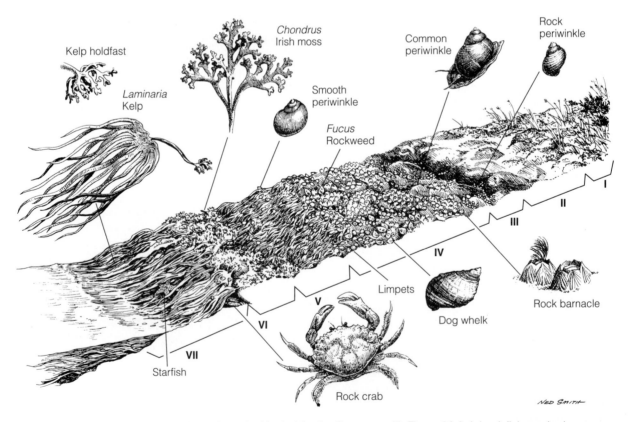

Figure 36.3 Zonation on a rocky shore along the North Atlantic. Compare with Figure 36.2. I, land: lichens, herbs, grasses; II, bare rock; III, black algae and rock periwinkle *(Littorina)* zone; IV, barnacle *(Balanus)* zone: barnacles, dog whelks, common periwinkles, mussels, limpets; V, fucoid zone: rockweed *(Fucus)* and smooth periwinkles; VI, Irish moss *(Chondrus)* zone; VII, kelp *(Laminaria)* zone.

gether often form a tight mat over the rocks. Here, well protected from waves in the dense growth, live infant starfish, sea urchins, brittle stars, and bryozoan sea mats or sea lace *(Membranipora)*.

The lowest part of the littoral zone, uncovered only at the spring tides and not even then if wave action is strong, is the **infralittoral fringe.** This zone, exposed for short periods of time, consists of forests of the large brown alga, *Laminaria*, one of the kelps, with a rich undergrowth of smaller plants and animals among the holdfasts.

Beyond the infralittoral fringe is the **sublittoral zone,** the open sea. This zone is principally neritic and benthic. It contains a wide variety of fauna, depending on the substrate, the presence of protruding rocks, gradients in turbulence, oxygen tensions, light, and temperature.

The pattern of life on rocky shores is heavily influenced by grazing, predation, competition, and larval settlement. Where wave action is heavy on New England intertidal rocky shores, periwinkles are rare, permitting a more vigorous growth of algae. Lack of grazing favors ephemeral algal species such as *Ulva* and *Enteromorpha*, whereas grazing allows the perennial *Fucus* to become established. On the New England coast the mussel *Mytilus edulis* outcompetes barnacles and algae; but predation by the starfish *Asterias* and the snail *Nucella lapillus* prevents dominance by mussels except on the most wave-beaten areas. A similar situation exists on the Washington State coast. Barnacles of several species tend to outcompete and displace algal species, but in turn the dominant mussel *M. californianus* destroys the barnacles by overgrowing them. However, where present the predatory starfish *Pisaster ochraceus* prevents the mussel from completely overgrowing barnacles. The end result of such interactions of the physical and the biotic is a patchy distribution of life across the rocky intertidal shore.

The ebbing tide leaves behind pools of water in rock crevices, in rocky basins, and in depressions (see Figure 36.1). They represent distinct habitats, which differ considerably from exposed rock and the open sea, and even differ among themselves. At low tide all pools are subject to wide and sudden fluctuations in temperature and salinity, changes most marked in shallow pools. Under the summer sun the temperature may rise above the maximum many organisms can tolerate. As water evaporates, especially in the shallower pools, salt crystals may appear around the edges. When rain or land drainage brings fresh water to the pools, salinity may decrease. In deep pools such fresh water tends to form a layer on top, developing a strong sa-

linity stratification in which the bottom layer and its inhabitants are little affected. If algal growth is considerable, oxygen will be high during the daylight hours but low at night, a situation that rarely occurs at sea. The rise of CO_2 at night lowers the pH.

Pools near low tide are influenced most briefly by the rise and fall of tides; those that lie near and above the high tide line are exposed the longest and undergo the widest fluctuations. Some may be recharged with seawater only by the splash of breaking waves or occasional high spring tides. Regardless of their position on the shore, most pools suddenly return to sea conditions on the rising tide and experience drastic and instantaneous changes in temperature, salinity, and pH. Life in the tidal pools must be able to withstand those fluctuations.

Function

Life on the rocky shore is both autotrophic and heterotrophic. Many organisms, such as barnacles, depend on tides to bring them food; others, such as periwinkles, graze on algal growth on the rocks. In fact, the functioning of the rocky seashore ecosystem involves complex interactions between physical and biotic aspects of the ecosystem. Like a salt marsh that daily receives energy from tidal flooding, rocky shores receive an energy subsidy indirectly from the waves (Figure 36.4). Wave-generated energy impinging on a rocky shore is 0.045 watts/cm^2, roughly twice that of solar radiation. Energy input is reflected in productivity. The kelp beds of the Pacific Northwest are more productive than the tropical rain forest. Part of this productivity relates to

Figure 36.4 Waves pounding on rocky shores provide an energy subsidy to intertidal life.

the much higher leaf surface area per square meter—about 20 m² of growing surface, compared to about 8 m² for a tropical forest. The highest standing crops of kelp, mussels, and other organisms are in areas receiving the heaviest wave action.

Intertidal organisms do not use wave energy directly. Rather, the force of the waves improves conditions for productivity. Heavy wave action reduces the activity of such predators of sessile intertidal invertebrates as starfish and sea urchins. Waves bring in a steady supply of nutrients and carry away products of metabolism. They keep the fronds of seaweeds in constant motion, moving them in and out of shadow and sunlight, allowing more even distribution of incident light and thus more efficient photosynthesis. By dislodging organisms, both plants and invertebrates, from the rocky substrate, waves open up space for colonization by algae and invertebrates and reduce strong interspecific competition. In effect disturbance, which influences community structure, is the root of intertidal productivity.

36.2   SANDY AND MUDDY SHORES

Sandy and muddy shores appear devoid of life at low tide, in sharp contrast to the life-studded rocky shore; but the sand and black mud are not as barren as they seem, for beneath them life lurks, waiting for the next high tide.

The sandy shore is a harsh environment; indeed, the matrix of this seaside environment is a product of the harsh and relentless weathering of rock, both inland and along the shore. Through eons the products of rock weathering are carried away by rivers and waves to be deposited as sand along the edge of the sea. The size of the sand particles deposited influences the nature of the sandy beach, water retention during low tide, and the ability of animals to burrow through it. Beaches with steep slopes are usually made up of larger sand grains and are subject to more wave action. Beaches exposed to high waves are generally flattened, for much of the material is carried away from the beach to deeper water and fine sand is left behind (Figure 36.5). Sand grains of all sizes, especially the finer particles in which capillary action is the greatest, are more or less cushioned by a film of water, reducing further wearing action. Sand's retention of water at low tide is one of the outstanding environmental features of the sandy shore.

In sheltered areas of the coast, the slope of the beach may be so gradual the surface appears to be

Figure 36.5 A long stretch of sandy beach washed by waves on the southern Australian coast. Although the beach appears barren, life is abundant beneath the sand.

flat. Because of the flatness the outgoing tidal currents are slow, leaving behind a residue of organic material settled from the water. In these situations mudflats develop.

Structure

Life on the sand is almost impossible. Sand provides no surface for attachment of seaweeds and their associated fauna; and the crabs, worms, and snails characteristic of rocky crevices find no protection there. Life, then, is forced to live beneath the sand.

Life on sandy and muddy beaches consists of epifauna and infauna. Most infauna either occupy permanent or semipermanent tubes within the sand or mud or are able to burrow rapidly into the substrate. Multicellular infauna obtain oxygen either by gaseous exchange with the water through their outer covering or by breathing through gills and elaborate respiratory siphons.

Within the sand and mud live vast numbers of **meiofauna** with a size range between 0.5 mm and 62 μm, including copepods, ostracods, nematodes, and gastrotrichs. Interstitial fauna are generally elongated forms with setae, spines, or tubercles greatly reduced. The great majority do not have pelagic larval stages. These animals feed mostly on algae, bacteria, and detritus. Interstitial life, best developed on the more sheltered beaches, shows seasonal variations, reaching maximum development in summer months.

Sandy beaches also exhibit zonation related to the tides (Figure 36.6), but you must discover it by dig-

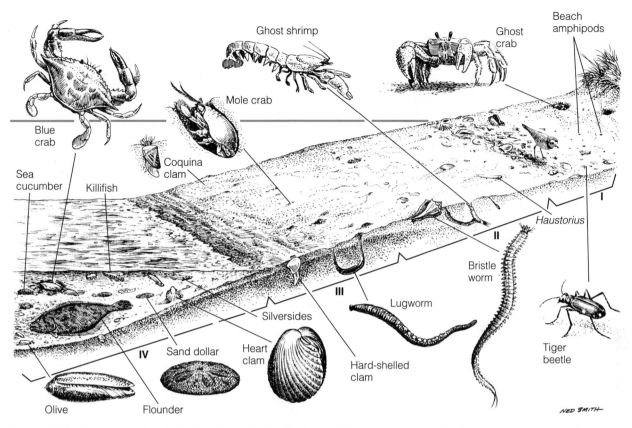

Figure 36.6 Life on a sandy beach along the mid-Atlantic coast. Although strong zonation is absent, organisms still change on a gradient from land to sea. I, supratidal zone: ghost crabs and sand fleas; II, flat beach zone: ghost shrimp, bristle worms, clams; III, intratidal zone: clams, lugworms, mole crabs; IV, subtidal zone: sand dollar, blue crab. The blue line indicates high tide level.

ging. Sandy and muddy shores divide roughly into supralittoral, littoral, and infralittoral zones, based on animal organisms, although a universal pattern similar to that of the rocky shore is lacking. Pale, sand-colored ghost crabs and beach hoppers occupy the upper beach, the supralittoral. The intertidal beach, the littoral, is a zone where true marine life appears. Although sandy shores lack the variety found on rocky shores, the populations of individual species of largely burrowing animals often are enormous. An array of animals, among them starfish and the related sand dollar, can be found above the low-tide line and in the infralittoral.

Organisms living within the sand and mud do not experience the same violent fluctuations in temperature as those on rocky shores. Although the surface temperature of the sand at midday may be 10° C or more higher than the returning seawater, the temperature a few centimeters below remains almost constant throughout the year. Nor is there a great fluctuation in salinity, even when fresh water runs over the surface of the sand. Below 25 cm, salinity is little affected.

Associated with these essentially herbivorous animals are predators, whether the tide is in or out. Near and below the low-tide line live predatory gastropods, which prey on bivalves beneath the sand. In the same area lurk predatory portunid crabs such as the blue crab and green crab, which feed on mole crabs, clams, and other organisms. They move back and forth with the tides. The incoming tides also bring other predators, such as killifish and silversides. As the tide recedes, gulls and shorebirds scurry across the sand and mudflats to hunt for food.

Function

For life to occupy the sandy shore, some organic matter has to accumulate. Most sandy beaches contain a certain amount of detritus from seaweeds, dead animals, feces, and material blown from inland. This organic matter accumulates within the sand, especially in sheltered areas. In fact, an inverse relationship exists between the turbulence of the water and the amount of organic matter on the beach, with accumulation reaching its maximum on the mudflats.

Organic matter clogs the spaces between the grains of sand and binds them together. As water moves down through the sand, it loses oxygen from both the respiration of bacteria and the oxidation of chemical substances, especially ferrous compounds. The point within the mud or sand at which water loses all its oxygen is a region of stagnation and oxygen deficiency. A layer of dark iron sulfides of variable depth marks it. On mudflats such conditions exist almost to the surface.

The energy base for sandy beach and mudflat fauna is organic matter. Much of it becomes available by bacterial decomposition, which goes on at the greatest rate at low tide. Bacteria concentrate around organic matter in the sand, where they escape the diluting effects of water. Each high tide dissolves and washes away into the sea the products of decomposition and brings in more organic matter. Therefore sandy beaches and mudflats are important sites for biogeochemical cycling, supplying offshore waters with phosphates, nitrogen, and other nutrients.

Energy flow in sandy beaches and mudflats differs from that in terrestrial and aquatic ecosystems, for the basic consumers are bacteria. In more usual energy-flow systems, bacteria act largely as reducers responsible for the conversion of dead organic matter into a nutrient form that can be used by producer organisms. In sandy beaches and mudflats, bacteria not only feed on detrital material and break down organic matter but also are a major source of food for higher-level consumers.

A number of deposit-feeding organisms ingest organic matter largely as a means of obtaining bacteria. Prominent among them are numerous nematodes and copepods (Harpacticoida), the polychaete worm *Nereis*, and the gastropod mollusks. Deposit feeders on sandy beaches obtain their food by burrowing through the sand and ingesting the substrate to feed on the organic matter it contains. Most common among them is the lugworm *Arenicola*, which is responsible for the conspicuous coiled and cone-shaped casts on the beach.

Other sandy beach animals are filter feeders, obtaining their food by sorting particles of organic matter from tidal water. Two of these "surf fishers," advancing and retreating with the tide, are the mole crab *Emerita* and the coquina clam *Donax*.

Except in the cleanest of sands, some primary production does take place in the intertidal zone. The major primary producers are the diatoms, confined mainly to fine-grained deposits of sand containing a high proportion of organic matter. Productivity is low;

one estimate places productivity of moderately exposed sandy beaches at 5 g C/m^2/yr. Production may be increased temporarily by phytoplankton carried in on the high tide and stranded on the surface. Again, the majority are diatoms. More important as producers are the sulfur bacteria in the black sulfide or reducing layer. These chemosynthetic bacteria use the energy released by oxidizing reduced compounds. Some mudflats may be covered with the algae *Enteromorpha* and *Ulva*, whose productivity can be substantial.

Because of their dependence on imported matter and their heterotrophic nature, perhaps we should not consider sandy beaches and mudflats as separate ecosystems. In effect, they are part of a larger coastal ecosystem involving the salt marsh, the estuary, and coastal waters (Figure 36.7). Sandy shores and mudflats act as sinks for energy and nutrients, because they draw their energy from organic matter that originates outside the area. Many nutrient cycles are only partially contained within the borders of the shore.

36.3 CORAL REEFS

Lying in the warm, shallow waters about tropical islands and continental land masses are colorful rich oases in nutrient-poor seas, the coral reefs (Figure 36.8). They are a unique accumulation of dead skeletal material built up by carbonate-secreting organisms, living coral (Cnidaria, Anthozoa), coralline red algae (Rhodophyta, Corallinaceae), green calcerous algae (*Halimeda*), foraminifera, and mollusks. Built only underwater at shallow depths, coral reefs need a stable foundation upon which to grow. Such foundations are provided by shallow continental shelves and submerged volcanos.

Coral reefs are of three types, with many gradations among them. (1) *Fringing reefs* grow seaward from the rocky shores of islands and continents. (2) *Barrier reefs* parallel shorelines of continents and islands and are separated from land by shallow lagoons. (3) *Atolls* are horseshoe-shaped rings of coral reefs and islands surrounding a lagoon, formed when a volanic mountain subsided. Such lagoons are about 40 m deep, usually connect to the open sea by breaks in the reef, and may hold small islands of patch reefs. Reefs build up to sea level.

Structure

Coral reefs are complex ecosystems that begin with the complexity of the corals themselves. Corals are modular animals, anemone-like cylindrical polyps, with prey-capturing tentacles surrounding the opening or

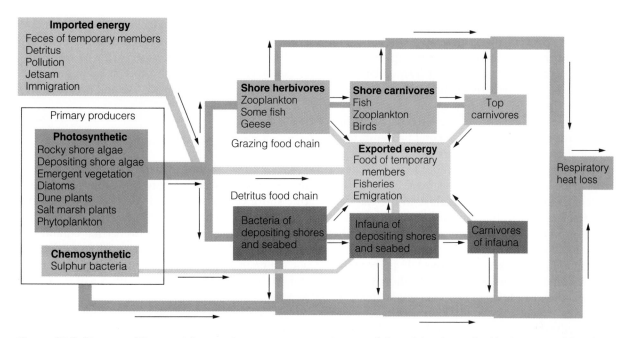

Figure 36.7 Diagram of the coastal ecosystem, a supraecosystem consisting of the shore, the fringing terrestrial regions, and the sublittoral zones. It connects two food webs: that of the rocky shore with its algae, herbivores, and zooplankton, and the detrital food webs involving the bacteria of the depositing shore and sublittoral muds and the dependent detritivores and carnivores. Coastal ecosystems are extremely productive; because energy imports exceed exports, the system is continuously gaining energy.

mouth. Most corals form sessile colonies supported on the tops of dead colonies and cease growth when they reach the surface of the water. There are close relationships between coral animals and algae. In the tissues of the gastrodermal layer live **zooxanthellae,** symbiotic, photosynthetically active, endozoic dinoflagellate algae upon which coral depend for most efficient growth. On the calcareous skeletons live still other kinds of algae, both the encrusting red and green coralline species and filamentous species, including turf algae. Also associated with coral growth are mollusks, such as giant clams *(Tridacna, Hippopus)*, echinoderms, crustaceans, polychaete worms, sponges, and a diverse array of fishes, both herbivorous and predatory.

The zonation and diversity of coral species depend on an interaction of depth, light, predation, competition, and disturbance. Light sets the depth at which the zooxanthellae can survive. Diversity is lowest at the crest near the surface, where only species such as massive pillar-shaped corals tolerant of intense or frequent disturbance by waves can survive. Diversity increases with depth to a maximum of about 20 m, a region occupied by brain, crustose, and delicate branched and fan corals; then it decreases as light attenuates, eliminating shade-intolerant species. Imposed upon this gradient are other abiotic and biotic disturbances that vary

Figure 36.8 A rich diversity of coral species, algae, and colorful fish occupy this reef in Fiji.

in intensity and decrease with depth. Growth rates of coral-containing photosynthetic zooxanthellae are highest in shallow depths; and a few species, especially the branching corals, can dominate the reef by overgrowing and shading the crustose corals and algae. Disturbances by wave action, storms, and grazing reduce the rate of competitive displacement among corals. Heavy grazing of overgrowing algae by sea urchins and fish, such as parrotfish, increases encrusting coralline algae. Light grazing allows rapidly growing filamentous and foliose algal species to eliminate crustose algae.

Intense disturbance or loss of species can have pronounced, even disastrous effects on coral reefs. In one population irruption a major coral predator, the crown-of-thorns starfish *Acanthaster planci*, ate nearly all of the corals on certain reefs in the western Pacific, destroying the reefs. Another disaster involved the most important predator of coral and algae in the Caribbean, the black sea urchin *Diadema antillarium*. Although it preyed mostly on newly settled young coral, the urchin did create algae-free settling areas for the larvae. An epizootic disease destroyed 95 percent of the *Diadema* population in 1983–1984, shortly after Hurricane Allen had scattered two dominant elkhorn coral species. The absence of black sea urchin allowed dense growths of microalgae to smother newly developing colonies of the corals. The result was the population collapse of these once dominant species along the north Jamaican coast.

Function

Coral are partially photosynthetic and partially heterotrophic. During the day zooxanthellae carry on photosynthesis and directly transfer organic material to coral tissue. At night coral polyps feed on zooplankton, securing phosphates and nitrates and other nutrients for the anthozoans and their symbiotic algae. Thus nutrients are recycled in place between the anthozoans and the algae. In addition, carbon dioxide concentrations in animal tissue enable the coral to extract calcium carbonate to build exoskeletons.

Adding to the productivity of the coral are crustose coralline algae, turf algae, macroalgae, sea grass, sponges, phytoplankton, and a large bacterial population. Coral reefs are among the most highly productive ecosystems on Earth. Net productivity ranges from 1500 to 5000 g C/m²/yr, compared to 15 to 50 g C/m²/yr for the surrounding ocean. Because the coralline community acts as a nutrient trap, offshore coral reefs are oases of productivity within a nutrient-poor sea.

This productivity and the varied habitats within the reef support a high diversity of life—thousands of kinds of invertebrates (some of which, such as sea urchins, feed on coral animals and algae), many kinds of herbivorous fish that graze on algae, and hundreds of predatory species. Some of these predators, such as

the puffers and filefish, are corallivores, feeding on coral polyps. Others lie in ambush for prey in coralline caverns. In addition, there is a wide array of symbionts, such as cleaning fish and crustaceans, that pick parasites and detritus from larger fish and invertebrates.

36.4 HUMAN IMPACT

Humans are strongly attracted to the seashore. Travel brochures show scenic, empty sandy beaches and pristine shore vegetation. Rarely do they show the real picture: crowded beaches backed by strands of hotels, boardwalks, and shops. Recreational and commercial development of seashores, along with intensive seasonal human use, has had a long-term impact on intertidal ecosystems, the severity of which will increase as human populations grow.

This use has had serious effects on intertidal wildlife, especially that of sandy shores. Beach-nesting birds such as the piping plover (*Charadrius melodus*) and the least tern (*Sterna antillarum*) are so disturbed by bathers and dune buggies that both species are in danger of extinction. Other terns and shore birds are subjected to competition for nest sites and to egg predation by rapidly growing populations of large gulls that are highly tolerant of humans and thrive on human garbage. Sea turtles and the horseshoe crab (*Limulus polyphemus*), dependent on sandy beaches for nesting sites, find themselves evicted and are declining rapidly for that reason.

Habitat destruction is only one aspect of human impact on intertidal ecosystems. Another is pollution. Seashore cottages use septic tanks that drain into sandy soil; commercial developments drain all sorts of wastes into the ground; and coastal cities and small towns pour raw sewage into shallow waters off the coast.

Each incoming tide brings into the beaches feces-contaminated water that makes beaches unhealthy for humans and wildlife alike. Tides also carry in old fishing lines, plastic debris and other wastes, and blobs of oil, all hazardous to humans and wildlife.

Oil is probably the most dramatic intertidal pollutant. Always subject to some degree of oil pollution from natural sources, intertidal life suffers most when oil spills off the coast reach the shores (see Chapter 35). The damage to intertidal life depends on the extent and intensity of the spill and the type of oil: heavy crude oil (involved in the Persian Gulf and Prince William Sound spills), diesel oil, or split oils.

Controlling oil pollution on shores is difficult because incoming tides wash more oil up onto the shore.

Millions of sea birds such as cormorants and diving ducks and thousands of marine mammals have been victims of oil pollution. A heavy coating of oil mats feathers, impairing the ability of birds to fly and swim, reduces the feathers' water repellency, and destroys the insulating effects of down. During cold weather a spot of oil as small as a button over a vital organ can induce a bird's death from hypothermia. As they preen to clean their feathers, birds ingest fatal amounts of oil. Marine mammals, especially seals and sea otters, fare no better. Oil destroys the insulating effects of fur, clogs ears and nostrils, and irritates eyes. Many such mammals succumb to hypothermia and dehydration.

Less visible to humans is the damage done to invertebrate and plant life of the shores. Particularly vulnerable to smothering oil pollution on sandy shores are interstitial life and sand crabs. Even after the sand appears to be scrubbed by tides, a layer of oil may still reside several meters below the surface. Some intertidal invertebrates, particularly mollusks, appear to be resistant to oil pollution, but their flesh becomes tainted, and the oil is passed along in the food chain. On rocky shores, barnacles resist oil pollution, but the grazing periwinkles, limpets, and whelks are particularly vulnerable. Oil works its way beneath their shells, causing these gastropods to lose their hold, and they are carried away by the tides. As oil eliminates these grazing invertebrates, the barren patches are colonized by algae and seaweeds, particularly the red algae. They too become encrusted with oil and are torn away by the waves. Aside from damage to particular species, oil pollution in the intertidal zone means a great reduction in species diversity, a simplification of the food web, and an increase in the populations of resistant species.

Human activity, especially dredging, sewage pollution, and overfishing, have degraded coral reefs. The most devastating human activity in reefs is overfishing, especially in the Philippines and the Caribbean. Overfishing removes both algal-feeding herbivorous fish and predatory fish that feed on the competitive algal-feeding sea urchins (*Diadema* spp.). Their removal upsets the complex food webs and interspecific relations on the reef and allows algae to overgrow the coral structures. The heavy growth of algae precludes the settlement of coral on the reef. Dynamiting reefs to capture fish for the marine aquarium trade destroys both the reefs and their fish populations.

Recovery of coral from disturbances is slow, often requiring 25 to 100 years. The rate of recovery depends upon the extent of disturbance, the nearness of a source of larvae, and favorable currents to sweep the larvae to substrates free of algal growth.

STUDY QUESTIONS

1. Describe the three major zones of the rocky shore.
2. What adaptations enable inhabitants of rocky shores to avoid desiccation, maintain position, and survive flooding?
3. What roles do predation and competition play in the rocky shore community?
4. C. M. Yonge, the British marine ecologist, called tide pools "microcosms of the sea" except for two significant environmental factors. What are they?
5. What are the major zones of life on sandy shores? How do they differ from those of the rocky shore?
6. How does life on a sandy shore survive in the harsh environment?
7. Contrast the energy source of a sandy or muddy shore with that of the rocky shore. What are the basic consumers on sandy shores?
8. What are coral reefs, and how do they form?
9. Explain how the symbiotic relationship between algae and anthozoans influences the distribution and functioning of corals.
10. Why are coral reefs so productive?
11. In what ways does human activity harm sandy and rocky shores?
*12. Make a list of endangered and threatened shore birds. How does their status relate to human use of the seashore?
*13. Explain how oil affects life on sandy shores and rocky shores. What is the difference?
*14. Report on the ecological and economic effects of some major oil spills, such as the *Amoco Cadiz* off the Brittany coast of France, the *Exxon Valdez* in Prince William Sound, Alaska, and the release of oil into the Persian Gulf during the Gulf War.

ESTUARIES, SALT MARSHES, AND MANGROVE FORESTS

OBJECTIVES

On completion of this chapter, you should be able to:

- Define an estuary and describe its characteristics.
- Relate freshwater and tidal inputs to the functioning of estuaries.
- Describe the major features of a salt marsh.
- Discuss the relationship of the salt marsh to the estuarine ecosystem.
- Discuss the unique features of a mangrove forest.
- Explain the importance of the mangrove forest to the tropical coastal ecosystem.
- Discuss the impact humans have had on the estuary, salt marsh, and mangrove forest.

Mangrove forests in Costa Rica support a variety of invertebrates, including the abundant mangrove crab (*Aratus pisonii*), which lives in the roots above the water line.

37.1 ESTUARIES

Waters of most streams and rivers eventually drain into the sea; and the place where this fresh water joins salt water is called an **estuary.** Estuaries are semienclosed parts of the coastal ocean where seawater is diluted and partially mixed with water coming from the land. Estuaries differ in size, shape, and volume of water flow, all influenced by the geology of the region in which they occur.

As the river reaches the encroaching sea, sediments drop in the quiet water. They accumulate to form deltas in the upper reaches of the mouth and shorten the estuary. When silt and mud accumulations become high enough to be exposed at low tide, tidal flats develop. These flats divide and braid the original channel of the estuary. At the same time, ocean currents and tides erode the coastline and deposit material on the seaward side of the estuary, also shortening the mouth. If more material is deposited than is carried away, barrier beaches, islands, and brackish lagoons appear.

Structure

Salinity, Temperature, and Nutrient Traps

The one-way flow of streams and rivers into the estuary meets the inflowing and outflowing tides. This meeting sets up a complex of currents that varies with the season, amount of rainfall, tidal oscillations, and winds. The interaction of fresh water and salt water influences the salinity of the estuarine environment.

Salinity varies vertically and horizontally, often within one tidal cycle (Figure 37.1). Salinity may be the same from top to bottom, or it may be completely stratified, with a layer of fresh water on top and a layer of dense salty water on the bottom. Salinity is homogeneous when currents, particularly eddy currents, are strong enough to mix the water from top to bottom. The salinity in some estuaries is homogeneous at low tide, but at flood tide a surface wedge of seawater moves upstream more rapidly than the bottom water. Salinity is then unstable, and density is inverted. The seawater on the surface tends to sink as lighter fresh water rises, and mixing takes place from the surface to the bottom. This phenomenon is known as **tidal overmixing.** Strong winds, too, tend to mix salt water with fresh water in some estuaries; but when the winds are still, the river water flows seaward on a shallow surface over an upstream movement of seawater, more gradually mixing with the salt.

Horizontally, the least saline waters are at the river mouth and the most saline at the sea (see Figure 37.1). Incoming and outgoing currents deflect this

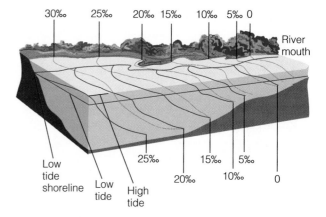

Figure 37.1 Vertical and horizontal stratification of salinity in an estuary at low and high tides. At high tide incoming seawater increases salinity toward the river mouth; at low tide salinity is reduced. Note also how salinity increases with depth, because lighter fresh water flows over denser salt water.

configuration. In all estuaries of the Northern Hemisphere, outward-flowing fresh water and inward-flowing seawater are deflected to the right because of Earth's rotation. As a result, salinity is higher on the left side.

The salinity of seawater is about 35‰; that of fresh water ranges from 0.065 to 0.30‰. Because the concentration of metallic ions carried by rivers varies from drainage to drainage, the salinity and chemistry of estuaries differ. The portion of dissolved salts in the estuarine waters remains about the same as that of seawater, but the concentration varies in a gradient from fresh water to sea.

Exceptions to these conditions exist in regions where evaporation from the estuary exceeds the inflow of fresh water from river discharge and rainfall (a negative estuary). This condition causes the salinity to increase in the upper end of the estuary, and horizontal stratification is reversed.

Temperatures in estuaries fluctuate considerably, both daily and seasonally. Waters are heated by the sun and inflowing and tidal currents. High tide on the mudflats may heat or cool the water, depending on the season. The upper layer of estuarine water may be cooler in winter and warmer in summer than the bottom, a condition that, as in a lake, will cause spring and autumn overturns.

Mixing waters of different salinities and temperatures creates a counterflow that works as a nutrient trap (Figure 37.2). Inflowing river waters more often than not impoverish rather than fertilize the estuary, except for phosphorous. Instead, nutrients and oxygen

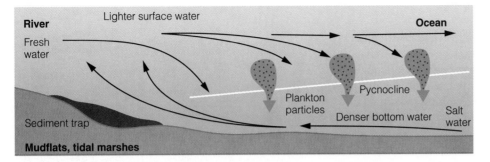

Figure 37.2 Circulation of fresh and salt water in an estuary creates a nutrient trap. A salty wedge of intruding seawater on the bottom produces a surface flow of lighter fresh water and a counterflow of heavier brackish water. This countercurrent traps nutrients, recirculating them toward the tidal marsh. The same countercurrent also sends phytoplankton up the estuary, repopulating the water.

are carried into the estuary by the tides. If vertical mixing takes place, these nutrients are not swept back out to sea but circulate up and down among organisms, water, and bottom sediments (Figure 37.3).

When nutrients are high in the upper estuary, they are taken up rapidly by tidal marshes and mudflats. These areas tend to trap particulate nitrogen and phosphorus, convert them to soluble forms, and export them back to open waters of the estuary. Plants on the tidal marshes and mudflats act as nutrient pumps between bottom sediment and surface water.

Estuarine Organisms

Organisms inhabiting the estuary face two problems—maintenance of position and adjustment to changing salinity. Most estuarine organisms are benthic. They attach to the bottom, bury themselves in the mud, or occupy crevices and crannies about sessile organisms. Mobile inhabitants are mostly crustaceans and fish, largely young of species that spawn offshore in high-salinity water.

Planktonic organisms are wholly at the mercy of the currents. Because the seaward movements of streamflow and ebb tide transport plankton out to sea, the rate of circulation or flushing time determines the size of the plankton population. If the circulation is too vigorous, the plankton population may be small. Phytoplankton in summer is mostly near the surface and in low-salinity water. In winter phytoplankton is more uniformly distributed. For plankton to become endemic in an estuary, reproduction and recruitment must balance loss from physical dispersal.

Salinity dictates the distribution of life in the estuary. Essentially, the organisms of the estuary are marine, able to withstand full seawater. Except for anadromous fish, no freshwater organisms live there. Some estuarine inhabitants cannot withstand lowered

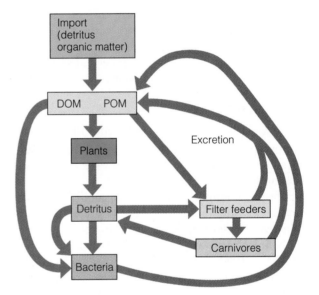

Figure 37.3 The estuary is mostly heterotrophic, depending on organic material from sea and land. This diagram shows how organic matter cycles in the estuary.

salinities, and these species decline along a salinity gradient. Sessile and slightly motile organisms have an optimum salinity range within which they grow best. When salinities vary on either side of this range, populations decline.

Size influences the range of motile species within estuarine waters, particularly fish. Some species, such as the striped bass (*Morone saxatilis*), spawn near the interface of fresh and low-salinity water (Figure 37.4). The larvae and young fish move downstream to more saline waters as they mature. Thus, for the striped bass the estuary serves as both a nursery and as a feeding ground for the young. Anadromous species such as the shad (*Alosa*) spawn in fresh water, but the young fish spend their first summer in the estuary, then move out

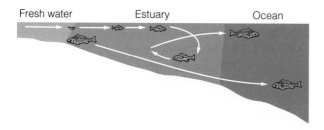

Figure 37.4 The relationship of a semianadromous fish, the striped bass, to the estuary. Adults live in the marine environment, but young fish grow up in the estuary.

to the open sea. Species such as the croaker (Sciaenidae) spawn at the mouth of the estuary, but the larvae are transported upstream to feed in plankton-rich low-salinity areas. Still others, such as the bluefish (*Pomatomus saltatrix*), move into the estuary to feed. In general, marine species drop out toward fresh water and are not replaced by freshwater forms. In fact, the mean number of species decreases progressively from the mouth of the estuary to upstream.

Salinity changes often affect larval forms more severely than adults. Larval veligers of the oyster drill *Thais* succumb to low salinity more easily than the adults. A sudden influx of fresh water, especially after hurricanes or a heavy rainfall, sharply lowers the salinity and causes a high mortality of oysters and their associates.

The oyster bed and oyster reef are the outstanding communities of the estuary. The oyster is the dominant organism about which life revolves. Oysters may be attached to every hard object in the intertidal zone, or they may form reefs, areas where clusters of living organisms grow cemented to the almost buried shells of past generations. Oyster reefs usually lie at right angles to tidal currents, which bring planktonic food, carry away wastes, and sweep the oysters clean of sediment and debris. Closely associated with oysters are encrusting organisms such as sponges, barnacles, and bryozoans, which attach themselves to oyster shells and depend on the oyster or algae for food.

Function

Estuarine food webs are both plankton-based and detritus-based. The producer component, particularly in the middle and lower estuary, consists of dinoflagellates and diatoms. The latter convert some of the carbon intake to higher-calorie fats and lipids rather than the low-energy carbohydrates typical of most green plants. This fat provides a high-energy food base for higher trophic levels.

Although inflowing water from rivers and coastal marshes carries nutrients into the estuary, phytoplankton production is regulated more by internal nutrient cycling than by external sources. This internal cycling involves excretion of mineralized nutrients by herbivorous zooplankton (see Chapter 23) and release of nutrients remineralized by invertebrates of the bottom sediments, by the roiling of sediments, and by steady-state exchanges between nutrients present in particulate and dissolved phases.

Nutrients accumulated over winter in temperate estuaries stimulate a winter-spring bloom. As the nutrients become depleted and zooplankton prey intensively on the phytoplankton, the bloom collapses. The phytoplankton falls to the bottom, where bivalves and other filter-feeding invertebrates feed upon it. In well-mixed estuaries, nutrients remineralized in the benthos return to the water column, stimulating a summer bloom.

In shallow estuarine waters, rooted aquatics such as widgeon grass and eelgrass (*Zostera marina*) assume major importance. These aquatic plants are complex systems supporting a large number of epiphytic and epizoic organisms. Such communities are important to certain vertebrate grazers, such as brant, Canada geese, the black swan in Australia, and sea turtles, and they provide a nursery ground for shrimp and bay scallops.

Human Impact

For centuries estuaries have carried the burden of human disturbance. Because they are semienclosed, natural harbors, located at the juncture of navigable rivers and the sea, estuaries have long been sites for cities and industry. As a result of this concentration of human activity, estuaries have become dumping grounds for untreated sewage, industrial effluents and wastes, and chronic oil pollution from ships and industry. Because rivers draining vast inland areas flow into them, estuaries receive a load of pollutants, from toxic chemicals, pesticides, and sewage to silt from mining, agriculture, construction, and lumbering. The commercial and recreational values of estuaries stimulate bulkheading, channel dredging, and filling to create waterfront industrial, recreational, and residential sites. Thermal power plants pour heated effluents into the water, raising the temperature near the shore. This warm water and the erosion of bottom sediments by its discharge reduce oxygen content of the estuarine water. All of these activities alter the flow of tidal currents, increase the anoxic conditions of the estuarine bottom, and in other ways change the nature of the estuary.

These intrusions have a pronounced ecological and economic effect on the estuary. Decreased oxygen on the bottom favors only a few benthic organisms, such as polychaete worms, and eliminates others, such as oysters. Excessive inputs of phosphorus stimulate the explosive growth of a few algae at the expense of others, as well as the spread of nonendemic plants. This reduction in species diversity simplifies the natural food web, affecting all life in the estuary.

Estuaries, rich sources of food and safe havens from predators, are major nurseries for commercially and recreationally important finfish and shellfish species on the Atlantic coast such as mullet, striped bass, flounder, shrimp, blue crab, Alaska crab, and clams. Pollution, salinity changes, turbidity, and loss of food have reduced the fishery resources. What remains is often contaminated with toxic bacteria and concentrations of toxic elements such as mercury, cadmium, and lead. These contaminants, as well as oil that taints the flesh of fish and shellfish, make them unfit for consumption.

37.2   SALT MARSHES

On the alluvial plains about the estuary and in the shelter of spits and offshore bars and islands exists a unique community, the salt marsh. Although salt marshes look like acres of waving grass, they are a complex of distinctive and clearly demarked plant associations. The reasons for this complex are tides and salinity.

The tides play the most significant role in plant segregation, for two times a day the salt-marsh plants on the outermost fringes are submerged in salty water and then exposed to full sun. Their roots extend into poorly drained, poorly aerated soil in which the soil solution contains varying concentrations of salt. Only plant species with a wide range of salt tolerances can survive such conditions.

Structure

Plants
From the edge of the sea to the high land, zones of vegetation distinctive in form and color develop, reflecting a microtopography that lifts the plants to various heights within and above high tide (Figure 21.2). Most conspicuous on the seaward edge of East Coast marshes and along tidal creeks are the deep green growths of *Spartina alterniflora*, which dominate the low marsh. Stiff, leafy, up to 3 m tall, and submerged in salt water at each high tide, salt marsh cordgrass forms a marginal strip between the open mud to the front

and the high marsh behind. No litter accumulates beneath the stand. Strong tidal currents sweep the floor of the *Spartina* clean, leaving only thick, black mud.

Spartina alterniflora is well adapted to grow on the intertidal flats of which it has sole possession. It has a high tolerance for salt water and is able to live in a semisubmerged state. It can live in a saline environment by selectively concentrating sodium chloride in its cells at a level higher than the surrounding salt water, thus maintaining its osmotic integrity. To rid itself of excessive salts, salt marsh cordgrass has special salt-secreting cells in the leaves. Water excreted with the salt evaporates, leaving behind sparkling crystals on the surface of leaves to be washed off by tidal water. To get air to its roots, buried in anaerobic mud, *Spartina alterniflora* has hollow tubes leading from the leaf to the root, through which oxygen diffuses.

Above and behind the low marsh is the high marsh, standing at the level of mean high water. At this level the tall *Spartina* gives way rather sharply to a short form of *Spartina alterniflora*, yellowish, almost chlorotic in appearance, contrasting with the tall, dark green form. This short form seemingly represents a phenotypic plastic response to environmental conditions of the high marsh. The high marsh has a higher salinity, a decreased input of nutrients, and an accumulation of toxic wastes, the result of lower tidal exchange rates. The shorter, more open canopy of the high marsh results in higher leaf and soil temperatures and a higher rate of evaporation than in the low marsh. These conditions, along with the decreased dominance of *Spartina*, allow other marsh plants to grow. Here are the fleshy, translucent glassworts (*Salicornia* spp.) (Figure 37.5) that turn bright red in fall, sea lavender (*Limonium carolinianum*), spearscale (*Atriplex patula*), and sea blite (*Suaeda maritima*).

Where the microelevation is about 5 cm above mean high water, short *Spartina alterniflora* and its associates are replaced by salt meadow cordgrass, *Spartina patens*, and an associate spikegrass or saltgrass, *Distichlis spicata*. Salt meadow cordgrass is a fine grass that grows so densely and forms such a tight mat that few other plants can grow with it (Figure 37.6).

As the microelevation rises several more centimeters above mean high tide and if there is some intrusion of fresh water, *Spartina* and *Distichlis* may be replaced by two species of black needlerush or black grass (*Juncus roemerianus* and *Juncus gerardi*), so called because their dark green color becomes almost black in the fall. Beyond the black grass and often replacing it is a shrubby growth of marsh elder (*Iva frutescens*) and groundsel (*Baccharis halimifolia*). These shrubs tend

Figure 37.5 Glasswort dominates highly saline areas on the salt marsh. The plant, which turns red in fall, is a major food of overwintering geese.

Figure 37.7 A tidal creek at low tide on the high marsh. Tall *Spartina* grows along the banks.

Figure 37.6 The distinctive cowlick sweep of salt meadow cordgrass next to a stand of saltgrass. Bayberry in the background marks the shrub zone.

to invade the high marsh where a slight rise in micro-elevation exists, but such invasions are often short-lived, as storm tides sweep in and kill the plants. On the upland fringe grow bayberry *(Myrica pensylvanica)* and the pink-flowering sea holly *(Hibiscus palustris)*.

Two conspicuous features of a salt marsh are the meandering creeks and the salt pans. The creeks form an intricate system of drainage channels that carry tidal waters back out to sea (Figure 37.7). The salt pans are circular to elliptical depressions. At high tide they are flooded. At low tide they remain filled with salt water. If they are shallow enough, the water may

evaporate completely, leaving an accumulating concentration of salt on the mud.

Salt pans support distinctive vegetation, which varies with the depth of the water and the salt concentration. Pools with a firm bottom and sufficient depth to retain tidal water support dense growths of widgeon grass *(Ruppia maritima)*, whose small black seeds are relished by waterfowl.

Shallow depressions in which water evaporates are covered with a heavy algal crust and crystalized salt. The edges of these salt flats may be invaded by *Salicornia*, *Distichlis*, or even short *Spartina alterniflora* (see Figure 37.5).

The exposed banks of tidal creeks that braid through the salt marshes support a dense population of mud algae, the diatoms and dinoflagellates, photosynthetically active all year. Photosynthesis is highest in summer during high tides and in winter during low tides, when the sun warms the mud. Some of the algae are washed out at ebb tide and become part of the estuarine plankton available to such filter feeders as oysters.

The salt marsh described is typical of the North American Atlantic coast. Many variations exist locally and latitudinally around the world. North America has several distinctive types. Arctic salt marshes have few species. East and Gulf coast marshes dominated by *Spartina* and *Juncus* reach their best development on the heavy silt deposits from North Carolina south. Pacific coast marshes, poorly developed, are dominated up to mean high-water level by *Spartina foliosa*, and above that by *Salicornia*. Salt marshes on the western coast of Europe and in Great Britain are dominated by *Salicornia* and *Puccinellia*. Those of northwestern Europe are dominated by *Spartina angelica*, a hybrid polyploid species that spontaneously formed from the native *Spartina townsendii*, and *Spartina alterniflora*, introduced from North America. Because of its vigorous growth, *S. angelica* is widely planted to stabilize and bind coastal mudflats.

Consumers

Animal life of the salt marsh, not noted for its diversity, is outstanding for its interesting organisms. Some of the inhabitants are permanent residents in sand and mud; others are seasonal visitors; and most are transients coming to feed at high and low tide.

Three dominant animals of the low marsh are the ribbed mussel (*Modiolus demissus*), fiddler crab (*Uca pugilator* and *U. pugnax*), and marsh periwinkle (*Littorina* spp.). The marsh periwinkle, related to the periwinkles of the rocky shore, moves up and down the stems of *Spartina* and onto the mud to feed on algae and detritus. Buried halfway in the mud is the ribbed mussel. At low tide the mussel is closed; at high tide the mussel opens to filter particles from the water, accepting some and rejecting others in a mucous ribbon known as pseudofeces.

Running across the marsh at low tide like a vast herd of tiny cattle are the fiddler crabs. Among the marsh animals they are the most adaptable. They have both lungs and gills. They can endure high tides and cold winters without oxygen. They have a salt-water control system that enables them to move from diluted seawater to briny pools. Omnivorous feeders, they eat detrital material, algae, and small animals. Fiddler crabs live in burrows, marked by mounds of freshly dug, marble-sized pellets. Like earthworms, burrowing crabs bring up nutrients to the surface.

Prominent about the base of *Spartina* and under debris are the sandhoppers (*Orchestia*). These detritus-feeding amphipods may be abundant and are important in the diet of some marsh birds.

Three conspicuous vertebrate residents of the low marsh of eastern North America are the diamond-backed terrapin (*Malaclemys terrapin*), the clapper rail (*Rallus longirostris*), and the seaside sparrow (*Ammospiza maritima*). The terrapin feeds on fiddler crabs, small mollusks, marine worms, and dead fish. The rail eats fiddler crabs and sandhoppers. The sparrow eats sandhoppers and other small invertebrates.

On the high marsh, animal life changes as suddenly as the vegetation. The small, coffee-bean-colored pulmonate snail *Melampus*, found by the thousands under the low grass, replaces the marsh periwinkle. Meadow mice (*Microtus*) have a maze of runways within the matted growth. Replacing the clapper rail and seaside sparrow on the high marsh are the willet (*Catoptrophorus semipalmatus*) and the seaside sharp-tailed sparrow (*Ammospiza caudacuta*).

Among the shrubby fringes of the marsh, dense growth of marsh elder, groundsel, and wax myrtle gives nesting cover for blackbirds and provides sites for heron rookeries. Remote stands of these shrubs support the nests of smaller herons and egrets, whereas tall dead pines and human-made structures support the nests of fish-eating osprey.

Low tide brings a host of predaceous animals onto the marsh to feed. Herons, egrets, gulls, terns, willets, ibis, raccoons, and others spread over the exposed marsh floor and muddy banks of tidal creeks to feed. At high tide the food web changes as the tide waters flood the marsh. Such fish as silversides (*Menidia menidia*), killifish (*Fundulus heteroclitus*), and four-spined stickleback (*Apeltes quadracus*), restricted to channel waters at low tide, spread over the marsh at high tide, as does the blue crab.

Function

The salt marsh is a detrital system; grazing herbivores play a minuscule role. In Georgia salt marshes, the salt-marsh grasshopper (*Orchelimum*), which eats plant tissues, and the plant hopper (*Prokelisis*), which sucks plant juices, use only about 3 to 4 percent of net production. Most of the detrital material from *Spartina* of the low marsh is washed out by the tide; that of the high marsh is decomposed in place by bacteria. Feeding on the detritus and the bacteria it supports are fiddler crabs and mussels. On a New England marsh, which has few fiddler crabs and mussels, most of the *Spartina* is broken up by winter ice instead of fiddler crabs. Much of the detrital material is carried out to the bay, where it is fed on by a major detritivore, the grass shrimp (*Palaemonetes pugio*). It breaks the coarse detritus into smaller pieces that are colonized by bacteria and diatoms. By its action the grass shrimp makes nutrients and biomass available for higher trophic levels.

The salt marsh, one of the most productive ecosystems, has become that way in part because of a tidal subsidy. Tidal flushing brings in new nutrients and sweeps out accumulated salts, metabolites, sulfides, and toxic wastes. It replaces anoxic interstitial water with oxygenated water. Added to this advantage is a tight internal nutrient cycling. Algal and bacterial populations turn over rapidly, and detrital material is fragmented and decomposed. Up to 47 percent of net primary production is respired by microbes; part is grazed by nematodes and microscopic benthic organisms, both of which are consumed by deposit feeders.

Depending on the nature of the salt marsh, between 11 and 66 percent of decomposed marsh grass is converted to microbial biomass. Belowground production enters an anaerobic food web of fermentation, dominated by sulfur-oxidizing bacteria that reduce sulfates and produce methane.

What happens to excess carbon production in a salt marsh is not well understood. Each salt marsh apparently differs in the way carbon is transformed in the food web and its route and amount of export. Some salt marshes are dependent on tidal exchanges and import more than they export, whereas others export more than they import. Some of the excess production goes into sediments; some may be transformed microbially in the water in the marsh and tidal creeks. A portion may be exported to the estuary physically as detritus, as bacteria, or as fish, crabs, and intertidal organisms in the food web.

The importance of salt marshes as a nutrient source and sink for the estuary, especially nitrogen, has been a question of long standing. In the salt marsh, denitrification often exceeds nitrogen fixation, and the marsh depends on inputs into the system. The amount of nitrogen cycled is determined by tidal input, physical and chemical exchanges with air and water, and biological fluxes. In some salt marshes, exemplified by the Great Sippewissett marsh in Massachusetts, groundwater inflow brings in nitrates, some of which percolate through the peat and are exported to the estuary as organic nitrogen, ammonium, and nitrates. Of the total influx, one-third is exported by denitrification and two-thirds by tides. The much larger salt marshes of Sapelo Island, Georgia, depend on inputs of nitrogen from the associated river and tides. Only in summer does the system appear to export any nitrogen. Mostly the marsh is a net sink for nitrogen.

Human Impact

Fifty-one percent of the population of the contiguous United States and 70 percent of all humanity live within 80 km of the coastlines. With so much humanity clustered about the coast for many reasons, it is obvious why coastal ecosystems are threatened and are disappearing rapidly. In spite of some efforts at regulation and acquisition at state and federal levels to slow the loss, coastal wetlands in the United States are still disappearing at a rate of 8000 ha a year. Commonly regarded as economic wastelands, salt marshes have been and are being ditched, drained, and filled for real estate development (everyone likes to live at the water's edge), industrial development, and agriculture. Reclamation of marshes for agriculture is most extensive in Europe, where the high marsh is enclosed within a sea wall and drained. Most of the marshland and tideland in Holland has been reclaimed in this fashion. Many coastal cities such as Boston, Amsterdam, and much of London have been built on filled-in marshes. Salt marshes close to urban and industrial developments become polluted with spillages of oil, which becomes easily trapped within the vegetation (Figure 37.8).

The most rapid loss of wetlands in the United States is taking place on the coast of Texas, Mississippi, and Louisiana. Two-thirds of the loss has been to deep water, the remaining third to urban development. At the present rate of destruction, about 125 km²/yr, most of the Louisiana coastal marshes will be gone in two decades. Canals laced through the marshes to accommodate oil-drilling rigs allow the intrusion of salt water and disrupt natural hydrology. Flood-control structures along the Mississippi interfere with the buildup of sediments in the delta areas necessary to maintain salt marshes along a sinking coastline. This sinking is further aggravated by extracting oil, gas, and groundwater, which allows the surface to subside.

Losses of coastal wetlands have a pronounced effect on the salt marsh and associated estuarine ecosystems. They are the nursery ground for commercial and recreational fisheries. There is, for example, a correlation between the expanse of coastal marsh and shrimp production in Gulf coastal waters. Oysters and blue crabs are marsh-dependent, and the decline of these important species is related to loss of salt marshes. Coastal marshes are major wintering grounds for waterfowl. One-half of the migratory waterfowl of the Mississippi Flyway depend on Gulf Coast wetlands,

Figure 37.8 Always targets for drainage, filling, and development, coastal marshes are often considered wastelands, as epitomized by this For Sale sign.

and the bulk of the snow goose population winters on the coastal marshes from the Chesapeake Bay to North Carolina. These geese may remove by grazing or uprooting nearly 60 percent of the belowground production of marsh vegetation. Forced concentration of these wintering migratory birds into shrinking salt-marsh habitats could jeopardize marsh vegetation and the future of the birds.

37.3   MANGROVE FORESTS

Replacing salt marshes on tidal flats in tropical regions are **mangrove forests** or **mangals,** which cover 60 to 75 percent of the coastline of the tropical regions. Mangals develop where wave action is absent, sediments accumulate, and the muds are anoxic. They extend landward to the highest vertical tidal range, where they may be only periodically flooded. The dominant plants are mangroves, which include eight families and 12 genera dominated by *Rhizophora, Avicennia, Bruguiera,* and *Sonneratia.* Growing with them are other salt-tolerant plants, mostly shrubs. Mangals reach their finest development and have the most species in the Indo-Malaysian region.

In growth form, mangroves range from short, prostrate forms to timber-size trees 30 m high. All mangroves have shallow, widely spreading roots, many with prop roots coming from trunk and limbs (Figure 37.9). Many species have root extensions called pneumatophores that take in oxygen for the roots. Their fleshy leaves, although tough, are often succulent and may have salt glands. Many mangrove species have a unique

Figure 37.9 Interior of a red mangrove stand. Note the prop roots.

method of reproduction, vivipary. Following fertilization, the embryos undergo uninterrupted development with no true resting seed. They grow into a seedling on the tree, drop to the water, and float upright until they reach water shallow enough for their roots to penetrate the mud.

Structure

The formation and appearance of mangrove forests are strongly influenced by the range and duration of tidal flooding. Mangals become established in areas where the lack of wave action allows fine sediments or mud to accumulate. The tangle of prop roots and pneumatophores further slows the movement of tidal waters, allowing more sediments to settle out. Land begins to march seaward, followed by colonizing mangroves.

A feature of this response is zonation, the changes in vegetation from seaward edge to true terrestrial environment. Although often used as an example, the mangals of the Americas, especially Florida, have the least pronounced zonation, largely because of the few species involved. The pioneering red mangrove (*Rhizophora mangle*) occupies the seaward edge and receives the deepest tidal flooding. Red mangroves are backed by black mangroves (*Avicennia germinanas*), shallowly flooded by high tides. The landward edge is dominated by white mangroves (*Laguncularia racemosa*), along with buttonwood (*Conocarpus erectus*), a nonmangrove species that acts as a transition to terrestrial vegetation.

Mangals of the Indo-Malaysian region, containing up to 40 species, have a more pronounced, although variable, zonation. The seaward fringe is dominated by one or several species of *Avicennia* and perhaps trees of the genus *Sonneratia* that do not grow well in the shade of other species. Behind the *Avicennia* is a zone of *Rhizophora*, characteristic mangrove species with prop roots and pneumatophores that grow in areas covered by daily high tides up to a point covered only by the highest spring tides. At and beyond the level of high spring tides is a broad zone of tall *Bruguiera*. The final and often indefinite mangrove zone is an association of small shrubs, mainly *Ceriops*.

Mangals are faunally rich, with a unique mix of terrestrial and marine life. Living and nesting in the upper branches are numerous species of birds, particularly herons and bitterns. As in the salt marsh, *Littorina* snails live on the prop roots and trunks of mangrove trees. Attached to the stems and prop roots are barnacles, and on the mud at the base of the roots are detritus-feeding snails. Fiddler crabs and tropical land crabs burrow into the mud during low tide and live on prop roots and high ground during high tide. In the Indo-Malaysian mangrove forests live mud skippers, fish of the genus *Periophthalmus*, with modified eyes set high on the head. They live in burrows in the mud and crawl about on the top of it. In many ways they act more like amphibians than fish. The sheltered waters about the roots provide a nursery and haven for the larvae and young of crabs, shrimp, and fish.

Function

Net productivity of mangrove forests is variable, ranging in Florida mangals from about 450 g C/m²/yr to 2700 g C/m²/yr. Productivity is influenced by tidal inflow and flushing, water chemistry, salinity, and soil nutrients, much as in the salt marsh. Highest rates of productivity occur in those mangrove forests under the influence of daily tides. For example, gross primary productivity of red mangroves, flooded daily by high tides, decreases with salinity, whereas gross primary productivity of white and black mangroves increases with increasing salinity. In areas of intermediate salinity, white mangroves have twice the productivity of red mangroves. Zonation of mangrove forests appears to reflect the optimal productivity niches of the species involved rather than physical or successional conditions.

Mangrove forests export a considerable portion of their production to the surrounding waters, largely as leaf fall and other detrital material. This material becomes the base of a detrital food web that supports an array of important commercial finfish and shellfish. In Florida, for example, these species include striped mullet, spotted seatrout, red drum, blue crab, and shrimp.

Human Impact

Mangrove forests have been exploited to some degree, mostly as firewood, since antiquity. Such exploitation probably destroyed the mangrove forests of the Red Sea and Persian Gulf. Large mangroves have been cut for fence posts, poles, and timber. In the Indo-Malaysian region, where mangroves are an important timber resource, silvicultural management systems have been developed for them. Pulp of certain species is used in the manufacture of rayon, lacquers, and cellulose acetate, and tannin is extracted from the bark. In the past, at least, bark and fruit have been used as a source of medicinals for treatment of rheumatism, boils, and eye infections. In addition, mangroves stabilize and protect the coast against erosion and shelter and support important commercial fisheries.

In spite of their value, mangrove forests have been and are being destroyed by filling and dredging for commercial development, marinas, and condominiums. They even are used as solid-waste dumps. In parts of the Pacific, especially the Philippines, mangals have been cleared for rice lands and for mariculture—raising fish, crabs, and prawns in brackish pools. The most massive destruction of mangals took place in Vietnam, where an estimated 100,000 ha were destroyed by herbicidal spraying during the Vietnam War. The mangals have never recovered.

STUDY QUESTIONS

1. What is an estuary?
2. What is the relationship of inflowing rivers and tides to salinity in the estuary?
3. How does tidal overmixing work?
4. How is the estuary a nutrient trap?
5. Why is the estuary so vulnerable to pollution?
6. What influences the major structural features of a salt marsh?
7. What accounts for the high productivity of the salt marsh?
8. What are the typical fauna of the salt marsh?
9. What are the unique characteristics and adaptations of mangroves?
10. Describe the zonation of mangals.
11. What is the ecological and economic importance of mangals?
12. What is the fate of many mangrove forests?
*13. There is growing concern over the condition of estuaries. What has prompted states to act on the problem? What are some of the solutions? Are these solutions universally accepted? Why?
*14. A battle is forming between those who would preserve coastal wetlands and those who would destroy them. Build up a set of arguments for both sides. What side in the long term has the strongest case?
*15. If you live or vacation along the coast, investigate the conditions of and attitudes toward salt marshes in that area. Of what importance are those salt marshes to wintering waterfowl? To the local economy?
*16. Investigate the importance of mangrove forests to a selected area—West Africa, Australia, Indo-Malaysia, the Philippines.

REFERENCES

CHAPTER 1 The Nature of Ecology

Bates, Marston. 1956. *The nature of natural history.* New York: Random House.

———. 1960. *The forest and the sea.* New York: Random House. Both of these classic books, available in many editions, provide a delightful introduction to ecology.

Brower, J. E., and J. H. Zar. 1984. *Field and laboratory methods for general ecology.* 2d ed. Dubuque, IA: Wm. Brown.

Cloud, P. E. 1988. Gaia modified. *Science* 240:1716.

Cox, G. W. 1985. *Laboratory manual for general ecology.* 5th ed. Dubuque, IA: Wm. Brown.

Egerton, F. N., ed. 1977. *History of American ecology.* New York: Arno Press.

Golley, F. B. 1993. *A history of the ecosystem concept in ecology: More than the sum of its parts.* New Haven: Yale University Press.

Hagen, J. B. 1992. *An entangled bank: The origins of ecosystem ecology.* New Brunswick, NJ: Rutgers University Press.

Kingsland, S. 1995. *Modeling nature.* Chicago: University of Chicago Press. A history of the development of population ecology.

Langenheim, J. H. 1966. Early history and progress of women ecologists: Emphasis on research contributions. *Annual Review of Ecology and Systematics* 27:1–53. Women were important in the development of ecology.

Lovelock, J. E. 1979. *Gaia: A new look at life on earth.* New York: Oxford University Press.

McIntosh, R. P. 1985. *The background of ecology: Concept and theory.* New York: Cambridge University Press. Excellent analysis of ecology in North America.

Sheall, J. (ed.). 1988. *Seventy-five years of ecology: The British Ecological Society.* Oxford: Blackwell Scientific Publishers. History of British ecology.

Worster, D. 1977. *Nature's economy.* San Francisco: Sierra Club Books. An engaging history of ecology with a different view from that of McIntosh.

CHAPTER 2 The Organism and Its Environment

Brown, J. H., and A. C. Gibson. 1983. *Biogeography.* St. Louis: C. V. Mosby. See Chapters 2 and 3.

Root, R. B. 1967. The niche exploitation of the blue-gray gnatcatcher. *Ecological Monographs* 37:317–350.

Root, T. 1988. *Atlas of wintering North American birds: An analysis of Christmas bird count data.* Chicago: University of Chicago Press.

Whittaker, R. H., S. A. Levi, and R. B. Root. 1973. Niche, habitat, and ecotope. *American Naturalist* 107:321–338.

Wiens, J. A. 1973. Pattern and process in grassland bird communities. *Ecological Monographs* 43:237–270.

Wiens, J. A., and J. T. Rotenberry. 1981. Habitat associations and community structure of birds in shrub-steppe environments. *Ecological Monographs* 51:21–41.

CHAPTER 3 Key Processes of Exchange

Anderson, J. M., and A. Macfadyen (eds.). 1976. *The role of terrestrial and aquatic organisms in decomposition processes.* Oxford: Blackwell Scientific Publications.

Mooney, H. A. 1986. Photosynthesis. Chapter 11 in M. J. Crawley, ed. *Plant ecology.* New York: Blackwell Scientific Publications. An excellent synthesis.

Swift, M. J., O. W. Heal, and J. M. Anderson. 1979. *Decomposition in terrestrial ecosystems.* Berkeley: University of California Press.

CHAPTER 4 Climate

Berry, R. G., and R. J. Corley. 1992. *Atmosphere, weather, and climate,* 5th ed. New York: Routledge.

Gates, D. M. 1962. *Energy exchange in the biosphere.* New York: Harper & Row.

Gates, D. M. 1972. *Man and his environment: Climate.* New York: Harper & Row. An excellent introduction.

Geiger, R. 1965. *Climate near the ground.* Cambridge, MA: Harvard University Press. A classic book on microclimate.

Landsberg, H. E. Man-made climatic changes. *Science* 170: 1265–1274.

Lee, R. 1978. *Forest microclimatology.* New York: Columbia University Press. Looks at climate within the forest.

Schaefer, V. J., and J. A. Day. 1981. *A field guide to the atmosphere.* Boston: Houghton Mifflin. A guide to all aspects of the ocean of air.

Schneider, S. H. 1989. *Global warming.* San Francisco: Sierra Club Books. A sound discussion of the topic.

Strahler, A. 1971. *The earth sciences.* New York: Harper & Row. Detailed presentation of climate, among other topics.

CHAPTER 5 Light

Bainbridge, R., G. C. Evans, and O. Rackham (eds.). 1966. *Light as an ecological factor.* Oxford, England: Blackwell Scientific Publications. Dated, but still an outstanding reference.

Caldwell, M. M., A. H. Teramura, and M. Tevini. 1989. The changing solar ultraviolet climate and the ecological consequences for higher plants. *TREE* 4:363–367.

Chazdon, R. L. 1988. Sunflecks and their importance to forest understory plants. *Advances in ecological research* 18:1–63.

Chazdon, R. L., and R. W. Pearcy. 1991. The importance of sunflecks for forest understory plants. *Bioscience* 41:760–765.

Dale, J. E. 1992. How do leaves grow? *Bioscience* 42:423–432.

Day, T. A., T. C. Vogelmann, and E. H. DeLucia. 1992. Are some plant life forms more effective than others in screening out ultraviolet radiation? *Oecologica* 92:513–516.

Grime, J. 1971. *Plant strategies and vegetative processes.* New York: Wiley. Excellent discussion of shade tolerance.

Kirk, J. T. O. 1983. *Light and photosynthesis in aquatic systems.* New York: Cambridge University Press.

CHAPTER 6 Temperature

These basic references on plant physiology cover relationships to temperature as well as moisture.

Etherington, J. R. 1982. *Environment and plant ecology*, 2nd ed. New York: Wiley.

Larcher, W. 1996. *Physiological plant ecology*, 3rd ed. New York: Springer-Verlag. Excellent reference on plant ecophysiology.

Long, S. P., and F. I. Woodward (eds.). 1988. *Plants and temperature*. Symposia of the Society for Experimental Biology, Number 42. Cambridge, England: The Company of Biologists Limited. Advanced and comprehensive.

Strain, B. R., and W. B. Billings (eds.). 1975. *Vegetation and the environment*. Vol. 6, *Handbook of vegetation science*. The Hague: W. Junk.

Turner, N. C., and P. J. Kramer (eds.). 1980. *Adaptation of plants to water and high temperature stress*. New York: Wiley.

These basic references on animal physiology have a strong emphasis on ecophysiology.

Heinrich, B. 1979. *Bumblebee economics*. Cambridge, MA: Harvard University Press. Excellent on energetics of bumblebees.

Heinrich, B. (ed.). 1981. *Insect thermoregulation*. New York: Wiley. Temperature regulation by insects, including heterothermy.

Heinrich, B. 1996. *The thermal warriors*. Cambridge: Harvard University Press. Semipopular account of myriad strategies insects use to heat their bodies.

Hill, R. W., and G. A. Wyse. 1989. *Animal physiology*. New York: Harper & Row. A comprehensive text strong on ecophysiology.

Lee, Jr., R. E. 1989. Insect cold-hardiness: To freeze or not to freeze. *Bioscience* 39:308–313. A good discussion of supercooling and cold hardening in insects.

Lyman, C. P., A. Malan, J. S. Willis, and L. C. H. Wang. 1982. *Hibernation and torpor in mammals and birds*. New York: Academic Press.

Schmidt-Neilsen, K. 1997. *Animal physiology: Adaptation and environment*, 5th ed. New York: Cambridge University Press. An excellent reference.

Storey, K. B., and J. M. Storey. 1996. Natural freezing survival in animals. *Annual Review of Ecology and Systematics* 27: 365–386.

CHAPTER 7 Moisture

Blom, C. W. P. M., and L. A. C. J. Voesenek. 1996. Flooding: The survival strategies of plants. *TREE* 11:290–295.

Feldman, L. J. 1988. The habits of roots. *Bioscience* 38:612–618.

Gilles, R. (ed.). 1979. *The mechanisms of osmoregulation in animals*. New York: Wiley.

Hill, R. W., and G. A. Wyse. 1989. *Animal physiology*. New York: Harper & Row. A comprehensive text strong on ecophysiology.

Kozlowski, T. T. 1982. Plant responses to flooding of soil. *Bioscience* 34:162–167.

Kozlowski, T. T. 1984. *Flooding and plant growth*. Orlando, FL: Academic Press.

Larcher, W. 1996. *Physiological plant ecology*, 3rd ed. New York: Springer-Verlag.

Maloly, C. M. O. (ed.). 1979. *Comparative physiology of osmoregulation in animals*. New York: Academic Press.

Schmidt-Neilsen, K. 1997. *Animal physiology: Adaptation and environment*, 5th ed. New York: Cambridge University Press. An excellent reference.

Schulze, E. D., R. H. Robichaux, J. Grace, P. W. Rundel, and J. R. Ehleringer. 1987. Plant water balance. *Bioscience* 37:30–37.

CHAPTER 8 Periodicity

Adkisson, P. L. 1966. Internal clocks and insect diapause. *Science* 154:234–241.

Aschoff, J., S. Daan, and G. A. Gross (eds.). 1982. *Vertebrate circadian rhythms*. New York: Springer-Verlag.

Beck, S. D. 1980. *Insect photoperiodism*, 2nd ed. New York: Academic Press. A good review.

Bioscience. 1983. Biological clocks. Special section in *Bioscience* 33:424–457.

Brady, J. 1982. *Biological timekeeping*. New York: Cambridge University Press.

Bünning, E. 1973. *The physiological clock*, 3rd ed. New York: Academic Press. A book by one of the pioneers in the field.

DeCoursey, P. J. 1960a. Daily light sensitivity rhythm in a rodent. *Science* 131:33–35.

DeCoursey, P. J. 1960b. Phase control of activity in a rodent. *Cold Spring Harbor Symposium on Quantitative Biology* 25:49–54.

DeCoursey, P. J. (ed.). 1976. *Biological rhythms in the marine environment*. Columbia: University of South Carolina Press.

Edmunds, L. N. 1988. *Cellular and molecular bases of biological clocks*. New York: Springer-Verlag.

Farner, D. S. 1985. Annual rhythms. *Annual Review of Physiology* 47:65–82.

Johnson, C. H., and J. W. Hastings. 1986. The elusive mechanisms of the circadian clock. *American Scientist* 74:29–36.

Leith, H. (ed.). 1974. *Phenology and seasonality modeling*. New York: Springer-Verlag.

Moore-Ede, M. C., F. M. Sulzman, and C. A. Fuller. 1982. *The clocks that time us*. Cambridge, MA: Harvard University Press. Circadian rhythms in humans.

Naylor, E. 1985. Tidally rhythmic behaviour of marine animals. *Symposium of Society of Experimental Biology* 39:63–93.

Palmer, J. D. 1976. *An introduction to biological rhythms*. San Diego: Academic Press.

Palmer, J. D. 1990. The rhythmic lives of crabs. *Bioscience* 40: 352–358.

Palmer, J. D. 1995. *The biological rhythms and clocks of intertidal animals*. New York: Oxford University Press.

Palmer, J. D. 1996. Time, tide, and living clocks of marine organisms. *American Scientist* 84:570–578.

Saunders, D. S. 1982. *Insect clocks*, 2nd ed. Elmsford, NY: Pergamon Press.

Takahashi, J. S., and M. Hoffman. 1995. Molecular biological clocks. *American Scientist* 83:158–165.

Winfree, A. T. 1986. *The timing of biological clocks*. Scientific American Library. New York: W. H. Freeman. A popular but challenging discussion of biological clocks.

CHAPTER 9 Nutrients

Aber, J. D., and J. M. Melillo. 1991. *Terrestrial ecosystems*. Philadelphia: Saunders. See Chapters 9–13.

Belovsky, G. E., and P. F. Jordan. 1981. Sodium dynamics and adaptations of a moose population. *Journal of Mammalogy* 63:613–621.

Chapin, F. S. III. 1980. The mineral nutrition of wild plants. *Annual Review of Ecology and Systematics* 11:233–260.

Cheatum, E. L., and C. W. Severinghaus. 1950. Variations in fertility of white-tailed deer related to range conditions. *Transactions of the North American Wildlife Conference* 15:170–189.

Dodd, R. H. 1973. Insect nutrition: Current developments and metabolic implications. *Annual Review of Entomology* 18:381–420.

Heslop-Harrison, Y. 1989. Carnivorous plants. Pages 92–106 in J. L. Gould and C. G. Gould (eds.), *Life at the edge: Readings from Scientific American*. New York: W. H. Freeman.

Hill, E. P. III. 1972. Litter size in Alabama cottontails as influenced by soil fertility. *Journal of Wildlife Management* 36:1199–1209.

Jones, R. L., and H. C. Hanson. 1985. *Biogeochemistry, mineral licks, and geophagy of North American ungulates*. Ames: Iowa State University Press.

Jones, R. L., and H. P. Weeks. 1985. Ca, Mg, and P in the annual diet of deer in south-central Indiana. *Journal of Wildlife Management* 49:129–133.

Schnell, D. E. 1976. *Carnivorous plants of the United States and Canada*. Winston-Salem, NC: John F. Blair, Publisher.

Swift, M. J., O. W. Heal, and J. M. Anderson. 1979. *Decomposition in terrestrial ecosystems*. Berkeley: University of California Press. A major reference.

Weeks, H. P. Jr., and C. M. Kirkpatrick. 1978. Salt preferences and sodium drive phenology in fox squirrels and woodchuck. *Journal of Mammalogy* 59:531–542.

Weir, J. S. 1972. Spatial distribution of elephants in an African national park in relation to environmental sodium. *Oikos* 23:1–13.

CHAPTER 10 Soil

Boul, S. W., F. D. Hole, R. J. McCracken, and R. J. Southward. 1997. *Soil genesis and classification*. Ames, IA: Iowa State University Press. Accessible description of soil genesis, soil groups, and classification.

Brady, N. C., and R. W. Weil. 1996. *The nature and properties of soils*, 11th ed. Upper Saddle River, NJ: Prentice Hall. Excellent basic text on soils.

Brown, J. H., and W. McDonald. 1995. Livestock grazing and conservation on southwestern rangelands. *Conservation Biology* 9:1644–1647.

Brown, L. R. 1984. Conserving soil. Pages 53–73 in L. Strake (ed.), *State of the world 1984*. New York: W. W. Norton.

Brown, L. R., and E. C. Wolf. 1984. *Soil erosion: Quiet crisis in the world economy*. Washington, DC: Worldwatch Institute.

Farb, P. 1959. *The living earth*. New York: Harper and Row. A small classic popular discussion of soil.

Fleischner, T. 1994. Ecological costs of livestock grazing in western North America. *Conservation Biology* 8:629–644.

Furley, P. A., and W. N. Newey. 1983. *Geography of the biosphere*. London: Butterworths. See Chapters 4 and 9 on soils.

Hodgson, J., and A. W. Illius. 1996. *The ecology and management of grazing systems*. Wallingford, Oxon UK: CAB International.

Jenny, H. 1980. *The soil resource*. New York: Springer-Verlag. A definitive reference source.

Killham, K. 1994. *Soil ecology*. New York: Cambridge University Press. Excellent overview of soil ecology with emphasis on soil microbes.

Levin, S. A., ed. 1993. Forum: Grazing theory and rangeland management. *Ecological Applications* 3:1–38.

Lutz, H. J., and R. F. Chandler. 1946. *Forest soils*. New York: Wiley. Dated, but still an excellent reference on native soils under forest growth.

Patton, T. R. 1996. *Soils: A new global view*. New Haven, CT: Yale University Press.

Pimentel, D., C. Harvey, P. Resosudarmo, et al. 1995. Environmental and economic costs of soil erosion and conservation benefits. *Science* 267:1117–1123.

Sears, P. B. 1947. *Deserts on the march*. Norman, OK: Oklahoma University Press. A frequently reprinted classic written after the Dust Bowl of the 1930s. Much of the book still holds true.

Worster, D. 1979. *The dust bowl*. New York: Oxford. An in-depth look at the Dust Bowl, past and present, and how it changed the plains.

CHAPTER 11 Properties of Populations

Begon, M., J. L. Harper, and C. R. Townsend. 1996. *Ecology: Individuals, populations, and communities*. New York: Blackwell Scientific Publishers. See Chapter 4 on unitary and modular populations.

Blower, J. G., L. M. Cook, and J. A. Bishop. 1981. *Estimating the size of animal populations*. London: George Allen & Unwin.

Brower, J. E., and J. H. Zar. 1982. *Field and laboratory methods for general ecology*, 3rd ed. Dubuque, IA. William C Brown.

Caughley, G. 1977. *Analysis of vertebrate populations*. New York: Wiley.

Krebs, C. J. 1989. *Ecological methods*. New York: Harper & Row.

Mueller-Dombois, D., and H. Ellenberg. 1974. *Aims and methods of vegetation ecology*. New York: Wiley.

Southwood, T. R. E. 1978. *Ecological methods*. London: Chapman and Hall.

CHAPTER 12 Life History Patterns

Alcock, J. 1995. *Animal behavior: An evolutionary approach*, 4th ed. Sunderland, MA: Sinauer Associates. Excellent treatment of topics covered in this chapter.

Andersson, M., and Y. Iwasa. 1996. Sexual selection. *Trends in Ecology and Evolution* 11:53–58.

Bajema, C. J. (ed.). 1984. *Evolution by sexual selection theory*. Benchmark Papers in Systemic and Evolutionary Biology. New York: Scientific and Academic Editions.

Boyce, M. S. 1984. Restitution of r and K selection as modes of density-dependent natural selection. *Annual Review of Ecology and Systematics* 15:427–447.

Buss, D. M. 1994. The strategies of human mating. *American Scientist* 82:238–249. Application of sexual selection theory to humans.

Clutton-Brock, T. H., F. E. Guinness, and S. D. Albion. 1982. *Red deer: Behavior and ecology of two sexes.* Chicago: University of Chicago Press. An outstanding study of reproductive behavior and success based on a long-term study of known individuals.

Cody, M. L. 1966. A general theory of clutch size. *Evolution* 20: 174–184.

Ellstrand, N. C. 1984. Multiple paternity within the fruits of wild radish *Raphanus sativus. American Naturalist* 123:819–828.

Freeman, C. L., K. T. Harper, and E. L. Charnov. 1980. Sex change in plants: Old and new observations and new hypotheses. *Oecologica* 47:222–232.

Grime, J. P. 1979. *Plant strategies and vegetative processes.* New York: Wiley.

Gubernich, D. J., and P. H. Klopher (eds.). 1981. *Parental care in mammals.* New York: Plenum.

Krebs, J. R., and N. D. Davies. 1981. *An introduction to behavioral ecology.* Oxford: Blackwell Scientific Publications. Excellent discussion of topics covered in this chapter.

Lack, D. 1954. *The natural regulation of animal numbers.* Oxford: Oxford University Press (Clarendon Press).

McGregor, P. K., J. R. Krebs, and C. M. Perrins. 1981. Song repertoires and lifetime reproductive success in the great tit *(Parus major). American Naturalist* 118:49–59. A study of reproductive success of known male and female birds.

Primack, R. B. 1979. Reproductive effort in annual and perennial species of *Plantago* (Plantaginaceae). *American Naturalist* 114:51–62.

Shapiro, D. Y. 1980. Serial sex changes after simultaneous removal of males from social groups of coral reef fish. *Science* 209: 1136–1137.

Small, M. F. 1992. Female choice in mating. *American Scientist* 80:142–151.

Thornhill, R., and J. Alcock. 1983. *The evolution of insect mating systems.* Cambridge, MA: Harvard University Press.

Willson, M. F. 1983. *Plant reproductive ecology.* New York: Wiley.

Wilson, E. O. 1980. *Sociobiology: The abridged edition.* Cambridge, MA: Harvard University Press.

Warner, R. R. 1988. Sex change and size-advantage model. *Trends in Ecology and Evolution* 3:133–136.

Wasser, S. K. (ed.). 1983. *Social behavior of female vertebrates.* New York: Academic Press.

CHAPTER 13 Population Growth

Begon, M., and M. Mortimer. 1995. *Population ecology: A unified study of plants and animals,* 2nd ed. Cambridge, MA. Blackwell Scientific Publications.

Burton, J. A., and B. Pearson. 1987. *Collins guide to rare mammals of the world.* London: Collins.

Caughley, G. 1977. *Analysis of vertebrate populations.* New York: Wiley.

Fitzgerald, S. 1989. *International wildlife trade: Whose business is it?* Washington, DC: World Wildlife Fund. Documents why some species go extinct.

Harper, J. L. 1977. *The population biology of plants.* London: Academic Press. The major reference on plant population biology.

Kaufman, L., and Mallory, K. (eds.) 1986. *The last extinction.* Cambridge, MA: MIT Press. Excellent overview.

Krebs, C. 1993. *Ecology: The experimental analysis of distribution and abundance.* New York: HarperCollins.

Mountfort, G. 1988. *Rare birds of the world.* London: Collins.

Shaw, J. 1985. *Introduction to wildlife management.* New York: McGraw-Hill. Discusses several concepts of carrying capacity.

Terbrough, J. 1988. *Where have all the birds gone?* Princeton, NJ: Princeton University Press. Effects of habitat fragmentation on neotropical birds.

CHAPTER 14 Intraspecific Population Regulation

Chepko-Sade, B. D., and Z. T. Halpin. 1987. *Mammalian dispersal patterns.* Chicago: University of Chicago Press.

Gaines, M. S., and L. R. McClenaghan, Jr. 1980. Dispersal in small mammals. *Annual Review of Ecology and Systematics* 11:163–169.

Greenwood, P. J., and P. H. Harvey. 1982. The natal and breeding dispersal in birds. *Annual Review of Ecology and Systematics* 13: 1–21.

Grime, J. P. 1979. *Plant strategies and vegetative processes.* New York: Wiley. Role of stress in plant populations.

Harestad, A. S., and F. L. Bunnell. 1979. Home range and body weight—A reevaluation. *Ecology* 60:390.

Harper, J. L. 1977. *Population biology of plants.* New York: Academic Press.

Houston, D. B. 1982. *The northern Yellowstone elk: Ecology and management.* New York: Macmillan. Excellent population study. Discusses the role of density-dependent mortality.

Jones, W. T. 1989. Dispersal distance and the range of nightly movements of Merriam's kangaroo rats. *Journal of Mammalogy* 20:29–34.

Krebs, J., and N. B. Davies (eds.). 1991. *Behavioral ecology: An evolutionary approach,* 3rd ed. Oxford, England: Blackwell Scientific Publications. Chapters on aspects of territoriality.

McCullough, D. R. 1981. Population dynamics of the Yellowstone grizzly. Pages 173–196 in C. W. Fowler and T. D. Smith (eds.), *Dynamics of large mammal populations.* New York: Wiley.

Mech, L. D. 1970. *The wolf: The ecology and behavior of an endangered species.* Garden City, NY: Doubleday.

Mloszewski, M. 1983. *The behavior and ecology of the African buffalo.* Cambridge: Cambridge University Press. Male and female social hierarchies in the buffalo.

Myers, K., C. S. Hale, R. Mykytowycz, and R. L. Hughes. 1971. The effects of varying density and space on sociality and health in animals. Pages 148–187 in A. E. Esser (ed.), *Behavior and environment: The use of space by animals and men.* New York: Plenum.

Nice, M. M. 1937. *Studies in the life history of the song sparrow.* Transactions of the Linnaean Society of New York IV. Reprint edition. New York: Dover Publications. A classic study in territoriality.

Searcy, W. A. 1982. The evolutionary effects of mate selection. *Annual Review of Ecology and Systematics* 13:57–85.

Sinclair, A. R. E. 1977. *The African buffalo: A study of resource limitations of populations.* Chicago: University of Chicago Press.

Smith, R. L. 1963. Some ecological notes on the grasshopper sparrow. *Wilson* Bulletin 75:159–165.

Smith, S. M. 1978. The underworld in a territorial adaptive strategy for floaters. *American Naturalist* 112:570–582.

Tamarin, R. H. (ed.). 1978. *Population regulation.* Benchmark Papers. Stroudsburg, PA: Dowden, Hutchinson, and Ross. Collection of important papers.

Zimen, E. 1981. *The wolf: A species in danger.* New York: Delacourt Press. Excellent behavioral study of the European wolf.

CHAPTER 15 Interspecific Competition

American Naturalist. 1983. Interspecific competition papers. Volume 122, No. 5 (November). A group of controversial articles on interspecific competition.

Brown, J. C., O. J. Reichman, and D. W. Davidson. 1979. Granivory in desert ecosystems. *Annual Review of Ecology and Systematics* 10:210–227.

Brown, J. H. 1995. *Macroecology.* Chicago: University of Chicago Press. See Chapter 3 for a discussion of species, niches, and communities.

Connell, J. H. 1983. On the prevalence and relative importance of interspecific competition: Evidence from field experiments. *American Naturalist* 122:661–696.

Heller, H. C., and D. Gates. 1971. Altitudinal zonation of chipmunks (*Eutamias*): Energy budgets. *Ecology* 52: 424–443.

Hutchinson, G. E. 1978. *An introduction to population ecology.* New Haven: Yale University Press. Builds a strong case for interspecific competition.

Pianka, E. R. 1994. *Evolutionary ecology,* 5th ed. New York: HarperCollins. Good discussion on competition.

Rice, E. L. 1984. *Allelopathy,* 2nd ed. Orlando, FL: Academic Press.

Schoener, T. W. 1983. Field experiments on interspecific competition. *American Naturalist* 122: 240–285.

Werner, E. E., and J. D. Hall. 1976. Niche shifts in sunfishes: Experimental evidence and significance. *Science* 191:404–406.

———. 1979. Foraging efficiency and habitat switching in competing sunfish. *Ecology* 60:256–264.

Wiens, J. A. 1977. On competition and variable environments. *American Scientist* 65:590–597.

CHAPTER 16 Predation

Curio, E. 1976. *The ethology of predation.* New York: Springer-Verlag. An interesting treatment of the behavioral side of predation.

Errington, P. L. 1946. Predation and vertebrate populations. *Quarterly Review of Biology* 22:144–177, 221–245. A classic study of predation.

———. 1963. *Muskrat populations.* Ames, IA: Iowa State University Press. A study of a prey species and its predators.

Fleming, T. H. 1988. *The short-tailed fruit bat: A study in plant-animal interactions.* Chicago: University of Chicago Press.

Fox, L. R. 1975. Cannibalism in natural populations. *Annual Review of Ecology and Systematics* 6:87–106.

Godfray, H. C. J. 1994. *Parasitoids: Behavioral and evolutionary ecology.* Princeton: Princeton University Press.

Holling, C. S. 1959. The components of predation as revealed by a study of small mammal predation of the European sawfly. *Canadian Entomologist* 91:293–320. Classic study of functional response.

———. 1966. The functional response of invertebrate predators to prey density. *Memoirs Entomological Society of Canada* 48:1–86. Further study of functional response.

Jackson, R. R. 1992. Eight-legged tricksters. *Bioscience* 42:590–596. Cannibalism in spiders.

Jedrzejewski, W., B. Jedrzejewski, and L. Szymura. 1995. Weasel population response, home range, and predation on rodents in a deciduous forest in Poland. *Ecology* 76:179–195.

Keith, L. B., J. R. Cary, O. J. Rongstad, and M. C. Brittingham. 1984. Demography and ecology of a declining snowshoe hare population. *Journal of Wildlife Management* 90:1–43.

Krebs, J. R., and N. B. Davies (eds.). 1984. *Behavioral ecology: An evolutionary approach,* 2nd ed. Oxford, England: Blackwell Scientific Publications. A theoretical approach.

Polis, G. 1981. The evolution and dynamics of intraspecific predation. *Annual Review of Ecology and Systematics* 12:225–251. Study of cannibalism.

Polis, G., and R. D. Holt. 1992. Intraguild predation: The dynamics of complex trophic interactions. *TREE* 7:151–154.

Sinclair, A. R. E., C. J. Krebs, J. M. N. Smith, and S. Boutin. 1988. Population biology of snowshoe hares. III: Nutrition, plant secondary compounds, and food limitation. *Journal of Animal Ecology* 57:787–806.

Spenser, C. N., B. McClelland, and J. A. Stanford. 1991. Shrimp stocking, salmon collapse, and eagle displacement. *Bioscience* 41:14–21.

Stephens, D. W., and J. W. Krebs. 1987. *Foraging theory.* Princeton: Princeton University Press. Theoretical look at optimal foraging.

Taylor, R. J. 1984. *Predation.* New York: Chapman and Hall. An overview of predator-prey dynamics.

Wagner, J. D., and D. H. Wise. 1996. Cannibalism regulates densities of young wolf spiders: Evidence from field and laboratory experiments. *Ecology* 77:639–652.

CHAPTER 17 Parasitism and Mutualism

Bacon, P. J. 1985. *Population dynamics of rabies in wildlife.* New York: Academic Press.

Barbour, A. G., and D. Fish. 1993. The biological and social phenomenon of Lyme disease. *Science* 260:1610–1616.

Boucher, D. H. (ed.). 1985. *The biology of mutualism.* Oxford, England: Oxford University Press. An excellent reference.

Boucher, D. H., S. James, and H. D. Keller. 1982. The ecology of mutualism. *Annual Review of Ecology and Systematics* 13: 315–347.

Burdon, J. J. 1987. *Diseases and plant population biology.* Cambridge, England: Cambridge University Press. Effect of diseases on individual plants and populations.

Buskirk, J. V., and R. S. Ostfeld. 1995. Controlling Lyme disease by modifying the density and species composition of tick hosts. *Ecological Applications* 5:1133–1140.

Dobson, A. P., and E. R. Carper. 1996. Infectious diseases and human population history. *Bioscience* 46:115–125.

Edwards, M. A., and U. McDonald. 1982. *Animal diseases in relation to animal conservation.* New York: Academic Press. Effects of disease on wild animal populations and human interactions.

Futuyma, D. J., and M. Slatkin (eds.). 1983. *Coevolution.* Sunderland, MA: Sinauer. Papers on processes and consequences of coevolution.

Garnett, G. P., and E. C. Holmes. 1996. The ecology of emergent infectious disease. *Bioscience* 46:127–135.

Garrett, L. 1994. *The coming plague: Newly emerging diseases in a world out of balance.* New York: Farrar, Straus, and Giroux.

Hamilton, W. D., and M. Zuk. 1982. Heritable true fitness and bright birds: A role for parasites? *Science* 218:384–387.

Hanzawa, F. M., A. J. Beattie, and D. C. Culvar. 1988. Directed dispersal: Demographic analysis of an ant-seed mutualism. *American Naturalist* 131:1–13.

Howe, H. F., and L. C. Westley. 1988. *Ecological relationships of plants and animals.* New York: Oxford University Press. An outstanding presentation of herbivory, pollination, and other relationships.

Lafferty, K. D., and A. K. Morris. 1996. Altered behavior of parasitized killifish increases susceptibility to predation by bird final hosts. *Ecology* 77:1390–1397.

McNeill, W. H. 1976. *Plagues and peoples.* New York: Doubleday. The role of infectious diseases in the course of history.

May, R. M. 1983. Parasitic infections as regulators of animal populations. *American Scientist* 71:36–45.

Moore, J. 1995. The behavior of parasitized animals. *Bioscience* 45:89–96.

Ostfeld, R. S., C. G. Jones, and J. O. Wolff. 1996. Of mice and mast. *Bioscience* 46:323–330.

Real, L. A. (ed.). 1983. *Pollination biology.* Orlando, FL: Academic Press. Recent research in pollination biology.

Real, L. A. 1996. Sustainability and the ecology of infectious disease. *Bioscience* 46:88–97.

Sheldon, B. C., and S. Verhulst. 1996. Ecological immunity: Costly parasite defenses and trade-offs in evolutionary ecology. *TREE* 11:317–321.

Spielman, A., M. I. Wilson, J. F. Levine, and J. Piesman. 1985. Ecology of *Ixodes domini*-borne human babesiosis and Lyme disease. *Annual Review of Entomology* 30:439–460.

Wakelin, D. 1997. Parasites and the immune system. *Bioscience* 47:32–40.

Wilson, E. O. 1975. *Sociobiology: The new synthesis.* Cambridge, MA: Harvard University Press. See Chapter 17, Social Symbiosis, for a detailed discussion of social parasitism.

Zuk, M. 1991. Parasites and bright birds: New data and new predictions. Pages 317–327 in J. E. Loge and M. Zuk (eds.), *Bird-parasite interactions.* Oxford: Oxford University Press.

CHAPTER 18   Human Interactions with Natural Populations

Baden, J. A., and D. Leal. 1990. *The Yellowstone primer.* San Francisco: Pacific Research Institute for Public Policy. An excellent review of the problems of Yellowstone Park, a microcosm of problems facing all parks and reserves.

Bricklemyer, E. C., Jr., S. Ludicello, and H. J. Hartmann. 1989. Discarded catch in U. S. commercial marine fisheries. Pages 259–295 in *Audubon Wildlife Report 1989/1990.* San Diego: Academic Press. Excellent review of current fishery regulations and the discard problem.

Caughley, G. 1976. Wildlife management and the dynamics of ungulate populations. *Applied Biology* 1:183–246. Good introduction to sustained yield in wildlife management.

Debach, P. 1974. *Biological control by natural enemies.* London: Cambridge University Press. Somewhat dated but an excellent introduction. Chapters on classic biological control successes.

DiSilvestro, R. L. 1989. *The endangered kingdom: The struggle to save America's wildlife.* New York: Wiley. Case histories of restoration and endangerment of selected wildlife species, as well as an assessment of wildlife management.

Dover, M. J., and B. A. Croft. 1986. Pesticide resistance and public policy. *Bioscience* 36:78–85. Discusses pesticide resistance and effective pest control.

Ellis, R. 1991. *Men and whales.* New York: Alfred Knopf. Excellent, well-illustrated history of whaling and the human-whale relationship.

Ellstrand, N. C., and C. A. Hoffman. Hybridization as an avenue of escape for engineered genes. *Bioscience* 40:438–442.

Fitzgerald, S. 1989. *International wildlife trade: Whose business is it?* Washington, DC: World Wildlife Fund. Why endangered species need protection; a review of illegal trade in wildlife species.

Gambell, R. 1976. Population biology and the management of whales. *Applied Biology* 1:237–343. A short history of whaling, population dynamics of species, and the failure of attempts to apply sustained yield management.

Harris, L. D. 1984. *The fragmented forest.* Chicago: University of Chicago Press. Integrates island biogeography and forest management in the preservation of biotic diversity.

Hoffman, C. A. 1990. Ecological risks of genetic engineering of crop plants. *Bioscience* 40:434–436.

Horn, D. J. 1988. *Ecological approach to pest management.* New York: The Guilford Press. Good review; many examples.

Huffaker, C. B., and R. L. Rabb (eds.). 1984. *Ecological entomology.* New York: John Wiley. Excellent chapters on evolutionary process in insects, natural control of insects, and application of ecology to insect population management.

Kleiman, D. G. 1989. Reintroduction of captive mammals for conservation. *Bioscience* 39:152–161. Authoritative discussion of guidelines for introducing endangered species into the wild.

Pimentel, D., H. Acquay, M. Biltonen, et al. 1992. Environmental and economic costs of pesticide use. *Bioscience* 42:750–760.

Prescott-Allen, C., and R. Prescott-Allen. 1986. *The first resource: Wild species in North American economy.* New Haven: Yale University Press.

Regier, H. A., and G. L. Baskerville. 1986. Sustainable development of regional ecosystems degraded by exploitive development. Pages 74–103 in W. E. Clark and R. E. Munn (eds.), *Sustainable development of the biosphere.* Cambridge, England: Cambridge University Press. Focuses on recovery of degraded forestry and fishery resource systems.

Regier, H. A., and W. L. Hartman. 1973. Lake Erie's fish community: 150 years of cultural stress. *Science* 180:1248–1255. History of Lake Erie's fisheries.

Small, G. 1976. *The blue whale.* New York: Columbia University Press. The authoritative depressing history of the blue whale.

Soule, M. E. (ed.). 1986. *Conservation biology: The science of scarcity and diversity.* Sunderland, MA: Sinauer Associates. Considers various aspects of conserving endangered populations.

Soule, M. E., and B. A. Wilcox (eds.). 1980. *Conservation biology: An evolutionary ecological perspective.* Sunderland, MA: Sinauer Associates. Excellent overview of the problems of conserving small populations, including captive propagation.

Thatcher, R. C., J. L. Searcy, J. E. Coster, and G. D. Hertel (eds.). 1986. *The southern pine beetle.* U. S. D. A. Forest Service Science and Education Tech. Bull. 1631. Washington, DC: U. S. Department of Agriculture. A colorful, accessible explanation of all aspects of IPM in action against the southern pine beetle.

Trefethen, J. B. 1975. *An American crusade for wildlife.* New York: Winchester Press. An excellent history of restoration of wildlife and the politics involved.

Walters, C. 1986. *Adaptive management of renewable resources.* New York: Macmillan. Advanced but challenging approaches to managing renewable resources.

Western, D., and M. C. Pearl. 1988. *Conservation for the twenty-first century.* New York: Oxford University Press. Focuses on problems in and approaches to wildlife conservation. Excellent reference.

CHAPTER 19 Population Genetics and Speciation

Bonnell, M. C., and R. K. Selander. 1974. Elephant seals. Genetic variation and near extinction. *Science* 184:908–909.

Chepko-Sade, B. D., and Z. T. Halpin (eds.). 1987. *Mammalian dispersal patterns: The effect of social structure on population genetics.* Chicago: University of Chicago Press.

Frankel, O. H., and M. E. Soule. 1981. *Conservation and evolution.* Cambridge, England: Cambridge University Press. A good introduction to population genetics of small populations.

Futuyma, D. J. 1984. *Evolutionary biology.* Sunderland, MA: Sinauer Associates. A superior treatment of the subject.

Grant, P. 1986. *Ecology and evolution of Darwin's finches.* Princeton, NJ: Princeton University Press. Updates the story of Darwin's finches. A modern field study in evolution. Should be read with Lack 1974.

Grant, V. 1971. *Plant speciation.* New York: Columbia University Press. A classic on all aspects of plant speciation.

———. 1985. *The evolutionary process: A critical review of evolutionary theory.* New York: Columbia University Press. An accessible discussion of natural selection and speciation.

Hanski, I. A., and M. E. Gilpin (eds.). 1996 *Metapopulation biology: Ecology, genetics, and evolution.* San Diego: Academic Press. Deals with various aspects of small populations.

Hartl, D. 1988. *A primer of population genetics.* Sunderland, MA: Sinauer Associates. An excellent introduction.

Highton, R. 1995. Speciation in eastern North American salamanders of the Genus *Plethodon. Annual Review of Ecology and Systematics* 26:579–600.

Lack, D. 1974. *Darwin's finches: An essay on the general biological theory of evolution.* London: Cambridge University Press. A classic in biology.

Lynch, M., J. Conery, and R. Burger. 1995. Mutation accumulation and the extinction of small populations. *American Naturalist* 146:489–518.

Mayr, E. 1963. *Animal species and evolution.* Cambridge, MA: Harvard University Press. A classic major work on animal speciation.

Otte, D., and J. Endler (ed.). 1989. *Speciation and its consequences.* Sunderland, MA: Sinauer Associates. A series of papers that provide a major reference source on the subject.

Rosensweig, M. 1995. *Species diversity in space and time.* New York: Cambridge University Press. See Chapters 5 and 6 for a clear discussion of speciation and extinction.

Schonewald-Cox, C. M., S. M. Chambers, B. Macbryde, and W. L. Thomas (eds.). 1983. *Genetics and conservation.* Menlo Park, CA: Benjamin Cummings. A reference manual for managing wild animal and plant populations.

Soule, M. (ed.). 1986a. *Conservation biology: The science of scarcity and diversity.* Sunderland, MA: Sinauer Associates. A source book on the ecology of small populations and extinction.

———. 1986b. *Viable populations for conservation.* Cambridge, England: Cambridge University Press. Discussion of minimum viable population concepts and management of small populations.

Thornhill, N. W. (ed.). 1993. *The natural history of inbreeding and outbreeding.* Chicago: University of Chicago Press. A collection of highly interesting papers covering the topic as well as some aspects of small populations and speciation.

Weiner, J. 1994. *The beak and the finch: A story of evolution in our time.* New York: Alfred A. Knopf. A superb book on species evolution, based on Grant's studies of Darwin's finches.

Western, D., and M. C. Pearl (eds.). 1989. *Conservation for the twenty-first century.* New York: Oxford University Press. Excellent overview of problems and possible solutions to the conservation of species and their habitat. A major reference.

White, M. J. D. 1978. *Modes of speciation.* San Francisco: Freeman. A major introduction to the speciation process.

CHAPTER 20 Community Structure

Huston, M. 1994. *Biological diversity: The coexistence of species on changing landscapes.* New York: Cambridge University Press.

Ricklefs, R. E., and D. Schluter (eds.). 1993. *Ecological communities: Historical and geographical perspectives.* Chicago: University of Chicago Press.

Wilson, E. O. 1992. *The diversity of life.* Cambridge, MA: Harvard University Press.

CHAPTER 21 Community Dynamics

Bond, W. J., and B. van Wilgen. 1996. *Fire and plants.* New York: Chapman and Hall. Plant population biology in relation to fire.

Bormann, F., and G. E. Likens. 1979. *Patterns and process in forested ecosystems.* New York: Springer-Verlag. A basic study of forest succession at Hubbard Brook, New Hampshire.

Braun, E. L. 1950. *Deciduous forests of eastern North America.* New York: McGraw-Hill.

Connell, J. H., and R. O. Slatyer. 1977. Mechanisms of succession in natural communities and their role in community stability and organization. *American Naturalist* 111:1119–1144.

Davis, M. B. 1983. Holocene vegetational history of the eastern United States. Pages 166–188 in H. E. Wright, Jr. (ed.), *Late quarternary environments of the United States.* Vol. II, The Holocene. Minneapolis: University of Minnesota Press.

Delacourt, P. A., and H. R. Delacourt. 1981. Vegetation maps for eastern North America, 40,000 yr BP to present. Pages 123–166 in R. Romans (ed.), *Geobotany*. New York: Plenum Press.

Forman, R. T. T. 1995. *Land mosaics: The ecology of landscapes and regions*. New York: Cambridge University Press. Ecology of heterogenous land areas; interaction between humans and natural processes on the land.

Galli, A. E., C. F. Leck, and R. T. T. Forman. 1976. Avian distribution patterns in forest islands of different sizes in central New Jersey. *Auk* 93:356–364.

Golley, F. (ed.). 1978. *Ecological succession*. Benchmark Papers. Stroudsburg, PA: Dowden, Hutchinson, and Ross. Succession as viewed over time.

Hanski, I. A., and M. E. Gilpin (eds.). 1997. *Metapopulation biology: Ecology, genetics, and evolution*. San Diego: Academic Press. Major reference on theory and application.

Harris, L. D. 1984. *The fragmented forest*. Chicago: University of Chicago Press. An important reference.

Huston, M. A. 1994. *Biological diversity: The coexistence of species on changing landscapes*. New York: Cambridge University Press. Chapter 9 is a good review of succession and landscape patterns.

Huston, M. A., and T. M. Smith. 1987. Plant succession: Life history and competition. *American Naturalist* 130:168–198.

Keever, C. 1950. Causes of succession in old fields of the Piedmont, North Carolina. *Ecological Monographs* 20:229–250.

MacArthur, R. H., and E. O. Wilson. 1967. *The theory of island biogeography*. Princeton: Princeton University Press. The original source.

Maser, C., and J. M. Trappe (eds.). 1984. *The seen and unseen world of the fallen tree*. USDA Forest Service General Technical Report PNW-164.

McCollough, D. R. (ed.). 1996. *Metapopulations and wildlife conservation*. Washington, DC: Island Press. Application of metapopulation theory to wildlife conservation.

Mooney, H. A., et al. (eds.). 1981. *Proceedings: Symposia on fire regimes and ecosystem properties*. General Technical Report WO-26. Washington, DC: USDA Forest Service. A major reference.

Olson, J. S. 1958. Rates of succession and soil changes in southern Lake Michigan sand dunes. *Botanical Gazette* 119:125–170.

Oosting, H. J. 1942. An ecological analysis of the plant communities of Piedmont, North Carolina. *American Midland Naturalist* 28:1–126.

Pickett, S. A., and P. White (eds.). 1984. *Natural disturbance: The patch dynamics perspective*. Orlando: Academic Press. The role of disturbance in succession and community diversity.

Ranney, J. W., M. C. Brunner, and J. B. Levenson. 1981. The importance of edge in the structure and dynamics of forest lands. Pages 67–91 in R. L. Burgess and D. M. Sharpe (eds.), *Forest island dynamics in man-dominated landscapes* (Ecological Studies No. 41). New York: Springer-Verlag.

Robbins, C. S., D. K. Dawson, and B. A. Dowell. 1989. Habitat area requirements of breeding forest birds of the Middle Atlantic States. *Wildlife Monographs* 103.

Sousa, W. P. 1979. Disturbance in marine intertidal boulder fields: The nonequilibrium maintenance of species diversity. *Ecology* 60:1225–1239.

———. 1984. Intertidal mosaics: Patch size, propagule availability, and spatially variable patterns of succession. *Ecology* 65:1918–1935.

Sprugle, D. G. 1976. Dynamic structure of wave-generated *Abies balsamea* forests in northeastern United States. *Journal of Ecology* 64:889–911.

Tilman, D. 1986. The resource-ratio hypothesis of plant succession. *American Naturalist* 116:362–369.

Walker, L. J., J. C. Zasada, and F. S. Chapin, III. 1986. The role of life history processes in primary succession on Alaskan flood plain. *Ecology* 67:1243–1253.

Watt, A. S. 1947. Pattern and process in the plant community. *Journal of Ecology* 35:1–22.

Watts, M. T. 1976. *Reading the American landscape*. New York: Macmillan. A delightful introduction to landscape ecology of the United States as influenced by humans.

West, D. C., H. H. Shugart, and D. B. Botkin (eds.). 1981. *Forest succession: Concepts and application*. New York: Springer-Verlag. Excellent reference on patterns and processes from temperate to tropical forests.

Whitcomb, R. F., J. F. Lynch, P. A. Opler, and C. S. Robbins. 1976. Island biogeography and conservation: Strategy and limitations. *Science* 193:1030–1032.

Whittaker, R. H. 1956. Vegetation of the Great Smoky Mountains. *Ecological Monographs* 26:1–80.

Williamson, M. 1981. *Island populations*. Oxford, England: Oxford University Press. Good summary of island biogeography theory.

Wilson, E. O. 1992. *The diversity of life*. Cambridge, MA: Harvard University Press. Excellent, accessible synthesis of biodiversity.

CHAPTER 22 Processes Shaping Communities

Austin, M. P., and T. M. Smith. 1989. A new model of the continuum concept. *Vegetatio* 83:35–47.

Crocker, R. L., and J. Major. 1955. Soil development in relation to vegetation and surface age at Glacier Bay, Alaska. *Journal of Ecology* 43:427–488.

Huston, M. 1994. *Biological diversity: The coexistence of species on changing landscapes*. New York: Cambridge University Press.

Huston, M., and T. M. Smith. 1987. Plant succession: Life history and competition. *American Naturalist* 130:168–198.

Pastor, J., R. J. Naiman, B. Dewey, and P. McInnes. 1988. Moose, microbes, and boreal forest. *Bioscience* 88:770–777.

Smith, T. M., and M. Huston. 1987. A theory of spatial and temporal dynamics of plant communities. *Vegetatio* 3:49–69.

CHAPTER 23 Production in Ecosystems

Aber, J. D., and J. M. Melillo. 1991. *Terrestrial ecosystems*. Philadelphia: Saunders College Publishing.

Coe, M. J., D. H. Cumming, and J. Phillipson. 1976. Biomass and production of large African herbivores in relation to rainfall and primary production. *Oecologica* 22:341–354.

Cooper, J. P. 1975. *Photosynthesis and productivity in different environments*. New York: Cambridge University Press.

Cowan, R. L. 1962. Physiology of nutrition of deer. Pages 1–8 in *Proceedings 1st National White-tailed Deer Disease Symposium*.

Gates, D. M. 1985. *Energy and ecology*. Sunderland, MA: Sinauer Associates. Energy as it relates to human affairs.

Leith, H. 1973. Primary production: Terrestrial ecosystems. *Human Ecology* 1:303–332.

Leith, H., and R. H. Whittaker (eds.). 1975. *Primary productivity in the biosphere.* Ecological Studies Vol. 14. New York: Springer-Verlag.

MacArthur, R. H., and J. H. Connell. 1966. *The biology of populations.* New York: John Wiley.

Meentemeyer, V. 1978. Macroclimate and lignin control of decomposition. *Ecology* 59:465–472.

National Academy of Science. 1975. *Productivity of world ecosystems.* Washington, DC: National Academy of Science. Good summary of world primary production.

Petrusewicz, K. (ed.). 1967. *Secondary productivity of terrestrial ecosystems.* Warsaw, Poland: Pantsworve Wydawnictwo Naukowe.

Phillipson, J. J. 1966. *Ecological energetics.* New York: St. Martin's Press. Although dated, this is still an excellent introduction to energy in ecosystems.

Transeau, E. N. 1926. The accumulation of energy by plants. *Ohio Journal of Science* 26:1–10.

Whittaker, R. H., and G. E. Likens. 1973. Carbon in the biota. In G. M. Woodwell and E. V. Pecan (eds.), *Carbon and the Biosphere Conference 72501.* Springfield, VA: National Technical Information Service.

Wiegert, R. C. (ed.). 1976. *Ecological energetics.* Benchmark Papers, Stroudsburg, PA: Dowden, Hutchinson & Ross. (Distributed by Academic Press, New York.) Collection of papers surveying the development of the concept.

CHAPTER 24 Trophic Structure

Anderson, J. M., and A. MacFadyen (eds.), 1976. *The role of terrestrial and aquatic organisms in the decomposition process.* Seventeenth Symposium, British Ecological Society, Oxford, England: Blackwell Scientific Publications.

Andrews, R. D., D. C. Coleman, J. E. Ellis, and J. S. Singh. 1975. Energy flow relationships in a shortgrass prairie ecosystem. *Proceedings 1st International Congress of Ecology* 22–28. The Hague: W. Junk Publishers.

Briand, F. 1983. Environmental control of food web structure. *Ecology* 64:253–263.

Cohen, J. E. 1989. Food webs and community structure. Pages 181–202 in J. Roughgarden, R. May, and S. Levin (eds.), *Perspectives in ecological theory.* Princeton, NJ: Princeton University Press.

Elton, C. 1927. *Animal ecology.* London: Sidgwick & Jackson. A classic, with original descriptions of food chains and ecological pyramids.

Fletcher, M., G. R. Gray, and J. G. Jones. 1987. *Ecology of microbial communities.* New York: Cambridge University Press. A good overview of ecology at the decomposer level.

Pimm, S. L. 1982. *Food webs.* London: Chapman and Hall. A detailed discussion of aspects of food web theory.

———. 1991. *The balance of nature?* Chicago: University of Chicago Press.

Swift, M. J., O. W. Heal, and J. M. Anderson. 1978. *Decomposition in terrestrial ecosystems.* Oxford: Blackwell Scientific Publications. An excellent review of the decomposition process.

Yodzis, P. 1989. *Introduction to theoretical ecology.* New York: Harper & Row.

Zaret, T. M., and R. T. Paine. 1973. Species introduction in a tropical lake. *Science* 182:449–455.

CHAPTER 25 Biogeochemical Cycles

Aber, J. D., K. N. Nadelhoffer, P. Steudler, and J. M. Melillo. 1989. Nitrogen saturation in northern forest ecosystems. *Bioscience* 39:378–386.

Alexander, M. (ed.). 1980. *Biological nitrogen fixation.* New York: Plenum. Ecology and physiology of nitrogen-fixing organisms.

Binkley, D., C. T. Driscoll, H. L. Allen, P. Schoeneberger, and D. McAvoy. 1989. *Acidic deposition and forest soils: Context and case studies in southeastern United States.* New York: Springer-Verlag. Best information on topic.

Bolin, B., E. T. Degens, S. Kempe, and P. Ketner (eds.). 1979. *The global carbon cycle.* SCOPE 13. New York: John Wiley. A comprehensive study of the carbon cycle.

Bormann, F. H., and G. E. Likens. 1979. *Pattern and process in a forested ecosystem.* New York: Springer-Verlag. A good discussion of nutrient cycling in the forest.

Dover, M. J., and B. A. Croft. 1986. Pesticide resistance and public policy. *Bioscience* 36:78–91. Problems created by pesticide resistance in insects.

MacKenzie, J. J., and M. T. El-Ashry. 1989. *Air pollution's toll on forests and crops.* New Haven: Yale University Press. Excellent global overview of the problem.

Pimentel, D., and C. A. Edwards. 1982. Pesticides and ecosystems. *Bioscience* 32:595–600. A summary.

Pimentel, D., and L. Levitan. 1986. Pesticides: Amounts applied and amounts reaching pests. *Bioscience* 36:86–91. Revealing review.

Pomeroy, L. R. 1974. *Cycles of essential elements.* Benchmark Papers in Ecology. Stroudsburg: PA: Dowden, Hutchinson, and Ross. Collection of important papers on mineral cycling.

Post, W. M., T. H. Peng, W. R. Emanuel, A. W. King, V. H. Dale, and D. L. DeAngelis. 1990. The global carbon cycle. *American Scientist* 78:310–326. An excellent updated review of the global carbon cycle—what we know and do not know about it.

Schulze, E. D., O. L. Lange, and O. Oren. 1989. *Forest decline and air pollution: A study of spruce* (Picea abies) *on acid soils.* New York: Springer-Verlag. Detailed studies of forest decline in Germany, with emphasis on the nutrient decline hypothesis.

Smith, W. H. 1990. *Air pollution and forests: Interaction between air contaminants and forest ecosystems,* 2nd ed. New York: Springer-Verlag. A thorough and indispensable reference source.

Sprent, J. I. 1988. *The ecology of the nitrogen cycle.* New York: Cambridge University Press. Various processes and magnitudes of the nitrogen cycle.

Wellburn, A. 1988. *Air pollution and acid rain: The biological impact.* Essex, England: Longman Scientific and Technical. (Co-published with J. Wiley, New York.) A basic introduction to the topic.

CHAPTER 26 Global Environmental Change

Amthor, J. S. 1995. Terrestrial higher-plant response to increasing atmospheric CO_2 in relation to the global carbon cycle. *Global Change Biology* 1:243–274.

Butcher, S. S., R. J. Charlson, G. H. Orians, and G. V. Wolfe (eds.). 1992. *Global biogeochemical cycles.* New York: Academic Press.

Edmonds, J. 1992. Why understanding the natural sinks and sources of CO_2 is important: A policy analysis perspective. *Water, Air and Soil Pollution* 64:11–21.

Gates, D. 1993. *Climate change and its biological consequences.* Sunderland, MA: Sinauer Associates. A highly accessible introduction to climate change and its effects, past and present.

Harte, J., and R. Shaw. 1995. Shifting dominance with a montane vegetation community: Results of a climate-warming experiment. *Science* 267:876–880.

Heimann, M. (ed.). 1993. *The global carbon cycle.* Berlin: Springer-Verlag.

Holdridge, L. R. 1947. Determination of world formulations from simple climatic data. *Science* 105:367–368.

Houghton, J. T., L. G. Meira Filho, B. A. Callander, N. Harris, A. Kattenberg, and K. Maskell (eds.). 1996. *Climate change 1995: The science of climate change.* Intergovernmental Panel on Climate Change. Cambridge, England: Cambridge University Press.

Houghton, R. A. 1995. Land-use change and the carbon cycle. *Global Change Biology* 1:275–287.

Peterjohn, W. T., J. M. Melillo, F. P. Bowles, and P. A. Steudler. 1993. Soil warming and trace gas fluxes: Experimental design and preliminary flux results. *Oecologia* 93:18–24.

Peters, R. L., and T. E. Lovejoy (eds.). 1992. *Global warming and biological diversity.* New Haven: Yale University Press. A sobering, readable examination of the potential effect of global warming on vegetation, soils, animals, parasites and diseases, and ecosystems.

Root, T. 1988. Energy constraints on avian distributions and abundances. *Ecology* 69:330–339.

Schneider, S. H. 1989. *Global warming.* San Francisco: Sierra Club Books. Highly readable examination of global warming and the debate engendered in politics and the media.

Smith, T. M., P. N. Halpin, H. H. Shugart, and C. Secrett. 1994. Global forests. In K. Strzpeck and J. Smith (eds.), *As climate changes: International impacts and implications.* Cambridge, England: Cambridge University Press.

Smith, T. M., R. Leemans, and H. H. Shugart. 1992. Sensitivity of terrestrial carbon storage to CO_2-induced climate change: Comparison of four scenarios based on general circulation models. *Climatic Change* 21:367–384.

Strain, B. R., and J. D. Cure (eds.). 1985. *Direct effects of increasing carbon dioxide on vegetation.* Washington, DC: U.S. Department of Energy Publication DOE/ER-0238.

Trabalka, J. R. (ed.). 1985. *Atmospheric carbon dioxide and the global carbon cycle.* Washington, DC: U.S. Department of Energy Report DOE/ER-0239.

Trabalka, J. R., and D. E. Reichle (eds.). 1994. *The changing carbon cycle: A global analysis.* New York: Springer-Verlag.

VEMAP. 1995. Vegetation/ecosystem modeling and anlaysis project: Comparing biogeography and biogeochemistry models in a continental-scale study of terrestrial ecosystem responses to climate change and CO_2 doubling. *Global Biogeochemical Cycles* 9:407–437.

Woodward, F. I. (ed.). 1992. *Global climate change: The ecological consequences.* London: Academic Press. Emphasis on climate change research.

CHAPTER 27 Grasslands and Savannas

Grasslands

Breymeyer, A., and G. Van Dyne (eds.). 1980. *Grasslands, systems analysis and man.* Cambridge University Press. Synthesis of International Biological Programme studies of grasslands; comprehensive.

Callenback, E. 1996. *Bring back the buffalo: A sustainable future for America's Great Plains.* Covelo, CA: Island Press. An ecological approach to the management of the shortgrass plains.

Coupland, R. T. (ed.). 1979. *Grassland ecosystems of the world: Analysis of grasslands and their uses.* Cambridge, England: Cambridge University Press. Good reference.

Duffy, E. 1974. *Grassland ecology and wildlife management.* London: Chapman and Hall. Examines ecology of tame grasslands.

Falt, J. H. 1976. Energetics of a suburban lawn ecosystem. *Ecology* 57:141–150. Ecological concepts applied to the lawn. An interesting study.

French, N. (ed.). 1979. *Perspectives on grassland ecology.* New York: Springer-Verlag. A good summary of grassland ecology.

Hodgson, J., and A. W. Illus (eds.). 1966. *Ecology and management of grazing systems.* New York: CAB International. Synthesis of grassland ecology and application of principles to management of domestic grassland and rangeland.

Manning, D. 1995. *Grassland: History, biology, politics, and promise of the American prairie.* New York: Penguin Books.

Reichman, O. J. 1987. *Konza prairie: Tallgrass natural history.* Lawrence, KS: University of Kansas Press. A good description of tallgrass prairie based on studies of one of the few remaining tracts.

Risser, P. G., et al. 1981. *The true prairie ecosystem.* Stroudsburg, PA: Dowden, Hutchinson, & Ross. Ecology of midgrass and tallgrass prairies.

Weaver, J. E. 1954. *North American prairie.* Lincoln, NE: Johnson. An old but important study of prairie vegetation by an outstanding botanist familiar with the original condition of the prairies.

Savannas

Bourliere, F. (ed.). 1983. *Tropical savannas: Ecosystems of the world* 13. Amsterdam: Elsevier Scientific Publishers. A major reference work on savannas.

Sarimiento, G. 1984. *Ecology of neotropical savannas.* Cambridge, MA: Harvard University Press. A major study of South American savannas.

Sinclair, A. R. E., and M. Norton-Griffiths (eds.). 1979. *Serengeti: Dynamics of an ecosystem.* A study of the African savanna ecosystem. Excellent.

Sinclair, A. R. E., and P. Arcese (eds.). 1995. *Serengeti II: Dynamics, management, and conservation of an ecosystem.* Chicago: University of Chicago Press. A masterful study of an ecosystem.

Tothill, J. C., and J. J. Mott (eds.). 1985. *Ecology and management of the world's ecosystems.* Canberra: Australian Academy of Science. Group of papers providing in-depth review of savannas and their management. An important reference.

CHAPTER 28 Shrublands and Deserts

Brown, G. W., Jr. (ed.). 1976–1977. *Desert biology,* 2 vol. New York: Academic Press. Basic information on biology and physical features of world deserts.

Castri, F. Di., and H. A. Mooney (eds.). 1973. *Mediterranean-type ecosystems*. Ecological Studies No. 7. New York: Springer-Verlag. An excellent basic introduction to all aspects of mediterranean ecosystems.

Castri, F. Di, D. W. Goodall, and R. L. Specht (eds.). 1981. *Mediterranean-type shrublands*. Ecosystems of the World No. 11. Amsterdam: Elsevier Scientific. A more comprehensive treatment of mediterranean ecosystems. A major reference.

Evenardi, M., I. Noy-Meir, and D. Goodall (eds.). 1985–1986. *Hot deserts and arid shrublands of the world*. Ecosystems of the World 12A and 12B. Amsterdam: Elsevier Scientific. A major reference work on world deserts.

Groves, R. H. 1981. *Australian vegetation*. Cambridge, England: Cambridge University Press. Concise descriptions of the diverse plant communities of Australia.

McKell, C. M. (ed.). 1983. *The biology and utilization of shrubs*. Orlando, FL: Academic Press. A major reference on shrubs worldwide.

Plumb, T. R., tech. coordinator. 1979. *Proceedings of the symposium on the ecology, management, and utilization of California oaks*. General Technical Report PSW-44. Berkeley, CA: Pacific Southwest Forest and Range Experiment Station, Forest Service, U.S. Dept. of Agriculture. Thorough coverage of this important and degraded mediterranean ecosystem.

Polunin, O., and M. Walters. 1985. *A guide to the vegetation of Britain and Europe*. Oxford, England: Oxford University Press. An outstanding survey of the plant communities of Europe as far as the Russian border; well-illustrated.

Specht, R. L. (ed.). 1979, 1981. *Heathlands and related shrublands*. Ecosystems of the World 9A and 9B. Amsterdam: Elsevier Scientific. A major reference on heathland communities worldwide.

Wagner, F. H. 1980. *Wildlife of the deserts*. New York: Harry W. Abrams. An excellent, well-illustrated introduction to desert animals' adaptation to arid environments.

CHAPTER 29   Tundra and Taiga

Tundra

Bliss, L. C., O. H. Heal, and J. J. Moore (eds.). 1981. *Tundra ecosystems: A comparative analysis*. New York: Cambridge University Press. A maor reference.

Brown, J., P. C. Miller, L. L. Tieszen, and F. L. Bunnell. 1980. *An arctic ecosystem: The coastal tundra at Barrow, Alaska*. Stroudsburg, PA: Dowden, Hutchinson & Ross.

Furley, P. A., and W. A. Newey. 1983. *Geography of the biosphere*. London: Butterworths. Chapter 10 gives a detailed discussion of the tundra biome.

Rosswall, T., and O. W. Heal (eds.). 1975. *Structure and function of tundra ecosystems*. Ecological Bulletins 20. Stockholm: Swedish Natural Sciences Research Council.

Smith, A. P., and T. P. Young. 1987. Tropical alpine plant ecology. *Annual Review of Ecology and Systematic* 18:137–158. An informative review paper.

Sonesson, M. (ed.). 1980. *Ecology of a subarctic mire*. Ecological Bulletins 30. Stockholm: Swedish Natural Sciences Research Council. A detailed study of structure and function.

Wielgolaski, F. E. (ed.). 1975. *Fennoscandian tundra ecosystems*. Part I, *Plants and microorganisms*; Part II, *Animals and systems analysis*. New York: Springer-Verlag.

Zwinger, A. H., and B. E. Willard. 1972. *Land above the trees*. New York: Harper & Row.

Taiga

Bonan, G. B., and H. H. Shugart. 1989. Environmental factors and ecological processes in boreal forests. *Annual Review of Ecology and Systematics* 20:1–18. An informative review paper.

Knystautas, A. 1987. *The natural history of the USSR*. New York: McGraw-Hill. Contains an interesting, well-illustrated description of the Siberian taiga.

Larsen, J. A. 1980. *The boreal ecosystem*. New York: Academic Press. A summary of what we know about the ecology of the boreal ecosystem.

————. 1989. *The northern forest border in Canada and Alaska*. Ecological Studies 70. New York: Springer-Verlag. An interesting description of the biotic communities and ecological relationships of the forest-tundra ecotone.

Polunin, O., and M. Walters. 1985. *A guide to the vegetation of Britain and Europe*. Oxford, England: Oxford University Press. Excellent description and guide to both tundra and taiga flora.

CHAPTER 30   Temperate Forests

Bormann, F. H., and G. E. Likens. 1979. *Pattern and process in a forest ecosystem*. New York: Springer-Verlag. Excellent discussion of forest ecosystem functions, with special reference to Hubbard Brook.

Brinson, M. M., B. L. Swift, R. C. Plantico, and J. S. Barclay. 1981. *Riparian ecosystems: Their ecology and status*. FWS/OBS-81/17. Kearneysville, WV: Eastern Energy and Land Use Team, U.S. Fish and Wildlife Service. General information on riparian ecosystems is difficult to find. This is an excellent introduction.

Curtis, J. T. 1959. *The vegetation of Wisconsin*. Madison: University of Wisconsin Press. A classic study of forest vegetation of the lake states.

Davis, M. B. (ed.). 1996. *Eastern old-growth forests: Prospects for rediscovery and recovery*. Covelo, CA: Island Press. First book on old-growth forests in eastern North America.

Duvigneaud, P. (ed.). 1971. *Productivity of forest ecosystems*. Paris: UNESCO. Somewhat dated, but an excellent reference on the subject.

Edmonds, R. L. (ed.). 1981. *Analysis of coniferous forest ecosystems in the western United States*. Stroudsburg, PA: Dowden, Hutchinson & Ross. Strong on function.

Franklin, J. F., and C. T. Dyrness. 1973. *Natural vegetation of Oregon and Washington*. USDA Forest Service Gen. Tech. Rept. PNW 8. Corvallis, OR: USDA Forest Service. Excellent description of old-growth forests and other vegetation.

Irland, L. C. 1982. *Wildland and woodlots: The story of New England's forests*. Hanover, NH: University Press of New England. Reviews the changing character and problems of New England's forests since colonial days.

Lang, G. E., W. A. Reiners, and L. H. Pike. 1980. Structure and biomass dynamics of epiphytic lichen communities of balsam fir forests in New Hampshire. *Ecology* 63:541–550.

Lansky, M. 1992. *Beyond the beauty strip: Saving what's left of our forests*. Gardiner, ME: Tilbury House. A detailed and critical examination of forestry practices in the boreal forests of the northeastern United States, applicable elsewhere.

Mather, G. A. S. 1990. *Global forest resources*. Portland, OR: Timber Press. Worldwide overview of forests, forest use, and management or lack thereof.

Norse, E. A. 1990. *Ancient forests of the Pacific Northwest*. Washington, DC: Island Press. Excellent description of the forest and its problems.

Packham, J. R., and D. J. L. Harding. 1982. *Ecology of woodland processes*. London: Edward Arnold. A concise, pleasant introduction to woodland ecology. Although it considers North American examples, it is Europe-oriented, which makes it an excellent introduction to European forest ecology.

Perlan, J. 1991. *A forest journey: The role of wood in the development of civilization*. Cambridge, MA: Harvard University Press. Destruction of the temperate and mediterranean forest and its impact on western civilization.

Peterken, G. F. 1996. *Natural woodlands: Ecology and conservation in north temperate regions*. New York: Cambridge University Press. How woods grow, die, and regenerate in the absence of human interference. An excellent book on old-growth forests.

Polunin, O., and M. Walters. 1985. *A guide to the vegetation of Britain and Europe*. New York: Oxford University Press. Outstanding well-illustrated guide to seminatural and natural vegetation, including detailed description of forests.

Reichle, D. E. (ed.). 1981. *Dynamic properties of forest ecosystems*. Cambridge, England: Cambridge University Press. The major reference source on functions of forest ecosystems throughout the world.

USDA Forest Service. 1980. *Environmental consequences of timber harvesting in Rocky Mountain coniferous forests*. USDA Forest Service Gen. Tech. Rept. INT-90. Ogden, UT: Intermt. For. and Range Exp. Stat. Technical, detailed assessment of timber harvesting in the mountains.

Whittaker, R. H., and G. M. Woodwell. 1969. Structure and function, production, and diversity of the oak-pine forest at Brookhaven, New York. *Journal of Ecology* 57:155–174.

CHAPTER 31   Tropical Forests

Ashton, P. S. 1988. Dipterocarp biology as a window to understanding of tropical forest structure. *Ann. Rev. Ecol. Syst.* 19: 347–370.

Bawa, K. S. 1990. Plant-pollinator interactions in tropical rain forests. *Ann. Rev. Ecol. Syst.* 21:399–422.

Buschbacker, R. J. 1986. Tropical deforestation and pasture development. *Bioscience* 36:22–28. Good review of the various aspects of the problem.

Colchester, M. 1989. *Pirates, squatters and poachers: The political ecology of the dispossession of the native peoples of Sarawak*. London: Survival International. Excellent review of the complex problems surrounding forestry, economic development, and welfare and rights of native peoples.

Collins, M. (ed.). 1990. *The last rain forests: A world conservation atlas*. New York: Oxford University Press. Authoritative, comprehensive world guide to all aspects of the tropical rain forest.

Denslow, J. S. 1987. Tropical rainforest gaps and tree species diversity. *Ann. Rev. Ecol. Syst.* 18:431–451.

Erwin, T. L. 1988. The tropical forest canopy: The heart of biotic diversity. In E. O. Wilson and F. M. Peters (eds.), *Biodiversity*. Washington, DC: National Academy Press.

Fleming, T. H., R. Breitwisch, and G. H. Whitesides. 1987. Patterns of tropical vertebrate frugivore diversity. *Ann. Rev. Ecol. Syst.* 18:71–90. Concise review of tropical frugivory.

Furtado, J. I., and K. Ruddle. 1986. The future of tropical forests. In N. Polunin (ed.), *Ecosystem theory and application*. New York: Wiley, pp. 145–171.

Golley, F. B. (ed.). 1983. *Tropical forest ecosystems: Structure and function*. Ecosystems of the World No. 14A. Amsterdam: Elsevier Scientific Publishing Company. An important reference.

Jansen, D. J. 1986. The future of tropical ecology. *Ann. Rev. Ecol. Syst.* 17:305–324. Discusses approaches needed to maintain the integrity of tropical ecosystems.

Jordan, C. F., and J. R. Kline. 1972. Mineral cycling: Some basic concepts and their application in a tropical rain forest. *Ann. Rev. Ecol. Syst.* 3:33–50.

Lathwell, D. J., and T. L. Grove. 1986. Soil-plant relations in the tropics. *Ann. Rev. Ecol. Syst.* 17:1–16.

Leigh, E. G., Jr. 1975. Structure and climate in tropical rain forest. *Ann. Rev. Ecol. Syst.* 6:67–86. Concise review of effects of climate on lowland and montane rain forests.

Martin, C. 1991. *The rainforests of West Africa*. New York: Birkhauser. A comprehensive treatment of the ecology and conservation of African rain forests.

Murphy, P. G., and A. E. Lugo. 1986. Ecology of tropical dry forests. *Ann. Rev. Ecol. Syst.* 17:67–88.

Myers, N. 1983. *A wealth of wild species*. Boulder, CO: Westview Press. Excellent reference on the value of tropical plants to human welfare.

Richards, P. W. 1996. *The tropical rain forest: An ecological study*, 2nd ed. New York: Cambridge University Press. A thoroughly revised edition of a classic.

Simpson, B. B., and J. Haffer. 1978. Spatial patterns in the Amazonian rain forest. *Ann. Rev. Ecol. Syst.* 9:497–518.

Sutton, S. L., T. C. Whitmore, and A. C. Chadwick (eds.). 1983. *Tropical rain forests: Ecology and management*. Oxford, England: Blackwell.

Tomlinson, P. B. 1987. Architecture of tropical plants. *Ann. Rev. Ecol. Syst.* 18:1–21.

Walter, H. 1971. *Ecology of tropical and subtropical vegetation*. Edinburgh: Oliver & Boyd.

Whitmore, T. C. 1984. *Tropical rainforests of the Far East*. Oxford: Clarendon Press. The authoritative reference on Southeast Asian rain forests.

Whitmore, T. C. 1990. *An introduction to tropical rain forests*. New York: Oxford University Press. An excellent reference on the structure and function of the rain forest.

Wilson, E. O., Jr. (ed.). 1988. *Biodiversity*. Washington, DC: National Academy Press. Contains information section on preserving biological diversity.

CHAPTER 32   Lakes and Ponds

Barnes, R. S. K., and K. H. Mann. 1980. *Fundamentals of aquatic ecosystems*. New York: Blackwell Scientific Publishers. A solid reference.

Brock, T. D. 1985. *A eutrophic lake: Lake Mendota, Wisconsin*. New York: Springer-Verlag. A study of the process of eutrophication.

Carpenter, S. A. 1980. Enrichment of Lake Wingra, Wisconsin, by submerged macrophyte decay. *Ecology* 61:1145–1155.

Carpenter, S. A., and J. F. Kitchell. 1984. Plankton community structure and limnetic primary production. *American Naturalist* 124:159–172.

Carpenter, S. A., J. F. Kitchell, and J. Hodgson. 1985. Cascading trophic interactions and lake productivity. *Bioscience* 35:634–639. Carpenter's three papers provide an excellent introduction to the functioning of lake ecosystems.

Hutchinson, G. E. 1957–1967. *A treatise on limnology.* Vol. 1, *Geography, physics, and chemistry.* Vol. 2, *Introduction to lake biology and limnoplankton.* New York: Wiley. A classic reference.

Likens, G. E. (ed.). 1985. *An ecosystem approach to aquatic ecology: Mirror Lake and environment.* New York: Springer-Verlag. An outstanding detailed study of a lake ecosystem and its interactions with the surrounding terrestrial communities.

Macan, T. T. 1970. *Biological studies of English lakes.* New York: Elsevier.

———. 1973. *Ponds and lakes.* New York: Crane, Russak. Macan's books are classic studies of English lakes.

———. 1974. *Freshwater ecology*, 2nd ed. New York: John Wiley & Sons. Strong on the physical aspects of freshwater ecosystems.

Maitland, P. S. 1978. *Biology of fresh waters.* New York: John Wiley & Sons. Although dated, still a good basic introduction.

Rich, P. H., and R. G. Wetzel. 1978. Detritus in the lake ecosystem. *American Naturalist* 112:57–71. Details the role of detritus in the economy of a lake ecosystem.

CHAPTER 33   Freshwater Wetlands

Cowardin, L. M., V. Carter, and E. C. Golet. 1979. *Classification of wetlands and deepwater habitats of the United States.* U.S. Department of Interior, Fish and Wildlife Service FWS/OBS-79/31. A revised classification of wetlands.

Dahl, T. E. 1990. *Wetland losses in the United States, 1780's to 1980's.* U.S. Department of Interior, Fish and Wildlife Service.

Dugan, P (ed.). 1993. *Wetlands in danger: A world conservation atlas.* New York: Oxford University Press. Survey of world wetland ecosystems and human impacts on them.

Ewel, K. C. 1990. Multiple demands on wetlands. *Bioscience* 40:660–666. Societal benefits of wetlands, with cypress swamps serving as a case study.

Ewel, K. C., and H. T. Odum (eds.). 1986. *Cypress swamps.* Gainesville: University Presses of Florida. In-depth studies of the structure, function, and management of cypress swamps in southern United States.

Good, R. E., D. F. Whigham, and R. L. Simpson (eds.). 1978. *Freshwater wetlands: Ecological processes and management potential.* New York: Academic. A review of functional aspects of wetlands and their management implications.

Gore, A. P. J. (ed.). 1983. *Mire, swamp, bog, fen, and moor.* Ecosystems of the world 4A and 4B. Amsterdam: Elsevier. A review of the world of freshwater wetlands.

Greesen, P. S., J. R. Clark, and J. E. Clark (eds.). 1979. *Wetland functions and values: The state of our understanding.* Minneapolis: American Water Resources Association. An earlier work still of value.

Lugo, A. E. 1990. *The forested wetlands.* Amsterdam: Elsevier. Excellent overview and discussion of structure and function.

Mitsch, W. J., and J. C. Gosslink. 1993. *Wetlands*, 2nd ed. New York: Van Nostrand Reinhold. A pioneering text and major reference.

Moore, P. D., and D. J. Bellemany. 1974. *Peatlands.* New York: Springer-Verlag. An outstanding introduction to the ecology and development of peatlands.

National Audubon Society. 1990. The last wetlands. *Audubon* 92(4). A highly informative issue devoted entirely to wetlands, their management and preservation.

Niering, W. A., and B. Hales. 1991. *Wetlands of North America.* Charlottesville, VA: Thomasson-Grant. An exceptionally illustrated survey of North American wetlands, both freshwater and salt.

Payne, N. F. 1992. *Techniques for wildlife habitat management of wetlands.* New York: McGraw-Hill. Although emphasis is on management, this book contains a wealth of information on wetland types, structure, and function.

Tiner, R. W. 1991. The concept of a hydrophyte for wetland identification. *Bioscience* 41:236–247. Reviews problems and means of identifying wetlands.

Van der Valk, A. (ed.). 1989. *Northern prairie wetlands.* Ames, IA: Iowa State University Press. Detailed studies of major wetlands rapidly disappearing.

Weller, M. W. 1981. *Freshwater wetlands: Ecology and wildlife management.* Minneapolis: University of Minnesota Press. An excellent nontechnical introduction.

CHAPTER 34   Streams and Rivers

Cummins, K. W. 1974. Structure and function of stream ecosystems. *Bioscience* 24:631–641.

Cummins, K. W. 1979. Feeding ecology of stream invertebrates. *Ann. Rev. Ecol. Syst.* 10:147–172. A good review of functional aspects of stream organisms.

Gore, J. A., and G. E. Petts. 1989. *Alternatives in regulated river management.* Boca Raton, FL: CRC Press. Focuses attention on ways to reduce ecological effects of river regulation.

Hynes, H. B. N. 1970. *The ecology of running water.* Toronto: University of Toronto Press. Dated, but a classic and valuable work; a major reference.

Meyer, J. L. 1990. A blackwater perspective on riverine ecosystems. *Bioscience* 40:643–651. A detailed look at stream ecosystem function, especially food webs.

Petts, G. E. 1984. *Impounded rivers: Perspectives for ecological management.* New York: Wiley. Detailed analysis of the effects of dams on the world's rivers, especially downstream.

Stanford, J. A., and A. P. Covich (eds.). 1988. Community structure and function in temperate and tropical streams. *J. North Amer. Benthol. Soc.* 7:261–529. A valuable special issue of the journal.

Vannote, R. L., G. W. Minshall, K. W. Cummins, J. R. Sedell, and C. E. Cushing. 1980. The river continuum concept. *Can. J. Fish. Aquat. Sci.* 37:130–137.

Ward, J. V. 1979. *The ecology of regulated streams.* New York: Plenum.

Whitten, B. A. (ed.). 1975. *River ecology.* Berkeley: University of California Press.

CHAPTER 35   Oceans

Boucher, G. 1985. Long-term monitoring of meiofauna densities after the *Amoco Cadiz* oil spill. *Mar. Pollution Bull.* 16:328–333.

Carson, R. 1961. *The sea around us.* New York: Oxford University Press. A classic book on the sea.

Fenchel, T. 1987. Marine plankton food chains. *Ann. Rev. Ecol. Syst.* 19:19–38.

Grassle, J. F. 1985. Hydrothermal vent animals: Distribution and biology. *Science* 229:713–717.

———. 1989. Species diversity in deep-sea communities. *Trends Ecol. Evol.* 4:12–15.

———. 1991. Deep-sea benthic diversity. *Bioscience* 41:464–469.

Gross, M. G. 1982. *Oceanography: A view of Earth*, 3rd ed. Englewood Cliffs, NJ: Prentice-Hall. An excellent reference on the physical aspects of the sea.

Hardy, A. 1971. *The open sea: Its natural history.* Boston: Houghton Mifflin. A classic introduction.

Hayman, R. M., and R. C. McDonald. 1985. The ecology of deep sea hot springs. *American Scientist* 73:441–449.

Kinne, O. (ed.). 1978. *Marine ecology*, 5 vol. A major and often technical reference source.

Marshall, N. B. 1980. *Deep-sea biology: Developments and perspectives.* New York: Garland STMP Press.

National Academy of Sciences. 1985. *Oil in the sea: Inputs, fates, and effects.* Washington, DC: National Academy Press.

Nelson-Smith, A. 1972. *Oil pollution and marine ecology.* London: Elek Science. Good, concise basic reference.

Nybakken, J. W. 1997. *Marine biology: An ecological approach*, 4th ed. Menlo Park, CA: Benjamin/Cummings. A solid reference on marine life and ecosystems.

Powell, M. A., and C. N. Somero. 1983. Blood components prevent sulfide poisoning of respiration of the hydrothermal vent tubeworm *Riftia pachyptila*. *Science* 219:297–299.

Rex, M. A. 1981. Community structure in the deep-sea benthos. *Ann. Rev. Ecol. Syst.* 12:331–353. A good review of deep-sea species diversity.

Steele, J. 1974. *The structure of a marine ecosystem.* Cambridge, MA: Harvard University Press.

CHAPTER 36 Intertidal Zones and Coral Reefs

Carson, R. 1955. *The edge of the sea.* Boston: Houghton Mifflin. A classic introduction.

Dayton, P. 1971. Competition, disturbance, and community organization: The provision and subsequent utilization of space in a rocky intertidal community. *Ecological Monographs* 45:137–159.

Eltringham, S. K. 1971. *Life in mud and sand.* New York: Crane, Russak. A concise, informative introduction.

Hiatt, R. W., and D. W. Strasburg. 1960. Ecological relationships of the fish fauna on coral reefs of the Marshall Islands. *Ecological Monographs* 30:66–120. An important, well-illustrated reference on coral reef fish.

Huston, M. 1985. Patterns of species diversity on coral reefs. *Ann. Rev. Ecol. Syst.* 16:149–177. An important reference on the roles of light and disturbance.

Jackson, J. B. C. 1991. Adaptation and diversity of reef corals. *Bioscience* 41:475–482. Relates patterns of species distribution to life history and disturbance.

Jones, O. A., and R. Endean (eds.). 1973, 1976. *Biology and geology of coral reefs.* Vols. II, III. New York: Academic Press.

Leigh, E. G., Jr. 1987. Wave energy and intertidal productivity. *Proc. Natl. Acad. Sci. USA* 84:1314.

Lessios, H. A. 1988. Mass mortality of *Diadema antillarum* in the Caribbean: What have we learned? *Ann. Rev. Ecol. Syst.* 19:371–393. Detailed review of the ecological impact of the die-off.

Lubchenco, J. 1978. Algal zonation in the New England rocky intertidal community: An experimental analysis. *Ecology* 61:333–344. An outstanding study and informative reference.

Mathieson, A. C., and P. H. Nienhuis (eds.). 1991. *Intertidal and littoral ecosystems. Ecosystems of the world 24.* Amsterdam: Elsevier. A detailed, broad survey of the intertidal and littoral zones of the world.

Moore, P. G., and R. Seed (eds.). 1986. *The ecology of rocky shores.* New York: Columbia University Press. Comprehensive, worldwide review of the rocky intertidal zone.

Newell, R. C. 1970. *Biology of intertidal animals.* New York: Elsevier. Adaptations of animals to the intertidal environment.

Nybakken, J. W. 1997. *Marine biology: An ecological approach*, 4th ed. Menlo Park, CA: Benjamin/Cummings. Informative chapters on rocky, sandy, and muddy shores and coral reefs.

Paine, R. T. 1969. The *Pisaster-Tegula* interaction: Prey patches, predator food preference, and intertidal community structure. *Ecology* 59:150–961. An outstanding paper on the role of predation in the rocky shore community.

Pomeroy, L. R., and E. J. Kuenzler. 1969. Phosphorus turnover by coral reef animals. In D. J. Nelson and F. E. Evans (eds.), *Symposium on radioecology conf.* 670503. Springfield, VA: National Technical Information Services. Pp. 478–483.

Reaka, M. J. (ed.). 1985. *Ecology of coral reefs.* Symposia series for undersea research, NOAA Underseas Research Program 3. Washington DC: US Department of Commerce.

Sale, P. F. 1980. The ecology of fishes on coral reefs. *Oceanogr. Mar. Biol. Ann. Rev.* 18:367–421. Sweeping review.

Stephenson, T. A., and A. Stephenson. 1973. *Live between the tidemarks on rocky shores.* San Francisco: Freeman. A detailed description of the structure of intertidal life around the world.

Underwood, A. J., E. J. Denley, and M. J. Moran. 1983. Experimental analyses of the structure and dynamics of midshore rocky intertidal communities in New South Wales. *Oecologica* 56:202–219.

Wellington, G. W. 1982. Depth zonation of corals in the Gulf of Panama: Control and facilitation by resident reef fishes. *Ecological Monographs* 52:223–241.

Wilson, R., and J. Q. Wilson. 1985. *Watching fishes: Life and behavior on coral reefs.* New York: Harper & Row. Written for a general audience, this well-illustrated book is accessible and informative.

Yonge, C. M. 1949. *The seashore.* London: Collins. A well-illustrated classic.

CHAPTER 37 Estuaries, Salt Marshes, and Mangrove Forests

Bertness, M. D. 1984. Ribbed mussels and *Spartina alterniflora* production on a New England marsh. *Ecology* 65:1794–1807. Relationship between consumer and producer in the salt marsh.

Bildstein, K. L., G. T. Bancroft, P. J. Dugan, et al. 1991. Approaches to the conservation of coastal wetlands in the Western Hemisphere. *Wilson Bulletin* 103:218–254. Excellent overview of problems.

Chapman, V. J. 1976. *Mangrove vegetation.* Leutershausen, Germany: J. Cramer. An authoritative reference on mangroves and mangrove forests.

Chapman, V. J (ed.). 1977. *Wet coastal ecosystems.* Amsterdam: Elsevier. Covers salt marshes and mangals of the world. A basic reference.

Clark, J. 1974. *Coastal ecosystems: Ecological considerations for the management of the coastal zone.* Washington, DC: Conservation Foundation.

Haines, B. L., and E. L. Dunn. 1985. Coastal marshes. In B. F. Chabot and H. A. Mooney (eds.), *Physiological ecology of North American plant communities.* New York: Chapman and Hall. Pp. 323–347.

Hopkinson, C. S., and J. P. Schubauer. 1984. Static and dynamic aspects of nitrogen cycling in the salt marsh graminoid *Spartina alterniflora. Ecology* 65:961–969.

Howarth, R. W., and J. Teal. 1979. Sulfate reduction in a New England salt marsh. *Limno. Oceanogr.* 24:999–1013.

Jefferies, R. L., and A. J. Davy (eds.). 1979. *Ecological processes in coastal environments.* Oxford: Blackwell.

Josselyn, M. 1983. The ecology of San Francisco tidal marshes: A community profile. U.S. Fish and Wildlife Service Office of Biological Services. FWS/OBS–82/83.

Ketchum, B. H. (ed.). 1983. *Estuaries and enclosed seas.* Ecosystems of the World 26. Amsterdam: Elsevier. Major reference on world estuarine structure and function.

Long, S. P., and C. F. Mason. 1983. *Salt marsh ecology.* New York: Chapman & Hall. A good, accessible introduction to the salt marsh ecosystem.

Lugo, A. E., and S. C. Snedaker. 1974. The ecology of mangroves. *Ann. Rev. Ecol. Syst.* 5:39–64.

McLusky, D. S. 1971. *Ecology of estuaries.* London: Heinemann Educational.

———. 1989. *The estuarine ecosystem,* 2nd ed. New York: Chapman & Hall. Clearly describes the structure and function of estuarine ecosystems.

Nixon, S. W., and C. A. Oviatt. 1973. Ecology of a New England salt marsh. *Ecological Monographs* 43:463–498.

Nybakken, J. W. 1997. *Marine biology: An ecological approach,* 4th ed. Menlo Park, CA: Benjamin/Cummings. Chapters on estuaries, salt marshes, and mangrove forests.

Odum, W. E., C. C. McIvor, and T. J. Smith III. 1982. The ecology of the mangroves of South Florida: A community profile. U.S. Fish and Wildlife Service Office of Biological Services. FWS/OBS 81/24.

Perkins, E. J. 1974. *The biology of estuaries and coastal waters.* New York: Academic Press.

Pomeroy, L. R., and R. G. Wiegert (eds.). 1981. *The ecology of a salt marsh.* New York: Springer-Verlag. Synthesis of a 20-year study of all aspects of a southern United States coastal marsh.

Stout, J. P. 1984. The ecology of irregularly flooded salt marshes of northeastern Gulf of Mexico: A community profile. U.S. Fish and Wildlife Service Biol. Rep. 85(7.1).

Teal, J. 1962. Energy flow in a salt marsh ecosystem of Georgia. *Ecology* 43:614–624.

Teal, J., and M. Teal. 1969. *Life and death of the salt marsh.* Boston: Little, Brown. A classic.

Valiela, I. 1984. *Marine biology processes.* New York: Springer-Verlag. An advanced synthesis.

Valiela, I., and J. M. Teal. 1979. The nitrogen budget of a salt marsh ecosystem. *Nature:* 47:337–371.

Valiela, I., J. M. Teal, and W. G. Denser. 1978. The nature of the growth forms in salt marsh grass *Spartina alterniflora. Am. Nat.* 112:461–470.

Wiley, M. 1976. *Estuarine processes.* New York: Academic Press.

Zedler, J., T. Winfield, and D. Mauriello. 1982. The ecology of southern California coastal marshes. A community profile. U.S. Fish and Wildlife Service Office of Biological Services. FWS/OBS 81/54.

A horizon Surface stratum of mineral soil, characterized by maximum accumulation of organic matter, maximum biological activity, and loss of such materials as iron, aluminum oxides, and clays.

abiotic Nonliving; the abiotic component of the environment includes soil, water, air, light, nutrients, and the like.

abrupt speciation Spontaneous rise of a new species, largely through polyploidy.

abundance The number of individuals of a species in a given area.

abyssal Relating to the bottom waters of oceans, usually below 1000 m.

acclimation Alteration of an individual's physiological rate or capacity to perform a function through long-term exposure to certain conditions.

acclimatization Changes in physiological state or tolerance that appear in a species after long exposure to different natural environments.

acid deposition Wet and dry atmospheric fallout with an extremely low pH, brought about when water vapor in the atmosphere combines with hydrogen sulfide and nitrous oxide vapors released by burning fossil fuels; the sulfuric and nitric acid in rain, fog, snow, gases, and particulate matter.

active transport Movement of ions and molecules across a cell membrane against a concentration gradient, involving an expenditure of energy, in a direction opposite to simple diffusion.

adaptation A genetically determined characteristic (behavioral, morphological, or physiological) that improves an organism's ability to survive and reproduce under prevailing environmental conditions.

adaptive radiation Evolution from a common ancestor of divergent forms adapted to distinct ways of life.

adiabatic cooling A decrease in air temperature when a rising parcel of warm air cools by expanding (which uses energy) rather than losing heat to the surrounding air; the rate of cooling is approximately 1° C/100 m for dry air and 0.6° C/100 m for moist air.

adiabatic lapse rate Rate at which a parcel of air loses temperature with elevation if no heat is gained from or lost to an external source.

adiabatic process A process in which heat is neither lost to nor gained from the outside.

aerenchyma Plant tissue with large air-filled intercellular spaces, usually found in roots and stems of aquatic and marsh plants.

aerobic Living or occurring only in the presence of free uncombined molecular oxygen either as a gas in the atmosphere or dissolved in water.

aestivation Dormancy in animals through a drought or dry season.

age-specific schedule of birth Average number of offspring produced per individual per unit time as a function of age class.

age structure The number or proportion of individuals in each age group within a population.

aggregate A group of soil particles adhering in a cluster; compare *ped.*

aggregative response Behavior of consumers who spend most time in food patches with the greatest density of prey.

aggressive mimicry Resemblance of a predator or parasite to a harmless species to deceive potential prey.

alfisol Soil characterized by an accumulation of iron and aluminum in the B horizon.

allele One of two or more alternative forms of a gene that occupies the same relative position or locus on homologous chromosomes.

allelopathy Effect of metabolic products of plants (excluding microorganisms) on the growth and development of other nearby plants.

allogenic succession Ecological change or development of species structure and community composition brought about by some external force, such as fire or storms.

allopatric Having different areas of geographical distribution; possessing nonoverlapping ranges.

allopatric speciation The separation of a population into two or more evolutionary units by some geographical barrier that causes reproductive isolation.

alluvial soil Soil developing from recent alluvium (material deposited by running water), exhibiting no horizon development, and typical of floodplains.

alpha diversity The variety of organisms occurring in a given place or habitat; compare *beta diversity, gamma diversity.*

altricial Condition among birds and mammals of being hatched or born usually blind and too weak to support their own weight.

altruism A form of behavior in which an individual increases the welfare of another at the expense of its own welfare.

ambient Surrounding, external, or unconfined in condition.

amensalism Relationship between two species in which one is inhibited or harmed, while the other (the amensal) is unaffected.

ammonification Breakdown of proteins and amino acids, especially by fungi and bacteria, with ammonia as the excretory by-product.

anaerobic Adapted to environmental conditions devoid of oxygen.

andisol Soil derived from volcanic ejecta, not highly weathered, with a dark upper layer.

annual grassland Grassland in California dominated by exotic annual grasses that reseed every year, replacing native perennial grasses.

anticyclone An area of high atmospheric pressure characterized by subsiding air and horizontal divergence of air near the surface in its central region.

apparent plants Large, easy to locate plants possessing quantitative defenses not easily mobilized at the point of attack, such as tannins.

aridisol Desert soil characterized by little organic matter and high base content.

arroyo A water-carved canyon in a desert.

asexual reproduction Any form of reproduction, such as budding, that does not involve the fusion of gametes.

assimilation Transformation or incorporation of a substance by organisms; absorption and conversion of energy and nutrients into constituents of an organism.

association A natural unit of vegetation characterized by a relatively uniform species composition and often dominated by a particular species.

atoll A ring-shaped coral reef that encloses or almost encloses a lagoon and is surrounded by open sea.

ATP Adenosine triphosphate; major energy-transferring molecules in all biological systems.

aufwuchs Community of plants and animals attached to or moving about on submerged surfaces; compare *periphyton*.

autogenic Self-generated.

autogenic succession Succession driven by environmental changes brought about by the organisms themselves.

autotrophic community Community whose energy source is photosynthesis, thus based on primary producers.

autotrophic succession Succession in a predominantly inorganic environment with early and continued dominance of green plants (autotrophs).

autotrophy Ability of an organism to produce organic material from inorganic chemicals and some source of energy.

available water capacity Supply of water available to plants in a well-drained soil.

B horizon Soil stratum beneath the A horizon, characterized by an accumulation of silica, clay, and iron and aluminum oxides and possessing blocky or prismatic structure.

basal metabolic rate The minimal amount of energy expenditure needed by an animal to maintain vital processes.

Batesian mimicry Resemblance of a palatable or harmless species, the mimic, to an unpalatable or dangerous species, the model.

bathyal Pertaining to anything, but especially organisms, in the deep sea, below the photic or lighted zone, and above 4000 m.

bathypelagic Lightless zone of the open ocean, lying above the abyssal or bottom water, usually above 4000 m.

behavioral ecology The study of the behavior of an organism in its natural habitat.

benthic zone The area of the sea bottom.

benthos Animals and plants living on the bottom of a lake or sea, from the high water mark to the greatest depth.

beta diversity Variety of organisms occupying different habitats over a region; regional diversity; compare *alpha diversity, gamma diversity*.

biennial Plant that requires two years to complete a life cycle, with vegetative growth the first year and reproductive growth (flowers and seeds) the second.

biochemical oxygen demand (BOD) A measure of the oxygen needed in a specified volume of water to decompose organic materials; the greater the amount of organic matter in water, the higher the BOD.

biodiversity A measure of the different kinds of organisms within a given region.

biogeochemical cycle Movement of elements or compounds through living organisms and the nonliving environment.

biological clock The internal mechanism of an organism that controls circadian rhythms without external time cues.

biological magnification Process by which pesticides and other substances become more concentrated in each link of the food chain.

biological species A group of potentially interbreeding populations reproductively isolated from all other populations.

bioluminescence Production of light by living organisms.

biomass Weight of living material, usually expressed as dry weight per unit area.

biome Major regional ecological community of plants and animals; usually corresponds to plant ecologists' and European ecologists' classification of plant formations and life zones.

biosphere Thin layer about Earth in which all living organisms exist.

biotic community Any assemblage of populations living in a prescribed area or physical habitat.

blanket mire Large area of upland dominated by sphagnum moss and dependent upon precipitation for a water supply; a moor.

bog Wetland ecosystem characterized by an accumulation of peat, acid conditions, and dominance of sphagnum moss.

bottleneck An evolutionary term for any stressful situation that greatly reduces a population.

brood parasitism Laying eggs in the nest of another species or in the nest of another individual of the same species.

browse The part of current leaf and twig growth of shrubs, woody vines, and trees available for animal consumption.

bryophyte Member of the division in the plant kingdom of nonflowering plants comprising mosses (Musci), liverworts (Hepaticae), and hornworts (Anthocerotae).

buffer A chemical solution that resists or dampens change in pH upon addition of acids or bases.

C horizon Soil stratum beneath the solum (A and B horizons), little affected by biological activity or soil-forming processes.

C₃ plant Any plant that produces as its first step in photosynthesis the three-carbon compound phosphoglyceric acid.

C₄ plant Any plant that produces as its first step in photosynthesis a four-carbon compound, malic or aspartic acid.

calcicole Plant susceptible to aluminum toxicity, acidity, and other factors influenced by the absence of calcium.

calcification Process of soil formation characterized by accumulation of calcium in lower horizons.

calcifuge Plant with a low calcium requirement that can live in soils with a pH of 4.0 or less.

caliche An alkaline, often rocklike salt deposit on the surface of soil in arid regions; it forms at the level where leached Ca salts from the upper soil horizons are precipitated.

calorie Amount of heat needed to raise 1 g of water 1° C, usually from 15° C to 16° C.

CAM plant (Crassulacean Acid Metabolism) Plant (cactus or other succulent) that separates the processes of carbon dioxide uptake and fixation when growing under arid conditions; it takes up gaseous carbon dioxide at night, when stomata are open, and uses it during the day, when stomata are closed.

cannibalism Killing and consuming one's own kind; intraspecific predation.

capillary water That portion of water in the soil held by capillary forces between soil particles.

carnivore Organism that feeds on animal tissue; taxonomically, a member of the order Carnivora (Mammalia).

carrying capacity (*K*) Number of individual organisms the resources of a given area can support, usually through the most unfavorable period of the year.

cation Part of a dissociated molecule carrying a positive electrical charge.

cation exchange capacity Ability of a soil particle to absorb positively charged ions.

chamaephyte Perennial shoot or bud from the surface of the ground to about 25 cm above the surface.

chaparral Vegetation consisting of broadleaved evergreen shrubs, found in regions of mediterranean-type climate.

chemical ecology Study of the nature and use of chemical substances produced by plants and animals.

chilling tolerance Ability of a plant to carry on photosynthesis within a range of +5° C to +10° C.

chromosome One of a group of threadlike structures of different lengths and sizes in the nuclei of cells of eukaryote organisms.

circadian rhythm Endogenous rhythm of physiological or behavioral activity of approximately 24 hours duration.

climate Long-term average pattern of local, regional, or global weather.

climax Stable end community of succession that is capable of self-perpetuation under prevailing environmental conditions.

climograph A diagram describing a locality based on the annual cycle of temperature and precipitation.

cline Gradual change in population characteristics over a geographical area, usually associated with changes in environmental conditions.

clone A population of genetically identical individuals resulting from asexual reproduction.

closed system A system that exchanges no energy with the surrounding environment.

coevolution Joint evolution of two or more noninterbreeding species that have a close ecological relationship; through reciprocal selective pressures, the evolution of one species in the relationship is partially dependent on the evolution of the other.

coexistence Two or more species living together in the same habitat, usually with some form of competitive interaction.

cohort A group of individuals of the same age.

cold resistance Ability of a plant to resist low temperature stress without injury.

collectors Feeding group of stream invertebrates that filter fine organic particles from flowing water or pick up particles from the stream bottom.

colluvium Mixed deposit of soil material and rock fragments accumulated near the base of a steep slope through soil creep, landslides, and local surface runoff.

commensalism Relationship between species that is beneficial to one, but neutral or of no benefit to the other.

community A group of interacting plants and animals inhabiting a given area.

community ecology Study of the living component of ecosystems; description and analysis of patterns and processes within the community.

compartment Major reservoir or component of an ecosystem.

compensation level Light intensity at which photosynthesis and respiration balance each other, so that net production is 0; in aquatic systems, usually the depth of light penetration at which oxygen utilized in respiration equals oxygen produced by photosynthesis.

competition Any interaction that is mutually detrimental to both participants, occurring between species that share limited resources.

competitive exclusion principle Hypothesis that when two or more species coexist using the same resource, one must displace or exclude the other.

competitive release Niche expansion in response to reduced interspecific competition.

conduction Direct transfer of heat from one substance to another.

consumer Any organism that lives on other organisms, dead or alive.

consumption efficiency Ratio of ingestion to production or energy available.

contest competition Competition in which a limited resource is shared only by dominant individuals; a relatively constant number of individuals survive, regardless of initial density.

continental shelf Gently seaward-sloping surface of a continent that extends to a depth of about 200 m.

continuum A gradient of environmental characteristics or changes in community composition.

convection Transfer of heat by the circulation of a liquid or gas.

convergent evolution Development of similar characteristics in different species living in different areas under similar environmental conditions.

coprophagy Feeding on feces.

Coriolis force Physical consequence of the law of conservation of angular momentum; as a result of Earth's rotation, a moving object veers to the right in the Northern Hemisphere and to the left in the Southern Hemisphere relative to Earth's surface.

countercurrent circulation An anatomical and physiological arrangement by which heat exchange takes place between outgoing warm arterial blood and cool venous blood returning to the body core; important in maintaining temperature homeostasis in many vertebrates.

covalence Sharing of a pair of electrons between two atoms.

critical daylength The period of daylight, specific for any given species, that triggers a long-day or a short-day response in organisms.

critical thermal maximum Temperature at which an animal's capacity to move is so reduced that it cannot escape from thermal conditions that will lead to death.

crown shyness Growth pattern of Asian tropical rain forest trees in which crowns are spaced about a meter apart, giving the impression of a jigsaw puzzle.

crude birth rate The number of young produced per unit of population.

crude density The number of individuals per unit area, compare *ecological density.*

cryptic coloration Coloration of organisms that makes them resemble or blend into their habitat or background.

cryptophyte Plant with overwintering buds buried in the ground on a bulb or rhizome.

cultural eutrophication Accelerated nutrient enrichment of aquatic ecosystems by a heavy influx of pollutants that causes major shifts in plant and animal life.

current Water movements that result in the horizontal transport of water masses.

cyclic replacement Succession in which the sequence of seral stages is repeated by imposition of some disturbance, so that the sere never arrives at a climax or stable sere.

day-neutral plant A plant that does not require any particular photoperiod to flower.

death rate Number of individuals in a population dying in a given time interval divided by the number alive at the midpoint of the time interval.

deciduous Of leaves, shed during a certain season (winter in temperate regions; dry season in the tropics); of trees, having deciduous parts.

decomposer Organism that obtains energy from the breakdown of dead organic matter to simpler substances; most precisely refers to bacteria and fungi.

decomposition Breakdown of complex organic substances into simpler ones.

deductive method In testing hypotheses, going from the specific to the general.

defensive mutualism Relationship in which one of the mutualists seems to protect the other from harm.

definitive host Host in which a parasite becomes an adult and reaches maturity.

deme Local population or interbreeding group within a larger population.

demography The statistical study of the size and structure of populations and changes within them.

denitrification Reduction of nitrates and nitrites to nitrogen by microorganisms.

density Size of a population in relation to a definite unit of space; see *crude density, ecological density*.

density dependence Regulation of population growth by mechanisms controlled by the size of the population; effect increases as population size increases.

density independence Being unaffected by population density; regulation of growth is not tied to population density.

dependent variable Variable *y*, the second of two numbers in an ordered pair (*x, y*); the set of all values taken on by the dependent variable is called the range of the function; compare *independent variable*.

desert grassland Grassland of hot, dry climates, with rainfall varying between 200 and 500 mm, dominated by bunchgrasses and widely interspersed with other desert vegetation.

desertification Process of desert expansion or formation as a consequence of climatic change, poor land management, or both.

deterministic model Mathematical model in which all relationships are fixed and a given input produces one exact prediction as an output.

detrital food chain Food chain in which detritivores consume detritus or litter, mostly from plants, with subsequent transfer of energy to various trophic levels; ties into the grazing food chain; compare *grazing food chain*.

detritivore Organism that feeds on dead organic matter; usually applies to detritus-feeding organisms other than bacteria and fungi.

detritus Fresh to partly decomposed plant and animal matter.

dew point Temperature at which condensation of water in the atmosphere begins.

diameter at breast height (dbh) Diameter of a tree measured at 1.4 m (4 feet, 6 inches) from ground level.

diapause A period of dormancy, usually seasonal, in the life cycle of an insect, in which growth and development cease and metabolism greatly decreases.

diffuse coevolution Coevolution involving the interactions of many organisms, in contrast to pair interactions.

diffuse competition Competition in which a species experiences interference from numerous other species that deplete the same resources.

diffusion Spontaneous movement of particles of gases or liquids from an area of high concentration to an area of low concentration.

dimorphism Existing in two structural forms, two color forms, two sexes, and the like.

dioecious Having male and female reproductive organs on separate plants; compare *monoecious*.

diploid Having chromosomes in homologous pairs, or twice the haploid number of chromosomes.

directional selection Selection favoring individuals at one extreme of the phenotype in a population.

disease Any deviation from a normal state of health.

dispersal Leaving an area of birth or activity for another area.

dispersion Distribution of organisms within a population over an area.

disruptive selection Selection in which two extreme phenotypes in the population leave more offspring than the intermediate phenotype, which has lower fitness.

distribution Arrangement of organisms within an area.

disturbance A discrete event in time that disrupts an ecosystem, community, or population, changing substrates and resource availability.

diversity Abundance of different species in a given location; species richness.

diversity index The mathematical expression of species richness of a given community or area.

doldrums Oceanic equatorial zone that has low pressure and light variable winds; moves seasonally north and south of the equator.

dominance In a community, control over environmental conditions influencing associated species by one or several species, plant or animal, enforced by number, density, or growth form; in a population, behavioral, hierarchical order that gives high-ranking individuals priority of access to essential resources; in genetics, ability of an allele to mask the expression of an alternative form of the same gene in a heterozygous condition.

dominant Population possessing ecological dominance in a given community and thereby governing type and abundance of other species in the community.

dormancy State of cessation of growth and suspended biological activity, during which life is maintained.

drought avoidance Ability of a plant to escape dry periods by becoming dormant or surviving as a seed.

drought resistance Sum of drought tolerance and drought avoidance.

drought tolerance Ability of plants to maintain physiological activity in spite of the lack of water or to survive the drying of tissues.

dynamic pool model Optimum yield model using growth, recruitment, mortality, and fishing intensity to predict yield.

dystrophic Term applied to a body of water with a high content of humic or organic matter, often with high littoral productivity and low plankton productivity.

E horizon Mineral horizon characterized by the loss of clay, iron, or aluminum and a concentration of quartz and other resistant minerals in sand and silt sizes; light in color.

early successional species Plant species characterized by high dispersal rates, ability to colonize disturbed sites, short life span, and shade intolerance.

easterlies A system of broad steady prevailing winds around Earth over the equatorial regions created by the westward deflection of air that follows the barometric pressure gradients from subtropical high to equatorial low; also called trade winds.

ecocline A geographical gradient of communities or ecosystems produced by responses of vegetation to environmental gradients of rainfall, temperature, nutrient concentrations, and other factors.

ecological density Density measured in terms of the number of individuals per area of available living space; compare *crude density*.

ecological efficiency Percentage of biomass produced by one trophic level that is incorporated into biomass of the next highest trophic level.

ecological pyramid A graphical representation of the trophic structure and function of an ecosystem.

ecological release Expansion of habitat or increase in food availability resulting from release of a species from interspecific competition.

ecology The study of relations between organisms and their natural environment, living and nonliving.

ecosystem The biotic community and its abiotic environment functioning as a system.

ecotone Transition zone between two structurally different communities; see also *edge*.

ecotype Subspecies or race adapted to a particular set of environmental conditions.

ectomycorrhizae Mutualistic association between fungi and roots in which the fungi form sheaths around the outside of the roots.

ectoparasite Parasite, such as a flea, that lives in the fur, feathers, or skin of the host.

ectothermy Determination of body temperature primarily by external thermal conditions.

edaphic Relating to soil.

edge Place where two or more vegetation types meet.

edge effect Response of organisms, animals in particular, to environmental conditions created by the edge.

effective population size The size of an ideal population that would undergo the same amount of random genetic drift as the actual population; sometimes used to measure the amount of inbreeding in a finite, randomly mating population.

egestion Elimination of undigested food material.

elaiosome Shiny, oil-containing, ant-attracting tissue on the seed coat of many plants.

emigration Movement of part of a population permanently out of an area.

endemic Restricted to a given region.

endoparasite Parasite that lives within the body of the host.

endothermic reaction Chemical reaction that gains energy from the environment.

endothermy Regulation of body temperature by internal heat production; allows maintenance of appreciable difference between body temperature and external temperature.

energy Capacity to do work.

entisols Embryonic mineral soils whose profile is just beginning to develop; common on recent floodplains and wind deposits, they lack distinct horizons.

entrainment Synchronization of an organism's activity cycle with environmental cycles.

entropy Transformation of matter and energy to a more random, more disorganized state.

environment Total surroundings of an organism, including other plants and animals and those of its own kind.

epidemic Rapid spread of a bacterial or viral disease in a human population; compare *epizootic*.

epifauna Benthic animals that live on or move over the surface of a substrate.

epiflora Benthic plants that live on the surface of a substrate.

epilimnion Warm, oxygen-rich upper layer of water in a lake or other body of water, usually seasonal.

epiphyte Plant that lives wholly on the surface of other plants, deriving support but not nutrients from them.

epizootic Rapid spread of a bacterial or viral disease in a dense population of animals.

equilibrium species Species whose population exists in equilibrium with resources and at a stable density.

equilibrium turnover rate Change in species composition per unit time when immigration equals extinction.

equitability Evenness of distribution of species abundance patterns; maximum equitability is the same number of individuals among all species in the community.

estivation Dormancy in animals during a period of drought or a dry season.

estuary A partially enclosed embayment where fresh water and seawater meet and mix.

euphotic zone Surface layer of water to the depth of light penetration where photosynthetic production equals respiration.

eutrophic Term applied to a body of water with high nutrient content and high productivity.

eutrophication Nutrient enrichment of a body of water; called cultural eutrophication when accelerated by introduction of massive amounts of nutrients from human activity.

eutrophy Condition of being nutrient-rich.

evaporation Loss of water vapor from soil or open water or another exposed surface.

evapotranspiration Sum of the loss of moisture by evaporation from land and water surfaces and by transpiration from plants.

evenness Degree of equitability in the distribution of individuals among a group of species; see *equitability*.

evolution Change in gene frequency through time resulting from natural selection and producing cumulative changes in characteristics of a population.

evolutionary ecology Integrated study of evolution, genetics, natural selection, and adaptations within an ecological context; evolutionary interpretation of population, community, and ecosystem ecology.

exothermic reaction Chemical reaction that releases heat to the environment.

exploitative competition Competition by a group or groups of organisms that reduces a resource to a point that adversely affects other organisms.

exponential growth (r) Instantaneous rate of population growth, expressed as proportional increase per unit of time.

extinction coefficient Point at which the intensity of light reaching a certain depth is insufficient for photosynthesis; ratio of intensity of light at a given depth to intensity at the surface.

F$_1$ generation The first generation of offspring from a cross between individuals homozygous for contrasting alleles; the F$_1$ is necessarily heterozygous.

F$_2$ generation Offspring produced by selfing or by allowing the F$_1$ generation to breed among themselves.

facilitation model A model of succession in which a community prepares or "facilitates" the way for a succeeding community.

facultative Able to adjust optimally to different environmental conditions.

fecundity Potential ability of an organism to produce eggs or young; rate of production of young by a female.

fell-field Area within a tundra characterized by stony debris and sparse vegetation.

fen Slightly acidic wetland dominated by sedges, in which peat accumulates.

fermentation Breakdown of carbohydrates and other organic matter under anaerobic conditions.

field capacity Amount of water held by soil against the force of gravity.

field study A controlled experimental study carried out in a natural environment rather than in the laboratory.

first law of thermodynamics Energy is neither created nor destroyed; in any transfer or transformation no gain or loss of total energy occurs.

first-level carnivores Organisms that feed on first-level consumers or plant eaters.

first-level consumers Organisms that feed on plants.

first trophic level Producers; organisms that fix energy that becomes the basic source of energy for consumers.

fitness Genetic contribution by an individual's descendants to future generations.

fixation Process in soil by which certain chemical elements essential for plant growth are converted from a soluble or exchangeable form to a less soluble or nonexchangeable form.

fixed quota Harvest removal of a certain percentage of a population, based on maximum sustained yield estimates.

floating reserve Individuals in a population of a territorial species that do not hold territories and remain unmated, but are available to refill territories vacated by death of an owner.

flux Flow of energy from a source to a sink or receiver.

foliage height diversity Measure of the degree of layering or vertical stratification of foliage in a forest.

food chain Movement of energy and nutrients from one feeding group of organisms to another in a series that begins with plants and ends with carnivores, detrital feeders, and decomposers.

food web Interlocking pattern formed by a series of interconnecting food chains.

foraging strategy Manner in which animals seek food and allocate their time and effort in obtaining it.

forb Herbaceous plant other than grass, sedge, or rush.

formation Classification of vegetation based on dominant life forms.

founder effect Effect of starting a population with a small number of colonists, which contain only a small and often biased sample of genetic variation of the parent population; a markedly different new population may arise.

fragmentation Reduction of a large habitat area into small, scattered remnants; reduction of leaves and other organic matter into smaller particles.

free-running cycle Length of a circadian rhythm in the absence of an external time cue.

frost pocket Depression in the landscape into which cold air drains, lowering the temperature relative to the surrounding area; such pockets often support their own characteristic group of cold-tolerant plants.

frugivore Organism that feeds on fruit.

functional response Change in rate of exploitation of a prey species by a predator in relation to changing prey density.

fundamental niche Total range of environmental conditions under which a species can survive.

Gaia hypothesis The idea that the biosphere is a self-regulating entity controlling the physical and chemical environment.

gamma diversity Differences among similar habitats in widely separated regions.

gap Opening made in a forest canopy by some small disturbance such as windfall; death of an individual tree or group of trees that influences the development of vegetation beneath.

garrigue Shrub woodland characteristic of limestone areas with low rainfall and thin, poor dry soils; widespread in the Mediterranean countries of southern Europe.

gaseous cycle A biogeochemical cycle with the main reservoir or pool of nutrients in the atmosphere and ocean.

gene Unit material of inheritance; more specifically, a small unit of a DNA molecule, coded for a specific protein to produce one of the many attributes of a species.

gene flow Exchange of genetic material between populations.

gene frequency Relative abundance of different alleles carried by an individual or a population; allele frequency.

gene pool The sum of all the genes of all individuals in a population.

genet A genetic individual that arises from a single fertilized egg.

genetic drift Random fluctuation in allele frequency over time, due to chance alone without any influence by natural selection; important in small populations.

genotype Genetic constitution of an organism.

genotypic frequency The proportion of various genotypes in a population; compare *gene frequency*.

geographic isolates Groups of populations that are semi-isolated from one another by some extrinsic barrier; compare *subspecies*.

geometric rate of increase (λ) Factor by which the size of a population increases over a period of time.

gleization A process in waterlogged soils in which iron, because of an inadequate supply of oxygen, is reduced to a ferrous compound, giving dull gray or bluish mottles and color to the horizons.

gley soil Soil developed under conditions of poor drainage, resulting in reduction of iron and other elements and in gray or bluish colors and mottles.

grazers Stream invertebrates that feed on algal coating on rocks and other substrates.

grazing food chain Food chain in which primary producers (green plants) are eaten by grazing herbivores, with subsequent energy transfers to other trophic levels.

greenhouse effect Selective energy absorption by carbon dioxide in the atmosphere, which allows short wavelength energy to pass through but absorbs longer wavelengths and reflects heat back to Earth.

greenhouse gas A gas that absorbs long-wave radiation and thus contributes to the greenhouse effect when present in the atmosphere; includes water vapor, carbon dioxide, methane, nitrous oxides, and ozone.

gross primary production Energy fixed per unit area by photosynthetic activity of plants before respiration; total energy flow at the secondary level is not gross production, but rather assimilation, because consumers use material already produced with respiratory losses.

gross reproductive rate Sum of the mean number of females born to each female age group.

groundwater Water that occurs below Earth's surface in pore spaces within bedrock and soil, free to move under the influence of gravity.

growth form Morphological category of plants, such as tree, shrub, or vine.

guild A group of populations that utilize a gradient of resources in a similar way.

gully erosion Form of surface erosion caused by torrents of water that bite deeply into topsoil and soft sediments.

gyre Circular motion of water in major ocean basins.

habitat Place where a plant or animal lives.

hadal That part of the ocean below 6000 m.

halophyte Terrestrial plant adapted morphologically or physiologically to grow in salt-rich soil.

haploid Having a single set of unpaired chromosomes in each cell nucleus.

Hardy-Weinberg law The proposition that genotypic ratios resulting from random mating remain unchanged from one generation to another, provided natural selection, genetic drift, and mutation are absent.

harvest effort Approach to harvesting populations by manipulating or controlling hunting efforts, by means such as setting seasons and bag limits.

heat dome Storage and reradiation of heat about and above urban areas, in which the temperature may be considerably higher than in the surrounding countryside.

hemicryptophyte Perennial shoots or buds close to the surface of the ground; often covered with litter.

hemiparasite Plant parasite that has chlorophyll, carries on photosynthesis, yet derives some nutrients from its host.

herbivore Organism that feeds on plant tissue.

herbivory Feeding on plants.

hermaphrodite Organism possessing the reproductive organs of both sexes.

heterogeneity State of being mixed in composition; can refer to genetic or environmental conditions.

heterotherm An organism that during part of its life history becomes either endothermic or ectothermic; hibernating endotherms become ectothermic, and foraging insects such as bees become endothermic during periods of activity; they are characterized by rapid, drastic, repeated changes in body temperature.

heterotrophic Requiring a supply of organic matter or food from the environment.

heterotrophic community Community that is dependent upon and supported by energy already fixed by the autotrophic community.

heterotrophic succession Succession that occurs on dead organic matter; detritivores feed in sequence, each group releasing nutrients used by the next group, until resources are exhausted.

heterozygous Containing two different alleles of a gene, one from each parent, at the corresponding loci of a pair of chromosomes.

hibernation Winter dormancy in animals, characterized by a great decrease in metabolism.

hierarchy A sequence of sets made up of smaller subsets.

histosol Soil characterized by high organic matter content.

home range Area over which an animal ranges throughout the year.

homeostasis Maintenance of nearly constant conditions in function of an organism or in interaction among individuals in a population.

homeotherm Animal with a fairly constant body temperature; also spelled homoiotherm and homotherm.

homeothermy Regulation of body temperature by physiological means.

homologous chromosomes Corresponding chromosomes from male and female parents that pair during meiosis.

homozygous Containing two identical alleles of a gene at the corresponding loci of a pair of chromosomes.

horizon Major zone or layer of soil, with its own particular structure and characteristics.

horse latitudes Subtropical latitudes coinciding with a major anticyclonic belt, characterized by generally settled weather and a light or moderate wind.

host Organism that provides food or other benefit to another organism of a different species; usually refers to an organism exploited by a parasite.

humus Organic material derived from partial decay of plant and animal matter.

hybrid Plant or animal resulting from a cross between genetically different parents.

hydrogen bonding A type of bond occurring between an atom of oxygen or nitrogen and a hydrogen atom joined to oxygen or nitrogen on another molecule; responsible for the properties of water.

hydroperiod In wetlands, the duration, frequency, depth, and season of flooding.

hydroscopic water Water held tightly by soil particles, so it is unavailable to plants.

hydrothermal vent Place on ocean floor where water, heated by molten rock, issues from fissures; vent water contains sulfides oxidized by chemosynthetic bacteria, providing support for carnivores and detritivores.

hyperthermia Rise in body temperature to reduce thermal differences between an animal and a hot environment, thus reducing the rate of heat flow into the body.

hypertrophic Condition of lakes that have received excessive amounts of nutrients, making them highly and unnaturally eutrophic; compare *eutrophic*.

hypervolume The multidimensional space of a species niche; compare *niche*.

hypha Filament of a fungus thallus or vegetative body.

hypolimnion Cold, oxygen-poor zone of a lake, below the thermocline.

hypothesis Proposed explanation for a phenomenon; we should be able to test it, accepting or rejecting it on the basis of experimentation.

igneous rock Rock that has crystallized after Earth's crust melts.

immigration Arrival of new individuals into a habitat or population.

immobilization Conversion of an element from inorganic to organic form in microbial or plant tissue, rendering the nutrient unavailable to other organisms.

importance value Sum of relative density, relative dominance, and relative frequency of a species in a community.

inbreeding Mating among close relatives.

inbreeding depression Detrimental effects of inbreeding.

inceptisol Mineral soil that has one or more horizons in which mineral materials have been weathered or removed and that is only beginning to develop a distinctive soil profile.

incipient lethal temperature Temperature at which a stated fraction of a population of animals (usually 50 percent) will die when brought rapidly to it from a different temperature.

inclusive fitness Sum of the fitness of an individual and the fitness of its relatives, weighted according to the degree of relationship.

independent variable Variable x, the first of two numbers of an ordered pair (x, y); the set of all values taken on by the independent variable is called the domain of the function; compare *dependent variable*.

individualistic concept The view, first proposed by H. A. Gleason, that vegetation is a continuous variable in a continuously changing environment; therefore no two vegetational communities are identical, and associations of species arise only from similarities in requirements.

induced edge Edge that results from some disturbance; adjoining vegetation types are successional, changing or disappearing with time, maintained only by periodic disturbances.

infauna Organisms living within a substrate.

infiltration Downward movement of water into the soil.

infralittoral fringe Region below the littoral region of the sea.

inherent edge Stable, permanent edge determined by long-term natural features and conditions.

inhibition model Model of succession proposing that the dominant vegetation occupying a site prevents colonization of that site by other plants of the next successional community.

instar Form of insect or other arthropod between successive molts.

integrated pest management Holistic approach to pest control that considers biological, ecological, economic, and social aspects; the object is to control pests before outbreaks can occur.

interference competition Competition in which access to a resource is limited by the presence of a competitor.

interior species Organisms that require large areas of habitat, even though their home ranges may be small.

intermediate host Host that harbors a developmental phase of a parasite; the infective stage or stages can develop only when the parasite is independent of its definitive host; compare *definitive host*.

internal cycling Movement or cycling of nutrients through components of ecosystems.

intersexual selection Choice of a mate, usually by the female.

interspecific Between individuals of different species.

intraguild predation Predation among species occupying the same trophic level and using a similar food resource.

intrasexual selection Competition among members of the same sex for a mate, most common among males and characterized by fighting and display.

intraspecific Between individuals of the same species.

intrinsic rate of increase The per capita rate of growth of a population that has reached a stable age distribution and is free of competition and other growth restraints.

inversion In genetics, reversal of part of a chromosome so that genes lie in reverse order; in meteorology, increase rather than decrease in air temperature with height, caused by radiational cooling of Earth (radiational inversion) or by compression and heating of subsiding air masses from high pressure areas (subsidence inversion).

island biogeography Study of distribution of organisms and community structure on islands.

isolating mechanism Any structural, behavioral, or physiological mechanism that blocks or inhibits gene exchange between two populations.

isotherm Line drawn on a map connecting points with the same temperature at a certain period of time.

iteroparity Having multiple broods over a lifetime.

K-selection Selection under carrying capacity conditions and a high level of competition.

keystone species A species whose activities have a significant role in determining community structure.

kin selection Differential reproduction among groups of closely related individuals.

kinetic energy Energy associated with motion; performs work at the expense of potential energy.

Krummholz Stunted form of trees characteristic of transition zone between alpine tundra and subalpine coniferous forest.

landscape ecology Study of structure, function, and change in a heterogeneous landscape composed of interacting ecosystems.

late successional species Long-lived, shade-tolerant plant species that supplant early successional species.

latent heat of fusion Amount of heat given up when a unit mass of a substance converts from a liquid to a solid state, or the amount of heat absorbed when a substance converts from the solid to liquid state.

laterization Soil-forming process in hot, humid climates, characterized by intense oxidation; results in loss of bases and in a deeply weathered soil composed of silica, sesquioxides of iron and aluminum, clays, and residual quartz.

law of tolerance The idea that organisms live within a range between maximum and minimum amounts of substances or conditions that limit their presence or success; compare *Liebig's law of the minimum.*

leaching Dissolving and washing of nutrients out of soil, litter, and organic matter.

leaf area index (LAI) Ratio of area of canopy foliage to ground area.

lek Communal courtship area males use to attract and mate with females.

lentic Pertaining to standing water, such as lakes and ponds.

Liebig's law of the minimum The idea that the growth of an individual or a population is limited by the lowest amount needed of an essential nutrient.

life expectancy The average number of years to be lived in the future by members of a population.

life table Tabulation of mortality and survivorship of a population; static, time-specific, or vertical life tables are based on a cross section of a population at a given time; dynamic, cohort, or horizontal life tables are based on a cohort followed throughout life.

life zone Major area of plant and animal life, equivalent to a biome; transcontinental region or belt characterized by particular plants and animals and distinguished by temperature differences; applies best to mountainous regions where temperature changes accompany changes in altitude.

light compensation point Depth of water or level of light at which photosynthesis and respiration balance each other.

light saturation point Amount of light at which plants achieve the maximum rate of photosynthesis.

limit cycle Stable oscillation in the population levels of a species, usually reflecting predator-prey interactions.

limiting resource Resource or environmental condition that limits the abundance and distribution of an organism.

limnetic Pertaining to or living in the open water of a pond or lake.

limnetic zone Shallow water zone of a lake or sea, in which light penetrates to the bottom.

lithosol Soil showing little or no evidence of soil development and consisting mainly of partly weathered rock fragments or nearly barren rock.

littoral zone Shallow water of a lake, in which light penetrates to the bottom, permitting submerged, floating, and emergent vegetative growth; also shore zone of tidal water between high water and low water marks.

locus Site on a chromosome occupied by a specific gene.

loess Soil developed from wind-deposited material.

logistic curve S-shaped curve of population growth that slows at first, steepens, and then flattens out at asymptote, determined by carrying capacity.

logistic equation Mathematical expression for the population growth curve in which rate of increase decreases linearly as population size increases.

long-day organism Plant or animal that requires long days—days with more than a certain minimum of daylight—to flower or come into reproductive condition.

lotic Pertaining to flowing water.

macromutation Mutation at the level of the chromosome.

macronutrients Essential nutrients plants and animals need in large amounts.

macroparasite Any of the parasitic worms, lice, fungi, and the like that have comparatively long generation time, spread by direct or indirect transmission, and may involve intermediate hosts or vectors.

mallee Sclerophyllous shrub community in Australia; most of the species are *Eucalyptus.*

mangal A mangrove swamp.

maquis Sclerophyllous shrub vegetation in the Mediterranean region.

marsh Wetland dominated by grassy vegetation such as cattails and sedges.

mating system Pattern of mating between individuals in a population.

mattoral Sclerophyllous shrub vegetation in regions of Chile with mediterranean climate.

maximum sustained yield The maximum rate at which individuals can be harvested from a population without reducing its size; recruitment balances harvesting.

mediterranean-type climate Semiarid climate characterized by a hot, dry summer and a wet, mild winter.

meiofauna Benthic organisms within the size range from 1 to 0.1 mm; interstitial fauna.

meiosis Two successive divisions by a gametic cell, with only one duplication of chromosomes, so the number of chromosomes in daughter cells is one-half the diploid number.

melatonin Special hormone in animals that serves to measure time; associated with the biological clock.

meristem Region in a plant containing actively or potentially dividing cells.

mesic Moderately moist.

mesopelagic Uppermost lightless pelagic zone.

mesophyll Specialized tissue located between the epidermal layers of a leaf; *palisade mesophyll* consists of cylindrical cells at right angles to upper epidermis and contains many chloroplasts; *spongy mesophyll* lies next to the lower epidermis and has interconnecting, irregularly shaped cells with large intercellular spaces.

metabolism The chemical reactions in cells responsible for breaking down molecules to provide energy (catabolism) and building more complex molecules from simpler molecules (anabolism).

metalimnion Transition zone in a lake between hypolimnion and epilimnion; region of rapid temperature decline.

metamorphic rock An aggregate of minerals formed when heat and pressure recrystallize rocks.

metamorphosis Abrupt transition between life stages.

metapopulation A population broken into sets of subpopulations held together by dispersal or movements of individuals among them.

metatrophic Having a moderate amount of nutrients; stage in a nutrient-poor lake becoming eutrophic.

micella Soil particle of clay and humus, carrying negative electrical charge at the surface.

microbial loop Feeding loop in which bacteria take up dissolved organic matter produced by plankton and nanoplankton consume the bacteria; adds several trophic levels to the plankton food chain.

microbivore Organism that feed on microbes, especially in the soil and litter.

microclimate Climate on a very local scale, which differs from the general climate of the area; influences the presence and distribution of organisms.

microflora Bacteria and certain fungi inhabiting the soil.

microhabitat That part of the general habitat utilized by an organism.

micromutation A mutation at the level of the gene; point mutation.

micronutrient Essential nutrient needed in very small quantities by plants and animals.

microparasite Any of the viruses, bacteria, and protozoans, characterized by small size, short generation time, and rapid multiplication.

migration Intentional, directional, usually seasonal movement of animals between two regions or habitats; involves departure and return of the same individual; a round-trip movement.

mimicry Resemblance of one organism to another or to an object in the environment, evolved to deceive predators.

mineralization Microbial breakdown of humus and other organic matter in soil to inorganic substances.

minimum viable population (MVP) Size of a population which, with a given probability, will ensure the existence of the population for a stated period of time.

mire Wetland characterized by an accumulation of peat.

mitosis Cell division involving chromosome duplication, resulting in two daughter cells with the full complement of chromosomes, genetically the same as the parent cell.

mixed-grass prairie Grassland in mid North America, characterized by great variation in precipitation and a mixture of largely cool season shortgrass and tallgrass species.

model In theoretical and systems ecology, an abstraction or simplification of a natural phenomenon, developed to predict a new phenomenon or to provide insight into existing ones; in mimetic association, the organism mimicked by a different organism.

moder Humus in which plant fragments and mineral particles form loose netlike structures held together by a chain of small arthropod droppings.

modular organism Organism that grows by repeated iteration of parts, such as branches or shoots of a plant; some parts may separate and become physically and physiologically independent.

mollisol Soil formed by calcification, characterized by accumulation of calcium carbonate in lower horizons and high organic content in upper horizons.

monoecious Having male and female reproductive organs separated in different floral structures on the same plant; compare *hermaphrodite, dioecious.*

monogamy In animals, mating and maintenance of a pair bond with only one member of the opposite sex at a time.

montane Related to mountains.

moor A blanket bog or peatland.

mor Humus in which unincorporated organic matter usually is matted or compacted or both and distinct from mineral soil; low in bases and acid in reaction.

morphological species Species described as monotypic, possessing little variation in color pattern, structure, proportion, and other features; the "field guide" species.

morphology Study of the form of organisms.

mortality rate The probability of dying; the ratio of number dying in a given time interval to the number alive at the beginning of the time interval.

mull Humus that contains appreciable amounts of mineral bases and forms a rich layer of forested soil, consisting of mixed organic and mineral matter; blends into the upper mineral layer without abrupt changes in soil characteristics.

Mullerian mimicry Resemblance of two or more conspicuously marked distasteful species, which increases predator avoidance.

mutation Transmissible changes in the structure of a gene or chromosome.

mutualism Relationship between two species in which both benefit.

mycelium Mass of hyphae that make up the vegetative portion of a fungus.

mycorrhizae Association of fungus with roots of higher plants, which improves the plants' uptake of nutrients from the soil.

myrmecochory Dispersal by ants.

myrmecochores Plants that possess ant-attracting substances on their seed coats.

nanoplankton Plankton with a size range from 2 to 20 μm.

natality Production of new individuals in a population.

natural selection Differential reproduction and survival of individuals that results in elimination of maladaptive traits from a population.

neap tide Tide of small range that occurs at the first and last quarters of the moon when Earth, moon, and sun are at right angles.

negative feedback Homeostatic control in which an increase in some substance or activity ultimately inhibits or reverses the direction of the processes leading to the increase.

nekton Aquatic animals that are able to move at will through the water.

neritic Marine environment embracing the regions where land masses extend outward as a continental shelf.

net production Accumulation of total biomass over a given period of time after respiration is deducted from gross production in plants and from assimilated energy in consumer organisms.

net reproductive rate Average number of female offspring produced by an average female during her lifetime.

neutrophilic Preferring a habitat that is neither acid nor alkaline.

niche Functional role of a species in the community, including activities and relationships.

niche breadth Range of a single niche dimension occupied by a population.

niche compression Restriction of the use of a resource such as food or space because of intense competition.

niche overlap Sharing of niche space by two or more species.

niche preemption Procurement by a species of a portion of available resources, leaving less for the next.

nitrification Breakdown of nitrogen-containing organic compounds into nitrates and nitrites.

nitrogen fixation Conversion of atmospheric nitrogen to forms usable by organisms.

nonshivering thermogenesis Production of metabolic heat by burning highly vascular brown fatty tissue capable of a high rate of oxygen consumption.

nucleotide A compound formed by the condensation of a nitrogenous base with a sugar and phosphoric acid; structural unit of DNA.

null hypothesis A statement of no difference between sets of values formulated for statistical testing.

numerical response Change in size of a population of predators in response to change in density of its prey.

nutrient Substance an organism requires for normal growth and activity.

nutrient cycle Pathway of an element or nutrient through the ecosystem, from assimilation by organisms to release by decomposition.

obligate Having no alternative in response to a particular condition or in way of life.

oceanic Referring to the regions of the sea with depths greater than 200 m that lie beyond the continental shelf.

old-growth forest Forest that has not been disturbed by humans for hundreds of years.

oligotrophic Term applied to a body of water low in nutrients and in productivity.

oligotrophy Nutrient-poor condition.

omnivore An animal that feeds on both plant and animal matter.

open system System with a constant input of energy.

opportunistic species Organisms able to exploit temporary habitats or conditions.

optimal foraging Tendency of animals to harvest food efficiently, selecting food sizes or food patches that supply maximum food intake for energy expended.

optimum yield Amount of material that can be removed from a population to produce maximum biomass on a sustained yield basis.

organismic concept of the community Idea that species, especially plant species, are integrated into an internally interdependent unit; upon maturity and death of the community, another identical plant community will replace it.

oscillation Regular fluctuation in a fixed cycle above or below some set point.

osmosis Movement of water molecules across a differentially permeable membrane in response to a concentration or pressure gradient.

osmotic potential The attraction of water across a membrane; the more concentrated a solution, the lower is its osmotic potential.

osmotic pressure Pressure needed to prevent passage of water or another solvent through a semipermeable membrane separating a solvent from a solution.

outbreeding Production of offspring by the fusion of distantly related gametes.

overdispersion Situation in which the distribution of organisms is random but clumped, with some areas empty and some heavily overpopulated; contagious distribution.

overturn Vertical mixing of layers in a body of water, brought about by seasonal changes in temperature.

oxisol Soil developed under humid semitropical and tropical conditions, characterized by silicates and hydrous oxides, clays, residual quartz, deficiency in bases, and low plant nutrients; formed by laterization.

paleoecology Study of ecology of past communities by means of the fossil record.

pampas Temperate South American grassland, dominated by bunchgrasses; much of the moister pampas are under cultivation.

parasitism Relationship between two species in which one benefits while the other is harmed (although not usually killed directly).

parasitoid Insect larva that kills its host by consuming the host's soft tissues before pupation or metamorphosis into an adult.

parthenogenesis Development of an individual from an egg that did not undergo fertilization.

peat Unconsolidated material consisting of undecomposed and only slightly decomposed organic matter under conditions of excessive moisture.

peatland Any ecosystem dominated by peat; compare *bog*, *mire*, and *fen*.

ped Soil particles held together in a cluster of various sizes.

pelagic Referring to the open sea.

percent base saturation The extent to which the exchange sites of soil particles are occupied by exchangeable base cations or by cations other than hydrogen and aluminum, expressed as percentage of total cation exchange capacity; compare *cation exchange capacity*.

percolation The movement of water downward and outward through subsurface soil, often continuing down to groundwater.

perennating tissue Underground organs or buds that store food for new shoots of the next growing season.

periphyton In freshwater ecosystems, organisms that are attached to submerged plant stems and leaves; see *aufwuchs*.

permafrost Permanently frozen soil in arctic regions.

permanent wilting point Point at which water potential in the soil and conductivity assume such low values that the plant is unable to extract sufficient water to survive and wilts permanently.

phanerophyte Tree, shrub, or vine that bears perennating buds on aerial shoots.

phenology Study of seasonal changes in plant and animal life and the relationship of these changes to weather and climate.

phenotype Physical expression of a characteristic of an organism, determined by both genetic constitution and environment.

phenotypic plasticity Ability to change form under different environmental conditions.

pheromone Chemical substance released by an animal that influences behavior of others of the same species.

photic zone Lighted water column of a lake or ocean, inhabited by plankton.

photoperiodism Response of plants and animals to changes in relative duration of light and dark.

photosynthate Energy-rich organic molecules produced during photosynthesis.

photosynthesis Use of light energy by plants to convert carbon dioxide and water into simple sugars.

photosynthetically active radiation (PAR) Those wavelengths in the radiation spectrum used by plants in photosynthesis.

phreatophyte Type of plant that habitually obtains its water supply from groundwater.

physiognomy Outward appearance of the landscape.

physiological ecology Study of the physiological functioning of organisms in relation to their environment.

phytochrome A protein pigment in plants involved in photoperiodic responses and other photoreactions.

phytoplankton Small, floating plant life in aquatic ecosystems; planktonic plants.

pioneer species Plants that invade disturbed sites or appear in early stages of succession.

plankton Small, floating or weakly swimming plants and animals in freshwater and marine ecosystems.

playas Low, generally flat basins in deserts that receive waters that rush down from higher elevations and water-cut canyons.

Pleistocene Geological epoch extending from about 2 million to 10,000 years ago, characterized by recurring glaciers; the Ice Age.

pneumatophore An erect respiratory root that protrudes above waterlogged soils; typical of baldcypress and mangroves.

podzolization Soil-forming process in which acid leaches the A horizon and iron, aluminum, silica, and clays accumulate in lower horizons.

poikilothermy Variation of body temperature with external conditions.

polyandry Mating of one female with several males.

polygamy Acquisition by an individual of two or more mates, none of which is mated to other individuals.

polygyny Mating of one male with several females.

polymorphism Occurrence of more than one distinct form of individuals in a population.

polyploid Having three or more times the haploid number of chromosomes.

population A group of individuals of the same species living in a given area at a given time.

population bottleneck See *bottleneck*.

population cycle Oscillation between periods of high and low population density.

population density The number of individuals in a population per unit area.

population dynamics Study of the factors that influence the number and density of populations in time and space.

population ecology Study of how populations grow, fluctuate, spread, and interact intraspecifically and interspecifically.

population genetics The study of changes in gene frequency and genotypes in populations.

population regulation Mechanisms or factors within a population that cause it to decrease when density is high and increase when density is low.

positive feedback Control in a system that reinforces a process in the same direction.

potential energy Energy available to do work.

potential evapotranspiration Amount of water that would be transpired under constantly optimal conditions of soil moisture and plant cover.

precocial In birds, hatched with down, open eyes, and ability to move about; in mammals, born with open eyes and ability to follow the mother, as fawns and calves can.

predation Relationship in which one living organism serves as a food source for another.

preferred temperature Range of temperatures within which poikilotherms function most efficiently.

primary production Production by green plants.

primary productivity Rate at which plants produce biomass.

primary succession Vegetational development starting on a new site never before colonized by life.

producer Green plant or chemosynthetic bacterium that converts light or chemical energy into organismal tissue.

production Amount of energy formed by an individual, population, or community per unit time.

productivity Rate of energy fixation or storage per unit time; not to be confused with production.

profundal zone Deep zone in aquatic ecosystems, below the limnetic zone.

pyramid of biomass Diagrammatic representation of biomass at different trophic levels in an ecosystem.

pyramid of energy Diagrammatic representation of the flow of energy through different trophic levels.

pyramid of numbers Diagrammatic representation of the number of individual organisms present at each trophic level in an ecosystem; the least useful pyramid.

quaking bog Bog characterized by a floating mat of peat and vegetation over water.

qualitative inhibitors In plants, chemical defense by toxic substances that interfere with a consumer's metabolism; quickly synthesized.

quantitative inhibitors In plants, chemical defense by substances that reduce digestibility or potential energy from food.

r-selection Selection under low population densities; favors high reproductive rates under conditions of low competition.

rain forest Permanently wet forest of the tropics; also the wet coniferous forest of the Pacific northwest of the United States.

rain shadow Dry area on the lee side of mountains.

raised bog A bog in which the accumulation of peat has raised the surface above both the surrounding landscape and the water table; it develops its own perched water table.

ramet An individual member of a plant clone.

random distribution Distribution lacking pattern or order; placement of each individual is independent of all other individuals.

rate of increase Factor by which a population changes over a given period of time; compare *exponential growth, geometric rate of increase, intrinsic rate of increase.*

realized niche Portion of fundamental niche space occupied by a population facing competition from populations of other species; environmental conditions under which a population survives and reproduces in nature.

recombination Exchange of genetic material by independent assortment of chromosomes and their genes during gamete production, allowing a random mix of different sets of genes at fertilization.

recruitment Addition of new individuals to a population by reproduction.

regolith Mantle of unconsolidated material below the soil, from which soil develops.

regular distribution A pattern in which individuals are more widely separated from each other than would be expected by chance; underdispersion.

relative humidity Water vapor content of air at a given temperature, expressed as a percentage of the water vapor needed for saturation at that temperature.

reproductive allocation Proportion of its resources that an organism expends on reproduction over a given period of time.

reproductive cost Decrease in survivorship or rate of growth when an individual increases its current allocation to reproduction; reflected in decreased potential for future reproduction.

reproductive effort Proportion of its resources an organism expends on reproduction.

reproductive isolation Separation of one population from another by inability to produce viable offspring when the two populations mate.

reproductive value Potential reproductive output of an individual at a particular age relative to that of a newborn individual at the same time.

reservoir Compartment of an ecosystem.

resource Environmental component used by a living organism.

resource allocation Apportioning the supply of a resource to specific uses.

respiration Metabolic assimilation of oxygen, accompanied by production of carbon dioxide and water, release of energy, and breakdown of organic compounds.

restoration ecology Study of the application of ecological theory to the restoration of highly disturbed sites.

rete A large network or discrete vascular bundle of intermingling small blood vessels carrying arterial and venous blood that acts as a heat exchanger in mammals and certain fish and sharks.

rhizobia Bacteria capable of living mutualistically with higher plants.

rhizome A horizontal underground stem that branches and gives rise to vegetative structures.

richness A component of species diversity; the number of species present in an area.

riffle Stretch of shallow, fast, rough water flowing between pools in a stream.

riparian woodland Woodland along the bank of a river or stream; riverbank forests are often called gallery forests.

root-to-shoot ratio Ratio of the weight of roots to the weight of shoots of a plant.

ruminant Ungulate with a three-chamber or four-chamber stomach; in the large first chamber or rumen bacteria ferment plant matter.

saprophage Organism that feeds on dead plant and animal matter; mainly bacteria, fungi, and invertebrates such as insect larvae.

saprophyte Plant that draws its nourishment from dead plant and animal matter, mostly the former.

saturation vapor pressure Maximum amount of water vapor a volume of air can hold at a given temperature.

savanna Tropical grassland, usually with scattered trees or shrubs.

scavenger Animal that feeds on dead animals or on animal products, such as dung.

sclerophyll Woody plant with hard, leathery, evergreen leaves that prevent moisture loss.

scramble competition Intraspecific competition in which limited resources are shared to the point that no individual survives.

scraper Any aquatic insect that feeds by scraping algae from a substrate.

search image Mental image formed in predators, enabling them to find more quickly and to concentrate on a common type of prey.

seasonality Recurrence of biological events with the seasons.

second law of thermodynamics In any energy transfer or transformation, part of the energy assumes a form that cannot be passed on any further.

second-level carnivores Organisms that feed on first-level carnivores or second-level consumers.

second-level consumers Organisms that feed on first-level consumers or herbivores; carnivores.

secondary producers Organisms that derive energy from consuming plant or animal tissue and breaking down assimilated carbon compounds.

secondary production Production by consumer organisms.

secondary substances Organic compounds that plants produce for chemical defense.

secondary succession Plant succession on disturbed sites that already support life.

sedimentary cycle Weathering of rock and leaching of its minerals, transport, deposition, and burial.

sedimentary rock Rock formed by the deposition and compression of mineral and rock particles.

selection Differential survival or reproduction of individuals in a population because of phenotypic differences among them.

selective pressure Any force acting on individuals in a population that determines which individuals leave more descendants than others; gives direction to the evolutionary process.

self-thinning Progressive decline in density of plants associated with the increasing size of individuals.

semelparity Having only a single reproductive effort in a lifetime over one short period of time.

semiarid Fairly dry in climate, with precipitation between 25 and 60 cm a year and with an evapotranspiration rate high enough that the potential loss of water to the environment exceeds inputs.

semispecies A group of organisms taxonomically intermediate between a race and a species with incomplete isolating mechanisms.

senescence Process of aging.

seral Following a series of stages.

sere The series of successional stages on a given site that lead to a terminal community.

serpentine soil Soil derived from ultrabasic rocks that are high in iron, magnesium, nickel, chromium, and cobalt and low in calcium, potassium, sodium, and aluminum; supports distinctive communities.

sessile Not free to move about; permanently attached to a substrate.

sex ratio The relative number of males to females in a population.

sex reversal A change in functioning so that a member of one sex behaves as the other.

sexual selection Selection by one sex for an individual of the other sex based on some specific characteristic or characteristics; usually takes place through courtship behavior.

shade-intolerant Growing and reproducing best under high light conditions; growing poorly and failing to reproduce under low light conditions.

shade-tolerant Able to grow and reproduce under low light conditions.

sheet erosion Transport of soil material from slopes by a thin, mobile sheet of water.

short-day organisms Plants and animals that come into reproductive condition under conditions of short days—days with less than a certain maximum length.

shortgrass plains Westernmost grasslands of the Great Plains, characterized by infrequent rainfall, low humidity, and high winds; dominated by shallow-rooted, sod-forming grasses.

shredders Stream invertebrates that feed on coarse particulate organic matter.

sibling species Species with similar appearance but unable to interbreed.

sigmoid curve S-shaped curve of logistic growth.

sink An unfilled, submarginal, or marginal habitat where the population can persist only by immigration from other habitats, because it experiences low reproduction or high mortality.

site Combination of biotic, climatic, and soil conditions that determine an area's capacity to produce vegetation.

snag Dead or partially dead tree at least 10.2 cm dbh and 1.8 m tall; important habitat for cavity-nesting birds and mammals.

social dominance Physical dominance of one individual over another, usually maintained by some manifestation of aggressive behavior.

social parasite Animal that uses other individuals or species to rear its young, such as the cowbird.

soil association A group of defined and named soil taxonomic units occurring together in an individual and characteristic pattern over a geographic region.

soil horizon Developmental layer in the soil with characteristic thickness, color, texture, structure, acidity, nutrient concentration, and the like.

soil profile Distinctive layering of horizons in the soil.

soil series Basic unit of soil classification, consisting of soils that are alike in all major profile characteristics except texture of the A horizon; soil series are usually named for the locality where the typical soil was first recorded.

soil structure Arrangement of soil particles and aggregates.

soil texture Relative proportions of the three particle sizes (sand, silt, and clay) in the soil.

soil type Lowest unit in the system of soil classification, consisting of soils that are alike in all characteristics, including texture of the A horizon.

solar constant Rate at which solar energy is received on a surface just outside of Earth's atmosphere; current value is 0.140 watt/cm^2.

speciation Separation of a population into two or more reproductively isolated populations.

species diversity Measurement that relates density of organisms of each type present in a habitat to the number of species in a habitat.

species richness Number of species in a given area.

specific heat Amount of energy that must be added or removed to raise or lower the temperature of a substance by a specific amount.

Sphagnum A genus of mosses that are most abundant in wet, acidic habitats; the dead cells rapidly fill with water, allowing the plant to hold many times its own weight in water.

spiraling Mechanism of retention of nutrients in flowing water ecosystems, involving the interdependent processes of nutrient recycling and downstream transport.

spodosol Soil characterized by the presence of a horizon in which organic matter and amorphous oxides of aluminum and iron have precipitated; includes podzolic soils.

spring tide A tide of greater than mean range that occurs every two weeks, when the moon is full or new; maximum spring tides occur when sun and moon are in the same plane as Earth; compare *neap tide*.

stabilizing selection Selection favoring the middle in the distribution of phenotypes.

stable age distribution Constant proportion of individuals of various age classes in a population through population changes.

stable equilibrium Ability of a system to return to a particular point if displaced by an outside force.

stable limit cycle A regular fluctuation in the abundance of predator and prey populations, when stabilizing and destabilizing interactions balance.

stand Unit of vegetation that is essentially homogeneous in all layers and differs from adjacent types qualitatively and quantitatively.

standard deviation Statistical measure defining the dispersion of values about the mean in a normal distribution.

standing crop Total amount of biomass per unit area at a given time.

static life table See *life table*.

stationary age distribution Special form of stable age distribution, in which the birth rate equals the death rate and age distribution remains fixed.

steppe Name given to Eurasian grasslands that extend from eastern Europe to western Siberia and China.

stomata Pores in the leaf or stem of a plant that allow gaseous exchange between the internal tissues and the atmosphere.

stratification Division of an aquatic or terrestrial community into distinguishable layers on the basis of temperature, moisture, light, vegetative structure, and other such factors, creating zones for different plant and animal types.

sublittoral zone Lower division of the sea, from about 40 m to below 200 m.

subsidence inversion Atmospheric inversion produced by sinking air.

subspecies Geographical unit of a species population, distinguishable by morphological, behavioral, or physiological characteristics.

succession Replacement of one community by another; often progresses to a stable terminal community called the climax.

successional sequence Pattern of colonization and extinction of plants on a given area over time; compare *sere*.

supercooling In ectotherms, lowering body temperature below freezing without freezing body tissue, by means of solutes, particularly glycerol.

supralittoral fringe The highest zone on the intertidal shore, bounded below by the upper limit of barnacles and above by the upper limit of *Littorina* snails.

surface runoff The excess water flowing across the surface of the ground when the soil becomes saturated during heavy rains.

surface tension Elastic film across the surface of a liquid, caused by the attractive forces between molecules at the surface of the liquid.

survivorship The probability that a representative newborn individual in a cohort will survive to various ages.

survivorship curve A graph describing the survival of a cohort of individuals in a population from birth to the maximum age reached by any one member of the cohort.

sustained yield Yield per unit time equal to production per unit time in an exploited population.

swamp Wooded wetland in which water is near or above ground level.

switching Changing the diet from a less abundant to a more abundant prey species; see *threshold of security*.

symbiosis Situation in which two dissimilar organisms live together in close association.

sympatric Living in the same area; usually refers to overlapping populations.

sympatric speciation Production of a new species within a population or the dispersal range of a population.

system Set or collection of interdependent parts or subsystems enclosed within a defined boundary; the outside environment may provide inputs and receive outputs.

systems ecology Application of general systems theory and methods to ecology, with emphasis on sets of compartments linked by fluxes of energy and nutrients.

taiga The northern circumpolar boreal forest.

tallgrass prairie A narrow belt of tall grasses dominated by big bluestem that ran north and south adjacent to the deciduous forest of eastern North America; presence maintained by fire; largely destroyed by cultivation.

temperate rain forest Forest in regions characterized by mild climate and heavy rainfall that produces lush vegetative growth; one example is the coniferous forest of the Pacific Northwest of North America.

territory Area defended by an animal; varies among animal species according to social behavior, social organization, and resource requirements.

thermal conductance Rate at which heat flows through a substance.

thermal neutral zone Among homeotherms, the range of temperatures at which metabolic rate does not vary with temperature.

thermal radiation Heat transfer by long-wave or infrared radiation.

thermal tolerance Range of temperatures in which an aquatic poikilotherm is most at home.

thermocline Layer in a thermally stratified body of water in which temperature changes rapidly relative to the remainder of the body.

thermogenesis Increase in production of metabolic heat to counteract the loss of heat to a colder environment.

therophyte Plant that survives unfavorable conditions in the form of a seed; annual or ephemeral plant.

thinning law, 3/2 Self-thinning plant populations, sown at sufficiently high densities, approach and follow a thinning line with a slope of roughly $-3/2$; therefore in a growing population, plant weight increases faster than density decreases to a point where the slope changes to -1.

threshold of security Point in local population density at which the predator turns its attention to other prey because of harvesting efficiency; the segment of prey population below the threshold is relatively secure from predation; see *switching*.

throughfall That part of precipitation that falls through vegetation to the ground.

tidal overmixing Mixing of fresh water and seawater when a tidal wedge of seawater moves upstream in a tidal river faster than fresh water moves seaward; seawater on the surface tends to sink as lighter fresh water rises to the surface.

tiller In grasses, a lateral shoot arising at ground level.

time lag Delay in a response to change.

time-specific life table See *life table*.

tolerance model Model that proposes that succession leads to a community composed of species most efficient in exploiting resources; colonists neither increase nor decrease the rate of recruitment or growth of later colonists.

topography Physical structure of the landscape.

torpor Temporary great reduction in an animal's respiration, with loss of motion and feeling; reduces energy expenditure in response to some unfavorable environmental condition, such as heat or cold.

trace element Element occurring and needed in small quantities; see *micronutrient*.

translocation Transport of materials within a plant; absorption of minerals from soil into roots and their movement throughout the plant.

transpiration Loss of water vapor from a plant to the outside atmosphere.

trophic Related to feeding.

trophic level Functional classification of organisms in an ecosystem according to feeding relationships, from first-level autotrophs through succeeding levels of herbivores and carnivores.

trophic structure Organization of a community based on the number of feeding or energy transfer levels.

trophogenic zone Upper layer of the water column in ponds, lakes, and oceans, in which light is sufficient for photosynthesis.

tropholytic zone Area in lakes and oceans below the compensation point.

tundra Area in an arctic or alpine (high mountain) region, characterized by bare ground, absence of trees, and growth of mosses, lichens, sedges, forbs, and low shrubs.

turgor The state in a plant cell in which the protoplast is exerting pressure on the cell wall because of intake of water by osmosis.

turnover rate Rate of species lost and others gained.

ultisol Low base soil associated with warm, humid climate and old terrain, taking on a reddish color from secondary iron oxides.

ungulate Any hoofed grazing mammal; usually refers to ruminants, such as cattle and deer.

unitary organism An organism, such as an arthropod or vertebrate, whose growth to adult form follows a determinate pathway, unlike modular organisms whose growth involves indeterminate repetition of units of structure.

upwelling Area where currents force water from deep within an ocean into the euphotic zone.

vapor pressure The amount of pressure water vapor exerts independent of dry air.

vector Organism that transmits a pathogen from one organism to another.

vegetative reproduction Asexual reproduction in plants by means of specialized multicellular organs, such as bulbs, corms, rhizomes, stems, and the like.

veld Extensive grasslands in the east of the interior of South Africa, largely confined to high terrain.

vertical stratification Layering of physical conditions and life in a community.

vertisol Mineral soil that contains more than 30 percent of swelling clays that expand when wet and contract when dry, associated with seasonal wet and dry environments.

vesicular arbuscular mycorrhizae (VAM) A form of endomycorrhizae in which the fungus enters and grows within the host's cells and extends widely into the surrounding soil.

viscosity Property of a fluid that resists the force that causes it to flow.

Wallace's line Biogeographic line between the islands of Borneo and the Celebes that marks the eastern boundary of many landlocked Eurasian organisms and the western boundary of the Oriental region.

water potential Measure of energy needed to move water molecules across a semipermeable membrane; water tends to move from areas of high or less negative potential to areas of low or more negative potential.

water-use efficiency Ratio of net primary production to transpiration of water by a plant.

watershed Entire region drained by a waterway into a lake or reservoir; total area above a given point on a stream that contributes water to the flow at that point; the topographic dividing line from which surface streams flow in two different directions.

weather The combination of temperature, humidity, precipitation, wind, cloudiness, and other atmospheric conditions at a specific place and time.

weathering Physical and chemical breakdown of rock and its components at and below Earth's surface.

weed Plant possessing a high rate of dispersal, occurring opportunistically on land or water disturbed by human activity, and competing for resources with cultivated plants; a plant growing in the wrong place.

wetfall Component of acid deposition that reaches Earth by some form of precipitation; wet deposition.

wetland A general term applied to open-water habitats and seasonally or permanently waterlogged land areas; defining the extent of a wetland is controversial because of conflicting land use demands.

wilting point Moisture content of soil at which plants wilt and fail to recover their turgidity when placed in a dark, humid atmosphere; measured by oven drying.

xeric Dry, especially in soil.

zero net growth isocline An isocline along which the population growth rate is zero.

zonation Characteristic distribution of vegetation along an environmental gradient; this gradient may form latitudinal, altitudinal, or horizontal belts within an ecosystem.

zoogeography Study of the distribution of animals.

zooplankton Floating or weakly swimming animals in freshwater and marine ecosystems; planktonic animals.

zooxanthellae One-celled dinoflagellates that live symbiotically in the tissues of various marine invertebrates; closely associated with corals.

TEXT/ILLUSTRATION CREDITS

Chapter 2   Figure 2.7: From R. M. Burns and B. H. Honkala (eds.), "Silvics of North America," vol. 2, *Hardwoods Ag. Handbook* 654 (1990): 60. (Washington DC: Forest Service USDA, 1990). **Figure 2.9a,b:** From J. Wiens, "Patterns and Process in Grassland Bird Communities," *Ecological Monographs* 43 (1973): 240, fig. 2. 1963 The Ecological Society of America. Reprinted by permission.

Chapter 4   Figure 4.1: Slim Films. Reprinted by permission. **Figure 4.4a–c:** Adapted from R. G. Barry and R. J. Chorley, *Atmosphere, Weather, and Climate,* 6th ed. (New York: Routledge, Chapman Hall, 1992), p. 25, fig. 1.22. © Routledge, Chapman Hall, Inc. **Figure 4.10:** From *This Great and Wide Sea* by Robert E. Coker. © 1947 The University of North Carolina Press, renewed 1977 by R. M. Coker. Used by permission of the publisher. **Figure 4.13:** Adapted from M. S. Schroeder and C. C. Buck, "Fire Weather," *Agricultural Handbook* 360 (1970): 29.

Chapter 5   Figure 5.1: Adapted from H. G. Haverson and J. L. Smith, "Solar Radiation as a Forest Management Tool," *USDA Forest Service Gen. Tech Report* PSW-33. **Figure 5.2:** From W. E. Reifsnyder and H. W. Lull, "Radiant Energy in Relation to Forests," *USDA Tech Bulletin* 1344 (1965): 21. **Figure 5.7:** Adapted from B. A. Hutchinson and D. R. Matt, "The Distribution of Solar Radiation Within a Deciduous Forest," *Ecological Monographs* 47 (1977): 205. © 1977 The Ecological Society of America. Reprinted by permission. **Figure 5.10a–c:** Adapted from C. K. Augspurger, "Light Requirements of Tropical Tree Seedlings: A Comparative Study of Growth and Survival," *Journal of Ecology* 72 (1982): 777–795. © 1982 The Ecological Society of America. Reprinted by permission.

Chapter 6   Figure 6.7, 6.9: Adapted from Peterson, et al., "Snake Thermal Ecology" in R. A. Siegel and J. T. Collins (eds.), *Snakes, Ecology, and Behavior* (New York: The McGraw-Hill Companies, 1993), p. 271, fig. 7.5b and p. 272, fig. 7.6c. **Figure 6.11:** Illustration by David E. Murrish in K. Schmidt-Nielsen, *Animal Physiology,* 3rd ed. (Englewood Cliffs, NJ: Prentice-Hall, 1970), p. 56.

Chapter 8   Figure 8.1: From N. Campbell et al., *Biology Concepts and Connections,* 2nd ed. (Menlo Park, CA: Benjamin/Cummings, 1997), fig. 37.9. © 1997 by The Benjamin/Cummings Publishing Company. **Figure 8.2:** From E. Bunning, "Circadian Rhythms and the Time Measurement in Photoperiodism," *Cold Spring Harbor Laboratory Symposium on Quantitative Biology* 25 (1960): 253. Reprinted by permission of Cold Spring Harbor Laboratory Press. **Figure 8.3:** Adapted from C. H. Johnson and J. W. Hastings, "The Elusive Mechanism of the Circadian Clock" *American Scientist* 74 (1986): 29–36. © 1986 Sigma Xi. Reprinted by permission. **Figure 8.6:** From J. D. Palmer, "The Rhythmic Lives of the Crabs," *Bioscience* 40 (1990): 353, fig. 2. © 1990 American Institute of Biological Sciences. Reprinted by permission.

Chapter 9   Figure 9.8: From H. A. Mooney and S. L. Gulman, "Environmental and Evolutionary Constraints on the Photosynthetic Characteristics of Higher Plants," *Bioscience* 32 (1982): 198–206. © 1982 American Institute of Biological Sciences. Reprinted by permission.

Chapter 10   Figure 10.4: From Soil Conservation Service.

Chapter 11   Figure 11.6a–c: Data from *United Nations: The Sex and Age of Population, The 1990 Revision of the United Nations Global Population Distribution, Estimates and Projections* (New York: United Nations, 1990). **Figure 11.7a:** From H. Hett and O. L. Loucks, "Age Structure Models of Balsam Fir and Eastern Hemlock," *Journal of Ecology* 64 (1976): 1035, fig. 1a. © 1976 Blackwell Scientific Publications. Reprinted by permission.

Chapter 12   Figure 12.5: From Baker et al, *American Naturalist* 128 (1986): 495. © 1977 University of Chicago Press. Reprinted by permission.

Chapter 13   Figure 13.1b: From R. R. Sharitz and J. R. McCormick, "Population Dynamics of Two Competing Plant Species," *Ecology* 54 (1973): 729, fig. 5. © 1973 The Ecological Society of America. Reprinted by permission. **Figure 13.5a:** From R. R. Sharitz and J. R. McCormick, "Population Dynamics of Two Competing Plant Species," *Ecology* 54 (1973): 730, tab. 2. Copyright © 1973 The Ecological Society of America. Reprinted by permission. **Figure 13.5b:** From J. Sarukhan and J. L. Harper, "Studies on Plant Demography," *Journal of Ecology* 64 (1976): 1063. © 1976 Blackwell Scientific Publications. Reprinted by permission. **Figure 13.9:** Data from the United States Census Bureau. **Figure 13.11:** Data from F. A. Pitelka, *Proceedings 18th Biology Colloquium,* pp. 79, 80. (Corvallis, OR: Oregon State University, 1957).

Chapter 14   Figure 14.2a: From M. C. Dash and A. R. Hota, "Density Effects on Survival Growth Rate, and Metamorphosis on *Rana tigrina* Tadpoles," *Ecology* 61 (1980): 1027, fig. 2. © The Ecological Society of America. Reprinted by permission. **Figure 14.3a:** From C. W. Fowler, "Density Dependence as Related to Life History Strategy," *Ecology* 62 (1981): 607, fig. 4. © 1981 The Ecological Society of America. Reprinted by permission. **Figure 14.5:** From T. Jones, "Dispersal Distance and Range of Nightly Movements in Merriam's Kangaroo Rats," *Journal of Mammalogy* 20 (1989): 31. © 1989 American Society of Mammalogists. Used by permission. **Figure 14.6:** From A. S. Harestad and F. L. Bunnell, "Home Range and Body Weight" *Ecology* 60 (1979): 390. © 1979 The Ecological Society of America. Reprinted by permission. **Figure 14.7:** From R. L. Smith, "Some Ecological Notes on the Grasshopper Sparrow," *Wilson Bulletin* 75 (1963): 159–163. **Box 14.1 Figure A:** From A. R. E. Sinclair, *The African Buffalo,* p. 140. (Chicago: University of Chicago Press, 1977). © 1977 University of Chicago Press. Reprinted by permission.

Chapter 15   Figure 15.3a–c: Reprinted with permission from ASLO, from Tilman et al., *Limnology and Oceanography,* vol. 27 (1981): 1025, 1027. **Figure 15.4:** Adapted from H. C. Heller and D. Gates, "Altitudinal Zonation of Chipmunks *(Eutamias)*: Energy Budgets," *Ecology* 52 (1971): 424, fig. 1. © 1971 The Ecological Society of America. Reprinted by permission. **Figure 15.5:** From N. A. Moran and T. G. Whigham, "Interspecific Competition Between Root-feeding," *Ecology* 71 (1990): 1056, fig. 5. © 1990 The Ecological Society of America. Reprinted by permission. **Figure 15.7:** Adapted from N. K. Wieland and F. A. Bazzaz, "Physiological Ecology of Three Codominant Successional Annuals," *Ecology* 56 (1975): 686, fig. 6. © 1975 The Ecological Society of America. Reprinted by permission. **Figure 15.9:** From E. R. Pianka, *Evolutionary Ecology,* 3rd ed. (New York: Harper & Row, 1983), fig. 7.4. © 1983 by

Eric E. Pianka. Reprinted by permission of Addison Wesley Longman, Inc. **Figure 15.11:** From P. Williamson "Feeding Ecology of the Red-Eyed Vireo *(Vireo olivaceous)* and Associated Foliage-Gleaning Birds," *Ecological Monographs* 41 (1971): 136, fig. 6. © 1971 The Ecological Society of America. Reprinted by permission.

Chapter 16   Figure 16.3a,b: From Jedrzejewski et al., *Ecology* 76 (1995): 192, fig. 11. © 1995 The Ecological Society of America. Reprinted by permission. **Figure 16.8:** From Jedrzejewski et al., *Ecology* 76 (1995): 190, fig. 10b. © 1995 The Ecological Society of America. Reprinted by permission. **Figure 16.9:** From Jedrzejewski et al., *Ecology* 76 (1995): 188, fig. 7. © 1995 The Ecological Society of America. Reprinted by permission. **Figure 16.12, 16.13:** From N. B. Davies, "Prey Selection and Social Behavior in Wagtails," *Journal of Animal Ecology* 46 (1977): 48, fig. 8. © 1977 Blackwell Science Ltd. Reprinted by permission.

Chapter 17   Figure 17.1: From Buskirk and Ostfeld, "Controlling Lyme Disease," *Ecological Applications* 5 (4) (1995): 1134, fig. 1. © 1995 The Ecological Society of America. Reprinted by permission.

Chapter 18   Figure 18.2a,b: From G. I. Murphy, "Vital Statistics of the Pacific Sardine and the Population Consequences," *Ecology* 48 (1967): 734. © 1967 The Ecological Society of America. Reprinted by permission. **Figure 18.3:** From C. J. Walters, *Adaptive Management of Renewable Resources* (New York: Macmillan, 1986), p. 37. © 1986 International Institute of Applied Systems Analysis. **Figure 18.8:** From P. DeBach, *Biological Control by Natural Enemies* (New York: Cambridge University Press, 1974), p. 4, fig. 2. © 1974 by Cambridge University Press. Reprinted by permission. **Figure 18.9:** Adapted from R. C. Thatcher, et al. (eds.), "The Southern Pine Beetle," *USDA Forest Service and Science and Educational Technology Bulletin* 1631.

Chapter 19   Figure 19.8: From O. H. Frankel and M. E. Soule, *Conservation and Evolution* (Cambridge, England: Cambridge University Press, 1981), p. 32, fig. 19.8. Reprinted by permission.

Chapter 20   Figure 20.4a,b: Data from R. H. McArthur and J. McArthur, *Ecology* 42: 594–598.

Chapter 21   Figure 21.8: From I. Hanski and M. E. Gilpin, *Metapopulation Biology: Ecology, Genetics, Evolution* (San Diego: Academic Press, 1977), fig. 2b, p. 49. Reprinted by permission. **Figure 21.11a,b:** Adapted from R. H. Whittaker, "Vegetation of the Great Smoky Mountains," *Ecological Monographs* 26 (1956): 1–80. Reprinted by permission of The Ecological Society of America. **Figure 21.12:** Data from E. L. Braun, *Deciduous Forests of Eastern North America* (Philadelphia: Blakison Co., 1950). **Figure 21.14:** Data from Sousa, *Ecology* 60: 1225–1239 and *Ecology* 65: 1918–1935. **Figure 21.21a–d:** Data from R. H. Whittaker, "Quaternary History and the Stability of Forest Communities," *Forest Succession: Concepts and Applications*, D. C. West et. al. (eds.) (New York: Springer-Verlag, 1981), p. 147.

Chapter 22   Figure 22.6: From Oosting, *American Midland Naturalist* (Notre Dame, IN: University of Notre Dame, 1942). Reprinted with permission. **Figure 22.7:** From M. P. Austin and T. M. Smith, "A New Model of the Continuum Concept," *Vegetatio* 83 (1989): 35–47. With kind permission of Kluwer Academic Publishers.

Chapter 23   Figure 23.3a,b: From H. Leith, "Primary Production Terrestrial Ecosystems," *Human Ecology* 1(1973): 303. Reprinted by permission of Plenum Publishing Corporation. **Figure 23.5:** Adapted from J. R. Etherington, *Environmental and Plant Ecology*, 2nd ed. (New York: John Wiley, 1975), p. 355. © John Wiley & Sons Limited. Reproduced with permission. **Figure 23.7:** Data from R. L. Cowan, "Physiology of Nutrition as Related to Deer," *Proceedings First National White-tailed Deer Symposium* (Athens, GA: University of Georgia, 1962), pp. 1–8. **Figure 23.8:** Data from M. J. Coe, D. H. Cumming, and J. Phillipson, "Biomass and Production of Large Herbivores in Relation to Rainfall and Primary Production," *Oecologia* 22 (1976): 341–354. **Table 23.1:** Data from R. H. Whittaker and G. E. Likens, "Carbon in the Biota," *Carbon and the Biosphere Conf.* 72501, G. M. Woodwell and E. V. Pecan (eds.) (Springfield, VA: National Technical Information Service), pp. 281–300.

Chapter 26   Figure 26.9: From VEMAP (Vegetation/Ecosystem Modeling and Analysis Project), "Comparing Biogeography and Biogeochemistry Models in a Continental Scale Study of Terrestrial Ecosystem Response to Climate Change and CO$_2$ Doubling," *Global Biogeochemical Cycles* 9 (1995): 407–437. © 1995 Global Biogeochemical Cycles. **Figure 26.11:** From J. Harte and R. Shaw, "Shifting Dominance with a Montane Vegetation Community: Results of a Climate-warming Experiment. *Science* 267 (1995): 876–880.

Chapter 27   Figure 27.5: Data from W. K. Lauenroth, "Grassland Primary Production," N. F. French (ed.), *Perspectives in Grassland Ecology* (New York: Springer-Verlag, 1979), p. 10.

Chapter 33   Figure 33.2: Adapted from J. R. Grosselink and R. E. Turner, "Role of Hydrology in Freshwater Wetland Ecosystems," in R. Good, et al. (eds.), *Freshwater Wetlands* (Orlando, FL: Academic Press, 1978), fig. 6, p. 73. **Figure 33.8:** Data from Whigham, et al. "Biomass and Primary Production of Freshwater Tidal Marshes," R. E. Good, D. F. Whigham, and R. L. Simpson (eds.), *Freshwater Wetlands* (New York: Academic Press, 1978), p. 12. **Figure 33.9:** Data from H. T. Odum and M. A. Heywood, "Decomposition of Intertidal Freshwater Marsh Plants," R. E. Good, D. F. Whigham, and R. L. Simpson (eds.), *Freshwater Wetlands* (New York: Academic Press, 1978), p. 92. **Figure 33.10:** From the United States Department of Agriculture. **Table 33.1:** Adapted from S. P. Shaw and C. G. Fredine, *Wetlands of the United States*, US Fish and Wildlife Circular 39.

Chapter 34   Figure 34.7: From Newbold et al, *Animal Naturalist* 120 (1982): 630, fig. 1. Reprinted by permission of University of Chicago Press.

Chapter 35   Figure 35.4: Data from J. E. Raymont, *Plankton and Productivity in Oceans* (New York: Pergamon Press, 1963), p. 547. **Figure 35.5:** Data from J. H. Steele, *The Structure of Marine Ecosystems* (Cambridge, MA: Harvard University Press, 1974). **Figure 35.6:** With permission from *Annual Review of Genetics*, vol. 9. © 1975 by Annual Reviews Inc.

Chapter 36   Figure 36.7: From S. K. Eltringham, *Life in Mud and Sand* (New York: Crane Russak, 1971), p. 203. Reprinted by permission of Hoddr & Stroughton.

PHOTOGRAPHS

Openers for Chapters 2, 11, 12, 15, 16, 34, and 37; Figures 4.6; 21.16: © Kennan Ward.

Openers for Chapters 21, 23, and 25; Figures 4.15a; 9.10; 21.13; 27.8b; 28.1; 28.6a,b; 29.8; 30.8; 30.9; 30.10; 31.5; 31.6a,b,c; 32.1c; 32.8b; 33.4; 36.1; 37.5; 37.6; 37.8: Courtesy of R. L. Smith.

Opener Backgrounds for Chapters 9 and 13; all Part Opener backgrounds: © 1997 Photodisc.

Part I Opener: © 1993 Joanne Lotter/Tom Stack and Associates.

Chapter 1   Opener: © David M. Dennis/Tom Stack and Associates. **1.1:** © Thomas Kitchin/Tom Stack and Associates.

Chapter 2 **2.5a:** © UPI/Corbis-Bettmann. **2.5b:** © Archive Photos.

Chapter 3 **Opener:** © Kerry T. Givens/Tom Stack and Associates.

Part II Opener: © Robert Winslow/Tom Stack and Associates.

Chapter 4 **Opener:** © Thomas Kitchin/Tom Stack and Associates. **4.15b:** © Richard H. Stewart/National Geographic Society.

Chapter 5 **Opener:** © Anna E. Zuckerman/Tom Stack and Associates. **5.6a:** © Thomas Kitchin/Tom Stack and Associates. **5.12a:** © John Shaw/Tom Stack and Associates. **5.12b:** © Rod Planck/Tom Stack and Associates.

Chapter 6 **Opener:** © Joe McDonald/Tom Stack and Associates.

Chapter 7 **Opener:** © Larry Tackett/Tom Stack and Associates. **7.7:** © Matt Bradley/Tom Stack and Associates.

Chapter 8 **Opener:** © Richard Allen Wood/Animals Animals. **Opener Background:** Courtesy of NASA.

Chapter 9 **Opener:** © Brian Parker/Tom Stack and Associates. **9.4a:** © Stanley L. Flegler/Visuals Unlimited. **9.4b:** © George J. Wilder/Visuals Unlimited. **Box 9.1a:** © 1993 David M. Dennis/Tom Stack and Associates. **Box 9.1b:** © John Gerlach/Tom Stack and Associates. **Box 9.1c:** © Gerald and Buff Corsi/Tom Stack and Associates.

Chapter 10 **Opener:** © Terry Donnelly/Tom Stack and Associates.

Part III Opener: © Rod Planck/Tom Stack and Associates

Chapter 12 **12.1:** © Roland Birke/OKAPIA/Photo Researchers, Inc. **12.3:** © John Gerlach/Tom Stack and Associates. **12.4:** © Rod Planck/Tom Stack and Associates. **12.6:** © Leonard Lee Rue III/Photo Researchers, Inc.

Chapter 13 **Opener:** © John Gerlach/Tom Stack and Associates.

Chapter 14 **Opener:** © Tom Stack/Tom Stack and Associates.

Chapter 16 **16.14:** © Merlin Tuttle/Bat Conservation International, Inc.

Chapter 17 **Opener:** © 1993 Eric A. Soder/Tom Stack and Associates. **17.3:** © Raymond Gehman/National Geographic Society.

Chapter 18 **Opener:** © Jim Nilsen/Tom Stack and Associates. **18.6:** © Wendy Shattil/Bob Rozinski/Tom Stack and Associates.

Chapter 19 **Opener:** © Tim Davis/Photo Researchers, Inc. **19.9a:** © Dan Sudia. **19.9b:** © W. Perry Conway/Tom Stack and Associates.

Part IV Opener: © Raymond K. Gehman/National Geographic Society.

Chapter 20 **Opener:** © Sharon Gerig/Tom Stack and Associates.

Chapter 21 **21.17b:** © Glenn M. Oliver/Visuals Unlimited.

Chapter 22 **Opener:** © Milton Rand/Tom Stack and Associates.

Chapter 24 **Opener:** © Kerry T. Given/Tom Stack and Associates.

Chapter 25 **Box 25.1a:** Courtesy of Hugh Morton.

Chapter 26 **Opener:** Courtesy of NASA. **Box 26.1a,b:** Courtesy of Peter Wolter. **Box 26.2a:** © Will Owens Photography.

Part V Opener: © Milton Rand/Tom Stack and Associates.

Part VI Opener: © Terry Donnelly/Tom Stack and Associates.

Chapter 27 **Opener:** © Brian Parker/Tom Stack and Associates. **27.2:** © Rod Planck/Tom Stack and Associates. **27.4:** © Jeff Foott/Tom Stack and Associates. **27.7:** © Rod Planck/Photo Researchers, Inc. **27.8a:** © Joe McDonald/Tom Stack and Associates. **27.8c:** © Barbara Gerlach/Tom Stack and Associates. **27.8d, 27.9:** Courtesy of T. M. Smith.

Chapter 28 **Opener:** © Joe McDonald/Tom Stack and Associates. **28.6c:** © Ray Ellis/Photo Researchers, Inc. **28.7:** © 1991 by David L. Brown/Tom Stack and Associates. **28.8a:** © E.R. Degginger/Animals Animals. **28.8b:** © Betty K. Bruce/Animals Animals. **28.9:** © Matt Bradley/Tom Stack and Associates.

Chapter 29 **Opener:** © Doug Sokell/Tom Stack and Associates. **29.2:** © Thomas Kitchin/Tom Stack and Associates. **29.3a:** © John Shaw/Tom Stack and Associates. **29.3b:** © Manfred Gottschalk/Tom Stack and Associates. **29.4:** © Peter Cummings/Tom Stack and Associates. **29.5:** © Thomas Kitchin/Tom Stack and Associates. **29.6a:** © Steve McCutcheon. **29.6b:** © Brian Parker/Tom Stack and Associates. **29.7:** © Biophoto.

Chapter 30 **Opener:** © Rod Planck/Tom Stack and Associates. **30.1a:** © Hans Reinhard/Bruce Coleman Inc. **30.1b:** © Milton Rand/Tom Stack and Associates. **30.1c:** © Don and Pat Valenti/Tom Stack and Associates. **30.1d:** © John Shaw/Tom Stack and Associates. **30.2:** © Raymond Gehman/National Geographic Society. **30.3:** © Greg Vaughn/Tom Stack and Associates. **30.6:** Courtesy of Westvaco. **30.7:** © 1991 J. Lotter/Tom Stack and Associates.

Chapter 31 **Opener:** © Dominique Braud/Tom Stack and Associates. **31.1:** © Inga Spence/Tom Stack and Associates. **31.2:** © Gregory G. Dimijian 1991/Photo Researchers, Inc. **31.4:** © Larry Tackett/Tom Stack and Associates. **31.6d:** © Inga Spence/Tom Stack and Associates.

Chapter 32 **Opener:** © Thomas Kitchin/Tom Stack and Associates. **32.1a:** © Jack S. Grove/Tom Stack and Associates. **32.1b:** © Doug Sokell/Tom Stack and Associates. **32.8a:** © Brian Parker/Tom Stack and Associates.

Chapter 33 **Opener:** © Larry Lipsky/Tom Stack and Associates. **33.3:** © Terry Donnelly/Tom Stack and Associates. **33.5:** © Mark Rollo/Photo Researchers, Inc. **33.11:** © Bob McKeever/Tom Stack and Associates.

Chapter 34 **34.1–34.3:** © John Shaw/Tom Stack and Associates. **34.9:** © Bob Pool/Tom Stack and Associates. **34.10:** © Joanne Lotter/Tom Stack and Associates.

Chapter 35 **Opener:** © Kerry T. Givens/Tom Stack and Associates **35.2:** © Flip Nicklin/Minden Pictures.

Chapter 36 **Opener:** © J. Lotter Gurling/Tom Stack and Associates. **36.4:** © Scott Blackman/Tom Stack and Associates. **36.5:** © Jack S. Grove/Tom Stack and Associates. **36.8:** © Tammy Peluso/Tom Stack and Associates.

Chapter 37 **37.7:** © Joe Arrington/Visuals Unlimited. **37.8:** © Greg Vaugh/Tom Stack and Associates. **37.9:** © Larry Lipsky/Tom Stack and Associates.

TRADEMARK ACKNOWLEDGMENTS

A-C Mirex is a registered trademark of Allied Chemical Corporation of New York, NY.

Dispel is a registered trademark of Vinings Industries, Inc. of Atlanta, GA.

Foray is a registered trademark of Novo-Nordisk A/S of Bagsvaerd, Denmark, owned by Abbott Laboratories of Abbott Park, IL.

Sevin is a registered trademark of Union Carbide Corporation of New York, NY.

Thuricide is a registered trademark of Bioferm Corporation of Wasco, CA.

INDEX